WITHDRAWN

OXFORD GEOLOGICAL SCIENCES SERIES NO 2

Series editors

P. Allen
E. R. Oxburgh
B. J. Skinner

OXFORD GEOLOGICAL SCIENCES SERIES

1. De Verle P. Harris: *Mineral resources appraisal: Mineral endowment, resources, and potential supply: concepts, methods, and cases*
2. J. J. Veevers (ed.): *Phanerozoic earth history of Australia*
3. Yang Zunyi (ed.): *The geology of China*

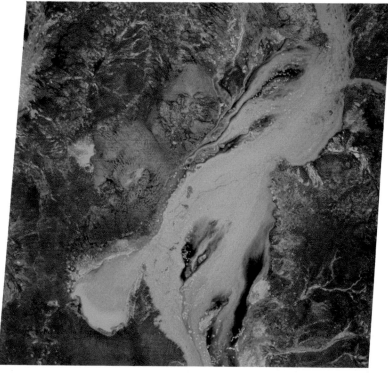

Innamincka depocentre of the Central-Eastern Lowlands: Cooper Creek, south-western Queensland, in its normal dry state, after a flood (above), and in its rare wet state, during a flood (below). The frequency of floods today matches that of the rare thalassocratic state of the Phanerozoic platform within its normal geocratic state. False-colour LANDSAT images (180 km across) from McCracken and Astley-Boden (1982), by permission of the authors.

Phanerozoic earth history of Australia

Edited by

J. J. VEEVERS

Contributions by

P. J. CONAGHAN
R. H. FLOOD
B. D. JOHNSON
J. G. JONES
K. L. McDONNELL
C. McA. POWELL
S. E. SHAW
J. A. TALENT
J. J. VEEVERS
S. Y. WASS
M. A. J. WILLIAMS
School of Earth Sciences, Macquarie University, North Ryde, N.S.W.

B. J. J. EMBLETON
CSIRO Division of Mineral Physics, North Ryde, N.S.W.

P. G. QUILTY
Antarctic Division, Department of Science and Technology, Kingston, Tasmania

CLARENDON PRESS · OXFORD · 1984

Oxford University Press, Walton Street, Oxford OX2 6DP
London New York Toronto
Delhi Bombay Calcutta Madras Karachi
Kuala Lumpur Singapore Hong Kong Tokyo
Nairobi Dar es Salaam Cape Town
Melbourne Auckland
and associated companies in
Beirut Berlin Ibadan Mexico City Nicosia

Oxford is a trade mark of Oxford University Press

Published in the United States
by Oxford University Press, New York

© J. J. Veevers and contributors listed on p. iii, 1984

All rights reserved. No part of this publication may be reproduced, stored in a retrieval system, or transmitted, in any form or by any means, electronic, mechanical, photocopying, recording, or otherwise, without the prior permission of Oxford University Press.

British Library Cataloguing in Publication Data
Phanerozoic earth history of Australia.—
(Oxford geological sciences series)
1. Geology, Stratigraphic 2. Geology—
Australia
I. Veevers, J.J.
551.7'00994   QE651
ISBN 0-19-854459-6

Library of Congress Cataloging in Publication Data
Main entry under title:
Phanerozoic earth history of Australia.
(Oxford geological sciences series)
Bibliography: p.
Includes index.
1. Geology—Australia. 2. Historical geology.
I. Veevers, J. J. II. Conaghan, P. J. III. Series.
QE340.P48 1984   551.7'0994   84-9679
ISBN 0-19-854459-6

Typeset by Cotswold Typesetting, Cheltenham.
Printed in Great Britain at the University Press, Oxford
by David Stanford, Printer to the University

PREFACE

The Australian continent represents a fortieth of the earth's surface, and a sixteenth of the continental lithosphere. Its land area is the same as that of the conterminous United States of America, but its margins are wider. From its equatorial border with south-east Asia in Timor and New Guinea, it extends 50° southward between the Indian and Pacific Oceans, its structure reflecting convergence of the Pacific Plate on the north-east and east, convergence with Sundaland on the north-west, and divergence from the rest of Gondwanaland on the south and west. In Australia, the most compact of continents, the present interplay of plate convergence on one side and divergence on the other is traceable through the entire Phanerozoic, to its beginning 575 Ma ago. The unity of its Phanerozoic history, treated at continental scale and related to the adjacent oceanic lithosphere, is matched by a unity of approach, expressed in a single tongue.

Of the Australian–New Guinean lithosphere, two parts are land, and the third part lies beneath the pericontinental sea of the margins and the epeiric sea between Australia and New Guinea. The geology of the submerged third of Australia, as well as that of the adjacent oceans, is a discovery of the past fifteen years, and post-dates previous Australian geologies, such as those of Brown, Campbell, and Crook (1968), and David and Browne (1950). Not only do the Australian divergent margins contain its chief resource of petroleum and gas, but most parts of them, thanks to a thin sediment cover, are more accessible to study of the pre-breakup stages of development than any others, and so are an important world resource of information on divergent margins. Petroleum and mineral resources are not part of the subject-matter of this book, but our account of the history of the interval during which were deposited all Australia's fossil fuels, most of its non-metallic minerals, and some of its metals provides the regional tectonic and depositional setting of these deposits.

This book is a review of the Phanerozoic earth history of Australia from a new perspective. Chapter I introduces the subject and, since chronology is the core of history, presents the time-scale of the Phanerozoic we employ. Australia's global setting now and in the past, the result of the growth of the eastern part of the Indo-Australian Plate out of Eastern Gondwanaland, is explored in Chapter II by studies of continental and oceanic palaeomagnetism, palaeoclimates and palaeoenvironments, and palaeobiogeography. The morphology and structure of Australia strike into Antarctica, and we document some aspects of this connection. New Zealand probably lies 'down-dip' from Australia, and connections are less clear. A review of the present anatomy and physiology of the Australian lithosphere in Chapter III is followed in Chapter IV by a morphotectonic analysis of the platform and margins. The land surface of Australia and New Guinea, the intervening Arafura Sea, and the continental margins express in their morphology the character and recency of the tectonic processes that formed them; for example, the cordilleran phase of the New Guinea Highlands and related basins can be retraced no further than the Pliocene, though the Eastern Highlands can be traced to the mid-Cretaceous, the western margin to the Mesozoic and late Palaeozoic, and the Musgrave Ranges of central Australia and other parts of the Great Western Plateau to the early Palaeozoic. We trace these features to their origin by uncovering successively older (and more obscure) layers in the related basins. Neotectonic studies of this kind are applicable to most parts of Australia at least back 100 Ma, because morphotectonic elements persist over this period, though they are now much modified in shape and activity. Features not directly accessible by neotectonic analysis are studied in Chapter V, and interpreted in the context of tectonic regimes, from the youngest to the oldest, that have marked the Phanerozoic development of Australia. We conclude, in Chapter VI, with a palaeogeographical cinematograph of the Australian Phanerozoic, some general comments on this history, and a list of pressing but unanswered questions.

This book is the fruit of collaboration of members of the Australian Plate Research Group of the School of Earth Sciences of Macquarie University, together with our associates Dr B. J. J. Embleton of the CSIRO Division of Mineral Physics, and Dr P. G. Quilty of the Antarctic Division of the Department of Science and Technology. Individual authorship, other than that of Veevers alone, is indicated in the list of contents and in the text.

Pre-publication information incorporated in the

book was kindly supplied by Associate Professor J. Roberts, Dr E. M. Truswell, and Professor C. C. von der Borch. Documentation includes published work up to early 1983. We thank the anonymous referees of our research papers that form the basis of much of this book, and the publishers who kindly granted permission to draw on this material: American Association of Petroleum Geologists, American Geophysical Union, Australian Museum, Australian Petroleum Exploration Association, Elsevier Scientific Publishing Company (*Earth and Planetary Science Letters, Tectonophysics*), *Journal of Geology, Journal of the Geological Society of Australia*, Dr W. Junk bv Publishers, *Nature*, Plenum Publishing Corporation, and the Southeast Asia Petroleum Exploration Society. Authors (and publishers) who consented to our incorporating extracts from their work are: American Association of Petroleum Geologists; Dr R. W. Day (*Tectonophysics*); Dr D. A. Falvey (*BMR Journal*); Mr D. M. Finlayson, Dr K. J. Muirhead, and Dr F. E. M. Lilley (*Tectonophysics*); Mr K. G. Grimes (Queensland Division, Geological Society of Australia); Dr D. E. Karig (*Earth and Planetary Science Letters*); Dr E. M. Kemp (*Palaeogeography, Palaeoecology, Palaeoclimatology*); Dr J. P. Kennett (Deep Sea Drilling Project); Dr E. C. Leitch (Journal of Geological Society of Australia); Dr X. Le Pichon (Elsevier); Dr B. McGowran (Consortium for Ocean Geosciences of Australian Universities); Dr T. J. Mount (Petroleum Exploration Society of Australia); Dr R. H. Pilger, Jr (*Journal of Geophysical Research*); Mr K. A. Plumb (*Earth-Science Reviews*); Dr W. D. Roots (*Earth and Planetary Science Letters*); Dr R. W. R. Rutland (Royal Society of New South Wales); the Times Atlas (copyright held by John Bartholomew & Son Limited, Edinburgh); and Prof. C. C. von der Borch (*Tectonophysics*).

The Australian Research Grants Scheme and the Australian Marine Sciences and Technologies Scheme provided grants to support much of our research.

I am grateful to Dr M. K. Horn, Cities Services Company, Tulsa, for arranging for me to take part in a field trip to the Southern Oklahoma Aulacogen, to Dr G. Bond, Lamont–Doherty Geological Observatory, for reviewing the account on hypsometry in Chapter IV, and to Dr J. G. Jones, for his unwearied guidance and counsel. Mr K. Rousell, Mr J. Cleasby, and Mr R. Bashford made the fair drawings, and Mrs A. M. Michal and Miss G. Keena did the wordprocessing. Miss R. E. Sefton typed the tables, and Mrs S. Roots made the computer runs of the oceanic reconstructions.

Macquarie University　　　　　　　　　　　　　J. J. V.
North Ryde   N.S.W.
1983

CONTENTS

I. Introduction   1
   1. General   1
   2. Phanerozoic time-scale   1
     (a) Magnetic-polarity scale (Cainozoic and late Mesozoic only)   1
     (b) Radiometric and biostratigraphical scales   3
       (i) Cainozoic and Mesozoic   3
       (ii) Palaeozoic   3
     (c) Explanation   3

II. Australia's global setting   5
   1. PRESENT GLOBAL SETTING   5
     (a) Introduction   5
     (b) Australian–Antarctic Depression   6
   2. PAST GLOBAL SETTINGS   11
     (a) Continental palaeomagnetism (by B. J. J. Embleton)   11
       (i) Introduction   11
       (ii) Australia's drift history   11
       (iii) Palaeolatitude of Australia   13
     (b) Oceanic palaeomagnetism (by B. D. Johnson and J. J. Veevers)   17
       (i) Introduction: growth of the Indo–Australian Plate   17
       (ii) 160 Ma: reconstruction of Eastern Gondwanaland   18
       (iii) Revised fit of Antarctica and Australia   21
       (iv) Revised fit of India and Antarctica/Australia   23
       (v) Revised pattern of spreading off the western margin of Australia   24
       (vi) Spreading from 160 to 128 Ma   25
       (vii) Spreading from 128 to 95 Ma   26
       (viii) Spreading from 95 to 81.7 Ma   27
       (ix) Reconstruction of Lord Howe Rise/New Zealand by closing the Tasman Sea Basin   28
       (x) Growth of the South-east Indian Ocean between Antarctica and Australia   34
       (xi) Growth of the Coral Sea between north-east Australia and the Papuan Peninsula   37
       (xii) Summary of plate motions   37
       (xiii) Independence of Australia between the opening and closing of oceanic straits   37
         (1) Western margin   37
         (2) Southern margin   38
         (3) North-western margin   38
     (c) Interaction of Australia and South-east Asia (by C. McA. Powell, B. D. Johnson, and J. J. Veevers)   38
       (i) Introduction   38
       (ii) Constraints   38

			(1)	Size of Greater India	38
			(2)	Position of south-east Asia	38
			(3)	Positions of Greater India and Australia since continental breakup 128 Ma ago	39
		(iii)		Possible reconstructions of the interaction between Sundaland, India, and Australia	41
3.	PALAEOCLIMATES AND PALAEOENVIRONMENTS				42
	(a) Quaternary environments (by M. A. J. Williams)				42
		(i)	Present-day morphoclimatic regime		42
		(ii)	Glacial aridity: Australia–New Guinea 18 ka ago		42
		(iii)	Warm, wet interglacial: Australia–New Guinea 9 ka ago		45
		(iv)	Conclusions		47
	(b) Phanerozoic climates and environments of Australia (by P. G. Quilty)				48
		(i)	Introduction		48
		(ii)	Cambrian		48
		(iii)	Ordovician		48
		(iv)	Silurian		48
		(v)	Devonian		49
		(vi)	Carboniferous		49
		(vii)	Permian		49
		(viii)	Triassic		49
		(ix)	Jurassic		50
		(x)	Cretaceous		50
		(xi)	Tertiary		50
			(1)	Late Palaeocene–early Eocene	50
			(2)	Middle–late Eocene	51
			(3)	Oligocene–middle Miocene	54
			(4)	Late Miocene and Pliocene	55
		(xii)	Summary		55
4.	AUSTRALIAN BIOGEOGRAPHY PAST AND PRESENT: DETERMINANTS AND IMPLICATIONS (by J. A. Talent)				57
	(a) The contemporary and Cainozoic biogeographic fabric				57
		(i)	Marine biota		58
		(ii)	Terrestrial biota		59
	(b) Chronology of land-bridges and island 'staging-points'				65
		(i)	The Moluccas–Lesser Sunda Islands 'inter-regional' zone		66
		(ii)	The Australia–Antarctica–South America dispersal route		68
		(iii)	Interaction with neighbours to the east: New Caledonia and New Zealand		69
	(c) The imprint of past biogeography on Australian biostratigraphy				70
		(i)	Global biogeographic differentiation		70
		(ii)	Provinciality within the Australian block		71
	(d) Biogeographic affinities of Australian pre-Cainozoic faunas and floras				72
		(i)	Ediacarian		73
		(ii)	Cambrian		73
		(iii)	Ordovician		75
		(iv)	Silurian		75

				(v)	Devonian	75
				(vi)	Carboniferous	76
				(vii)	Permian	78
				(viii)	Triassic	78
				(ix)	Jurassic	79
				(x)	Cretaceous	80
				(xi)	Cainozoic	81
		(e)	Résumé of main biogeographic events and relationships			81
		(f)	Bibliography of Australian biogeography and palaeobiogeography			82

III. Lithospheric structure — 94

1. GROSS ANATOMY — 94
   (a) Structural differences between the Phanerozoic and Precambrian terrains — 94
       (i) Deep structure — 94
       (ii) Shallow structure — 97
   (b) Crustal and upper mantle stratigraphy beneath eastern Australia (by S. Y. Wass) — 98
       (i) Mantle — 98
       (ii) Lower crust — 99
       (iii) Upper crust — 100
2. PHYSIOLOGY — 100
   (a) Seismicity — 100
   (b) Dextral shear within the eastern Indo-Australian Plate (by J. J. Veevers and C. McA. Powell) — 102
3. REFLECTION OF EARTH MOTIONS IN THE CAINOZOIC VOLCANICITY OF EASTERN AUSTRALIA — 103

IV. Morphotectonics of the Australian platform and margins — 106

1. DIVISION OF THE AUSTRALIAN LITHOSPHERE INTO PLATFORM AND MARGINS — 106
2. MORPHOTECTONICS OF THE CONVERGENT NORTHERN MARGIN, FOCUSED ON THE NEW GUINEA HIGHLANDS — 107
   (a) Morphotectonics — 107
   (b) Summary history — 114
3. MORPHOTECTONICS OF THE PLATFORM REGIONS, FOCUSED ON THE HIGHLANDS — 115
   (a) Eastern Highlands (by J. G. Jones and J. J. Veevers) — 115
       (i) Cainozoic history of the South-east Highlands — 115
           (1) Summary — 115
           (2) Introduction — 115
           (3) Gross morphology and current tectonic activity of South-east Australia — 115
           (4) Basalts of the South-east Highlands — 117
           (5) Sediments of the Murray Basin — 117
           (6) Sedimentation signatures of the Gippsland, Bass, and Otway Basins — 118
           (7) Cainozoic tectonism in South-east Australia: a model — 119

		(8)	Late Cainozoic tectonic cycles of South-east Australia	119
			(a) Pliocene–Quaternary cycle	119
			(b) Late Oligocene–Miocene cycle	121
		(9)	Inception of the Highlands	123
		(10)	Outline of a Cainozoic history of the Highlands	123
		(11)	Discussion	123
	(ii)	Cainozoic history of the North-east Highlands		124
	(iii)	Mesozoic origins and antecedents of the Eastern Highlands		125
		(1)	Summary	125
		(2)	Introduction	126
		(3)	Eastern Highlands defined	126
		(4)	North-east Australia 200 to 90 Ma ago	127
			(a) Surat–Maryborough transect	127
			(b) Carpentaria–Papua transect	127
			(c) Orogen and foreland basin	129
		(5)	South-east Australia 200 to 95 Ma ago	131
			(a) Early and Middle Jurassic rocks	131
			(b) Late Jurassic and Early Cretaceous rocks	132
			(c) Discussion	133
		(6)	Tectonic setting of Eastern Australia 200 to 90 Ma ago	136
			(a) North-east sector: 140 to 90 Ma	136
			(b) North-east sector: 200 to 140 Ma	136
			(c) South-east sector: 200 to 140 Ma	136
			(d) South-east sector: 140 to 90 Ma	136
			(e) Synthesis: 200 to 90 Ma	136
		(7)	Eastern Highlands terrain 95/90 to 65 Ma ago	137
		(8)	Eastern Highlands: a dynamic epeirogen	138
		(9)	Eastern Highlands and continental margin	138
		(10)	Late Mesozoic development of the Pacific Borderland	139
		(11)	Birth of the Eastern Highlands	142
(b)	History of the South Australian Highlands			143
	(i)	Introduction		143
	(ii)	Cainozoic		143
	(iii)	Later Cretaceous		148
	(iv)	Jurassic through Cenomanian		148
	(v)	Late Triassic–Early Jurassic		148
	(vi)	Late Carboniferous–mid-Triassic		148
	(vii)	Discussion		148
(c)	History of the Great Western Plateau			149
	(i)	Western Shield		149
		(1)	Topography and sedimentation in the Cainozoic	149
		(2)	Late Cretaceous	153
		(3)	Late Carboniferous to mid-Cretaceous	154
		(4)	Cambrian to Carboniferous	154
		(5)	Discussion	155

	(ii)	Kimberley Block, Pine Creek Inlier, Arnhem Block	155
		(1) Introduction	155
		(2) Cainozoic patterns of erosion and deposition	156
		(3) Late Jurassic to end of Cretaceous	159
		(4) Late Carboniferous to Middle Jurassic	159
		(5) Late Devonian to Early Carboniferous	159
		(6) Silurian to Middle Devonian	159
		(7) Cambrian and Ordovician	160
		(8) Conclusion	160
	(iii)	Central Australia: Amadeus Transverse Zone	160
		(1) Introduction	160
		(2) Cainozoic record	160
		(3) Late Carboniferous to Cretaceous	161
		(4) Devonian and Carboniferous	161
		(5) Latest Proterozoic/earliest Cambrian to Silurian	167
(d)		Summary: the antiquity of the highlands of Australia	167
4.		MORPHOTECTONICS OF THE DIVERGENT OR RIFTED MARGINS	168
(a)		Development of rifted margins: a conceptual framework	168
	(i)	Life cycle	168
	(ii)	Falvey's (1974) model for the recognition of developmental stages	171
(b)		Southern margin	174
	(i)	Introduction	174
	(ii)	Age of breakup of the southern and south-eastern margins	175
	(iii)	Ceduna Depocentre	176
	(iv)	East of the Ceduna Depocentre	179
	(v)	West of the Ceduna Depocentre	181
	(vi)	Summary	181
(c)		Western and north-western margins	182
	(i)	Introduction	182
	(ii)	Physiography and structure	183
	(iii)	Age of breakup	183
	(iv)	Boundary between oceanic and continental crust	186
	(v)	Stages of development	189
		(1) Failed-arm stage	189
		(2) Infrarift stage	190
		(3) Rifted-arch stage	191
		(4) Rim-basin stage	198
		(5) Open-margin stage	199
	(vi)	Comparison of rifted-arch and later stages with the conceptual morphotectonic model	200
(d)		Eastern margin	201
	(i)	Introduction	201
	(ii)	South-eastern margin	201
	(iii)	North-eastern margin	204
	(iv)	Comparison with the western margin	205
	(v)	Development of the eastern margin	205

5. QUANTITATIVE ESTIMATES OF THE VERTICAL MOTIONS OF THE PLATFORM SINCE THE JURASSIC — 210
    (a) Introduction — 210
    (b) Summary — 211
    (c) Bond's method — 211
    (d) Past sea levels — 212
    (e) Australian hypsometry — 213
    (f) Local anomalies of continent level — 215
        (i) Since the middle Miocene — 215
        (ii) Between the Eocene and Miocene — 216
        (iii) Between the later Cretaceous and mid-Eocene — 216
        (iv) Between the Aptian–Albian and later Cretaceous — 217
        (v) Between the Late Jurassic and the Aptian–Albian — 218
    (g) Discussion — 218

V. Australia's Phanerozoic history — 222
1. INTRODUCTION — 222
2. POTOROO REGIME — 222
    (a) Preamble — 222
    (b) Late Innamincka Regime — 222
    (c) Cenomanian Interregnum and later Cretaceous deposition — 222
    (d) Later Cretaceous lacuna on the platform — 228
    (e) Cainozoic resumption of platform deposition — 228
        (i) Cycles in the Cainozoic — 229
    (f) Summary — 230
        (i) Innamincka prelude — 230
        (ii) Interregnum — 230
        (iii) Potoroo Regime — 230
    (g) Australian–Antarctic Depression in Antarctica (by J. J. Veevers and P. G. Quilty) — 232
        (i) Late Cretaceous and Cainozoic history — 232
        (ii) Earliest history — 234
        (iii) Conclusions — 234
3. INNAMINCKA REGIME — 235
    (a) Introduction — 235
    (b) Mid-Carboniferous lacuna — 235
    (c) Gondwanan style of deposition — 235
        (i) Glacigene sediments — 235
        (ii) Coal measures — 235
        (iii) Carbonate — 238
        (iv) Evaporites — 238
    (d) Late Carboniferous jump of the magmatic arc and related basins — 238
    (e) Early Innamincka stage — 239
        (i) Review — 239
        (ii) Papuan Basin analogue and a foreland basin model for the Bowen–Sydney Basin (by J. G. Jones, P. J. Conaghan, K. L. McDonnell, R. H. Flood, and S. E. Shaw) — 243

		(1)	Introduction	243
		(2)	Modern Papuan Basin	245
		(3)	Late Permian Bowen Basin	247
		(4)	Late Permian Sydney Basin	249
		(5)	Tectonics of the Papuan Basin	251
		(6)	Tectonics of the Bowen–Sydney Basin	257
		(7)	A foreland basin model	258
		(8)	Bowen–Sydney Basin: mid-Permian to mid-Triassic part of the early Innamincka Regime	259
		(9)	Appendix: Permo-Triassic time-scale and palynostratigraphy	261
	(iii)	Epicratonic basins		262
(f)	Middle Innamincka stage: Late Triassic/earliest Jurassic			262
(g)	Late Innamincka stage			265
	(i)	Early and Middle Jurassic		265
	(ii)	Late Jurassic		265
	(iii)	Early Cretaceous		266
	(iv)	Cenomanian Interregnum		269

4. ULURU AND ADELAIDEAN REGIMES — 270
   (a) Introduction — 270
   (b) Pre-Adelaidean: final consolidation of the Precambrian craton — 270
   (c) Adelaidean: rift-valley stage — 270
   (d) Epi-Adelaidean: regional shear (by J. J. Veevers and C. McA. Powell) — 278
   (e) Cambrian: plate divergence on the east and north-west, followed by plate convergence on the east (by J. J. Veevers and C. McA. Powell) — 282
       (i) Breakup and volcanism — 282
       (ii) Depositional events — 283
       (iii) Pattern of divergence — 284
       (iv) Ordian transgression — 284
       (v) Middle and Late Cambrian convergence — 284
           (1) Templetonian: start of convergence — 284
           (2) Late Cambrian deformation and plutonism — 284
           (3) Reconstructions — 289
   (f) Ordovician to earliest Silurian: marginal sea and island arc (by C. McA. Powell) — 290
       (i) Tasman Orogen of south-eastern Australia — 290
           (1) Major facies distribution — 290
               (a) Western littoral zone — 291
               (b) Central deep-water zone — 292
               (c) Central-eastern mafic volcanic belt — 292
               (d) Eastern turbidite zone — 292
           (2) Palaeocurrents and provenance — 293
           (3) Modern analogue — 295
               (a) Morphotectonic comparison — 295
               (b) Deformation pattern — 297
           (4) Nature of the basement — 298
               (a) Distribution and composition of granitic batholiths — 298
               (b) Oceanic or continental crust? — 302

				(5)	Tectonic models	303
		(ii)	Tasman Orogen of north-eastern Australia			305
		(iii)	Platform and north-western margin			307
	(g)	Silurian to mid-Devonian – dextral transtensional margin (by C. McA. Powell)				309
		(i)	Tasman Orogen of south-eastern Australia			309
			(1)	Major facies distribution		309
			(2)	Recognition of sedimentary basins		310
			(3)	Main sedimentary basins		310
			(4)	Early Silurian		311
			(5)	Mid-Silurian to mid-Devonian		315
				(a)	Wenlock and Ludlow	315
				(b)	Pridoli and Lochkovian	317
				(c)	Early Devonian	319
				(d)	Middle Devonian	320
		(ii)	Major facies distribution elsewhere in the Tasman Orogen			321
		(iii)	Tectonic models			323
			(1)	Early Silurian transform margin		323
			(2)	Mid-Silurian to mid-Devonian dextral transtensional phase		324
		(iv)	Australian platform			327
	(h)	Late Devonian and Early Carboniferous: continental magmatic arc along the eastern edge of the Lachlan Fold Belt (by C. McA. Powell)				329
		(i)	Tasman Orogen of south-eastern Australia			329
			(1)	Major facies distribution in the north-eastern Lachlan Fold Belt		329
			(2)	Correlation of the Lambian Facies in south-eastern Australia		329
			(3)	Deposition as sheet-like bodies or in grabens?		331
			(4)	Frasnian palaeogeography		332
			(5)	Famennian-Tournaisian palaeogeography		333
		(ii)	Tasman Orogen elsewhere			334
			(1)	New England Fold Belt in New South Wales		334
			(2)	Queensland sector		335
		(iii)	Tectonic models			337
		(iv)	Australian platform			337
	(i)	Comparative tectonics of the transverse structural zones of Australia and North America (by J. J. Veevers and C. McA. Powell)				340
		(i)	Introduction			340
		(ii)	Amadeus Transverse Zone			340
		(iii)	Southern Oklahoma Aulacogen			344
		(iv)	Comparisons			344
			(1)	Southern Oklahoma Aulacogen		344
			(2)	Ancestral Rocky Mountains		346
		(v)	North–south shortening			347
		(vi)	Discussion			347
	(j)	Termination of the Uluru Regime: the mid-Carboniferous lacuna (by C. McA. Powell and J. J. Veevers)				348

VI.	**Synopsis** (by J. J. Veevers, J. G. Jones, C. McA. Powell and J. A. Talent)	351
	1. INTRODUCTION	351
	2. TIMETABLE OF EVENTS	351
	(a) Narrative	351
	(b) Pattern of plate development	354
	(i) Plate divergence	354
	(ii) Plate convergence	354
	3. CINEMATOGRAPH	354
	4. DISCUSSION	363
	5. PRESSING UNANSWERED QUESTIONS	364

References   365

Indexes
   AUTHOR INDEX   403
   SUBJECT INDEX   413

I. INTRODUCTION

1. GENERAL

Geological and geophysical exploration has reached the stage of knowing fairly well the uppermost few kilometres of the Australian landmass, and the salient features of the offshore margins and oceanic perimeter. Historical analysis depends on precise dating, and during the last decade palynological methods have increased their resolving power for dating the formerly intractable non-marine formations that abound in the latter half of the Phanerozoic. Furthermore, the proliferation of radiometric dates of igneous rocks, exemplified by the 400 odd for the Cainozoic basalts of eastern Australia, means that attempts at correlating igneous and sedimentary events rest on an increasingly reliable basis.

Our work is founded on actualism. Fold belts were once highlands, and sedimentary basins lowlands. We believe that fold belts and basins can be understood best through understanding highlands and lowlands that exist today, with which they can be compared and contrasted, not just by sketches and cartoons but by scaled maps and cross-sections, and quantitative time–space diagrams. We find that linked studies of palaeo-currents and sediment composition are the key to understanding the morphology and dynamics of ancient landscapes, and hence the tectonics. Actualistic analogues are drawn *inter alia* from modern New Guinea, Indonesia, and Africa/Arabia. Where suitable modern analogues are unavailable, we turn to the comparative tectonics of ancient systems, such as the transverse tectonic zones of North America and Australia.

This study begins and ends with the modern surface of Australia, the only living expression we have of Australian tectonic regimes. Since the modern is the standard, we work back from here, and only in the concluding cinematographic review do we follow the convenience of chronological narrative by telling events from the oldest (and dimmest) to the present day.

2. PHANEROZOIC TIME-SCALE

The first task in writing this book was to compile a time-scale to ensure uniformity in the citation of numerical equivalents of the biostratigraphical and magnetic-polarity scales, and ages cited, except where noted to the contrary, relate to Fig. 1. Had we started later, we could have saved ourselves the trouble, because during the final stages of writing, a spate of comprehensive time scales (Harland *et al*. 1982; Odin 1982; Palmer 1983, part of which is derived from the preceding two) have appeared, and another, *Geochronology and the geologic record*, by the Geological Society of London, is in press. In the event, we are gratified to find that all but a few of our numerical estimates coincide with the newly published scales. The only substantial difference is about the Jurassic–Cretaceous boundary: we cite it as 129 Ma, Harland *et al*. (1982) and Palmer (1983) as 144 Ma.

The time-scale has three components: (1) a biostratigraphical scale, shown in the columns on the right-hand side; (2) a numerical scale, in millions of years (Ma), derived from measurements of radioactive decay, on the left-hand side; and (3) a magnetic-polarity scale, with normal intervals shown by solid black, and reversed intervals by white, derived from seafloor-spreading magnetic anomalies, on the far left.

The biostratigraphical and magnetic scales are correlated against the radiometric scale expressed in equal intervals of millions of years.

In the text, we cite ages in whole numbers of millions of years and, for the Cainozoic, decimals. This does not imply blindness to the analytical (let alone, geological) errors. Rounding the numbers, while helping to express the imprecision, would lead to confusion. Therefore, each number must be understood as having a halo of error.

(a) Magnetic-polarity scale (Cainozoic and late Mesozoic only)

For the Cainozoic and Late Cretaceous (A1 to A34), the scale is from American Association of Petroleum Geologists (1981) derived from La Brecque *et al*. (1977), and updated by Mankinen and Dalrymple (1979) to include the new constants used in K–Ar dating. On the other side of the Cretaceous normal polarity interval (KN), part of which is distinguished by an oblique outline, the scale is from Larson and Hilde (1975) (M0–M25) and Cande *et al*. (1978) (M26–M29). The decrease in dipole magnetic field intensity from 140 to 160 Ma (Cande *et al*. 1978), and the older

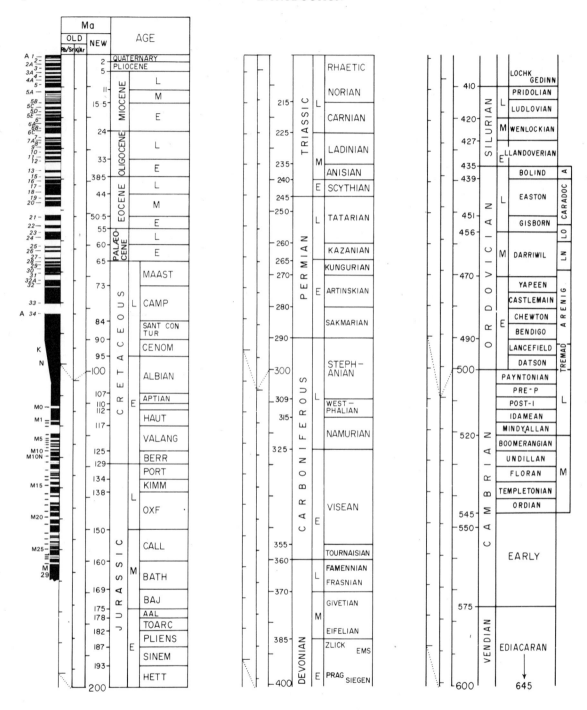

Fig. 1 Phanerozoic time-scale.

part of the land-based magnetic polarity scale (McElhinny 1978) are not shown.

(b) Radiometric and biostratigraphical scales

(i) Cainozoic and Mesozoic

1. Constants used for calculation of K–Ar and Rb–Sr ages:

	Old	New (IUGS)
$^{40}K/K$	1.19×10^{-4} mol/mol	1.167×10^{-4} mol/mol
$\lambda_\beta^{40}K$	$4.72 \times 10^{-10} a^{-1}$	$4.962 \times 10^{-10} a^{-1}$
$\lambda_\epsilon + \lambda'_\epsilon {}^{40}K$	$0.585 \times 10^{-10} a^{-1}$	$0.581 \times 10^{-10} a^{-1}$
^{87}Rb	$1.39 \times 10^{-11} a^{-1}$	$1.42 \times 10^{-11} a^{-1}$

From Steiger and Jager (1978) and Dalrymple (1979). Scale of K–Ar ages converted from old to new constants by multiplying by factors of 1.0268 to 1.0166 according to the table given by Dalrymple (1979), and of Rb–Sr ages converted from old to new constants by multiplying by the simple ratio of 1.39/1.42 ($= 0.979$). Dotted lines at each round hundred Ma show the discrepancy between the various constants.

Estimates of the analytical precision that exceed a few per cent are given in the text.

2. Cainozoic (0–65 Ma) part of scale from Abele (1976, table 8.1) converted to new constants by the table of Dalrymple (1979).
3. Cretaceous to Oxfordian (65–138 Ma) part of scale from Odin (1978).
4. Oxfordian to Triassic (138–200 Ma) part of scale from Van Hinte (1976) converted to new constants.
5. Triassic (200–245 Ma) part of scale from Webb (1981).

(ii) Palaeozoic

1. Based on Crook and Powell (1976), with the constants of Armstrong and McDowell (1975) and Armstrong (1978b):

$^{40}K/K$ 1.18×10^{-4} mol/mol
$\lambda_\beta^{40}K$ $4.905 \times 10^{-10} a^{-1}$
$\lambda_\epsilon + \lambda'_\epsilon {}^{40}K$ $0.575 \times 10^{-10} a^{-1}$
^{87}Rb $1.42 \times 10^{-11} a^{-1}$

The Rb constant is the same as the new IUGS constant, and the K/Ar constants are effectively the same.

2. The period boundaries are modified from those of Armstrong (1978b, table 6) by placing the Permian/Triassic boundary at 245 Ma (Webb 1981), the Carboniferous/Devonian boundary at 360 Ma (Roberts 1981), the Devonian/Silurian boundary at 410 Ma (Owen and Wyborn 1979), the Silurian/Ordovician boundary at 435 Ma (Lanphere *et al.* 1977), and the Ordovician/Cambrian boundary at 500 Ma, a compromise from Webby *et al.* (1981) between the two extremes suggested by Ross *et al.* (1978). Boundaries within periods have been adjusted from these dates using the scale given by Crook and Powell (1976), except those within the Permian, in which the Kungurian and later stages are adjusted according to Waterhouse (1978) (see Table 10), the Carboniferous, from Roberts (1981), the Ordovician, from Churkin *et al.* (1977) and Webby and Packham (1982), and the Cambrian, from Cook (1982), with the Middle and Late Cambrian expanded linearly to take account of the 500 Ma Ordovician/Cambrian boundary.

3. The Cambrian/Precambrian (Vendian) boundary is at 575 Ma, and the Vendian extends back to 680 Ma (Walter 1972), and the Ediacaran to 645 Ma (Cowie and Cribb 1978) or the Ediacarian to 670 Ma (Cloud and Glaessner 1982). While sympathetic to Cloud's (in Cloud and Glaessner 1982) inclusion of the Ediacarian in the Phanerozoic, we observe the original definition of the Phanerozoic as starting in the Cambrian.

(c) Explanation

Subdivisions not shown are:

		Ma
Cainozoic		0–65
Quaternary		0–2
Tertiary	Neogene	2–38.5
	Paleogene	38.5–65
Mesozoic		65–245
Palaeozoic		245–575
Precambrian		>575.

Abbreviations:

[General]
L   Late
M   Middle
E   Early

[Specific (in stratigraphical order)]

MAAST	Maastrichtian
CAMP	Campanian
SANT	Santonian
CON	Coniacian
TUR	Turonian
CENOM	Cenomanian
BARR	Barremian
HAUT	Hauterivian

VALANG	Valanginian	EASTON	Eastonian			
BERR	Berriasian	GISBORN	Gisbornian			
PORT	Portlandian	DARRIWIL	Darriwilian			
KIMM	Kimmeridgian	YAPEEN	Yapeenian			
OXF	Oxfordian	CASTLEMAIN	Castlemainian			
CALL	Callovian	CHEWTON	Chewtonian			
BATH	Bathonian	BENDIGO	Bendigonian			
BAJ	Bajocian	LANCEFIELD	Lancefieldian			
AAL	Aalenian	DATSON	Datsonian			
TOARC	Toarcian	PRE-P	Pre-Payntonian			
PLIENS	Pliensbachian	POST-I	Post-Idamean.			
SINEM	Sinemurian					
HETT	Hettangian					

In those figures that require extreme reduction, the following abbreviations are used.

ZLICK	Zlichovian				
EMS	Emsian				
PRAG	Pragian	Cz	Cainozoic	ᴦ	Triassic
SIEGEN	Siegenian	Q	Quaternary	P	Permian
LOCHK	Lochkovian	Pl	Pliocene	C	Carboniferous
GEDINN	Gedinnian	M	Miocene	D	Devonian
A	Ashgill	Ol	Oligocene	S	Silurian
LO	Llandeilo	E	Eocene	O	Ordovician
LN	Llanvirn	Pa	Palaeocene	Є	Cambrian
TREMAD	Tremadoc	K	Cretaceous	p-Є	Precambrian
BOLIND	Bolindian	J	Jurassic		

II. AUSTRALIA'S GLOBAL SETTING

1. PRESENT GLOBAL SETTING

(a) Introduction

Australia occupies much of the eastern part of the Indo-Australian Plate (Fig. 2), so-called, in preference to either Indian or Australian Plate, because it is occupied jointly today by India and Australia; as shown below, from 128 Ma to 44 Ma, India was the only continent on the plate we call the Indian Plate, as, from 95 Ma to 44 Ma, Australia was the only continent on the Australian Plate; and the coalescence of the Indian and Australian Plates by the cessation of spreading along their divergent plate boundary 44 Ma ago formed the Indo-Australian Plate. Today, the Indo-Australian Plate has a consuming boundary on the north and east, and an extensional boundary on the south and west.

The continental lithosphere of Australia–New Guinea is girt by the lithosphere of divergent oceans on its western, southern, and eastern margins, and of convergent oceans on its northern margin from Timor to the Papuan Peninsula (Figs. 3 and 4). The convergent boundary is marked by high relief (3 km above sea level in Timor and 5 km in New Guinea), intense earth movement, seismicity, and scattered volcanicity; south of this boundary, as befits its intra-plate position, Australia is low-lying (relief of 2.2 km in the south-east, little more than 1 km elsewhere), and quiescent except for diffuse seismicity in the south-west, in the central-west at 22° S, 127° E, in the south about longitude 138° E, and in the south-east, and two areas of volcanicity, one in north-east Queensland, the other on the southern margin about longitude 141° E. Outside these areas, earth movement is presumably very slow.

The submarine features of the Australian divergent margins have a certain bilateral symmetry about the meridian half-way across Australia (Fig. 3): complexes of marginal plateaux and abyssal plains backed by a

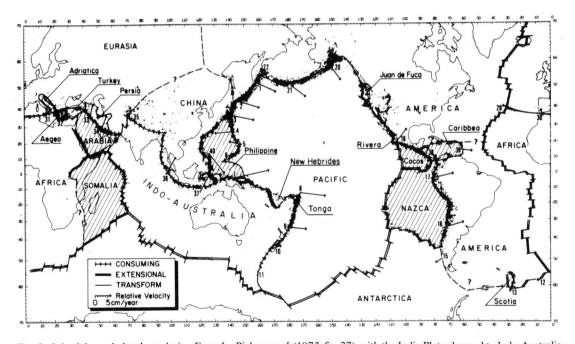

Fig. 2. Seismicity and plate boundaries. From Le Pichon *et al.* (1973, fig. 27), with the India Plate changed to Indo-Australia.

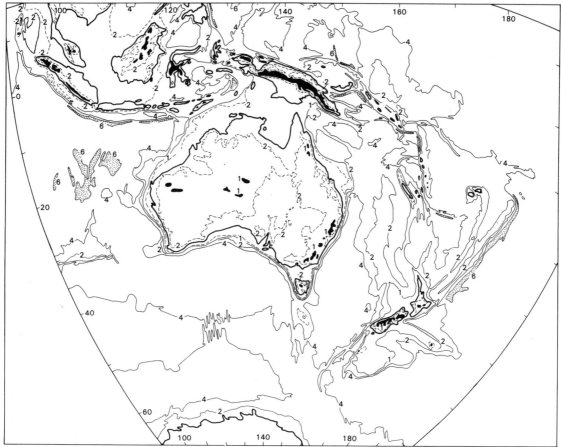

Fig. 3. Morphology (contours in km) of Australia and neighbouring continents and oceans. From American Association of Petroleum Geologists (1978). Lambert azimuthal equal-area projection.

broad shelf in the north, and narrow margins expanding into long appendages (Naturaliste Plateau and Tasmania/South Tasman Rise) in the south. This symmetrical pattern extends into the arrangement of seafloor-spreading magnetic anomalies (Fig. 4), in that the azimuth of the anomaly set off the north-west is reflected by that off the north-east, as is that off the south-west by that off the south-east, while the set south of Australia is crossed at right angles by the line of symmetry. The symmetry does not extend beyond the geometry, and the ages of these reflected anomaly sets are totally different.

The Eastern Highlands are a relic of an early phase of interaction with the Pacific Plate, as the ridges and oceanic basins of the adjacent South-west Pacific are the more modern ones (Fig. 4).

Plate divergence along the north-western and western margins started 160 Ma and 128 Ma ago, and led to a complex morphology of marginal plateaux and intervening abyssal plains. Since 95 Ma ago, the original pattern of spreading has changed twice, so that these margins are no longer related to the present spreading ridge, in contrast to the southern margin.

Plate divergence along the southern margin is reflected in its simple, low-lying morphology. Very slow divergence started 95 Ma ago, and rapid divergence 44 Ma ago, but with an unchanged pattern of spreading. The southern margin and its Antarctic counterpart are therefore related to the present spreading ridge, and a morphological depression along the ridge can be traced north and south into the continents.

(b) Australian–Antarctic Depression

The Australian–Antarctic Depression (Fig. 5) (Veevers 1982b) straddles the South-east Indian Ocean Ridge (SEIR) and extends into Australia and Antarctica. The

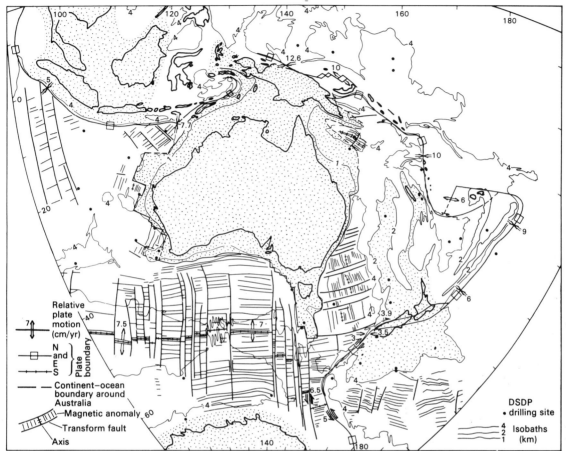

Fig. 4. Plate-Tectonic map of Australia and neighbouring continents and oceans. From American Association of Petroleum Geologists (1981). Same projection as Fig. 3. Continent-backed lithosphere of Sundaland, Australia–New Guinea, New Zealand–Lord Howe Rise, and Antarctica distinguished by stippling. Oceanic lithosphere left clear, with 2 and 4 km isobaths. Continent–ocean boundary round Australia marked with heavy lines. Convergent and strike-slip plate boundaries on north and east indicated by boxes along full line, and divergent boundary on south by full line with sea-floor-spreading magnetic anomalies detailed in Figs. 28 to 30, and 33, and modelled in Fig. 26. Double lines in New Zealand indicate zone of deformation.

region of Wilkes Land of East Antarctica, the South-east Indian Ocean, and southern Australia has two principal morphological aspects. The dominant latitudinal aspect is the ocean basin flanked by continents, all symmetrically disposed about the east-trending SEIR, and brought about by seafloor spreading during the past 95 Ma. The subordinate longitudinal aspect is a depression, some 15×10^6 km² in area, disposed about a north-trending axis of symmetry that passes indifferently across Antarctica, the South-east Indian Ocean, and Australia (double line in Fig. 5). Most clearly expressed in the crest of the SEIR at the Australian–Antarctic discordance (Fig. 6; Weissel and Hayes 1974), this axis is traceable through the flanks of the ridge, shown by the shape of the 7 and 29 Ma isochrons, and the Australian (and, presumably, the Antarctic) continent–ocean boundary (COB) to a distance of about 1000 km in Australia and 1500 km in Antarctica.

As shown in Figure 5, the depression is flanked (wiggly line) on the east by the Eastern Highlands (EH), Tasmania, the South Tasman Rise (STR), a high point on the SEIR at long 150° E, the submarine ridge surmounted by the Balleny Islands (BI), and the Trans-antarctic Mountains (TAM); on the west, it is flanked less distinctly by the sub-ice Gamburtsev Mountains (GM), a long spur delimited by the 0 m contour that encloses Mount Sandow (MS), a high point on the SEIR at long. 95° E, and a spur to the north-east, and the south-east flank of the Great Western Plateau

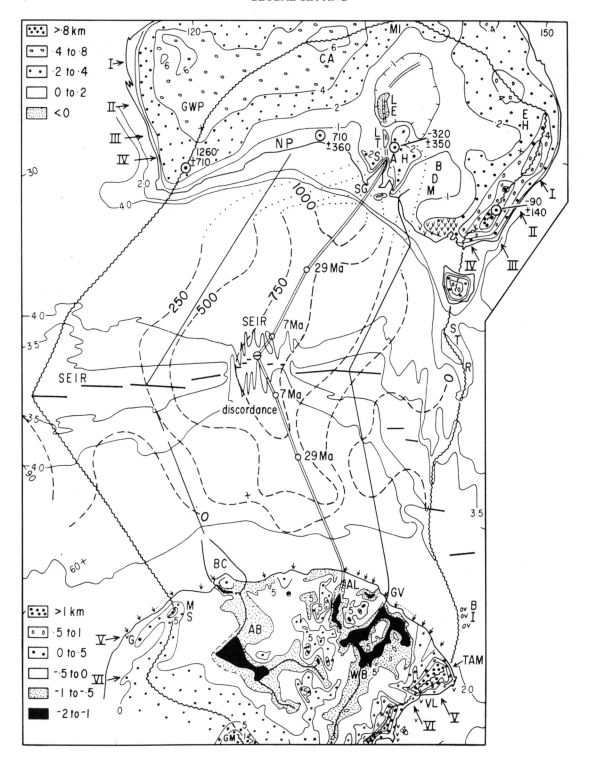

(GWP); and on the north by central Australia (CA), and the Mount Isa block (MI). The axis, shown by a double line, extends through Lakes Eyre (LE) and Torrens (LT), Spencer Gulf (SG) within the South Australian Highlands (SAH), across the flank of the SEIR, with datum points at the 7 and 29 Ma isochrons (Weissel and Hayes 1974), to Adelie Land (AL), which contains a trough 2.23 km below sea level between highlands (Steed and Drewry 1982). Two subsidiary axes (single lines) can be defined: on the west, an axis extends from the Nullarbor Plain (NP) across the SEIR to the Budd Coast (BC) and the Aurora sub-ice basin (AB); on the east, another axis passes through the Murray-Darling basin (MDB) to George V Land (GV) and the Wilkes sub-ice basin (WB). Modern (<5 Ma) volcanics (V's) are confined, except at the crest of the SEIR and at Gaussberg (G) (A. Gleadow *in* Sheraton and Cundari 1980), to the eastern flank of the depression, in South-east Australia (Aziz-ur-Rahman and McDougall 1972), the Balleny Islands (Kyle *et al.* 1979), and Victoria Land (VL) (Armstrong 1978a).

Profiles of the depression (Fig. 6) have a first-order trough shape, with secondary crests and troughs flanking the median axis on the continent. A tendency to greater amplitude and shorter wavelength of relief for the eastern boundary compared to the western is also characteristic of the continental profiles. The Antarctic profiles have an amplitude much greater than that of the others, but without its maximum cover of 4 km of ice, which has appeared only since the mid-Tertiary (Drewry 1976), the deepest parts of the depression in Antarctica would rebound isostatically about 1 km ($4 km \times (0.9/3.3)$), shown by the broken lines in Fig. 6, to bring the axis to or above sea level, and the highlands about the axis a kilometre or so higher. The crest of the Transantarctic Mountains, much of which is barely covered by ice, would rebound much less. With these adjustments, the discrepancies between Antarctica and Australia are reduced, but the Antarctic profiles remain with a much greater relief. The outcrop that rises above sea-level along parts of the coast of East Antarctica (shown by short arrows in Fig. 5) is probably due to its unloaded state in contrast to the loaded state of the depressed interior (Warren 1965), and has no counterpart in Australia.

Weissel and Hayes (1974) noted the correspondence of the Australian-Antarctic discordance with a negative satellite free-air gravity anomaly (FAA), which they inferred to be due to downward convecting currents in the asthenosphere. The extension of this correspondence northward and southward (Fig. 7) through the residual depth anomalies (Fig. 5) to cover the entire depression is taken to indicate dynamic control of the depression. We show in Chapter V how this dynamic state has persisted back at least to the Mesozoic.

Wellman (1982) compared continental and oceanic values of altitude anomalies: 'A lack of systematic variation across the region indicates that altitude anomalies are unlikely to be due to asthenospheric forces on the base of the lithosphere as these would be continuous across lithosphere boundaries', but this is contradicted by the four values of altitude anomalies across southern Australia (C, A, E, and F in fig. 2 of Wellman 1982), shown in Fig. 5. The trend of these altitude anomalies follows that of the oceanic anomalies, with the minimum of -320 ± 350 m near the axis of the Australian-Antarctic Depression, flanked by higher values on the east and west, so confirming the conclusion that the Depression is determined by dynamic asthenospheric forces on the base of the lithosphere.

Fig. 5. Morphology of the Australian-Antarctic Depression. Australia shown by contours (km) of $1° \times 1°$ area mean rock altitude (Wellman 1979b) to afford direct comparison with the ice-smoothed topography of Antarctica; South-east Indian Ocean by the 3.5 and 4.0 km isobaths (Efimova 1974) and the crest of the South-east Indian Ocean ridge (SEIR); and Wilkes Land, Antarctica, by sub-ice bedrock contours (km) (Drewry 1975a; Johnson *et al.* 1980b; Masolov *et al.* 1981; Steed and Drewry 1982; Allen and Whitworth 1970; Walker 1966). Residual depth anomalies (m) of the south-east Indian Ocean (broken lines) from Watts and Daly (1981), interpolated by dotted lines to the continental values across southern Australia (Wellman 1982). Coastal outcrop in Wilkes Land west of the Transantarctic Mountains (Bentley 1962) shown by short arrows. Modern (<5 Ma) volcanics (Aziz-ur-Rahman and McDougall 1972; Armstrong 1978a; Kyle *et al.* 1979) shown by Vs. Axis of depression shown by double lines, subsidiary axes by single lines, and flank crests by wiggly lines. Circles along axis indicate datum points along isochrons. Arrow tips indicate location of profiles of Fig. 6. Abbreviations explained in text. Lambert equal-area projection.

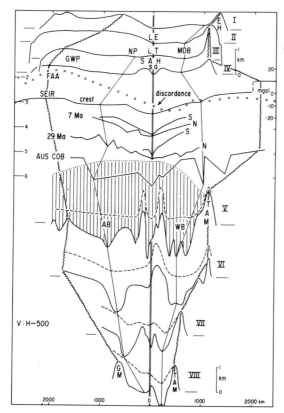

Fig. 6. Profiles of the modern Australian–Antarctic Depression, located on Figs. 5 and 153, aligned along the main axis (double line) with subsidiary axes shown by full lines, and the perimeter by wiggly lines. On continental profiles (I–VIII), sea level is marked by horizontal lines, and the vertical scale by a bar on IV and VIII. The oceanic profiles show the crest of the South-east Indian Ocean ridge (SEIR) with the satellite free-air gravity anomaly (open circles, Anderson et al. 1973), the depth of the 7 and 29 Ma isochrons (from Weissel and Hayes 1974), and the continent–ocean boundary (COB) of southern Australia (from Veevers 1981, 1982a, extended from Cooney et al. 1975, fig. 3, and Kennett et al. 1975, pp. 226, 272). In Antarctica (profile V only), the ice sheet between the surface (dotted line) and bedrock (full line) is shown by vertical lines. The bedrock surface adjusted to its pre-ice elevation after isostatic rebound is shown by broken line. Abbreviations given in text.

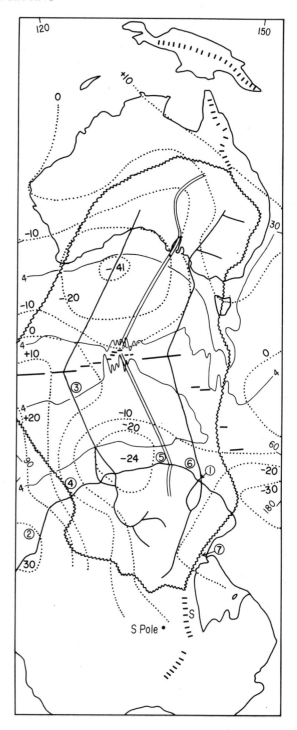

Fig. 7. Map of the extended region showing the correspondence of negative satellite free-air gravity anomalies (dotted lines) with the Australian–Antarctic Depression. Gravity anomalies, in mGal, from Anderson et al. (1973) to lat. 70° S, extended southward from Gaposchkin and Lambeck (1971). Morphology outlined by the perimeter (wiggly line), main axis (double line), minor axes (full lines), and the 4 km isobath. New Guinea Highlands and extensions of Eastern Highlands of Australia and Transantarctic Mountains shown by lines of bars. Circled numbers indicate occurrences of *in situ* (1) and reworked (2-7) microfossils. S=Scott Glacier. Lambert equal-area projection.

2. PAST GLOBAL SETTINGS

(a) Continental palaeomagnetism (by B. J. J. Embleton)

(i) Introduction

Methods that provide quantitative estimates of location through time are required to understand the geological evolution of Australia in a global context. Analyses of palaeomagnetic data together with evidence from radiometric geochronology have enabled us to establish continental associations and their drift history during the Phanerozoic.

The palaeomagnetic method comprises several techniques, which cover sampling, measurement, and analysis, and has been described by a number of authors, for example Irving (1964), Collinson *et al.* (1967), McElhinny (1973), and most recently by Collinson (1983). Improvements in the precision of measuring and equipment sensitivity have been achieved through the development of cryogenic rock magnetometers (Goree and Fuller 1976). Refined techniques for the analysis of raw palaeomagnetic data (for example Zidjerveld 1967; Kirschvink 1980; Schmidt 1982) have also played an important role in achieving objective resolution of complex, multicomponent magnetizations, and the identification of secondary magnetizations, referred to as overprints.

The results of palaeomagnetic studies are often presented in a form independent of sampling locality by means of the palaeomagnetic pole. This is a time-averaged estimate of the geomagnetic pole according to the axial-geocentric dipole-field model. An apparent polar-wander path (APWP) represents a chronologically ordered set of poles; when obtained for a single land mass, it describes the motion of the land mass with respect to the Earth's axis of rotation.

(ii) Australia's drift history

The APWP for Australia has been constructed from palaeomagnetic information contained in a review by Embleton (1981) with supplementary results for the late Precambrian and Early Palaeozoic from McWilliams and McElhinny (1980) and Klootwijk (1980), for the mid-Palaeozoic by Goleby (1980), for the Mesozoic by Schmidt and Embleton (1981) and Schmidt (1982), and for the Cainozoic by Embleton and McElhinny (1982). Throughout most of the Phanerozoic, the palaeomagnetic pole positions form distinct groups which provide a convenient, natural hierarchy for the purpose of determining pole averages. The results are listed in Table 1 together with an estimate of the time range to which they apply. The APWP shown in Fig. 8 is for the time interval 600 Ma

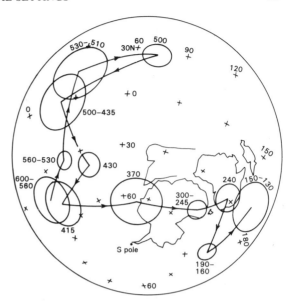

Fig. 8. Apparent polar wander with respect to Australia from the late Precambrian to the Late Jurassic, plotted on a reconstruction of eastern Gondwanaland. Linear polar projection. Ages in Ma.

to 130 Ma, during which eastern Gondwanaland (Australia, India, Antarctica) remained intact. The following description of the continental drift of the three land masses recognizes the fact that the present Pacific margins of Australia and Antarctica should not be regarded as integral parts of the shield regions for the whole of pre-Cretaceous time. It follows therefore, that in order to interpolate from a particular locality to the whole of a land mass, we must be able to assert confidently that there has been no relative motion between the two geological entities since that respective time. In the case of Australia, the tectonic framework of the Tasman Fold Belt is yet to be investigated further in terms of assessing its integrity with the main platform. Similar work in Antarctica (for example Watts and Bramall 1981) has shown that its Pacific margin (western Antarctica) comprises a number of displaced terrains.

From the late Precambrian (600 Ma) to the end of the Silurian (415 Ma), the pole track defines an approximate great-circle loop with respect to Australia. From the Silurian through the Late Carboniferous (300 Ma), the South Pole and Australia underwent relative convergence as the pole tracked across

TABLE 1. *Summary of palaeomagnetic results for the Phanerozoic of Australia*

| Ma | Pole position | | | Palaeolat. | Palaeodec. |
	Latitude	Longitude	A95	(For locality 24° S, 134° E)	
600–560	42.2° S	350.6° E	12.0°	16° N	207°
560–530	31.6° S	16.5° E	4.5°	8° N	230°
530–510	14.8° N	30.4° E	15.0°	18° N	279°
500	27.0° N	72.0° E	7.5°	11° S	307°
500–435	0.2° S	21.8° E	15.0°	20° N	260°
430	39.1° S	34.4° E	7.0°	8° S	231°
415	48.4° S	356.3° E	12.0°	8° N	207°
370	62.8° S	76.8° E	14.0°	36° S	208°
300–245	47.7° S	137.4° E	5.5°	66° S	174°
240	30.0° S	147.0° E	8.0°	77° S	120°
190–160	51.0° S	182.0° E	4.0°	45° S	139°
150–130	20.0° S	164.0° E	13.0°	62° S	88°
128*	33.0° S	162.0° E	–	64° S	117°
115*	42.0° S	161.0° E	–	61° S	135°
105	53.0° S	158.0° E	6.5°	56° S	154°
90	55.0° S	147.0° E	5.0°	58° S	166°
80*	51.0° S	134.0° E	–	63° S	180°
65*	56.0° S	120.0° E	–	56° S	194°
53*	66.0° S	118.0° E	–	47° S	189°
45*	67.0° S	114.0° E	–	45° S	191°
25*	77.0° S	110.0° E	–	36° S	186°
5*	87.0° S	86.0° E	–	26° S	182°
Present	90.0° S	–	–	24° S	180°

*Interpolated pole position based on estimates of Tertiary APW from Embleton (1981) and Embleton and McElhinny (1982).

A95 is the semi-angle of the cone of confidence about the mean calculated at the 95 per cent probability level (P = 0.05) for a Fisher distribution (Fisher 1953).

Antarctica. Thus between 600 Ma and 300 Ma, apparent polar wander with respect to Australia may be described in three phases:

(i) the main Australian platform experienced counter-clockwise rotation of about 90° between 600 Ma and 500 Ma, followed by
(ii) a similar amount of clockwise rotation between 500 Ma and approximately 400 Ma, and
(iii) Australia drifted directly into south polar latitudes between 400 Ma and 300 Ma, more or less maintaining its azimuthal orientation.

The component of drift experienced during the Early Palaeozoic (600–415 Ma) may be modelled using an Euler pole near Australia.

Throughout Permo-Carboniferous and Early Triassic times (300–225 Ma), the South Pole remained positioned close to, or within, the Australian continent. During the remainder of the Triassic, the Jurassic, and Early Cretaceous (see also Fig. 9) the South Pole defined a loop eastwards to the position of present day New Zealand, then northwesterly into the Coral Sea region whence it tracked approximately southerly to reach a position south-east of Tasmania

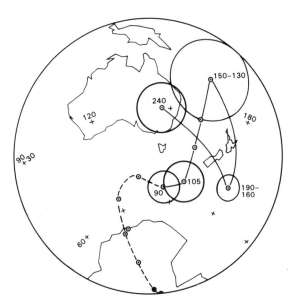

Fig. 9. Mesozoic and Cainozoic APW for Australia. The dashed part of the APWP, based on an analysis of laterites, is poorly calibrated. Ages in Ma.

by mid-Cretaceous time. This Mesozoic APW loop is found also from Africa (Hargraves and Onstott 1980) and South America (Schult *et al.* 1981). The time interval (300–95 Ma) represented here is of course significant in terms of the breakup history of Gondwanaland. Although Antarctica remained juxtaposed with Australia until about 95 Ma, India had begun its northward drift, having separated from Antarctica and Australia at approximately 130 Ma.

Since the mid-Cretaceous (95 Ma) the South Pole moved westerly, south of the Bight, from longitude 150° to 120°, and, during the Tertiary, the pole tracked southwards more or less directly towards its present location.

Some problems remain to be solved. For example, refinement of the palaeomagnetic record for the Tertiary with suitable calibration from radiometric studies is required. The form of the APWP has been based on an analysis of results from laterites and weathered profiles (Embleton and McElhinny 1982). The approach is one which makes use of the fact that secular variation of the geomagnetic field is generally averaged within a single rock sample. On a regional scale, crustal weathering and laterite-forming processes appear to have occurred as protracted events throughout the Tertiary and although the age of magnetization is not precisely known, the results taken as a whole define the average form of the APWP for this period.

The configuration of the geomagnetic field during the mid-Palaeozoic is also poorly understood. This corresponds to the period of Australia's drift history when its motion transforms from essentially one of changing azimuthal orientation to a migration towards the South Pole. A considerable volume of palaeomagnetic data is available from the Tasman Fold Belt (Goleby 1980) but whether it relates generally to the main Australian platform or specifically to discrete tectonic units within the fold belt, for example, as described by Embleton *et al.* (1974), is yet to be resolved. A dearth of information for this period from the other members of Gondwanaland rules out unequivocal models of APW.

The drift history of the continents since the breakup of Gondwanaland has also been modelled from the results of seafloor spreading, as shown in the next section. Continental distributions for specific anomaly times have provided a framework against which we have been able to check the consistency of results from studies of continental palaeomagnetism. Thus Embleton and McElhinny (1982) have demonstrated that the palaeomagnetic results for the southern hemisphere continents and India are consistent with the drift history as determined from Norton and Sclater's (1979) analysis of marine magnetic anomalies. This provides us with an independent assessment of the correctness of our magnetic field model used for interpreting palaeomagnetic data. In a similar fashion, reconstructions of the supercontinent based on matching the shape of the continental margins and geological trends, for example, Bullard *et al.* (1965) and Smith and Hallam (1970), have been used as a backdrop for comparing results from different continents. The general form of these reconstructions has been supported by the seafloor-spreading data. We are therefore confident in applying the axial-geocentric dipole-field model to the interpretation of directional data for times older than the oldest seafloor.

Comparisons of continental-margin morphology as a basis for reconstruction may not yield unique solutions nor yet be susceptible to further detailed constraint from consideration of regional geological information. Continental palaeomagnetism offers the potential to provide quantitative assessments of the effects of subtle rearrangements of land masses; see, for example, reconstructions by Smith and Hallam (1970), Norton and Sclater (1979), and Powell *et al.* (1980d). Analyses carried out with the specific objective of assessing the goodness of fit of different Gondwanaland reconstructions (Embleton *et al.* 1980 and Thompson and Clark 1982) have been unsuccessful in demonstrating the superiority of any single reconstruction. The reason is that the palaeomagnetic data are not sufficiently precise to provide the degree of resolution sought.

(iii) Palaeolatitude of Australia

The field model used for interpreting palaeomagnetic data relates the inclination of magnetization (I) to the latitude λ according to the dipole formula:

$$\tan I = 2\tan \lambda.$$

This model is the appropriate one for describing the time-averaged behaviour of the Earth's magnetic field so that a palaeomagnetic pole position may be regarded as an analogue of the palaeogeographic pole, that is, defining the Earth's ancient axis of rotation. Hence, the palaeomagnetic equator and the palaeogeographic equator coincide, as do the palaeomagnetic and palaeogeographic latitudes. An estimate of the palaeoinclination, measured from a rock section for a particular site, will therefore yield the latitude in which those rocks formed. The declination value defines the meridian through the site, thereby fixing the azimuthal orientation. Palaeolatitude diagrams may then be constructed. Alternatively, when a considerable data

GLOBAL SETTING

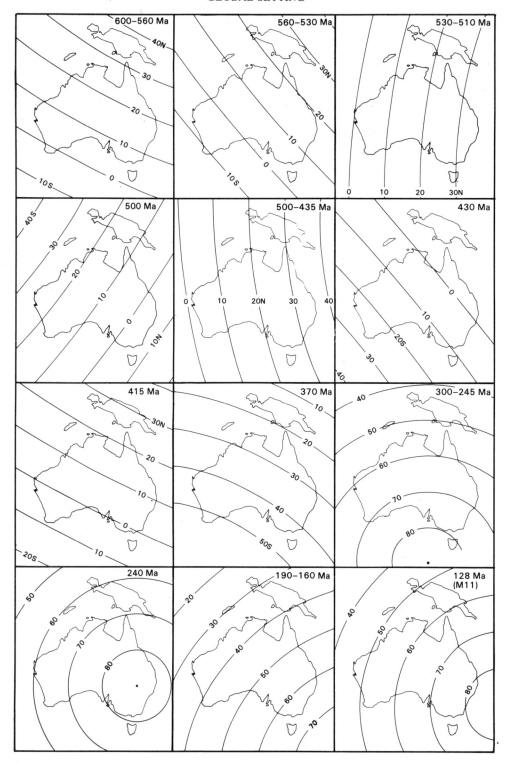

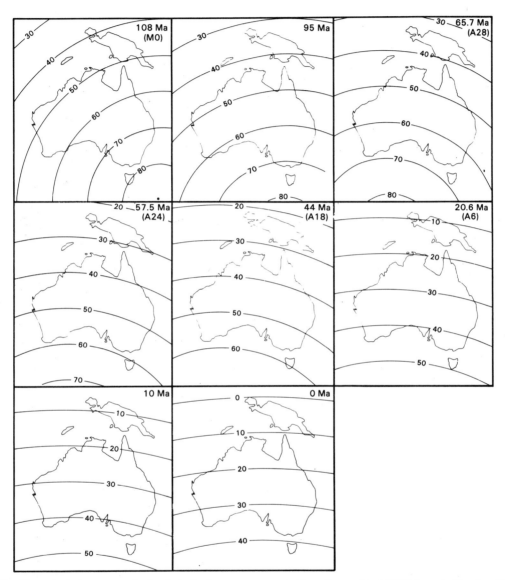

Fig. 10. Chronological sequence of palaeolatitude diagrams constructed from the results of APW shown in Figs. 8 and 9. Pole position data listed in Table 1.

base is available, as is the case for Australia, averaged palaeomagnetic poles are transformed to yield the palaeolatitudes.

We have already described APW relative to Australia. Conversely, the pole may be regarded as fixed and the continent in relative motion, thus describing the palaeolatitude in chronological sequence. The APWP contains all of the information required in order to describe a continent's drift history, and this remains the commonest way to present palaeomagnetic results. However, the APWP is often not visually descriptive in terms of palaeolatitudes. A sequence of palaeolatitude diagrams therefore have been prepared based on the calibrated APWP, and appear in Fig. 10. The maps for the period 600 Ma through to 160 Ma correspond to the quasi-static intervals listed in Table 1, that is, times for which the continent may be considered to have remained approximately stationary with respect to the pole. For the period after 160 Ma, the palaeomagnetic data have been interpolated to yield pole positions that correlate with specific marine magnetic anomaly times. These estimates are also listed in Table 1. A reference locality within Australia, close to the geographical centre (24° S, 134° E), has been chosen in order to relate the change of latitude with time. This is shown in Fig. 11.

The palaeolatitude history of Australia was reviewed by Embleton (1973), and the present analysis, based on a considerable upgrade of that data base, confirms the general conclusion reached at that time, although with some improvement in detail. Probably the most contentious issue remaining is whether the late Precambrian glaciation occurred in low or high palaeolatitudes. Low-inclination magnetizations have been recorded from sediments genetically related to glaciations (for example Tarling 1974; Morris 1977;

and McWilliams and McElhinny 1980). However, none of the components has been unequivocally demonstrated to be primary in origin. To the contrary, Stupavsky *et al.* (1982) have shown that magnetizations reported as primary are, in fact, overprints. Until the thermal history of the rocks has been thoroughly investigated, it would seem reasonable to exercise caution in the interpretation of such data. Models of apparent polar wander suggest that Australia did occupy polar regions during the late Proterozoic (for example Klootwijk 1980 and Embleton 1981), but the calibration of the pole track requires further study.

For an interval of 200 Ma, including the latest Precambrian, Cambrian, Ordovician, and Silurian, Australia occupied equatorial latitudes. There is abundant evidence from the geological climatic indicators to support this conclusion (Quilty, below). Its western margin was adjacent to Greater India, the northern margin probably lay in New Guinea in much the same way as at present, the southern margin was juxtaposed with the Antarctic shield, and Australia's Pacific margin was undergoing a process of accretion. This general continental association was maintained through the following 100 Ma as Australia drifted into high latitudes. The effects of the Early Permian glaciation are clearly evident in the geological record. The continent remained in polar latitudes until the Early Triassic. During the Late Triassic and Jurassic, there was a relatively brief period (~50 Ma) when Australia lay between approximately 35° S and 65° S, prior to a return to polar latitudes during the Cretaceous. At this time the continental reconstruction that had pertained throughout the Palaeozoic and Mesozoic began to change in response to the late Mesozoic phase of seafloor spreading. By 130 Ma ago, eastern Gondwanaland started to disintegrate, with India separating from Antarctica/Australia (Markl 1974a; Larson 1977).

During the Late Cretaceous the South Pole migrated across Wilkes Land in East Antarctica with Australia drifting along a small circle path concentric about the pole. The eastern and southern continental margins of Australia were created at this time by seafloor spreading forming the Tasman Sea and the southeastern Indian Ocean as the Lord Howe Rise and Antarctica broke away. Early in the Tertiary Period seafloor spreading ceased in the Tasman Sea (Weissel and Hayes 1977), and Australia commenced its northward drift as the South Indian Ocean Ridge became more active (Cande and Mutter 1982). Relative drift between Australia and India ceased during the Eocene (Norton and Molnar 1977), though this did not affect Australia's motion northwards throughout the remainder of the Cainozoic.

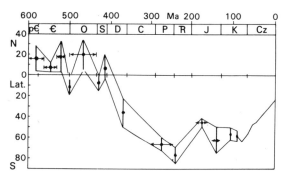

Fig. 11. The changing palaeolatitude of Australia with time. A central geographical location (24° S, 134° E) was chosen as a reference point. The arrows indicate the time intervals, and the vertical lines the errors.

(b) Oceanic palaeomagnetism (by B. D. Johnson and J. J. Veevers)

(i) Introduction: growth of the Indo-Australian Plate

On all sides but the north, the Australian continental block is flanked by a divergent oceanic lithosphere marked by seafloor-spreading magnetic anomalies (Fig. 4). Calibrated by deep-sea drilling, the magnetic anomalies range in age from Late Jurassic in the north-west, Early Cretaceous in the west, Late Cretaceous in the south and south-east, and Palaeocene in the north-east. The magnetic anomalies record the growth of the oceanic lithosphere between the former neighbours of Australia in Eastern Gondwanaland: with reference to directions from Australia today, Greater India on the west, Antarctica on the south, Lord Howe Rise/New Zealand Plateau on the east, and the Papuan Peninsula on the north-east. Eastern Gondwanaland (Figs. 12D, 13, 14) was reconstructed by rotations about the finite poles listed in Table 2, and reconstructions of the stages of growth of that part of the Indo-Australian Plate round Australia (Figs. 15–26) were made by rotating the continents about the stage poles listed in Table 3. Each map, except Figs. 16 to 20 which show the separation of India, is centred on an Australia fixed within the frame of Fig. 4, and shows the synthetic seafloor-spreading isochrons. The maps are presented in sequence from the oldest at 160 Ma – a

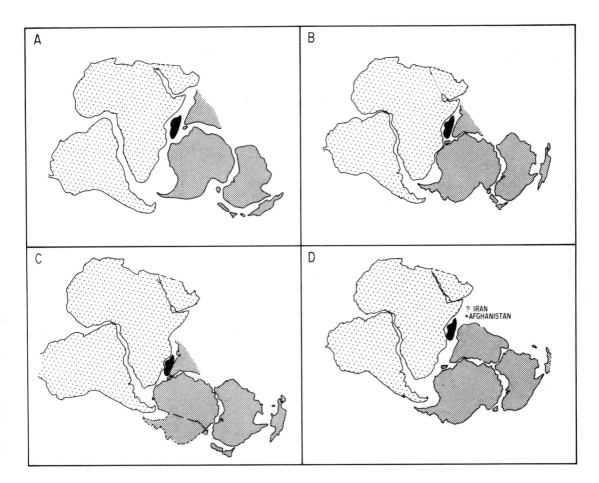

Fig. 12. Reconstructions of East and West Gondwanaland. Coastline shown only. East Gondwanaland (close stipple), West Gondwanaland (open stipple), and Madagascar (solid). A—Du Toit (1937). Sketch only. B—Smith and Hallam (1970). Lambert equal-area projection centred at 5° S, 35° E. C—Tarling (1972). Stereographic projection. Dashed line indicates postulated line of shear across Antarctica. D—Powell et al. (1980d). Same projection as B.

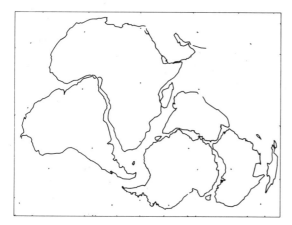

Fig. 13. Gondwanaland reconstruction (>160 Ma) of Veevers *et al.* (1980), modified in subsequent figures by loose fits of India and Australia to Antarctica.

reconstruction of part of Eastern Gondwanaland – to the present day, through steps of 30 Ma or shorter, and can be viewed as a set of cinematographic or time-lapse pictures of the growth of the eastern part of the Indo-Australian Plate. Palaeolatitudes of each stage are derived from the studies of continental palaeomagnetism described by Embleton in the previous section. Details of the reconstructions, and detailed maps showing the match of the synthetic isochrons with the observed isochrons of the magnetic anomalies and deep-sea drilling sites (Figs. 27–33) follow the set of reconstructions.

The reconstructions were computed using an interactive version, developed by Johnson, of R. L. Parker's HYPERMAP program. The maps have the same centre (35° S, 135° E) and projection (Lambert Azimuthal Equal-Area) as the Plate-Tectonic Map of the Circum-Pacific region (AAPG 1981), and Wellman and McCracken's (1979) Plate-Tectonics map in the BMR Earth Science Atlas of Australia.

(ii) 160 Ma: reconstruction of Eastern Gondwanaland (Figs. 12, 13, 14)

A 160 Ma reconstruction of Gondwanaland (Figs. 12D, 13), from Veevers *et al.* (1980), shows the assembly before breakup of South America and Africa, constituting Western Gondwanaland, linked through Madagascar with India, Antarctica, and Australia/Lord Howe Rise/New Zealand, constituting Eastern Gondwanaland. Western Gondwanaland separated from Eastern Gondwanaland about 160 Ma ago, at the same time as spreading started off north-west Australia; India, in the form of Greater India (Veevers *et al.* 1975) separated from Antarctica/Australia 128 Ma ago. The reconstruction of 160 Ma (Fig. 14) is therefore a pre-breakup fit. New information from the South-east Indian Ocean, involving a reinterpretation of magnetic anomalies (Cande and Mutter 1982) and a more precise location of the continent–ocean boundary (COB)

TABLE 2. *Finite poles and angles, relative to Australia, for the reconstruction of Eastern Gondwanaland (≥160 Ma), by closing India, Antarctica, Lord Howe Rise (LHR)/New Zealand (NZ), and the Papuan Peninsula to Australia. Positive values indicate north latitudes, east longitudes, and anti-clockwise rotations when viewed from above the pole*

Plate	Pole			References and Remarks
	Lat.°	Long.°	Angle°	
India	11.5	−170.2	64.76	Resolution of India to Antarctica (−1.5, 12.4, −89.7) and Antarctica to Australia (10.3, 32.7, 28)
Antarctica	10.3	32.7	28.0	Weissel *et al.* (1977), table III, pole 7, and Konig's (1981) closing angle of 28°
LHR/NZ	−14.0	142.0	−25.0	Weissel and Hayes (1977), table 1 and p. 82, with an additional 2° of rotation
Papuan Peninsula	−1	127	−5.54	Weissel and Watts (1979), table 1; closing angle of −5.54° to unite 62 Ma (old constants) isochrons

TABLE 3. *Stage poles and angles, relative to Australia, for the motion of India, Antarctica, Lord Howe Rise (LHR)/New Zealand (NZ), and the Papuan Peninsula from Australia. Positive values indicate north latitudes, east longitudes, and anticlockwise rotations*

Pole	Plate	Magnetic anomalies from	Magnetic anomalies to	Age (Ma) from	Age (Ma) to	Lat.°	Long.°	Angle°	Reference and remarks	Fig.
a	East Gondwanaland off NW margin	>M25	M11	160	128	−28	19	17.8	Pole calculated here from data in Heirtzler *et al.* (1978) and Powell (1978)	14
b1	India	M11	>M3	128	118	−56.5	−2.7	5.6	Amended from Johnson *et al.* (1980a). Total angle $(17.8°) \times (104/330) = 5.6°$	15
b2	India	>M3	<M0	118	108	−56.5	−2.7	5.2	Amended from Johnson *et al.* (1980a). Total angle $(17.8°) \times (96/330) = 5.2°$	16
b3	India	<M0 KN		108	95	−56.5	−2.7	7.0	Amended from Johnson *et al.* (1980a). Total angle $(17.8°) \times (130/330) = 7.0°$	17
c1	India		A34	95	81.7	−2	8	13.0	Amended from Johnson *et al.* (1980a):40/100 of total angle from 95 to 65.7 Ma	18
d	LHR/NZ		A34	95	81.7	−14.0	142.0	2.0	Extrapolated from fit pole of Weissel and Hayes (1977)	19
e1	Antarctica		A34	95	81.7	12.7	22.2	−1.17	Weissel *et al.* (1977), table III: pole 7* − 6 = e, total angle = −4.5°. Rotation is $(95 - 81.7)/(95 - 44) \times -4.5° = (13.3/51) \times -4.5° = -1.17°$. *Pole 7 with rotation of −28°	
c2	India	A34	A28	81.7	65.7	−2	8	19.5	Amended from Johnson *et al.* (1980a): 60/100 of total angle from 95 to 65.7 Ma	20
f	LHR/NZ	A34	A28	81.7	65.7	−17.0	142.0	17.24	Weissel and Hayes (1977), table 1: resolution of poles at A33 and A29, angle from A28 to fit	
e2	Antarctica	A34	A28	81.7	65.7	12.7	22.2	−1.41	Weissel *et al.* (1977), table III: pole e, rotation $(16/51) \times -4.5° = -1.41$	
g	LHR/NZ	A28	A24	65.7	57.5	−8.1	141.6	7.3	Weissel and Hayes (1977), table I: resolution of poles at A29 and A26, angle pro-rated for A28 to A24	21
h	Papuan Peninsula	A28	A24	65.7	57.5	−1	127	5.54	Weissel and Watts (1979): rotation of 5.54° units 62 Ma (old) isochrons	
i1	India	A28	A24	65.7	57.5	4	−3	10.93	Sclater and Fisher (1974): pro-rated rotation angle	
e3	Antarctica	A28	A24	65.7	57.5	12.7	22.2	−0.72	Weissel *et al.* (1977), table III: pole e, rotation = $(8.2/51) \times -4.5° = -0.72$	
e4	Antarctica	A24	A18	57.5	44.0	12.7	22.2	−1.19	Weissel *et al.* (1977), table III: pole e, rotation = $(13.5/51) \times -4.5° = -1.19$	22
i2	India	A24	A18	57.5	44.0	4	−3	10.67	Sclater and Fisher (1974): pro-rated rotation angle	
j	Antarctica	A18	A13	44.0	36.8	−0.3	34.8	−3.2	Weissel *et al.* (1977), table III: pole 6−5	23
k	Antarctica	A13	A6	36.8	20.6	9.3	33.6	−8.5	Weissel *et al.* (1977), table III: pole 5−2	24
l	Antarctica	A6	A5	20.6	10.0	19.2	32.7	−5.3	Weissel *et al.* (1977), table III: pole 2−1	25
m	Antarctica	A5	A1	10.0	0	9.7	36.5	−6.75	Weissel *et al.* (1977), table III: pole 1	26

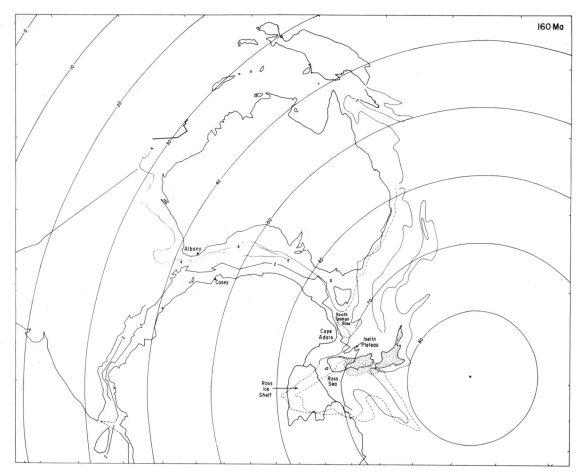

Fig. 14. 160 Ma: Greater India, including a rotated Sri Lanka, Australia, Antarctica, and New Zealand/Lord Howe Rise assembled in Eastern Gondwanaland. Western margin of Australia and north-eastern margin of Greater India defined by the 4 km isobath, and north-western margin of Australia by line of breakup at 160 Ma (estimated COB). The margins between India and Antarctica outlined by the 2 km isobath; between Australia and Antarctica by the 2 and 4 km isobaths, including the overlapping South Tasman Rise (dotted line of 2 km isobath) shown in its present position with respect to the rest of Australia; between south-eastern Australia (2 km isobath—broken line) and Lord Howe Rise/New Zealand by the 2 km isobath, and between north-eastern Australia and the Papuan Peninsula by the 2 km isobath. New Caledonia Basin shown in present open position. Palinspastic, pre-late Cainozoic configuration of New Zealand (stipple) and New Zealand Plateau (broken line) from Bradshaw *et al.* (1981). Note broad overlap of 2 km isobath (dotted line) over the Ross Sea and Ice Shelf. Palaeolatitudes from Embleton's palaeomagnetic determinations given above.

(Talwani *et al.* 1979), has led us to revise the fit of Antarctica and Australia, and correspondingly that of Antarctica and India. We defer carrying these revisions through the entire Gondwanaland fit for a later occasion. With these qualifications, Fig. 13 suffices as a general picture of Gondwanaland, the eastern part of which is updated in Fig. 14.

Except in the present-day map of Fig. 26, New Zealand is shown in two configurations: (1) the present coastline and 2 km isobath by full lines, and (2) a palinspastic outline, from Bradshaw *et al.* (1981), in broken lines, delineating the Alpine Fault Zone and south of it the displaced shoreline and 2 km isobath. Walcott (1978) argues that the lateral displacement and bending of formerly continuous linear features can be accounted for by extrapolating the present rate of shear back to the time of A5 or A6, to about 15 Ma ago. Concerning the overlap of New Zealand, in

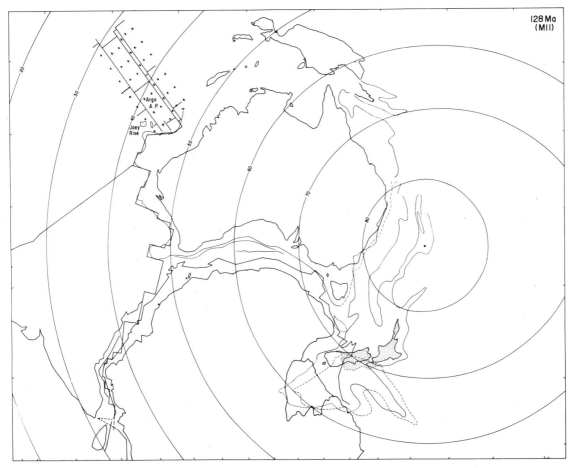

Fig. 15. 128 Ma (M11): seafloor spreading in the Argo Abyssal Plain, involving a ridge-jump to Australia in the southwestern-most compartment, and generation of the epilith of the Joey Rise, and appearance of the suture between Greater India and Australia/Antarctica. Active ridge crest shown by heavy line, and abandoned ridge crest by a line with bars.

particular the Campbell Plateau, and West Antarctica, we offer no contribution to the solution of this problem except to show the overlap within the frame of our maps. The overlap is inherent in the class of reconstructions that close the Tasman Sea from Weissel and Hayes's (1977) poles, and nothing short of a major reconstruction of New Zealand or Antarctica or both suffices to remove the overlap. The problem is discussed at length by Molnar *et al.* (1975), Weissel *et al.* (1977), Barron and Harrison (1979), and Stock and Molnar (1982).

Spreading in the marginal seas north of New Zealand (Fig. 26) is from Wellman and McCracken (1979) and Malahoff *et al.* (1982). Coleman (1980) summarizes the development of the South-west Pacific during the last 100 Ma, and Packham (1982) the evolution of the India–Pacific plate boundaries.

(iii) Revised fit of Antarctica and Australia (Fig. 14)

Talwani *et al.* (1979) located the COB off southern Australia at 130° E and off the adjacent part of Antarctica at the landward edge of the set of seafloor-spreading magnetic anomalies, at an approximate depth of 4 km, but did not show the revised fit of Antarctica and Australia beyond a single profile. Konig (1981) determined a new closing pole (1.5° N, 37.0° E) and rotation angle (28°) from the trend and extent of the oldest seafloor magnetic anomaly, A22, and Cande and Mutter (1982) have reinterpreted this anomaly as A34 or older, as detailed below in our account of the

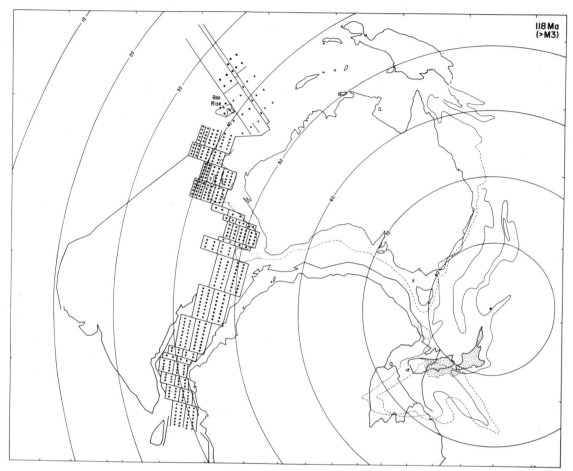

Fig. 16. 118 Ma (~M3): continued spreading in the Argo Abyssal Plain, involving generation of the Roo Rise epilith, and the first phase of spreading between Greater India and Australia/Antarctica, involving ridge-jumps (abandoned ridge crests shown by bars) and the generation of the epiliths of the Sonne Ridge and the Wallaby Plateau. Frame of map shifted westward to include all of India.

southern margin. For the new fit shown here, we use Weissel *et al.*'s (1977) pole of 10.3° N, 32.7° E (cf. Griffiths's (1974) pole of 10.8° N, 32.3° E) and Konig's (1981) closing angle of 28°, which is 2° smaller than that of Weissel *et al.* This fit differs from that of Griffiths (1974) and Weissel *et al.* (1977) only in being looser by 2°. Because the pole is about 90° distant, the loosening of the fit by 2° amounts to 210 km on the east, between the South Tasman Rise and the Iselin Plateau, and 220 km on the west between Albany and Casey. The slight overlap of the 2 km isobaths of the South Tasman Rise (shown by dotted lines in Fig. 14) and of the margin off north-east Victoria Land (Cape Adare) is attributable to southwestward extension of the entire South Tasman Rise isthmus during the later stages of opening. The depth of the suture ranges from 4 km and deeper in the Great Australian Bight and on adjacent Antarctica, as shown in Fig. 18, to 2 km between Tasmania and Victoria Land, and to between 2 and 3 km off Antarctica west of 120° E. The Balleny Islands and Seamounts, and the Adare Seamounts off Victoria Land (not shown) overlap Tasmania and the South Tasman Rise in the fit, and hence are inferred to be subsequent oceanic volcanic upgrowths or epiliths. Similar, but unspecified, fits are given by Deighton *et al.* (1976), Falvey and Mutter (1981), and Veevers (1981). Cande and Mutter's (1982) interpretation of A22 as A34 or older, which is bounded by the Creta-

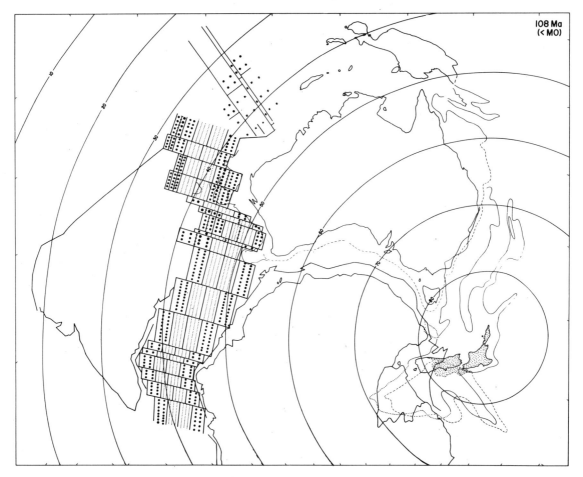

Fig. 17. 108 Ma (~M0): continued spreading between Greater India and Australia/Antarctica, involving ridge-jumps and generation of most of the Zenith Seamount.

ceous long normal polarity interval (KN), changes the inception of spreading from 54 Ma to 82 Ma or older; in Chapter IV we argue that spreading started at 95 Ma ago.

(iv) Revised fit of India and Antarctica/Australia

To conform with the loose fit of Antarctica and Australia, we have loosened the fit of India to Antarctica from the Smith and Hallam (1970) fit, which entailed an overlap of the 2 km isobaths off Enderby Land and off the Coromandel Coast, so that the 2 km isobaths are now separated in the fit by 50 to 150 km along a median line equivalent to a smoothed 4 km isobath of the margins from which the very thick post-breakup sediment has been removed (Fig. 27). Sri Lanka is rotated to a position south of the tip of India (Veevers *et al.* 1980). The minimum northern extent of Greater India is given by its boundary with the Cape Range Fracture Zone (Veevers and Powell 1979).

The finite poles for this revised reconstruction of Eastern Gondwanaland (Fig. 14) are given in Table 2. In an attempt to estimate the relative motion of the Pacific and Indo-Australian Plates, Stock and Molnar (1982) examined the uncertainties in the relative positions of the Indo-Australian, Antarctic, Lord Howe, and Pacific Plates since the Late Cretaceous. The poles that produce the best fit differ little from those used here.

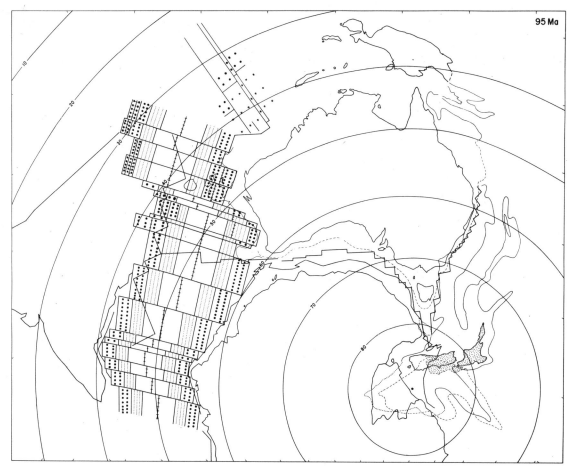

Fig. 18. 95 Ma, in the Cretaceous long normal polarity interval (KN): continued spreading between Greater India and Australia/Antarctica, with abandonment of old ridge and reorganization of spreading about a new ridge, which extends east of a triple junction between Australia and Antarctica, at the same time as the inception of a spreading ridge between southeast Australia and the Lord Howe Rise/New Zealand.

(v) Revised pattern of spreading off the western margin of Australia

The changed positions of Antarctica and India with respect to Australia entail a change in the rotation parameters of India and Australia. The northward track of India relative to Australia since 90 Ma ago is tightly constrained (Johnson *et al.* 1980a), so that the new pole of rotation for India and Australia must satisfy the following constraints: (a) the new fit of India to Antarctica/Australia; (b) the direction of spreading parallel to the Zenith–Wallaby and Cape Range Fracture Zones, and (c) India's northwestward track must include a common position on its northward track.

The pole (b) is determined as 56.5° S, 2.7° W, and the rotation angle as 17.8°. Rotation of India from its position 65.7 Ma ago (A28) to its initial position on the northwestward track is from the old pole of 2° S, 8° E, and an angle of 32.5°, an increase of 3.26° from the previously determined angle of 29.24° (Johnson *et al.* 1980a). Fig. 28, showing the details of the reconstruction off the western margin, updates figs. 3 and 4 of Johnson *et al.* (1980a), and for direct comparison the ages are expressed in the old K–Ar constants. The changed pole of rotation makes only small changes, mainly in the curvature of the fracture zones. The age of the change of spreading regime, from northwestward to northward, is measured from the new initial

PAST GLOBAL SETTINGS

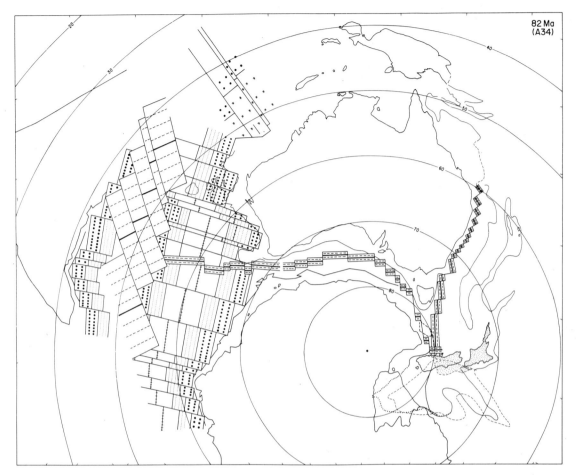

Fig. 19. 82 Ma (A34): rapid spreading east of India, and slow spreading between Australia and Antarctica, and between Australia and the Lord Howe Rise/New Zealand.

position of the northward-spreading ridge. This is shown in Fig. 29, which updates fig. 5 of Johnson *et al.* (1980a), and again the ages are expressed in the old constants. Larson *et al.*'s (1978) identification of A33 (76.48 Ma) and A34 (79.65 Ma) indicates a spreading half-rate of 5.84 cm/year. From A34 to the initial ridge position is 725 km, which at this rate involved 12.42 Ma of spreading, so that the northward spreading started at 79.65 + 12.42 Ma = 92.07 Ma. In the new constants, this is 94.37 Ma, which we round to 95 Ma.

(vi) *Spreading from 160 to 128 Ma (Fig. 15)*

As an extension of Falvey's (1972) and Larson's (1975) work, Heirtzler *et al.* (1978) and Powell (1978) shqwed the pattern of spreading in the Argo Abyssal Plain region between M25 and M9 (Fig. 107). Spreading is estimated to have started 160 Ma ago, as argued in Chapter IV. Powell's (1978) map showed four straight fracture zones, which imply an equatorial pole, which we calculate as 28° S, 19° E (pole a). The oldest identified anomaly in all four spreading compartments, M24, was rotated by 1.8° back to the approximate position of the COB and forward by 7.1° to the estimated 128 Ma (M11) position, and the COB reflected through $2 \times (7.1° + 1.8°) = 17.8°$ to its 128 Ma position. M11 was then moved in the south-western compartment to its observed position, to take account of a long ridge-jump in this compartment between M22 and M13 (Powell 1978) (Fig. 107).

From the persistence of marine facies in north-western Australia from the early Palaeozoic, as dis-

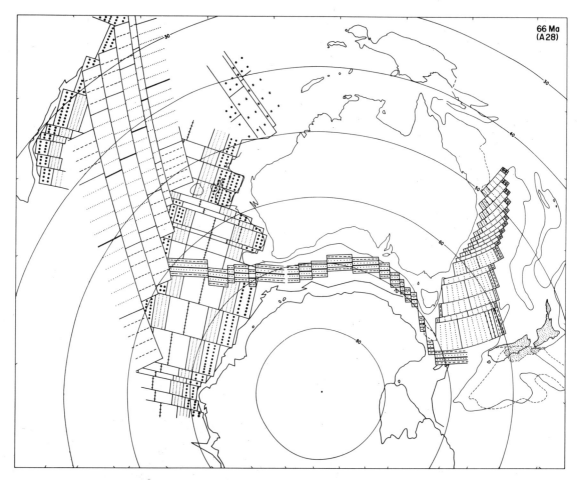

Fig. 20. 66 Ma (A28): continued rapid spreading east of India, and slow spreading south of Australia. Rapid spreading in Tasman Basin, involving ridge-jumps to Australia and asymmetrical spreading.

cussed in Chapter IV, Veevers and Heirtzler (1974) argued that whatever continental lithosphere was broken off the north-western margin 160 Ma ago must have been narrow to have allowed ingress of the sea from Tethys. Larson (1975) nominated the Tarim Block of Asia as a possibility. We refer to this plate as 'off the north-west margin'.

(vii) Spreading from 128 to 95 Ma (Figs. 16–18)

Spreading in the Argo Abyssal Plain continued to at least M9, and the Roo Rise (Fig. 16) was generated between M11 (128 Ma) and M9 (124 Ma) (Heirztler *et al.* 1978). The Joey Rise was generated earlier between M24 and M22. Details of an inferred triple-ridge junction north-west of the Joey Rise that operated after 128 Ma ago are given by Powell (1978) but are not shown here.

Markl (1974a), Larson *et al.* (1979), and Johnson *et al.* (1980a) showed that spreading between India and Antarctica–Australia started in the Early Cretaceous. The time of breakup is estimated as 128 Ma, as detailed in Chapter IV, and this stage of spreading lasted to a radical reorganization of spreading at 95 Ma, as described above. Ridge-jumping about an abandoned spreading axis 117 Ma old in the Cuvier Abyssal Plain, marked today by the Sonne Ridge (von Stackelberg *et al.* 1980), and about spreading axes north-west of Perth at various times is detailed by Johnson *et al.* (1980a, fig. 7) and is shown in Fig. 28.

The interval 110.9 Ma (M0) to 95 Ma and beyond to

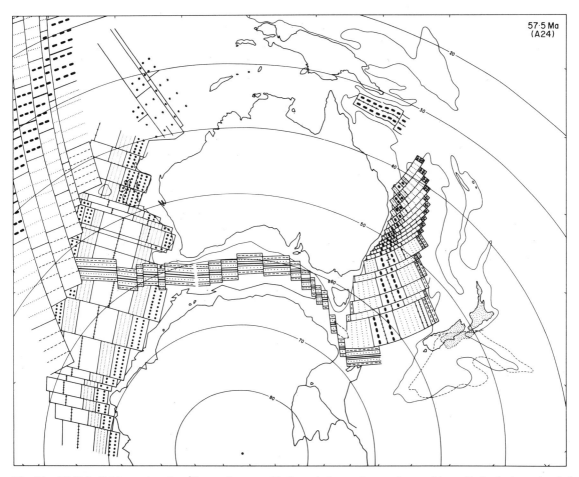

Fig. 21. 57.5 Ma (A24): continued rapid spreading east of India; end of spreading south-east of Australia; beginning and end of spreading between north-east Australia and the Papuan Peninsula; and continued slow spreading between Australia and Antarctica. Frame of map shifted eastward to show New Zealand.

81.7 Ma (A34) corresponds to the Cretaceous long normal polarity interval (KN), and the only control on the ages interpolated between magnetic anomalies is from DSDP site 257 and from a core (E54–7) that contains Late Cretaceous fossils (Quilty 1973). In the Johnson *et al.* (1980a) reconstruction, based on the former tight fit of Antarctica and Australia, E54–7 lay too far north; the loose fit of the present reconstruction puts it in a better position.

Spreading from 128 to 95 Ma was about a single pole (b), with a total rotation of 17.8°, and the three stages shown, at 118, 108, and 95 Ma have been drawn by apportioning the rotation angle for each stage according to spreading rates. Fig. 18 shows the terminal position of the spreading ridge (marked by bars) at 95 Ma, and the initial position of the new ridge, which extended eastward from a triple junction to pass between Antarctica and Australia. Another ridge system developed simultaneously off south-east Australia.

(viii) Spreading from 95 to 81.7 Ma (Fig. 19)

Following the reorganization of the spreading ridge at 95 Ma, India moved northward from pole c, and the old 95 Ma spreading ridge was abandoned on the Australian Plate.

Two new plates, Antarctica and Lord Howe Rise/New Zealand, came into being 95 Ma ago, with the inception of spreading from Australia.

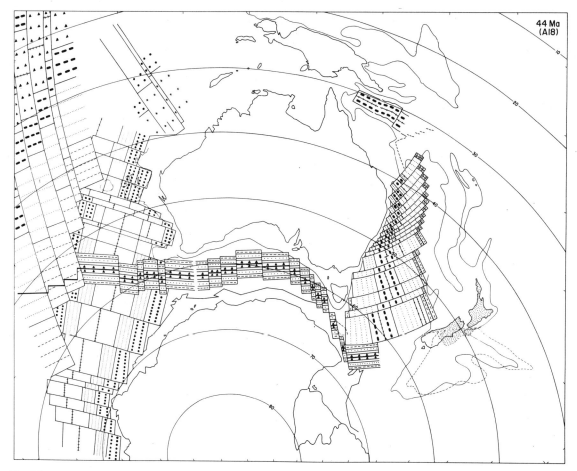

Fig. 22. 44 Ma (A18): final phase of spreading south-east of India, and continued slow spreading between Australia and Antarctica; westward propagation of South-east Indian Ocean Ridge.

(ix) Reconstruction of Lord Howe Rise/New Zealand by closing the Tasman Sea Basin

Hayes and Ringis (1973), revised by Weissel and Hayes (1977) and Weissel *et al.* (1977), mapped seafloor-spreading magnetic anomalies and fracture zones in the Tasman Sea Basin, and interpreted the growth of the basin 'in terms of a simple two-plate spreading system active between about 82 and 60 m.y. ago'.

We note the following departures from a simple spreading system. (a) The spreading pattern is asymmetrical in two ways. First (Fig. 30C), the spreading axis (A24) lies, except in compartment 5–6, west of the mid-point of the Tasman Sea Basin between the 2 km isobaths of Australia and the Lord Howe Rise, taken to mark the continent–ocean boundary. Secondly, as shown in Fig. 31, between A24 and A33, the length of seafloor generated on the east limb was only 60 to 91 per cent of that on the west. Fig. 31 was constructed by accepting Weissel and Hayes's (1977) mapped anomalies, and measuring on either side the distances between the extinct spreading axis, mapped as A24, and A28, between A28 and A32, and between A32 and A33. The only synthetic profile of the anomalies appears in Hayes and Ringis (1973, fig. 2), and the physical model of normally and reversely polarized blocks from which the synthetic profile was generated is not shown. Weissel and Watts (1979) found that the spreading ridge in the Coral Sea Basin became extinct at the same time as that in the Tasman Sea Basin, shortly after the generation of A24, at 56.00 Ma (= 57.44 Ma, in the new constants). Weissel and Watts's (1979, figs. 3, 6) model of the growth of the

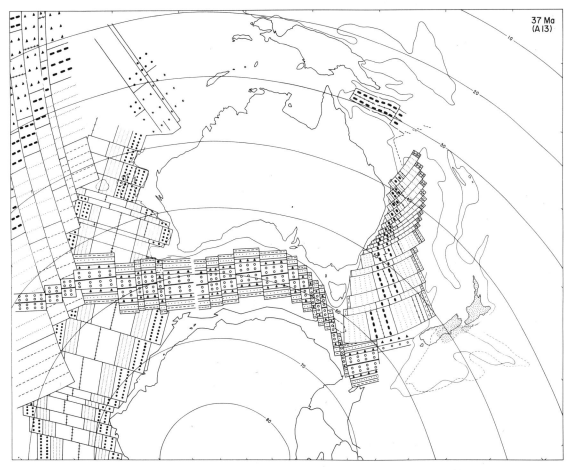

Fig. 23. 37 Ma (A13): rapid spreading between Australia and Antarctica and to the west.

Coral Sea Basin includes a block model, which shows that they mapped the peaks of the anomalies on either side of the axis, as Hayes and Ringis (1973) did for the Tasman Sea Basin, whereas, strictly speaking, the peaks above the one edge of the blocks on one limb must correspond with troughs above the reflected edge on the other. In the Coral Sea, the block model is symmetrical but because of this artefact the mapped anomalies imply asymmetrical spreading by an amount equal to the peak-to-trough distance of the anomalies or width of the normally polarized blocks. This is 10 km only, and can be ignored. In the Tasman Basin, the error is likely to be correspondingly small, and again can be ignored, so that the asymmetry shown by the outer pairs of anomalies is probably real. The asymmetry between A28 and the axis is sensitive, however, to the precise location of the axis. Comparison of Hayes and Ringis's (1973) synthetic profile for the Tasman Basin with that of Weissel and Watts (1979) for the Coral Sea shows that the same point on the magnetic anomalies was taken to mark the axis, and the only uncertainty resides in the picking of this point on the observed profiles of the Tasman Sea. This uncertainty may account for some of the apparent asymmetry for the interval A24 (axis) to A28. We therefore conclude that the asymmetry indicated by the mapped magnetic anomalies in the Tasman Sea is real, as shown in Fig. 31, except that in the interval from the axis to A28, with an extreme asymmetry of 60 per cent in compartment 7–8, it may be exaggerated due to the difficulty of picking the axis. Some confidence that this interval is not exceptional, except in

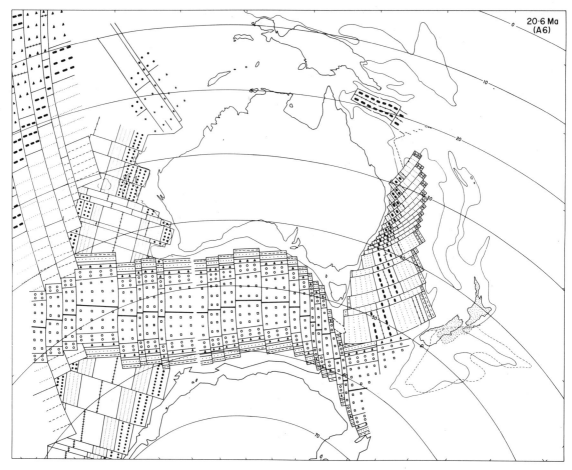

Fig. 24. 20.6 Ma (A6): continued rapid spreading between Australia and Antarctica, and to the west.

compartment 7–8, is provided by the linearity of the spreading rates for each limb seen in Fig. 31.

A second plot (Fig. 32), of the asymmetry against the distance from the pole, shows another linear relationship, again with the exception of compartment 7–8. Here the asymmetry from the axis to A28 on either limb increases with the distance from the pole.

(b) A second departure from a simple spreading system is found in the trend of the magnetic anomalies, starting at A32, in the compartment between fracture zones 6 and 7 (Weissel and Hayes 1977, fig. 1). Here the anomalies strike the Australian margin at an angle of 45°. As pointed out by Roots (1976), this pattern must entail a shorter set of spreading segments between A32 and the margin; and accommodating the seafloor generated before A32 must involve ridge-jumping to Australia so that the older seafloor is left on the east limb.

These departures from the simple kind of spreading system found in the broad oceans suggest to us that the tectonic setting of the Tasman Sea is not that of a simple rifted ocean but, as Hayes and Ringis (1973) pointed out originally, that of a back-arc basin. As detailed in Chapter IV, many features can be explained by viewing this spreading episode in the context of a Pacific Ocean back-arc in place of an Atlantic-type splitting of a supercontinent.

Previous authors (Hayes and Ringis 1973; Weissel and Hayes 1977; Weissel *et al.* 1977; and Shaw, 1978) have concentrated on the details of the spreading pattern; Weissel and Hayes (1977) made reconstructions back to a fit position but only for the area north of

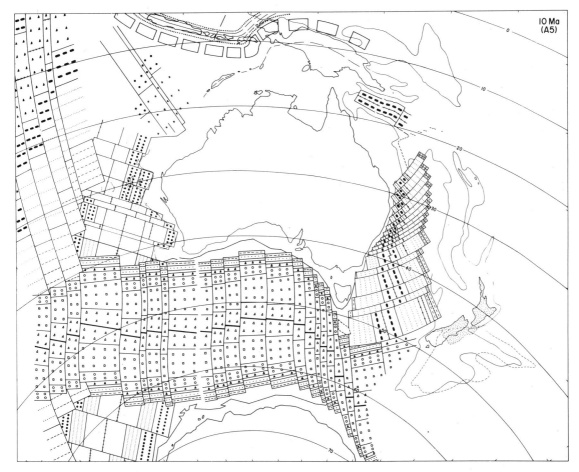

Fig. 25. 10 Ma (A5): continued rapid spreading between Australia and Antarctica, and to the west. Final adjustment of distended South Tasman Rise. Appearance in north of leading edge of Sundaland (dotted pattern) and Pacific Plates with an outer non-volcanic arc (blocks) and an inner arc (broken lines).

42° S on Lord Howe Rise, and north of 44° S on Australia. Shaw's (1978) oldest reconstruction is to A33 only. Here we show the closure of the Tasman Sea Basin along its entire length. Whereas the pattern of closure in the oceans treated already, the Eastern and South-eastern Indian Oceans, is clear, the pattern in the Tasman Sea is obscure, and we concentrate on making a reconstruction that is constrained principally by the shape of the adjacent margins.

The geometrical constraints on the extent of seafloor spreading in the Tasman Sea Basin are the South-east Australian COB, well marked by the steep and narrow slope, and the adjacent but ill-defined COB on the western side of the Lord Howe Rise. Off South-east Australia, the COB must lie beneath the steep continental slope, between the 2 km and 4 km isobaths, which are only 20 to 30 km apart. The only published map of the COB on the western side of the Lord Howe Rise, by Karner and Watts (1982, fig. 3a), is based on seismic and magnetic data, but no details are provided. Their COB corresponds to the 2 km isobath at about 25° S and 30° to 32° S, the 3 km isobath at 29° S, and the 4 km isobath and deeper from 32° to 40° S. This area of a postulated deep COB is crossed at 36° S by a seismic profile published by Willcox *et al.* (1980, fig. 6B), who interpret a change of crustal structure, from fault-block terrain (continental) to rugged crystalline basement, which we take to be oceanic crust, at a

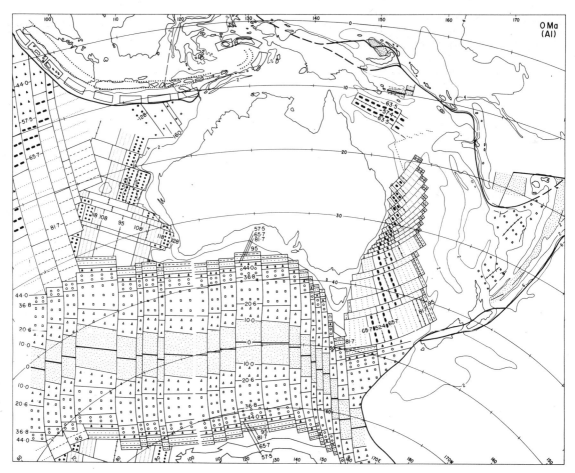

Fig. 26. 0 Ma (A1): continued rapid spreading between Australia and Antarctica. Back-arc spreading along Pacific Borderland, arc–continent collision north of Australia, and plate consumption along the Java Trench. Heavy line indicates convergent or transform boundary. Parallel dotted lines indicate Indonesian volcanic arc.

water depth of 1.9 km. So here, as off Australia, we adopt the 2 km isobath as the best estimate of the position of the COB.

The spreading pattern is constrained by the seafloor-spreading magnetic anomalies, which are calibrated by the age of the oceanic lithosphere at DSDP site 283. Using stage poles derived from Weissel and Hayes's (1977, table 1) finite poles and their mapped anomalies, we made the reconstructions shown in Fig. 30. The position of fit, according to Weissel and Hayes, by a rotation of half the closing angle of 23° on either side of the axis about their pole 5, overlaps Australia. Incidentally, the angle of rotation of this fit, with an extrapolated spreading rate from A32 (72.8 Ma) to A33 (78.41 Ma), implies breakup at 82.00 Ma. (Note the different dates of the magnetic anomalies in Weissel and Hayes 1977, from those of La Brecque et al. 1977, updated in new constants, by Mankinen and Dalrymple 1979). We remove the overlap by postulating ridge-jumping to Australia, which preserves the seafloor-spreading pattern and the total allowable amount of opening between the adjacent COBs. Independent evidence, in particular from the Strzelecki/Gippsland Basin, as detailed in Chapter IV, suggests that breakup took place at 95 Ma. We therefore postulate (a) that an early stage of spreading took place between 95 and 82 Ma, by a rotation of 2° about the closing pole, and (b) that the oceanic lithosphere

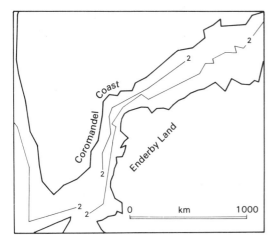

Fig. 27. New fit of India to Antarctica, by a rotation of −89.7° from a pole at 2.5° S, 12.4° E, showing coastlines and 2 km isobaths. This is an enlarged part of the pre-breakup reconstruction, though with a different map centre.

generated during this early interval of spreading was accommodated next to the Lord Howe Rise by a ridge-jump to Australia before A33, except in compartment 4–5, which requires no jump.

In Fig. 30A, the axis, shown by circles, has been rotated on either side about the closing pole of Weissel and Hayes (1977) by half the angle of closure ($\frac{1}{2} \times 23°$) to the crosses on the flanks, and additionally by 1° on either side, for a total of 25°, to the triangles. The rotated points in compartments 1–3, 6–7, 7–8, and immediately north of 8 overlap Australia. In Fig. 30B, these points are rotated so that they lie on the Australian 2 km isobath, and the equivalent points on the eastern flank are rotated an equal amount. In all those compartments with adequate magnetic anomalies, except one, the rotated points that represent a rotation of 25° match the 2 km isobath of the Lord Howe Rise, so that this is taken to be the angle of closure. The exception is in compartment 6–7, where the triangles overlap the 2 km isobath by 150 to 200 km. Finally, in Fig. 30C, we show for the entire area the eastern part of each compartment that would have been generated before the ridge jumped to Australia to generate the seafloor asymmetrically disposed about the final axis near A24. South of 34° S, the jump took place at about 82 Ma; at 34° S, in compartment 6–6′, at about 79 Ma; and north of 34° S, at about 72 Ma, as seen from information in compartment 7–8. Further north there is no relevant information, and what is

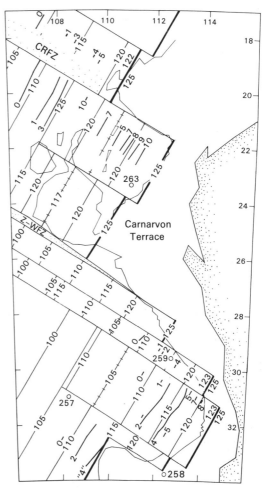

Fig. 28. Synthetic isochrons (Ma, in old constants) compared with observed M-series magnetic anomalies, DSDP sites, and 4 km isobaths in the Perth and Cuvier Abyssal Plains, updating figs. 3 and 4 of Johnson *et al.* (1980a). CRFZ=Cape Range Fracture Zone; Z–WFZ=Zenith–Wallaby Fracture Zone. Ages expressed in old constants here and in Fig. 29 to provide continuity with ages in Johnson *et al.* (1980a).

shown in Fig. 30C is extrapolated from compartment 7–8.

This reconstruction observes (a) the geometrical constraint of the amount of oceanic lithosphere between the 2 km isobaths, except in compartment 6–7; (b) the further geometrical constraint of the direction of spreading given by Weissel and Hayes's (1977) rotation poles; and (c) the position of the magnetic anomalies mapped by Weissel and Hayes

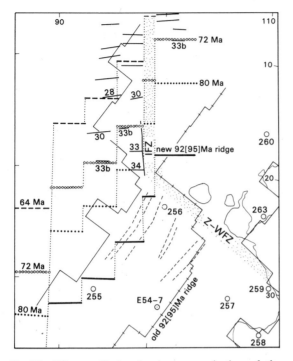

Fig. 29. Wharton Basin, showing reorganization of the spreading pattern at 92 Ma (=95 Ma in new constants), updating fig. 5 of Johnson *et al.* (1980a). Full offset lines show 125 Ma isochrons along margins of India and Australia on either side of old 92 Ma spreading ridge. New 92 Ma spreading ridge shown by heavy full line offset along dotted lines, 80 Ma isochron (=A34) by heavy dotted line, 72 Ma isochron (=A33b) by small open circles, and 64 Ma isochron (=A28) by heavy dashed line. Deep-sea drilling sites shown by large open circles. Light broken lines indicate magnetic lineations of the Broken Ridge set from Markl (1974a). Note the continuation of the distributed fractures between Zenith–Wallaby Fracture Zone (Z–WFZ) and DSDP site 259 into the Investigator Fracture Zone (IFZ), all stippled, as suggested by Larson *et al.* (1978). E54-7 is the reconstructed position of a core that penetrated Cenomanian sediment on the Kerguelen Ridge (Quilty 1973).

(1977). The reconstruction implies that the Dampier Ridge is oceanic, contrary to the views of Willcox *et al.* (1980). On Lord Howe Rise itself, the reconstruction implies than an area south-east of Lord Howe Island, and areas on the western edge of the northern part of the Rise are oceanic. Willcox *et al.* (1980, figs. 2, 6A) show seismic profiles across the area south-east of Lord Howe Island at 30.7° S and 31° S, and interpret this part of the Lord Howe Rise as part of the western province of the Lord Howe Rise, characterized by horsts and grabens of continental crust. The area of overlap to the north includes the Middleton Chain of seamounts and islands, including Lord Howe Island, the oldest known rocks of which are 7 Ma old (McDougall *et al.* 1981). Weissel and Hayes (1977) point out, however, the likelihood that this volcanic chain has modified and extended westward the original rifted margin of the Lord Howe Rise.

Shaw (1978) determined new finite poles from information including new magnetic data in the northern Tasman Sea Basin, and found poles more southerly than those of Weissel and Hayes (1977). From such poles, the overlap in our reconstruction would tend to be reduced. We have not used Shaw's (1978) poles, however, because the magnetic anomaly interpretations on which the new poles are based are unavailable, except for a double line across the northern Tasman Sea, which shows the great difficulty of correlating anomalies in this area.

In conclusion, our reconstruction (Fig. 30C) satisfies the major constraints of the geometry of the adjacent margins, and the pattern of seafloor spreading; in the process, it requires the Dampier Ridge and part of the Lord Howe Rise to be oceanic. The first stage of opening, between 95 and 82 Ma, is by a rotation of 2° from the closing pole, at a half-rate of 4 mm/year at 40° S, and the second stage, between 82 and 57.5 Ma, by a rotation of 23°, at half-rates of 2.1 cm/year on the eastern limb, and 2.7 cm/year on the western limb.

(x) Growth of the South-east Indian Ocean between Antarctica and Australia

The growth of the South-east Indian Ocean is based on two phases of spreading (Cande and Mutter 1982): the first, at a very slow rate, from breakup (95 Ma) to A18 (44 Ma), and the second, at a rapid rate, from A18 to the present. Weissel and Hayes (1971, 1972), who mapped the magnetic anomalies, showed that spreading from A6 to A14 in a zone between 128° and 138° E was only 71 per cent on the south limb of what it was on the north limb. Other irregularities from 'normal' spreading are suggested by the recent extensive aeromagnetic survey of the Australian–Antarctic discordance (Vogt *et al.* 1979). Until the details of this survey are published, it would be premature to attempt a detailed model of the spreading history. Our reconstruction therefore starts from the closure, as detailed already, and accounts for each of the individual stages in spreading from Weissel *et al.*'s (1977) poles. A synthetic ridge, equal to the line of suture east of Albany (118° E) was rotated forward, and, west of Albany, the present ridge, from McKenzie

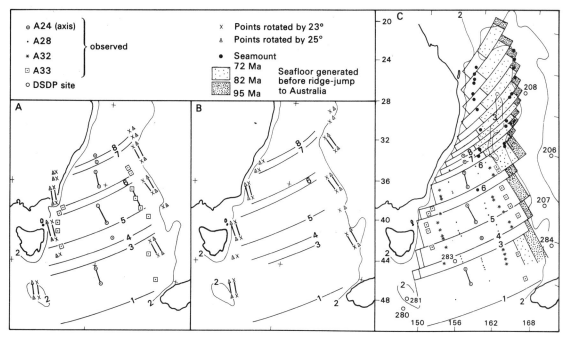

Fig. 30. Geometry of the reconstruction between the 2 km isobaths. A. Spreading axis (~A24) shown by circles, rotated on either side by half the closing angle (½×23°) of Weissel and Hayes (1977) to the crosses, and then an additional 1° on either side to the triangles. Squares indicate observed A33. Numbered fracture zones from Weissel and Hayes (1977). B. Rotation of triangles in each compartment to lie along the 2 km isobath of Australia, and corresponding rotation of other points. C. Synthetic COBs along eastern Australian margin and western Lord Howe Rise, with stippled area showing seafloor generated in early stage of spreading before a postulated ridge-jump to Australia. Dampier Ridge outlined by 3 km isobath. Tasmantid and Middleton chains of islands and seamounts shown by filled circles. Lord Howe Island is the southernmost of the Middleton chain. The oldest material recovered by deep-sea drilling on the Lord Howe Rise is rhyolite 95 Ma old at DSDP site 207 (McDougall and van der Lingen 1974). In the Tasman Basin, DSDP site 283 penetrated oceanic basalt overlain by 66 Ma old sediment, consistent with the determination of A30 (Weissel and Hayes 1977).

and Sclater (1971), was rotated backward. Fig. 33 shows the synthetic isochrons compared with the observed ones from the magnetic anomalies and the DSDP sites. The observed ages at sites 265, 266, and 267B confirm the magnetic isochrons. Sites 268 and 269 provide minimum ages only, as probably does 280, which bottomed in an intrusion. Site 282, west of Tasmania, recovered 45 Ma sediment above basalt; this is equivalent to A18, and suggests that 282 belongs in the compartment to the west. Discrepancies are attributable to the asymmetrical spreading and to local ridge-jumps, in particular in the compartments south of Tasmania. Some of the synthetic fracture zones in the area east of 118° E, generated from the closing suture, do not match the observed fractures, and would require individual adjustment, but the set of transform faults south of Tasmania, each with a substantial offset, is well modelled.

The synthetic South-east Indian Ocean spreading ridge was extended westward to a triple junction with the reorganized ridge in the Eastern Indian Ocean (Fig. 18). Johnson et al.'s (1976, 1980a) reconstructions entailed a separation of Antarctica from Australia that started no earlier than 54 Ma, so that India separated from a combined Antarctica/Australia before 54 Ma. With the recognition of an earlier though very slow separation of Antarctica from Australia, India's motion with respect to Antarctica is now described by the resolved pole of its separation from Australia (−2°, 8°, 13.0°) plus the separation of Antarctica from Australia (12.7°, 22.2°, −1.17°). This resolved pole is −0.93°, 9.27°, 14.1°, which, due to the very slow spreading between Antarctica and Australia, is barely distinguishable from the India-from-Australia pole of −2°, 8°, 13.0°. The additional spreading of India from Antarctica and Australia was taken up by a

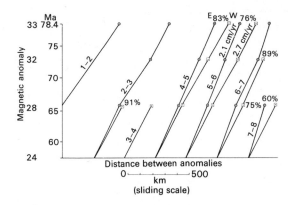
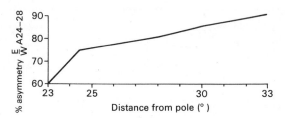

Fig. 31. Spreading rates on either limb (circles on east, squares on west) of the Tasman Sea spreading system for each compartment between fracture zones 1 to 8. Asymmetry expressed in ratio of distance in east limb to that in west limb. Average spreading half-rates shown for compartment 5–6.

Fig. 32. Asymmetry of distance from A24 (axis) to A28 plotted against distance (in degrees) from pole.

slightly faster rate of spreading in the Eastern Indian Ocean, probably until about 44 Ma, when spreading east of the Ninetyeast Ridge ceased. This age is a revision of the 32 Ma age that Johnson *et al.* (1976, fig. 2F) calculated from a separation of Antarctica from Australia at 54 Ma. Independent evidence that spreading east of the Ninetyeast Ridge ceased earlier is provided by Liu *et al.* (1982), who find beneath the

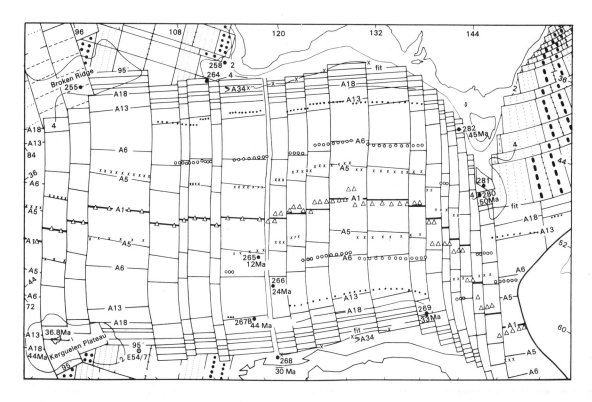

Fig. 33. Synthetic isochrons compared with the observed isochrons of the magnetic anomalies. Observed isochrons from Wellman and McCracken (1979) except for the oldest isochron (≥A34), from Cande and Mutter (1982).

TABLE 4. *Ages of spreading about the poles listed in Table 3, relative to Australia, showing the fragmentation of Eastern Gondwanaland (India, Antarctica, Australia) through the intermediate India, Antarctic/Australian, and Australian Plates to the present Indo-Australian and Antarctic Plates. Wiggly lines indicate inception of spreading, and double lines cessation. The Australian Plate existed between the dispersal of Antarctica 95 Ma ago to the coalescence of the Indian Plate 44 Ma ago*

Plate	Stage Ma												
	160	128	118	108	95	81.7	65.7	57.5	44.0	36.8	20.6	10.0	0
Off NW margin	a				?						Indo-		
India		b1	b2	b3	c1	c2	i_1	i_2			Australian		
LHR/NZ						d	f	g					
Papuan Peninsula								h			Plate		
Antarctica						e1	e2	e3	e4	j	k	l	m

(Eastern Gondwanaland spans early columns; Antarctic Plate labelled at right.)

Nicobar Fan at 3.5° S a fossil spreading ridge marked by A20 (46 Ma). In the timetable of the plate motions involved in the fragmentation of Eastern Gondwanaland (Table 4), the intermediate Indian Plate is therefore shown as lasting to the demise of the spreading ridge east of Ninetyeast Ridge about 44 Ma.

The position of core E54–7, shown in the 95 Ma reconstruction of Fig. 29, comes to lie, without adjustment, on the southern side of the new spreading system. Our reconstruction implies that the southeastern part of Broken Ridge and the north-eastern part of Kerguelen Plateau were generated during the early phase of very slow spreading.

(xi) Growth of the Coral Sea between north-east Australia and the Papuan Peninsula

Weissel and Watts (1979) identified in the Coral Sea seafloor-spreading magnetic anomalies A25, A26, and perhaps A27 on either side of an extinct spreading axis a little younger than A24, dated as 56.00 Ma (old K–Ar constants), equivalent to 57.44 Ma in the new constants. At DSDP site 287, near the postulated axis (Fig. 4), Andrews *et al.* (1975) recovered sediment as old as 55 Ma (early Eocene), confirming the anomaly identification. Weissel and Watts (1979) determined a pole of rotation (h) from the curvature of a well-mapped fracture zone, and we calculate an angle of rotation of 5.54° to unite the outermost isochrons of 62 Ma (old constants), equivalent to 63.5 Ma (new constants). Following Taylor and Falvey (1977), Weissel and Watts (1979) drew the continent–continent transform fault on the western side of the Coral Sea through the Moresby Trough and the Bismarck–Lagaip fault zones in New Guinea.

(xii) Summary of plate motions

In summary, the divergent plate motions (Table 4) involve:

(1) fragmentation of Eastern Gondwanaland by separation from Australia
   (a) of the plate 'off NW margin of Australia' at 160 Ma
   (b) of the Indian Plate at 128 Ma
   (c) of the Lord Howe Rise/New Zealand Plate and of the Antarctic Plate at 95 Ma, and
   (d) of the Papuan Peninsula Plate at 65.7 Ma, followed by

(2) Coalescence of the Indo-Australian Plate by the demise of the spreading ridges between Australia and
   (a) India at about 44 Ma
   (b) the Lord Howe Rise/New Zealand and the Papuan Peninsula at 57.5 Ma, leaving

(3) two divergent plates today:
   (a) the Indo-Australian Plate, and
   (b) the Antarctic Plate.

(xiii) Independence of Australia between the opening and closing of oceanic straits

(1) Western margin

For some 10 Ma after breakup, till about 118 Ma ago, India and Australia remained connected along the divergent continent–continent transform fault of the

Cape Range Fracture Zone (Veevers and Powell 1979), and along the eastern part of the Zenith–Wallaby Fracture Zone. The trailing edge of India unkeyed from the Exmouth Plateau 123 Ma ago (= 120 Ma in old constants), and from the Carnarvon Terrace (Fig. 28) 122 Ma ago (= 119 Ma in old constants) but an oceanic strait did not open between the continents until 95 to 100 Ma ago with the end of growth of the epiliths of the Wallaby Plateau and Zenith Seamount (Figs. 17 and 18); and circulation of oceanic water between Australia and India could not have started therefore before about 95 Ma. The replacement of detrital sediment by carbonate along the western margin of Australia at this time, detailed in Chapter IV, correlates with the inception of oceanic circulation.

(2) Southern margin

The protuberance of Tasmania and the South Tasman Rise did not unkey from its Antarctic counterpart until about 37 Ma ago (Fig. 23), in the early Oligocene, and it was not until a little later, at about 30 Ma ago, in the late Oligocene, that the Circum-Antarctic Current was able to flow unimpeded between Australia and Antarctica (Weissel and Hayes 1972; Kennett et al. 1975; Deighton et al. 1976; Kemp 1978). This means that Australia did not become independent of its last Gondwanaland neighbour until as recently as 30 Ma ago.

(3) North-western margin

Independence was short-lived, because by 10 Ma ago the Sunda Arc and Sundaland had encroached the opposite flank of Australia from the north-west (Fig. 25), as described below.

(c) Interaction of Australia and South-east Asia (by C. McA. Powell, B. D. Johnson, and J. J. Veevers)

(i) Introduction

This account of Australia's interaction with the Sundaland salient of South-east Asia starts with an analysis of the relative motions of India and Sundaland, which provides a constraint on Sundaland's westerly motion and hence a constraint on its interaction with Australia. The source is Powell et al. (1980b), updated with the new fits of India, Australia, and Antarctica.

The oceanfloor path of the Gondwanaland fragment, Greater India, can be traced from its original position alongside Western Australia 128 Ma ago to its present position in Asia (Fig. 34). This oceanfloor path shows that the continental margin of Greater India that lay alongside Western Australia moved northwestward relative to Australia from 128 Ma to 95 Ma ago (Larson et al. 1979; Johnson et al. 1980a, updated in the preceding section), and then moved north along the Investigator Fracture Zone, at longitude 100° E. The easternmost 500 km of this oceanfloor path is now truncated by the Java Trench, western Sumatra, and the Andaman–Nicobar region, which consequently must have been further east until the trailing edge of Greater India had passed. Reconstructions (Fig. 35) show that the eastern margin of Greater India was in the present latitudes of Sumatra and Burma from 60 Ma ago (Palaeocene) until 15 Ma ago (mid-Miocene), so that South-east Asia could not have encroached on Greater India's oceanfloor path until 15 Ma ago. The leading edge of Australia, which arrived at the present latitude of Java about 15 Ma ago, narrowly missed a continent-to-continent collision with the trailing edge of the westward-moving South-east Asia, and the resulting relative motion of the Banda Arcs has been dominated by a large sinistral shear (Fig. 36) (Powell and Johnson 1980; Powell et al. 1980c).

(ii) Constraints

(1) Size of Greater India

The size of Greater India depends on the fits of India, Australia, and Antarctica in Gondwanaland, and the improved fits described above are used here to update Powell et al. (1980b) and Powell and Johnson (1980).

The northern edge of Greater India in Gondwanaland can be reconstructed from two lines of evidence: (1) unravelling the telescoping of Indian crust in the Himalayan–Tibetan region, and (2) analysing the facies distribution on the West Australian continental margin. Powell (1979) discussed several possibilities for the northern edge of Greater India, and concluded that it stretched against Australia at least as far north as latitude 25° S. It is possible, but direct evidence is lacking (Veevers and Powell 1979), that Greater India extended as far north as the Cape Range Fracture Zone (Johnson et al. 1980a). In our modelling, we have taken the view that Greater India extended as far north along the western margin of Australia as latitude 20° S (Powell and Johnson 1980; and Fig. 34). This is consistent with the Palaeozoic and Mesozoic facies trend along the western margin from marine in the north to non-marine in the south.

(2) Position of South-east Asia

Palaeomagnetic work (McElhinny et al. 1974; Haile and Tarling 1975; Haile et al. 1977) imposes two constraints on the position of South-east Asia. First, the major continental blocks have occupied approximately their present latitude since 85 Ma ago (Santonian)

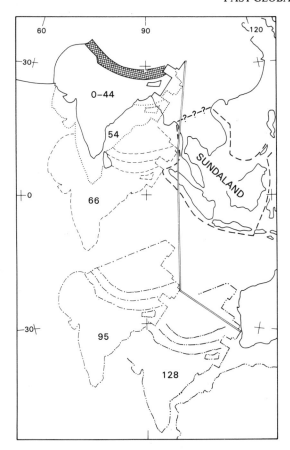

Fig. 34. Interaction of Greater India and Sundaland, presented in two steps. This figure gives the amount that Sundaland overlaps the wake of Greater India by showing the seafloor track of the easternmost part of Greater India with respect to the present position of Australia and Sundaland, whereas the correction for the palaeolatitudinal motion of India and Australia is given in Fig. 35. Timing of Sundaland's westward motion into its present position by showing the palaeolatitudinal position of Greater India with respect to Sundaland, assumed to have maintained its present latitude at least since the Cretaceous. This figure shows the present position (full line) of India, Australia, and south-east Asia (Sundaland—bounded by heavy dashed line), and positions relative to Australia of Greater India 54, 66, 95, and 128 Ma ago. Double line shows track of easternmost tip of Greater India with respect to Australia, which is held fixed in this diagram. Note that there has been no relative motion between India and Australia during the last 44 Ma. The northwestern margin of Greater India is marked by the Hindu Kush–Baluchistan Ophiolite belt, and the northern margin by the Indus–Tsangpo Ophiolite line. The shaded area along the northern margin of India is the doubled Indian crust between the Himalayan front and the Ophiolite line. The Tethyan margin of Greater India before collision is extended by the reflection of the Himalayan front about the Indus–Tsangpo line (Powell and Johnson 1980, fig. 1). The 128 and 95 Ma reconstructions show additionally a line that indicates the maximum northern extent of Greater India as suggested by Powell and Johnson (1980). The eastern margin of Greater India is the counterpart of the continental margin of Australia/Antarctica carried through since breakup without deformation.

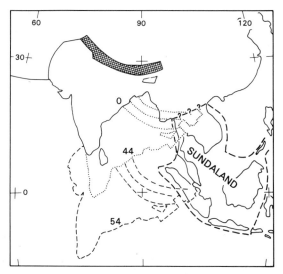

Fig. 35. Present position (full line) of India and South-east Asia (Sundaland). Boundaries of Greater India as in Fig. 34. Reconstructed position of India 44 and 54 Ma ago derived from seafloor spreading data, as in Fig. 34, corrected for the apparent polar wander path of India (Klootwijk and Peirce 1979).

while rotating a net 40° to 50° anticlockwise (Powell and Johnson 1980, figs. 3 and 5). Secondly, the directions of palaeomagnetization from sites in Indo-China, Malaysia, and Kalimantan show approximately the same declination at similar times, suggesting that much of the older continental nuclei of South-east Asia have acted as a single continental block, called Sundaland (Powell and Johnson 1980), at least since 85 Ma ago. The margins of Sundaland (Fig. 34, details in Powell and Johnson 1980) exclude the island arcs and marginal seas, some of which may have grown by extension away from the continental nucleus.

(3) Positions of Greater India and Australia since continental breakup 128 Ma ago

The relative positions of India and Australia since their Early Cretaceous breakup can be determined by

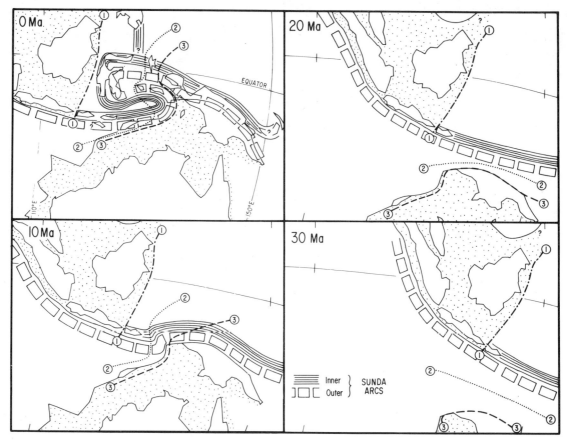

Fig. 36. Formation of the Banda Arcs by interactions between Sundaland and Australia (stippled). No attempt has been made at a detailed reconstruction of the evolution of the Sunda Arcs into their present S-shape, and uncertainties in the timing of collision and the rate of movement of Sundaland require further palaeomagnetic and geological data for resolution. For example, Haile (1978) shows that western Sulawesi was associated with Kalimantan in the Cretaceous. A possible evolution of three zoogeographic lines is indicated: (1) Wallace's Line, of faunal diversity contrast; (2) Weber's Line, of faunal balance; and (3) Lydekker's Line (Powell *et al.* 1980c). Note that Weber's Line is discordant to the tectonic grain, as discussed below by Talent.

removing successively older stages of oceanic crust between the continents according to the rotation poles given in the preceding section. These stage reconstructions can then be oriented with respect to the earth's rotation axis by applying a uniform polar-wander correction. Klootwijk and Peirce (1979) have recently derived an independent polar-wander curve for India, and have resolved several of the uncertainties in the Australian polar-wander curve. We have used this curve for our reconstructions (Fig. 35).

The Indian polar-wander curve modifies, but does not change significantly, the reconstructed positions of India and Australia shown by Powell and Johnson (1980, fig. 5). In essence, the northern, leading edge of Greater India reached the present latitude of Sundaland around 60 Ma ago, and the trailing edge did not pass the region until 15 Ma ago. Australia was well south of the eastward extension of the arcs bordering the southern margin of Sundaland (Sunda Arcs) until the latter half of the Cainozoic; Australia crossed the projected line of the Sunda Arcs about 15 Ma ago, narrowly avoiding collision with the trailing southeastern edge of Sundaland.

(iii) Possible reconstructions of the interaction between Sundaland, India, and Australia

The oceanfloor data combined with continental palaeomagnetism provide precise positions for the dispersion of India, Australia, and Antarctica, but the position of these continents relative to Sundaland cannot be constrained uniquely. For instance, we do not know when the 40° to 50° of counter-clockwise rotation of Sundaland occurred other than it is post-Santonian (85 Ma). There may have been more than 40° to 50° counter-clockwise rotation at some intermediate stage, with more recent clockwise rotation to give the net total of 40° to 50°, as indeed some data suggest (Ninkovitch 1976).

We are also uncertain about the extent of what can be considered as a coherent continental block for the purposes of reconstruction. Holcombe (1977) suggests that Sundaland is cut by a maze of faults, and if his suggestions are valid, the shape of Sundaland has been greatly distorted during the Cainozoic. We can, however, conclude that if any of the continental nuclei within Sundaland were more widely dispersed than at present during the Palaeogene, then the constraint placed by Greater India on the most westerly position implies that the eastern part of Sundaland lay even farther east than we have shown (Powell and Johnson 1980). Our model is conservative on the possible easterly positions of South-east Asia, but by 15 Ma ago Sundaland must have had essentially the present shape.

The tight constraint imposed by the eastern edge of Greater India on the north-western part of Sundaland accords with the geological evolution of the Andaman-Nicobar region – a system that has developed since the mid-Miocene by marginal sea spreading away from the Malay Peninsula (Curray *et al.* 1979). Prior to 15 Ma ago, the Andaman and Nicobar island chains lay alongside the Shan Scarp, in continuity with the Indo-Burman Ranges, which have also expanded east-west in the late Cainozoic.

The close approach of the Australian continent to the south-eastern edge of Sundaland in the mid-Miocene provides a framework in which the shape of the Banda Arcs can be understood. We envisage that prior to Miocene a southward-facing double arc (outer non-volcanic, and inner volcanic, arcs) extended eastward from Java to connect with similar arcs in the South-west Pacific (Fig. 36). When the leading edge of Australia crossed the line of the arcs in the mid-Miocene, subduction was terminated in the New Guinea segment and transferred northward to other consuming plate boundaries. Because of the irregular shape of its northern margin, the Australian continent did not reach the Sunda Arc system in the vicinity of Timor until 5 Ma ago (Pliocene), which was the major time of uplift and deformation in the region (Carter *et al.* 1976; Veevers *et al.* 1978).

The Australian continent has penetrated some 800 km north since the mid-Miocene, a distance equal to the longitudinal amplitude of the Banda Arcs. We think that the present shape of the Banda Arcs may thus be caused by the interaction of a large sinistral shear between the west-moving Sundaland Plate and the northward-moving Australian Plate, in the fashion of a large drag fold (Carey's (1958) oroclinotath) that grew in longitudinal amplitude with time (Fig. 36; see Powell *et al.* 1980c). If this model is correct, then large Neogene counter-clockwise rotations may be expected in the eastern Banda islands, such as Seram. Haile's (1979) palaeomagnetic evidence of a 74° counter-clockwise rotation for Seram since the extrusion of basalt 7.6 Ma ago confirms the model. Furthermore, the oceanfloor in the Banda Sea north of Timor may be quite old, having been attached to the trailing edge of Sundaland since it was formed, and not Neogene as suggested by some authors (for example Carter *et al.* 1976).

In our view, the kinematics, at least in outline, of the continent-arc collision on Australia's north-western flank are clear. What remains obscure are the details of the northern limit of the Australian lithosphere and of the southern limit of the allochthonous Asian and Pacific terrain. The boundaries sketched on Figs. 25 and 26 are therefore controversial in detail, and are shown for comprehensiveness only.

Two models have been used to explain details of the Banda Arcs. Hamilton (1979) argues that the inner volcanic arc and the outer non-volcanic arc are the product of a subduction system; the Timor, Aru, and Seram Troughs are interpreted as the trace of the subduction zone that generated the volcanoes in the inner arc, and the outer non-volcanic arc as a subduction *melange*. In contrast, Audley-Charles *et al.* (1979) argue that the outer arc is a zone of Pliocene collision between Asian overthrusted rock and Australian basement; the subduction trench is interpreted as lying between the inner and outer arcs, so that the outer trough and the outer arc are underlain by Australian continental crust, a view which we share. Further palaeomagnetic data from the Banda Arcs, beyond the reconnaissance by Chamalaun (1977a, b), would seem to be the most fruitful approach to the resolution of the conflicting models. Discussion of the problem is provided by Bowin *et al.* (1980), Chamalaun and Grady (1978), Crostella and Powell (1975), Curray *et* al. (1977), Johnston and Bowin (1981), Norvick (1979), and von der Borch (1979); in our description of the northern margin (Chapter IV), we examine the application of these models to the geology of Timor.

3. PALAEOCLIMATES AND PALAEO-ENVIRONMENTS

(a) Quaternary environments (by M. A. J. Williams)

As shown by palaeomagnetic evidence, and following its separation from Antarctica, Australia began to move northwards from 95 Ma ago into warmer subtropical latitudes dominated by semi-permanent anticyclones over land and sea. The post-Eocene disintegration of the drainage network in the western half of Australia was one outcome of Australia's equatorward drift, as was the inception and expansion of the arid zone throughout inland Australia. A second important consequence of the post-breakup history of Australia and Antarctica was the creation of a cold circumpolar current in the newly developed Southeast Indian Ocean. From the late Eocene onwards sea-surface temperatures fell by 10–15°C in the Southern Ocean (Shackleton and Kennett 1975). Thermally isolated from warmer northern seas, Antarctica grew progressively colder. Mountain glaciers formed, coalesced, and eventually grew into major ice-caps. With the development of the West Antarctic ice-cap during the mid-Miocene, Antarctica became the land we know today: the coldest, driest continent in the world. Australia has the dubious distinction of being the hottest and second driest continent.

From about 3 Ma onwards, and particularly after the sudden growth of continental ice-caps in the Northern Hemisphere towards 2.5 Ma (Shackleton and Opdyke 1977), the northern ice became a decisive factor modulating global climates. As ice-caps waxed and waned, sea levels rose and fell. Glacial maxima were times of lowest sea level, minimum temperature, and peak intertropical aridity. Conversely, interglacials were warmer at high latitudes and wetter at low latitudes. During the past 1.8 Ma of Quaternary time there were at least 17 distinct glacial–interglacial cycles, each of about 100 ka duration (Fink and Kukla 1977). Interglacials were short-lived, occupying about 10 per cent of each cycle. The rest of the cycle was taken up by a saw-toothed build-up towards glacial maximum and sea-level minimum (Broecker and Van Donk 1970). The amplitude of near-surface air-temperature change from glacial to interglacial was on average about 5–10°C, often within a time interval of only 10 ka. The contrast between the 30 Ma involved in the 10°C cooling of the Southern Ocean and the 10 ka involved in the 5–10°C postglacial warming of Australia and New Guinea suggests that the rate of Quaternary climatic change may have been up to a thousand times faster than Tertiary climatic oscillations of broadly comparable magnitude.

(i) Present-day morphoclimatic regime

Key elements of the present climate are shown in Fig. 37A. High summer rainfall is confined to the tropical north and north-east of Australia, which is also characterized by very high seasonal runoff. Despite a six- to eight-month dry season, the northern rivers discharge more water into the sea than all but a few of the southern rivers. Sediment yields are also high in the seasonally wet tropics of Australia (Douglas 1967), and many of the northern coastal plains are actively prograding behind a protective barrier of coastal mangroves. In parts of northern Australia land is encroaching on the sea at rates of 20–30 cm/yr (Clarke et al. 1979). Well-developed platform reefs off the coast of tropical Australia and New Guinea point to generally warm sea-surface temperatures and to efficient mangrove trapping of intertidal muds minimizing turbidity off shore.

Rainfall increases with elevation in central New Guinea and eastern Australia so that lowland areas in the rainshadow of major uplands are prone to drought. The Australian arid zone, with the significant exception of the Murray Basin, is a region of internal drainage. At the distal end of the major seasonal rivers that ferry quartz sand from their head-waters to the arid depressions there are well-defined concentrations of linear dunes. About 40 per cent of the present land area of Australia is mantled by wind-blown sand, as opposed to only 20 per cent of the much drier Sahara.

(ii) Glacial aridity: Australia–New Guinea 18 ka ago

Judging from isotopic estimates of ice volume change and the best available sea-level curves spanning the last 0.7 Ma, it seems that at least the last seven glacial maxima were broadly similar in terms of global ice volume, sea-level lowering, and intertropical aridity – the latter reflected in aeolian dust input to oceans, evident in well-dated deep-sea cores (Bowles 1975). The last glacial maximum is dated between 25 ka and 17 ka, with 18 ka as the time-spike adopted by the project members of CLIMAP (McIntyre et al. 1976), although somewhat earlier time-spikes would probably be equally appropriate.

The information given in Fig. 37B is based on a variety of sources, including consensus conclusions embodied in recent reviews (Williams 1975b; Bowler et

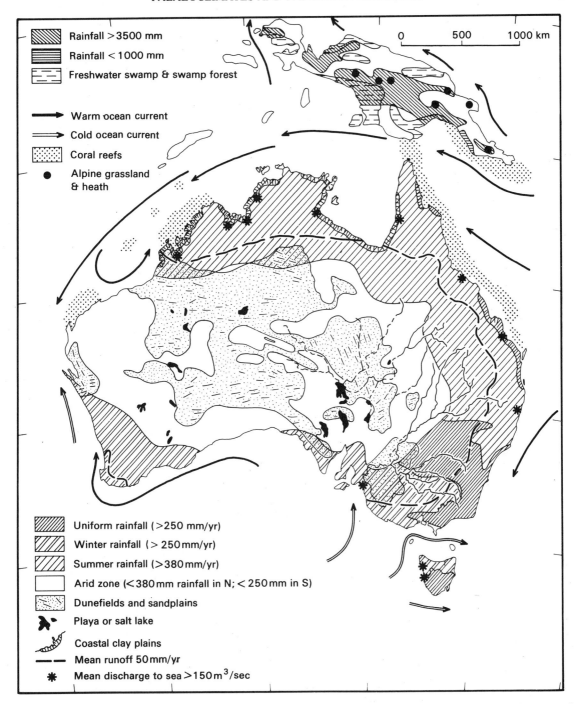

Fig. 37A. Morphoclimatic map of Australia–New Guinea at 0 ka. Isohyets and rainfall zones after Commonwealth Bureau of Meteorology (1971) and Brookfield and Hart (1966). Sandplains and dune orientations after Mabbutt and Sullivan (1970). Coral reefs, coastal clay plains, river discharge, and annual runoff after Davies (1977). Ocean currents from *Times Concise Atlas* (1972). Swamp vegetation after Flenley (1979) and Soil Map of New Guinea (1967). Alpine vegetation in New Guinea after Hope (1976).

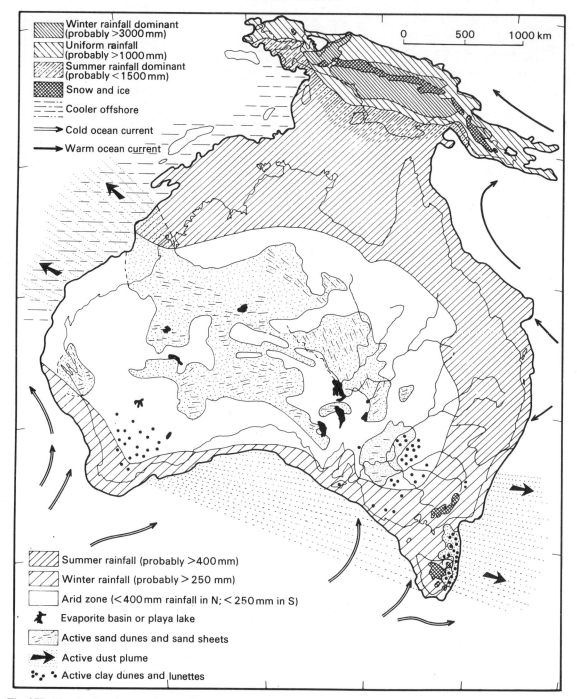

Fig. 37B. Morphoclimatic map of Australia–New Guinea at 18 ka. Rainfall zones deduced from Atlas of Australian Resources (1970) and from Brookfield and Hart (1966). Lunettes and dust plumes after Bowler (1973) and Colhoun (pers. comm.). Conservative limits of snow and ice after Galloway (1965) and Hope and Hope (1976). 18 ka shoreline and likely maximum land area here defined by 200 m submarine contour. Dunefields and sandplains after Mabbutt and Sullivan (1970). Ocean currents and cooler waters offshore deduced from McIntyre (1976), CLIMANZ (1983), and Pittock *et al.* (1978).

al. 1976; Rognon and Williams 1977; Galloway and Kemp 1980; Walker and Singh 1981; CLIMANZ 1983, and summarized by Salinger 1981).

It is unlikely that the 18 ka sea level was 200 metres lower than present. An estimate of 150 metres seems closer to the truth, so that the land area shown in Fig. 37B is a maximum value. Tasmania and New Guinea were certainly joined to mainland Australia, and much of the Sahul Shelf was dry land. Lower sea-surface temperatures to the south, west, and north-west influenced rainfall. The incidence of tropical cyclones was drastically curtailed in the more continental tropical north of Australia. Linear dunes were active out on to the now submerged continental shelf as far north as Derby in north-western Australia (Jennings 1975). Pollen spectra from Lynch's Crater on the Atherton Tableland of north-east Queensland are consistent with an effective rainfall one-third of present at this time (Kershaw 1978).

Between about 30–25 ka, when southern lakes like the Willandra lakes began to dry out, and 18 ka, when most of the lakes were dry, the Australian arid zone attained its greatest extent (Bowler 1976). East of the present 800 mm isohyet in Tasmania, now-fixed dunes were then active, as they were on Kangaroo Island, as well as in the south-western and south-eastern sectors of Australia well beyond the present arid zone. Summers were dry, hot, and windy. Rivers flowing inland from the Eastern Highlands were highly seasonal bed-load streams. Huge quantities of sand were transported into the desert by rivers flowing across the Channel Country and the Riverine Plain (Bowler 1978; Veevers and Rundle 1979; Rust 1981). Each major seasonal river had its downwind string of source-bordering dunes, including the upland Shoalhaven south of Braidwood, New South Wales. Desert dust was blown out to sea, and some high-level dust finally came to rest on the ice of central Antarctica (Petit et al. 1981). Calcareous dusts were blown off the exposed continental shelf and from the playa lake floors of the arid inland, some to accumulate as piedmont clays mantling Devonian orthoquartzites 500 km east of the lower Darling river.

In the far south-east, conditions were even more rigorous. Northward movement of the Antarctic Polar Front from about 50 ka onwards (Heusser 1981) culminated in ice buildup in Tasmania and the Snowy Mountains. The area of snow and ice shown on Fig. 37B is probably conservative (Galloway 1965; Davies 1974). Pollen spectra analysed by Dr E. A. Colhoun offer additional insight (CLIMANZ 1983). In northeast Tasmania grassland and herbaceous taxa increase at the expense of woodland, suggesting lower precipitation and possibly temperature. In western Tasmania an alpine flora at lower elevations suggests a 5°C temperature decrease relative to present. A change from eucalypt forest to savanna grassland in southeastern Tasmania coincides with the increase in aeolian activity noted earlier. At the peak of the last glaciation Tasmania was distinctly colder and windier as well as marginally drier in the east, beyond the ice-caps.

In New Guinea diminished precipitation was not likely to have been a major factor limiting plant growth except perhaps in the southern lowlands, which even today experience several months of mild drought before the onset of the summer rains. The New Guinea highlands today have a slight tendency to receive more rain in winter than in summer. With 18 ka temperatures 5 to 10°C lower, winter rainfall may have been more significant, with most of the precipitation above 3000 metres falling as sleet and snow. Pollen spectra from a number of upland sites which span the last 20 ka indicate a substantial depression of tree-line (and associated vegetation belts) equivalent to an 18 ka temperature drop of 5 to 8°C below present values (Hope 1976; Bowler et al. 1976; Flenley 1979, pp. 99–100).

Quaternary glacial maxima in Australia–New Guinea were cold, dry, and windy relative to today. Over most of the Australian mainland away from the Eastern Highlands, aridity was the dominant factor controlling erosional and depositional processes, as well as the nature and distribution of plant and animal life. Where rainfall was not a major factor limiting plant growth, as in New Guinea, Tasmania, and parts of southern and eastern Australia, lower glacial temperatures were. The seven thousand years or so leading up to 18 ka were thus a time of severe environmental stress for the Australasian biota, culminating in megafaunal extinctions. Prehistoric fires further modified the habitats, and may also have helped to accelerate hillslope erosion.

(iii) Warm, wet interglacial: Australia–New Guinea 9 ka ago

In contrast to the glacial aridity of 18 ka, the 9 ka post-glacial spike illustrated in Fig. 37C was a time of mild and humid climate in Australia and New Guinea, with temperatures slightly warmer than today. Throughout the Quaternary, interglacials were times when global ice volumes were low, sea levels were at (or even slightly above) present level, and world climates were broadly similar to those of today.

After about 18–17 ka, the sea rose, rapidly at first, then more slowly. Present level was reached towards 7–6 ka in Australia–New Guinea (Thom and Chappell

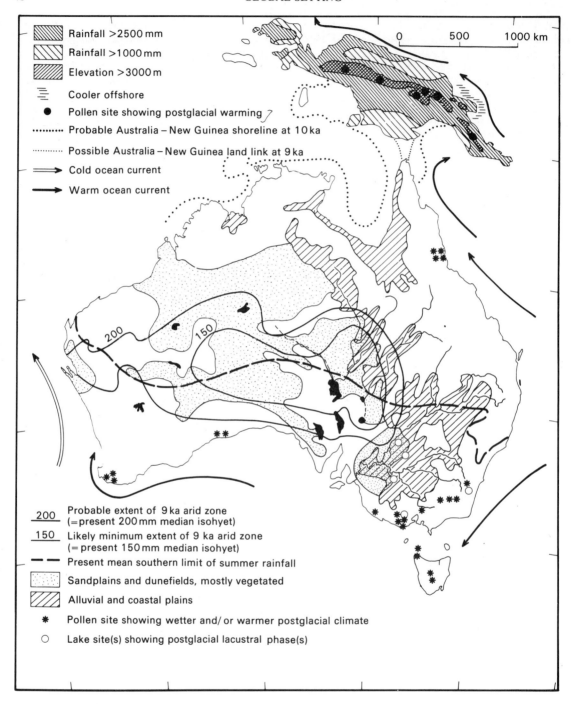

Fig. 37C. Morphoclimatic map of Australia–New Guinea at 9 ka. Arid-zone limits deduced from isohyets in Atlas of Australian Resources (1970). New Guinea rainfall deduced from Brookfield and Hart (1966). Summer rainfall limits after Commonwealth Bureau of Meteorology (1971). Coastal and alluvial plains, dunefields, and sandsheets after Mabbutt and Sullivan (1970). Pollen sites and lakes after Walker and Singh (1981) and CLIMANZ (1983). 10 ka Australia–New Guinea partial shoreline after Chappell (1976). 9 ka land link deduced from Jennings (1971). Cooler offshore zone deduced from Aharon (CLIMANZ 1983). Ocean currents deduced from *Times Concise Atlas* (1972).

1975), after which the available evidence does not support eustatic fluctuations of more than about one metre, discounting local and regional tectonic and isostatic effects. The land link with New Guinea was probably severed soon after 9 ka, by which time the Bassian plain that once joined Tasmania to the mainland was also submerged (Jennings 1971). Flooding of the Sahul Shelf was accompanied by an influx of warm sea water into the Gulf of Carpentaria as the equatorial ocean current resumed its flow westwards through the Torres Strait. Postglacial warming of the tropical oceans was very evident towards 13–11 ka, as was a rapid increase in summer rainfall over northern Australia. Several factors contributed to this rainfall increase, including a general reduction in continentality, marine flooding of the Great Barrier Reef, intensification of the summer monsoon, an increase in tropical cyclone activity, and weakening of the anticyclones inland. The net effect was an expansion of forest at the expense of grassland, stabilization of hitherto active dunes, and a general contraction of the Australian desert.

Lake levels rose again in the far south, but many lakes in the present semi-arid zone did not receive much runoff, and so failed to attain their 40–30 ka preglacial shoreline levels. Pollen evidence from Lake Frome shows that summer rains reached as far south as 30–31° S during the early to middle Holocene, but after about 4–5 ka the summer rains became more erratic in southern Australia, and the levels of many lakes fell (Singh 1981).

Ice retreat from Tasmania and from all of New Guinea except the Carstenz Mts. was accompanied by a temperature increase of 5–10 °C, culminating in a 1000 metre upward shift in vegetation belts by about 9–7 ka. Rain forest or eucalypt woodland replaced alpine grassland or heath, glacial moraines became at least partially vegetated, and periglacial solifluction mantles became fixed and stable. Renewed periglacial activity and hillslope instability from about 5–4 ka onwards in upland Australia and New Guinea reflect a return to colder, less equable conditions during the late Holocene.

Establishment of the long summer wet-season in tropical Australia–New Guinea towards 9 ± 2 ka had several effects. Rivers draining the Eastern Highlands became far less seasonal, and began to carry and deposit a suspension load of silt and clay. In the Channel Country the thin surface mantle of clay represents the overbank deposits of Holocene high-sinuosity, suspension-load streams that are superimposed on the sandsheets of the Pleistocene braided-river system (Veevers and Rundle 1979). The same fining-upward alluvial sequence is evident in late Quaternary distributary channels of the Riverine Plain (Bowler 1978).

An interesting possible by-product of a stronger northwesterly monsoon towards 9 ka is the possible 1.5–2 °C sea-surface cooling detected isotopically by Dr P. Aharon (CLIMANZ 1983) off the Huon Peninsula in New Guinea. A similar effect is evident off the east coast of Saudi Arabia during the early Holocene, and is attributed by Duplessy (1982) to monsoon-induced cold upwelling offshore.

(iv) Conclusions

(1) The contrast between the morphoclimatic regimes in Australia–New Guinea at 18 ka and at 9 ka epitomizes the difference between glacial and interglacial environments in this region throughout the Quaternary, during which time there were at least 17 glacial–interglacial cycles, each of about 100 ka duration. During the present postglacial interval, peak temperatures were reached towards 9 ka ago, which is about 3 ka before the sea attained its present level. Pollen spectra from eastern Australia and New Guinea suggest that towards 9 ka mean temperatures were up to 1 °C higher than they are today.

(2) Glacial maxima were characterized by sea levels as much as 150–200 metres lower than present, resulting in the merging of New Guinea, Tasmania, and other islands with the Australian mainland. Temperatures were 5–10 °C lower, tree-lines were depressed, and summer rainfall was very much curtailed. Now-vegetated dunes were active throughout a much expanded arid zone. The overall climate was more continental: colder in winter, hot and windy in summer. Perennial ice-caps were a feature of the uplands of New Guinea, Tasmania, and the Snowy Mountains.

(3) Interglacial climates were warm and wet relative to glacial maxima. The ice melted, the sea rose to present levels, and summer rainfall was again high. Desert dunes became vegetated, montane forests encroached on alpine grasslands, and tropical rain forests expanded once more. Lake levels rose in the wetter parts of southern Australia, and mean temperatures rose rapidly from about 11 ka onwards, as did annual rainfall.

(4) The contraction and expansion of the Australian arid zone in sympathy with the rise and fall of the sea level is consistent with a periodic oscillation from glacial aridity to interglacial forest expansion throughout the intertropical zone (Street 1981).

(b) Phanerozoic climates and environments of Australia (by P. G. Quilty)

(i) Introduction

The methods of interpreting past climates are described by Frakes (1979). Care must be taken not to interpret too much from the study of rocks and their contained fossils. For example, non-marine sediments deposited since the mid-Cretaceous are thin and sporadic over large areas of continental Australia. The sediments represent fluvial or lacustrine accumulations in local drainage basins, and in places contain evidence of vegetation near a river channel. This should not be taken as firm evidence against aridity, because even in arid environments ephemeral streams may be bordered by trees, or pass into local lake systems. The Sudd in Africa is a classic example of a humid locality in an arid region (Williams and Adamson 1973).

The present marked differentiation of climatic zones between the equator and poles differs from many past patterns. For example, the Late Cretaceous and Palaeocene were times of low differentiation of climatic zones, and the Maastrichtian rocks of Antarctica contain warm-water foraminifera that suggest warm marine conditions at high latitude. Thus there are grounds for error in assuming that climatic zones have remained constant and that continents have simply moved through them under the influences of continental dispersion.

Much of the material concerning Mesozoic and Cainozoic climates is drawn from Quilty (1984). A parallel review of Australian Phanerozoic climates, seen after this account was written, is by Frakes and Rich (1982).

(ii) Cambrian (575–500 Ma)

During the Cambrian, Australia was warm to hot and arid (Cook 1982), consistent with the palaeolatitude as indicated by palaeomagnetism (Embleton, above). Data on the Early Cambrian are sparse but the analogy of South Australian archaeocyathids with modern colonial corals has been used to infer warm-water conditions off an arid hinterland. Interpretations in the Ordian (early Middle Cambrian) and later are based on features of sediments, such as phosphorites, pseudomorphs after halite (Southgate 1982), and gypsum and other evaporites. The extent of aridity increased from Early Cambrian to a maximum craton-wide in the Ordian, contracted again to the area of South Australia during the rest of the Middle Cambrian, and expanded again in latest Cambrian, but not as much as in the early Middle Cambrian.

(iii) Ordovician (500–435 Ma)

The Ordovician (Webby 1976, 1978) began with sedimentation in the Canning Basin and in a large area on the east, the two areas being separated by land of moderate relief. Warm conditions are indicated by carbonate in the Georgina Basin, and halite pseudomorphs in the Daly River Basin and Amadeus Basin. Through the latest Early and the Middle Ordovician, Australia was separated by the Larapintine Sea into a humid equatorial zone of low relief on the north and an arid zone of high relief on the south. In the Middle Ordovician, the north-west was cooler than the east, probably because of a counter-clockwise gyre in the sea to the west of Australia which circulated cooler water originating in the glaciated area over the South Polar region, then centred in north Africa (Frakes 1979). In contrast, eastern Australia was further north, and was bathed by a warm current from a tropical source. Mingling of the nutrient-rich cooler (but still warm) western water and warmer eastern water is suggested by phosphorites in central Australia.

The Ordovician generally was warm as indicated by colonial corals and the high diversity in organisms, especially molluscs and graptolites. Non-biological support for this interpretation includes halite pseudomorphs and desiccation cracks.

By the Late Ordovician, the Larapintine Sea had contracted, deposition in north-western Australia ceased, and the Late Ordovician is thus interpreted from eastern Australia, where warm-water carbonate sediments had become restricted to the shelf areas of Tasmania.

(iv) Silurian (435–410 Ma)

Probably Silurian aeolian, deltaic, and lacustrine sediments that have the potential to tell us much about environments and climate are known from several areas of Australia, but precise time-control is usually lacking. This contrasts with the high precision of correlation frequently deduced from pelagic elements in the marine sequences of the Tasman Mobile Belt of eastern Australia (Talent et al. 1975). Among these sequences from eastern Victoria to northern Queensland are numerous occurrences of Early and Late Silurian limestones, many biostromal, that by the abundance and diversity of rugose and tabulate corals can be taken to be consistent with warm water. The Mereenie Sandstone of central Australia, partly Devonian in age, is a record of predominantly aeolian and fluvial environments (Wells et al. 1970); within it may be a record of aridity during some or all of

Silurian time on the craton. Better age-control is available for the thick Late Silurian (Ludlovian) carbonate and evaporite sequence of the Dirk Hartog Limestone in the Carnarvon Basin of Western Australia (Playford *et al.* 1975). In summary, the evidence for Silurian time is for warm waters on both sides of the continent and for areas with evaporitic and probably aeolian environments.

(v) Devonian (410–360 Ma)

Similar environments persisted into the Devonian, as witnessed by the thick evaporitic Carribuddy Formation (probably Early Devonian, Wells 1980) of the Canning Basin and similar strata in the Bonaparte Gulf Basin, and evaporites in the Middle Devonian of the Adavale Basin, the Mereenie Sandstone of the Amadeus Basin, the Craven's Peak Beds of south-west Queensland, and the Grampians Group of western Victoria, the last a great thickness of fluvial sandstones and 'redbeds' of age seemingly somewhere in the range Late Silurian–Early Devonian. Redbeds underlying, intercalated with, or conformably overlying Early Devonian marine sequences have been recorded from Ravine, Taemas, and Wee Jasper, in New South Wales. The Early and Middle Devonian marine sequences of eastern Australia are notable for the high species-diversity of some units, such as the Buchan and Taemas Groups and Garra Formation of south-eastern Australia (the last with more than 550 species; principally J. R. Farrell, B. D. Johnson, and A. C. Lenz, in preparation), the richness of the coral faunas, and the presence of small but indubitable bioherms especially in the north Queensland sequences (Wyatt and Jell 1980). All of this is consistent with warm waters. The latest Middle Devonian to Early Carboniferous sequence of the northern Canning Basin is also richly fossiliferous; the main, Devonian, part is a spectacular reef complex, one of the best examples of reef development in the geological record (Playford 1980). This must be good evidence for warm waters. Redbeds are abundant in the widespread Late Devonian non-marine sequences of eastern Australia and persist into the Early Carboniferous of the Mansfield to Mount Wellington area of Victoria. They presumably signify warm humid climates, but, as elsewhere in Australia, they have not been subjected to the close scrutiny necessary to determine their precise environmental significance. Coaly material (thin layers of vitrinite) in the Late Devonian piedmont deposits of the Amadeus Basin constitutes the oldest-known coal in Australia.

(vi) Carboniferous (360–290 Ma)

A unique occurrence of Tournaisian reef limestone in the Bonaparte Gulf Basin (Roberts and Veevers 1973) indicates the persistence of warm water in the northwest. A relatively high diversity of faunas persisted through the Tournaisian and Visean (Talent, below) to be followed by a rapid drop in diversity that seems to be connected with the southward movement of the Australian block indicated by the palaeomagnetic data (Embleton, above). Thin, rare coal started accumulating in eastern Australia from the Visean. Tillites and fluvio-glacial sediments in the Late Carboniferous of New South Wales testify to the onset of refrigeration that became widespread in the Early Permian (Crowell and Frakes 1971a, b, 1975).

(vii) Permian (290–245 Ma)

Glaciation is indicated by striated and faceted boulders, tillites, glacial pavements, varves, glendonites (calcite or siderite pseudomorphs after glauberite), and dropstones (Crowell and Frakes 1971a, b, 1975), associated cold-water benthic marine invertebrates (Dickins 1978), and oxygen-isotope studies (Rao and Green 1982). The peak of the glaciation in the late Stephanian and early Sakmarian was much colder and more extensive than the only other known Phanerozoic glaciation during the Quaternary. The causes of the glaciations include the poleward migration of Australia and the concomitant development in eastern Australia of the Kanimblan highlands.

The Late Carboniferous alpine glaciation intensified to a continent ice-sheet glaciation in the late Stephanian and early Sakmarian (Fig. 155), which waned through the rest of the Permian (Crowell and Frakes 1971a, b, 1975) to a termination in the Kazanian. Eastern Australia remained colder than the west during the Artinskian (Dickins 1978). Towards the end of the Permian, marine invertebrates in the north-west indicate an emphatic change to warmer conditions, probably reflected in eastern Australia by the main interval of coal deposition. These and other widespread coals were deposited after the peak of the glaciation, and indicate high water-tables and poor drainage in the basins.

(viii) Triassic (245–200 Ma)

Frakes and Rich (1982) infer from the lack of Early Triassic coal deposits that this interval was relatively dry. Dolby and Balme (1976) differentiated two microfloras in the Middle and Late Triassic. The Onslow Microflora contains both Gondwanan and European elements, and was taken to indicate temperate rainforests. The Ipswich Microflora, coeval with the Onslow Microflora, is less diverse and contains no European elements, and was taken by Dolby and Balme to indicate colder conditions. Townrow (1964) suggested a cool-temperate climate for the Rhaeto-Liassic in Tasmania. However, it must be emphasized that the Tasmanian results are based on the study of coal-measure floras, which may be of very local significance. Douglas and Ferguson (1976) suggested a cool-temperate climate for the Victorian Triassic. Widespread Late Triassic coal measures in eastern Australia indicate a return to high water-tables in the basins.

(ix) Jurassic (200–129 Ma)

The climate during most of the Jurassic was less differentiated than now. There is no evidence of any polar ice (Hallam 1975, Quilty 1978b). It seems that the cosmopolitanism of the biota in the Early and Middle Jurassic indicates uniform conditions world-wide, although some differentiation is still possible.

Townrow (1964) suggested a temperate climate for Tasmania in the earliest Jurassic. Bowen (1961) gave palaeotemperature measurements of 18.5–29.2 (average 24.4) °C for belemnites from the Middle Jurassic Newmarracarra Limestone in Western Australia. These results have been questioned because the area was deeply weathered in the mid-Tertiary. However, Bowen's results are in general agreement with the result obtained by detailed palynology by Filatoff (1975), who suggested that the Early Jurassic of the Perth Basin was warm to hot and arid, and that the Middle/Late Jurassic was warm to hot and wet; other evidence of warm marine conditions (large foraminifera, corals, limestone) is lacking.

From oxygen-isotope studies of Late Jurassic belemnites from New Zealand, Clayton and Stevens (1967) and Stevens and Clayton (1971) found a minimum palaeotemperature of 15 °C.

(x) Cretaceous (129–65 Ma)

As in the Jurassic, evidence of warm seas is lacking, as shown by the low diversity of the Early Cretaceous planktic foraminiferal faunas (Scheibnerova 1976). Waldman (1971) concluded that the fish fauna in the Early Cretaceous Strzelecki Group at Koonwarra, 120 km south-east of Melbourne, indicates mass mortality ('winterkill') by anoxic winter conditions beneath an ice cover in a lake. Gill (1972) summarized the oxygen-isotope data of Dorman and Gill (1959) and Dorman (1968), to show a temperature gradient from cool in the south-east to warmer in the north-west, consistent with the distribution of reptiles.

In the later Cretaceous, the foraminiferal fauna of the western margin indicates an open warm sea; a hint of 'tropical' large foraminifera to the north-east is provided by Glaessner's (1960) record of *Pseudorbitoides israelskyi* from Papua. India had cleared south-west Australia but Antarctica and Australia were effectively united, so that a northerly current probably passed along the western margin (Gordon 1973). Douglas and Ferguson (1976), from palynological and foraminiferal analyses, suggest that the south-east was cool until the Santonian, with rising temperature and humidity to the end of the Cretaceous. Oxygen-isotope studies of belemnites (Stevens and Clayton 1971) suggest that the sea-water of New Zealand was markedly warmer than at present in the Turonian-Coniacian, was only slightly warmer than at present in the Santonian, and became warmer again in the Campanian and Maastrichtian. Webb and Neall (1972) report warm-water foraminifera from Taylor Valley, Antarctica, so that the warm current along the east Australian margin must have carried warm water to a high palaeolatitude. Sporadic coal swamps in the Gippsland, Bass, Otway, and Great Australian Bight Basins indicate that the water-table along the southern margin was locally high.

(xi) Tertiary (65–2 Ma)

The environments of the Tertiary are described for the depositional cycles (Fig. 38) (Quilty 1977, and others; see Fig. 151) that correlate with eustatic highstands of the sea, and concomitant tectonism (McGowran 1979; Jones and Veevers 1982). Here we attempt to extract the climatic components from the record; for their interplay with tectonic events, see McGowran (1979), who describes also the extra-tropical excursions of large foraminifera.

(1) Late Palaeocene–early Eocene

The Palaeocene contains cosmopolitan marine faunas that suggest low differentiation of polar to equatorial environments. Palynological evidence (Kemp 1978) suggests temperate conditions with high rainfall in the south-east and south-west, and probably in the interior, as confirmed by the dating of deep weathering profiles in south-western Queensland as Palaeocene and Eocene (Fig. 38) (Idnurm and Senior 1978). The tentative atmospheric circulation is shown in Fig. 39. Oceanic circulation (Edwards 1975; Deighton *et al.* 1976) included a warm northerly directed current along the western margin and a southerly directed offshoot of the Pacific equatorial current between the east coast and Lord Howe Rise–New Zealand, as indicated by the high (18–20 °C) surface-water temperature on the

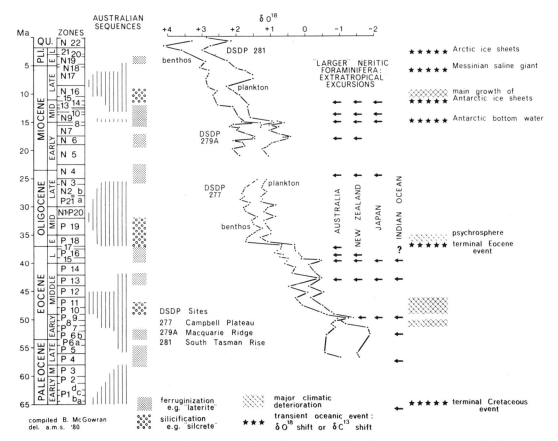

Fig. 38. A synopsis of Cainozoic palaeoclimatic data, from McGowran (1981, fig. 21) after a previous version (McGowran 1979, fig. 6). 'The Australian stratigraphic record occurs as four packets of strata shown schematically under "Australian Sequences" (vertical ruling represents hiatuses).

'For most of the Cainozoic (before Mid Miocene) δO^{18} (proportion of oxygen isotope) shifts reflect temperature changes (heavier=cooler); from the late Mid Miocene, the ice effect increases. The curves are based on benthonic and planktonic Foraminifera from oceanic carbonates (DSDP=Deep Sea Drilling Project of the US National Science Foundation). Three major climatic "steps" in the overall climatic deterioration are identified at the boundaries of the Early/Mid Eocene, Eocene/Oligocene, and Mid/Late Miocene. Six particularly important oceanic changes, identified by shifts in stable isotope values, are shown as transient oceanic events. The flights of arrows identify extratropical excursions by larger benthonic foraminifera which inhabit primarily the tropical neritic environment. There is a satisfactory correlation in several cases of this neritic biogeographic data with "warm peaks" in the isotope curves based on oceanic material. Until the Early Eocene (Zone P7) climate was warm at high latitudes. During the subsequent decline, two periods of "restoration" are indicated in Mid to Late Eocene and Early to Mid Miocene, but "climate" fluctuated rapidly during those times. Chemically formed duricrusts are listed as a series of rapid changes. This is (a) based on limited evidence for age, and (b) extended generously on the now impressive accumulation of cross-correlated evidence that Cainozoic climatic change was rapid, far-reaching and bipolar. Indeed, Tertiary climatic shifts occurred on scales of 10^5 years.'

Campbell Plateau found by Shackleton and Kennett (1975), and shown in Figs. 38 and 40.

(2) Middle–late Eocene

Large foraminifera are common in the shallow marine sediments of Western Australia and New Guinea and occur to the south-east as far as Esperance (Cockbain 1967). These indicate warm waters, confirmed by Cockbain's (1969) record of warm-water dasycladacean algae at Esperance and by Hos's (1975) analysis of palynomorphs which suggested a warm, humid climate. A humid climate is indicated also by terri-

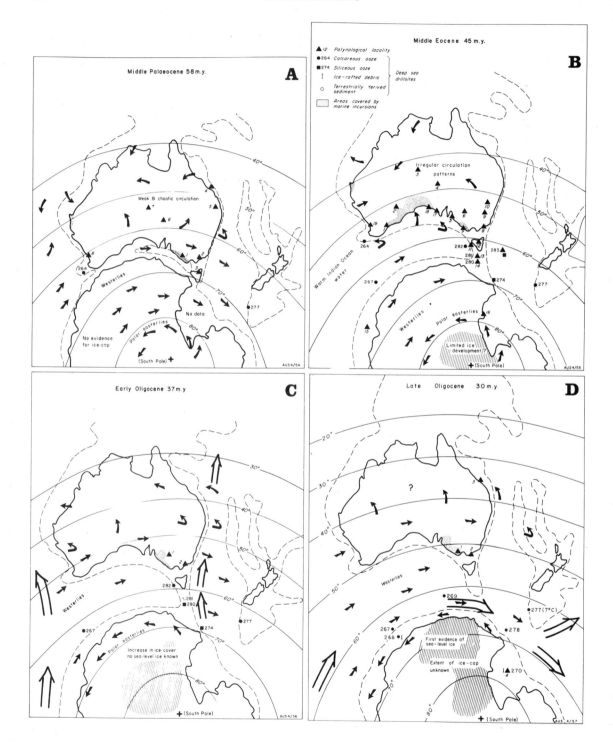

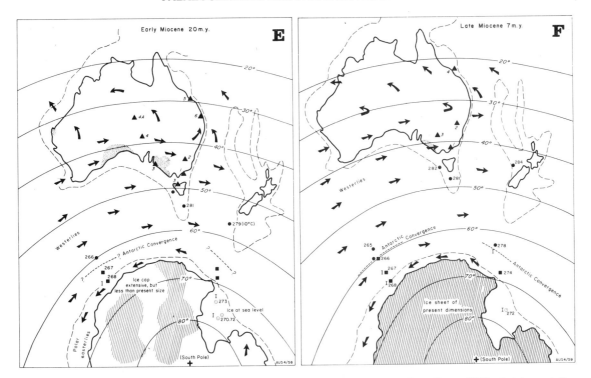

Fig. 39. Palaeogeographical reconstruction of the South-east Indian Ocean region (from Kemp 1978), showing tentative atmospheric circulation (heavy arrows). The broad arrows in C and D indicate bottom-water circulation (from Fig. 133). The position of Antarctica is assumed to be fixed in its present position, with the result that the palaeolatitude differs slightly from that given by Embleton above. Likewise, the fit of Australia and Antarctica and the time of breakup differ in detail from those given by Johnson and Veevers above (Figs. 18–26). The broken lines approximate the present position of the 3 km isobath; the northern margin is left vague. Further explanation is given in B.
A. Middle Palaeocene. B. Middle Eocene, with palynological and deep-sea sedimentological data spanning the middle–late Eocene interval. C. Early Oligocene. D. Late Oligocene. E. Early Miocene. Palynological and deep-sea sedimentological data spanning the early–middle Miocene interval. F. Late Miocene.

genous sediments in the Eucla Basin, Bremer Basin, onshore Carnarvon Basin, and in the area of Perth, and by lignite in southern Australia and oil shale in Queensland, fostered by high water-tables in graben valleys. An effective drainage system was active, the last time in which this seems to be true. Davies et al. (1977) have suggested that there was a world-wide middle Eocene interval of higher sedimentation and runoff, and, in the late Eocene, the beginning of an arid, low runoff phase. The common differentiation in Australia of middle Eocene terrigenous from late Eocene carbonate marine sediment agrees with this.

In south-eastern Australia, drainage systems supplied terrigenous sediment to be redistributed by the sea as marine conditions became general across the southern margin. The characteristics of several species of foraminifera (unpublished information) would suggest that a temperature gradient from higher in the west to lower in the east may have prevailed. This is in conflict with Edwards's (1975) suggested gyre development with its westerly flowing current in the Great Australian Bight. The South Tasman Rise was still an effective deterrent to current flow south of Tasmania. Because no general circulation existed south of Australia, the east and west coasts probably were washed by the same currents as in the Palaeocene and Late Cretaceous.

Oxygen-isotope data are available from Dorman (1966) and Shackleton and Kennett (1975). Hornibrook (1971), Dorman (1966), and evidence from foraminifera from Australia all suggest warm conditions in the late Eocene, after a cooler early and middle Eocene, in contrast to the data in Shackleton and Kennett (1975), which imply a general cooling of surface waters over the Campbell Plateau from the late Palaeocene through late Eocene. The pattern of Australia and New Zealand in the Eocene being considerably warmer than the waters of the Campbell Plateau is in keeping with the palynological evidence of Kemp and Barrett (1975). The general change in surface waters of the Campbell Plateau is reflected in the deeper waters also. There was still no significant difference in temperature between surface and bottom waters. Thus the later substantial cool current caused by major glaciation on Antarctica was lacking, and *Nothofagus*-dominated vegetation was probably still widespread on Antarctica (Kemp and Barrett 1975).

Wopfner *et al.* (1974) suggest increasing seasonal aridity to explain the development in the Eocene, and possibly continuing into the Oligocene, of the widespread silcrete of the interior. Kemp (1978) (Fig. 39) cites evidence for widespread rain forest as indicating high humidity at the Eocene palynological sites (all in the southern half of Australia), with some seasonal effects in the interior.

(3) Oligocene–middle Miocene

By the start of the Oligocene, the South Tasman Rise had cleared north-eastern Victoria Land (Fig. 23), and oceanic circulation between Australia and Antarctica (Circum-Antarctic Current) developed fully by the end of the Oligocene, as shown by the open arrows in Fig. 39, and by Fig. 133. From deep-sea drilling at Macquarie Ridge and the Campbell Plateau, Shackleton and Kennett (1975) found a marked cooling of surface waters which then remained approximately constant throughout the Oligocene, early Miocene, and early middle Miocene, after which was another marked cooling (Fig. 38). Also prominent from Shackleton and Kennett's data is the continuing decline in temperature of deeper water, even as the surface remained approximately constant. The early Oligocene temperature decrease is in agreement with the interpretation of Edwards (1968), based on calcareous nannoplankton.

To these data must be added the information summarized by Hayes and Frakes (1975), Kemp (1975), and Kemp and Barrett (1975) on the onset of Antarctic glaciation, based on analysis of ice-rafted detritus, sedimentation rates, and palynology. Glaciation seems to have become widespread in the late Oligocene and to have removed effectively any vegetation from Antarctica. The last phases of vegetation lived in an 'alpine temperate-type of glacial environment' (Kemp 1975). In early and middle Miocene times, glaciation expanded and covered essentially all of Antarctica but did not develop extensive ice shelves. The rapid deterioration in climate and the extent of the late Oligocene ice-cap probably meant the extinction of the Antarctic terrestrial biota and probably also the destruction of most evidence of the earlier biota.

According to Kemp (1978), the low diversity of the pollen spectrum in south-eastern Australia correlates with the lowered temperatures suggested by the oxygen-isotope data. 'The observation that some coals in the early Oligocene are dominated by pollen of the *Dacrydium franklinii* type may also reflect relatively low temperatures. Abundant rainfall is suggested by the high volume of coals and lignites – it seems highly probable, from Figs. 6 and 7 [our Figs. 39C and 39D], that this south-east region would have come under the influence of rain-bearing westerly winds throughout the year. It is also likely that precipitation over Australia as a whole would have been reduced from that of the Eocene, with the increase in Antarctic glacial activity, but there are no data available to support the suggestion that arid regions developed in the north and north-west.'

The Oligocene in central Australia is represented by the continued generation of silcrete (Wopfner *et al.* 1974; Callen and Tedford 1976) and in Western Australia by laterite (Johnstone *et al.* 1973; Schmidt and Embleton 1976), both suggesting seasonal humidity. Davies *et al.* (1977) suggest world-wide aridity in the Oligocene; this may be reflected in the near lack of terrigenous sediment carried by the drainage in Australia during the Oligocene–middle Miocene, perhaps even less than would be expected from the present drainage. In the late Oligocene and early Miocene in particular, marine sediments are dominantly biogenic carbonate, to a degree much greater than at any time before or since, suggesting that little sediment from the land reached the sea except in New Guinea. The sea round most of Australia was warm; Australia had moved a considerable distance northward from Antarctica, the Circum-Antarctic Current was fully developed, and planktic foraminiferal faunas from the north-west and west, New Guinea, and Queensland have tropical affinities; the south-eastern faunas are less diverse, and indicate a lower water-temperature except in the late early Miocene when large foraminifera reached as far south as Tasmania (Quilty 1972).

By the middle Miocene, the climate had changed dramatically. The sediment that contains the vertebrate faunas of the Lake Frome–Lake Eyre area shows evidence of large lakes and wooded areas with eucalypts (Stirton *et al.* 1967; Callen 1977), suggesting more hospitable conditions there than at the present. Callen (1977) further suggested seasonal aridity and low relief. The fossils recorded by Lloyd (1968a, b) from northern Australia also suggest a more humid phase.

Kemp (1978) concluded that palynological data for the Miocene suggest a dominance of rainforest in those regions for which such data are available. In the east and southeast, there is little evidence of open vegetation throughout the epoch, although the pollen spectra show that some of the major elements of modern open forest, such as *Acacia*, Compositae and Gramineae, were already established. Closed forest, dominated by *Nothofagus brassi* species and by gymnosperms, appears to have extended from western Victoria to coastal Queensland, and inland to western New South Wales. In central Australia however, at localities near Lake Frome in central South Australia, and at Ti Tree in the southern part of the Northern Territory, there is evidence of more open grassland vegetation in what is broadly a middle Miocene interval. Even there, however, rainforest trees apparently persisted along watercourses, suggesting that conditions remained relatively humid.

The picture of continuing high precipitation in the southeast accords with the circulation patterns postulated in Figs. 8 and 9 [our Figs. 39E and 39F]. Throughout the Miocene this region lay in latitudes in which the influence of westerly wind systems would have been felt throughout the year. In the early Miocene in particular, waters adjacent to the continent were relatively warm, and there were transgressive shallow seas in the south and east, so that heavy precipitation would have been a likely result. The inland, middle Miocene locality near Lake Frome may have been marginal to the westerly wind belt, and may thus have experienced some seasonal variation in precipitation, a climatic pattern which accords with the presence of grassland.

(4) Late Miocene and Pliocene

Again, according to Kemp (1978)

The general increase in aridity which is recorded in the Murray Basin, and probably later in the Gippsland Basin, was no doubt associated with continued northwards motion of Australia into drier climatic belts, and with a precipitation decrease related to increasing Antarctic ice build-up. It is tempting to relate the disappearance of *Nothofagus brassi* species from regions west of the Great Dividing Range to the cooling event of the latest Miocene. A dramatic decrease in ocean temperatures, plus an increase in the volume of Antarctic ice, would have led to a widespread precipitation decrease; in inland areas such as western New South Wales this may have been critical enough to make survival of this rainforest type no longer possible.

Shackleton and Kennett (1975) show that on the South Tasman Rise a marked decrease in sea-water temperature started in the late Miocene, 7 Ma ago, interpreted as reflecting the increase of glacial activity on Antarctica. The effects of this change become most obvious in the late Pliocene when the major worldwide glacial period seems to have begun. The correlation in this time interval between sea level and glacial activity is obvious; earlier sea-level fluctuations were probably tectonically controlled.

Dorman's (1966) oxygen-isotope data from southeastern Australian molluscs also show late Pliocene cooling. Stratigraphic data from Singleton *et al.* (1976) show that this change occurred in approximately late Pliocene (N20) time. The change is marked not only by oxygen-isotope changes but also by decreased planktic-foraminiferal diversity. Comparable information is available in north-western Australia. Quilty (1974) mentioned briefly results based on studies of planktic foraminifera from North Tryal Rocks Well. *Globorotalia cultrata* (d'Orbigny) is sinistrally coiled in the late Miocene, changes to dextral (indicating warming) in the earliest Pliocene – almost exactly the reverse of the pattern expected from Shackleton and Kennett's curve – and changes back to sinistral (implying cooling) late in the interval N19/20. Immediately after the last instance of cooling seen in North Tryal Rocks–1, a marked decrease in the planktic-foraminiferal content of the foraminiferal faunas was taken as indicating a significant shallowing of the water, due to the progradation of the edge of the continental shelf or to a glacio-eustatic fall in sea level, or both.

For earlier time intervals, particularly the late Eocene, late Oligocene, and middle Miocene, larger foraminifera have been taken as indicating warm water, by analogy with the distribution of modern forms. In the late Miocene–Pliocene, little evidence of larger foraminifera in most areas studied (Fig. 38) suggests generally cooler conditions. This evidence agrees with the oxygen-isotope studies but is not binding because depositional facies do not appear to have been favourable for larger foraminifera.

In summary, though the data for this time interval are sparse, it seems that the late Miocene–Pliocene was generally a time of increasing aridity and of generally cooler sea-water temperatures, decreasing markedly at the end of the interval with the onset of a Quaternary glacial interval described above by Williams.

(xii) Summary

Fig. 40 shows the variation through the Phanerozoic of latitude, evaporites, glaciation, temperature, coal seams, and rainfall in Australia, compared with the global pattern of temperature and rainfall (dotted lines), as summarized by Frakes and Rich (1982). The latitude refers to Alice Springs, from which the continent extends on all sides for about 2000 km or 20°. The latitude shown is therefore only a central value of an envelope up to 40° wide. Covering such an area, Australia, from one extremity to the other, has notable differences in climate today and presumably also in the

56   GLOBAL SETTING

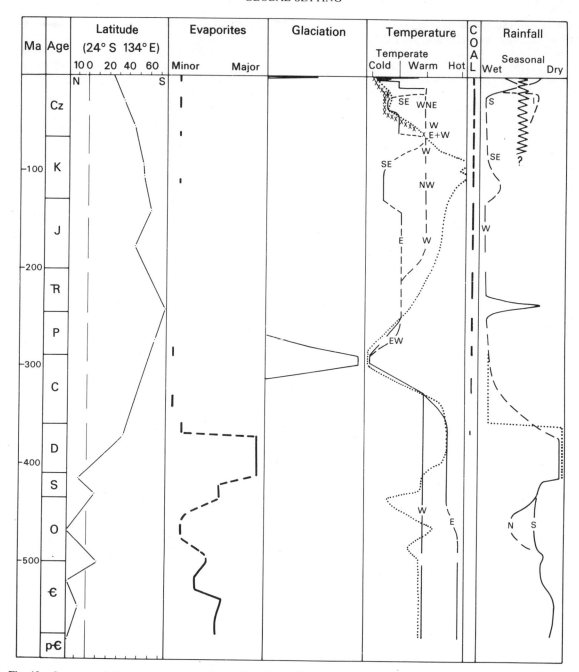

Fig. 40. Summary of changing latitude (Embleton, Fig. 11) and climate in the Phanerozoic.
  (1) Evaporites, from Wells (1980). The scale ranges from major occurrences of thick-bedded halite in the Devonian (740 m in the Carribuddy Formation, McLarty–1 Well, Canning Basin, and 472 m in the Boree Salt Member, Boree–1, Adavale Basin), through intermediate occurrences in the Cambrian (exceeding 200 m of halite and anhydrite in the Amadeus Basin) and Silurian (36 m of halite in the Carnarvon Basin), to minor occurrences: halite pseudomorphs (Ordovician of Amadeus and Daly River Basins, Late Devonian of Amadeus Basin); patches, veins, and interbeds of anhydrite (Sakmarian of the Browse Basin, Early Carboniferous of the Canning Basin (Anderson Formation),

past. The correlations that follow are of events that transcend the individual regions. Correlations with the independently determined latitude are as follows. (1) Evaporites (with the notable exception of the Sakmarian anhydrite in the Browse Basin) and the related aridity correlate with latitudes lower than 55°, and the major (Palaeozoic) occurrences with latitudes lower than 40°. (2) Temperature correlates negatively with latitude, with an offset of 40 Ma between the early Sakmarian glacial maximum and the Permo-Triassic maximum latitude. An obvious mismatch is the cold Quaternary and low latitude, to be expected from the wide latitudinal extent of this glaciation. (3) Higher rainfall correlates with latitudes greater than 30°, and seasonal rainfall with latitudes less than 50°.

Frakes and Rich (1982) point out that the Phanerozoic climates of Australia reflect Australia's changing latitude and the changing global climatic state. Their summary of Phanerozoic global climates is shown in Fig. 40 by the dotted lines in the columns for temperature and rainfall. The globe was warm from the Cambrian to the end of the Devonian, with rapid oscillations of climate in the Ordovician; aridity in the Devonian changed to humidity during the Carboniferous. Glaciation affected a large part of Gondwanaland during the second half of the Carboniferous and the first quarter of the Permian. By the mid-Permian, the globe became relatively warm. 'The earth apparently reached its thermal maximum during the Mesozoic Era, between the middle Triassic and the late Cretaceous. Oxygen isotopes define two peaks of warmth in the Cretaceous–Albian (100 Ma) and Coniacian–Santonian (85 Ma), the latter synchronous with the highest peak of the Phanerozoic sea level curve of Vail *et al.* (1977). The latest Cretaceous began the long and intermittent cooling which resulted ultimately in the Pleistocene glaciations. Large temperature drops occurred in the oceans at the end of the Eocene (39 Ma) and in the middle Miocene (13 Ma), the latter apparently being the time during which most Antarctic ice accumulated. The last two million years has been characterized by advances and retreats of ice sheets in both hemispheres and by cooling of the oceans on a cyclicity of about 100,000 years. The last interval of significant ice growth ended about 18,000 years ago.'

The Australian record follows the global pattern except that Australia remained cool to warm during the Mesozoic global thermal maximum, probably due to Australia's fairly high Mesozoic latitude.

4. AUSTRALIAN BIOGEOGRAPHY PAST AND PRESENT: DETERMINANTS AND IMPLICATIONS (by J. A. Talent)

In this section we examine cursorily the distinctiveness of the contemporary Australian fauna and flora in relation to global biogeography and in turn will consider the chronology of formation and destruction of island 'staging-points' and land-bridges that have been determinants of migration to and from Australia, the sequence of past faunas and floras and their palaeobiogeographic relationships and, as well, we will comment briefly on how patterns of past biogeography have hampered or facilitated biostratigraphic correlation. References cited here are grouped in the bibliography that starts on p. 82.

(a) The contemporary and Cainozoic biogeographic fabric

Australia's terrestrial and to a lesser degree its shallow marine floras and faunas are striking for the comparatively high level of endemism at family-group and higher levels, and for the appreciable number of groups that are not known from the northern hemisphere but which have disjunct distributions

---

Palaeocene and late Oligocene of the Capricorn Basin), and of gypsum (Early Cretaceous Rolling Downs Group of Eromanga Basin; ?Quaternary Whitula Formation of the Birdsville Basin); and halite and gypsum in the modern playa lakes that range in area from the smallest salt-pan to Lake Eyre (Dulhunty 1982). At Lake Eyre, the modern lake sediments, according to Wells (1980), are only 4.5 m thick, and only 0.5 m of this is salt.

(2) Glaciation includes the Quaternary (Williams, above), the Late Carboniferous–Permian, with a peak in the early Sakmarian (Crowell and Frakes 1975), and (off-scale, about 700 Ma ago) the late Adelaidean (Plumb *et al.* 1981; Rutland *et al.* 1981).

(3) Temperature and rainfall from sources cited in the text. Global trends (dotted lines) from Frakes and Rich (1982). The cardinal points indicate the regions, and I the interior. Seasonal rainfall in the interior, indicated by the zigzag, implies local aridity. The line of crosses is the oxygen-isotope temperature curve for the surface waters of the Campbell Plateau (DSDP 277) (late Palaeocene–Oligocene), Macquarie Ridge (DSDP 279) (early Miocene), and South Tasman Rise (DSDP 281) (middle Miocene–present).

(4) Coal is sporadic from the earliest record, in the Late Devonian of the Amadeus Basin (Playford *et al.* 1976a), to the latest record, in the late Miocene of Victoria (Abele *et al.* 1976) and in the Quaternary of the Papuan Basin (Ridd 1976). Thick coal measures first appear in the Sakmarian.

across the southern hemisphere crustal blocks. It is a complex legacy that has involved passive dispersal or isolation by rifting and dispersal of continental blocks, active dispersal ('island-hopping') of organisms across formerly narrow late Mesozoic and Cainozoic seaways, immigration from the north, particularly as a consequence of Neogene interplay with south-east Asia and, as well, the strong imprint of evolutionary/selective pressures connected with the progressive climatic change that occurred as Australia migrated northwards through the 'horse latitudes' into the tropics (Beard 1977).

Biogeography has been approached in at least five ways:

(1) Inferring antiquity and origins subjectively from present-day geographic patterns such as diversity gradients and distributions of primitive forms, supplementing this with geological evidence where available. In the absence of harder evidence, this can be a useful first approximation. That it is not constraining and can be misleading has been stressed by Lange (1982).

(2) Classifying elements according to their inferred geographic origins, such as Palaeoaustral, Malayo-Pacific, Neoaustral, and so on. This has been used effectively by Fleming (1975) to convey a broad picture of the origins and evolution of the New Zealand flora and fauna.

(3) Endeavouring to use a hierarchy of larger biogeographic units – realms, regions, provinces – preferably with computer assistance rather than by 'eyeballing'. Though useful, such categories are in large measure mental constructs (Talent, in press), units varying according to the state of knowledge and the relative vagilities of the groups being considered.

(4) Using cladistic analysis, again preferably computer-assisted, to produce cladograms or alternative hypotheses (see Ashlock 1974 for review). This has been used compellingly for southern hemisphere chironomid midges (Brundin 1966) and dipterans (Hennig 1966); there is still an insufficiency of such examples for one to be able to rely solely on this method.

(5) Using co-evolutionary systems such as the intricate patterns of plant–phytophagous insect–parasitoid insect–hyperparasitoid insect relationships. This approach is obviously more constraining than distribution patterns of ecologically unrelated organisms, but as yet has not been sufficiently exploited. Its potential is exemplified by a study (Schlinger 1974) of insects ecologically associated with *Nothofagus*.

Because of the need to provide an overview, the following discussion is unabashedly a medley of all five of these approaches.

(i) Marine biota

There is fairly general agreement (Hedley 1904; Whitley 1932; Bennett and Pope 1953, 1960; Hedgpeth 1957; Knox 1960, 1963, 1975, 1980; Briggs 1974; Dartnall 1974) that four or perhaps five contemporary marine provinces can be discriminated usefully (Fig. 41A) for faunas and perhaps floras from the shelf waters around Australia and New Guinea. One tract, extending from the vicinity of Shark Bay northwards around to Sandy Cape (Fraser Island), forms part of the vast Indo-West Pacific tropical province or region. Classification and subdivision of the latter is left open as there is no consensus on how to proceed with the bewildering array of species, the gross deficiencies in data, and the wide geographic range of so many species. Numerous suggestions have been made that range from utilizing a single province (for example Ekman 1953; Hedgpeth 1957; Valentine 1973), as was done by Forbes (1856), to recognizing as many as 18 subdivisions (Powell 1957). It is beyond the scope of this contribution to become involved with this question, but there appears to be sufficient endemism of fishes and echinoderms (Whitley 1948; Clark 1946) in the area between Shark Bay and Torres Strait to retain Hedley's Dampierian Province – here used as a subprovince – for this tract.

The connection between the arrangement of marine shelf provinces and the oceanic circulation pattern is obvious; the latter, and accordingly the former, are modified as the distributions and configuration of crustal blocks change through time (Luyendyk *et al.* 1974). It is anticipated therefore that there must have been profound changes in the nature and even in the number of marine zoogeographic provinces around the shores of the Australia–New Guinea block during the Cretaceous and Cainozoic, i.e. since the beginning of the Potoroo Regime (discussed elsewhere in this book); this enormous problem has yet to be broached quantitatively. Many authors, however, (for example Jenkins 1975; Chaproniere 1980) have noted the ingression into shelf faunas of southern Australia of warm-water elements such as the foraminifers *Lepidocyclina, Cycloclypeus, Marginopora,* and *Flosculinella* and the crabs *Calappilia* and a tropical section of *Nucia* during the late early and early middle Miocene; the same event – an increase in Indo-West Pacific taxa – is apparent in the molluscan faunas (T. A. Darragh, personal communication). The development and decline of this warmer interval (see also previous section on palaeoclimates) contrasts with the overall global pattern of cooling through the Late Cainozoic. Presumably it is to be connected with the speed of northward drift of Austrlia–New Guinea being for a time more rapid than the equatorward contraction of

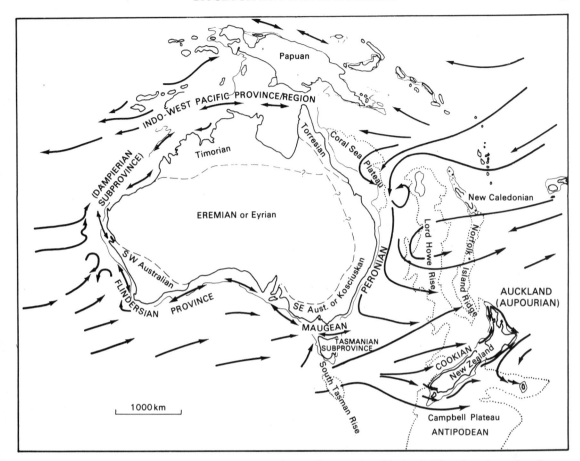

Fig. 41A. Generalized biogeographical subdivision of Australia and adjacent regions. Marine shelf faunal provinces (from Bennett and Pope 1953 and Knox 1963) are capitalized. Terrestrial regions or provinces, with only initial capitals, are a blend of phytogeographic (Tate 1889; Cameron 1935; Burbidge 1960; Beard 1981) and zoogeographic units (Spencer 1896; Kikkawa and Pearse 1969; Horton 1973). Consensus on classification has not been reached, boundaries of units proposed varying according to groups studied and problems arising from overlaps in taxa replacements. A suite of fluvifaunal provinces that has been proposed (McMichael and Hiscock 1958), being connected with present and past drainage systems, differs markedly from the pattern presented here. It will be seen that in a general way the division into marine zoogeographic provinces is connected with the pattern of oceanic circulation. Data for the movement of surface waters is generalized from various sources (principally Wyrtcki 1960; Wyrtcki *et al.* 1971; Phillips 1981). Seasonal changes in direction are shown by double-headed arrows. It should be noted that circulation in the South-east Indian Ocean is weak and variable.

the marine temperature zonation. Alternatively, it may be due to a peculiar marine circulation pattern engendered by some configuration of blocks that occurred just prior to interplay between Asia and Australia–New Guinea.

(ii) Terrestrial biota

Cretaceous–Cainozoic tectonic history has been used to help interpret distribution patterns of souther hemisphere plants (Raven and Axelrod 1972, 1974; Barlow 1981), passeriform birds (Cracraft 1973; Rich 1975), marsupials (Tedford 1974), vertebrates in general (Cracraft 1974), and some groups of non-marine invertebrates (Main 1981); shorter contributions on various groups are to be found in symposia volumes edited by Keast (1981a), Whitmore (1981a), and Rich and Thompson (1982). For all groups, it is fairly easy to recognize families and genera that probably originated in Australia–New Guinea or that, because of their present disjunct distributions, presumably arose

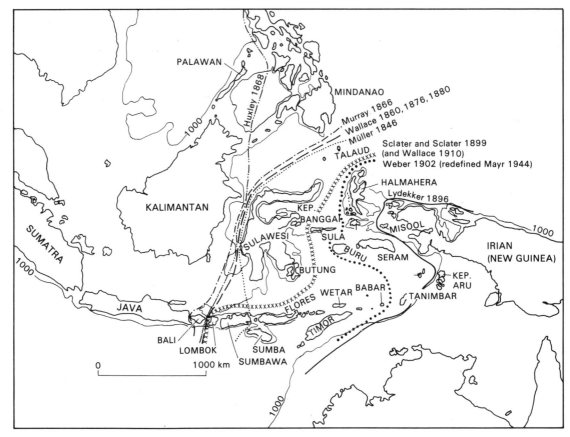

Fig. 41B. A selection of the zoogeographic limits that have been proposed in the Moluccas–Lesser Sunda Islands 'inter-regional zone' (based on Simpson 1977, fig. 1).

in Gondwanaland before fragmentation. It is also easy to discriminate many groups whose entry from the north must be recent, in most cases presumably post-mid-Miocene because of their distribution patterns: limited penetration from the north; conspicuously low diversity compared with Eurasia, especially South-east Asia; and little if any differentiation at generic or specific levels compared with regions further west. Conversely, high diversity in Australia coupled with limited penetration westwards to or towards Asia is strongly suggestive of origin in Gondwanaland, if not specifically on the Australian block. This approach has been used in developing Table 5. One is not constrained, however, to interpret data in this way. In the absence of detailed cladistic analysis from which dispersal trackways might be postulated, a more conservative stance may be taken. Whatever the reality of the table as an expression of origins and interchange, substantial colonization by biota from the north is obvious. The question then arises: What biota passed into extinction, what were displaced as a consequence of competitive interaction with the invaders? Was there substantial extinction as a consequence of Australia drifting northwards too rapidly for evolutionary radiation to establish community equilibrium before the onset of invasion of Oriental elements? The pre-Miocene fossil record, as presently known, provides no significant evidence on these questions.

Marsupials and monotremes are highly characteristic of the Australia–New Guinea block. The evolutionary history of the Australian marsupials and their relationships to South American marsupials have been reviewed recently (Archer and Bartholomai 1978; Archer 1981), and there is now one record from the former Antarctic land-route between them: a c. 40 Ma polydolopid marsupial from Seymour Island

TABLE 5. *Postulated origins and Cainozoic dispersals of a selection of non-marine plants and animals relative to the Huxley/Wallace and Lydekker lines (Fig. 41B)*

Groups of probable Gondwanaland origin that have not dispersed westwards from the Australia–New Guinea block	Groups of Gondwanaland origin that appear to have crossed the Huxley/Wallace Line dispersing westwards *from* the Australia–New Guinea block	Groups that appear to have dispersed eastwards from Asia across Lydekker's Line *to* Australia–New Guinea (pre- or post-collision not implied)
PLANTAE		
Haemodoraceae (kangaroo paws)	Podocarpaceae (plum-pines)	Cupressaceae (cypresses)
Xanthorrhoeaceae (grass trees)	Auraucariaceae (araucarias)	Magnoliaceae (magnolias)
Myoporaceae (emu bushes)	Winteraceae (winter's bark)	Dipterocarpaceae
Tremandraceae (pink-bells)	Epacridaceae (Aust. 'heaths')	Meliaceae (mahoganies)
Himantandraceae	Proteaceae (proteas, grevilleas)	Fagaceae (beeches) – except *Nothofagus*
Aizoaceae (mesembryanthemums)	Casuarinaceae (she-oaks)	Ericaceae (heaths)
Eucryphiaceae (eucryphias)	Myrtaceae (myrtles, eucalyptus)	Loranthaceae (mistletoes) [in part]
Corynocarpaceae	Stylidiaceae (trigger plants)	Flacourtiaceae
Brunoniaceae	Goodeniaceae (Goodenias and fan-flowers)	Icacinaceae
	Pittosporaceae	Poaceae (grasses)
	Stackhousiaceae (candles)	Geraniaceae (geraniums, pelargoniums)
	Monimiaceae	Sapotaceae
	Centrolepidaceae	Celastraceae (spindle trees)
	Restionaceae (twine-rushes)	Pandanaceae (screw-pines)
		Lauraceae
		Olacaceae
		Annonaceae (sweet-sop and sour-sop)
		Dioscoreaceae (yams)
		Apocynaceae
		Cucurbitaceae (gourds and pumpkins)
		Asteraceae (composites)
BIVALVIA		
Hyriidae		
GASTROPODA		
Athoracophoridae, Cystopeltidae		
HIRUDINEA		
		Haemadipsidae
ONYCHOPHORA		
Peripatopsidae		
ARACHNIDA		
Scorpionida (scorpions)		
Bothriuridae		
Scorpionidae-Urodacinae		
Pseudoscorpionida (pseudo-scorpions)		
Synphyronidae		
Araneida (spiders)		
Actinopodidae, Migidae, Toxopidae,	?Desidae	numerous
Cycloctenidae, Hickmaniidae,		
Gradungulidae, Archaeidae,		
Hadrotarsidae, Symphytognathidae		
CRUSTACEA		
Anaspidacea		
Anaspididae, Koonungidae		
Isopoda		
Phreatoicidea (also in India)		
Decapoda		
Parastacidae		Potamidae

TABLE 5 – *continued*

Groups of probable Gondwanaland origin that have not dispersed westwards from the Australia–New Guinea block	Groups of Gondwanaland origin that appear to have crossed the Huxley/Wallace Line dispersing westwards *from* the Australia–New Guinea block	Groups that appear to have dispersed eastwards from Asia across Lydekker's Line *to* Australia–New Guinea (pre- or post-collision not implied)
INSECTA		
Odonata (dragon-flies, damsel-flies)		
Synthemiidae, Hemiphlebiidae, Lestoideaidae, Petaluridae		Protoneuridae, Megapodagrionidae, Chlorolestidae
Isoptera (termites)		
Mastotermitidae		
Plecoptera (stone-flies)		
Eustheniidae, Austroperlidae, Gripopterygidae		
Embioptera		
Australembiidae		
Phthiraptera (lice)		
Boopiidae		Oligotomidae
Hemiptera (leafhoppers, bugs)		
Tettigarctidae, Peloridiidae, Eurymelidae, Hyocephalidae		numerous, e.g. Machaerotidae, Scutellaridae, Plataspidae
Neuroptera (lacewings)		
Stilbopterygidae, Nymphidae		
Coleoptera (beetles)		numerous
Belidae		
Mecoptera (scorpion-flies)		
Choristidae, Nannochoristidae		
Siphonaptera (fleas)		
Stephanocircidae, Macropsyllidae, Rhopalopsyllidae	?Pygiopsyllidae	Ischnopsyllidae
Diptera (flies)		
numerous Nematocera and Orthorrapha have disjunct distributions between Australia, New Zealand and South America		numerous
Trichoptera (caddis-flies)		
Philanisidae, Plectrotarsidae, Philorheithridae, Tasimiidae		Calamoceratidae, Leptoceridae
Lepidoptera (moths, butterflies)		
Anthelidae, Cyclotomidae, Carthaeidae, Lasiocampidae		numerous, e.g. Papilionidae, Aegeriidae, Pieridae, Sphingidae, Tineodidae, Amatidae
Hymenoptera (wasps, bees, ants)		
Formicidae-Myrmeciinae	Pergidae	Tenthredinidae, Argidae, Formicidae-Dorylinae, and Pseudomyrmecinae
PISCES		
Ceratodontidae (lung-fish)	Osteoglossidae	
AMPHIBIA		
?Microhylidae-Astrophryinae		Ranidae-Platymantinae
Myobatrachidae		
REPTILIA		
Chelyidae (Australian 'tortoises')		Carettochelyidae
Gekkonidae-Diplodactylinae		Scincidae (skinks)
Pygopodidae (legless lizards)		Varanidae (monitors)
		?Elapidae
		Colubridae (water and tree-snakes)
		Boidae (pythons)

TABLE 5 – continued

Groups of probable Gondwanaland origin that have not dispersed westwards from the Australia–New Guinea block	Groups of Gondwanaland origin that appear to have crossed the Huxley/Wallace Line dispersing westwards *from* the Australia–New Guinea block	Groups that appear to have dispersed eastwards from Asia across Lydekker's Line *to* Australia–New Guinea (pre- or post-collision not implied)
AVES		
Casuariidae (cassowaries)	Megapodiidae (megapodes)	Podicipedidae (grebes)
Dromaiidae (emus)	Psittacidae (parrots)	Falconidae (falcons)
Spheniscidae (penguins)	Columbiidae (pigeons)-?in part	Threskiornithidae (ibises, spoon-bills)
Pedionomidae (plains wanderers)	Acanthizinae (Aust. warblers)	Ciconiidae (storks)
Menuridae (lyre-birds)	Artamidae (wood-swallows)	Anatidae (ducks)
Atrichornithidae (scrub-birds)		Phasianidae (pheasants, quail)
Pachycephalinae (thickheads)		Gruidae (cranes)
Falcunculinae (shrike-tits)		Cuculidae (cuckoos)
Cinclosomatiinae (quail-thrushes)		Otidae (bustards)
Malurinae (Aust. wrens)		Rallidae (rails)
Neosittidae (Aust. nuthatches)		Jacanidae (jacunas)
Climacteridae (tree-creepers)		Strigiidae (hawk-owls)
Pardalotidae (pardalotes)		Caprimulgidae (nightjars)
Ephthianuridae (Aust. chats)		Coraciidae (rollers)
Meliphagidae (honeyeaters)		Alcedinidae (kingfishers)
Cracticidae (Aust. magpies)		Bucerotidae (hornbills)
Grallinidae (magpie-larks)		Pittidae (pittas)
Corcoracidae (Aust. choughs)		Alaudidae (larks)
Ptilinorhynchidae (bower-birds)		Hirundinidae (swallows)
Paradisaeidae (birds of paradise)		Campephaginidae (cuckoo-shrikes)
Aegothelidae (owlet-nightjars)		Turdinae (thrushes)
?Podargidae (frogmouths)		Timaliinae (babblers)
		Motacillidae (wagtails)
		Zosteropidae (white-eyes)
		Sturnidae (starlings)
		Laniidae (shrikes)
		Corvidae (crows, jays)
MAMMALIA		
Monotremata (spiny ant-eaters, platypuses)		Muridae (rats, mice)
Dasyuroidea (marsupial cats, mice, and ant-eaters)		Chiroptera (bats)
		Canidae (dogs – 'dingo')
Perameloidea (bandicoots)		Hominidae
Phalangeroidea (kangaroos, possums, wombats)		
Notoryctidae (marsupial moles)		

(Woodburne and Zinsmeister 1982). The oldest fossil marsupials from Australia are of late Oligocene age, the oldest monotremes mid-Miocene; their evolutionary history in Australia is assumed to extend back until some time in the Cretaceous. The earliest record of Chiroptera is also mid-Miocene; again earlier entry is likely, but endemism (only two endemic genera) is not high. The earliest known rodents are of early Pliocene age; they are conspicuously absent from older faunas rich in small vertebrates. It is therefore assumed their entry into Australia commenced between 4 and 5 Ma ago.

The freshwater fish fauna of Australia is dull compared with the rich faunas that evolved in tropical Africa and South America. Only two primary freshwater groups are represented, the Ceratodontidae (lungfish) and the Osteoglossidae, the former being a relic of a group that had wide distribution during the Mesozoic. There is not the diversity of endemics that one might expect from a long history of isolation. The phylogeny of amphibians is not sufficiently precise for a clear statement to be made on the biogeographic implications of most groups, but the Myobatrachidae are endemic to Australia–New Guinea and South America and perhaps have always been so.

The Carettochelyidae, now restricted to New Guinea, were formerly widespread in Laurasia. Among lizards restricted to Australia are the Diplodactylinae, a group of geckos whose closest relationship is with either the pantropical Gekkoninae or the south and central American Sphaerodactylinae, showing they could have come from either the north or the south. It is popularly assumed that the varanoid lizards ('goannas') are distinctively Australian, but the family, though now restricted to Africa, Australasia, and southern Asia, formerly ranged over Laurasia. It is known from the Late Cretaceous of North America (Hoffstetter 1968) and could have originated in the northern hemisphere. Though the greatest diversity of elapid snakes (cobras, tiger snakes, copperheads, and their ilk) is in Australia, they too appear to have been formerly more widespread in Laurasia and seem to have originated somewhere there.

The Australian bird record goes back to the Early Cretaceous (Talent et al. 1966) but, apart from ratites from the Late Cainozoic (Rich 1979), it is rather scrappy. The high level of endemism in the contemporary avifauna – 30 per cent for the families, 90 per cent for the genera, and 95 per cent for the species (Keast 1981b) – is symptomatic of the evolutionary radiations that have taken place in isolation on the Australia–New Guinea block. The ratite birds, the former Asian occurrences of ostriches and *Eleutherornis* aside, are clearly a Gondwanan group (Cracraft 1974) as they occur or have been found fossil on all southern continents and most micro-continents from South America and the Canary Islands (extinct) to Africa, Madagascar (extinct), Australia (emus, cassowaries, and the extinct Dromornithidae – Rich 1979), New Zealand, and New Caledonia (the extinct *Sylviornis*). In all cases, because of their flightlessness, low dispersal capability is assumed; their origin and dispersal must have predated the breakup of Gondwanaland, or at least before appreciable rafting, block from block, had occurred. Several of the other non-passeriform groups could well have originated in Gondwanaland if not in fact in the Australia–New Guinea block (Rich 1975); the same obviously applies to a number of passeriform families and subfamilies (Table 5).

The small number of families of spiders endemic to Australia–New Guinea is understandable because so many of them are dispersed by 'ballooning', but this of course does not apply to the large spiders and forest litter-dwellers where patterns are being deciphered that can be explained, at least in part, by global tectonics (Forster 1975; Platnick 1976; Main 1981). The litter-dwelling Cycloctenidae, Gradungulidae, and Toxopidae, for instance, are endemic to Australia and New Zealand; their origins therefore presumably predate the separation of New Zealand from Australia–New Guinea.

Considering that most groups of insects are active fliers and readily dispersed by air currents, it is perhaps surprising how many families and subfamilies are restricted to Australia or have strong taxonomic connections with New Zealand and South America reflecting former connection or juxtaposition. The selection shown in Table 5 does not being out the relative richness in endemic lineages derived from 'primitive' stocks and the tendency of these to be concentrated in the south rather than the north.

As far as the terrestrial flora is concerned, the majority of families represented in Australia, more than 200 of them, occur widely around the globe and might be referred to as quasi-cosmopolitan. About 34 per cent (just over 100) of the families of angiosperms of the world do not occur on the Australia–New Guinea block; this figure is increased to about 40 per cent if only Australia proper is considered. Only about 10 per cent of the families represented in Australia can be described as being Gondwanan, having been dispersed across Gondwanaland before its fragmentation or having originated on the Australian block after parting company with Antarctica, without subsequent dispersal having obscured their origins. But these include the numerically prominent Proteaceae, Myrtaceae (with about 500 species of *Eucalyptus*), Casuarinaceae as well as the distinctive southern gymnosperms (Araucariaceae, Podocarpaceae) and *Nothofagus*. The gymnosperms and *Nothofagus* are very much a relict component confined to moist habitats, usually closed forests, and showing relatively little evolutionary modification since the Mesozoic. The others have diversified into many habitats with often remarkable adaptations to aridity, nutrient deficient soils, fire, and high climatic variability. Scleromorphy – small rigid leaves, small plant size, short internodes – is especially characteristic of this Gondwanan group. This response to poor soils exaptated them (in the old sense of pre-adaptation) for the mediterranean climate that became widespread during the Cainozoic (Beadle 1954, 1966, 1981; Loveless 1961; Johnson and Briggs 1981). It is almost certainly this adaptation to low-nutrient soils that limited penetration westwards through the generally richer soils of Malesia (van Steenis 1979). Migration from south-east Asia has been much more important than migration westwards from Australia through Malesia. This finds expression in the low level of generic endemism in Australia's tropical zone compared with that of the temperate zone (14 per cent vs. 46.6 per cent according to Burbidge 1960). Most of the non-endemic Indo-Malayan genera, about 360 in all (Burbidge 1960), have only one or very few species in Australia, testifying to the relative recentness of their migration into this region; some of these are shown in Table 5. Not all groups that penetrated from the north were strictly cosmopolitan or Indo-Malayan elements. There are a small number of Indo-Gondwanan elements hypothesized as having been rafted on the Indian block to Asia, their derivatives subsequently being able to re-enter Australia (Specht 1981; Nelson 1981); among such are two genera of Loranthaceae (mistletoes), *Decaisnina* and *Dendrophthoe*. Loranthacean pollen makes its first appearance in Australia during the Eocene.

Palynology is helping establish a sequence of first appearances for a number of groups (Stover and Partridge 1973; Martin 1978, 1981; Kemp 1978). It has already demonstrated that some of the supposed cosmopolitans or groups assumed to have Laurasian origins had arrived well before the mid-Miocene collision between Asia and Australia–New Guinea. For instance, the Santalaceae (sandalwood family) and Aquifoliaceae (hollies) were present in the Paleocene; the Loranthaceae (mistletoes), Sparganiaceae (burr-reeds), and Euphorbiaceae were present in the Eocene, and the Poaceae (grasses) by about 32 Ma (Wellman and McDougall 1974), about the middle of the Oligocene; the record for the Poaceae is based on macrofossils (von Ettingshausen 1888), not on pollen, and is not completely reliable. Should these be interpreted as minimal dates? Take the case of the Mimosaceae with almost 700 species in 10 genera (including 650 in the incredibly diverse genus *Acacia*) and the Asteraceae (Compositae) with about 705 species in 104 genera. One may be tempted, if there were no palynological data, to infer that entry had taken place long before the Miocene, but this is the age of the oldest pollen identifiable as belonging to *Acacia* (Cookson 1954), and the age of the oldest asteracean pollen. Beadle (1981) has suggested that the distribution of the Gummiferae, the most primitive section of *Acacia*, is in accord with a Gondwanaland origin for the genus and that Australia may already have had a few species of the group at the time of fragmentation; there is, however, no fossil evidence to support this. Similarly the diversity of grasses, some 1240 species in 210 genera, and the strong Gondwanan element (disjunct distributions across the southern continents) have led Clifford and Simon (1981) to question the efficiency of long-distance dispersal as explanation and to speculate that there may have been grasses much earlier than suggested by the fossil record, even before separation of various parts of Gondwanaland. The greater antiquity in Australia is acceptable, but distribution across Gondwanaland before the final separation of Africa from the rest, even if this were a latest Cretaceous event (Rand and Mabesoone 1982), is difficult to reconcile with the present global data for the origin of grasses being an early Cainozoic (?Eocene) event. Perhaps the grasses have greater antiquity than presently assumed and perhaps the Australian palynological record has been too biased towards forest habitats. However that may be, the present data are in accord with the Mimosaceae and Asteraceae arriving just about or soon after the Asia–Australia collision. Some cosmopolitan groups, none the less, such as the Malvaceae (mallows) and Cyperaceae (reeds and sedges), have an Eocene record in Australia proving establishment long before the Miocene interplay between the Asian and Australia–New Guinea continental blocks. The elevation of mountain systems as a consequence of this interplay can be assumed to have generated a dispersal route from Malaya to New Guinea for cool temperate plants, though with some largish steps in the route.

More than half the land area of Australia is continuously or seasonally arid (the Eremian Province). Its flora, about 363 genera of seed plants, is composite, consisting of about 102 endemics and approximately equal numbers of cosmopolitan genera, genera represented in temperate Australia and genera from adjacent tropical lowlands (Burbidge 1960). It is widely accepted as having been derived from older adjacent plant communities as arid conditions overtook the continent from the north concurrent with Australia's northward drift, at least 15 Ma ago (Beard 1976). The development of the arid conditions may go back into the Oligocene (Barlow 1981; Carolin 1981) and the palaeoclimatic scenario may be more complicated (Bowler 1981). Moreover, the tropical lowland component may be a recent colonizer of the arid zone (Carolin 1981). It is appropriate at this stage to consider more closely the routes by which colonization has taken place.

(b) Chronology of land-bridges and island 'staging-points'

Theoretically (in the absence of active dispersal), the degree of relationship of biota made disjunct by dispersal of crustal blocks should be according to the chronology of block-dispersal (Fig. 42A). An approach to such a pattern is to be anticipated for

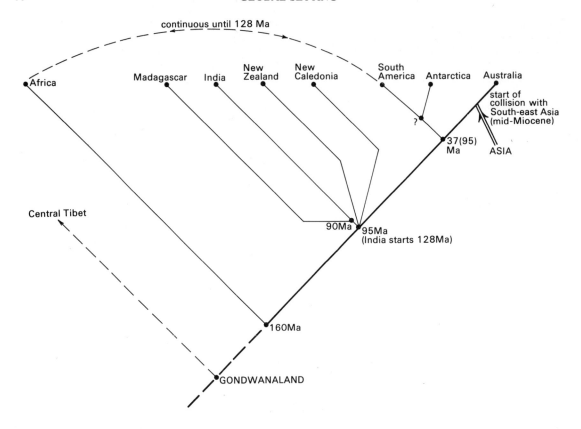

Fig. 42A. Dispersal of Gondwanaland continental and microcontinental fragments, as shown by seafloor spreading, with Australia as prime focus and with spacing of branches from the main axis being proportional to elapsed time. Figures at the nodes are from elsewhere in the book; they represent the latest dates, on present information, for effective separation. Dates of rifting and/or minimal separation may be substantially earlier, as was the case of Australia from Antarctica. Theoretically, if biotic dispersal had been purely passive, the degree of relationship of biota now disjunct between these fragments should have been according to this pattern. This, however, is subordinate to (in fact overwhelmed by) the effects of other factors: 'island-hopping', waif-dispersal, secondary development of land-bridges; protracted maintenance of land-bridge or island-chain filter-zones—as was the case in the Eocene-Oligocene between Australia and Antarctica.

groups having low vagility, such as freshwater fishes, chironomid midges, and litter-dwelling spiders. For most organisms, however, patterns due to gross tectonic dispersal are masked by the effects of regional and global climatic history in conjunction with dispersal opportunities and capabilities that, insignificant in the short term, become highly significant in the context of geological time. Staging-points ('islands') for dispersal are manifold, ranging from patches of mountain habitats to submarine pinnacles or platforms of appropriate depth for particular organisms.

(i) The Moluccas–Lesser Sunda Islands 'inter-regional zone'

Biogeography in the region north and west of Australia has been bedevilled by a search for a single boundary between the Oriental and Australian zoogeographic regions, or, as Mayr (1944) has so graphically put it, to find 'the most convenient place where to replace one color by another'. Seven discrete lines (Fig. 41B) and numerous modifications of them have been proposed as regional zoogeographic boundaries in the region between Kalimantan and West Irian (Mayr 1944; Simpson 1977). Only two of these, Lydekker's Line

(1896) and Huxley's Line (1868), a modification of the original Wallace's Line (1860-80), appear to have some meaning, especially if made to correspond with the edges of the Sahul and Sunda Shelves respectively. Attempting to draw a line of 'faunal balance' between these lines has been shown (for example Holloway and Jardine 1968) to be a futile exercise because it varies according to the taxonomic group studied. A single boundary for the kingdom Animalia as a whole has not been and probably cannot be established; and even if it were possible one wonders what deeper meaning it would have. Simpson (1977) suggests, prudently I believe, to keep the zoogeographic boundaries of the Oriental and Australian regions at Huxley's and Lydekker's Lines and not assign the intervening islands to any region, subregion, or province in order that the study of faunas on these islands, their compositions, affinities, histories, and ecologies might not be inhibited by classificatory strait-jackets.

What can we make out about the question of former land connections and past and present island staging-points? It should be noted before we proceed that the Lombok and Makassar Straits are of insufficient width to have caused noticeable differentiation in shallow marine faunas. The original Wallace's Line (1860-80) and indeed all of the lines proposed by various workers in the region between Kalimantan-Java-Bali and Australia-New Guinea refer only to terrestrial biota; they have no meaning for marine faunas which belong to a single Indo-Pacific Province. Likewise for the plants; the whole region from Sumatra and the Malay peninsula through to New Guinea and the Bismarck archipelago has been embraced within a single phytogeographic province, the Malesian Province (Whitmore 1975).

Even cursory examination of Indonesia would suggest two broad pathways for interregional migration: from Sulawesi through the Moluccas to New Guinea and from Java through the Lesser Sunda Islands to north-western Australia. For how long have these been effective migration routes?

The palaeomagnetic evidence (Haile 1978; Sasajima *et al.* 1980) suggests western Sulawesi having been close to Kalimantan since Cretaceous times and therefore effectively part of the Asian aggregate during the Cainozoic, whereas some at least of eastern Sulawesi seems to have been formerly part of the Australian region of Gondwanaland. Collision between the Sula-Banggai micro-continent and Sulawesi commenced in the mid Miocene about 15 Ma (Kundig 1956) bringing part at least of east Sulawesi above sea level and shallowing and perhaps exposing various parts of Sula-Banggai by the late Miocene. Collision between Sula-Banggai and the Inner Banda Arc in the vicinity of Seram and Ambon was appreciably later: close to the Mio-Pliocene boundary, zones N18-N19. Precise dating of emergence of Seram and Buru is uncertain, beyond a generalized Pliocene age, though early Pliocene is possible. The emergences of Obi and Misool seem however to have been later events, presumably Pleistocene though with the possibility of some Pliocene land (Audley-Charles 1981). In summary, though collisions may have commenced in the mid-Miocene, it was probably not until the Pliocene that a series of staging-points approaching comparability with what we see today between Sulawesi and New Guinea came into being.

The Lesser Sunda Islands between Java and Timor superficially appears a better route. Generation of a chain of islands had commenced by the late Miocene; by the end of the Miocene a continuous chain extended from Java to the Australian continental margin at Timor that would have allowed migration from east and west (van Bemmelen 1949; Audley-Charles and Hooijer 1973; Audley-Charles 1981). With such constraints in mind, it is unlikely that the entry of rats and mice into Australia from Asia will be pushed back earlier than the end of the Miocene or beginning of the Pliocene. The fossil record for vertebrates along this route is lacking or deficient especially as regards bats and birds, but pre-Pliocene migration of these highly vagile groups is to have been expected. Nevertheless, despite an overemphasis on the most highly vagile groups – the birds, bats, and butterflies – in biogeographic studies of the region (Talent 1981), even these show appreciable levels of endemism (instance the 38 per cent for birds of Sulawesi and progressive decline in diversity of the Sundaic bird component eastwards from 87 per cent on Bali to 63 per cent on Flores). Interestingly, there is less fall-off in diversity of the seemingly less vagile amphibians and reptiles (Mertens 1930; Mayr 1944; Inger 1966). Extended stretches of water have a remarkable isolating effect on most birds, the widespread 'tramps' aside (Diamond 1974).

Though the Lombok and Makassar Straits have been taken by many to mark the eastern limit of a distinctive Sundaic terrestrial vertebrate fauna, exclusively Asian in origin, the terrestrial vertebrates of the Philippines, Sulawesi, and the Lesser Sunda Islands nevertheless are predominantly Asian in affinities. In fact Lombok Strait does not mark a transition or disjuncture any more decisive than between other pairs of neighbouring islands in the Lesser Sunda Chain. Sulawesi, however, is different, largely because of the relatively high level of endemism of its terrestrial fauna; this has led one author (Cranbrook 1981) to suggest that if zoogeographical lines are drawn, one should be drawn circumscribing Sulawesi!

It has been suggested (Audley-Charles 1981) that there may have been land connection between Kalimantan and Sulawesi at some time during the Pliocene or Pleistocene, but if so it must have been short-lived or incomplete as truly freshwater fishes did not cross the Straits of Makassar eastwards; all immigrants from Kalimantan or elsewhere to the west of Wallace's line almost certainly had to cross a marine barrier. Barriers to migration from the east remained more profound than those from the west throughout the Neogene; this is clear from the overwhelming dominance of Asian over Papuasian elements in the terrestrial faunas of Sulawesi. Elephants and stegodont proboscideans were able to reach Sulawesi and stegodonts migrated as far as Flores, Sumba, and Timor during the Plio–Pleistocene (Hooijer 1967). Elephants are capable of swimming, and if we assume stegodonts were too, their dispersal would not have required continuous land-bridges. Timor, interestingly, has yielded giant varanids like *Varanus komodoensis* and giant rats from latest Pleistocene or Holocene strata (Hooijer 1957, 1972; Cranbrook 1981 – q.v. for bibliography). None of these crossed the Timor Sea, despite its narrowness during Pleistocene glacial intervals.

Finally a note of caution. As has just been argued, neither Wallace's, Huxley's, nor any other postulated biogeographic boundary in the region has fundamental tectonic significance. This should be borne in mind, for there has been a tendency to imply the reverse; in some cases, for example in Raven and Axelrod's seminal paper (1972, p. 1382), the generation of Wallace's Line has been explicitly connected with the Late Cainozoic collision between the Australian and Asian plates. As was outlined above, the palaeomagnetic evidence indicates that West Sulawesi was juxtaposed or even sutured with Kalimantan by Cretaceous times, long before this. Any explanation for the floral and faunal history and contrast across Wallace's Line at the Makassar Strait must therefore be approached by considering the emergence–submergence and marine circulation history of the region as well as the relative vagilities of the organisms involved.

(ii) The Australia–Antarctica–South America dispersal route

As described in an earlier chapter, the separation of Australia from Antarctica occurred during the mid-Cretaceous. The date of separation almost coincides with the origin of the angiosperms and marsupials, or occurred soon thereafter. Had the separation produced a broad marine barrier so long ago it is difficult to account for the strong floral and faunal connections between Australia and South America, especially for angiosperms and marsupials; one would not expect so many families and even genera with disjunct distributions across the southern Pacific to have had these disjunctures for 100 Ma or more. That the separation was not great until well into Cainozoic times is more consistent with the evidence; an area of ligature, a strait or narrow sea-way with perhaps a few island staging-points is hypothesized as having been maintained in the area now known as the South Tasman Rise until perhaps Oligocene times.

Marine conditions indeed penetrated into the Otway Basin from the west during the Early Cretaceous, but oceanic influence remained weak through the Cretaceous into the Early Cainozoic as indicated by the benthic foraminiferal faunas with agglutinated and essentially non-calcareous tests, organic-walled but not calcareous phytoplankton, and low diversity macrofaunas (Taylor 1964, 1965; Harris 1965; McGowran 1970; McGowran *et al.* 1971). A dramatic change in conditions commenced about 47 Ma in the mid-Eocene when truly oceanic influence became apparent, spreading eastwards diachronously. This has been associated with the onset of active separation of Australia and Antarctica. Faunal and floral interplay between Australia and Antarctica may have terminated at that time, but it seems more likely that dispersal between the two continental blocks may have been maintained via the South Tasman Rise until perhaps as recently as 37 Ma ago in the early Oligocene – in other words, so long as the two continental blocks were sliding past each other along a transform.

The rupture of the land route between Antarctica and South America was appreciably later. It has been argued (for example Kennett 1977–80) that opening of the Drake Passage occurred in late Oligocene or early Miocene times. Boltovsky (1980), however, argues that this event and generation of the Circumpolar Antarctic Current and its branch, the Malvin Current, did not take place until latest Miocene times or very close to the Miocene–Pliocene boundary about 5 Ma ago.

The time of development of the Antarctic ice-sheet is also important in this discussion. Some workers (for example Mercer and Sutter 1982) have concluded from the evidence of marine regression, long-distance ice-rafting, and pronounced cooling of the ocean surface that the major expansion of the Antarctic ice-sheet took place about 6 to 6.5 Ma ago, but it is clear there was a large mass of ice well before this, perhaps as early as 9 Ma (Ciesielski *et al.* 1982), or even earlier in

West Antarctica as evidenced by subglacially erupted lavas of various ages between 14 and 27 Ma (Kerr 1982). Other workers (for example Woodruff and Douglas, quoted in Kerr) have concluded from oceanographic evidence that the 'transition from a relatively unglaciated world to one similar to today's' occurred between 16.5 and 13 Ma with greatest change between approximately 14.8 and 14 Ma. Though the debate is far from settled, it is clear that the trans-Antarctic land dispersal route between Australia and South America had ceased to have practical significance some time before and perhaps long before pronounced separation of South America and West Antarctica had taken place.

Despite the evidence for land connection or virtual connection between Australia, Antarctica, and South America persisting well into the Cainozoic, and the evidence of marsupials, plants, insects, and some birds (ratites and parrots) having probably dispersed along the Antarctic route, it is curious that none of the ancient placental groups already present in the Palaeocene of South America, the edentates, condylarths, notoungulates, litopterns, trigonostylopoids, or xenungulates got through the 'filter system' to Australia, presumably due to persistent water gaps and/or profound climatic barriers having developed before the end of the Cretaceous.

(iii) Interaction with neighbours to the east: New Caledonia and New Zealand

The chronology of events in the rifting and separation eastwards of New Zealand, New Caledonia, the Lord Howe Rise–Chesterfield Plateau, and Norfolk Ridge is discussed elsewhere. The timing of subsidence of the two rises is poorly known, but is thought to have been early among these events as provincial contrast between the Australian and New Zealand marine faunas set in during the Late Cretaceous and had become pronounced by the Campanian; this is assumed to have been connected with the disappearance of shallow-water intermediate staging-points. Following the cessation of spreading in the Tasman and New Caledonia Troughs about 60 Ma (Hayes and Ringis 1973; Weissel *et al.* 1977), New Caledonia and New Zealand appear to have remained more or less in their present positions relative to Australia.

New Caledonia was sufficiently isolated for the terrestrial fauna and flora to develop such a high level of endemicity (*c.* 90 per cent) for it to be regarded as a distinct though small biogeographic province or region (Figs. 41A and 42A; Holloway 1979). It has particular interest because of the poor representation of more advanced groups of Cainozoic origin and the large number of relict groups of Monimiaceae, Araliaceae, Cunoniaceae, Sapindaceae, Escalloniaceae, Rutaceae, and palms (Raven and Axelrod 1972) as well as having notable diversity of Winteraceae and the Gondwanan Araucariaceae (15 species of *Araucaria* and 5 of *Agathis*) (Whitmore 1981*a*). Less than 5 per cent of its flora (total *c.* 3000 species) is thought to have been of Australian origin. Between 45 and 50 per cent are of Old World Origin and about 10 per cent of Malesian origin (Balgooy 1971); these have immigrated by 'island-hopping' through the island arcs from New Guinea. The marine biota do not show a level of endemicity of genera and species comparable to that of the terrestrial biota, due no doubt to the nature of the oceanic circulation pattern (Fig. 41A) and the abundance of island staging-points and shallow platforms throughout the Indo-West Pacific province or region.

The New Zealand terrestrial and marine biota are characterized by greater endemicity than New Caledonia, but the land flora is surprisingly less diverse (2000 vs. 3000 species) despite the much greater area. Separation from Australia and Antarctica was clearly too early for mammals (apart from Chiroptera) and many characteristically Australian groups of plants and animals to reach either of these areas overland. The palynological record (Mildenhall 1980) documents the arrival of several Australian plant groups and their subsequent extinction: Casuarinaceae (Palaeocene–early Pleistocene), *Acacia* (early Pliocene–middle Pleistocene), *Eucalyptus* (Miocene–early Pleistocene), the proteaceans *Banksia* (late Eocene), *Isopogon* (Oligocene), *Symphyonema* (Miocene–early Pleistocene), and many others. Other groups may have entered via New Caledonia, for example the proteacean *Beauprea* (mid Oligocene–early Pleistocene; extant on New Caledonia), the winteracean *Zygogynum* (?Oligocene–Pliocene; extant on New Caledonia); the olacacean *Anacolosa* (from Asia; Palaeocene–Eocene; not known pre-Eocene in Australia).

The width of the Tasman Sea (*c.* 2000 km) is two or three times wider than the majority of larvae of molluscs and echinoids can be expected to cross with normal rates of oceanic circulation (Talent, in press). The high level of provinciality of New Zealand marine faunas that developed in the Late Cretaceous–Cainozoic, following opening of the Tasman Sea, is therefore to be expected. Immigration was easier down the island arcs from the north, with shorter distances between staging-points, but in general this would have been against the marine temperature gradient. Significant invasions of warm-water Indo-West Pacific forms took place, as they did in southern Australia, during warmer intervals in the Eocene and early Miocene. Throughout the Cainozoic there was a trickle of Australian molluscs getting across the Tasman Sea, but never sufficient to modify appreciably the provincial aspect of the New Zealand shelf faunas (Fleming 1975). There does not appear to be any evidence of New Zealand biota crossing the Tasman Sea westwards to Australia. This would have to be against the general atmospheric and oceanic circulation pattern (cf. Fig. 41A). On the other hand, the scatter of reefs between New Caledonia and eastern Australia must have facilitated migration of Indo-West Pacific species from the vicinity of New Caledonia and the New Hebrides to Lord Howe Island and Australia along the East Australian Current. Just what species may have migrated in this way has yet to be probed. This probability aside, interaction of the Australian and New Guinea biota with our now distant neighbours to the east has become a 'one-way street', with dispersal of terrestrial and marine biota from the tropical (New Guinea) part of the Australian block down the island arcs to New Caledonia and New Zealand, supplemented by a subordinate amount of trans-Tasman waif-dispersal, especially of insects and spiders but also of occasional land plants and marine biota. This pattern must have persisted since substantial separation developed in the Late Cretaceous.

One should be wary of discounting New Zealand and New Caledonia as centres of origin from which occasional forms may have dispersed widely. Witness the case of the Trigoniidae, one of the important groups of Mesozoic bivalves. Fleming (1964) has analysed their fossil record and shown them to have originated from myophoriid ancestors in the Chile–New Zealand region of Gondwanaland, diversifying and dispersing to New Caledonia, Timor, Japan, and even the Arctic during the Late Triassic with only a few reaching Europe by latest Triassic; near-global distribution occurred during the Jurassic. The last trigoniids occur as 'living fossils' off south-eastern Australia.

(c) The imprint of past biogeography on Australian biostratigraphy

Accumulating data and refining correlations, basin to basin and with internationally agreed stratotype sections, seem to be without end! Problems of differing lithological suites and attendant community or 'biofacies' differences are proverbial. Not always so clearly formulated are the impediments to precise correlation that arise from the ever-changing global and regional patterns of biogeography.

(i) Global biogeographic differentiation

Global biogeographic history has an element of cyclicity about it (Fig. 42B). Provinciality has increased and declined repeatedly from high levels of provinciality, such as in the Late Cambrian–Early Ordovician, Early Devonian, and Late Carboniferous–Permian, to remarkably high levels of cosmopolitanism, as in the Late Devonian–Early Carboniferous, the Early Triassic, and the Early to Middle Jurassic. In the absence of quantitative data tied to an agreed or consistent basis of definitions of provinces, regions and realms, it is not clear how much of this pattern can be ascribed to changing orientations and patterns of aggregation and disaggregation of crustal blocks and the consequent changes in oceanic circulation patterns and, in the case of terrestrial biota, how much can be ascribed to the climatic changes brought about, for instance, by generation of major mountain belts. There is circumstantial evidence of major conjunctions and dispersals of crustal blocks having profound biogeographic consequences. Examples are the increase in provinciality after the Early Carboniferous collision of Africa and Europe that terminated the circum-global equatorial current (Ross 1979a), and the generation of the circum-global, cold-water circulation after the separation of Antarctica from South America. How much of the increase and decrease in provinciality may be ascribed to other fundamental

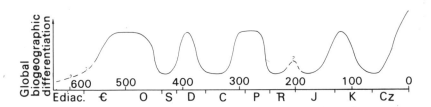

Fig. 42B. Highly generalized pattern of waxing and waning of provinciality through time.

causes, such as change in inclination of the Earth's spin-axis, is not clear. The possible factors are legion. What concerns us here is the ways in which such perturbations in provinciality have hampered or facilitated precise correlation of Australian faunas and floras with stratotype sections elsewhere in the world, or correlation with faunas and floras (and thus of sedimentary and tectonic events) of adjacent crustal blocks and even widely separated regions of the Australian block.

During intervals of low provinciality (high cosmopolitanism or low endemism), there is increased ease of establishing long-distance correlations. Take, for instance, the conodonts. For the Late Silurian (Link and Druce 1972), Late Devonian (Druce 1969; 1976), and Early Triassic (McTavish 1973a) when little provincialism is apparent, great success has been obtained in establishing refined intercontinental correlations. Admittedly, at these times the conodonts underwent major radiations as well as having wide distribution. By contrast, during the Early Devonian and the Late Carboniferous–Permian, both epochs of high provinciality, intercontinental correlation is much more difficult; this can be connected to increased provinciality for conodonts in particular (Telford 1972, 1979; Klapper and Johnson 1980; Fåhraeus 1976) and for marine faunas as a whole. Brachiopods and rugose corals are especially prominent in Australian Middle Ordovician to Permian communities and were the traditional basis for most attempts at intercontinental correlation before the advent of conodont investigations. The high level of provincialism for eastern Australian Early Devonian brachiopod and coral faunas (Boucot *et al.* 1967, 1969; Talent *et al.* 1972; Hill 1957) severely hampered accurate correlation with the standard Devonian sections in Europe. Likewise with the Permian, the relatively low diversity and high endemism of Gondwanan faunas long hindered precise correlation with the standard sections in the Soviet Union and ultimately lead to proposals for suites of zones and stages for the southern cold-water faunas based on New Zealand and Australian sequences (Waterhouse 1967a, 1969, 1970, 1976; Runnegar 1969; Runnegar and McClung 1975).

Perhaps significantly, correlation by means of ammonoids has been easy during intervals of low provinciality, particularly the Late Devonian where authors have commented on the striking similarity of the ammonoid faunas of the Canning Basin to contemporaneous faunas from Europe and North Africa (for example Teichert 1941; Glenister 1958). The latest monograph on Canning Basin ammonoids concerned the Famennian (Petersen 1975); of 23 species identified with certainty at the specific level, 7 were new, 1 previously described from North America, and 15 (65 per cent) were referred to forms previously described from western Europe. By contrast the Permian ammonoid fauna of Western Australia (Glenister and Furnish 1961; Glenister *et al.* 1973; Cockbain 1980) has 12 forms identified at the specific level, and all are endemics except for one, *Metalegoceras australe* (Smith), known elsewhere only from Oman and Timor. The Early Permian ammonoid faunas of Timor (Haniel 1915; Smith 1927; Gerth 1950) are rich, but only one species, the same *M. australe*, has been found in the Canning, Carnarvon, or Perth Basins of Western Australia! Most of the Timor sequences, however, are younger than the ammonoid-bearing Permian sequences of Western Australia, all of which are early Kungurian or older. There are similarities and differences, when other elements of the Permian faunas of Timor are compared with those of eastern or western Australia; there are generic resemblances, but, at the specific level (the basis on which modern provinces are delineated), endemism in the three areas is very high. It should be pointed out that the relationship of Timor and the Vogelkop region of West Irian (Archbold *et al.* 1982) to Australia during the Permian remains an open question; both may be exotic terrains.

(ii) Provinciality within the Australian block

Because of latitudinal and therefore climatic spread, all present-day continents, the Antarctic shelf faunas possibly excepted, have such profound differentiation of marine and terrestrial faunas and floras that authors repeatedly propose systems of named biogeographic regions and provinces (cf. Figs. 41A, 43A–I). As was discussed earlier, these usually have broad and therefore nettlesome transitional belts (the biogeographic equivalents of the ecotones of ecologists); the boundaries and even numbers of subdivisions vary according to the group or groups under consideration and according to the classificatory procedures adopted. The size and especially the latitudinal extent of the Australian and New Zealand continental blocks, however, is sufficient to cause differentiation into suites of marine shelf provinces at the present day. One should not therefore be surprised if provincial differentiation to a similar degree had occurred in the past.

Incidentally, most exercises in palaeobiogeographic analysis have been based on genera and therefore have resulted in much simpler biogeographic patterns than occur today; the resulting 'provinces' equate more with present-day biogeographic realms. On the basis of general practice, Kauffman (1973) proposed that provinces should have 25–50 per cent endemism of species, subprovinces 10-25 per cent, regions 50–75 per cent, and realms more than 75 per cent, but pointed out that many workers, among them Woodward (1851–6), Hedgpeth (1957), and Coomans (1962), have opted for a minimum of 50 per cent endemism for province-discrimination. How do past faunas from the Australian block fare when these criteria are applied?

Exceptionally rich faunas have been monographed from the Late Devonian, Carboniferous, and Permian of western Australia; the contrast between those faunas and contemporaneous faunas from eastern Australia has long been recognized. Roberts (1971) suggested that the profound differences in the Early Carboniferous brachiopod faunas implied profound isolation, but Runnegar and Campbell (1976) stressed the contrast in lithological suites (and inferred environmental differences) between east and west as being of equal importance. The same reservations can be advanced for the Late Devonian and Permian when lithological differences between east and west were considerable.

The sample of invertebrates so far described from the Permian of Australia (c. 500), Timor, New Zealand, and India must now be adequate for fairly valid biogeographic inferences to be drawn. Of the 339 species of sponges, corals, bryozoans, brachiopods, bivalves, cephalopods, trilobites, and crinoids described from the Permian of western Australia (principally Thomas 1958, 1971; Coleman 1957; Campbell 1965; Dickins 1956, 1957, 1963; Glenister and Furnish 1961; Glenister et al. 1973; Teichert 1949; earlier literature in Townley 1970), only 6 (a mere 1.8 per cent) are confidently identified as occurring on both sides of Australia, and only 13 (3.8 per cent) as occurring on Timor as well. More similarity is indicated for the foraminiferal faunas from eastern and western Australia (Crespin 1958; Scheibnerova 1982) and for the bryozoans (Wass 1968) than for other invertebrate groups. Some of the Timor contrast is due to the dearth of Early Permian on the island. A strict application of the modern bases for defining biogeographic entities would require the three areas to be considered as belonging to three faunal realms, but a more careful stage-by-stage comparison is needed. Perhaps in the past, in the quest for correlations, we have focused too much on the similarities of species to those described from elsewhere and paid too little attention to the differences, so important in biogeography.

A similar exposition can be made comparing faunas of the Late Devonian (principally Veevers 1959a, 1959b; Roberts 1971; McKellar 1970; Maxwell 1951; Grey 1978) and Early Carboniferous (principally Campbell 1957, 1961; Campbell and Engel 1963; Campbell and McKelvey 1972; Maxwell 1961; Roberts 1963, 1964, 1971; Thomas 1971) of eastern and western Australia. The results are the same: almost total contrast at the specific level. Differences in the lithological suites must be reiterated as a partial explanation, but the contrast is so profound that other causes must be adduced. There do not appear to have been significant palaeolatitudinal differences between the eastern and western faunas during the Devonian, Carboniferous, and Permian – no more than 5–10°. Perhaps a fundamental control was dissimilarity of situation relative to the global pattern of oceanic circulation, isolating faunas on either side of the Australian block in much the same way as the contemporary oceanic circulation pattern isolates the Flindersian Province of western Australia from the Peronian Province of eastern Australia (Fig. 41A).

(d) Biogeographic affinities of Australian pre-Cainozoic faunas and floras

As was mentioned earlier, palaeontologists analysing faunas globally have generally done so in terms of 'provinces', proposing these on the basis of generic presences or absences. The resulting entities, though useful generalizations in themselves, do not align well with contemporary hierarchies of realms, regions, provinces, and subprovinces favoured by many but not all biologists, and based in the first instance on species distributions (cf. Kauffman 1973; Ross 1974). In most cases, a palaeontological province corresponds roughly with a biologist's realm or region. Some palaeontologists nevertheless have attempted to apply a present-day style of hierarchy to the past; two notable examples are Kauffman's (1973) analysis of Cretaceous bivalves, and Boucot's (1975) overview of Silurian and Devonian brachiopod biogeography.

In the following attempt to sketch a panorama of the changing patterns of Australian biogeography through time, a province–region–realm nomenclature will be used for brevity, though not always with rigour as very little of the palaeontological record has been analysed quantitatively. The result should be regarded as impressionistic and often over-generalized compared with patterns the neontologists can give us. But they too have conceptual and procedural problems and not all of them are enamoured of biogeographic nomenclature. In some respects the patchiness of the preserved record is to the palaeontologist's advantage, enabling him or her to avoid some of the neontologists' problems, though with a concomitant loss of rigour. Most of the trends presented here have been based on the distribution of one, two, or three groups of organisms; no holistic quantitative analyses seem yet to have been attempted for any era or epoch. Additionally there is under-representation of work on microfossils as the main thrust of micropalaeontology has been towards establishing increasingly refined correlations with only minor concern for provinciality in shallow marine faunas; the Cretaceous works of Scheibnerova (1971) and Haig (1979) are notable exceptions.

(i) Ediacarian

The remarkable fauna of soft-bodied cnidarians, annelids, primitive arthropods, Problematica, and trace fossils originally described from the Pound Quartzite at Ediacara in South Australia (Sprigg 1947; Glaessner 1958, 1962, 1969; Glaessner and Daily 1959; Glaessner and Wade 1966; Wade 1972; Jenkins and Gehling 1978) has now been shown to have had wide distribution at low latitudes and to have had elements spread almost world-wide (Cloud and Glaessner 1982). The most remarkable of these other occurrences is in the Valdai sediments of the White Sea area of northeast Europe where there are some 10 species and 10 genera that are represented at Ediacara; this is symptomatic of a surprising degree of cosmopolitanism, the more remarkable from being so early in the evolution of the Metazoa. Information none the less seems to be too sparse for patterns of Ediacarian provinciality to be detected yet.

(ii) Cambrian

A similar lack of global provincial differentiation seems to apply to the Tommotian (earliest Cambrian) shelly faunas. Zhuravleva (1968) has pointed to minor provincialism in the Archaeocyatha, but Tommotian archaeocyathan faunas are not well known outside the Soviet Union. Post-Tommotian Archaeocyatha are better known and show a distinct global pattern of two regions and about five provinces. The richest Australian faunas, from the Flinders Ranges of South Australia, are regarded as a discrete province. Its closest affinities are with faunas described from Antarctica (Fig. 43A; Hill 1965), which is what would be anticipated from the structural continuity between Australia and East Antarctica at that time.

With the entry of trilobites, a clear pattern of zoogeographic differentiation into two regions or realms can be made out for the remainder of the Early Cambrian: the Redlichiid including Australia, Antarctica, China, and southern Asia, and the Olenellid including North and South America and north-western Europe (Cowie 1971). A third intermediate realm has been proposed (Kobayashi 1972) but seems not to have gained recognition yet. Reanalysis and refinement of Early Cambrian trilobite zoogeography is needed for rationalization of the trilobite and archaeocyathan schemes. There seems a need for numerical analysis in similar fashion to that carried out for the Middle Cambrian (Jell 1974) wherein three zoogeographic regions or realms have been discriminated; these will have nomenclatorial priority where appropriate over a suite of four names proposed subsequently for Late Cambrian entities by Palmer (1979). In Jell's analysis, Australia's Middle Cambrian trilobite faunas cluster with those of north and south China, Siberia, northern India, New Zealand, and Antarctica into the single Tolchuticook realm or region. South America tends to cluster with North American faunas; Jell, quoting palaeomagnetic data, suggests this was due to a 30° separation of eastern and western Gondwanaland. Ziegler *et al.* (1981), however, show the two parts of Gondwanaland sutured in the Late Cambrian and explain the contrast with the rest of Gondwanaland as being a product of oceanic circulation. If the zoogeographic regions or realms based on trilobites for each of the Cambrian epochs have the same meaning, then there would be an overall increase in provinciality from perhaps a single region or realm in the Ediacarian and Tommotian to two, three, then four as one proceeds in time through the Early to Middle and Late Cambrian.

Apart from Jell's analysis of the Middle Cambrian trilobites, there has been no recent biogeographic overview of the Australian Cambrian faunas. A voluminous literature has been published on Cambrian, especially Middle and Late Cambrian, trilobites (for key to earlier literature see Shergold 1973, 1982; Öpik 1978, 1982). The high level of provinciality at the species-level is manifest; it is only with agnostids that there is a significant proportion of species (c. 20 per cent) known from elsewhere in the world. This relative cosmopolitanism, taken to reflect a pelagic mode of life (Öpik 1978), has facilitated development of refined correlations with Eurasian and North American sequences. A late Middle Cambrian inarticulate brachiopod fauna from the south island of New Zealand has been shown to have four out of five species in common with Australia (Henderson and MacKinnon 1981); the two areas obviously formed parts of the same biogeographic province at that time. The Cambrian molluscan faunas are of special interest (for example Runnegar and Jell 1976; Pojeta and Shergold 1977) but their biogeographic significance has yet to be deciphered.

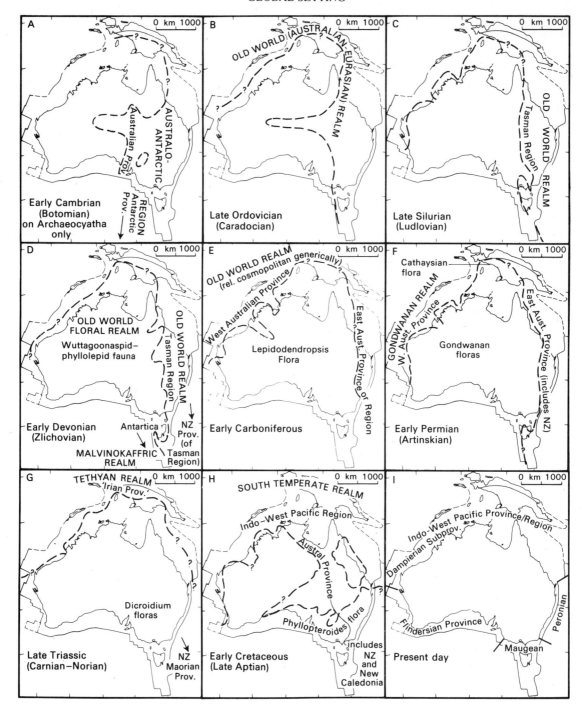

Fig. 43. Résumé of biogeographic terms that have been used in the Australia–New Guinea area, highlighting the current need to develop consistency through time.

(iii) Ordovician

Numerical analysis of Ordovician trilobites (Whittington and Hughes 1972) and brachiopods (Williams 1973) indicates that, for marine benthic faunas, there were four or five zoogeographic regions or realms in the Early Ordovician; provinciality declined rapidly during the Late Ordovician. It has been stated or implied by a number of workers that during the late Ashgillian (Hirnantian) and Early Silurian there was very little endemism: a fundamentally cosmopolitan fauna extended globally. This may be generally true, but is in part an artifact of poor knowledge of the rarely preserved and difficult-to-correlate southern cold-water faunas that arose with the onset of the Late Ordovician glaciation and inaugurated the distinctive Malvinokaffric Realm that was to persist through the Silurian into the Devonian.

What is known of Australian Ordovician shelly faunas would appear to be generally consistent with this analysis. The Early Ordovician trilobite (Shergold 1975; Legg 1976), cephalopod (Teichert and Glenister 1954; Pojeta and Gilbert-Tomlinson 1977), brachiopod (Laurie 1980), and even conodont faunas (Jones 1971; Druce and Jones 1971; McTavish 1973b) have a high level of specific endemism. Middle and Late Ordovician trilobite (Webby 1974), mollusc (Teichert and Glenister 1953; Pojeta and Gilbert-Tomlinson 1977; Wade 1977), and brachiopod faunas (Percival 1978, 1979) are likewise highly endemic at the specific level but the proportion of cosmopolitan to endemic genera seems to have increased up the column (cf. Banks and Burrett 1980). Traditionally, Ordovician correlations in Australia have focused on graptolite evidence wherever available (Thomas 1960; Webby *et al.* 1981) but, despite problems of endemism, increasing use is being made of shelly faunas (see Banks and Burrett 1980, Webby *et al.* 1981, and Webby and Packham 1982 for keys to literature on Ordovician faunas and sequences).

(iv) Silurian

Knowledge of global biogeography for the Silurian would seem to resemble a quagmire. The principal reasons for this are the patchy record of modern work on Llandoverian and Wenlockian coral faunas (apart from eastern North America, Europe, and Central Asia), lack of integration of work on trilobites, rugosans, tabulates, and brachiopods, and workers having restricted analysis to presence–absence studies of genera. Additionally, there are problems that relate to global tectonics (Pickett 1975). As a broad generalization, it would seem that generic-level biogeographic differentiation was not especially pronounced for Tabulata during the Llandoverian (Leleshus 1970) or later Silurian; it was more obvious for the Rugosa (Hill 1959, 1981; Ivanovsky 1965; Kaljo and Klaamann 1973), and was relatively pronounced by Late Silurian times when, despite earlier assumptions about rugose coral faunas of that age being nearly cosmopolitan, clear differentiation of an Eastern North American Realm can be made out (Oliver 1977). A similar pattern has been presented for brachiopods, but with recognition of the cold-water, essentially coral-free Malvinokaffric Realm (Boucot 1975) that persisted from Early Silurian to the Middle Devonian in the southern polar regions.

Australia (Fig. 43C) has been regarded as a separate biogeographic region of the Old World Realm, having greatest affinities with faunas from the Urals area of Eurasia and the Cordilleran area of North America (Boucot 1975). Its palaeontology however has tended to be neglected (reviews in Talent *et al.* 1975; Pickett 1982). With three notable exceptions (Chatterton and Campbell 1980; Sherwin 1971; Strusz 1980, 1982), little work has been done on Australian Silurian trilobites or brachiopods in recent years – too little on which to base sophisticated biogeographic conclusions. Of 19 Llandoverian rugosans from New South Wales that have been identified to specific level, only two are known from outside Australia; provincial affinities again are doubtful (McLean 1977). Similarly for the Late Silurian corals (literature in Hill 1978), specific endemism is high for the tabulates and near total for the rugosans, but generic endemism is not great. To summarize: though fine work has been done on intercontinental correlation utilizing graptolites (Sherwin 1974) and conodonts (Link and Druce 1972), the corpus of recent work on Australian shelly faunas is insufficient for a confident statement to be made about changes in zoogeographic relationships during the Silurian.

(v) Devonian

Global trends in provinciality are fairly well known: increasing through the Early Devonian to a maximum at about Zlichovian time and then declining to a minimum in the Late Devonian (Boucot *et al.* 1969; Boucot 1975; Oliver 1977). There were three well-delineated faunal realms in the first half of the Devonian: the Eastern North American, the cold-water Malvinokaffric, and the vast Old World Realm. Among subdivisions of the last of these is the Tasman Region (Fig. 43D), proposed to express the distinctiveness of the Early Devonian faunas from eastern Australia and New Zealand.

Provinciality of brachiopods and bivalves became so pronounced late in the Early Devonian that a discrete New Zealand Province has been proposed for faunas of approximately Zlichovian age from the Reefton area of the South Island. The latter include a number of elements, such as *Pleurothyrella* and *Australocoelia*, otherwise characteristic of the Malvinokaffric Realm that extended over the seaways of much of Gondwanaland, including Antarctica. The transitional character of these faunas thus accords with the plate-tectonic reconstructions for the Devonian in which New Zealand is juxtaposed in one way or another with the Marie Byrd Land region of West Antarctica (Talent *et al.* 1972).

Provinciality in corals and brachiopods declined in Australia as well as globally during the Middle Devonian until a nadir of provinciality was reached in the Frasnian when almost all genera were cosmopolitan though species contrasts are still obvious (Veevers 1959a, 1959b; Hill and Jell 1970; Roberts 1971). The lowest provinciality for Australian Late Devonian ammonoids appears to have been in the Famennian (Petersen 1975). The Late Devonian conodont faunas from Australia are remarkably rich and cosmopolitan, facilitating intricate zonation and accurate correlation with classical sections in western Europe.

It has been suggested (Young 1981) that five Early Devonian vertebrate provinces or regions can be delineated, the Australian faunas constituting one of these. The late Early Devonian (approximately Zlichovian) fishes are highly endemic (for example White 1978; Young 1980) but, as with the marine invertebrates and seemingly the plants as well, zoogeographic differentiation declined in the later half of the Devonian. Young has suggested this may have been due to marine regression and generation of land connection between Gondwanaland and Euramerica; such a connection, however, should have generated increased provincialism in marine faunas.

The global pattern of Silurian land floras is obscure, but Early Devonian floras appear to be differentiated in somewhat similar fashion to the marine faunas: a Malvinokaffric flora in South America, southern Africa, and Antarctica, and an Old World flora, of which the *Baragwanathia* flora (latest Silurian–Pragian; Table 6) of Australia is a subset. It is known from numerous localities in south-eastern Australia. The described species (for example Lang and Cookson 1935) are all endemic, but most genera are known from elsewhere in the Old World Realm; even the especially prominent *Baragwanathia* has been found in Canada (F. Hueber, oral communication). A small number of new general however have yet to be published (J. Timms in preparation). Somewhat younger floras that lack *Baragwanathia* (for example Lang and Cookson 1930) are still poorly known. At the very close of the Early Devonian, the lycopod *Leptophloeum australe* (McCoy) entered the record and remained prominent through the remainder of the Devonian (Gould 1976). Almost all genera associated with *L. australe* (Table 6) are Old World types; *Protolepidodendron scharianum* was first described from Europe, and *L. australe* may be identical with the North America *L. rhombicum* Dawson. By the Late Devonian, there was a single floral realm world-wide with obvious ecological segregation and with possible threefold provincial or regional differentiation. The Australian Late Devonian floras have closest affiliation with the floras of Kazakhstan, China, and Japan that were dominated by lycopods, especially *Leptophloeum,* a genus now known in Siberia (Edwards 1973).

(vi) Carboniferous

A high level of cosmopolitanism persisted from the Late Devonian into the Early Carboniferous (Tournasian and Visean) for Australian terrestrial floras (Table 6; Morris 1980) and marine faunas (Campbell and McKellar 1969; Jenkins 1974; Runnegar and Campbell 1976) with several forms specifically identical or very close to forms from Eurasia and even North America (Roberts 1971). Provinciality (Fig. 43E) appears to have been more noteworthy among corals than among brachiopods. Hill (1948, 1973) suggested the eastern Australian coral faunas were sufficiently distinctive compared with those of the rest of the world to justify grouping as a discrete Australian province or region characterized by Aphrophyllidae and richness in species of *Lithostrotion*. Roberts (1971) contended that the faunal differences between Early Carboniferous brachiopod faunas from eastern and western Australia were sufficiently profound as to signify two zoogeographic provinces, relating this to geographic isolation; Runnegar and Campbell (1976), however, have stressed environmental contrasts as perhaps equally important determinants.

Both marine and terrestrial provinciality increased globally after the Visean. This has been connected with collision between Gondwanaland and the Euramerican continental aggregate stopping circum-global warm-water circulation (Ross 1979b). However that may be, palaeomagnetic data seem to be consistent with a rapid southward shift of about 20° of latitude in the position of the Australian region of Gondwanaland between late Visean and some time in the Westphalian first suggested by Irving (1964; cf. Embleton – this chapter).

TABLE 6. *The sequence of Palaeozoic and Mesozoic terrestrial mega-floras*

Flora	Age	Salient associated genera
Unnamed flora	Late Cretaceous	*Ginkgoites,* undescribed conifers and angiosperms
Phyllopteroides flora	Early Cretaceous	*Phyllopteroides, Equisetum, Cladophlebis, Ptilophyllum, Taeniopteris, Pagiophyllum, Nathorstionella, Xylopteris, Otozamites, Nilssonia, Gleichenites, Sphenopteris, Podozamites, Pachypteris, Osmundacaulis, Lycopodites, Coniopteris, Adiantites, Alamatus, Amanda, Hausmannia. Ginkgoites, Baiera, Thinnfeldia, Brachyphyllum, Hydrocotylophyllum, Lappacarpus, Bellarinea, Hepaticites*
Otozamites-Ptilophyllum flora	Jurassic	*Otozamites, Ptilophyllum, Lycopodites, Agathis, Phlebopteris, Equisetum, Dictyophyllum, Coniopteris, Cladophlebis, Sagenopteris, Taeniopteris, Rissikia, Elatocladus, Pachypteris, Podozamites, Allocladus, Mataia, Araucaria, Araucarites, Pagiophyllum, Marattia, Osmundocaulis, Hausmannia, Nilssonia, Pentoxylon, Gleichenites, Isoetites, Nathorstia, Cupressinoxylon, Mesembrioxylon*
Dicroidium floras	Triassic	*Dicroidium, Marchantites, Cyclomeia, Skilliostrobus, Scoresbya, Cyclostrobus, Rissikia, Cladophlebis, Lepidopteris, Dejersya, Lobifolia, Pteruchus, Karibocarbon, Kurtziana, Anthrophyopsis, Johnstonia, Tetrapilon, Phyllotheca, Rienitsia, Taeniopteris, Voltziopsis, Asterotheca, Dictyophyllum, Ginkgo, Hoegia, Nilssonia, Pseudoctenis. Umkomasia, Phoenicopsis, Xylopteris, Pachypteris, Pterophyllum, Yabeiella, Todites, Chriopteris, Microphyllopteris, Pilophorosperma, Ginkgoites, Sphenobaiera, Neocalamites, Strzeleckia, Linguifolium*
'Thinnfeldia' callipteroides flora	latest Late Permian	*'Thinnfeldia', Schizoneura, Voltziopsis,* occasional *Glossopteris*
Glossopteris flora	Permian	*Glossopteris, Gangamopteris, Selaginella, Trizygia, Elatocladus, Phyllotheca, Umbellaphyllites, Raniganjia, Eretomonia, Palaeosmunda, Alethopteris, Sphenopteris, Partha, Dundedoonia, Austroglossa, Dictyopteridium, Isodictyopteridium, Plumsteadia, Vertebraria, Walkomiella, Neomariopteris, Noeggerathiopsis, Nummulospermum, Lelstotheca, Cyclodendron*
Botrychiopsis flora	Late Carboniferous (Stephanian) to very Early Permian	*Botrychiopsis, Aphlebia*
Otopteris argentinica and *Sphenopteridium* floras	Late Carboniferous (Namurian–Westphalian)	*Pseudorhacopteris, Aphlebia, Sigillaria, Stellotheca, Botrychiopsis, Cardiopteris, Cyclostigma, Triphyllopteris, Adiantites, Sphenopteridium, Austroclepsis, Lepidodendron, Stigmaria, Calamites* or *Archaeocalamites*
Lepidodendropsis flora	Early Carboniferous to Late Carboniferous (Westphalian)	*Lepidodendropsis, Stigmaria, Archaeocalamites, Rhodea*
Leptophloeum flora	latest Early Devonian to ?earliest Carboniferous	*Leptophloeum, Protolepidodendron, Astrocaulis, Barinophyton, Archaeopteris, Taeniocrada, ?Lepidosigillaria, Cordaites, Schizopodium*
Unnamed 'post-*Baragwanathia*' flora	Early Devonian (post-Pragian) – ?Middle Devonian	*Zosterophyllum, Sporongites, 'Hostinella'*
Baragwanathia flora	latest Silurian to Early Devonian (Pragian)	*Baragwanathia, Yarravia, Zosterophyllum, Hedeia, Salopella, Pluricaulis, Stenocaulon, Sporongites, Pachytheca, Hostinella*

Main data sources: Gould (1976, 1981), Douglas (1969), Retallack (1980), Morris (1975, 1980), Holmes (1982), Holmes and Ash (1979).

This is obviously sufficient to account for all or most of the floral, faunal, and climatic change that happened in Australia during this interval. Eastern Australian marine faunas dropped sharply in diversity, the *'Otopteris' argentinica* flora rapidly replaced the near-cosmopolitan *Lepidodendropsis* flora and, during the Westphalian, varved shales and tillites appeared (White 1968) and became increasingly important elements in the stratigraphic column. Despite frequent reference to the well-documented Late Carboniferous glaciation in eastern Australia and documentation of the low diversity marine faunas (Campbell 1962) resembling those of Argentina, obviously connected with emergence of a cold-water Gondwanan Realm at this time, one still reads that Late Carboniferous 'southern high latitude faunas have yet to be described' (Ziegler et al. 1981, p. 251)!

(vii) Permian

The global pattern of profound climatic and biogeographic differentiation persisted from the Late Carboniferous through most of Permian time (Waterhouse and Bonham-Carter 1975; Runnegar and Campbell 1976; Runnegar 1972, 1979). Most of Australia lay within the cold-water Gondwanan Realm, but the northern edge of what subsequently became the Australian block intruded into the warm-water Tethyan Realm. Faunas from the Bonaparte Gulf region as well as from Timor, Irian, and the Northland region of New Zealand have Tethyan faunas or show strong Tethyan influence as do the post-Sakmarian faunas from the north-western part of the Indo-Pakistan block.

The provincial contrast between faunas from eastern and western Australia (Fig. 43F) was discussed earlier. There is much closer similarity between the eastern Australian faunas and contemporary faunas from New Zealand (Waterhouse 1963, 1964, 1967b, 1968, 1969, 1980a, 1980b). That warm-water elements, specifically the rugose corals *Waagenophyllum* and *Wentzelella* and the fusulinids *Yabeina, Verbeekina,* and *Neoschwagerina*, extended into the Northland region of New Zealand some 15° or so further south than their penetration into Timor on the western flank of the Australian block was presumably due to the influence of an oceanic gyre bringing warm or tropical waters down the east-facing shores of Gondwanaland in much the same way as the East Australian Current brings warm water to the Peronian Province of eastern Australia today.

During the Late Permian, there was a world-wide tendency towards marine regression or increasing 'continentality' (Schopf 1974; Simberloff 1974), global climatic differentiation decreased, and the cold-water Gondwanan Realm contracted to just Australia and New Zealand (Dickins 1978) before finally disappearing during the latest Permian – as did the Boreal Realm in the northern hemisphere.

The Late Carboniferous *Botrychiopsis* flora was replaced in the very early Permian by the *Gangamopteris-Glossopteris* flora (Table 6) with *Gangamopteris*, a swamp-dwelling gymnosperm, tending to predominate in earlier floras. The entry of the latter coincided with the entry of an abundance of *Protohaploxypinus* pollen; similar pollen has been extracted from glossopterid microsporophylls. Retallack (1980) has likened the flora at that time to the present-day boreal taiga and the subsequent *Glossopteris*-dominated floras he has categorized as deciduous swamp forests. The *Glossopteris* flora seems to have extended over all of Australia except for a temporary extension of the Cathaysian Realm (*Gigantopteris* flora) into West Irian at about 5° S, 135°50′ E (Jongmans 1940; Lacey 1975).

The significance of the *Gangamopteris-Glossopteris* flora in the Permian sequences of Australia, South America, southern Africa, Madagascar, India, Antarctica, and New Zealand, and especially the now disjunct distributions of so many of its genera and even species has been commented on by numerous workers (for example Wegener 1924; Rigby 1972; Plumstead 1973). Decades before formulation of the new global tectonics, these distributions forming a single phytogeographic realm or even province were taken by many workers to be compelling evidence that in the Late Palaeozoic the southern continents had been aggregated into a super-continent: Gondwanaland. Interestingly, the microfloras are considerably more diverse than might be expected from the known macrofloras. Gould (1982) suggests that the limited diversity of form in *Glossopteris* and *Gangamopteris* may be deceptive, suggesting a parallel with present-day *Eucalyptus* where the range of leaf-form is small relative to the great number of species – about 500.

The flora of the latest Permian, immediately postdating the coal measures of the Sydney Basin, has been termed the *'Thinnfeldia' callipteroides* flora by Retallack (1980); it was not diverse (Table 6) and was short-lived. The nominate species possibly gave rise to *Dicroidium,* a pivotal genus in the Triassic floras (Retallack 1980).

(viii) Triassic

The Early Triassic was a time of strikingly low diversity of marine invertebrates (Kummel 1973). Ammonoids and conodonts are especially prominent in Early

Triassic faunas; many species achieved near-global distribution, thus making intercontinental correlation a matter of relative ease when either group is represented. Bivalves too were very common, but the diversity was low; several species of pteriod bivalves attained extraordinarily wide geographic ranges. Significant marine Early Triassic faunas are known from only two areas on the Australian mainland: in the southern Perth Basin (Dickins and McTavish 1963; McTavish and Dickins 1974; McTavish 1973a) and in the Gympie-Maryborough area of south-east Queensland (Fleming 1966; Runnegar 1969). Both have low-diversity faunas with key ammonoid genera, thus enabling correlation to be made with ease and precision, but a measure of provincialism in eastern Australia is indicated by all five specifically identifiable ammonoids from Gympie being new.

For better appreciation of provinciality in marine faunas during the Triassic, it is necessary to look farther afield. Molluscan faunas from the Late Triassic (Carnic-Noric) of the Jimi River area of New Guinea, south of the north New Guinea Tertiary suture and therefore assumed to be part of the Australian Triassic sedimentary prism, are profoundly provincial (Skwarko 1967). These faunas have no species in common with contemporaneous faunas from Misool (Krijnen 1931), Seram (Krumbeck 1923), or Timor (Krumbeck 1924); available evidence therefore is consistent with some sort of biogeographic boundary, provincial or regional, between the latter and northern New Guinea during the later Triassic (Fig. 43G). There is no relationship either to the contemporaneous faunas from the Maorian Province of New Zealand and New Caledonia, characterized by archaic endemics. Fossiliferous marine Triassic strata, incidentally, are not known from Antarctica, southern Australia, and South Africa, so the Maorian Province may be all that remains of a distinctive Triassic Austral realm (Fleming 1975).

In summary, the marine Triassic faunas of the Australian block, and of blocks assumed to have been aggregated with it during the Triassic, imply possibly two faunal realms: a very restricted cold or cool-water Maorian or Austral Realm in the New Zealand-New Caledonia area, and the remainder forming part of the warm-water Tethyan Realm, though with possibly considerable biogeographic differentiation between the benthic molluscan faunas of the northern or New Guinea margin and areas to the north-west.

Non-marine faunas from the Triassic of Australia include remarkable labyrinthodonts (Cosgriff 1974; Warren 1981), some showing relationships to South American forms. The spectrum of relationships to labyrinthodonts described from elsewhere on the globe is broad, indicating ease of intercontinental or global dispersal at that time. Fish faunas mainly from the Sydney Basin (for example Wade 1935, 1940) were diverse with some relationship to Madagascar and South Africa, but their overall biogeographic affinities, whether Laurasian or Gondwanan, are not sharply defined. Numerous Triassic insect faunas have been described mainly from the eastern states (Riek 1970), but their biogeographic significance is uncertain. Conchostracan faunas from eastern and western Australia (Tasch and Jones 1979) were also diverse; the occurrence of a few species on both sides of the continent testifies to ease of dispersal and absence of profound climatic contrast between east and west.

The Triassic floras (Table 6) contrast sharply with the preceding Permian floras, being dominated by pteridosperms, prominent among which is *Dicroidium*, but including sphenopsids, cycads, ginkgos, ferns, and conifers, several of the genera occurring in both the northern and southern hemispheres. An interesting exercise in discrimination of plant communities has been undertaken (Retallack 1977, 1980); the presence of analogues of present-day mangrove scrub, coastal woodlands, heath, coniferous forest, and so on have been suggested. Biogeographic differentiation within the Australian floras is not apparent, though admittedly the Triassic floral record, palynological records apart, is restricted to occurrences in eastern Australia and Tasmania.

(ix) Jurassic

With few exceptions (Arkell and Playford 1954; Brunnschweiler 1960, 1963; Skwarko 1974), Australian marine Jurassic faunas are not especially diverse nor important biogeographically; one has therefore to look to adjacent areas for better understanding of what shelf faunas must have been like in the Australian region during the Jurassic. Early Jurassic (Liassic) faunas were quasi-cosmopolitan, but there was global increase in provinciality during the Middle and Late Jurassic; differentiation into three zoogeographic regions or realms is well documented (Arkell 1956). This global tendency apart, there does not appear to have been significant finer-scale provincial developments within the Australia-New Zealand sector of Gondwanaland (Stevens 1977, 1980). Close faunal similarity and, by implication, ease of migration was maintained with Iran, India-Pakistan, south-east Asia, west Antarctica, and southern South America, all of which were parts of the Tethyan Realm.

Australian Jurassic floras are dominated by conifers and are basically cosmopolitan in aspect, but the appearance in relative abundance of Podocarpaceae and Araucariaceae (Townrow 1969; Vakhrameev 1972; Gould 1981, 1982) is an expression of the generation of a new Gondwana flora. It has been argued (Gould 1982) that the palaeolatitudes suggested from palaeomagnetic data (Smith *et al.* 1973) are inconsistent with the palaeobotanical evidence; a 30–60° S position is suggested rather than the 45–80° S reconstructed from palaeomagnetic data. A similar case has been carefully documented for fossil forests from the Early Cretaceous of Antarctica; again growth characteristics are those of trees growing in warm temperate areas and not like those of modern high-latitude trees (Jefferson 1982).

(x) Cretaceous

Provinciality in Cretaceous bivalves has been carefully reviewed, using quantitative data, by Kauffman (1973). He has shown that a distinctive Indo-Pacific Region with 50 per cent endemism and two clearly defined provinces is discernible in the Early Cretaceous, the one that concerns us, the Austral Province, consisting of Australia, New Caledonia, New Zealand, New Guinea, and some associated islands. Provincial contrast with the East African Province or subprovince about Mozambique–Madagascar accords with the tectonic isolation of Africa from Australia–New Zealand–Antarctica that had already occurred. Both provinces experienced rapid decline in provinciality commencing in the Aptian, decreasing to about 30 per cent in the Late Cretaceous. In New Zealand, however, endemism increased in the Late Cretaceous, becoming pronounced by the Campanian; this too is compatible with the tectonic isolation, in this case from spreading in the Coral Sea–Tasman Sea area bringing into being a wide deep-water barrier between the Australian and New Zealand blocks. It followed soon after the onset of widespread transgression in New Zealand and is consistent with the 95 Ma separation of the Lord Howe Ridge from eastern Australia (suggested elsewhere in this volume) rather than the somewhat later separation suggested by earlier authors (Hayes and Ringis 1973; Weissel *et al.* 1977; Crook and Belbin 1978).

Australian Early Cretaceous, especially Aptian and Albian ammonoid, bivalve, and foraminiferal faunas are well documented (Cox 1961; Skwarko 1963, 1966, 1967, 1970; Day 1969, 1974; Crespin 1963; Scheibnerova 1976; Haig 1980, 1981, 1982), and are provincially distinctive. Though the ammonoids are not generically distinctive, two of the Albian genera, *Labeceras* and *Myloceras*, are known elsewhere only from Mozambique and Madagascar. The bivalves are much more discriminative with several endemic genera: *Austrotrigonia, Nototrigonia, Tatella, Maranoana, Tancretella, Fissilunula, Barcoona, Cyrenopsis, Eyrena,* and *Pseudavicula* (Day 1969). Endemism was less apparent during the Late Cretaceous, though the bivalves *Climacotrigonia, Actinotrigonia, Entolium* (*Ctenopleurium*), and *Fissiluna* indicate a continuing distinctiveness (Kauffmann 1979). Less is known of Late Cretaceous marine faunas because of the great reduction in development compared with the Early Cretaceous. Important Late Cretaceous foraminiferal and ammonoid faunas have been described from Bathurst Island, the Carnarvon and Otway Basins, and New Guinea (Wright 1963; Belford 1960; Owen 1973; Taylor 1964; Brunnschweiler 1966). There was not significant biogeographic differentiation within the Cretacecus faunas of Australia *sensu stricto*. Foraminiferal faunas from the northern or New Guinea margin (Haig 1981), however, show that during the Early Cretaceous the Australian Block projected into the warm waters of the Tethyan Realm; some of this influence extended into north Queensland. There was no separation into contrasting eastern and western Australian faunas during the Early Cretaceous (Fig. 43H) as was the case for the Devonian and Carboniferous. The Neocomian faunas are not sufficiently well known, but the intermittent trans-continental marine connections across Arnhem Land (Skwarko 1966) and possibly from the Great Artesian Basin to the Canning Basin in the Aptian–Albian (when faunas are better known), may have helped counteract tendencies towards east–west provincial differentiation.

An extraordinary rich insect fauna from the Early Cretaceous (Neocomian–Aptian) from Koonwarra, Victoria, is notable for apparent complete endemism and remarkable similarity of many species and genera to contemporary forms (E. F. Riek, personal communication). This apparent precocity and endemism is surely a reflection of spotty knowledge of insect evolution and palaeobiogeography, and has no implications as regards isolation and palaeoclimate of the Australian block during Cretaceous times. On the other hand, scraps of information on dinosaurs from beds of similar age elsewhere in Victoria for example and from Queensland (Molnar 1980) suggest fairly high cosmopolitanism and thus relative ease of communication between Laurasia and Gondwanaland until early in the Cretaceous.

Angiosperms apparently arose in the Barremian in western Gondwanaland (Brenner 1976), and became widely distributed during the Early Cretaceous (Raven and Axelrod 1974). Their migration into the Australian region, however, seems not to have commenced until the early Albian (Dettmann 1981), perhaps by a subtropical route from Africa through Madagascar and India (Raven and Axelrod 1974). By the middle of the Cretaceous their pollen had become more abundant than that of ferns and gymnosperms; some of the modern angiosperm families were already in existence by that time and many more were to enter the record during the Late Cretaceous.

Extensive Early Cretaceous floras are known from Queensland (Walkom 1919) and Victoria (Douglas 1969, 1974); the similarities of the latter to the Early Cretaceous floras of Alexander Island, West Antarctica, is remarkable (T. Jefferson, oral communication), so much so that similar latitude and climate (cool temperate) and ease of interplay must be inferred. Late Cretaceous macrofloras are not well known (Table 6), but the microfloras are becoming well documented (Dettman 1981); these, even from northern Australia, have been interpreted as temperate, not tropical (Burger in Norvick and Burger 1975).

(xi) Cainozoic

The composition and novelty of Australia's contemporary and Cainozoic marine and terrestrial faunas and floras have been discussed in earlier sections of this account. Special attention was given to the chronology of land-bridges and island 'staging-points' and when these were most effectual.

Next to nothing is known of Palaeogene terrestrial faunas, but a sophisticated picture of the succession of early Miocene (or latest Oligocene) to recent mammalian faunas is being developed (Archer 1981) and something of the avifaunas is being deciphered (Rich and Thompson 1982). Difficulty has been experienced in interpreting the Cainozoic floras from leaf-form, but a new era of intense and critical investigation has commenced with special attention to physiognomic analysis. The most interesting example of this has been the demonstration that the early middle Eocene flora from Maslin Bay near Adelaide is closely comparable with the present-day vine-forests of eastern Australia between 20° and 25° south latitude – climatically equivalent to 10–15° farther north than Maslin Bay today (Christophel and Blackburn 1978; Blackburn 1981). Despite frequent difficulties of identifying spores and pollens with specific elements of macroflora, a broad outline of the evolution of the Australian flora is being delineated from palynology (Martin 1978; Kemp 1978; Singh 1981, 1982; Singh *et al.* 1981) though there is a great lacuna in information for the northern half of the continent. As a result of about 30 years of intensive labour by many workers on foraminifers, a finely subdivided zonal scheme has been developed and applied to onshore and offshore Cainozoic sequences; a salient achievement has been to couple this information with oxygen-isotope measurements to generate and repeatedly refine temperature curves for the waters around the Australian block (Kennett and Shackleton 1976; McGowran 1978, 1979). We have now entered a phase where Cainozoic palaeoclimatic events are being correlated with increasing precision, especially in the marine context.

Biogeographic change in Australia and adjoining areas during the Cainozoic can be summarized as having involved the following elements:

(1) A continuous influx of terrestrial fauna and flora from the huge biotic reservoir in south-east Asia, dispersing through the Moluccas–Lesser Sunda Islands inter-regional zone to the New Guinea area (Papuan zoogeographic province; part of Malesian phytogeographic province) of the Australian block, percolating southwards into Australia proper (and also filtering southeastwards towards New Caledonia and New Zealand) to interact with:

(2) A corpus of what for convenience might be termed autochthonous groups, archaic in themselves (cf. Table 5) and now relict (such as podocarps and araucarians), or that have originated in Gondwanaland before substantial dispersion of its constituent blocks (for example the Proteaceae, Winteraceae, and ratites) (cf. Fig. 42A). Additionally there has been:

(3) A subordinate but continuous process of waif-dispersal of marine and terrestrial biota across the Tasman Sea from temperate Australia to New Zealand. This is well documented for several elements of the marine biota and inferred for various terrestrial biota, for example several groups of insects, spiders, birds, and plants. Reciprocal migration of terrestrial biota in the reverse direction is not documented, but the oceanic circulation pattern and the existence of intermediate staging-points suggest a possibility of this from the New Caledonia–New Hebrides part of the Indo-West Pacific marine province (or region).

(e) Résumé of main biogeographic events and relationships

(1) In the context of what seems to have been a general global increase in provinciality through Cambrian time, Australian archaeocyathan and trilobite faunas remained provincial with greatest similarity to faunas from Antarctica, China, and south-west Siberia.

(2) A relatively high level of global provincial differentiation is reflected in the high level of endemism of Australian Early and Middle Ordovician faunas. The global decline in provinciality of Caradocian faunas is apparent, but the Australian picture for the Ashgillian and Early Silurian is not well known.

(3) Marine provinciality increased from the Late Silurian reaching an acme in the late Early Devonian (approximately Zlichovian); endemism then declined to low levels from late Middle Devonian (Givetian) through to Early Carboniferous. Provinciality of non-marine faunas is apparent, but terrestrial floras were relatively cosmopolitan.

(4) Rapid cooling and increase in provinciality in the Late Carboniferous was connected with Australia's rapid southward movement. Australia became part of the emerging Gondwanan marine and terrestrial realms.

(5) During the Permian, Australian marine faunas formed part of the cold-water Gondwanan Realm, having strong Tethyan influence in the north-west. Terrestrial floras were an integral part of the *Gangamopteris-Glossopteris* flora, apart from a brief incursion of the temperate or tropical *Gigantopteris* flora on the northern (West Irian) margin.

(6) Diversity and provinciality of global faunas appears to have decreased rapidly at the close of the Permian. During the Triassic, the cold-water Gondwanan Realm disappeared, or became restricted to New Zealand and New Caledonia as the Maorian Province; the remaining faunas about Australia, though Tethyan, exhibit provincial differentiation. Floras dominated by pteridosperms including *Dicroidium* spread rapidly and persisted with little fundamental change through the Triassic.

(7) Provinciality of marine shelf faunas remained low through the Early Jurassic, but increased progressively through the remainder of the period to peak in the earliest Cretaceous when Australia and its immediate neighbours (still attached) formed a discrete Austral province within the emerging Indo-Pacific Realm. Jurassic floras of Australia were broadly cosmopolitan.

(8) Commencing in the Aptian, marine provinciality declined rapidly through the remainder of the Cretaceous. Angiosperms made their appearance in terrestrial floras in the Albian and rapidly rose to dominance.

(9) From Palaeocene to early early Miocene there was an overall increase in provinciality of marine benthos, with transient incursions of tropical elements into Australian waters, for example of larger foraminifers, especially in the late Eocene. Marsupials, parrots, and ratites diversified. Numerous other vertebrate, invertebrate, and plant groups arrived seriatim from the north (and some from the south), presumably by 'island-hopping', for example grasses, cypresses, chenopods, bur-reeds and Symplocaceae. Curiously there was no faunal influx of placental groups via the Antarctic route from South America.

(10) During the late early and early middle Miocene there was a notable influx of Indo-West Pacific taxa into marine shelf faunas of southern Australia. Consequent on the interplay of Australia and SE Asian continental margins, colonization from the Oriental Realm accelerated. Muridae (rats and mice) and several groups of plants, for example Asteraceae, arrived. Aridity increased.

(11) From late middle Miocene to present there appears to have been increasing provinciality of southern Australian shallow marine faunas; northern faunas formed part of the Indo-West Pacific Province or Region. The Eremian (or Eyrian) Province expanded due to increasing aridity, concurrent with contraction of the south-west and south-east Australian Provinces. Rain-forest areas fragmented and became restricted to the east coast and highland regions. There was a great increase in grasslands, chenopods, and in *Eucalyptus* and *Acacia* as canopy-dominants.

(12) Marsupial and ratite diversity decreased during the Quaternary. There was increased incidence of fire, possibly connected with the arrival of man, and consequent further increase in importance of *Eucalyptus* and *Acacia* at the expense of more fire-sensitive groups.

(f) Bibliography of Australian biogeography and palaeobiogeography

These references overlap only slightly those listed at the back. They constitute the core of the literature of Australian biogeography and palaeobiogeography, and, for utility, are retained as an entity.

Archbold, N. W., Pigram, C. J., Ratman, N. and Hakim, S. (1982). Indonesian Permian brachiopod fauna and Gondwana–South-east Asia relationships. *Nature* **296**, 556–8.

Archer, M. (1981). A review of the origins and radiations of Australian mammals. In *Ecological biogeography of Australia* (ed. A. Keast), pp. 1435–88. Junk, The Hague.

—— and Bartholomai, A. (1978). Tertiary mammals of Australia: a synoptic view. *Alcheringa* **2**, 1–19.

Arkell, W. J. (1956). *Jurassic geology of the world*. Oliver and Boyd, Edinburgh and London. 806 pp.

—— and Playford, P. E. (1954). The Bajocian ammonites of Western Australia. *R. Soc. Lond., Philos. Trans., Ser. B.* **237**, 547–605, pls. 27–40.

Ashlock, P. D. (1974). The uses of cladistics. *Annu. Rev. Ecol. Syst.* **5**, 81–99.

Audley-Charles, M. G. (1981). Geological history of the region of Wallace's line. In *Wallace's line and plate tectonics* (ed. T. C. Whitmore), pp. 24–35. Clarendon, Oxford.

—— and Hooijer, D. A. (1973). Relation of Pleistocene migrations of pygmy stegodonts to island arc tectonics in eastern Indonesia. *Nature* **241**, 197–8.

Balgooy, M. M. J. van (1971). Plant geography of the Pacific. *Blumea Supp.* VI. 22 pp.

Banks, M. R. and Burrett, C. F. (1980). A preliminary Ordovician biostratigraphy of Tasmania. *Geol. Soc. Aust., J.* **26**, 363–76.

Barlow, B. A. (1981). The Australian flora: its origin and evolution. In *Flora of Australia*, vol. 1, pp. 25–75. Bureau of Flora and Fauna, Canberra.

Beadle, N. C. W. (1954). Soil phosphate and the delimitation of plant communities in eastern Australia. *Ecol.* **35**, 370–5.

—— (1966). Soil phosphate and its role in molding segments of the Australian flora and vegetation with special reference to xeromorphy and sclerophylly. *Ecol.* **47**, 991–1007.

—— (1981). Origins of the Australian angiosperm flora. In *Ecological biogeography in Australia* (ed. A. Keast), pp. 407–26. Junk, The Hague.

Beard, J. S. (1976). The evolution of Australian desert plants. In *The evolution of desert biota* (ed. D. W. Goodall), pp. 52–63. Univ. Texas, Austin.

—— (1977). Tertiary evolution of the Australian flora in the light of latitudinal movements of the continent. *J. Biogeogr.* **4**, 111–18.

—— (1981). The history of the phytogeographic region concept in Australia. In *Ecological biogeography in Australia* (ed. A. Keast), pp. 335–53. Junk, The Hague.

Belford, D. J. (1960). Upper Cretaceous Foraminifera from the Toolonga Calcilutite and Gingin Chalk, Western Australia. *Aust., Bur. Miner. Resour., Geol. Geophys., Bull.* **57**, 127 pp., 35 pls.

Bemmelen, R. W. van (1949). *The geology of Indonesia. Vol. 1A: General geology of Indonesia and adjacent archipelagoes*, pp. 1–732. Govt. Printing Office, The Hague.

Bennett, I. and Pope, E. C. (1953). Intertidal zonation of the exposed rocky shores of Victoria, together with a rearrangement of the biogeographical provinces of temperate Australian shores. *Aust. J. Mar. Freshw. Res.* **4**, 105–59, 6 pls.

—— —— (1960. Intertidal zonation of the exposed rocky shores of Tasmania and its relationship with the rest of Australia. *Aust. J. Mar. Freshw. Res.* **11**, 182–221, 7 pls.

Blackburn, D. T. (1981). Tertiary megafossil flora of Maslin Bay, South Australia: numerical taxonomic study of selected leaves. *Alcheringa* **5**, 9–28.

Boltovsky, E. (1980). The age of the Drake Passage. *Alcheringa* **4**, 289–97.

Boucot, A. J. (1975). *Evolution and extinction rate controls.* Elsevier, Amsterdam. 427 pp.

——, Johnson, J. G. and Talent, J. A. (1967). Lower and Middle Devonian faunal provinces based on Brachiopoda. In *International Symposium on the Devonian System* (ed. D. H. Oswald), II, pp. 1239–54. Alberta Soc. Petrol. Geol., Calgary.

—— —— —— (1969). Early Devonian brachiopod zoogeography. *Geol. Soc. Am., Spec. Pap.* **119**, 113 pp., 20 pls.

Bowler, J. M. (1981). Age, origin and landform expression of aridity in Australia. In *Evolution of the flora and fauna of arid Australia* (ed. W. R. Barker, P. J. M. Greenslade, and P. R. Baverstock), pp. 35–46. Peacock Pubs., Frewville, South Australia.

Brenner, G. J. (1976). Middle Cretaceous floral provinces and early migrations of angiosperms. In *Origin and evolution of angiosperms* (ed. C. B. Beck), pp. 23–47. Columbia Univ., New York.

Briggs, J. C. (1974). *Marine zoogeography.* McGraw-Hill, New York. 475 pp.

Brundin, L. (1966). Transantarctic relationships and their significance, as evidenced by chironomid midges, with a monograph of the subfamily Podonominae, Aphrotaeniinae and the austral Heptagiae. *Kgl. Sv. Ventenskapsakad. Handl.* (4) **11**, 1–472.

Brunnschweiler, R. O. (1960). Marine fossils from the Upper Jurassic and the Lower Cretaceous of Dampier Peninsular, Western Australia. *Aust., Bur. Miner. Resour., Geol. Geophys., Bull.* **59**, 53 pp., 3 pls.

—— (1963). A review of the sequence of *Buchia* species in the Jurassic of Australasia. *R. Soc. Vic., Proc.* **76**, 163–8.

—— (1966). Upper Cretaceous ammonites from the Carnarvon Basin of Western Australia. 1: The heteromorph Lytoceratina. *Aust., Bur. Miner. Resour., Geol. Geophys., Bull.* **58**. 58 pp., 8 pls.

Burbidge, N. (1960). The phytogeography of the Australian region. *Aust. J. Bot.* **8**, 75–211.

Cameron, L. M. (1935). The regional distribution of vegetation in New South Wales. *Aust. J. Geog.* **2**(5), 18–32.

Campbell, K. S. W. (1957). A Lower Carboniferous brachiopod-coral fauna from New South Wales. *J. Paleontol.* **31,** 34–98, pls. 11–17.

—— (1961). Carboniferous fossils from the Kuttung rocks of New South Wales. *Palaeontology* **4,** 428–74, pls. 53–63.

—— (1962). Marine fossils from the Carboniferous glacial rocks of New South Wales. *J. Paleontol.* **36,** 38–52.

—— (1965). Australian Permian terebratuloids. *Aust., Bur. Miner. Resour., Geol. Geophys., Bull.* **68.** 113 pp., 17 pls.

—— and Engel, B. A. (1963). The faunas of the Tournaisian Tulcumba Sandstone and its members in the Werrie and Belvue Synclines, New South Wales. *Geol. Soc. Aust., J.* **10,** 55–122, pls. 1–9.

—— and McKellar, R. G. (1969). Eastern Australian Carboniferous invertebrates: sequence and affinities. In *Stratigraphy and Palaeontology. Essays in honour of Dorothy Hill* (ed. K. S. W. Campbell), pp. 77–119. Aust. Nat. Univ. Press, Canberra.

—— and McKelvey, B. C. (1972). The geology of the Barrington district, N.S.W. *Pac. Geol.* **5,** 7–48.

Carolin, R. C. (1981). The development of the biogeography of arid Australia. In *Evolution of the flora and fauna of arid Australia* (ed. W. R. Barker, P. J. M. Greenslade, and P. R. Baverstock), pp. 119–24. Peacock Pubs., Frewville, South Australia.

Chaproniere, G. C. H. (1980). Influence of plate tectonics on the distribution of late Paleogene to early Neogene larger foraminiferids in the Australasian region. *Palaeogeogr., Palaeoclimat., Palaeoecol.* **31,** 299–317.

Chatterton, B. D. E. and Campbell, K. S. W. (1980). Silurian trilobites from near Canberra and some related forms from the Yass Basin. *Palaeontogr., Abt. A* **167,** 77–119, 5 pls.

Christofel, D. C. and Blackburn, D. T. (1978). Tertiary megafossil flora of Maslin Bay, South Australia: preliminary report. *Alcheringa* **2,** 311–9.

Ciesielski, P. F., Ledbetter, M. T. and Ellwood, B. B. (1982). The development of Antarctic glaciation and the Neogene paleo-environment of the Maurice Ewing Bank. *Mar. Geol.* **46,** 1–51.

Clark, H. L. (1946). The echinoderm fauna of Australia: its composition and origin. *Carnegie Inst., Pub.* **566,** 1–567.

Clifford, H. T. and Simon, B. K. (1981). The biogeography of Australian grasses. In *Ecological biogeography of Australia* (ed. A. Keast), pp. 537–54. Junk, The Hague.

Cloud, P. and Glaessner, M. F. (1982). The Ediacarian Period and System: the Metazoa inherit the Earth. *Science* **218,** 783–92.

Cockbain, A. E. (1980). Permian ammonoids from the Carnarvon Basin – a review. *W. Aust., Geol. Surv., Annu. Rep.* **1979,** 100–5.

Coleman, P. J. (1957). Permian Productacea of Western Australia. *Aust., Bur. Miner. Resour., Geol. Geophys., Bull.* **40.** 147 pp., 21 pls.

Cookson, I. C. (1954). The Cainozoic occurrence of *Acacia* in Australia. *Aust. J. Bot.* **2,** 52–9.

Coomans, H. E. (1962). The marine mollusk fauna of the Virginian area as a basis for defining zoogeographical provinces. *Beaufortia* **9**(98), 83–104.

Cosgriff, J. W. (1974). Lower Triassic Temnospondylii of Tasmania. *Geol. Soc. Am., Spec. Pap.* **149,** 134 pp.

Cowie, J. W. (1971). Lower Cambrian faunal provinces. In *Faunal provinces in space and time* (ed. F. A. Middlemiss, P. F. Rawson and G. Newall), pp. 31–46. Seel House, Liverpool.

Cox, L. R. (1961). The molluscan fauna and probable Lower Cretaceous age of the Nanutarra Formation of Western Australia. *Aust., Bur. Miner. Resour., Geol. Geophys., Bull.* **61,** 39 pp., 7 pls.

Cracraft, J. (1973). Continental drift, palaeoclimatology and the biogeography of birds. *J. Zool. Lond.* **169,** 455–545.

—— (1974). Continental drift and vertebrate distribution. *Annu. Rev. Ecol. System* **5,** 215–61.

Cranbrook, The Earl of (1981). The vertebrate faunas. In *Wallace's line and plate tectonics* (ed. T. C. Whitmore), pp. 57–69. Oxford, Clarendon.

Crespin, I. (1958). Permian Foraminifera of Australia. *Aust., Bur. Miner. Resour., Geol. Geophys., Bull.* **48.** 207 pp., 33 pls.

—— (1963). Lower Cretaceous arenaceous Foraminifera of Australia. *Aust., Bur. Miner. Resour., Geol. Geophys., Bull.* **66.** 110 pp., 18 pls.

Crook, K. A. W. and Belbin, L. (1978). The southwest Pacific area during the last 90 million years. *Geol. Soc. Aust., J.* **25,** 23–40.

Dartnall, A. J. (1974). Littoral biogeography. In *Biogeography and ecology in Tasmania* (ed. W. D. Williams), pp. 171–94. Junk, The Hague.

Day, R. W. (1969). The Lower Cretaceous of the Great Artesian Basin. In *Stratigraphy and palaeontology: essays in honour of Dorothy Hill* (ed. K. S. W. Campbell), pp. 140–73. Austr. Nat. Univ., Canberra.

—— (1974). Aptian ammonites from the Eromanga and Surat Basins, Queensland. *Qld, Geol. Surv. Publ.* **360.** 19 pp., 8 pls.

Dettmann, M. E. (1981). The Cretaceous flora. In

Ecological biogeography of Australia (ed. A. Keast), pp. 355-75. Junk, The Hague.

Diamond, J. M. (1974). Colonization of exploded volcanic islands by birds: the supertramp strategy. *Science* **184**, 803-6.

Dickins, J. M. (1956). Permian pelecypods from the Carnarvon Basin, Western Australia. *Aust., Bur. Miner. Resour., Geol. Geophys., Bull.* **29**. 42 pp., 6 pls.

—— (1957). Lower Permian pelecypods and gastropods from the Carnarvon Basin, Western Australia. *Aust., Bur. Miner. Resour., Geol. Geophys., Bull.* **41**. 55 pp., 6 pls.

—— (1963). Permian pelecypods and gastropods from Western Australia. *Aust., Bur. Miner. Resour., Geol. Geophys., Bull.* **63**. 150 pp., 26 pls.

—— (1978). Climate of the Permian in Australia: the invertebrate faunas. *Palaeogeogr., Palaeoclimatol., Palaeoecol.* **23**, 33-46.

—— and McTavish, R. A. (1963). Lower Triassic marine fossils from the Beagle Ridge (BMR 10) bore, Perth Basin, Western Australia. *Geol. Soc. Aust., J.* **10**, 123-40.

Douglas, J. G. (1969). The Mesozoic floras of Victoria Parts 1 and 2. *Vic., Geol. Surv., Mem.* **28**. 310 pp.

—— (1974). The Mesozoic floras of Victoria Part 3. *Vic., Geol. Surv., Mem.* **29**. 185 pp.

Druce, E. C. (1969). Devonian and Carboniferous conodonts from the Bonaparte Gulf Basin, northern Australia. *Aust., Bur. Miner. Resour., Geol. Geophys., Bull.* **98**. 157 pp., 43 pls.

—— (1976). Conodont biostratigraphy of the Upper Devonian reef complexes of the Canning Basin, Western Australia. *Aust., Bur. Miner. Resour., Geol. Geophys., Bull.* **158**. 303 pp., 97 pls.

—— and Jones, P. J. (1971). Cambro-Ordovician conodonts from the Burke River Structural Belt. *Aust., Bur. Miner. Resour., Geol. Geophys., Bull.* **110**. 158 pp., 20 pls.

Edwards, D. (1973). Devonian floras. In *Atlas of palaeobiology* (ed. A. Hallam), pp. 105-15. Elsevier, Amsterdam.

Ekman, S. (1953). *Zoogeography of the sea.* Sidgwick and Jackson, London. 417 pp.

Ettingshausen, C. von (1888). Contributions to the Tertiary flora of Australia. *N.S.W., Geol. Surv., Mem. Palaeontol.* **2**. 192 pp., 15 pls.

Fåhraeus, L. E. (1976). Possible Early Devonian conodontophorid provinces. *Palaeogeogr., Palaeoclimat., Palaeoecol.* **19**, 201-17.

Fleming, C. A. (1964). History of the bivalve family Trigoniidae in the southwest Pacific. *Aust. J. Sci.* **26**, 196-204.

—— (1975). The geological history of New Zealand and its biota. In *Biogeography and ecology in New Zealand* (ed. G. Kuschel), pp. 1-86. Junk, The Hague.

Fleming, P. J. G. (1966). Eotriassic marine bivalves from the Maryborough Basin, south-east Queensland. *Qld, Geol. Surv., Publ.* **333**, 17-29, pls. 7-9.

Forbes, E. (1856). Map of the distribution of marine life. In *The physical atlas of natural phenomena*, plate 31. W. and A. K. Johnston, Edinburgh.

Forster, R. R. (1975). The spiders and harvestman. In *Biogeography and ecology of New Zealand* (ed. G. Kuschel), pp. 439-505. Junk, The Hague.

Gerth, H. (1950). Die Ammonoideen des Perms von Timor und ihre Bedeutung für die stratigraphische Gliederung der Permformation. *Neues Jahrb. Mineral. Geol. Palaeontol. Abh.* **91**, 233-320.

Glaessner, M. F. (1958). New fossils from the base of the Cambrian in South Australia. *R. Soc. S. Aust., Trans.* **81**, 185-8, 1 pl.

—— (1962). Pre-cambrian fossils. *Biol. Rev.* **37**, 467-94.

—— (1969). Trace fossils from the Precambrian and basal Cambrian. *Lethaia* **2**, 369-93.

—— and Daily, B. (1959). The geology and Late Precambrian fauna of the Ediacara Fossil Reserve. *S. Aust. Mus., Rec.* **13**, 369-401.

—— and Wade, Mary (1966). The Late Precambrian fossils from Ediacara, South Australia. *Palaeontology* **9**(4), 599-628, 7 pls.

Glenister, B. F. (1958). Upper Devonian ammonoids from the *Manticoceras* zone, Fitzroy Basin, Western Australia. *J. Paleontol.* **32**, 58-96.

—— and Furnish, W. (1961). The Permian ammonoids of Australia. *J. Paleontol.* **35**, 673-736, pls. 78-86.

—— Windle, D. L. and Furnish, W. M. (1973). Australasian Metalegoceratidae (Lower Permian ammonoids). *J. Paleontol.* **47**, 1031-43, pls. 1-5.

Gould, R. E. (1976). The succession of Australian pre-Tertiary megafossil floras. *Bot. Rev.* **41**, 453-83.

—— (1981). The coal-forming flora of the Walloon Coal Measures. *Coal Geol.* (Austr.) **1**(3), 83-105.

—— (1982). Early Australasian vegetation history. In *A history of Australasian vegetation* (ed. J. M. B. Smith), pp. 32-43. McGraw-Hill, Sydney.

Grey, K. (1978). Devonian atrypid brachiopods from the reef complexes of the Canning Basin. *W. Aust., Geol. Surv., Rep.* **5**. 60 pp., 5 pls.

Haig, D. W. (1979). Global distribution patterns for mid-Cretaceous foraminiferids. *J. foraminiferal Res.* **9**, 29-40.

—— (1980). Early Cretaceous textulariine foramini-

ferids from Queensland. *Palaeontogr., Abt. A* **170**, 87–138, 11 pls.
—— (1981). Mid-Cretaceous foraminiferids from the Wahgi Valley, central highlands of Papua New Guinea. *Micropaleontology* **27**, 337–51.
—— (1982). Early Cretaceous milioline and rotaliine benthic foraminiferids from Queensland. *Palaeontogr., Abt. A* **177**, 1–88, 15 pls.
Haile, N. S. (1978). Reconnaissance palaeomagnetic results from Sulawesi, Indonesia, and their bearing on palaeomagnetic reconstructions. *Tectonophysics* **46**, 77–85.
Haniel, C. A. (1915). Die Cephalopoden der Dyas von Timor. *Paläontologie von Timor* Lief. 3, Abh. **6**. 153 pp. 11 pls.
Harris, W. K. (1965). Basal Tertiary microfloras from the Princetown area, Victoria, Australia. *Palaeontogr., Abt. B* **115**, 75–106.
Hayes, D. E. and Ringis, J. (1973). Sea-floor spreading in the Tasman Sea. *Nature Phys. Sci.* **243**, 454–8.
Hedgpeth, J. W. (1957). Marine biogeography. In *Treatise on marine ecology and palaeoecology,* vol. 1 (ed. J. W. Hedgpeth). *Geol. Soc. Am., Mem.* **67**, 359–82.
Hedley, C. (1904). The effect of the Bassian Isthmus on the marine fauna. *Linn. Soc. N.S.W., Proc.* **28**, 876–83.
Henderson, R. A. and MacKinnon, D. I. (1981). New Cambrian inarticulate Brachiopoda from Australasia and the age of the Tasman Formation. *Alcheringa* **5**, 289–309.
Hennig, W. (1966). The Diptera fauna of New Zealand as a problem in systematics and zoogeography. *Pac. Insects Monogr.* **9**, 1–81.
Hill, D. (1948). The distribution and sequence of Carboniferous coral faunas. *Geol. Mag.* **85**, 121–48.
—— (1957). The sequence and distribution of Upper Palaeozoic coral faunas. *Aust. J. Sci.* **19**, 42–61.
—— (1959). Distribution and sequence of Silurian coral faunas. *R. Soc. N.S.W., J. Proc.* **92**, 151–73.
—— (1965). Archaeocyatha from Antarctica and a review of the phylum. *Trans-Antarctic Expedition 1955–58, Sci. Rept.* No. 10 (Geol. No. 3). 151 pp., 12 pls.
—— (1973). Lower Carboniferous corals. In *Atlas of palaeobiogeography* (ed. A. Hallam), pp. 133–42. Elsevier, Amsterdam.
—— (1978). Bibliography and index of Australian Palaeozoic corals. *Qld Univ., Dep. Geol., Publ.* **8**(4). 38 pp.
—— (1981). Part F, Coelenterata, Supplement 1, Rugosa and Tabulata. In *Treatise on Invertebrate Paleontology* (ed. C. Teichert). Geol. Soc. Am., Boulder, and Univ. Kansas, Lawrence. 762 pp.
—— and Jell, J. S. (1970). Devonian corals from the Canning Basin, Australia. *W. Aust., Geol. Surv., Bull.* **121**. 158 pp.
Hoffstetter, R. (1968). Présence de Varanidae (Reptilia, Sauria) dans le Miocène de Catalogne. Considérations sur l'histoire de la famille. *Bull. Mus. nat. Hist. natur.* **40**, 1051–64.
Holloway, J. D. (1979). *A survey of the Lepidoptera, biogeography and ecology of New Caledonia.* Junk, The Hague. 588 pp.
—— and Jardine, N. (1968). Two approaches to zoogeography: a study based on the distribution of butterflies, birds and bats in the Indo-Australian area. *Linn. Soc. Lond., Proc.* **179**, 153–88.
Holmes, W. B. K. (1982). The Middle Triassic flora from Benelong near Dubbo, central-western New South Wales. *Alcheringa* **6**, 1–33.
—— and Ash, S. R. (1979). An Early Triassic megafossil flora from the Lorne Basin, New South Wales. *Linn. Soc. N.S.W., Proc.* **103**, 47–70.
Hooijer, D. A. (1957). Three new giant prehistoric rats from Flores, Lesser Sunda Islands. *Geologische Mededeelingen, Leiden* **35**, 229–314.
—— (1967). Indo-Australian insular elephants. *Genetica* **38**, 143–62.
—— (1972). *Varanus* (Reptilia Sauria) from the Pleistocene of Timor. *Zoologische Mededeelingen, Leiden* **47**, 445–80.
Horton, D. (1973). The concept of zoogeographic subregions. *Syst. Zool.* **22**, 191–5.
Huxley, T. H. (1868). On the classification and distribution of the Alectoromorphae and Heteromorphae. *Linn. Soc. Lond., Proc.* **1868**, 294–319.
Inger, R. F. (1966). The systematics and zoogeography of the Amphibia of Borneo. *Fieldiana: Zool.* **45**, 1–268.
Irving, E. (1964). *Paleomagnetism and its application to geological and geophysical problems.* Wiley, New York. 399 pp.
Ivanovsky, A. B. (1965). Stratificheskiy i paleobiogeograficheskiy obzor rugoz Ordovika i Silura. Akad. Nauk SSSR, Sibirskoe otd., Inst. Geol. Geofiz., Nauka (Moscow). 119 pp.
Jefferson, T. H. (1982). Fossil forests from the Lower Cretaceous of Alexander Island, Antarctica. *Palaeontology* **25**, 681–708.
Jell, P. A. (1974). Faunal provinces and possible planetary reconstruction of the Middle Cambrian. *J. Geol.* **82**, 319–50.
Jenkins, R. J. F. (1975). The fossil crab *Ommato-*

carcinus corioensis (Cresswell) and a review of related Australasian species. *Natl Mus. Vict. Mem.* **35**, 33-62, pls. 4-8.

—— and Gehling, J. G. (1978). A review of the frond-like fossils of the Ediacara assemblage. *S. Aust. Mus., Rec.* **17**, 347-59.

Jenkins, T. B. H. (1974). Lower Carboniferous conodont biostratigraphy of New South Wales. *Palaeontology* **17**, 909-24.

Johnson, L. A. S. and Briggs, B. G. (1981). Three old southern families – Myrtaceae, Proteaceae and Restionaceae. In *Ecological biogeography of Australia* (ed. A. Keast), pp. 427-69. Junk, The Hague.

Jones, P. J. (1971). Lower Ordovician conodonts from the Bonaparte Gulf Basin and the Daly River Basin, northwestern Australia. *Aust., Bur. Miner. Resour., Geol. Geophys., Bull.* **117**. 80 pp., 9 pls.

Jongmans, W. J. (1940). Beiträge zur Kenntnis der Karbonflora von Niederlandisch *Neu-Guinea*. *Meded. geol. Sticht.* **1938-9**, 263-74.

Kaljo, D. and Klaamann, E. (1973). Ordovician and Silurian corals. In *Atlas of palaeobiogeography* (ed. A. Hallam), pp. 37-45. Elsevier, Amsterdam.

Kauffman, E. G. (1973). Cretaceous Bivalvia. In *Atlas of palaeobiogeography* (ed. A. Hallam), pp. 353-83. Elsevier, Amsterdam.

—— (1979). Cretaceous. In *Treatise on invertebrate paleontology,* Part A, pp. 418-87. Geol. Soc. Amer., Boulder, and Univ. Kansas, Lawrence.

Keast, A. (ed.) (1981a). *Ecological biogeography of Australia.* 2142 pp., 3 vols. Junk, The Hague.

—— (1981b). The evolutionary biogeography of Australian birds. In *Ecological biogeography of Australia* (ed. A. Keast), pp. 1585-635. Junk, The Hague.

Kemp, E. M. (1978). Tertiary climatic evolution and vegetation history in the southeast Indian Ocean region. *Palaeogeog., Palaeoclimatol., Palaeoecol.* **24**, 169-208.

Kennett, J. P. (1977). Cenozoic evolution of Antarctic glaciation, the circum-Antarctic ocean and their impact on global paleoceanography. *J. Geophys. Res.* **82**, 3843-60.

—— (1978). The development of planktonic biogeography in the Southern Ocean during the Cenozoic. *Mar. Micropalaeont.* **3**, 301-45.

—— (1980). Paleoceanographic and biogeographic evolution of the Southern Ocean during the Cenozoic. *Palaeogeogr., Palaeoclimatol., Palaeoecol.* **31**, 123-52.

—— and Shackleton, N. J. (1976). Oxygen isotope evidence for the development of the psychrosphere 38 Ma ago. *Nature* **160**, 513-5.

Kerr, R. A. (1982). New evidence fuels Antarctic ice debate. *Science* **216**, 973-4.

Kikkawa, J. and Pearse, K. (1969). Geographical distribution of land birds in Australia – a numerical analysis. *Aust. J. Zool.* **17**, 821-40.

Klapper, G. and Johnson, J. G. (1980). Endemism and dispersal of Devonian conodonts. *J. Paleontol.* **54**, 400-55.

Knox, G. A. (1960). Littoral ecology and biogeography of the southern oceans. *R. Soc. Lond., Proc., Ser. B.* **152**(949), 577-624.

—— (1963). Littoral biogeography and intertidal ecology of the Australasian coasts. *Annu. Rev. Oceanogr. Mar. Biol.* **1**, 341-404.

—— (1975). Marine benthic ecology and biogeography. In *Biogeography and ecology in New Zealand* (ed. G. Kuschel), pp. 353-403. Junk, The Hague.

—— (1980). Plate tectonics and the evolution of intertidal and shallow-water benthic biotic distribution patterns of the southwest Pacific *Palaeogeogr., Palaeoclimatol., Palaeoecol.* **31**, 267-97.

Kobayashi, T. (1972). Three faunal provinces in the Early Cambrian Period. *Japan Academy, Proc.* **48**(4), 242-7.

Krijnen, W. F. (1931). Palaeozoic and Mesozoic Gastropoda, Lamellibranchiata and Scaphopoda. *Leidsche Geologische Mededeelingen* **5**, 164-205.

Krumbeck, L. (1923). Die Brachiopoden, Lamellibranchiaten und Gastropoden aus der oberen Trias der Insel Seran (Mittel-Seran). *Palaeontogr.* Suppl. 4, pt. **3**, 185-246, pls. 12-15.

—— (1924). Die Brachiopoden, Lamellibranchiaten und Gastropoden der Trias von Timor. 2. *Paläontologie von Timor* **13**, 143-417, pls. 179-198.

Kummel, B. (1973). Lower Triassic (Scythian) molluscs. In *Atlas of Palaeobiogeography* (ed. A. Hallam), pp. 225-33. Elsevier. The Hague.

Kundig, E. (1956). Geology and ophiolite problems of east Celebes. *Verhandelingen van het nederlandsch Geologisch-Mijnbouwkundig Genootschap* **16**, 210-35.

Lacey, W. S. (1975). Some problems of 'mixed' floras in the Permian of Gondwanaland. In *Gondwana geology* (ed. K. S. W. Campbell), pp. 125-34. Austr. Nat. Univ., Canberra.

Lang, W. H. and Cookson, I. C. (1930). Some fossil plants of early Devonian type from the Walhalla Series, Victoria, Australia. *R. Soc. Lond., Philos. Trans., Ser. B.* **219**, 133-63.

—— —— (1935). On a flora, including vascular land

plants, associated with *Monograptus,* in rocks of Silurian age, from Victoria, Australia. *R. Soc. Lond., Philos. Trans., Ser. B.* **224**, 421–49.

Lange, R. T. (1982). Australian Tertiary vegetation. In *A history of Australasian vegetation* (ed. J. M. B. Smith), pp. 44–89. McGraw-Hill, Sydney.

Laurie, J. R. (1980). Early Ordovician orthide brachiopods from southern Tasmania. *Alcheringa* **4**, 11–24.

Legg, D. P. (1976). Ordovician trilobites and graptolites from the Canning Basin, Western Australia. *Geol. et Palaeont.* **10**, 1–58.

Leleshus, V. L. (1970). Paleozoogerafiya ordovika, silura i rannego devona po tabulyatomorfnym korallam i granitsy siluriyskoy sistemy. *Izv. Akad. Nauk SSSR ser. geol.* **9**, 184–92. (English translation in *Int. Geol. Rev.* **13**(3), 427–34.)

Link, A. G. and Druce, E. C. (1972). Ludlovian and Gedinnian conodont stratigraphy of the Yass Basin, New South Wales. *Aust., Bur. Miner. Resour., Geol. Geophys., Bull.* **134**. 136 pp.

Loveless, A. R. (1961). A nutritional interpretation of sclerophylly based on differences in the chemical composition of sclerophyllous and mesophytic leaves. *Ann. Bot. (Lond.)* n.s. **25**, 168–84.

Luyendyk, B. P., Forsyth, D. and Phillips, J. D. (1974). Experimental approach to the paleocirculation of the oceanic surface waters. *Geol. Soc. Am., Bull.* **93**, 2649–64.

Lydekker, R. (1896). *A geographical history of mammals.* Cambridge University Press. 400 pp.

McClung, G. (1975). Late Palaeozoic glacial faunas of Australia. In *Gondwana geology* (ed. K. S. W. Campbell), pp. 381–90. Aust. Nat. Univ., Canberra.

McGowran, B. (1970). Late Paleocene in the Otway Basin: biostratigraphy and age of key microfaunas. *R. Soc. S. Aust., Trans.* **94**, 1–14.

—— (1978). Stratigraphic record of Early Tertiary oceanic and continental events in the Indian Ocean region. *Mar. Geol.* **26**, 1–39.

—— (1979). The Tertiary of Australia: foraminiferal overview. *Mar. Micropalaeontol.* **4**, 235–64.

——, Lindsay, J. M. and Harris, W. K. (1971). Attempted reconciliation of Tertiary biostratigraphic systems. In *The Otway Basin of southeastern Australia* (ed. H. Wopfner and J. G. Douglas), pp. 273–81. *S. Aust. and Vict., Geol. Survs., Spec. Bull.*

McKellar, R. G. (1970). The Devonian productoid brachiopod faunas of Queensland. *Qld, Geol. Surv., Publ.* 342 (Palaeont. Pap. 18). 40 pp., 12 pls.

McLean, R. A. (1977). Biostratigraphy and zoogeographic affinities of the Lower Silurian corals of New South Wales, Australia. *Bur. Rech. geol. et min. Mem.* **89**, 102–7.

McMichael, D. F. and Hiscock, I. D. (1958). A monograph of the freshwater mussels (Mollusca: Pelecypoda) of the Australian region. *Aust. J. Mar. Freshw. Res.* **9**, 372–507.

McTavish, R. A. (1973a). Triassic conodonts from Western Australia. *Neues Jahrb. Geol. Paläont. Abh.* **143**, 275–303.

—— (1973b). Prionodontacean conodonts from the Emanuel Formation (Lower Ordovician) of Western Australia. *Geol. et Palaeont.* **7**, 27–51, 3 pls.

—— and Dickins, J. M. (1974). The age of the Kockatea Shale (Lower Triassic), Perth Basin. *Geol. Soc. Aust., J.* **21**, 195–201.

Main, B. Y. (1981). A comparative account of the biogeography of terrestrial invertebrates in Australia: some generalizations. In *Ecological biogeography of Australia* (ed. A. Keast), pp. 1055–77. Junk, The Hague.

Martin, H. A. (1978). Evolution of the Australian flora and vegetation through the Tertiary: evidence from pollen. *Alcheringa* **2**(3), 181–202.

—— (1981). The Tertiary flora. In *Ecological biogeography of Australia* (ed. A. Keast), pp. 393–406. Junk, The Hague.

Maxwell, W. G. H. (1951). Upper Devonian and Middle Carboniferous brachiopods of Queensland. *Qld Univ., Dep. Geol., Pap.* **3**(14), 1–27, 4 pls.

—— (1961). Lower Carboniferous brachiopod faunas from Cannindah, Queensland. *J. Paleontol.* **35**, 82–103, 2 pls.

Mayr, E. (1944). Wallace's line in the light of recent zoogeographic studies. *Quart. Rev. Biol.* **19**, 1–14.

Mercer, J. H. and Sutter, J. F. (1982). Late Miocene-earliest Pliocene glaciation in southern Argentina: implications for global ice-sheet history. *Palaeogeogr., Palaeoclimatol., Palaeoecol.* **38**, 185–206.

Mertens, R. (1930). Die Amphibien und Reptilien der Inseln Bali, Lombok, Sumbawa und Flores. *Abhandlungen hrsg. von der Senckenbergischen Naturforschenden Gessellschaft*, 115–344 (*non vidi*).

Mildenhall, D. C. (1980). New Zealand Late Cretaceous and Cenozoic plant biogeography: a contribution. *Palaeogeogr., Palaeoclimatol., Palaeoecol.* **31**, 197–233.

Molnar, R. E. (1980). Australian late Mesozoic terrestrial tetrapods: some implications. *Mém. Soc. géol. Fr.* (n.s) **59**, 131–43.

Morris, N. (1975). The *Rhacopteris* flora in New South Wales. In *Gondwana Geology* (ed. K. S. W. Campbell), pp. 99–108. Aust. Nat. Univ. Press, Canberra.

—— (1980). Carboniferous floral succession in eastern Australia. In *A guide to the Sydney Basin* (ed. C. Herbert and R. Helby). *N.S.W., Geol. Surv., Bull.* **26**, 350–8.

Müller, S. (1846). Über den Character der Thierwelt auf den Inseln des indischen Archipels. *Arch. Naturg.* **12**, 109–28.

Murray, A. (1866). *The geographical distribution of mammals.* Day and Son, London (*non vidi*).

Nelson, E. C. (1981). Phytogeography of southern Australia. In *Ecological biogeography of Australia* (ed. A. Keast), pp. 733–59. Junk, The Hague.

Norvick, M. S. and Burger, D. (1975). Palynology of the Cenomanian of Bathurst Island, Northern Territory, Australia. *Aust., Bur. Miner. Resour., Geol. Geophys., Bull.* **151**. 169 pp., 34 pls.

Oliver, W. A. (1977). Biogeography of Late Silurian and Devonian rugose corals. *Palaeogeogr., Palaeoclimatol., Palaeoecol.* **25**, 85–135.

Öpik, A. A. (1978). Middle Cambrian agnostids: systematics and biostratigraphy. vol. 1 (Text), vol. 2 (Plates). *Aust., Bur. Miner. Resour., Geol. Geophys., Bull.* **172**. 252 pp., 64 pls.

—— (1982). Dolichometopid trilobites of Queensland, Northern Territory and New South Wales. *Aust., Bur. Miner. Resour., Geol. Geophys., Bull.* **175**. 85 pp., 32 pls.

Owen, M. (1973). Upper Cretaceous planktonic Foraminifera from Papua New Guinea. *Aust., Bur. Miner. Resour., Geol. Geophys., Bull.* **140**. 47–65, pls. 14–20.

Palmer, A. R. (1979). Cambrian. In *Treatise on invertebrate paleontology, part A* (ed. C. Teichert *et al.*), pp. 119–35. Geol. Soc. Am., Boulder, and Univ. Kansas. Lawrence.

Percival, I. G. (1978). Inarticulate brachiopods from the Late Ordovician of New South Wales, and their palaeoecological significance. *Alcheringa* **2**, 117–41.

—— (1979). Ordovician plectambonitacean brachiopods from New South Wales. *Alcheringa* **3**, 91–116.

Petersen, M. S. (1975). Upper Devonian (Famennian) ammonoids from the Canning Basin, Western Australia. *Palaeontol. Soc., Mem.* **8** (Supplement to *J. Paleontol.* **49**, No. 5). 55 pp.

Phillips, B. F. (1981). Circulation of southeastern Indian Ocean and the planktonic life of western rock lobster. *Oceanogr. Mar. Biol. Ann. Rev.* **19**, 11–39.

Pickett, J. W. (1975). Continental reconstructions and the distribution of coral faunas during the Silurian. *R. Soc. N.S.W., J. Proc.* **103**, 147–56.

—— (ed.) (1982). The Silurian system in New South Wales. *N.S.W., Geol. Surv. Bull.* **29**. 264 pp., 5 pls.

Platnick, N. (1976). Drifting spiders or continents?: Vicariance biogeography of the spider subfamily Laroniinae (Araneae: Gnaphosidae). *Syst. Zool.* **25**(2), 101–9.

Plumstead, E. P. (1973). The late Palaeozoic *Glossopteris* flora. In *Atlas of Palaeobiogeography* (ed. A. Hallam), pp. 187–205. Elsevier, Amsterdam.

Pojeta, J. and Gilbert-Tomlinson, J. (1977). Australian Ordovician pelecypod molluscs. *Aust., Bur. Miner. Resour., Geol. Geophys., Bull.* **174**. 93 pp., 29 pls.

—— —— and Shergold, J. H. (1977). Cambrian and Ordovician molluscs from northern Australia. *Aust., Bur. Miner. Resour., Geol. Geophys., Bull.* **171**. 81 pp., 27 pls.

Powell, A. W. B. (1957). Marine provinces of the Indo-West Pacific. *Proc. 8th Pacific Sci. Congr.* **3**, 359–62.

Rand, H. M. and Mabesoone, J. M. (1982). Northeastern Brazil and the final separation of South America and Africa. *Palaeogeogr., Palaeoclimatol., Palaeoecol.* **38**, 163–83.

Raven, P. H. and Axelrod, D. I. (1972). Plate tectonics and Australasian palaeobiogeography. *Science* **176**, 1379–86.

—— —— (1974). Angiosperm biogeography and past continental movements. *Ann. Missouri Bot. Gardens* **61**, 539–673.

Retallack, G. J. (1977). Reconstructing Triassic vegetation of eastern Australasia: a new approach for the biostratigraphy of Gondwanaland. *Alcheringa* **1**, 247–77.

—— (1980). Late Carboniferous to Middle Triassic mega fossil floras from the Sydney Basin. In *A guide to the Sydney Basin* (ed. C. Herbert and R. Helby). *N.S.W., Geol. Surv., Bull.* **26**, 384–430.

Rich, P. V. (1975). Antarctic dispersal routes, wandering continents, and the origin of Australia's non-passeriform avifauna. *Natl. Mus. Vic., Mem.* **36**, 63–125.

—— (1979). The Dromornithidae, a family of large extinct ground birds endemic to Australia. *Aust., Bur. Miner. Resour., Geol. Geophys., Bull.* **184**, 193 pp.

—— and Thompson, E. M. (eds.) (1982). *The fossil vertebrate record of Australasia.* Monash Univ. Offset Printing Unit, Clayton (Australia). 759 pp.

Riek, E. F. (1970). Fossil history. In *The insects of Australia*, pp. 168–86. Melbourne Univ., Melbourne.

Rigby, J. F. (1972). The Gondwana palaeobotanical province at the end of the Palaeozoic. *Proc. 24th Int. Geol. Congr.*, Montreal **7**, 324–30.

Roberts, J. (1963). A Lower Carboniferous fauna from Lewinsbrook, New South Wales. *R. Soc.*

N.S. W., J. Proc. **97**, 1–29.

—— (1964). Lower Carboniferous brachiopods from Greenhills, New South Wales. *Geol. Soc. Aust., J.* **11**, 173–94, 6 pls.

—— (1971). Devonian and Carboniferous brachiopods from the Bonaparte Gulf Basin, northwestern Australia. *Aust., Bur. Miner. Resour., Geol. Geophys., Bull.* **122**, 319 pp., 59 pls.

Ross, C. A. (1974). Paleogeography and provinciality. In *Palaeogeographic provinces and provinciality* (ed. C. A. Ross). *Soc. Econ. Paleontol. Mineral., Spec. Publ.* **21**, 1–17.

—— (1979a). Late Paleozoic collision of North and South America. *Geology* **7**, 41–4.

—— (1979b). Carboniferous. In *Treatise on invertebrate palaeontology,* Part A, Introduction (ed. C. Teichert *et al.*), pp. 254–90. Geol. Soc. Amer., Boulder, and Univ. Kansas, Lawrence.

Runnegar, B. (1969). The Permian faunal succession in eastern Australia. *Geol. Soc. Aust., Spec. Pub.* **2**, 73–98.

—— (1972). Late Palaeozoic Bivalvia from South America: provincial affinities and age. *An. Acad. Brasil. Cienc.,* **44** (Suplemento), 295–312.

—— (1979). Marine fossil invertebrates of Gondwanaland: palaeogeographic implications. *4th Intl. Gondwana Symp., 1977, Calcutta,* pp. 144–59. Hindustan, Delhi.

—— and Campbell, K. S. W. (1976). Late Palaeozoic faunas of Australia. *Earth-Sci. Rev.* **12**, 235–57.

—— and Jell, P. A. (1976). Australian Middle Cambrian molluscs and their bearing on early molluscan evolution. *Alcheringa* **1**, 109–38.

—— and McClung, G. (1975). A Permian time scale for Gondwanaland. In *Gondwana Geology* (ed. K. S. W. Campbell), pp. 425–41. Aust. Nat. Univ., Canberra.

Sasajima, S., Nishimura, S., Hirooka, K., Otofuji, Y., van Leeuwen, T. and Hehuwat, F. (1980). Palaeomagnetic studies combined with fission-track datings on the western arc of Sulawesi, east Indonesia. *Tectonophysics* **64**, 163–72.

Scheibnerova, V. (1971). Foraminifera and their Mesozoic biogeoprovinces. *N.S.W., Geol. Surv., Rec.* **13**(3), 135–74.

—— (1976). Cretaceous Foraminifera of the Great Artesian Basin. *N.S.W., Geol. Surv., Mem. Palaeont.* **17**. 277 pp., 77 pls.

—— (1982). Permian Foraminifera of the Sydney Basin. *N.S.W., Geol. Surv., Mem. Palaeont.* **19**. 125 pp.

Schlinger, E. I. (1974). Continental drift, *Nothofagus,* and some ecologically associated insects. *Annu. Rev. Entomol.* **19**, 323–43.

Schopf, T. J. M. (1974). Permo-Triassic extinctions: relation to sea-floor spreading. *J. Geol.* **82**, 129–43.

Sclater, W. L. and Sclater, P. L. (1899). *The geography of mammals.* Kegan Paul, Trench, Trübner, London. 355 pp.

Shergold, J. H. (1973). Bibliography and index of Australian Cambrian trilobites. *Aust., Bur. Miner. Resour., Geol. Geophys., Bull.* **140**, 67–84.

—— (1975). Late Cambrian and Early Ordovician trilobites from the Burke River Structural Belt, western Queensland. *Aust., Bur. Miner. Resour., Geol. Geophys., Bull.* **153**. 251 pp., 58 pls.

—— (1982). Idamean (Late Cambrian) trilobites, Burke River Structural Belt, western Queensland. *Aust., Bur. Miner. Resour., Geol. Geophys., Bull.* **187**. 69 pp., 17 pls.

Sherwin, L. (1971). Trilobites of the subfamily Phacopinae from New South Wales. *N.S.W., Geol. Surv., Rec.* **13**(2), 83–99, 8 pls.

—— (1974). Llandovery graptolites from the Forbes district, New South Wales. *Spec. Pap. Palaeontology* **13**, 149–75, pls. 10–12.

Simberloff, D. S. (1974). Permo-Triassic extinctions: effects of area on biotic equilibrium. *J. Geol.* **82**, 267–74.

Simpson, G. G. (1977). Too many lines; the limits of the Oriental and Australian zoogeographic regions. *Am. Philos. Soc., Proc.* **121**, 107–20.

Singh, G. (1981). Late Quaternary pollen records and seasonal palaeoclimates of Lake Frome, South Australia. In *Athalassic Salt Lakes* (ed. W. D. Williams). Junk, The Hague.

—— (1982). Quaternary upheaval: vegetation of Australasia during the Quaternary. In *A history of Australasian vegetation* (ed. J. M. B. Smith), pp. 90–108. McGraw-Hill, Sydney.

——, Kershaw, A. P. and Clark, R. L. (1981). Quaternary vegetation and fire history in Australia. In *Fire and the Australian biota* (ed. A. M. Gill *et al.*), pp. 23–54. Austr. Acad. Sci., Canberra.

Skwarko, S. K. (1963). Australian Mesozoic trigoniids. *Aust., Bur. Miner. Resour., Geol. Geophys., Bull.* **67**. 42 pp., 6 pls.

—— (1966). Cretaceous stratigraphy and palaeontology of the Northern Territory. *Aust., Bur. Miner. Resour., Geol. Geophys., Bull.* **73**. 133 pp., 15 pls.

—— (1967). Mesozoic Mollusca from Australia and New Guinea. *Aust., Bur. Miner. Resour., Geol. Geophys., Bull.* **75**. 101 pp., 12 pls.

—— (1970). Bibliography of the Mesozoic palaeontology of Australia and eastern New Guinea. *Aust., Bur. Miner. Resour., Geol. Geophys., Bull.* **108**, 237–79.

—— (1974). Jurassic fossils of Western Australia. 1: Bajocian Bivalvia of the Newmarracarra Limestone and the Kojarene Sandstone. *Aust., Bur. Miner. Resour., Geol. Geophys., Bull.* **150**, 57–109, pls. 21–36.

Smith, A. G., Briden, J. C. and Drewry, G. E. (1973). Phanerozoic world maps. *Spec. Pap. Palaeontology* **12**, 1–42.

Smith, J. P. (1927). Permian ammonoids of Timor. *Mijn. Nederlandsch-Indië Jaarb.*, Jaarg. 55, Verh. **1**, 1–91, 16 pls.

Specht, R. L. (1981). Evolution of the Australian flora. In *Ecological biogeography of Australia* (ed. A. Keast), pp. 783–805. Junk, The Hague.

Spencer, B. (1896). *Report on the work of the Horn Scientific Expedition to Central Australia.* Pt. 1, Summary. Dulau, London. 220 pp.

Sprigg, R. C. (1947). Early Cambrian (?) jellyfishes from the Flinders Ranges, S. Australia. *R. Soc. S. Aust., Trans.* **71**, 212–24, 4 pls.

Steenis, G. G. G. J. van (1979). Plant-geography of east Malesia. *Linn. Soc. Lond., Bot. J.* **79**, 97–178.

Stevens, G. R. (1977). Mesozoic biogeography of the south-west Pacific and its relationship to plate tectonics. In *International symposium on Geodynamics in south-west Pacific, Noumea (New Caledonia)*, 27 Aug. to 2 Sept. 1976, pp. 309–26, Editions Technip, Paris.

—— (1980). Southwest Pacific faunal palaeogeography in Mesozoic and Cenozoic times: a review. *Palaeogeogr., Palaeoclimatol., Palaeoecol.* **31**, 153–96.

Stover, L. E. and Partridge, A. D. (1973). Tertiary and Late Cretaceous spores and pollen from the Gippsland Basin southeastern Australia. *R. Soc. Vic., Proc.* **85**, 237–86.

Strusz, D. L. (1980). The Family Encrinuridae and related trilobite families, with a description of Silurian species from southeastern Australia. *Palaeontogr., Abt. A* **168**, 1–68.

—— (1982). Wenlock brachiopods from Canberra, Australia. *Alcheringa* **6**, 105–42.

Talent, J. A. (1981). Birds, bats, butterflies and biogeography in the Indo-Australian region. In *Status of environmental studies in India* (ed. N. G. K. Nair), pp. 5–6. Cent. Earth Sci. Studies, Trivandrum.

—— in press. Provinciality as a means for qualified resolution of separation of continental blocks in the past: preliminary exemplification from Western Pacific borderlands. XIV Pacific Sci. Congr., Khabarovsk, 1979.

——, Berry, W. B. N. and Boucot, A. J. (1975). Correlation of the Silurian rocks of Australia, New Zealand and New Guinea. *Geol. Soc. Am., Spec. Pap.* **150**. 108 pp.

——, Campbell, K. S. W., Davoren, P. J., Pickett, J. W. and Telford, P. G. (1972). Provincialism and Australian Early Devonian faunas. *Geol. Soc. Aust., J.* **19**, 81–97.

——, Duncan, P. M. and Handby, P. L. (1966). Early Cretaceous feathers from Victoria. *Emu* **66**, 81–6, pls. 4, 5.

Tasch, P. and Jones, P. J. (1979). Carboniferous, Permian and Triassic conchostracans of Australia. *Aust., Bur. Miner. Resour., Geol. Geophys., Bull.* **185**. 54 pp., 7 pls.

Tate, R. (1889). On the influence of physiographic changes in the distribution of life in Australia. *Rep. Austr. Assoc. Adv. Sci.* **1**, 312–25.

Taylor, D. J. (1964). Foraminifera and the stratigraphy of the western Victorian Cretaceous sediments. *R. Soc. Vic., Proc.* **77**, 535–602.

—— (1965). Preservation, composition and significance of Victorian Lower Tertiary 'Cyclammina faunas'. *R. Soc. Vic., Proc.* **78**, 143–60.

Tedford, R. H. (1974). Marsupials and the new palaeogeography. In *Palaeogeographic provinces and provinciality* (ed. C. A. Ross). *Soc. Econ. Paleontol. Min. Spec. Pubs.* **21**, 109–26.

Teichert, C. (1941). Upper Devonian goniatite succession of Western Australia. *Am. J. Sci.* **239**, 148–53.

—— (1949). Permian crinoid *Calceolispongia*. *Geol. Soc. Am., Mem.* **34**, 1–126, 5 pls.

—— and Glenister, B. F. (1953). Ordovician and Silurian cephalopods from Tasmania. *Bull. Am. Paleontol.* **34**(144), 187–299, pls. 1–6.

—— —— (1954). Early Ordovician cephalopod fauna from northwestern Australia. *Bull. Am. Paleontol.* **35**(150). 110 pp., 10 pls.

Telford, P. G. (1972). Conodonts. In *Provincialism and Australian Early Devonian faunas* (J. A. Talent et al.). *Geol. Soc. Aust., J.* **19**, 81–97.

—— (1979). Devonian conodont distribution – provinces or communities? In *Historial biogeography, plate tectonics, and the changing environment* (ed. J. Gray and A. J. Boucot), 201–13. Oregon State Univ., Corvallis.

Thomas, D. E. (1960). The zonal distribution of Australian graptolites. *R. Soc. N.S.W., J. Proc.* **94**, 33–84.

Thomas, G. A. (1958). The Permian Orthotetacea of Western Australia. *Aust., Bur. Miner. Resour., Geol. Geophys., Bull.* **39**. 115 pp., 22 pls.

—— (1971). Carboniferous and early Permian

brachiopods from western and northern Australia. *Aust., Bur. Miner. Resour., Geol. Geophys., Bull.* **56.** 276 pp., 31 pls.

Townley, K. A. (1970). Bibliography of Australian Permian invertebrates. *Aust., Bur. Miner. Resour., Geol. Geophys., Bull.* **116,** 157–72.

Townrow, J. A. (1969). Some Lower Mesozoic Podocarpaceae and Araucariaceae. In *Gondwana stratigraphy.* IUGS Symposium, Buenois Aires. Paris, Unesco, 159–84.

Vakhrameev, V. A. (1972). Mesozoic floras of the southern hemisphere and their relationship to the floras of the northern continents (in Russian). *Paleont. Zh.* **1972**(3), 146–61. (Translated in *Paleont. J.* **6,** 409–21).

Valentine, J. W. (1973). *Evolutionary paleoecology of the marine biosphere.* Prentice-Hall, Englewood Cliffs, NJ. 511 pp.

Veevers, J. J. (1959a). Devonian brachiopods from the Fitzroy Basin, Western Australia. *Aust., Bur. Miner. Resour., Geol. Geophys., Bull.* **45.** 137 pp., 18 pls.

—— (1959b). Devonian and Carboniferous brachiopods from north western Australia. *Aust., Bur. Miner. Resour., Geol. Geophys., Bull.* **55.** 34 pp., 4 pls.

Wade, M. (1969). Medusae from uppermost Precambrian sandstones, central Australia. *Palaeontology* **12,** 351–65, pls. 68, 69.

—— (1972). Hydrozoa and Scyphozoa and other medusoids from the Precambrian Ediacara fauna, South Australia. *Palaeontology* **15,** 197–225, pls. 40–43.

—— (1977). Georginidae, new family of actinoceroid cephalopods, Middle Ordovician, Australia. *Qld Mus., Mem.* **18,** 1–15.

Wade, R. T. (1935). *The Triassic fishes of Brookvale, New South Wales.* Brit. Mus. (Nat. Hist.). 110 pp., 10 pls.

—— (1940). Australian Triassic fishes. I. The Triassic fishes of St. Peters, Sydney, New South Wales. II. The relationships of the Australian Triassic fishes to each other and to other bony fishes. *R. Soc. N.S.W., J. Proc.* **74,** 377–96, 1 pl.

Walkom, A. B. (1919). Mesozoic floras of Queensland. Parts 3 and 4. The floras of the Burrum and Styx River Series. *Qld, Geol. Surv., Publ.* **263,** 1–77.

Wallace, A. R. (1860). On the zoological geography of the Malay Archipelago. *Linn. Soc. Lond., Zool. J.* **4,** 172–84.

—— (1869). *The Malay Archipelago.* Macmillan, London. 2 vols., 515 pp.

—— (1876). *The geographical distribution of animals with a study of the relations of living and extinct faunas as elucidating the past changes of the earth's surface.* Macmillan, London. 2 vols., 1139 pp.

—— (1880). *Island life: or the phenomena and causes of insular faunas and floras, including a revision and attempted solution of the problems of geological climates.* Macmillan, London. 526 pp.

—— (1910). *The world of life.* Chapman and Hall, London. 441 pp.

Warren, A. (1981). A horned member of the labyrinthodont superfamily Brachyopoidea from the Early Triassic of Queensland. *Alcheringa* **5,** 273–88.

Wass, R. E. (1968). Permian Polyzoa from the Bowen Basin, Queensland. *Aust., Bur. Miner. Resour., Geol. Geophys., Bull.* **90.** 138 pp., 18 pls.

Waterhouse, J. B. (1963). Permian gastropods of New Zealand, Pts. 1–3. *N.Z. J. Geol. Geophys.* **6,** 88–112, 115–54, 587–622.

—— (1964). Permian brachiopods from New Zealand. *N.Z., Geol. Surv., Palaeontol. Bull.* **35,** 1–287.

—— (1967a). Proposal of series and stages for the Permian in New Zealand. *N.Z. J. Geol. Geophys.* **12,** 713–33.

—— (1967b). Upper Permian (Tatarian) brachiopods from New Zealand. *N.Z. J. Geol. Geophys.* **10,** 74–118.

—— (1968). The classification and descriptions of Permian Spiriferida from New Zealand. *Palaeontogr., Abt. A* **129.** 94 pp.

—— (1969). The Permian bivalve genera *Myonia, Megadesmus, Vacunella* and their allies, and their occurrence in New Zealand. *N.Z., Geol. Surv., Palaeontol. Bull.* **41,** 1–141.

—— (1970). Correlation of marine Permian faunas for Gondwana. *Proc. Pap. Second Gondwana Symp., S.Afr.* (ed. S. H. Haughton), pp. 381–94.

—— (1976). World correlations for Permian marine faunas. *Qld Univ., Dep. Geol., Pap.* **7**(2), 1–232.

—— (1980a). Permian bivalves from New Zealand. *R. Soc. N.Z., J.* **10,** 97–133.

—— (1980b). Scaphopod, gastropod and rostroconch species from the Permian of New Zealand. *R. Soc. N.Z., J.* **10,** 195–214.

—— and Bonham-Carter, G. (1975). Global distribution and character of Permian biomes based on brachiopod assemblages. *Can. J. Earth Sci.* **12,** 1085–146.

Webby, B. D. (1974). Upper Ordovician trilobites from central New South Wales. *Palaeontology* **17,** 203–52, pls. 28–34.

—— and Packham, G. H. (1982). Stratigraphy and regional setting of the Cliefden Caves Limestone Group (Late Ordovician), central-west New South

Wales. *Geol. Soc. Aust., J.* **29**, 297-317.

——, VandenBerg, A. H. M., Cooper, R. A., Banks, M. R., Burrett, C. F., Henderson, R. A., Clarkson, P. D., Hughes, C. P., Laurie, J., Stait, B., Thomson, M. R. A., and Webers, G. F. (1981). The Ordovician System in Australia, New Zealand and Antarctica. *Intl. Union. Geol. Sci., Publ.* **6.** 68 pp., 1 pl.

Weber, M. (1902). *Der Indo-australische Archipel und die Geschichte seiner Tierwelt.* Jena. 46 pp. (*non vidi*).

Wegener, A. (1924). *The origin of continents and oceans* (English trans., 3rd edition by J. Skerl). Methuen, London. 212 pp.

Weissel, J. K., Hayes, D. E. and Herron, E. M. (1977). Plate tectonics synthesis: the displacements between Australia, New Zealand and Antarctica since the Late Cretaceous. *Mar. Geol.* **25**, 231-77.

Wellman, P. and McDougall, I. (1974). Potassium-argon ages on the Cainozoic volcanic rocks of New South Wales. *Geol. Soc. Aust., J.* **21**, 247-72.

White, A. H. (1968). The glacial origin of Carboniferous conglomerates west of Barraba, New South Wales. *Geol. Soc. Am., Bull.* **79**, 675-86.

White, E. I. (1978). The larger arthrodiran fishes from the area of the Burrinjuck Dam, N.S.W. *Lond., Zool. Soc., Trans.* **34**, 149-262.

Whitley, G. P. (1932). Marine zoogeographical regions of Australasia. *Aust. Nat.* **8**, 166-7.

—— (1948). A list of the fishes of Western Australia. *West. Aust. Fish. Dept., Fish. Bull.* **2.** 35 pp.

Whitmore, T. C. (1975). *Tropical rain forests of the Far East.* Clarendon Press, Oxford. 282 pp.

—— (1981a). *Wallace's line and plate tectonics* (ed. T. C. Whitmore). Clarendon Press, Oxford. 91 pp.

—— (1981b). Wallace's line and some other plants. In *Wallace's line and plate tectonics* (ed. T. C. Whitmore), pp. 70-80, Clarendon Press, Oxford.

Whittington, H. B. and Hughes, C. P. (1972). Ordovician geography and faunal provinces deduced from trilobite distribution. *R. Soc. Lond., Philos. Trans., Ser. B.* **263**, 235-78.

Williams, A. (1973). Distribution of brachiopod assemblages in relation to Ordovician palaeogeography. In *Organisms and continents through time* (ed. N. F. Hughes), pp. 241-69. *Spec. Pap. Palaeontol.* 12.

Woodburne, M. O. and Zinsmeister, W. J. (1982). Fossil land mammal from Antarctica. *Science* **218**, 284-6.

Woodward, S. P. (1851-1856). *A manual of the Mollusca or a rudimentary Treatise of Recent and Fossil shells.* J. Weale, London. 486 pp., 25 pls.

Wright, C. W. (1963). Cretaceous ammonoids from Bathurst Island, northern Australia. *Palaeontology* **6**, 597-614, pls. 81-9.

Wyrtki, K. (1960). The surface circulation in the Coral and Tasman Seas. *CSIRO Tech. Pap.* 8.

——, Bennett, E. B. and Rochford, D. J. (1971). *Oceanographical atlas of the International Indian Ocean Expedition.* National Science Foundation, Washington. 531 pp.

Young, G. C. (1980). A new Early Devonian placoderm from New South Wales, Australia, with a discussion of placoderm phylogeny. *Palaeontogr., Abt. A* **167**, 10-76.

—— (1981). Biogeography of Devonian vertebrates. *Alcheringa* **5**, 225-43.

Zhuravleva, I. T. (1968). Biogeografiya i geokhronologiya rannego Kembriya po arkheotsiatam. *XXIII Sessiya Mezhdunarodnogo geologicheskogo Kongressa,* 1968, Doklady Sovetskikh geologiv, 3. Problemy Paleontologii, 33-44.

Ziegler, A. M., Bambach, R. K., Parrish, J. T., Barrett, S. F., Gierlowski, E. H., Parker, W. C., Raymond, A. and Sepkoski, J. J. (1981). Paleozoic biogeography and climatology. In *Paleobotany, paleoecology and evolution* vol. 2 (ed. K. J. Niklas), pp. 231-66. Praeger, New York.

III. LITHOSPHERIC STRUCTURE

This review of the lithospheric structure of Australia concentrates on three features: (1) the gross anatomy of the lithosphere, in particular the difference in deep structure between the Precambrian terrain to the west of the Tasman Line and the Phanerozoic terrain to the east; (2) the physiology of modern earth motions of the lithosphere, as expressed in seismicity, stress measurements, and faults and folds; and (3) the reflection of these motions in the Cainozoic volcanicity of eastern Australia.

1. GROSS ANATOMY

Australia is divided by the Tasman Line (Figs. 44, 45) into a western terrain of exposed Precambrian blocks and fold belts overlain by thin Phanerozoic basins, and an eastern terrain of exposed Phanerozoic fold belts overlain by younger Phanerozoic basins. The exposed Precambrian terrain (Plumb 1979; Rutland 1981, 1982) comprises (1) the Archean (3.5 to 2.5 Ga old) Western Shield of the Pilbara and Yilgarn Blocks, (2) the older Proterozoic Kimberley and Gawler Blocks, Curnamona Craton, North Australian Craton, and North-east Australian Orogens, all embedded in (3) younger Proterozoic mobile belts. Except western Tasmania and the Adelaide region, not consolidated until about 0.5 Ga ago, the Precambrian terrain was consolidated by about 1.0 Ga ago, and then overlain by (4) the platform cover of the Adelaidean succession. Towards the end of the Proterozoic, the Petermann Ranges Orogeny produced thrusts, nappes, and uplift in north-western and central Australia, metamorphism and uplift in south-western Australia and Tasmania, and granite intrusions in Western Australia, as a prelude to Phanerozoic continental breakup by plate divergence along the Tasman Line, and in the northwest.

A regional gravity survey of Australia, with a density of stations of 1 per 130 km^2, was completed in 1976 (BMR 1976). Elongate gravity anomalies of wavelength 20 to 100 km mainly reflect density variations in the non-sedimentary part of the crust, and were used by Wellman (1976) to define structural trends of the crust, whether exposed or not, over the entire mainland, as shown in Fig. 45. Wellman (1976) defined areas of subparallel elongate anomalies, and the major areas correspond with the crustal subdivisions described by Plumb (1979), and identified in Fig. 45. The boundary between the Phanerozoic Tasman Fold Belt and the Precambrian terrain to the west – the Tasman Line – is emphasized by heavy bars in Fig. 45. We postpone to the account of the Adelaidean–Cambrian and Ordovician history in Chapter V a detailed description of the Tasman Line, and also the implications of a postulated Precambrian source rock for the granitoids of south-eastern Australia.

(a) Structural differences between the Phanerozoic and Precambrian terrains

(i) Deep Structure

In a review of the geophysical differences in the lithosphere between Phanerozoic and Precambrian Australia, Finlayson (1982) found regional differences in seismic-velocity structure at depths greater than 200 km. Cleary et al. (1972) (Fig. 46A) demonstrated differences in teleseismic travel-time residuals from recording stations sited across the southern Tasman Fold Belt and on the Precambrian shield areas to the west. Their recordings of the Cannikin (Aleutians) atomic bomb blast at angular distances of about 90° indicate residual differences of about 1.1 seconds between sites in eastern Australia and sites on the Precambrian craton, with the earlier arrivals being recorded in the Precambrian. According to Finlayson (1982), differences due to structure in the upper 60 km can be estimated from known crustal structure to be, at most, about 0.20 second, so that a substantial proportion of the residuals must be attributed to structure deeper than 60 km. Deeper structure is less constrained, but attributing the teleseismic travel-time residuals to the interval above a depth of 200 km results in average velocities that are not observed, and a distribution of residuals within a greater depth range (60 to 500 km) produces average velocities which are more in accord with observations.

Another method that reveals differences in deep structure between the Phanerozoic and Precambrian terrains is electromagnetic sounding. Lilley et al. (1981) demonstrated from long-period sounding a major conductivity increase at a depth of about 200 km in south-eastern Australia and at about 500 km in central Australia (Fig. 46B), thus strengthening the

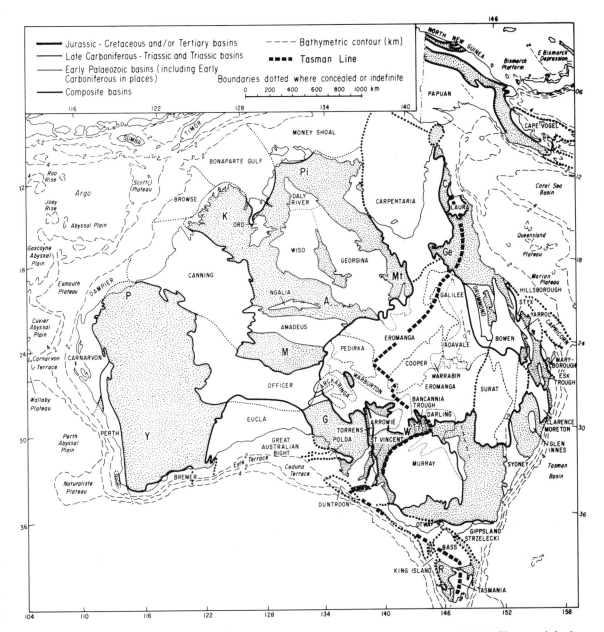

Fig. 44. Exposed Precambrian and Phanerozoic fold belts and Precambrian blocks and basins (stippled), Phanerozoic basins (clear) (Veevers *et al.* 1982), and the Tasman Line that divides the continent into Precambrian terrain on the west and Phanerozoic terrain on the east. Within the Precambrian terrain, the major exposed blocks (Rutland 1976) are: A, Arunta Block; C, Coen Inlier; G, Gawler Block; Ge, Georgetown Inlier; K, Kimberley Block; Mt, Mount Isa Block; T, Tyennan Geanticline; W, Willyama Block; WS, Western Shield; Y, Yilgarn Block.

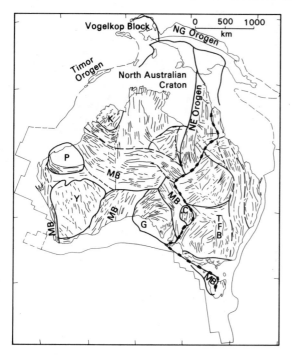

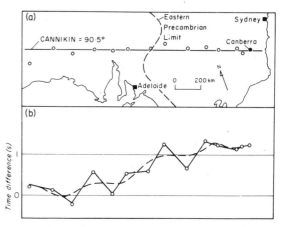

Fig. 45. Gravity trends in mainland Australia (from Wellman 1976) and major crustal subdivisions (from Plumb 1979). In increasing age, these are the Timor Orogen (Neogene), New Guinea Orogen (Mesozoic–Cainozoic), Vogelkop Block (Palaeozoic), Tasman Fold Belt (Palaeozoic–Mesozoic), the various Precambrian mobile belts (MB) that were cratonized from about 1.4 to 0.6 Ga ago; the North-east Orogens, parts stabilized since 1.4 Ga ago, others reactivated in the Palaeozoic; the Curnamona and Gawler Cratons, stabilized about 1.4 Ga ago; the North Australian Craton (including the Kimberley Block), fairly stable since 1.7 Ga ago; and the Yilgarn and Pilbara Blocks, stabilized by 2.5 Ga ago. Tasman Line shown by heavy bars.

Fig. 46A. (a) Location of stations in south-eastern Australia recording the Cannikin atomic explosion in the Aleutian Islands (after Cleary et al.1972). The line through the stations is at an angular distance of 90.5° from the explosion site. The eastern limit of the Precambrian craton is indicated by a dashed line. (b) Relative residual travel-time differences between stations. An arbitrary residual zero point has been taken on the Precambrian craton. Open circle represents observed difference; dashed line is the smoothed three station running average. From Finlayson (1982).

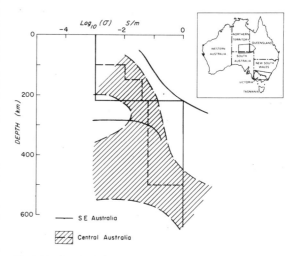

Fig. 46B. Conductivity-model ranges for the mantle under south-eastern and central Australia (after Lilley et al. 1981). The logarithm of the conductivity (σ) is represented. The solid and dashed lines represent the preferred models for Phanerozoic and Precambrian Australia respectively; the shaded and hachured areas represent the limits of models which fit the data. From Finlayson (1982).

case for rapid variations in the physical properties between Phanerozoic and Precambrian Australia at depths greater than 200 km. Shorter-period (magnetotelluric) sounding indicates a similar trend, with sub-crustal conductivities increasing at depths less than 100 km under the Phanerozoic terrain (Murray Basin, Vozoff et al. 1975; Eromanga Basin, cited in Finlayson 1982), but not until depths of 200 to 250 km in the Precambrian terrain of the Officer Basin (Jupp et al. 1979) and about 150 km under the North Australian Craton (cited in Finlayson 1982). An intermediate depth of 125 km to a low-velocity zone beneath the southern New Guinea platform was found by Brooks's (1969) study of the dispersion of surface seismic waves.

(ii) Shallow structure

Rutland (1973, 1976, 1981, 1982) divided the Australian continental crust into three major provinces corresponding to crustal evolution during three chelogenic cycles, defined as global thermal cycles during which large areas of continental crust were stabilized as shield areas, with each characterized by a peak of metamorphism and granitoid intrusion of short duration but wide extent. The Archean cycle is represented by the Pilbara granite–greenstone belt complex, in which the principal granites are older than 3.0 Ga and the greenstones about 3.45 Ga, and the Yilgarn Complex, in which the principal granites are about 3.0 to 2.5 Ga, and the greenstones in between. The Proterozoic chelogenic cycle is bounded by major phases of basic dyke emplacement at about 2.4 and 1.0 Ga, with a peak of granitoid intrusion about 1.9 to 1.7 Ga ago. 'Eastern Australia is the product of an, as yet, incomplete chelogenic cycle, encompassing an assemblage of Late Proterozoic and Phanerozoic fold belts, which display varying degrees of morphotectonic rejuvenation in the Neogene' (Rutland 1982).

The regional variation in shallower seismic structure appears to be strongest between the Archean terrain of the Pilbara Block and either Proterozoic or Phanerozoic terrain (Fig. 46C) (Rutland 1982). The Pilbara profile apparently lacks the relatively high-velocity layers present in the lower crust of the Proterozoic and Phanerozoic terrains. The lower crust of the Pilbara Block is interpreted as felsic granulite (Drummond 1981), and, as detailed by Wass in the next section, the lower crust of the Phanerozoic terrain is attributed to rocks of basaltic to ultramafic composition in the granulite facies.

In the Proterozoic terrain of south-central Australia and western New South Wales, a suite of garnet clinopyroxenite, eclogite, and granulite xenoliths (Ferguson et al. 1979) is inferred to wedge out into the eastern Phanerozoic terrain, and differs from xenoliths in the Phanerozoic terrain in containing quartz, kyanite, and sphene (Wass, below). Ferguson et al. (1979) estimate that this suite comes from depths of 35 to 60 km, which overlaps the depths of 40 to 48 km of the high-velocity lower crust in the Proterozoic terrain from Mt Isa to Tennant Creek (Fig. 46C). The Proterozoic areas sampled by xenoliths and by seismic sounding are 1500 km apart, but if the results represent the whole terrain, then they show that the lower crust

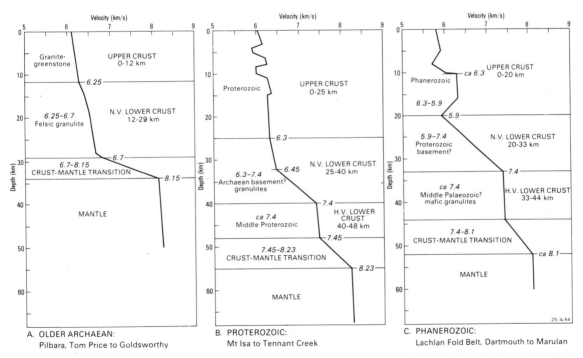

Fig. 46C. Velocity-depth functions for Phanerozoic, Proterozoic, and Archaean provinces in Australia, indicating their possible interpretation in terms of crustal structure and age. N.V.=normal velocity; H.V.=high velocity. From Rutland (1982).

the composition of the lower crust of eastern Australia is reviewed in the next section.

(b) Crustal and upper mantle stratigraphy beneath eastern Australia (by S. Y. Wass)

The Cainozoic basaltic rocks of eastern Australia (Fig. 47) contain abundant xenoliths of upper mantle and crustal rocks entrained from all stratigraphic levels above the locus of magma generation. This xenolith population provides unique access to rocks not exposed at the surface in eastern Australia. Theoretical calculations (Mercier 1979) and experimental evidence (Kushiro *et al.* 1976) suggest that the transit time of lavas to the surface after entrainment of dense upper mantle and lower crustal xenoliths does not exceed 60 hours. Therefore, samples of deep country rock are usually unaltered by reaction with the magma. An idealized stratigraphic column of the upper mantle and lower crustal rocks for south-eastern Australia is presented in Fig. 48.

(i) Mantle

The most abundant xenolith type (Wass and Irving 1976) is the spinel lherzolite, characterized by the four-

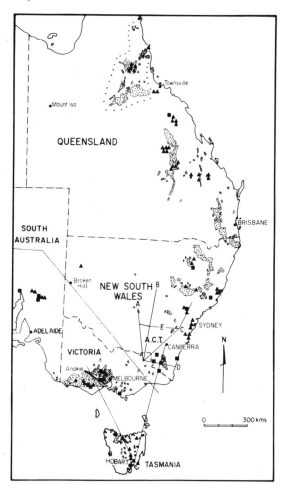

Fig. 47. Distribution of lower crustal mafic xenoliths in eastern Australia (based on Sutherland and Hollis 1982). Square symbols denote garnet-bearing, triangular symbols garnet-free, assemblages. Stippled areas enclose dominantly Cainozoic basaltic rocks. Straight lines represent seismic traverses (Finlayson *et al.* 1980).

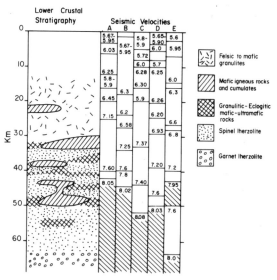

Fig. 48. Schematic representation of the lower crustal stratigraphy beneath south-eastern Australia (Griffin *et al.* 1984). Adjacent seismic velocities for comparison are from Finlayson (1982) for the Lachlan Fold Belt (traverses shown on Fig. 47). Shaded areas of seismic-velocity columns represent mantle velocities. Seismic velocities in km/s.

of the Proterozoic terrain has a similar velocity structure to that of the Phanerozoic, but a different composition.

In summary, the Precambrian (Archaean and Proterozoic) terrain is underlain by a thick lithosphere, possibly as much as several hundred kilometres thick, with a lower crust of felsic granulite (Archean) to eclogite (Proterozoic), whereas the Phanerozoic terrain on the east is underlain by a thin lithosphere, perhaps only 100 km thick, with a lower crust of mafic to ultramafic granulite. The evidence from xenoliths of phase assemblage of olivine, orthopyroxene, clinopyroxene (Cr-bearing diopsides), and spinel. Rare garnet-bearing lherzolites (Ferguson *et al.* 1977) sample deeper material. Both of these xenolith types represent the main rock types constituting the upper mantle beneath eastern Australia. Spinel lherzolite forms the uppermost mantle. Garnet lherzolite appears with increasing pressure and thus underlies the spinel lherzolite (Fig. 48). The lherzolites show a wide variety of microstructures ranging from polygonal, relatively strain-free fabrics to strongly deformed microstructures. These different microstructures record different strain environments within the mantle and therefore may provide clues concerning major deep-seated tectonic processes which have affected the overlying crust at a given time.

Not only to these upper mantle xenoliths record movement, but they also commonly have an imprint of geochemical processes which affect the mantle. Many large xenoliths show contact relationships indicating that the upper mantle is, in detail, heterogeneous. The most common rock type found in contact with lherzolite is an ultramafic rock containing Al-augite as the clinopyroxene in contrast to the Cr-diopside of the lherzolite.

Some of the Al-augite-bearing rocks have a chemical composition which suggests that they are frozen basaltic liquids which have not escaped from the upper mantle. Others, however, do not represent reasonable melt compositions, but are cumulates from basaltic liquids. One mechanism put forward to explain the origin of these high-pressure cumulates (Irving 1980) is flow crystallization of liquidus phases on mantle conduit walls as the magma flows upwards. Xenoliths of Al-augite rock types adjacent to spinel lherzolite are samples of such mantle wall–rock contacts. Disaggregation of such conduit cumulates may be the origin of many of the megacrysts (large single crystals of, for example, Al-augite, spinel, orthopyroxene, olivine) which are common in the alkali basaltic rocks of eastern Australia.

Another group of rarer mantle xenoliths is composed largely of amphibole and apatite (Wass 1979), and some spinel lherzolites show alteration to or addition of these minerals. The amphibole/apatite xenoliths are rich in volatiles and elements not found in significant concentrations in spinel lherzolites, such as Cl, P_2O_5, CO_2 and rare earth elements. They probably represent small amounts of volatile-rich liquids released from deeper in the mantle. The presence of similar mantle rock types and metasomatic upper mantle events has been linked with volume expansion of upper mantle material and may be a causal factor in crustal uplift associated with tensional regimes (Lloyd and Bailey 1975).

(ii) Lower crust

Granulitic xenoliths (for example Ferguson *et al.* 1979; Wilkinson 1974; Irving 1974; Lovering and White 1969; Wass and Hollis 1983; and author's unpublished data) represent the only direct access we have in eastern Australia to deep old crust. The mineralogy and geochemistry of these xenoliths suggest that the lower crust beneath eastern Australia is basaltic to ultramafic in character (Figs. 47, 48). Although these rocks have now been metamorphosed to the granulite and eclogite facies, they probably formed by deep intrusion or underplating of continental-type basaltic magmas during successive igneous episodes spanning the time of evolution of the eastern Australian crust. Geothermometry and geobarometry on these xenoliths indicate that the lower crust may extend as deep as 60 km beneath western Victoria and that the geothermal gradient was high at the time of entrainment. Ferguson *et al.* (1979) have described a suite of garnet-clinopyroxenite, eclogite, and granulite xenoliths from south central Australia and western New South Wales. These rocks commonly contain quartz, kyanite, and sphene, which are absent from the other xenoliths reported here. The xenoliths, which have come up through Precambrian terrain, are thus distinguishable from the xenoliths in eastern Australia by their higher content of silica and alumina. Volcanics that have come up through the only obviously Precambrian terrain in eastern Australia, the Georgetown Inlier west of Townsville, contain high-pressure xenoliths of undefined origin.

A Cainozoic analogy is provided by some southern and central Queensland basaltic provinces. Ewart *et al.* (1980) show that deep crustal intrusion of plutonic equivalents of surface volcanics have effected crustal thickening of about 60 km in this region.

This interpretation of a generally basaltic character for the lower crust is supported by geophysical evidence (Finlayson *et al.* 1980) which shows that seismic velocities gradually increase downwards in the lower crust (Fig. 48). The Moho appears not to be a sharp break in seismic velocity, but to be the culmination of a downward increase in seismic velocity.

In summary, xenolith evidence suggests the lower crust is a complex of granulitic and eclogitic rocks successively intruded as basaltic magma around the crust/mantle boundary and above. This results in an interlayered mafic/ultramafic complex at the Moho (Fig. 48) and a generally mafic lower crust.

(iii) Upper crust

The upper crust is more rarely sampled than the mantle and lower crust but xenoliths have still provided information on hidden underlying rocks. The Boomi Creek sill (30° S, 151° E) (Wilkinson and Taylor 1980) contains xenoliths of a middle-level differentiated tholeiitic intrusion. Xenoliths of clastic upper crustal rocks are found in the Mogo diatreme (Emerson and Wass 1980) and these suggest that a sequence like that of the Hill End Trough may extend beneath the Sydney Basin.

2. PHYSIOLOGY

(a) Seismicity

Seismicity divides the region into two parts (Fig. 49): (1) The northern margin of Australia, along the boundary of the Indo-Australia Plate and the Eurasian and Pacific Plates, and peppered with epicentres; and (2) the almost aseismic intra-plate Australian platform and its margins. Bounded on the north-west and north-east by Benioff zones, and as part of the Meso-zoic–Cainozoic Alpine–Himalayan and Circum-Pacific orogens, the northern margin is a zone of intense seismicity, volcanicity, and earth movement, with rates of differential vertical movement of thousands of metres per million years; relative motion between the Indo-Australian and Pacific Plates is oblique to their northern boundary (Figs. 2, 4), and produces sinistral shear in New Guinea. As part of the interior of the Indo-Australia Plate, the Australian platform is a region of low seismic and volcanic activity, with rates of vertical movement of tens of metres per million years, two orders of magnitude less than those of the northern margin. The major activity along the northern margin, vividly reflected in the present morphology, is described in Chapter IV; here we concentrate on the subdued modern movements of the Australian platform, which, not obvious in the present morphology, require special studies to be detected.

All the recorded earthquakes in Australia come from within the crust at depths less than 40 km. Denham *et al.* (1975) divided the platform and adjacent western, southern, and eastern margins into three regions of earthquake activity: (1) the eastern region has a diffuse distribution of earthquakes over a wide area that corresponds with the South-eastern Highlands, with clusters in the vicinity of Wilsons Promontory (39° S, 146° E) and Gunning (35° S, 149° E), and two groups of epicentres in the Tasman Sea, at lat. 40° S. (2) The central region, from Adelaide to the Simpson Desert (26° S, 137° E), contains clusters in the area immediately north of Spencer Gulf, and in the Simpson Desert, notable for six earthquakes with magnitude greater than six. (3) The western region contains scattered activity except in a zone 90 km east of Perth, and in the eastern Canning Basin (22° S, 127° E).

The analysis of earthquake focal mechanisms, measurements of stress by overcoring techniques in mines, tunnels, and shallow drill-holes, observations of rock movements that follow excavations, and interpretations of the origin of folds and joints show that the upper crust is in a state of substantial horizontal compression, and that the horizontal pressures are commonly much greater than the overburden pressure (Denham 1979; Denham *et al.* 1979). The direction of principal horizontal stress ranges from N–S to NE–SW in the Sydney Basin (Shepherd and Huntington 1981; Gray 1982); farther south, the direction is 120° (ESE) (Denham *et al.* 1981), and NNW–SSE in Victoria (Barton 1981); it is N–S in the Simpson Desert, and E–W near Perth. Denham *et al.* (1979) discussed the evidence thus:

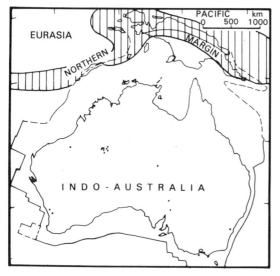

Fig. 49A. Seismicity. Individual earthquake epicentres on mainland Australia, 1964–77, magnitude 5.0–7.4, focal depth ≤70 km, and zone of intense activity (ruled) of northern margin. From American Association of Petroleum Geologists (1981). Only 15 events with M >5.0 were recorded from the Australian mainland, in contrast to the hundreds recorded north of lat. 12° S (shown individually in B), along a seismic boundary that marks the northern margin.

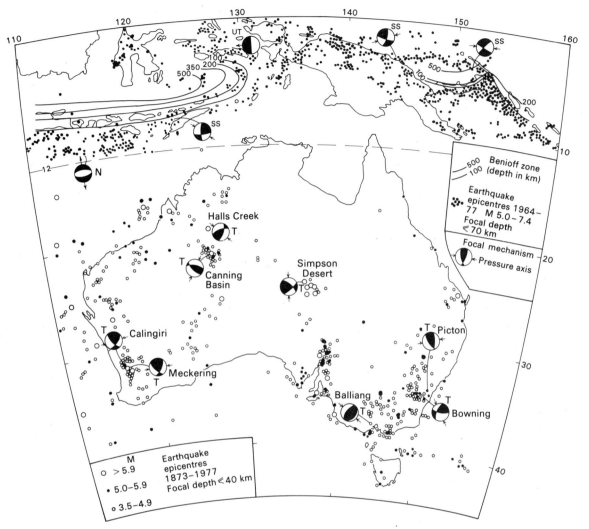

Fig. 49B. Earthquake epicentres. (1) North of lat. 12° S, 1964–77, magnitude 5.0 to 7.4, focal depth ≤70 km. From American Association of Petroleum Geologists (1981). Deeper-focus earthquakes indicated by contours on Benioff zones, that on the north-east from Wellman and McCracken (1979), that on the north-west from Hamilton (1979). Focal mechanisms of selected shallow earthquakes from Indonesia (Cardwell and Isacks 1978), New Guinea (Johnson 1979), and mainland Australia (Denham et al. 1979). All the selected Australian mechanisms represent thrust-fault (T) interpretations; elsewhere the mechanisms include strike-slip (SS), which is general for the northern plate boundary, underthrusting (UT), and normal (N). (2) South of lat. 12° S, 1873–1977, magnitude 3.5 to 7.4, focal depth ≤70 km. From Denham (1979).

There is now good evidence from several parts of the continent that compressive stresses are present throughout the upper crust. However, the simple models derived from standard plate-tectonic concepts that predict a consistent direction for stress throughout the whole continent do not apply. Within specific regions, such as the southwest Yilgarn Block and central New South Wales, the evidence from earthquake focal mechanisms, overcoring measurements, and ground deformations all give consistent results. However, the boundaries of these specific regions are difficult to define, and stress directions can change abruptly from region to region. For example, the 1973 Picton and 1977 Bowning earthquakes [Fig. 49B] which were close together, gave pressure axes almost at right angles to each other, although both mechanisms indicated that compressive forces close to horizontal caused the earthquakes. Similarly, the 1970

Canning Basin and the 1978 Halls Creek earthquakes [Fig. 49B] show thrust fault solutions, but with pressure axes in different directions.

The reasons for the apparent inconsistencies in the directions of the pressure axes obtained from the earthquakes are not evident. However, there seem to be two main possibilities. The first is that the earthquakes occur along pre-existing zones of weakness in the crust. These zones would then determine the attitudes of the fault planes and hence the true regional stress pattern would not be revealed – as suggested by McKenzie (1969); the second possibility is that the stress field in the crust is distorted by lateral inhomogeneities. These would be expected at craton boundaries, and at the edges of similar large crustal features.

The results of the overcoring measurements are reasonably consistent. However, the data are scattered thinly, and the problem remains of ascertaining whether or not the rock tested is properly coupled to the crust.

Wellman (1981) analysed recent crustal movement from repeat surveying in south-eastern and south-western Australia.

Both the pattern of crustal movement and the isostatic compensation process appear to be controlled by crustal strength. In southeastern NSW, the crustal movement rate and direction vary smoothly with position, and earthquakes of magnitude greater than five are distributed uniformly . . . In this region, an analysis of gravity anomalies (Wellman 1979b) showed that areas of $0.5° \times 0.5°$ have a standard deviation of their mean heights of only 80 m from the height for perfect isostatic equilibrium. Hence this is an area of weak crust composed of small blocks near isostatic equilibrium. It is thought that stresses and strains are propagated linearly through the blocks, the strain being released uniformly. In southwestern WA the crustal movement and seismicity seem to be localised along zones of weakness, and stress and strain can change in direction by 90° over a short distance. In this region only areas of $2° \times 2°$ seem to be in close to isostatic equilibrium. These findings are consistent with a strong crust, broken by widely spaced fractures, the strength and size of the blocks making the strain release geographically discontinuous. In this interpretation, the N–S stress and strain at Goomalling near Meckering [Fig. 49B] is regarded as an edge effect resulting from the interaction of several blocks. The stronger crust in western, as compared to eastern Australia is likely to be related to lower lithospheric temperatures, as indicated by lower heat flow (Cull & Denham 1979).

A more coherent pattern of stress, expressed as shear, is indicated by Neogene faults and folds in the eastern Indo-Australian Plate.

(b) Dextral shear within the eastern Indo-Australia Plate (by J. J. Veevers and C. McA. Powell)

The fault and fold pattern in Neogene sediments on either side of Australia is consistent with dextral or anticlockwise motion of Australia relative to the rest

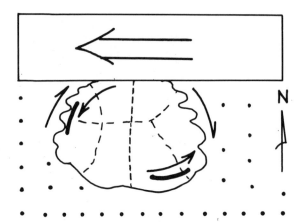

Fig. 50A. Mechanical analogue of motions within the plate driven by motion along its edge.

of the enclosing plate, caused by the sinistral shear of the Pacific and Indo-Australia Plates in New Guinea. A mechanical analogue of this (Fig. 50A) is the anticlockwise rotation imparted by a moving slab to a mosaic of blocks set in a less rigid matrix. The inferred horizontal motions in Australia, summed over the past 20 Ma, are very small, perhaps of the order of a kilo-

Fig. 50B. Broad arrows show modern sinistral shear motion in New Guinea and modern dextral shear in New Zealand, and small arrows the modern small-scale dextral motion in the Gippsland Basin (Fig. 50C), and the Carnarvon–Dampier Basin (Fig. 50D). Base map from Fig. 4.

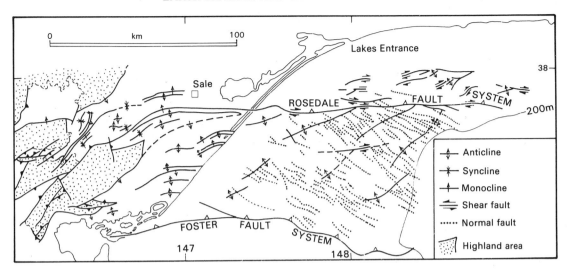

Fig. 50C. Gippsland Basin, showing late Eocene to Recent shear faults and *en échelon* folds superimposed on Late Jurassic to middle Eocene normal faults. From Threlfall *et al.* (1976), Abele *et al.* (1976), and Davidson (1980).

metre, in contrast to the hundreds of kilometres or more of relative motion in the same period between the Pacific and Indo-Australia Plates. To apply the analogue, two conditions must be satisfied: (1) synchronous motion by the slab (Pacific Plate) and anticlockwise rotation of the mosaic of blocks (Australia), and (2) the same sense of (dextral) motion throughout the mosaic (Australia), along azimuths determined by the local structural grain.

Evans (1981, 1982) drew attention to the origin by shearing of the *en échelon* folds that act as reservoir structures for petroleum in the Carnarvon and Gippsland Basins. Fig. 50B shows the large-scale sinistral shear in New Guinea (Carey 1970, 1976; Grund 1976), and the large-scale dextral or anticlockwise shear in New Zealand, and the inferred small-scale dextral shear in the Gippsland Basin (Fig. 50C) and the Carnarvon Basin (Fig. 50D). In northern New Guinea, the movements started in the late Miocene, and continue to the present day (Grund 1976). In New Zealand, the present shearing can be extrapolated back 10 to 20 Ma to the middle or late Miocene to account for the total deformation across the fault zone (Walcott 1978); what part, if any, is attributable to the postulated small-scale anticlockwise rotation of the Indo-Australian Plate is overshadowed by the major plate motion. The deformation in the Gippsland Basin involves Late Jurassic to middle Eocene dextral-divergent rotation expressed in normal faults, followed by late Eocene to Recent dextral-convergent rotation expressed by shear faults and *en échelon* folds (Davidson 1980), with pronounced shearing during the late Eocene and Oligocene, and late Miocene (Threlfall *et al.* 1976). In the Carnarvon Basin, the folds are expressed at the surface in a middle Miocene limestone (Crank 1973; Thomas and Smith 1974) and hence are late Miocene or younger, and, as shown by emerged warped marine terraces on Cape Range (van de Graaff *et al.* 1977a) and to the south (Denman and van de Graaff 1978), continue growing to the present day. All these movements therefore involve shearing during the past 10 Ma. Evans (1981, 1982) also noted similar folds in the continental interior in the Eromanga/Cooper and Eromanga/Pedirka Basins. This evidence suggests that the conditions of synchronous motion at the northern plate boundary and within the plate, and the same sense of motion along appropriate azimuths, are met.

The late Eocene to modern system of compressional-dextral shear in the Gippsland Basin is traceable back to the Late Jurassic to middle Eocene dextral-divergent shear (Davidson 1980), showing the longevity of this shear system in south-east Australia.

3. REFLECTION OF EARTH MOTIONS IN THE CAINOZOIC VOLCANICITY OF EASTERN AUSTRALIA

As with seismicity and earth movements, so with volcanicity the northern margin has the lion's share,

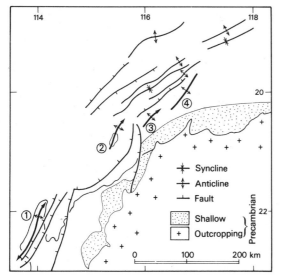

Fig. 50D. Carnarvon–Dampier Basin, showing *en échelon* folds of the Cape Range (1), Barrow Island (2), Mermaid Nose (3), and Enderby Trend (4). From Crostella and Barter (1980), Thomas (1978), and Crank (1973).

leaving only four widely scattered Quaternary volcanic centres for the entire Australian platform. Quaternary volcanicity is confined on the mainland to south-west Victoria and the adjacent part of South Australia, and south-east and northern Queensland (area enclosed by dotted lines in Fig. 47); to the north, volcanicity extends through the Maer Island province in Torres Strait (Fig. 53C) to the volcanic rim of the Pacific in New Guinea and New Britain. In south-west Victoria and South Australia, basalt flows and ash were erupted as recently as 1410 ± 90 years ago in the latest manifestation of volcanism that started in the Pliocene; the region has a high heat flow (Cull and Denham 1979). Basalt cones, scoria cones, maars or tuff rings, and flows are the main forms, and the lava ranges from nepheline basanite to alkali olivine basalt and olivine tholeiite (Douglas and Ferguson 1976). In Queensland, the volcanism is likewise mafic in the small field on the coast near Bundaberg (lat. 25° S) (Wellman 1978) and in the large northern field that stretches along the Eastern Highlands from 21° to 15° S. The northern field (Stephenson *et al.* 1980) of flows and cones last erupted about 13 ka ago, and ranges from olivine tholeiite to strongly alkaline types. In the Maer Island Province (Fig. 53C), basalt flows and pyroclastic cones have erupted through Pleistocene or Holocene carbonate reefs (Stephenson *et al.* 1980).

In New Guinea (Johnson 1979), active volcanoes in the Bismarck Volcanic Arc (tholeiitic basalt, low-silica andesite, less common high-silica andesite and dacite), in the eastern part of the New Guinea Highlands (tholeiitic basalt and andesite, trachybasalt and trachyandesite), and in the Vogelkop in the west, lie along the zone of convergence between the Indo-Australian and Pacific plates, and are described further in Chapter IV.

Wellman and McDougall (1974b) distinguished three main types of igneous province in the Cainozoic of eastern Australia: central volcanoes, lava fields, and a single mafic, high-potassium province. The ages and distribution of the central volcanoes suggests migration of the locus of volcanism southward at 66 ± 5 mm/year, from Cape Hillsborough (lat. 21° S) 33 Ma ago to Mount Macedon (lat. 37° S) 6 Ma ago; they related this migration to the movement of the Indo-Australian plate relative to two hot-spot sources in the underlying asthenosphere. Sutherland (1981) extended this model to the basaltic lava provinces, and postulated six volcanic trails, with the modern basalt fields of Victoria–South Australia occupying the present position on the central volcano trail, but, as noted by Pilger (1982), requiring several hot spots south of Tasmania, evidence of which is lacking. The volcanic pattern appears to be incompatible with a 'simple, discrete, deep-hotspot hypothesis', and Pilger (1982) presented an alternative that hinges on (1) the progressive cessation of both central volcano and lava field activity that began 35 Ma ago at lat. 20° S, and extends to the present, near 37° S, and (2) the long period of igneous activity before cessation (Fig. 51). He found that

Paleostress and contemporary stress indicators as well as the volcanic pattern support a model in which the trace forms due to intraplate extension normal to the trend of the trace, resulting in pressure release and melting at the base of the lithosphere, beginning in the Late Cretaceous. This stress field persisted until 35 Ma when progressive reorientation of the stress field occurred. Cessation of igneous activity reflected onset of compression normal to the trend of the trace. As a consequence, a migrating stress node is recorded in the progressive pattern of extinction of volcanic activity along the trace. . . . If the melting responsible for the Highlands igneous activity occurred as a result of a reduction of pressure due to extension at the base of the lithosphere, then the north–south trend of the province and the paleostress indicators suggest that east–west extension could have been the predominant stress state over the entire length of the province prior to 30 m.y. ago. Beginning at 30 Ma, the predominant stress orientation changed, with east–west compression beginning in the north and progressively migrating south. Thus the progressive cessation of igneous activity represents a southward migrating stress node separating significantly different stress fields (Sass and

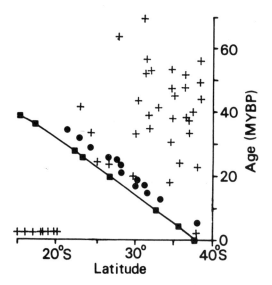

Fig. 51. Age-versus-latitude plot of igneous rocks of the Eastern Australian Highlands province, modified after Wellman and McDougall (1974b). Pluses indicate lava field, and solid circles central volcanoes. Progressive cessation of activity, beginning at 22° S latitude at 30 Ma, is well shown. Note the long period of igneous activity prior to cessation. Connected solid squares indicate the projected locus of a hot-spot trace. Note the close parallelism of the volcanic centre and termination trend with the predicted trend. From Pilger (1982).

Lachenbruch, 1979). . . . Heat flow patterns in eastern Australia appear to be compatible with the intraplate stress model (Sass and Lachenbruch 1979). East–west compression has apparently had a cooling effect, indicated by lower measured heat flow values along the northern part of the trace.

This would mean that the present state of horizontal compressive stress in eastern Australia, described above, started no earlier than 30 Ma ago.

Another pattern discernible in the Cainozoic volcanoes of eastern Australia is a cycle in the age distribution of the total sample of dated material, with modes in the Pliocene and Miocene. As we relate in the analysis of the Eastern Highlands (Chapter IV), these modes correlate with cycles in the flanking basins of transgressive-regressive-lacuna that entail uplift of the Highlands, in both the south (Jones and Veevers 1982) and north (Grimes 1980). This pattern arises from the same data as the latitudinal pattern found by Wellman and McDougall (1974b), and elaborated by Pilger (1982), and suggests to us that areas of volcanic activity indicate uplift, regardless of their position within the Eastern Highlands province. For example, the northern Queensland basalt, which is not accounted for by Wellman and McDougall (1974b) or Pilger (1982), correlates with Pliocene uplift as do the coeval basalts of western Victoria.

IV. MORPHOTECTONICS OF THE AUSTRALIAN PLATFORM AND MARGINS

1. DIVISION OF THE AUSTRALIAN LITHOSPHERE INTO PLATFORM AND MARGINS

The continental lithosphere of the eastern part of the Indo-Australia Plate occupied by Australia (Fig. 52A) has an emergent or shallowly submerged surface, here termed the platform, continuous from southern New Guinea through the epicontinental Arafura Sea and Gulf of Carpentaria to the mainland and across Bass Strait to Tasmania, bounded by the continental margins, mountainous in the north, and submerged beneath a pericontinental sea on all other sides (Fig. 52B). The present shoreline, except round the Arafura Sea and Bass Strait, is not ephemeral; at continental scale it has persisted in much the same position for most of the past 300 Ma, as shown by the superposition of past shorelines (Fig. 52C). That is to say, for at least the second half of its Phanerozoic history, the Australian platform has maintained its present outline. The most complete evidence of the fixed position of the shoreline comes from the west and north-west, where, with only the few exceptions in the Canning Basin, the shorelines back to the Permian lie within a narrow zone. In the north, except for a wide southern excursion in the Aptian–Albian and narrower excursions in the later Cretaceous, early Miocene, and today, the shoreline has lain near the southern flank of the New Guinea Highlands. On the east, the shoreline has remained near its present position since the Early Permian, and on the south since the later Cretaceous. In concrete terms, the boundary between the platform and the margins is manifested in the wedge-out of the marine sediments of the margin against the persistent land of the platform. The shoreline, equally in the past as today, is not only a morphological boundary but a tectonic one too; it is the hinge-line about which the western, southern, and eastern margins have subsided after breakup, and the northern margin in New Guinea has risen. Before breakup along the western, southern, and eastern margins, the shoreline, at least back to the Permian, marked the hinge of a subsiding zone that preceded the modern margins. Our account of the morphotectonics of Australia accordingly follows these divisions.

A radical excursion of the shoreline during the Aptian–Albian (Fig. 52C), in which a eustatic highstand of the sea combined with an active subsidence of part of the land, suggests the morphotectonic longevity of the three elements of the platform. Fig. 52B shows the subdivision by the 200 m contour – the median height of the land and continental shelf (Fig. 137) – into the Great Western Plateau, Central-Eastern Lowlands, and Eastern Highlands, as distinguished by David and Browne (1950) and Gentilli and Browne (1963). The Central-Eastern Lowlands extend

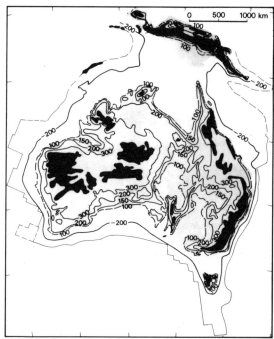

Fig. 52A. Morphology, presented on a base map widely used throughout this book, on the same projection (Lambert Conformal Conic) as the Tectonic Map of Australia and New Guinea (Geological Society of Australia 1971). Smoothed topographical contours (m), 200 m bathymetric contour, and continent–ocean boundary. Main contours from American Association of Petroleum Geologists (1978) and Times Atlas (1975). Note that the clear area round Lake Eyre is below sea level. Ground above 500 m in solid black.

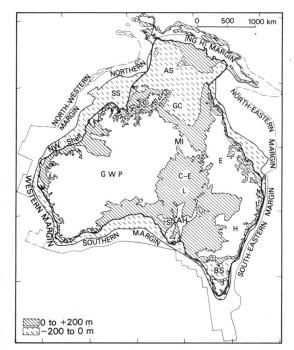
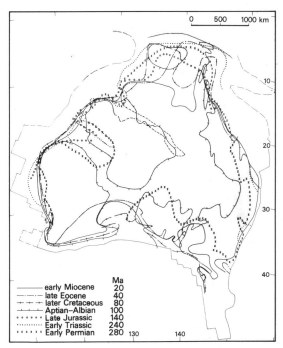

Fig. 52B. Division of the land surface into areas above (clear) and below (shaded) the median height of 200 m of the land and the continental shelf (broken lines), defining the Great Western Plateau (GWP), Central-Eastern Lowlands (C-EL), and Eastern Highlands (EH), of David and Browne (1950) and Gentilli and Browne (1963). Boundary between platform and margins denoted by heavy broken line. AL Arnhem Land, AS Arafura Sea, BS Bass Strait, GC Gulf of Carpentaria, MI Mount Isa, NGH New Guinea Highlands, SAH South Australian Highlands, SS Sahul Shelf, T Tasmania.

Fig. 52C. Shorelines since the Early Permian. With the notable exception of the Aptian–Albian epicontinental seas and the wide range of shoreline positions between the New Guinea orogen and the mainland, the shoreline has oscillated within narrow limits about its present position. Information from this chapter and from Chapter V.

2. MORPHOTECTONICS OF THE CONVERGENT NORTHERN MARGIN, FOCUSED ON THE NEW GUINEA HIGHLANDS

(a) Morphotectonics

northward past a saddle near Mt Isa into the Gulf of Carpentaria, and southward into the southern margin on either side of the South Australian Highlands. The Great Western Plateau has an outlier in Arnhem Land, and the Eastern Highlands an outlier, past Bass Strait, in Tasmania. The Aptian–Albian shoreline (Fig. 52C) blocks out the platform into the same elements to suggest that the high ground of the Great Western Plateau and Eastern Highlands, and the intervening Central-Eastern Lowlands and South Australian Highlands are not ephemeral but had morphotectonic forebears in the Cretaceous. Our morphotectonic analysis of the platform is focused on the history of the high ground. And our analysis of the entire continent starts with the highest ground of all along the northern convergent margin.

In its present morphology, volcanicity, seismicity, and rapid vertical motions, the northern margin bears the marks of vigorous growth, as distinct from the other margins, along the west, south, and east, which have long since been senescent. As sketched in Fig. 36, and perceived by Wegener (1929), as quoted by Johnson (1979), 'From the southeast came the huge Australian block with its front thickened like an anvil: New Guinea (folded to form a high-altitude mountain range) plus shelf; this forced itself between the chains of the southernmost Sunda Islands and the Bismarck Archipelago.'

This interaction between the continental lithosphere of the eastern Indo-Australian Plate with the oceanic

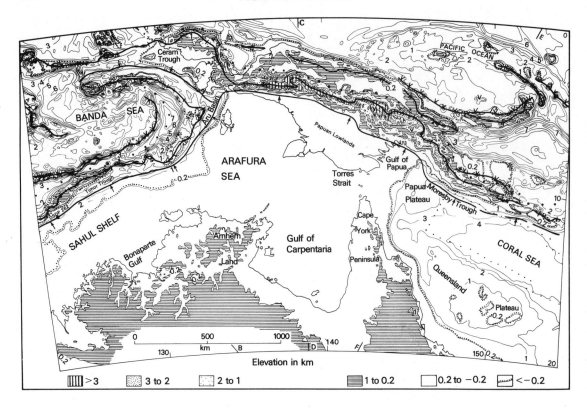

Fig. 53A. Morphology of the northern margin, showing the protrusion of the New Guinea salient into the oceanic realm of the Pacific on the north-east, and the Banda Sea on the west. Contours from American Association of Petroleum Geologists (1978). Morphotectonic features indicated (symbols in Fig. 53B) are (1) the northern limit of the foreland basin, coincident with the frontal thrust (shown by arrows that are not vectors), (2) the crest of the orogen, (3) the boundary between continental lithosphere and Cainozoic oceanic lithosphere, (4) the Northern Ranges of New Guinea, and (5) active volcanic arcs. Banda Arcs from Audley-Charles (1974).

lithosphere is expressed today in the features shown in Fig. 53: (1) a foreland basin and frontal thrust coupled to an orogen traceable for 7000 km through the archipelago of the troughs and high islands of the Outer Banda arc and the loop of east Sulawesi and the Vogelkop to the central New Guinea cordillera, and then southeastward through the Owen Stanley Range of the Papuan Peninsula and the foreland basin of the Moresby Trough; (2) 100 to 200 km oceanward of the crest of the orogen, the boundary between the continental lithosphere and Cainozoic oceanic lithosphere – the edge of Wegener's (1929) Australian block; this boundary lies along the axis of the North New Guinea Basin inboard of (3) the Northern Ranges Province in central New Guinea; the final features are (4) volcanic arcs on the west (Inner Banda Arc) and east (Eastern Papuan Province, Bismarck Volcanic Arc).

The three widely spaced profiles in Fig. 53B show the common structure of the northern margin: an epeiric or epicontinental basin up to 1250 km wide that past a foreswell deepens or contains thicker sediment at the toe of the frontal thrust to form a narrow foreland basin bounded by an equally narrow orogen, together only 300 to 400 km wide. At the line of profile EF, the epicontinental sea passes eastward into the deep oceanic lithosphere of the Coral Sea. The common size and shape of the rim orogen and foreland basin, regardless of their position in the system – whether in the central cordillera or peripheral archipelagoes – contrast with the shape of the bordering oceanic lithosphere: the uniform rim of the continental lithosphere passes into an almost chaotic oceanic lithosphere in which the most regular elements discernible are those moulded by the continental anvil: the

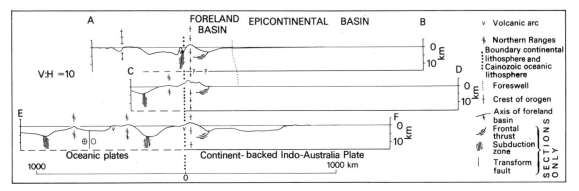

Fig. 53B. Profiles across the northern margin, located in Fig. 53A. Continent-backed lithosphere indicated by full line at base of each section, and ocean-backed lithosphere by broken line. Note in EF left-lateral transform fault indicated by ⊕ (motion out from page).

volcanic Inner Banda Arc and the western half of the Bismarck Volcanic Arc on the periphery, and the Northern Ranges Province in the centre.

The complex structure of small oceanic plates east of New Guinea (Johnson 1979) (Fig. 53C) is probably due in part to the shearing motion between the large Indo-Australian and Pacific Plates (the direction of relative motion is shown in Figs. 2 and 4). This means that sinistral strike-slip motion, observed on the ground along the Sorong Fault of the Vogelkop and in northern New Guinea (Grund 1976), and interpreted from studies of first motions of earthquakes along the northern boundary of the Bismarck Plate (Johnson 1979), probably exceeds convergence. In Chapter II, we followed Carey (1958) in recognizing sinistral shear as the source of the drag fold of the Seram–eastern Sulawesi–Vogelkop (Fig. 36). Zones of epicentres of earthquakes shallower than 125 km outline the Bismarck and Solomon Sea Plates (Fig. 49B) and, on the west, the Eurasia Plate. A more diffuse zone across north New Guinea lies up to 350 km from the Pacific Plate boundary, and probably reflects the dissipation of shear motion over a wide zone along the northern margin. Deeper earthquakes define (a) Benioff zones

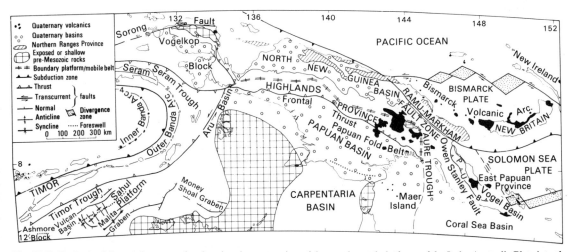

Fig. 53C. Geology of the northern margin, showing the protrusion of the continental platform of the Indo-Australia Plate into the oceanic lithosphere of the Pacific, Bismarck, and Solomon Sea Plates, all bordering the western part of the Pacific Ocean, and of the Eurasia Plate (Sundaland) on the north-west. Modified from Geological Society of Australia (1971), American Association of Petroleum Geologists (1981), Johnson (1979), Grund (1976), and Ridd (1976). Details of Sahul Shelf from Fig. 82.

beneath the Bismarck Plate and Eurasia Plate, and (b) a diffuse zone along the Highlands.

Active volcanoes (Figs. 53A, and 54, top left) are restricted to the Inner Banda Arc, Bismarck Volcanic Arc, the East Papuan Province, and two widely spaced occurrences in the Fly-Highlands Province (Johnson 1979), which contains many extinct volcanoes, as does the Maer Island basaltic province (Fig. 53C). The only active volcano is in the Vogelkop. The Fly-Highlands Province is the only modern occurrence of volcanism not situated in or near oceanic terrain. According to Johnson et al. (1978), the commonly accepted compositional divisions of circum-oceanic rocks in arc-trench systems – tholeiitic, calc-alkalic, and shoshonitic – have little significance for Papua New Guinea: in the mainland province, calc-alkalic rocks form compositional continua with shoshonitic rocks, and, in the south Bismarck arc, continua with tholeiites.

The depositional systems of the northern margin comprise the wide epicontinental basin of the Arafura Sea and Gulf of Carpentaria and Sahul Shelf: detrital mud in depressions, such as the Bonaparte Depression (Fig. 81); quartz sand along the Australian coast, as in the Karumba Basin round the Gulf of Carpentaria; and reef carbonate and glauconitic sand in the shallow parts of the Sahul Shelf, including its seaward edge, and reef carbonate on the structural ridge of Torres Strait. In the centre, the epicontinental basin terminates northward against a foreswell, on the other side of which the foreland basin passes through the swampy deltaic plain of the Papuan Lowlands to the frontal thrust. The foreland basin here contains up to 2.5 km of Pliocene–Quaternary volcanogenic mudstone, sandstone, and conglomerate, with coal seams; marine limestone is also present. In the west, at DSDP Site 262, in the axis of the foreland basin of the Timor Trough, immediately before the frontal thrust, about 300 m of Quaternary and late Pliocene bathyal radiolarian and clayey coccolith and foraminiferal ooze have accumulated above a Pliocene shallow marine calcarenite. As in the Papuan foreland basin, so here much modern sediment is being redeposited from the uplifted submarine foothills as the tectonic front of the thrust advances southward. On the east, at the head of the Gulf of Papua, 3 km of Quaternary–Pliocene mudstone has been pierced by mud diapirs (Tallis 1975). The principal source of the foreland basin sediment is the orogen, which, as we relate below, started rising to its present elevation of 5 km at the close of the Miocene. Between the orogen and the Northern Ranges Province, intermontane basins accumulated up to 5 km of Pliocene–Quaternary sediment, mainly detrital, deltaic to turbiditic, mud and sand. With the complete emergence of the northern ranges in the Pleistocene, regressive reef carbonate and shoreline sand were deposited round the edges, and deltaic sand and mud in the intermontane basins as the shoreline regressed to its present position (Grund 1976). The final uplift was rapid, as indicated by deep-water Pliocene sediments at an elevation of 1.8 km on the north flank of the Finisterre Ranges (Grund 1976).

The rapid pace of modern activity along the northern margin is indicated by the rates of differential vertical motion between the orogen and foredeep: 1800 m/Ma in the New Guinea Highlands (Kubor Block) and adjacent foreland basin during the past 5 Ma (Jenkins 1974) and 2000 m/Ma across Timor and the Timor Trough during the past 3 Ma (Veevers et al. 1978). As mentioned already, deep-water Pliocene sediment now at an elevation of 1.8 km in the Northern Ranges Province of New Guinea (Grund 1976) implies a minimum net uplift rate of 600 m/Ma. Chappell (1974) found the rate of uplift on the Huon peninsula to be about 3000 mm/ka over the past 230 ka, and Chappell and Veeh (1978) found the north coast of Timor to have risen during the past 120 ka at 500 mm/ka for the eastern part of the north coast of Timor, only 20 to 40 mm/ka for the western part of the north coast, and 470 mm/ka for the south coast of the adjacent extinct volcanic island of Atauro, on the Inner Volcanic Arc. Except for the western north coast of Timor, these rates are one or two orders of magnitude greater than those measured on the intra-plate mainland, such as the uplift rate of 70 mm/ka for the past 700 ka in the Naracoorte area of south-eastern South Australia (Idnurm and Cook 1980), or the subsidence rate during the Cainozoic of 7 m/Ma for the Innamincka depocentre, with a maximum of 50 m/Ma in the Cenomanian or 150 m/Ma including isostatic loading by sediment (Fig. 144). In terms of vertical motion, the northern margin today is lively, and the rest of Australia moribund. The propensity of New Guinea to rise may reflect the strong positive geoidal anomaly in the region, pointed to by Johnson (1979), as the propensity of southern Australia to subside reflects the negative anomaly of the Australian–Antarctic Depression (Chapter II).

The eastern New Guinea Highlands are marked today by volcanoes, and flanked by the intermontane North New Guinea Basin on the north, and by the Fly Lowlands of the Papuan Foreland Basin on the south (Fig. 54). This geography has developed since the beginning of the Pliocene (5 Ma), with the outward advance of volcanogenic alluvium over the sea that initially lapped the flanks of the highlands. The

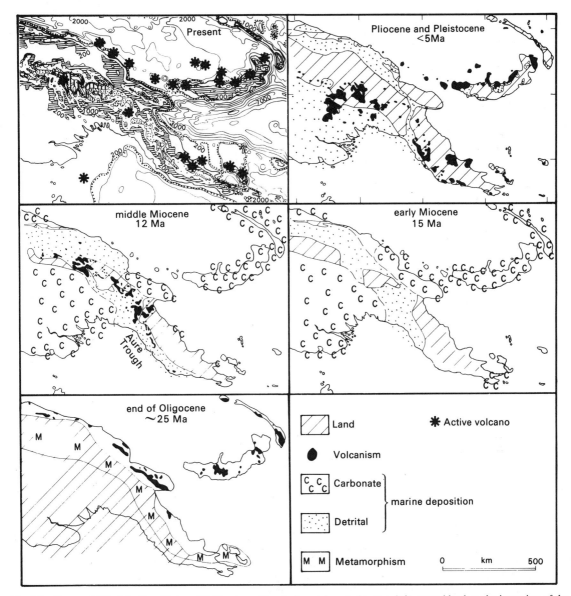

Fig. 54. Eastern half of the New Guinea Highlands and bordering regions today (top left), traced back to the inception of the cordillera in the Pliocene (top right), with archipelagic forebears in the Miocene (middle frames), and emergent metamorphic terrain in the Oligocene (lower left). After Dow (1977).

highlands grew out of a Miocene archipelago surrounded by shallow carbonate banks, and set in a deep marine trough, the Aure Trough, which accumulated at least 5 km of turbiditic mud and sand derived from the adjacent metamorphic and volcanic terrain (Dow 1977). Incidentally, at the same time, a sea covered a zone across northern Australia (Fig. 139).

The archipelagic trough in turn grew out of an Oligocene metamorphic belt, possibly including the Papuan Ultramafic Belt, abutted by the ancestral Bismarck Volcanic Arc. According to Page (1976), most radiometric dates of the metamorphic rocks fall between 20 and 27 Ma, about the Oligocene/Miocene boundary. Also at the beginning of the Miocene, volcanic activity ceased throughout the Bismarck Volcanic Province (until the Pliocene), and was followed by a short period of erosion and then limestone deposition (Dow 1977). We take this abrupt cessation of volcanism in the earliest Miocene (c. 23 Ma ago) as the best estimate of the age of the collision between the volcanic arc and the continent.

Hamilton (1979) interpreted the pre-middle-Miocene mobile zone north of the continental platform (Fig. 53C) as a *melange*, with three parts.

On the south is the belt of subduction melange. The southern part of this melange represents the crumpled terrigenous clastic sediments of the upper Mesozoic and Paleogene continental slope of New Guinea, and the rest is polymict melange containing many exotic elements. In the middle is a belt containing small to enormous masses of ophiolite. Farthest north is another broad belt, formed largely of submarine mafic and intermediate volcanic rocks and their intrusive and volcanigenic–sedimentary equivalents. These three belts are regarded as parts of an island arc that migrated southward during the Late Cretaceous and Paleogene and collided with the previously stable northern margin of New Guinea during Miocene time.

The precise boundary of the allochthon in the mobile belt against the autochthon of the Kubor Range Anticline is obscure. From our study of the Eastern Highlands of Australia and their antecedents, given in the next section, we concur with Hamilton (1979, p. 238) in finding that 'the Mesozoic volcanic rocks in at least the Kubor Anticline belong to the eastern Australian tectonic and magmatic belts', and trace them northward to the southern edge of a zone of ultramafic rocks.

Davies and Smith (1971) interpreted the geology of the Papuan Peninsula along the same lines, except that the Papuan Ultramafic Belt was emplaced in the early Eocene (55–50 Ma), shortly after the opening of the Coral Sea Basin by seafloor spreading (63–57 Ma) (Chapter II); associated metamorphic rocks were not widely exposed until the late Oligocene/early Miocene, and their radiometric ages probably reflect this cooling event. Hamilton (1979) interpreted the initial eruption in the middle Miocene of calc-alkalic magmatic rocks in the archipelago and platform as indicating reversal of the polarity of subduction towards the south. The opposing view, that no polarity reversal took place after the collision, is summarized by Johnson (1979), who regards the Fly-Highlands volcanic province as due to 'partial melting in mantle sources which had been modified by southward subduction in the Cretaceous'.

Visser and Hermes (1962), Harrison (1969), and Hermes (1974) found a history of western New Guinea (Irian Jaya) similar to that of the east: on the continental platform, open-water deposition in the Jurassic and Cretaceous, and shallow-water limestone on the platform and pelagic limestone and volcanics in the mobile belt (Hermes's eugeosyncline) in the Palaeogene. Volcanic activity ceased at the beginning of the Miocene, and thick flysch was depositied in a deep trough. Then starting in the early Pliocene, the New Guinea Highlands started their rapid emergence, which continues today. In the later account of eastern Australia. we retrace the forebears of the New Guinea Highlands, as of the Eastern Highlands, to the Early Jurassic. And, in developing a modern analogue of the Permo–Triassic Bowen–Sydney Basin, we examine the development of the modern New Guinea foreland basin.

At the western end of the rim orogen, by the mid-Pliocene or 3 Ma ago, Timor had started to rise and the Timor Trough foreland basin to subside as a topographic wave started to migrate 80 km to the southeast (Veevers *et al.* 1978; Johnston and Bowin 1981), as a result of complex interaction with the Eurasia Plate involving a mid-Pliocene (3 Ma) collision of the northern continental margin of Australia with a volcanic arc at the edge of the Eurasia Plate. New Guinea collided with the arc some 20 Ma earlier because it protruded deeply into the ocean whereas Timor lay southwestward along the recessive part of the margin, as sketched in Fig. 36. Also sketched in this figure is the evolution of biogeographical boundaries in the zone of collision, adumbrated by Talent in Chapter II.

Authors agree on the Pliocene collision but disagree on the structure and mode of origin of Timor (Fig. 55). Carter *et al.* (1976) regard the superstructure of Timor as (1) an allochthon of five thrust sheets generated by the early Pliocene at the leading edge of the Eurasia Plate and emplaced in the mid-Pliocene on (2) a para-autochthonous and autochthonous Permian to Palaeocene 'Australian' facies above (3) the deep continental crust of the northern margin. Subduction ceased in the Pliocene, as shown by the extinction of volcanism in that part of the inner arc opposite Timor. The subcrop of the steeply dipping plate boundary is along the north coast of Timor but, strictly speaking, the southern edge of the Eurasian Plate extends to the outcrop of thrust 1. Before being emplaced in the mid-Pliocene, the allochthon was overlain by an olistostrome in turn overlain by an early Pliocene bathyal lutite. Subsequent uplift after the emplacement of the allochthon caused the olistostrome to slump southward across southern Timor to the axis of the Timor Trough, so that material originally deposited on the

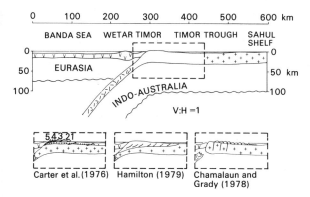

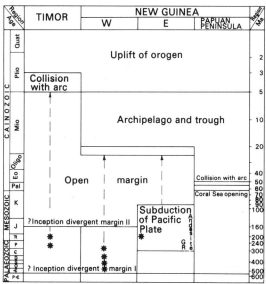

Fig. 55A. Cross-section of the northern margin at Timor (above), adapted from Hamilton (1979), and three structural models of Timor (below). The model of Carter *et al.* (1976) shows an allochthon of five thrust sheets that rests on a Permian and Mesozoic autochthon or para-autochthon that passes beneath the Timor Trough into coeval autochthonous sediment. The model of Hamilton (1979) has Timor as entirely allochthonous above deep basement. And Chamalaun and Grady's (1978) model shows a detached downgoing slab of the Indo-Australia oceanic plate and a rebounded north coast of Timor with allochthonous sediment, originally deposited in the arc-trench gap, shed southward to the Timor Trough.

Fig. 55B. Phanerozoic events along the northern margin, arranged on a logarithmic time-scale. Asterisks represent the oldest recorded marine fossils. GR=granite.

Eurasian Plate has now reached the axis of the Timor Trough.

According to Hamilton (1979), Timor, and the rest of the Outer Bands Arc, 'represents the ramping onto continental crust of material pushed in front of an advancing island arc. The subcrop of the primary plate boundary, beneath the surficial debris wedge, is approximately along the inner edge of the outer-arc ridge.' But, as in Carter *et al.*'s (1976) model, the plate boundary extends, in a strict sense, to the southern edge of the wedge in the Timor Trough. This interpretation differs from that of Carter *et al.* (1976) only in the thickness and extent of the allochthonous material, so that the entire rock record exposed on Timor is part of the Asian accretionary prism.

In company with other authors, Chamalaun and Grady (1978) envisage subduction of the Indo-Australian Plate beneath the Eurasian Plate and the accumulation of arc-trench gap sediments above the plate boundary. With the arrival and initial downwarp of the northern continental margin of Australia at the trench in the late Miocene, the buoyancy of the continental lithosphere became equal to the pull on the subducted oceanic slab resulting in the detachment of the slab. Released from its sinker, the northern margin rose rapidly along steeply dipping faults to expose metamorphic basement on the north coast. The sediment originally deposited at the arc-trench gap slid under gravity to the Timor Trough. This is the only allochthonous material on Timor. Chamalaun and Grady (1978) support this interpretation by evidence from a palaeomagnetic reconnaissance (Chamalaun 1977a, b) that the directions of magnetization of two Permian formations, one recognized by Carter *et al.* (1976) as autochthonous, the other as allochthonous, are indistinguishable, and moreover that they agree with that of the Permian of mainland Australia, suggesting no palaeomagnetically detectable change in the relative position of Australia and Timor since the Permian. Chamalaun and Grady (1978) comment on the apparent discrepancy of the warm-water fauna found in the various Permian limestones of Timor in the context of a coeval glaciation in mainland Australia, explained (Carter *et al.* 1976) by requiring the limestones to have originated along the distant tropical shelf of Tethys. As shown by the brachiopod faunas, most of the limestones are Late Permian (Grant 1976), and post-date the last evidence, in the Kazanian, of glaciation in Australia. Moreover, Permian limestone in the basal 50 m of Sahul Shoals-1, on the indubitably autochthonous terrain at the edge of the Sahul Shelf, only 150 km from the south coast of Timor (Fig. 106), contains brachiopod and bryozoan fossils of Late Permian age with close affinities to those of the Basleo Beds of Timor (Campbell 1970). Likewise the Cretaceous–Palaeocene bathyal manganiferous claystone of southern Timor is compatible with coeval deep-water (slope) deposits of the North-west Shelf (Powell 1976b; Apthorpe 1979).

The final word must go to Brunnschweiler (1978), who compared Timor with the Swiss Alps:

It took a century and a half and the work of hundreds of geologists to unravel the story of the Alpine Orogene, and even today some of its chapters remain hotly disputed. Timor may well be an even more difficult case, not only on account of its remoteness, but also because one suspects that what one sees is merely the top part of a very big pile of nappes which has just begun to emerge from the sea.

(b) Summary history

These connections suggest that Timor lay along the north-western (Tethyan) margin of Australia (Fig. 114) before a sliver of continental lithosphere was rifted off to the north-west in the phase of plate divergence that started in the Middle/Late Jurassic. Timor may then have been part of the continental rim, which later subsided to form the bathyal part of the north-western margin. The subsequent events described above relate to the later change from divergent to convergent motion along the north-western margin.

The earlier history of New Guinea is also consistent with its lying along the southern margin of Tethys but the record of Permian emplacement of granite and of Late Triassic to Cretaceous andesitic volcanism in eastern New Guinea indicates its connection with eastern Australia as part of the convergent Pacific margin, as described below in our account of the Eastern Highlands.

The chronology of events of the northern margin (Fig. 55B) starts with a postulated inception of Australia's north-western margin and possibly also the adjacent northern margin by the generation of the Tethyan Ocean, reflected in the first appearance of marine facies in north-western Australia during the Cambrian, as described later (Fig. 111). No evidence of this is found in Timor or eastern New Guinea until the Permian and Triassic, but in western New Guinea the oldest known marine fossils are Early to Middle Ordovician (470 Ma) cephalopods from the southern slope of the Central Range near the eastern border of Irian Jaya, reworked into a Callovian mudstone, probably from the exposed Kariem Formation, which itself has not yielded any fossils (Kobayashi and Burton 1971). Silurian, Devonian, and Permo-Carboniferous marine fossils are known from pebbles and boulders in rivers that drain the southern flank of the Central Range (Visser and Hermes 1962). Autochthonous Silurian graptolites are found in the Vogelkop (Visser and Hermes 1962), whose palinspastic position, however, is uncertain. Nevertheless, all these tantalizing occurrences suggest that the continental platform of western New Guinea was frequently covered by the sea at least from the Ordovician, and that, in common with north-western Australia, the sea may have been part of the Tethyan Ocean. The oldest known marine fossils in Timor are Permian. As mentioned already, they have warm-water Tethyan affinities; Carter *et al.* (1976) ascribed them to the warm northern margin of Tethys, and emplaced on Timor during the Pliocene collision. But, as Chamalaun and Grady (1978) point out, they are also relatable to the warm-water marine fauna of north-western Australia through the occurrence in Sahul Shoals-1, only 150 km from the coast of Timor, of a Permian limestone with brachiopods and bryozoans. The oldest marine fossils recorded from the eastern New Guinea platform (Kuta Formation) are Late Triassic (205 Ma) (Skwarko *et al.* 1976). These first occurrences of marine fossils (Fig. 55B) are the earliest indication of the pericontinental sea, but the possibility of the Cambrian (575 Ma) inception of the northern margin as a divergent margin to the Tethys/Pacific Ocean, is also indicated on the figure.

The next events are the intrusion and cooling of the Kubor Granodiorite about 240 Ma ago (Page 1976), or older granite to the south about 300 Ma ago, and the Late Triassic to mid-Cretaceous (215–90 Ma) volcanic activity and volcanogenic deposition on either side of the Kubor Massif (Fig. 70), all interpreted as indicating southward to westward subduction of the Pacific Plate beneath eastern New Guinea.

A second phase of seafloor spreading off north-western Australia at the end of the Middle Jurassic (160 Ma) (Fig. 14), this time by the Indian Ocean opening near the old margin, possibly extended north-eastward through Timor and western New Guinea to form a renewed divergent margin, but did not extend to eastern New Guinea, which remained a convergent margin until the mid-Cretaceous (90 Ma ago).

In the modern era, the Papuan Peninsula was rotated away from north-eastern Australia by the opening of the Coral Sea Basin (63–57 Ma) (Fig. 21), and then collided with a volcanic arc in the early Eocene (55–50 Ma). The Australian salient in New Guinea collided with a volcanic arc by consumption of the north-dipping oceanic part of the Indo-Australia

Plate in the early Miocene (25–20 Ma), and the western shoulder in Timor in the early Pliocene (5–3 Ma). Finally, the New Guinea Highlands started to emerge at the beginning of the Pliocene, 5 Ma ago, and Timor in the mid-Pliocene, 3 Ma ago.

3. MORPHOTECTONICS OF THE PLATFORM REGIONS, FOCUSED ON THE HIGHLANDS

(a) Eastern Highlands (by J. G. Jones and J. J. Veevers)

Recent stratigraphical analysis of the Cainozoic basins that flank the Eastern Highlands and of the Highlands basalts (Grimes 1980; Jones and Veevers 1982), and an extension of this analysis back into the Mesozoic (Jones and Veevers 1983) provide the basis for this account of the history of the Eastern Highlands.

(i) Cainozoic history of the South-east Highlands

(1) Summary

We have found that the Cainozoic basalt chronology of the South-east Highlands of Australia correlates with depositional cycles in the flanking basins, in particular the Murray and Gippsland Basins, and gives rise to the hypothesis that periods of more intense volcanism correspond with uplift of the Highlands and concomitant subsidence of the flanking basins. Conversely, periods of less intense volcanism correspond with settling of the Highlands and concomitant rising of the flanking basins. In terms of current extent, the South-east Highlands came into existence between 60 and 50 Ma ago with the inception and rapid alluviation of the Murray Basin. Within the Cainozoic, the Highlands have undergone at least three episodes of uplift with concurrent basalt volcanism, linked with episodes of marine transgression followed by regression in the flanking basins.

(2) Introduction

Until the last decade, reconstruction of the history of the South-east Highlands was severely hampered by an almost total lack of age constraints from the Highlands themselves, and by a very incomplete knowledge of the stratigraphy of the flanking basins. For the flanking basins, that deficiency has been remedied largely as an outcome of petroleum- and groundwater-exploration, and an intensive programme of radiometric dating of basalt flow remnants by Wellman and McDougall (1974a, b) has provided a wealth of dated events in the Highlands.

The focus of this account is the correlation we perceive between the basalt chronology of the South-east Highlands and the cycles of marine transgression/regression in the flanking basins. Although current eustatic models may adequately account for the basinal cycles, they can provide no clue to their correlation with basaltic volcanism in the uplands, nor to the evident fluctuations in sediment supply from those uplands to the flanking basins. The tectonic scheme we propose encompasses all three. We do not reject eustatic explanations for the stratigraphy of the flanking basins, believing that our evidence supports a synchronism of local tectonism and global eustasy, a theme which is beyond the scope of this analysis. Nor do we reject climatic influences on sediment supply. We do believe, however, that our tectonic scheme is a simple and sufficient explanation for the available information.

(3) Gross morphology and current tectonic activity of South-east Australia

In plan, the morphology of South-east Australia is dominated by two entities: the South-east Highlands and the Murray–Darling Basin (Fig. 56A). The dominantly erosional domain of the South-east Highlands occupies eastern New South Wales and southern Victoria, and the dominantly depositional domain of the Murray–Darling Basin occupies western New South Wales, northern Victoria, and south-eastern South Australia. The topographic contours of the Highlands define a crest that runs north-north-east to south-south-west parallel to the continental margin in New South Wales, and east to west in Victoria, roughly paralleled by the north coast of Bass Strait.

The Highlands culminate at around 2000 m, at the elbow between the New South Wales and Victorian segments. North-east of this elbow they extend with notably fluctuating altitude beyond the Queensland border; westwards they decline to vanishing point at Victoria's western border. By contrast, the Murray–Darling plain lies almost wholly below 200 m (Fig. 56A).

At present, sediments from the South-east Highlands are delivered westwards and northwards into the riverine plains of the Murray–Darling Basin, and eastwards and southwards into coastal estuaries flooded by the Holocene transgression.

In terms of seismicity the south-east is one of the most active regions in mainland Australia (Fig. 49B). The seismic activity occurs largely in the central and southern Highlands of New South Wales, though in Victoria it expands beyond the Highlands' perimeter, northwards into the Riverina and southwards into Bass Strait. Evidence of Quaternary ground movements is reported in both the Highlands and the flanking basins

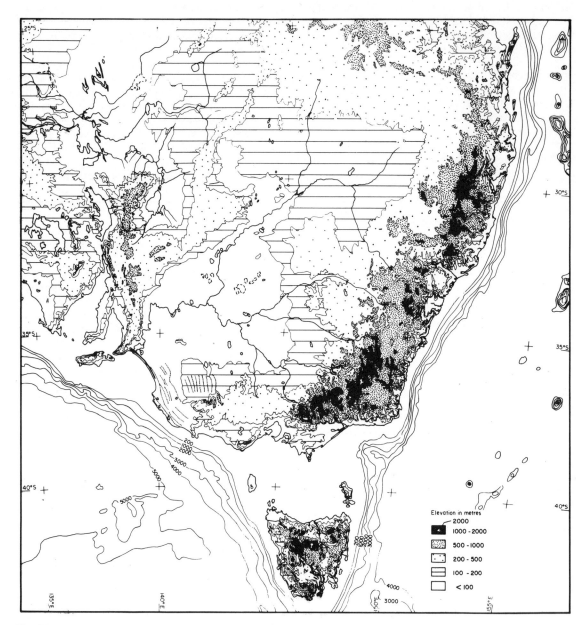

Fig. 56A. Morphology of South-east Australia, showing the South-east Highlands giving way on the west to the Murray–Darling Basin. Land contours from Times Atlas of the World (1975), submarine contours from BMR Earth Science Atlas (1979). Stranded beach ridges (dotted lines) about lat. 37° S, long. 140° E from Hills (1975, p. 280) and Idnurm and Cook (1980). Figs. 56 to 61 from Jones and Veevers (1982).

(Lawrence et al. 1976), though measures of displacement are rare, except on fossil strandlines. Warped and tilted Quaternary strandlines rise to 40 m above current sea-level in southern Victoria (Lawrence et al. 1976), and to as much as 100 m in South Australia just beyond the western end of the Highlands (Idnurm and Cook 1980). The western end of the Highlands in western Victoria has been the locus of widespread basaltic volcanism over the last 5 Ma (Joyce 1975), and this region can be considered currently active, though feebly so, the youngest dated volcanism being about 1500 years ago (Lawrence et al. 1976).

(4) Basalts of the South-east Highlands

The basalts of western Victoria are the most recent of a scatter of Cainozoic basalt flows along the length of the South-east Highlands (Fig. 56B). Within the present topography, these basalts occupy a range of situations from valley floor to ridge crest, but at their time of emplacement most appear to have filled valleys in a landscape of significant relief. Thus, in and around the Snowy River Valley, 90 km south of Kosciusko, basalt flows about 40 Ma old (late Eocene;

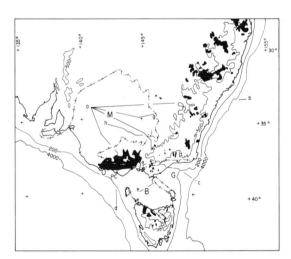

Fig. 56B. Distribution of Late Cretaceous and Cainozoic volcanic rocks, and Cainozoic sedimentary basins of South-east Australia. Distribution of volcanics from Wellman and McDougall (1974a), Wellman (1974), McDougall and Wellman (1976), and Sutherland (1978). Outlines of Murray (M), Gippsland (G), Bass (B), and Otway (O) sedimentary basins from Geological Society of Australia (1971). Morphology indicated by 500 m land contour and 200 m and 4000 m submarine contours. Lines ab, ac, ad, ag, ef, and hi show location of cross-sections of Figs. 59A and 61. PC=Port Campbell. Renmark is situated at a.

Wellman 1974) occur over an altitudinal range of 600 m (Hills 1938, 1975). If post-basalt warping and faulting are locally negligible, as it appears, it can be inferred that the local landscape 40 Ma ago had at least 600 m of relief. On the southern slopes of Barrington Tops, a southerly outlier of the Northern Tablelands, basalts about 50 Ma old (mid-Eocene; Wellman and McDougall 1974a) outcrop from 900 m to the plateau surface at 1500 m (Galloway 1967) where they skirt outcrops of Palaeozoic rocks, indicating a local relief of at least 600 m 50 Ma ago.

The Cainozoic basalts of the South-east Highlands have been the subject of an intensive programme of radiometric dating by Wellman and McDougall (1974a), and their results for New South Wales and Victoria, together with a few more recent dates (including those from Tasmania), are summarized in Fig. 57. Dating shows the basalts to span the Cainozoic. The primary pattern is one of increasing numbers of dates towards the present, which we attribute to the increasing probability of preservation with decreasing age. Superimposed on this trend are secondary maxima and minima, notably a gap in the distribution 7-9 Ma ago inclusive (late Miocene) and a paucity of dates around 28 Ma ago (late Oligocene). These secondary variations might simply reflect uneven sampling, but our analysis of the stratigraphy of the Murray Basin suggests a quite different significance.

(5) Sediments of the Murray Basin

The Murray Basin is an almost circular basin draining the Victorian Highlands in the south and the New South Wales Highlands in the east (Figs. 56A, B, 58). Rivers from these sources have constructed a broad crescent of alluvium around the eastern perimeter of the basin – the riverine plain or Riverina – gently declining westwards to the flat Mallee Plain of sand-dunes and scattered lakes.

The Mallee is less than 100 m above present sea-level and its flat expanse is a direct inheritance of the Murray Basin's most recent inundation by the sea. That incursion broke across the low south-west sill of the basin about 5 Ma ago, to be met by a waxing influx of silt, sand, and gravel to the riverine plain from the Highlands' streams (Fig. 58).

This Pliocene–Quaternary cycle of transgression/regression is the most recent and best substantiated of several such cycles within the Murray Basin. The preceding cycle spans the late Oligocene and Miocene, and a still older cycle of uncertain but comparable duration spans the mid-Eocene to late Oligocene (Fig. 57).

We are struck by the degree of correspondence

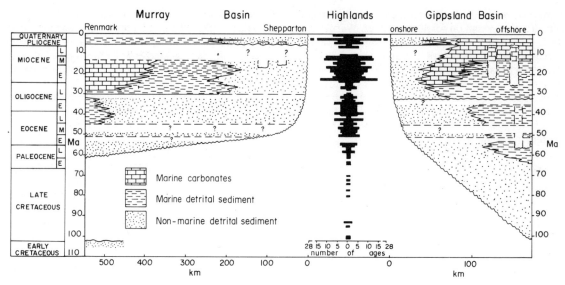

Fig. 57. Time–space diagram of the Murray and Gippsland Basins, and histogram of the 290 published ages of volcanic rock, dominantly basalts, of New South Wales, Victoria, and Tasmania. The histogram is two-sided, with values reflected on either side of zero, and comprises K–Ar ages grouped in 1 Ma classes, with a mode of 28 from 2 to 3 Ma ago. Ages expressed in new IUGS constants (Dalrymple, 1979) from determinations given by Wellman (1974) and Wellman and McDougall (1974a), for which the average precision is about 2 per cent (2 standard deviations), supplemented by Abele and Page (1974), Aziz-ur-Rahman and McDougall (1972), Carr and Facer (1980), Cooper et al. (1963), Cromer (1980), Dulhunty (1971, 1972, 1973), Dulhunty and McDougall (1966), Dury et al. (1969), Facer and Carr (1979), Harding (1969), Leaman (1976), McClenaghan et al. (1981), McDougall et al. (1966), McDougall and Gill (1975), Mc Dougall and Leggo (1965), Mc Dougall and Wellman (1976), Mc Dougall and Wilkinson (1967), Singleton et al. (1976), Stipp and McDougall (1968), Sutherland et al. (1973), Wellman et al. (1969), Wellman et al. (1970), and Young and Bishop (1980). Time–space diagrams made from information in Abele et al. (1976), James and Evans (1971), Lawrence (1975), Lindsay and Bonnett (1973), Macumber (1978), McGowran (1977, 1978, 1979), Partridge (1976), and Woolley (1978). In the Gippsland Basin all the non-marine detrital sediment, except in the Quaternary–Pliocene, belongs to the Latrobe Group; the marine detrital sediment includes the late Oligocene to middle Miocene Lakes Entrance Formation, and the marine carbonates include the Gippsland Limestone. Details of interfingering boundaries are well constrained in the Murray Basin, less well-constrained in the Gippsland Basin. Channels are shown by wavy lines. In both basins, the interval 0 to 2 Ma ago is shown as a lacuna; in the Gippsland Basin the late Pliocene Haunted Hill Gravels are now incised by streams along the piedmont downs, and much of the offshore surface sediment we believe to be transient; likewise, we believe that the surface reworked sediment of the Murray Basin has a low potential for preservation. Note the variable scale of kilometres.

between the peaks and saddles of the Highlands' basalt-age histogram and the transgression/regression/lacuna signature of the Murray Basin (Fig. 57), and this leads us to review the signatures of basins along the southern flanks of the Highlands.

(6) Sedimentation signatures of the Gippsland, Bass, and Otway Basins

The thick Cainozoic fill of the Gippsland, Bass, and Otway Basins contrasts with the thin fill of the Murray Basin and the New South Wales continental margin (Fig. 59A). Thanks to a uniform biostratigraphical zonation, summarized by Abele et al. (1976, table 8.1), fairly precise correlation of the basins is possible.

We recognize three (and possibly four) Cainozoic cycles of transgression/regression/lacuna in the basins that flank the South-east Highlands, and, since the cycles follow earlier ones, denote them by the letters W to Z (Fig. 59B). They correspond to cycles of wider extent called sequences 1 to 4 by McGowran (1979) and cycles 1 to 4 by Quilty (1977), and X and Y to tectono-eustatic intervals of transgressive shorelines related by Pitman (1978) to the rate of sea-level changes due to fluctuating seafloor-spreading rates. Z and Y correlate with peaks in the basalt-age histogram, and X probably correlates with a poorly defined peak. The few dates older than 50 Ma rule out a correlation with W, beyond the fact that the initial concentration of

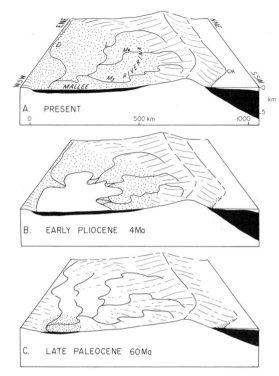

Fig. 58. Schematic block diagrams, looking northeastward, of South-east Australia, with the South-east Highlands flanked by the Gippsland and Murray Basins. Position of the late Palaeocene shoreline north of Cape Howe not known. D, Darling; L, Lachlan; Me, Murrumbidgee; and My, Murray Rivers; CH, Cape Howe.

ages at about 65 Ma corresponds with the onset of cycle W. Cycle Z is not yet complete. We attribute the mismatches in correlation of sedimentation signatures and the basalt histogram mainly to errors in calibration of the radiometric and biostratigraphic time-scales, and to inadequate sampling.

(7) Cainozoic tectonism in South-east Australia: a model

The stratigraphy of the Cainozoic sedimentary basins of South-east Australia displays recurrent cycles of transgression followed by regression. These cycles in turn appear to be linked to phases of more, then less, intense volcanic activity in the Highlands. We interpret these relationships in terms of recurrent cycles of activity in South-east Australia in the Cainozoic. After a state of rest (Fig. 60A), the cycle commences with uplift of the highlands and concomitant subsidence of the lowlands (Fig. 60B). Transgression is an instantaneous response where the sea has access to a lowland; but prograding sediment from rejuvenated highlands may eventually turn transgression into regression (Fig. 60C, D). Volcanism is seen to be concurrent with this phase of uplift/subsidence. Transition to a passive state is marked by a phase of 'recovery', in which highlands settle and lowlands rise, with a concomitant reduction in gross relief, together with a decline in volcanic activity (Fig. 60F). This phase is marked by valley alluviation or rapidly declining erosion in the highlands, and erosion or much reduced clastic inflow within the hitherto depositional realm of the lowlands.

(8) Late Cainozoic tectonic cycles of South-east Australia

(a) Pliocene–Quaternary cycle. Of the Cainozoic tectonic cycles that we believe have shaped South-east Australia, the most recent is that of the last 5 Ma (Pliocene–Quaternary). In terms of our model, its history, as expressed in the stratigraphy of the Murray Basin, may be read as follows (Figs. 57, 58, 60, 61). After a late Miocene interlude of passivity or quiescence, uplift of the South-east Highlands resumed with concomitant subsidence of the Murray Basin. The Basin, which may have experienced general subaerial exposure in the late Miocene (Mologa Surface of Macumber 1978; Williams and Drury 1980) as it does today, was perhaps almost instantaneously inundated to the limits of the modern Mallee (Bookpurnong Beds of Fig. 61, section a–g). Here the transgression was halted and reversed by a fast-expanding, piedmont plain of coarse sand and gravel (Calivil Sand of Fig. 61) from the freshly uplifted Highlands, which rapidly overtopped the entrenched channels of the degraded Mologa piedmont. Basalt from Highlands vents in central and western Victoria spilled down rejuvenated valleys to interleave with the fresh flux of quartz gravel and sand (Macumber 1978; Hills 1975, p. 311). Later in the Pliocene, a second phase of uplift/subsidence caused transgressive reworking of the Calivil Sand to form the lower Parilla Sand (Fig. 61; Macumber 1978). Subsequent regression, which we take to express shallowing of the Basin in the recovery phase of the cycle (Fig. 60F), is recorded in the sequence of Quaternary beach ridges in upper Parilla Sand; these extend from the middle of the Murray Basin to the present Coorong coast of South Australia (Fig. 55) and decline in height from 100 m to sea level (Idnurm and Cook 1980). Kalf and Woolley (1977) and Woolley (1978, 1980) attributed a change from Pliocene white clay and quartz sand and gravel to Pleistocene variegated clay and thin polymictic sand and gravel in New South Wales to decreasing rainfall. We regard such a change

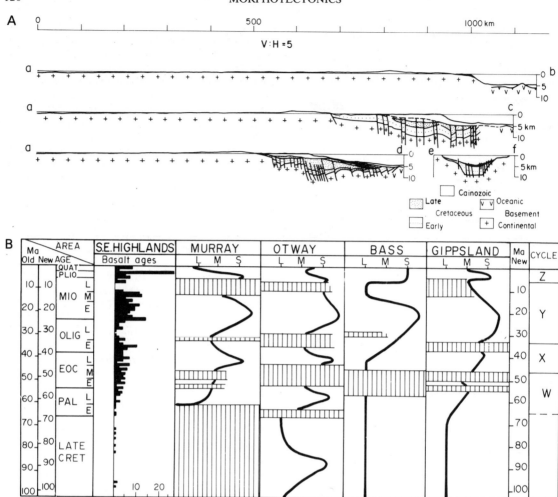

Fig. 59. A. Cross-sections, located in Figure 56B; a–b, a–c, and a–d from Veevers (1982a), e–f from Brown (1976) and Robinson (1974). B. Time–facies diagrams of the Gippsland, Bass, Otway, and Murray Basins and histogram of ages of volcanic rocks, from Fig. 57. K–Ar time-scale given with the new IUGS constants (both sides) and with old constants on far left only. Depositional environments range from L (land, non-marine, sub-aerial) through M (mixed non-marine and shallow marine) to S (shallow marine). Lacunae designated by vertical lines. The columns represent the offshore Gippsland Basin, the south-east part of the Bass Basin, the Port Campbell area of the Otway Basin, and the Renmark area of the Murray Basin. Sources for Murray and Gippsland Basins as in Fig. 57; for Bass: Brown (1976), Robinson (1974), and Quilty (1972, 1980a); and for Otway: Abele *et al.* (1976) and McGowran (1977, 1978, 1979). Transgression/regression/lacuna cycles W to Z, corresponding to McGowran's (1979) sequences 1 to 4 and Quilty's (1977) sedimentation cycles 1 to 4, are shown on the right-hand side.

as secondary to the primary decrease in grain size from the Calivil Sand to the Shepparton Formation, which, with Woolley (1978), we attribute to waning uplift.

A very similar story can be read in sediments spanning the last 5 Ma in the Gippsland Basin. Here a late Miocene interlude of low clastic input, characterized along the coast by glauconitic silt and mud and phosphate nodules (Carter 1978, 1980) and by limestone offshore, was terminated by renewed clastic influx (transgressive/regressive sequence of Jemmy's Point Formation and Wurruk Sand Member: Abele *et al.* 1976; Carter 1980) presumably expressing uplift of the Highlands. A brief mid-Pliocene relapse, as in the Murray Basin, was followed by a new pulse of uplift/subsidence and a further influx of silt, sand, and gravel (Haunted Hill Gravels; correlative of Shepparton

is an unconformity clearly evident in seismic profiles between 32°30′ S and 35° S (Davies 1975, 1979) 'in the same depth range as the early to middle Pliocene unconformity' of the Gippsland Basin.

We have taken the Miocene–Pliocene boundary (5 Ma age) as a mean for onset of this most recent first-order tectonic cycle in South-east Australia, as expressed in the sedimentary record of the Murray and Gippsland Basins. For the Highlands, currently available basalt dates (Fig. 57) indicate a resumption of volcanism at 7 Ma ago after an interval of dormancy from 10–7 Ma ago, 2 Ma (at face value) in advance of resumption of sedimentation in the flanking basins. The peak in the basalt histogram at 2 to 3 Ma ago may correspond with the late Pliocene pulse of the cycle described above.

(b) Late Oligocene–Miocene cycle. In accordance with our model, we interpret the cluster of basalt dates in the range 28–10 Ma ago (Fig. 57), followed by the 10–7 Ma ago lacuna, as a manifestation of the next preceding first-order cycle of uplift, followed by settling of the Highlands (though it was probably composed of several secondary or sub-cycles). In the Murray Basin this phase witnessed a dramatic incursion of the sea (Ettrick Marl and Geera Clay of Fig. 61) at around 30 Ma ago, far more extensive than any previous. From mid-Oligocene to mid-Miocene, this incursion was held close to the limits of the Pliocene transgression by upbuilding sands and silts from the resurgent Highlands, though a dwindling clastic input was reflected farther from the shore by increasingly calcareous sediments (Winnambool Formation and Duddo Limestone of Fig. 61). Some time in the mid- to late Miocene, the recovery phase set in, involving cessation of sedimentation in the west of the basin, erosional loss in the east with entrenchment of the streams, weathering of the interfluves of the piedmont fans (Mologa Surface of Macumber 1978), and cessation of basaltic volcanism in the Highlands.

Again, a very similar story can be read for the same time span in the Gippsland Basin. An early Oligocene interval of negligible clastic input, and perhaps of net loss, in the offshore Basin (Gurnard Formation: Abele *et al.* 1976) was followed by a notable transgression (base of Lakes Entrance Formation) some time between 35 and 30 Ma ago, the onset of the cycle. Declining clastic influx following the initial pulse of deposition (Childers Formation) is reflected, as in the Murray Basin, by increasingly calcareous sediments (transition from Lakes Entrance Formation to Gippsland Limestone), a widespread latest Miocene phosphatic nodule bed (Carter 1978, 1980) recording the

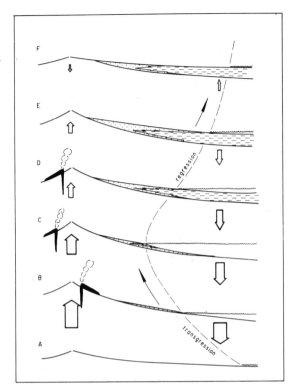

Fig. 60. Model linking tectonic activity of the highlands, expressed by volcanism and uplift, and the transgression/regression/lacuna cycle in the flanking sedimentary basin. Sediment symbols as for Fig. 57. Arrows depict direction and rate of ground movement. Surface relief due to ground movement is reduced by erosion of the highlands and deposition in the lowlands.

Formation of Murray Basin in Fig. 61: Abele *et al.* 1976; Bolger 1980). A Quaternary phase of recovery has resulted in entrenchment of modern streams in the Pliocene piedmont alluvium (piedmont downs of Talent 1969; Jenkin 1976) and elevation of strandline terraces and beach ridges (Jenkin 1968, 1976), as in the south-west Murray Basin.

If the Pliocene–Quaternary sedimentation signatures of the Murray and Gippsland Basins are a record of a first-order cycle of uplift followed by settling of the South-east Highlands, one might expect similar signatures along the length of the footslopes. The depositional domain on the eastern footslopes is almost wholly submerged on the New South Wales continental shelf, and its stratigraphy is known only in seismic profiles unconstrained by drilling. The most conspicuous internal element of the shelf stratigraphy

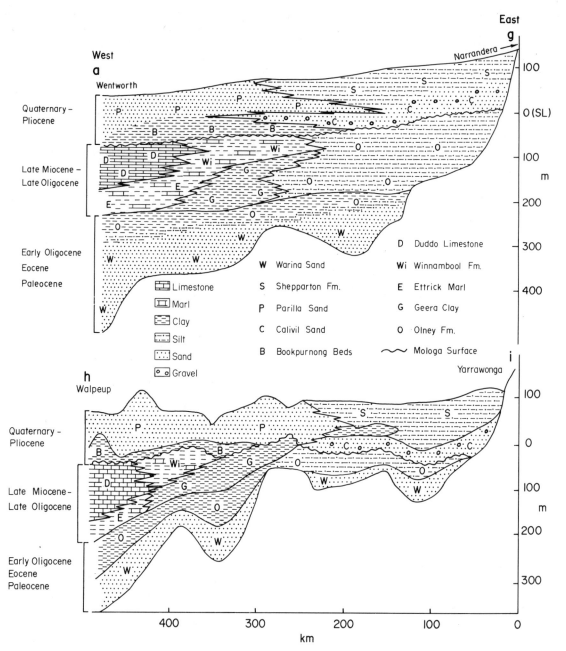

Fig. 61. Cross-sections of the Murray Basin, located in Fig. 56B. a–g: Wentworth–Narrandera, modified from Woolley (1978, fig. 2). h–i: Walpeup–Yarrawonga, from Lawrence (1975, fig. 14). Miocene–Pliocene boundary indicated by Mologa Surface; early–late Oligocene boundary indicated by heavy, near-horizontal line in a–g only.

low point of supply. Late Miocene recovery is expressed by a break in deposition onshore (Mallett 1978), and by channelling of earlier Miocene sediments, both onshore (Carter 1980) and offshore (Threlfall *et al.* 1976), presumably at the same time as the incision and weathering of the Mologa Surface of the Murray Basin.

(9) Inception of the Highlands

Forty to fifty Ma ago, local relief in the South-east Highlands was greater than 500 m, judging from the varying elevation of basalts of the same age. Moreover, the present valleys of inland streams from the Highlands are traceable for up to 100 km beneath the Murray Basin fill, where they contain sediments as old as Eocene (Macumber 1978; Woolley 1978). This evidence suggests that 40 or 50 Ma ago the Highlands were more extensive, and if anything higher, than they are now.

The gross configuration and sediment-movement pattern of the Murray Basin, as recorded in its sedimentary fill and basement morphology (Macumber 1978), can also be read back to 50 Ma ago. Beyond this time, the extent of the basin diminishes rapidly to vanishing point at about 60 Ma ago (late Palaeocene; Fig. 57). On the Gippsland flank of the Highlands the present pattern of sedimentation goes back to 100 Ma ago, that continuity being expressed in the Latrobe Group (fig. 57).

In the Late Cretaceous the region occupied by the Murray Basin in the Cainozoic was an erosional domain, and in this respect the Late Cretaceous highlands of South-east Australia covered at least twice the area of the Cainozoic highlands. Thus, if extent is thought critical to their definition, the South-east Highlands came into existence between 60 and 50 Ma ago with the inception and rapid alluviation of the Murray Basin (Figs. 57, 58).

(10) Outline of a Cainozoic history of the Highlands

We envisage that at the beginning of the Cainozoic the presently onshore area of South-east Australia was an essentially erosional domain culminating in a highland crest close to, if not coincident with, the crest of the present South-east Highlands. Early in the Palaeocene this extensive upland began to contract rapidly, as alluvial sediment lapped up its north-west flank to initiate the Murray Basin.

We suggest that within the Cainozoic the Highlands have experienced three or more major episodes of uplift followed by settling, with concurrent though erratically distributed basaltic volcanism, linked to episodes of transgression followed by regression in the flanking basins. Of these cycles of renewal, the most clearly defined are the most recent ones; one spanning the late Oligocene and Miocene, and one the Pliocene–Quaternary.

(11) Discussion

The morphotectonic cycle on which we build our interpretation of the history of Australia's South-east Highlands is a development of King's (1967) 'cymatogenic' cycle; our 'recovery' phase and links to volcanism are significant additions. Our cyclic chronology closely resembles that of King (1967, table IX), though based substantially on information unavailable to him. King's morphotectonic cycle linked unconformities in flanking basins with erosion surfaces in the uplands (King, 1967, fig. 78), and in the South-east Highlands of Australia he saw a landscape compounded of three such Cainozoic cyclic surfaces (King 1967, figs. 134, 135).

Interpretations of the history of the South-east Highlands in terms of cyclic denudation go back to the turn of the century, and prior to King were largely couched within the framework of W. M. Davis's Geographical Cycle (Young 1978). Their chronologies, in the absence of radiometric and stratigraphic constraints, were either imported or invented (Young 1978). Such poorly constrained denudation chronologies were bound to be vulnerable to a programme of extensive radiometric dating of Highlands basalts, and since the initial publications of Wellman (1974) and Wellman and McDougall (1974a) there has been a strong impulse to review the Highlands' history afresh (Young 1978; Francis *et al.* 1979; Wellman 1979a).

Nowhere are the conventional chronologies more vulnerable than for that portion of the Highlands within the limits of the Sydney Basin, where near-horizontal, relatively resistant sheet sandstones of Permian and Triassic age form a series of base-levels for pediplanation. From his review of denudation chronologies for this region, Young (1978) concluded that 'a cyclic origin for the major features of the region can no longer be taken for granted' – a conclusion we extend to the South-east Highlands generally. However, we do not advocate the abandonment of cyclic concepts, but insist that their value be judged by their capacity to incorporate the evidence.

A review of latest Cainozoic volcanism and tectonics in the Victorian Highlands led Joyce (1975) to conclude that 'the cycle of stability followed by uplift and then volcanism which took place in southeastern Australia between the Pliocene and the Holocene was probably only the most recent of a number of similar cycles which followed the formation of Australia as a separate continent'. Quilty (1977, 1980b) recognized sedimentation cycles in the Cainozoic of Western Australia which he suggests are continent-wide and which for the late Cainozoic correspond closely with those recognized by us (Quilty, 1980b, figs. 3, 7), as do the Cainozoic 'sequences' of McGowran (1979, esp. fig. 2), framed on the basis of a review of the Tertiary foraminiferal biostratigraphy of the Australian continent and environs. Ollier (1979), after stating that 'Cyclical theories ... do not fit the Australian scene very well', outlined an essentially similar Australia-wide chronology of marine transgression/regression, concluding that 'All these changes of sea level should be conducive to rejuvenation and formation of new cycles of erosion on the Davisian scheme, and to some extent some sort of cyclical landscape development did occur in the Tertiary.' Quilty's (1980b) and McGowran's (1978) cycles are probably global, as are the basinal sea-level changes we describe, but the exploration of this possibility is beyond the scope of this account. The important point is that the sea-level rises correlate with uplift of South-eastern Australia, and not with a static or subsiding land as expected from present sea-level theory.

That tectonic cycles of the kind we propose should leave their imprint not only in the sediments of the flank basins but also in the landscape of the Highlands seems inescapable, and must provide a crucial test of the hypothesis. But there is no reason to suppose that surface lowering in the Highlands during the Cainozoic ever advanced substantially towards the end-point of a general peneplain or pediplain, as earlier writers assumed. The surfaces to be reconstructed within the Highlands will not be plains, except in places within the perimeter of the present foothills. Their reconstruction must involve the upslope tracing, in large part by extrapolation, of the unconformities of the flank basins, which themselves represent surfaces of long exposure (commonly characterized by duricrusts) within a plains or shelf environment.

(ii) Cainozoic history of the North-east Highlands

In a review of the Cainozoic geology of the northern (tropical) part of Queensland, Grimes (1980) proposed a cyclic scheme of landscape evolution very similar to our own, both in concept and chronology. For Queensland in the Cainozoic, Grimes recognized 'three major cycles of geological activity ... Each cycle commenced with a period of epeirogenic movement which uplifted areas to become erosional sources of sediment, and downwarped basins to receive these sediments. There was accompanying volcanism in some areas. The erosion and deposition continued into the subsequent phase of tectonic stability until the relief was reduced to a planation surface which was then deeply weathered. The available data suggest that, with some exceptions, each of the cycles commenced and developed synchronously throughout all the regions.' As shown in Fig. 64, his first cycle corresponds to our cycles W and X, his second cycle to Y, and his third cycle to Z.

The regions recognized by Grimes (1980) are shown in Fig. 62, and the Cainozoic basins, volcanics, and structure in Fig. 63. The cycles, condensed from Grimes (1980), are compared with a histogram of volcanic ages in Fig. 64, which has the same format as Fig. 59B above to facilitate direct comparison. With only half as many radiometric ages as are available from the South-east Highlands, and without the refined stratigraphical dating of the South-east Australian sediments afforded by the search for petroleum and ground-water, the North-east

Fig. 62. Location of the Cainozoic regions of North-east Australia (from Grimes 1980) shown, except the Burnett and Brisbane regions, in the correlation columns of Fig. 64. Circled dots show the exploration wells at Anchor Cay (AC), Aquarius (A), and Capricorn (C).

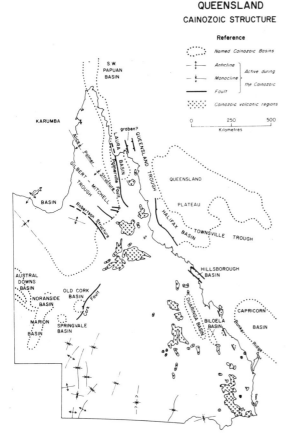

Fig. 63. Cainozoic basins, volcanics, and structure (from Grimes 1980).

Highlands information is less reliable than that from the south-east. The dearth of information offshore, from only three exploration wells in two widely spaced areas, is particularly limiting.

The cycles of deposition and lacuna, from the youngest to the oldest, are (3) Pliocene–Quaternary marine deposition in the Northern Coastal and Fitzroy regions, and non-marine deposition elsewhere, correlated with voluminous basaltic eruption in the Burdekin region; (2) late Oligocene through middle Miocene marine deposition in the North Coastal and Fitzroy regions, probable early and middle Miocene deposition during a marine incursion in the Inland region (Lloyd 1968a), and non-marine deposition in other parts of this and the other regions, correlated with late Oligocene and early Miocene volcanic eruption in south-east Queensland. A middle to late Miocene lacuna in deposition corresponds with a lacuna in volcanism; (1) following a later Cretaceous lacuna, marked in the Inland region by the deeply weathered Morney Profile, Palaeocene and Eocene non-marine deposition everywhere except marine deposition in the later part of the Eocene in the Northern Coastal region, all succeeded by an early to complete Oligocene lacuna marked by deep weathering. The remaining ages of volcanism range from 40 to 63 Ma, but are too few to indicate correlation – positive or negative – with the depositional cycle. A lacuna in the Eyre Formation of the Inland region (Wopfner et al. 1974), continuous with the rest of the Birdsville Basin, may indicate a separation of Grimes's (1980) cycle 1 into the two cycles found in South-east Australia.

Within the limitations of the narrow data-base from North-east Australia, comparison of Fig. 64 with Fig. 59B above shows tentative correlation of the cycles, which are thus recognized as extending to the entire Eastern Highlands and adjacent basins.

In the next section, we trace the entire Eastern Highlands to their origin and antecedents in the Mesozoic.

(iii) Mesozoic origins and antecedents of the Eastern Highlands

(1) Summary

Australia's Eastern Highlands are a conspicuous manifestation of a tectonic regime which, as we have shown above, goes back at least 65 Ma. This review of the Mesozoic stratigraphy of Eastern Australia gives evidence of a very different regime before 95 Ma, related to the presence of a plate boundary close to the present east coast of the continent.

During the prior regime, cratonic sedimentation in Eastern Australia was dominated by labile sediment from an andesitic orogen coincident with the coast north of Brisbane during the Cretaceous and further offshore in the Jurassic. While the plate boundary north of Brisbane appears to have been simply convergent, that south to Bass Strait may have experienced prolonged oblique-slip, manifested in the Jurassic by alkaline volcanism within the South-east Highlands terrain.

Following a Cenomanian (95–90 Ma) phase of transition during which the Eastern Australia plate boundary may have resembled that margining western North America at present, the plate boundary migrated away from mainland Australia, as is manifest in the subsequent dominance of quartzose sedimentation on the craton, and the fission-track and palaeomagnetic evidence of rapidly falling geotherms

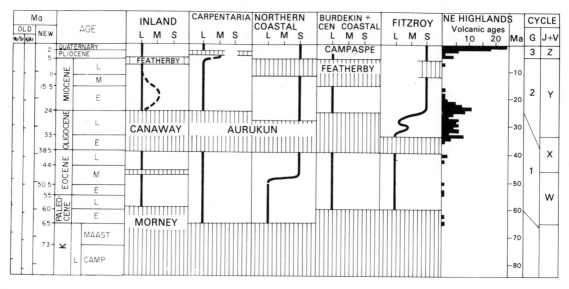

Fig. 64. Time–facies diagrams of the regions (except Burnett and Brisbane) located in Fig. 62, and histogram of ages of volcanic (mainly basaltic) rocks. Depositional environments range from L (land, non-marine) through M (mixed non-marine and shallow marine) to S (shallow marine). Lacunae designated by vertical lines. Facies information from Grimes (1980), with emphasis on the marine facies of the eastern flank in the Papuan and Capricorn Basins known from oil exploration wells. The broken line in the column of the Inland region portrays the marine ingression indicated by foraminifers of Miocene (or younger) age in the Austral Downs Limestone (Lloyd, 1968a). Deeply weathered surfaces indicated by capital letters. K: Kendall Surface. Volcanic ages, in new constants, from Exon *et al.* (1970), Griffin and McDougall (1975), McDougall and Slessar (1972), Sutherland *et al.* (1978), Webb and McDougall (1967a), Webb *et al.* (1967), Wellman (1978), Wellman and McDougall (1974b), and Wyatt and Webb (1970).

in the Late Cretaceous. The Eastern Highlands were initiated around 90 Ma ago, and the crestline subsequently migrated west from an initial location at the present coastline.

The geography and history of the Eastern Highlands are inconsistent with concepts of continental margin development based on analogues outside the Pacific realm. The Highlands are an intrinsic element of a continent formerly fronting the Pacific Ocean, but now abutting a back-arc basin.

(2) Introduction
The Eastern Highlands are the most conspicuous linear morphotectonic element of mainland Australia (Fig. 65A: inset). Without resolution of their development, initiation, and antecedents, there can be no satisfactory account of the evolution of the Pacific borderland of the continent.

The foregoing analysis of the history of the Southeast (New South Wales and Victorian) Highlands led us to the conclusion that for South-east Australia the present gross morphology, and the tectonic regime it expresses, extend back at least to the beginning of the Cainozoic (65 Ma), if not well into the Late Cretaceous

(65 to 95 Ma). Grimes (1980) reached a similar conclusion for North-east Australia. To date the inception of the Eastern Highlands and flank basins we need to characterize the preceding regime and to determine the time of transition.

(3) Eastern Highlands defined
The gross morphology of the Pacific borderland of mainland Australia (Fig. 65A) may be divided into two sectors meeting in the easterly salient of the continent around the latitude of Brisbane: a north-east sector in the state of Queensland running north-west from Brisbane, and a south-east sector running south-west through New South Wales.

Within eastern New South Wales the crest of the Highlands (most clearly defined by the 0.5 km contour) runs roughly north-east to south-west, parallel to the coast and continental margin, around 150 km inland (Fig. 65A). At its culmination in Mt. Kosciusko it attains 2 km, but for most of its length in New South Wales it is about 1 km. The eastern fall is much more abrupt than the western, as reflected in the contour spacing.

Within eastern Queensland the geomorphology is

much more diffuse (Fig. 65A). The Highlands are coastal as far north as Brisbane. North of Brisbane, the watershed between coastal and inland streams, commonly little more than 0.5 km above sea level, swings inland as much as 400 km in a broad arc, reapproaching the coast at the base of the Cape York Peninsula. From New South Wales to Queensland, this inland swing of the watershed is matched by a commensurate salient in a greatly broadened continental shelf (Fig. 65A), with Highlands and shelf separated by a more sinuous coastline.

In both New South Wales and Queensland the coastline may conveniently be taken as the present eastern boundary of the Highlands, marking at continental scale the boundary between a grossly erosional and a grossly depositional realm, at least for the late Cainozoic. By the same criterion the western boundary is the eastern limit of Cainozoic interior basins: the Murray, Birdsville, and Karumba (Fig. 65A), all delineated by the 0.2 km contour. Thus defined, the Queensland sector of the Highlands, though considerably lower in elevation, is notably more expansive.

In the Cainozoic the Eastern Highlands have been a terrain of patchy, sporadic basaltic volcanism, as shown in Fig. 65A. Apparent maxima in volcanic activity are related to phases of uplift in the highlands accompanied by subsidence of the flanking basins.

(4) North-east Australia 200 to 90 Ma ago

(a) Surat–Maryborough transect. Only around the Brisbane salient are sediments of Jurassic–Cretaceous age preserved across the width of the Highlands terrain (Figs. 65B and 66). Here the Surat and Maryborough basin successions on the fringes of the Highlands span the Lower Jurassic to mid-Cretaceous (Albian: 100 Ma), whereas the median Moreton Basin succession, the sole preserved link between late Mesozoic successions west and east of the divide, is truncated at the Middle Jurassic (Fig. 67).

The Surat Basin succession (Exon 1976; Exon and Senior 1976; Exon 1980; Exon and Burger 1981) consists of 2 km of sandstone, siltstone, and mudstone that rest unconformably on the Permo-Triassic succession of the Bowen–Sydney Basin (Fig. 68B). The Surat succession is a record of apparently continuous sedimentation from the onset of the Jurassic to the Albian. The Jurassic succession of the Basin consists at formation scale of alternating intervals of dominant sandstone (aquifers of the Great Artesian Basin) and dominant siltstone–mudstone (aquicludes). The sandstones of the sandstone formations are typically quartzose, and the sandstone interbeds of the siltstone–mudstone formations are typically labile, containing fragments of mafic to felsic volcanic rock and feldspar (Allen and Houston 1964). Coal and bentonite are typical accessories of the siltstone–mudstone formations. The Cretaceous succession is almost wholly of labile siltstone–mudstone. The quartz sandstone facies is typically fluvial, whereas the labile siltstone–mudstone facies ranges from fluvio-lacustrine, through paralic to shallow marine.

Permo-Triassic rocks forming the eastern slopes of the Highlands presently separate the Jurassic–Cretaceous succession of the Surat Basin from that of the coastal Maryborough Basin 150 km north-east (Ellis 1968; Ellis and Whitaker 1976; Day *et al.* 1974). This latter succession differs from that of the Surat Basin most notably in its much greater thickness (4.5 km v. 2 km), the presence of volcanics (intermediate flows and pyroclastics with coeval intrusives), and of conglomerates with volcanic clasts.

The contrasts between the Surat and Maryborough successions are most simply explained in terms of relative proximity to a shared volcanic source, which Ellis (1968) located along the eastern margin of the Maryborough Basin and which Ericson (1976) has suggested may lie beneath the Bunker Ridge 70 km offshore. Flows and coarse pyroclastics from that source are preserved in the Maryborough succession, while the Surat succession contains the more distantly flushed volcanic sands and silts, and deposits of fine airborne volcanic ash in the form of bentonites.

Such an explanation necessitates a degree of fluvial interconnection, if not substantial continuity, between the two successions through the Jurassic and Early Cretaceous. Nothing in the stratigraphy of the north-east Surat Basin discounts their former continuity, most formations persisting from the axis of the structural basin to their north-east erosional limits with little or no loss of thickness (Exon 1980, especially fig. 5B). Moreover, there is near continuity of Lower and Middle Jurassic strata from the Surat to Maryborough Basin via the Moreton and Nambour Basins. However, local stratigraphic overlap of Lower–Middle Jurassic by Upper Jurassic–Lower Cretaceous with basement onlap along the western edge of the Maryborough Basin (Ellis 1968; Ellis and Whitaker 1976) suggests some temporary intervening erosional relief.

These relations are summarized in a time–space diagram of the Eastern Highlands terrain for the last 200 Ma (Fig. 69) in a transect from the Coonamble lobe of the Surat Basin north-east to the Maryborough Basin (Figs. 65B, 68B).

(b) Carpentaria–Papua transect. The western flanks of the Eastern Highlands in north Queensland are

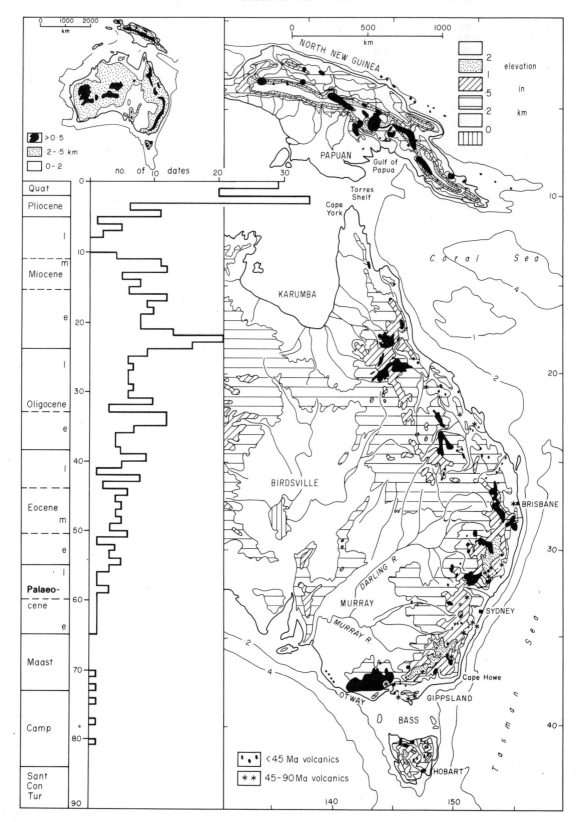

underlain in part by the Jurassic–Lower Cretaceous sediments of the Carpentaria Basin, continuous with those of the Surat Basin via the Eromanga Basin across the Euroka Arch (Fig. 65A, B: Smart *et al.* 1980). The Carpentaria Basin succession is comparable in thickness to the Coonamble lobe of the Surat Basin, which it further resembles in being essentially quartzose sandstone facies in the Jurassic and labile siltstone–mudstone facies in the Cretaceous (Smart *et al.* 1980).

North of 15° S the Eastern Highlands vanish, and the Carpentaria Basin laps thinly across the Peninsula to merge east of Cape York with the Southern Papuan Basin beneath the Torres Shelf (Fig. 65). North-east of Cape York, the Jurassic–Lower Cretaceous succession extends unbroken from the Torres Shelf, beneath the Papuan Platform, to the flanks of the Kubor Anticline at the crest of the New Guinea Highlands (Fig. 68A). It thickens from a few hundred metres at Cape York to values comparable to the Surat maximum beneath the head of the Gulf of Papua (2–3 km: Fig. 68A, B), and to a greater but uncertain value beneath the foothills. Beyond the crest of the Kubor Anticline, where continuity of the succession is broken by a 50 km inlier of Permo-Triassic felsic volcanics, granites, and metamorphics that resemble the rocks separating the Surat and Maryborough Basin successions, exposure of Jurassic–Early Cretaceous rocks persists for another 100 km across the faulted Bismarck Range (Fig. 68A).

Though basement rocks at the crest of the Kubor Anticline were exposed in the Early and Middle Jurassic (Fig. 70: Bain *et al.* 1975), there is no more reason to doubt the substantial continuity of the North New Guinea and Papuan basins through the Jurassic and Early Cretaceous than in the case of the Maryborough and Surat basins: and their proximal–distal relationship to volcanism is equally manifest. The proximal character of the Jurassic–Lower Cretaceous succession of the North New Guinea Basin is evident in the presence of substantial mafic to silicic lavas, coarse pyroclastics and derived conglomerates, and dominant labile sandstones (Fig. 70: Bain *et al.* 1975; Dow 1977). In contrast, the coeval succession of the foothills (north-east) sector of the Papuan Basin is dominated by labile siltstone–mudstone, and coarse pyroclastics and lavas are insignificant. Further south-west, from the head of the Gulf of Papua to Cape York, craton-derived quartz sands and silts form a substantial part of the succession and dominate the Jurassic (Fig. 70: Ridd 1976; Burns and Bein 1980), though Smart and Rasidi (1979) report Early/Middle Jurassic 'tuff and tuffaceous sandstone' from Cape York.

As in the Surat and Maryborough basins, there is a clear correlation between sediment composition and environmental 'wetness'. The characteristically quartzose Jurassic from the Gulf of Carpentaria to the head of the Gulf of Papua is fluvial to paralic, whereas the open-marine facies is conspicuous in the labile Jurassic further north-east and in the ubiquitously labile Cretaceous.

(c) Orogen and foreland basin. The highlands of Papua New Guinea and Irian Jaya are an orogen that rims the northern edge of the Australian craton, impounding the Papuan foreland basin and the Karumba epicratonic basin (Fig. 65A). The Karumba traps the wash of quartz sand, silt, and mud from the craton, whereas the flood of gravel, lithic sand, and silt from the orogen, conspicuously volcanic in Papua New Guinea, accumulates as the Papuan piedmont. The Jurassic–Early Cretaceous stratigraphy of the Carpentaria–Papua transect is most simply interpreted as the relic of an ancestral configuration involving homologous morphotectonic elements: Carpentaria epicratonic basin, Mesozoic Papuan foreland basin, and Mesozoic New Guinea orogen. We envisage a comparable configuration for the Surat–Maryborough transect, from the Eromanga epicratonic basin, across the Surat foreland basin, to the Maryborough orogen.

The Surat Basin, as defined by its present erosional limits, extends no further north than 25° S (Fig. 65B). But the physical continuity of the Eromanga and Carpentaria epicratonic basins and their similar facies at similar times are consistent with the former continuity of the Surat and Papuan basins which adjoin them. Moreover the volcanics of the Maryborough are attributed to a Jurassic–Early Cretaceous

---

Fig. 65A. Morphology of eastern half of Australia and New Guinea, from Times Atlas (1975); and distribution of volcanics, 0 to 45 Ma old (solid black) and 45 to 90 Ma old (asterisks), from Stephenson *et al.* (1980) and references cited below. Inset, top left, shows morphology of mainland Australia (Gentilli and Browne 1963). Inset on left shows frequency-distribution of K–Ar dates of Eastern Australian volcanics, mainly basalt, 0 to 90 Ma old, using the new constants, from sources given mainly in Jones and Veevers (1982), Wellman and McDougall (1974a), and Wellman (1974); and additionally in Exon *et al.* (1970), Griffin and McDougall (1975), Hamilton (1981), McDougall and Slessar (1972), Sutherland *et al.* (1978), Webb and McDougall (1967a), Webb *et al.* (1967), and Wyatt and Webb (1970). Fig. 65 and Figs. 67-73, from Jones and Veevers (1983).

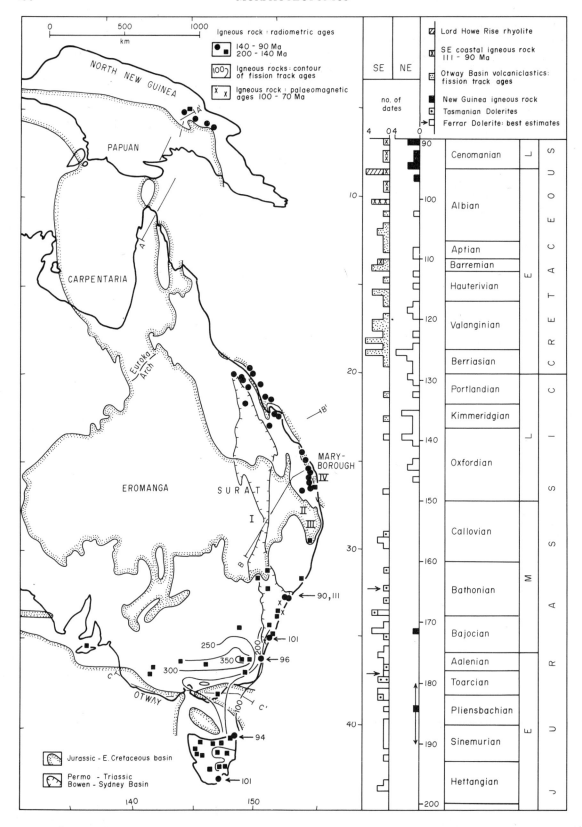

volcanic arc, the relics of which outcrop in similarly proximal facies in the Proserpine–Whitsunday Island area (Fig. 65B: Ellis 1968; Clarke et al. 1971) and which may have extended the length of the present east coast of Queensland (Henderson 1980). What are probably the distal products of this arc are encountered throughout the Surat and Eromanga Basins, particularly in the Aptian–Albian–Cenomanian (110–90 Ma: Exon and Senior 1976; Veevers and Evans 1973).

The gross configuration of the Queensland sector of the Highlands terrain in the Jurassic–Early Cretaceous recalls the character and context of the Permo-Triassic Bowen–Sydney Basin (Fig. 65B: Jones and McDonnell 1981; Conaghan et al. 1982; Jones et al., chapter V). The Bowen–Sydney Basin developed as a foreland basin between the Gondwanaland craton and an eastern, Pacific rim-orogen, with labile debris from the orogen and quartz from the craton. The ratio labiles:quartz waxed and waned with volcanic activity in the orogen to define at least 3 major tectonic cycles in the Permian through Triassic (previous references and McDonnell et al. in preparation). Similar cycles and sub-cycles can be defined within the Surat Basin succession (Fig. 69), commencing with an influx of labiles from the eastern orogen and terminating with resurgence of quartz from the craton.

For the Sydney Basin this model rests on detailed palaeocurrent and petrographic analysis, coupled with close correlation of the Basin stratigraphy with dated magmatic/metamorphic events in the orogen. Data of these kinds from the Bowen Basin are much more scanty, but consistent with the model, as are the patchy palaeocurrent data from the Surat Basin, which are almost confined to the quartzose facies in which orientations range from north-west through north-east to south-east (Fig. 66, inset: Playford and Cornelius 1967; Swindon 1960; Exon 1976; Martin 1981; Arditto 1982): that is, from orogenwards to orogen-parallel, but not cratonwards.

The close association of coal with labile sediments including tuffs in the Surat and coeval basins in eastern Australia is also characteristic of the Bowen–Sydney Basin. In the context of a foreland basin model, this association reflects optimal conditions for peat accumulation and preservation through high groundwater levels and rapid sedimentation resulting from synchronous persistent basin subsidence and orogen uplift during phases of volcanic/tectonic activity.

A detailed application of these concepts to the Jurassic–Early Cretaceous basins of eastern Australia is beyond the scope of this account. However, the independent development of concepts of cyclical sedimentation for the Surat and Eromanga Basins from Whitehouse (1954) to Exon and Burger (1981) requires brief comment. Exon and Burger (1981) divide the Surat succession into cycles commencing with dominant sandstone, commonly quartzose, and terminating with dominant labile siltstone–mudstone. These they attribute to base-level changes resulting from global sea-level changes, commencing with rapid fall and terminating with slow rise: a model that accounts for trends in grain size and sedimentary structures, but not in sediment composition. In essence their scheme involves eustatic influence on an inert craton. In the context of a model involving a cratonic foreland periodically subsiding in synchrony with a rising, volcanically active orogen (cf. Jones and Veevers 1982), we focus on cycles commencing with labile sediments derived from the orogen and deposited in 'wet' fluvio-lacustrine, paralic, and shallow marine conditions; and terminating with quartzose sediments, derived from the craton, and deposited in 'dry' fluvial conditions. Our cycles correlate with global sea-level motions to a degree that suggests global links between tectonism and eustasy (cf. Jones and Veevers 1982).

(5) South-east Australia 200 to 95 Ma ago

(a) Early and Middle Jurassic rocks. Only two small occurrences of Early through Middle Jurassic sediments are known in South-east Australia, south of

---

Fig. 65B. Distribution of dated igneous (mainly volcanic) rocks 90 to 200 Ma old, and cited coeval sedimentary basins; contours of fission-track ages of apatite from igneous rocks in South-east Australia from Morley et al. (1981) and magnetic overprints 100-70 Ma old from Schmidt and Embleton (1981). Arrows indicate igneous rocks 101-90 Ma old along South-east coast (Carr and Facer 1980; Evernden and Richards 1962; Gleadow and Duddy 1981; Hamilton 1981; McDougall and Roksandic 1974). Lines show location of the sections and time–space diagrams of Figs. 68 to 71. Insert on right shows frequency-distribution of K–Ar dates and, from the Otway Basin, concordant fission-track dates on zircon, sphene, and apatite in volcaniclastics with errors <12 Ma (Gleadow and Duddy 1981), all using the new constants; sources of K–Ar dates are Dulhunty (1972, 1976), Dulhunty and McDougall (1966), Flint et al. (1976), McDougall and Leggo (1965), McDougall and Wellman (1976), Wellman et al. (1970), Schmidt and McDougall (1977), Carr and Facer (1980), and Hamilton (1981), and, for DSDP Site 207 on the Lord Howe Rise, McDougall and van der Lingen (1974). Arrows indicate the best estimates of the age of the Ferrar Supergroup of the Transantarctic Mountains (Kyle et al. 1981). Emplacement age of diatremes of the Sydney Basin from enclosed fossils (Crawford et al. 1980, and Helby and Morgan 1979).

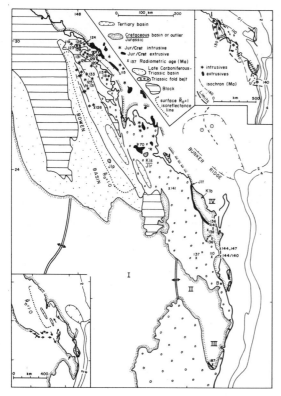

Fig. 66. East-central Australia, emphasizing the occurrence of Jurassic and Cretaceous igneous rocks. Cainozoic rocks, except for sedimentary basins, ignored. From Tectonic Map of Australia and New Guinea (Geological Society of Australia, 1971), augmented by

(a) Tertiary basins: Grimes (1980), Ericson (1976), Hentsridge and Missen (1981), Gray (1976), Green and Bateman (1981).

(b) Jurassic–Cretaceous basins or outliers: (1) outliers south-west of Rockhampton (R), Stanwell Coal Measures (Kls) and Razorback Beds (Jlr): Playford and Cornelius (1967), Skwarko (1968); (2) outlier south-west of the Duaringa Basin, on the Blackdown Tableland, Precipice Sandstone (Jlp): Malone et al. (1969); (3) Maryborough Basin from the Early Jurassic Tiaro Sandstone (Jlt) to the Albian Burrum Coal Measures (Klb), and its northeastward extension beneath the Capricorn Basin: Ellis (1976), Ericson (1976); (4) Styx Basin: Malone et al. (1969); (5) Moreton Basin: Day et al. (1974); (6) Surat and Eromanga Basins: Exon and Senior (1976).

(c) Jurassic–Cretaceous igneous rocks with radiometric dates (Ma) and inset, top right: Evernden and Richards (1962), Webb and McDougall (1964, 1968), Clarke et al. (1971), Paine et al. (1974), Green and Webb (1974); Mt. Hedlow trachyte (minimum age of 70 Ma) and probable ages in the Port Clinton area (Double Mountain and Peninsula Range Volcanics, Pyri Pyri, and Bayford Granites) from Murray (1975) and Kirkegaard et al. (1970); probable ages in northern Bowen Basin from Dickins and Malone (1973).

(d) Surface $R_0=1$ isoreflectance line from Shibaoka et al. (1973), Cook (1975), and Staines and Koppe (1980).

(e) Palaeocurrent directions in basal Jurassic sandstone (inset, lower left) from Martin (1981), Swindon (1960), and Playford and Cornelius (1967).

32°30′ S; viz. Early Jurassic in diatremes near Sydney (Crawford et al. 1980) and mid-Jurassic in the western Otway Basin as dated by associated basalt flows (Douglas and Ferguson 1976; see also Wopfner and Douglas 1971; Ellenor 1976). However, this is an interval of widespread magmatism in this region (Fig. 65B). Scattered throughout the South-east Highlands are dominantly mafic necks, dykes, and sills with ages in the interval 200–140 Ma (Figure 65B: McDougall and Wellman 1976, and other sources cited in caption). Their distribution within the Highlands is as dispersed as the Cainozoic basalts, and if they have the same tectonic significance, they may record equally widespread mild uplift in the Early and Middle Jurassic.

Several orders of magnitude more voluminous than the intrusives just described are the Early–Middle Jurassic sills of Tasmania (Fig. 65B: Schmidt and McDougall 1977). In palinspastic context (Veevers 1982b), these are part of a flood basalt province that runs the length of the Transantarctic Mountains (Kyle et al. 1981).

(b) Late Jurassic and Early Cretaceous rocks. A thick succession of Late Jurassic through Early Cretaceous strata spans the length of the Otway and Gippsland basins of Victoria (Figs. 65B, 68, 71), largely concealed beneath a comparably thick Late Cretaceous–Cainozoic cover. Within the limits of the Gippsland Basin, the succession is more than 3 km thick and is dominated by labile (intermediate volcanic) sandstones with subordinate conglomerate and mudstone and minor bentonite and coal (Douglas and Ferguson 1976). The equally thick coeval succession of the Otway Basin differs in that the labile sediments are dominantly siltstone–mudstone and in the presence of a widespread and variably thick (up to 1.5 km) unit of quartz sandstone (Wopfner and Douglas 1971; Ellenor 1976; Douglas and Ferguson 1976). These facies and their trends recall those of the Surat–Maryborough and Carpentaria–Papua transects (cf. Figs. 69, 70, 71). However, the Otway–Gippsland succession differs in being wholly non-marine and in its confinement within an east–west linear basin bounded on its northern edge

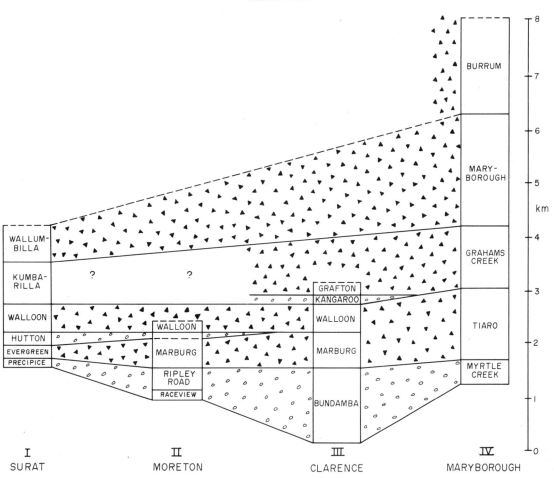

Fig. 67. Stratigraphical columns of the Surat, Moreton, Clarence, and Maryborough Basins, from Day *et al.* (1974), using their correlations, with additions. Located on Fig. 65B by Roman numerals.

by coeval uplands coincident with the present Victorian Highlands (Wopfner and Douglas 1971; Douglas and Ferguson 1976).

Gleadow and Duddy (1981) regard their fission-track dates on detrital minerals from the Otway rocks (Fig. 65B) as evidence that 'volcanism was effectively synchronous with deposition' and they conclude that 'the Early Cretaceous was one of the most significant volcanic episodes in the history of SE Australia'.

Prior to 100 Ma, evidence of Cretaceous magmatism is confined to the Otway–Gippsland succession. However, the interval 100–90 Ma is marked by the emplacement of several small, mafic-intermediate intrusions in coastal locations from a little north of Sydney to south of Hobart (Fig. 65B).

(c) Discussion. For the Late Jurassic–Early Cretaceous interval, the Otway–Gippsland succession suggests tectonic links with coeval rocks in the north-east of the continent as does the coastal location, if not composition, of igneous rocks of this age (Fig. 65B). However, the Early–Middle Jurassic interval is one of strong contrasts: in the north-east widespread sedimentation, partially volcanogenic but with a dearth of igneous rocks; in the south-east widespread mafic volcanism but almost no evidence of sedimentation.

There is indirect evidence of significant sedimentation beyond the limits of the Otway–Gippsland Basins in the interval 130–95 Ma. Coal ranks in the Permo-Triassic Sydney Basin (Fig. 65B) at the present Highlands surface indicate a loss of at least 1 km of

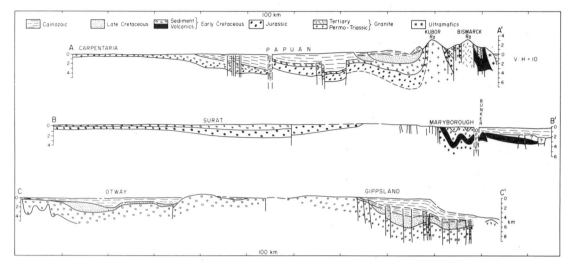

Fig. 68. Cross-sections, at the same scale, of: AA′, Carpentaria–Papua–New Guinea Basins (Bain *et al.* 1975; Burns and Bein 1980; Smart and Rasidi 1979); BB′, Surat–Maryborough Basins (Ericson 1976; Ellis 1968; Exon 1976; Bembrick 1976); CC′, Otway–Gippsland Basins (Veevers 1982a; Abele 1979; Douglas and Ferguson 1976; Wopfner and Douglas 1971). Located on Fig. 65B.

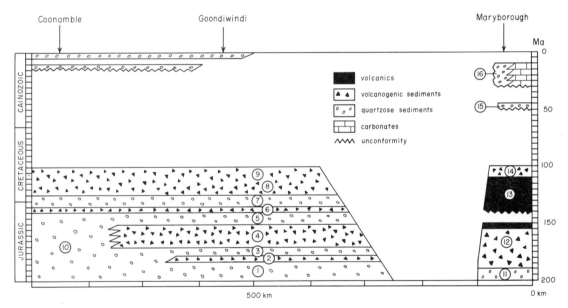

Fig. 69. Time–space diagram of the Surat and Maryborough Basins, located on the same section line as BB′ of Fig. 68, showing distribution of preserved lithofacies and of lacunae (shown blank). Circled numbers indicate named stratigraphic units: 1, Precipice Sandstone; 2, Evergreen Formation; 3, Hutton Sandstone; 4, Injune Creek Group; 5, Gubberamunda Sandstone; 6, Orallo Formation; 7, Mooga Sandstone; 8, Bungil Formation; 9, Rolling Downs Group; 10, Pilliga Sandstone; 11, Myrtle Creek Sandstone; 12, Tiaro Coal Measures; 13, Grahams Creek Formation; 14, Maryborough Formation and Burrum Coal Measures; 15, Fairymead Beds; 16, Elliott Formation. Sources: Allen (1976), Allen and Houston (1964), Arditto (1982), Bourke (1980), Cranfield *et al.* (1976), Ellis (1968, 1976), Ellis and Whitaker (1976), Exon (1980), Exon and Burger (1981), Exon and Senior (1976), Martin (1980), McKellar (1980a, b), Robertson (1979).

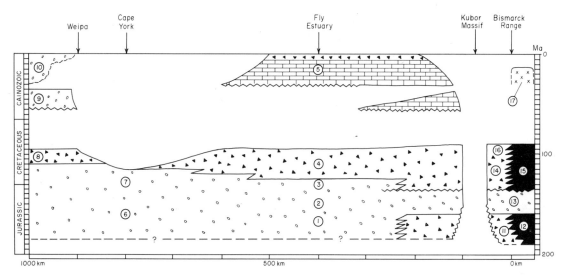

Fig. 70. Time–space diagram of the Carpentaria, Papuan, and New Guinea Basins along the same line as Fig. 68, AA'. Symbols as for Fig. 69. Circled numbers: 1, 'Lower Sand'; 2, 'Lower Shale'; 3, 'Upper Sand'; 4, 'Upper Shale'; 5, Danai Limestone; 6, Helby Beds; 7, Gilbert River Formation; 8, Rolling Downs Group; 9, Bulimba Formation; 10, Wyaaba Formation; 11, Balimbu Greywacke; 12, Mongum Volcanics; 13, Maril Shale; 14, Kondaku Tuff; 15, Kumbruf Volcanics; 16, Chim Formation; 17, Bismarck Granite. Sources; Bain *et al.* (1975), Brown *et al.* (1979), Burns and Bein (1980), Conybeare and Jessop (1972), Dow (1977), Powell *et al.* (1976), Smart *et al.* (1980), Smart and Rasidi (1979).

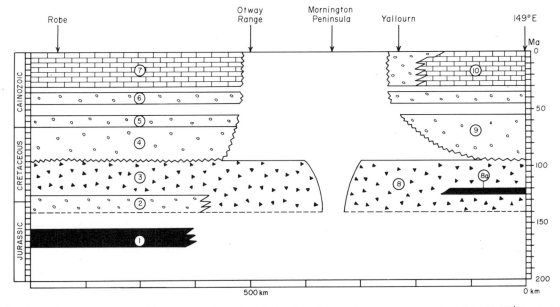

Fig. 71. Time–space diagram of the Otway and Gippsland Basins along the same line as the cross-section of Fig. 68, CC'. Symbols as for Fig. 69. Circled numbers: 1, Casterton Beds; 2, Pretty Hill Sandstone; 3, Eumeralla Formation; 4, Sherbrook Group; 5, Wangerrip Group; 6, Nirranda Group; 7, Heytesbury Group; 8, Strzelecki Group; 8a, Duck Bay Basalt; 9, Latrobe Group; 10, Seaspray Group. Sources: Brown (1976), Douglas and Ferguson (1976), Ellenor (1976), Gleadow and Duddy (1981), Hocking (1972), Wopfner and Douglas (1971).

former cover (Facer *et al.* 1980; Middleton and Schmidt 1982), and Cretaceous overprint magnetizations in the same region suggest the loss of at least 2 km of cover since the mid-Cretaceous (Schmidt and Embleton 1981). If the south-east was an erosional realm during the Early and Middle Jurassic, as parallels with the Cainozoic suggest, then this lost sediment is likely to have been Late Triassic or Late Jurassic through Early Cretaceous in age.

(6) Tectonic setting of Eastern Australia 200 to 90 Ma ago

(a) North-east Sector: 140 to 90 Ma. The Surat Basin has already been designated a Jurassic through Early Cretaceous foreland basin between the Australian craton and a volcanic arc or orogen roughly coincident with the present Queensland coast. For the interval 140–90 Ma that reconstruction rests both on regional stratigraphy and on coastal igneous geology (Figs. 65B, 66, 69). Extrusive igneous rocks of this age in the Maryborough Basin are described (Ellis 1968; Ellis and Whitaker 1976) as dominantly andesite with early basalt and late rhyolite: coeval intrusives inland range from gabbro to granite (Webb and McDougall 1968). Lavas of this age interval in the New Guinea Highlands are described as ranging from basalt through andesite to rhyodacite (Bain *et al.* 1975), and Cenomanian intrusives as granodiorite and basalt (Page 1976). According to the comprehensive review by Gill (1981), '*all* active orogenic andesite volcanoes occur within 500 km of a boundary where two plates now converge'. Assuming that the Early Cretaceous andesites of Queensland and New Guinea are orogenic as defined by Gill, they denote proximity to a convergent plate boundary, perhaps north-east of the magmatic arc (cf. Henderson 1980).

(b) North-east sector: 200 to 140 Ma. For the interval 200–140 Ma, the regional stratigraphy of the North-east sector suggests a closely similar tectonic setting, but one in which the influence of the volcanic arc extends a shorter distance over the craton (Fig. 69). In contrast to the succeeding interval, however, there is a dearth of igneous rocks (Fig. 65B): in this interval, basalt occurs in the New Guinea Highlands (Bain *et al.* 1975) and 'trachyandesite' (Ellis 1968) in the Maryborough Basin. Despite this difference, we are confident of the stratigraphic unity of the Jurassic through Early Cretaceous in the North-east Sector. We conclude that in the interval 200–140 Ma, the volcanic arc lay a little north-east of its successor (presently offshore in Queensland), a reconstruction consistent with the contrasting extent of volcanogenic cover on the craton in these intervals.

(c) South-east sector: 200 to 140 Ma. In the absence of sediments, igneous rocks assume a still more important role in deciphering the tectonic character of the South-east sector in the interval 200–140 Ma. Falling in this interval are mafic intrusives of the South-east Highlands terrain (Fig. 65B), wide in extent but slight in amount. Their small volume and dominantly alkaline character contrast with the large volume and tholeiitic character of the mafic intrusives and extrusives of the Tasmania–Transantarctic Mountains province (McDougall and Wellman 1976; Schmidt and McDougall 1977; Kyle *et al.* 1981).

On the basis of unusually high $^{87}Sr/^{86}Sr$ ratios and their setting as intrusions into and flows above a Permo-Triassic foreland basin succession separating the East Antarctic craton from the West Antarctic orogen, the Antarctic rocks have been tentatively related to Jurassic extension, analogous to back-arc spreading, on a plate boundary marked by Triassic compression, plate divergence succeeding plate convergence (Elliot 1976; Kyle *et al.* 1981). The same conclusions apply to the coeval and comagmatic Tasmanian dolerites; but the change to an alkaline province across Bass Strait indicates a change to a significantly different but unresolved plate tectonic setting.

(d) South-east sector: 140 to 90 Ma. The tectonic setting of the South-east sector in the interval 140–90 Ma is as equivocal as for the preceding interval, but it is clearly different. From 140 to 100 Ma, evidence is confined to the Otway–Gippsland succession which has generally been regarded as the fill of a newly founded rift-valley (Douglas and Ferguson 1976, and references therein). The dominant lithic component of its largely labile sands and silts was denoted andesitic on petrographic criteria by Edwards and Baker (1943). However, Gleadow and Duddy (1981) characterize the volcanic component of the Otway sediments as mildly alkaline, in contrast to the calc-alkaline designation of coeval igneous rocks in Queensland (Webb and McDougall 1968).

(e) Synthesis: 200 to 90 Ma. From 200 to 90 Ma a plate boundary was located parallel to and, at most, several hundred kilometres east of the present east coast of Australia. It extended northwards to the east of Kubor (Papua New Guinea) and southwards east of Tasmania and thence along the Pacific perimeter of the Transantarctic Mountains terrain. The character of this boundary varied along its length, as do modern plate

boundaries at the same scale. Changes of character took place (1) in sectors marked by conspicuous change of azimuth (for example the Brisbane salient), and (2) with time, as at 140 Ma, in response to changing stress patterns at continental, if not global, scale.

Throughout the interval 200 to 90 Ma, the plate boundary north of about 30° S had a convergent character and probably exhibited orogenic volcanism throughout its length, though subsequent dilation of the Coral Sea and subsidence of its margins have obscured the proximal products between 10° and 20° S. At 140 Ma the volcanic arc moved cratonward perhaps several hundred kilometres, to a position coincident with the present shoreline.

From 200–90 Ma we envisage an orogen in this sector, fluctuating in aspect in time as the Sunda orogen has varied with the oscillating sea-level of the Quaternary: from a volcanic mountain range at the margin of a dry continent, to a line of mountainous volcanic islands backed by a widespread shallow sea.

We can only speculate on the chemical character of the 200–140 Ma magmatic arc in the North-east sector, but petrographic similarities between Early–Middle Jurassic and Late Jurassic–Early Cretaceous volcanics in the Maryborough Basin (Ellis 1968) suggest calc-alkaline affinities for both. If so, between 200 and 140 Ma an Eastern Australian plate boundary was characterized by calc-alkaline magmatism north of 30° S, alkaline magmatism between 30° S and 40° S, and tholeiitic magmatism still further south.

We have postulated that the Jurassic–Early Cretaceous plate boundary ran parallel to the present coast. The gross trends of that coast for 10° of latitude north and south of Brisbane diverge by 60°. Such radical changes in azimuth on modern plate boundaries are commonly sites of change of character; as between Kyushu and south-west Honshu, Japan, from simple convergence with calc-alkaline magmatism to oblique transcurrence with alkaline magmatism (Gill 1981).

Around 140 Ma the plate boundary south of 30° S changed character, presumably in response to a regional change in the pattern of stress. The previously extensive alkaline magmatism of the South-east Highlands terrain ceased, perhaps accompanied by general settling and more widespread sedimentation. At this time the Otway–Gippsland Basins formed as a latitudinal rift progressively filled from the eastern end over the next 45 Ma by the products of mildly alkaline intermediate volcanism. Comparison of facies of the Otway–Gippsland transect with those of the Surat–Maryborough and Carpentaria–Papua transects suggests that the volcanic source is unlikely to have been more distant than the present continental margin. Whether it was one of a number along the plate boundary to 40° S or a solitary complex at the Gippsland triple junction is unresolved. The coastal intrusions in the South-east sector belonging to the interval 100–90 Ma (Fig. 65B) might suggest the former: but we regard them as the vanguard of a new tectonic regime in Eastern Australia.

(7) Eastern Highlands terrain 95/90 to 65 Ma ago

Nowhere in the Eastern Highlands terrain, and beyond to the limits of the Carpentaria and Eromanga basins, is there a rock record of the Late Cretaceous after the Cenomanian (90 Ma: for example Figs. 69, 70). Throughout this vast region, Jurassic, Lower Cretaceous, and Cenomanian strata, where not exposed, are overlain by Cainozoic sediments, uniformly non-labile. The extensively exposed Cenomanian labile sandstones of the Eromanga Basin are characterized by deep weathering profiles, the most massive of which (Morney Profile, up to 90 m thick) appears to have formed in the interval 90–60 Ma (Idnurm and Senior 1978; Grimes 1980).

Post-Cenomanian Cretaceous strata do occur in the Papuan foreland basin but west of the transects of Figs. 68A and 70 (Ridd 1976). In the south of the continent, the Gippsland and Otway basins contain thick successions (3 km max.) of Late Cretaceous strata: fluvial, paralic, and shallow marine sediments derived from contiguous high ground (Fig. 68C: Wopfner and Douglas 1971; Douglas and Ferguson 1976; Ellenor 1976). These Otway–Gippsland successions rest unconformably on Early Cretaceous rocks, but pass conformably into the Cainozoic, with which they are further allied in their quartzose, exclusively non-labile, composition.

In modern Eastern Australia (Fig. 65), relatively short, steeply graded streams draining the eastern flanks of the Highlands flow north-east through south-east into the Coral and Tasman seas to deposit quartz sand, silt, and mud within coastal estuaries and on the continental shelf, whereas the long, gently graded streams draining the western flanks flow south-west into Lake Eyre or the Southern Ocean and deposit their sediment largely within the interior lowlands. For the coastal flank of the Victorian Highlands at least, the Otway and Gippsland Basin successions indicate morphotectonic continuity back to 95 Ma (Jones and Veevers 1982); and in the absence of evidence to the contrary, we speculate that the same is true of coastal New South Wales. For the inland flanks of the New South Wales and Victorian Highlands, the Murray

Basin succession indicates morphotectonic continuity back to the opening of the Cainozoic (Jones and Veevers 1982), prior to which we speculate that streams draining the western flanks of a more extensive Highlands deposited their sediment in the Ceduna depocentre beneath the present southern continental margin (Veevers 1982b). Like the Murray, the Birdsville succession spans the Cainozoic (Wopfner et al. 1974; Grimes 1980). But whereas the Cenomanian is a lacuna in the precincts of the Murray Basin along with the rest of the Late Cretaceous, in the precincts of the Birdsville Basin it is represented by thick and extensive labile sandstones of the precursor Eromanga Basin.

In our view, the modern morphotectonic regime in Eastern Australia can be retraced with confidence to the beginning of the Cainozoic, and in southern Victoria to the mid-Cretaceous (95 Ma). We speculate that the lineaments of the current regime took shape over Eastern Australia in the interval 95–90 Ma (Cenomanian). The Cenomanian separates the Late Cretaceous–Cainozoic regime from its Jurassic–Early Cretaceous precursor.

(8) Eastern Highlands: a dynamic epeirogen

The Eastern Highlands are a broad swell spotted by basalts of slight volume. Grimes (1980) and Jones and Veevers (1982) detect at least three Cainozoic cycles of uplift with renewed outflow of lava, and the Otway Basin succession records another in the interval 90–65 Ma (Ellenor 1976; Douglas and Ferguson 1976).

Since Taylor (1911), several authors have considered the possible horizontal mobility of the Highlands crest, and in particular evidence of inland migration with time. In the late Cainozoic, for which the signs must be freshest and least equivocal, Bishop (1982) sees no evidence in central and southern New South Wales: but Hills (1975) claims 10–20 km of inland migration for a segment of the divide in Victoria.

All of the seventeen recorded sites of old basalts (90–45 Ma: Fig. 65A) lie east of the present divide (cf. Wellman and McDougall 1974a, fig. 7), the greatest separation occurring in the north-east sector, where the divide is furthest inland. Grimes (1980) and Jones and Veevers (1982) have reiterated the coincidence of basalts and highlands and linked both to uplift-cum-volcanism. In the context of this hypothesis, the pattern of basalt ages indicates that the Highlands crest has at least doubled its distance from the coast in the last 90 Ma. Analogy with uplift predisposes us to see divide migration as intermittent: we speculate that it may have slowed with time, hence the inconspicuous migration of the Late Cainozoic.

(9) Eastern Highlands and continental margin

Despite acknowledged discrepancies, particularly for the Tasman continental margin, there has been a strong tendency to view the late Mesozoic–Cainozoic development of the Pacific borderland in terms of concepts of continental margin development based on types or analogues drawn from regions outside the Pacific (for example Falvey 1974; Falvey and Mutter 1981).

In New South Wales the cross-profile asymmetry of the Highlands and their parallelism and relative proximity to the continental margin suggest a back-worn, degraded rift-valley shoulder (for example Ollier 1982). Yet this margin is 'notable for its marked lack of rift and infra-rift sedimentary basins' (Falvey and Mutter 1981); and a recent geophysical reconnaissance of the Lord Howe Rise, its pre-drift adjunct, yielded 'no evidence of a single sediment-filled depression . . . such as occurs on the "Atlantic-type" southern margin of Australia' (Willcox et al. 1980; cf. Jongsma and Mutter 1978).

The concept of the Eastern Highlands as a degraded rift-valley shoulder Ollier (1982) applies not only in New South Wales but also in Queensland. Ollier focuses on the coast-facing escarpment that runs between the coast and the divide for virtually the entire length of the Highlands. He sees the escarpment as forming through retreat of the western scarp of a pre-drift rift-valley. But this ignores the fact that, though the escarpment is on average about the same distance from the coast in Queensland as in New South Wales, it is up to five times the distance from the continental margin. In New South Wales the proximity and parallelism of Highlands and continental margin, despite the absence of evidence for a pre-drift rift-valley, indicate some link in their history. No such link can be assumed in Queensland, where the Highlands and continental margin are neither proximal nor parallel.

Like the Highlands morphology and volcanology, the pattern of post-Carboniferous regional metamorphism in Eastern Australia, as indicated by coal rank, shows a clear spatial relationship to the present coastline, but not to the continental margin. Outcrops of coal of bituminous rank in Eastern Australia are confined to within 200 km of the coast where vertical rank gradients also tend to be higher than elsewhere (Cook 1975). Ranks are high in the southern Sydney and Maryborough basins, but highest in the northern Bowen Basin 600 km from the continental margin (Fig. 66). According to Cook (1975), 'the high rank gradients are presumably associated with high geothermal gradients and therefore high rates of heat flow which in the case of the Sydney and Bowen Basins probably relates to their location on or marginal to a substrate which was part of an orogen'.

Recently the significance of coal rank has been reviewed in the context of evidence for a Late Cretaceous thermal event in south-eastern Australia. Apatite fission-track dates from the South-east Highlands (Fig. 65B), declining coastwards to 81 Ma, are interpreted by Morley et al. (1981) to give 'clear evidence of a thermal and tectonic event in SE Australia associated with the initiation of the Tasman Rift'. We regard that thermal event as a profound cooling of the present coastal zone of Eastern Australia at the close of the 140–95/90 Ma regime. We attach the same significance to the Cretaceous (100–70 Ma approx.) magnetic overprint in Highlands rocks of the Sydney Basin, which, according to Schmidt and Embleton (1981), 'appears to indicate rapid lowering of the ambient crustal temperature either by lowering the regional geothermal gradient or by rapid erosion, or both'.

(10) Late Mesozoic development of the Pacific Borderland

The palaeogeography of Eastern Australia at the culmination of Jurassic volcanism, as measured by the widest expanse of labile (volcanogenic) sediment, is shown in Fig. 72A. In the north, an andesitic volcanic cordillera, parallel to but several hundred kilometres east of the present coastline sheds its detritus westwards in tongues of river-borne sand and silt which extend over 1000 km into the craton. In the south is an erosional domain culminating in uplands coextensive with the Cainozoic South-east Highlands and similarly characterized by widespread basalt flows of small volume. In contrast, Tasmania is the site of basaltic flood magmatism.

Fig. 72B depicts the last few million years of the 200–140 Ma cycle. Quartz sands from the cratonic perimeter of the Eromanga and Surat basins, previously overwhelmed by labile sediment, blanket the Eromanga Basin and lap the inactive piedmont of the eastern orogen. Volcanism in the south-east, as in the north-east, has declined to a low level of activity.

Fig. 72C depicts a fresh advance of labile sediment across the Surat Basin, following inception of andesitic volcanism in an arc parallel to but several hundred kilometres west of its precursor. In the southern uplands, at the latitude of Melbourne, the Gippsland–Otway rift valley has come into being and is filling with river-borne labile sediment from a volcanic complex at its eastern end.

The latest Jurassic–earliest Cretaceous saw a lapse in volcanic activity, recorded by a phase of widespread quartzose fluvial sedimentation. But by 125 Ma the first pulse of a fresh flood of labile sediment was pressing into the craton from a resurgent orogen, reaching its utmost extent at 90 Ma. Between 125 and 95 Ma the advance of volcanogenic sand and silt across the alluvial plains of the Surat and Eromanga basins was punctuated by progressively more extensive inundations of the sea, culminating in the Albian when the continent experienced its most extensive marine transgression since the Cambrian (Veevers and Evans 1973; Morgan 1980).

Fig. 72D shows Eastern Australia at peak transgression around 100 Ma. A shallow sea extends to the southern and western limits of the Surat and Eromanga Basins, and transforms the volcanic arc into a chain of islands, like the Sunda arc today, or New Guinea in the Miocene (Dow 1977). As in the north-east of the continent, volcanic activity continues at the eastern end of the Gippsland–Otway rift valley but sedimentation is persistently fluvial.

Fig. 72E shows a critical moment in the history of the Pacific borderland; the end of the interval of transition between the Jurassic–Early Cretaceous regime and the Late Cretaceous–Cainozoic regime. In the north-east a New Guinea-style volcanic cordillera, at the end of its active life, feeds a vast piedmont plain, analogous in form to the Fly–Digoel shelf south of the New Guinea Highlands, but much larger. In the southeast the almost equally long-lived volcanic complex feeding the Gippsland–Otway rift valley has died out by 95 Ma, as witnessed by the onset of quartzose sedimentation. On either side of Tasmania, shallow arms of the sea creep northerly along incipient rifts that ultimately dilate into oceans, the eastern (Tasman) arm margined by a few small mafic-intermediate volcanoes.

With the death of the north-east orogen at 90 Ma came the last of the intervals of voluminous and extensive labile sedimentation that recur through the late Palaeozoic and Mesozoic of Eastern Australia. Henceforth detrital sediment is quartzose, and, within the limits of the present shoreline, an order of magnitude less voluminous than for the Jurassic–Early Cretaceous. Some time in the interval 95–65 Ma – probably around 95 Ma (Veevers 1982b) – the south-west quadrant becomes for the first time the principal exit for Eastern Australian drainage and the principal point of entry for subsequent marine transgressions (Jones and Veevers 1982) but never again on the scale of the Aptian transgression.

Eastern Australia at around 80 Ma is depicted in Fig. 72F. A broad belt of highlands, a precursor of the Cainozoic Eastern Highlands, runs the length of the

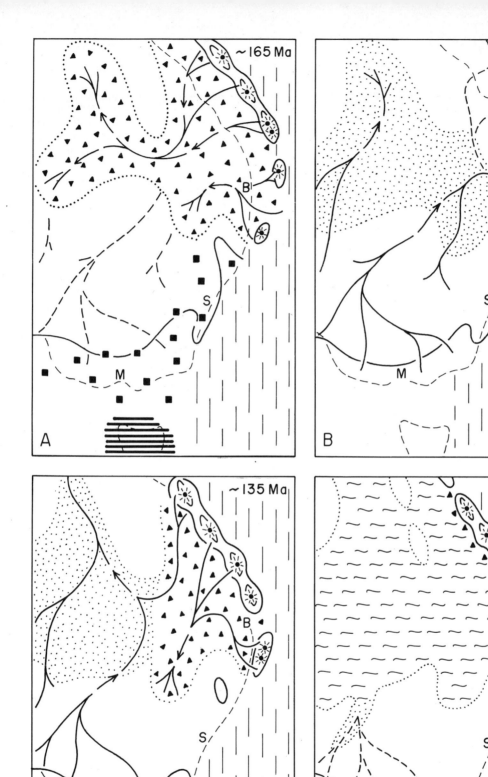

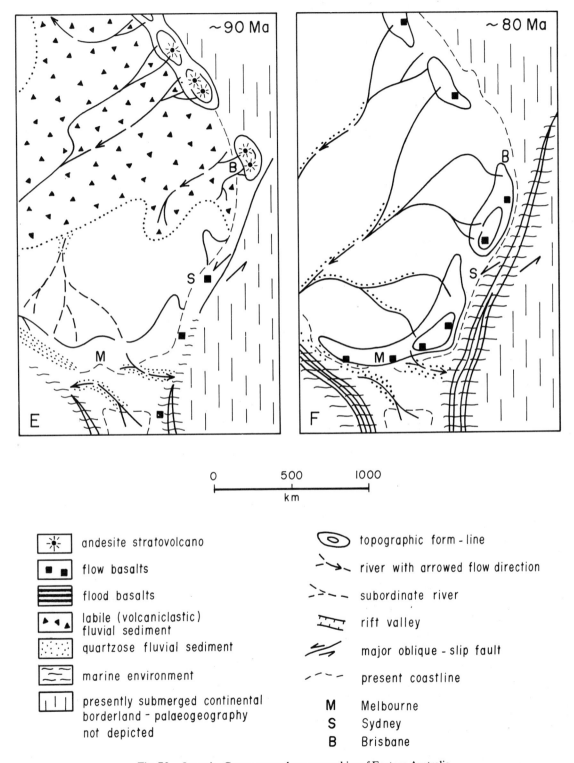

Fig. 72. Jurassic–Cretaceous palaeogeographies of Eastern Australia.

Pacific borderland, cresting 100–200 km east of the modern divide and spotted by basaltic vents and flows. Streams from the western flanks flow south-west to the Ceduna Depocentre (Veevers 1982b) across a continental expanse which for the first time for more than 100 Ma is wholly an erosional domain subject to progressively deeper weathering. Rifting has propagated along the eastern flanks of the highlands to the latitude of Brisbane, accompanied by a postulated marine incursion.

(11) Birth of the Eastern Highlands

The Eastern Highlands are an erosional domain spanning 25° of latitude, extending inland more than 400 km from the line of the present east coast which their crestline parallels, and further characterized by scattered basalt lava flows of small volume. In terms of these characteristics, the Eastern Highlands have a history that goes back 90 Ma.

The importance of the Cenomanian (95–90 Ma) as a turning-point in the Phanerozoic history of Eastern Australia cannot be overemphasized. Before 90 Ma that history was punctuated, if not dominated, by the activity of an eastern rim orogen. From 90 Ma to the present is an aftermath, the relatively placid response of a craton no longer flanked by a plate boundary.

The present-day pattern of tectonic elements of the Pacific borderland of North America between about 30° N and 50° N provides a prima-facie analogue for the configuration of tectonic elements in Eastern Australia at 90 Ma (Fig. 73). Here activity on the plate boundary margining North America has markedly different expression north and south of the salient of Cape Mendocino (cf. the Brisbane salient). To the north its principal expression onshore is the andesitic volcanism of the High Cascades (cf. the Cenomanian volcanic cordillera of north-east Australia). To the

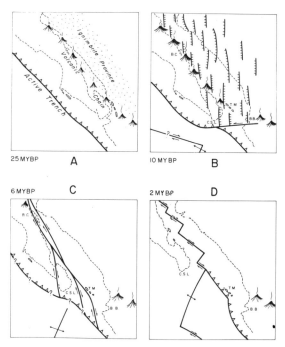

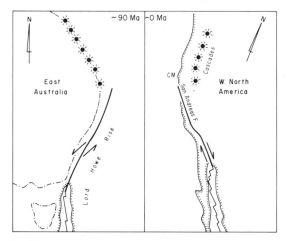

Fig. 73. Comparative tectonic configurations of Pacific borderlands of Eastern Australia in Cenomanian and western North America at present. Drawn at same scale. CM: Cape Mendocino.

Fig. 74. The proto-Gulf of California. From Karig and Jensky (1972). 'Development of the southern Gulf of Californian region, based on the assumption that the proto-Gulf was a subduction-related extensional zone. The displacements assume a constant North America–Pacific motion for the past 30 my, the validity of the magnetic data of Larson (1972), and reasonable subduction rates. A. Before the proto-Gulf developed and during the formation of both an andesitic-volcanic chain and an ignimbrite plateau. B. During the east–west extensional opening of the proto-Gulf. Normal faults are shown diagramatically. Volcanism associated with extension is not shown. C. During the interval between extensional pulses. Strike-slip faults parallel the proto-Gulf and join the subduction zone and spreading zone off Baja California in an unknown manner. The subduction zone west of Baja California is probably still active. D. During inter-plate extension, when the Pacific–Farallon spreading center and subduction zone have disappeared. The displacement direction is now northwest and a transform-spreading center pattern appears in the Gulf.'

south its principal onshore expression is the activity of the San Andreas Fault, which separates mainland North America from the California Borderland and peninsula (cf. the ?Cenomanian oblique-slip zone, coincident with the present continental margin of South-east Australia, separating mainland Australia from the northern Lord Howe Rise). At its southern end, the San Andreas Fault transforms to the oblique-slip zone of the Gulf of California (Figs. 73 and 74), an embryonic ocean basin (cf. the ?Cenomanian chasm between Tasmania and the southern Lord Howe Rise).

In the Late Cretaceous–Cainozoic aftermath, the Pacific borderland of Australia has behaved, not as a continent abutting an Atlantic-style ocean, but as one that formerly fronted the Pacific Ocean and now abuts a back-arc basin.

(b) History of the South Australian Highlands

(i) Introduction

Lying centrally within the Australian–Antarctic Depression (Fig. 75A), the South Australian Highlands (extended from Gentilli and Browne's 1963 definition to include the Gawler Ranges) have the form of a cross, with a sigmoidal meridional arm through Kangaroo Island, the Mt. Lofty Ranges, and the Flinders Ranges, and an east–west arm from the Barrier Ranges to the Gawler Ranges. The meridional trend is reinforced by the intramontane and marginal depressions of Spencer Gulf–Lake Torrens, Gulf St Vincent, and Lake Frome. The region (Fig. 75B) is underlain on the west by the Early Proterozoic Gawler Block, on the east by the Early Proterozoic Willyama Block, and on the meridional axis by the Stuart Shelf of comparatively little deformed Late Proterozoic or Adelaidean sediments and the Adelaide Fold Belt of Adelaidean and Cambrian sediments intruded on the east by early Palaeozoic granite. Today, the South Australian Highlands are close-pressed by sediment-filled depressions. On the north-west is the Lake Eyre drainage basin with the distal mud of Lake Eyre and the proximal sand of the Channel Country of Cooper Creek, and on the north-east, between the Flinders Ranges and the Barrier Ranges, is Lake Frome with a thin layer of clay that interleaves in the foothills with thick scarp deposits of gravel, sand, and clay, in the same fashion as the Pleistocene Hindmarsh Clay of the Adelaide Plains passes into the piedmont deposits of the Mt. Lofty Ranges (Parkin 1969, p. 222). On the east is the Murray–Darling drainage basin in which the westward-flowing Murray River is diverted southward by the Highlands. On the west is the arheic Nullarbor Plain of Tertiary limestone, and on the south the continental margin with a thin cover of modern carbonate.

If the Highlands were invisible today, we could infer their presence by the onlap of modern sediment, and this is the main method followed below in tracing their history, with care to discriminate between (a) an eroded feather-edge of strata brought about by disruption of a continuous sheet subsequently uplifted to form the Highlands, and (b) an onlapping wedge of strata originally deposited against the flank of the Highlands, with attendant facies changes.

Fig. 75B shows the South Australian Highlands ringed by Cainozoic basins (Eucla, Birdsville, Murray, and the southern margin), by Late Jurassic–Early Cretaceous basins (Eucla, Eromanga, Renmark, and Duntroon), and by Late Carboniferous–Permian basins (Arckaringa, Pedirka, Cooper, Renmark, Troubridge, and Denman). The evidence presented below favours the view that most of these are original depositional basins that were deposited against the flanks of the ancestral Highlands and subsequently eroded to varying degrees.

(ii) Cainozoic

The Birdsville Basin started with the deposition of the quartzose sheet sand of the Eyre Formation in the Palaeocene, after a lacuna that spanned most of the Late Cretaceous. According to Wopfner et al. (1974) 'Epeirogenic movements in the Paleocene are thought to have been responsible for the sudden onset of the deposition of the Eyre Formation. Uplift of the source area, in particular the Olary Block and the Barrier Ranges/Cobar region, would have created the gradients necessary for the establishment of major drainage systems [shown in Fig. 75B] ... The abundant pebbles of silicified wood and agate in the basal units of the Eyre Formation clearly identify the source. They could only have been derived from the Jurassic sandstone along the southern margin of the Great Artesian Basin ... The Southwestern Tablelands [south and south-west of Lake Eyre] seem to have been an area of lacustrine deposition, the fine quiet-water clay and silt providing an excellent medium for the preservation of leaves and other plant macrofossils.'

This configuration of fine lacustrine sediment downstream from deposits of mainly sand – the regional slope having been southwestward (Grimes 1980) – anticipated the present facies distribution of mud in Lake Eyre (Dulhunty 1982) and sand upstream in the Channel Country (Veevers and Rundle 1979), and is compatible with the geography found today. A second

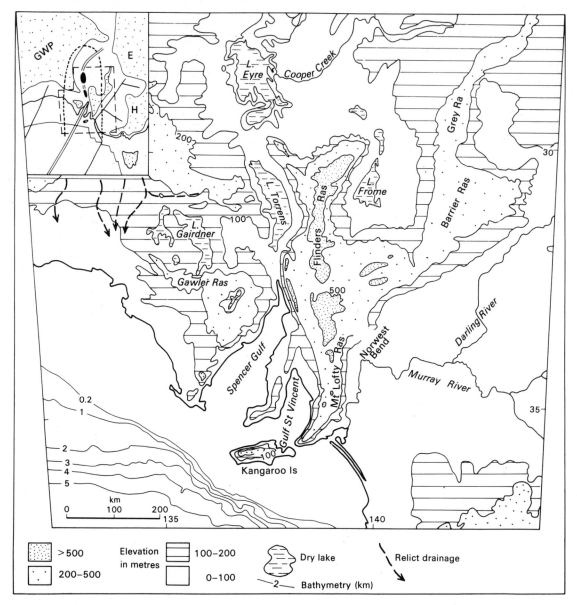

Fig. 75A. Morphology of the South Australian Highlands within the Central-Eastern Lowlands. Land contours from Times Atlas of the World (1975), submarine contours from BMR Earth Science Atlas (1979), relict drainage in west from Pitt (1979) and Barnes and Pitt (1976). Inset shows location of South Australian Highlands (stippled) within the Central-Eastern Lowlands part of the Australian–Antarctic Depression (axis shown by double lines) flanked by the subsidiary axes of the Nullarbor Plain on the west and of the Murray and Darling Rivers on the east, all enclosed by the Great Western Plateau (GWP) and the Eastern Highlands (EH). Central earthquake region (BMR, 1979) shown by dotted line.

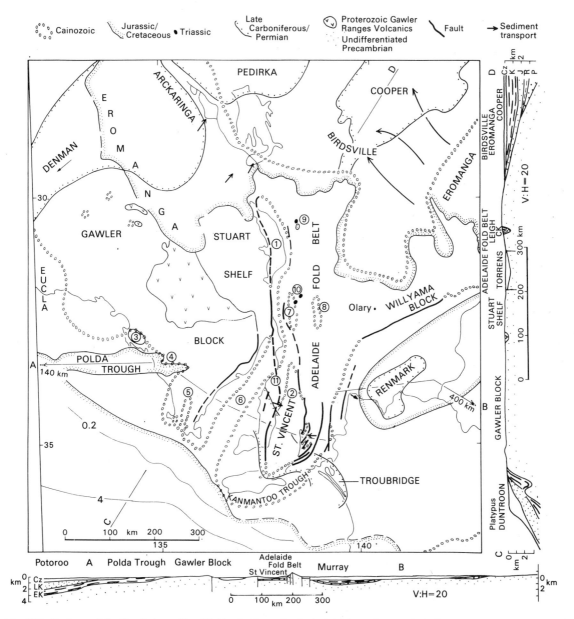

Fig. 75B. Geology of the South Australian Highlands (from Geological Society of Australia 1971) bounded by the Murray, Renmark, Troubridge, Duntroon, Eucla, and Birdsville/Eromanga/Cooper Basins, and cut across by the large intramontane basins (Eocene and younger) of Lake Torrens (1) and Gulf St Vincent (2), and the smaller basins at Robinson (3), Polda (4), Cummins (5), Spencer Gulf (6), Willochra (7), and Walloway (8); older intramontane basins are the Late Triassic–Early Jurassic basins at Leigh Creek (9), unconformably overlain by Late Jurassic/Early Cretaceous sediment, and Boolcunda/Springfield (10). The Torrens lineament (11) bounds the west side of the St Vincent and Torrens Basins. Late Carboniferous and Permian sediment is shown in Fig. 77.

Cross-section AB, with horizontal scale reduced from that of the map, extends outside limits of map into the Great Australian Bight Basin on the west, and to the edge of the Murray Basin and the western flank of the Eastern Highlands on the east. Cross-section CD has a horizontal scale only slightly reduced from that of the map. In both sections, the crosses represent pre-Late Carboniferous basement.

pulse of deposition started in the middle Eocene, and extended into the late Eocene.

The Murray Basin started likewise in the Palaeocene after the Late Cretaceous lacuna. According to Twidale *et al.* (1978), 'Tertiary sedimentation in the western Murray Basin was initiated by uplift of the ancestral Mt Lofty Ranges along rejuvenated Delamerian fault-trends, probably in the early Paleocene ... The oldest Tertiary sediments in the area are palynologically dated middle Palaeocene and are of fluvio-lacustrine origin ... In a borehole at Waikerie, quartz sands, gravels and carbonaceous clays of this age were intersected ... further southeastwards into the basin, and away from the marginal ranges, carbonaceous silts ... are of the same age.' This trend from coarse to fine sediment is shown by an arrow in Fig. 75B.

Williams and Goode (1978) postulate 'that prior to the uplift of the Mount Lofty–Flinders Ranges [in the late Miocene] an ancient Murray River in South Australia extended across the present line of the ranges to a major delta in Spencer Gulf'. In reply to a discussion of their paper, Goode and Williams (1980) question Twidale *et al.*'s (1978) interpretation of an early Palaeocene uplift of the ancestral Mount Lofty Ranges. 'The Paleocene deposits in the western Renmark–Morgan area have not been shown to be of piedmont origin, nor have any palaeocurrent directions indicating derivation from the west been determined.' Furthermore, from a recent drilling campaign, they find no piedmont deposits close to the ranges. Important as this point may be, it is secondary to the chief evidence here and in the Birdsville Basin of the Late Cretaceous lacuna being terminated in the Palaeocene by the widespread simultaneous deposition of fluvial sediments shed from the ancestral Eastern Highlands, reflecting the sudden damming of the main drainage by the uplift of the South Australian Highlands. The rare occurrence of gravel distant from the main eastern source is consistent with proximity to the secondary source of the postulated ancestral South Australian Highlands. Of course, the postulated dam may have impeded but not blocked the flow, and local and intermittent spillways, as required by Williams and Goode (1978) to supply sediment to a delta in Spencer Gulf, could be expected.

The evidence on the south-west side of the Highlands is of the same kind. The Pidinga Formation is a sheet of mainly marine sand that transgressed northeastward from the Great Australian Bight and Duntroon Basins in the Palaeocene (Fraser and Tilbury 1979) to reach the north-east Eucla Basin (Lindsay and Harris 1975) and the southern and northwestern parts of the Gawler Block in the middle Eocene. The transgression lapped over ground that rose to the north-east, but how high it was cannot be told.

Altogether, the evidence from the surrounding Cainozoic basins suggests that an ancestral South Australian Highlands existed in the Palaeocene. Wopfner *et al.* (1974) mention 'an apparent renewal of epeirogenic movements' during the middle Eocene in the southern Birdsville Basin, and this second cycle of deposition in this and the other flanking basins, following an early Eocene lacuna, coincides with the inception of deposition of the intramontane basins (1 to 8 of Fig. 75B), whose time relations are shown in Fig. 76. Piedmont deposits here show that at least from the middle Eocene the ancestral Highlands had enough local relief to supply coarse sediment to the intramontane basins. According to Parkin (1969), 'These basins [1, 2, 6–8 of Fig. 75B] are developed parallel to the ancient Torrens Crush Zone. They follow a complex system of grabens which lie immediately west of the tilted blocks or horsts of the Highland Chain. Similar grabens east of the chain are found in the Frome Embayment and beneath the riverine plains in Victoria and New South Wales. Steeply dipping marginal faults separate the horsts and grabens. Development of the rift is recorded within the basins by thick sequences of clay, sand and gravel.'

The basal, Eocene, part of the Torrens Basin adjacent to the Torrens Fault comprises non-marine mudstone with occasional gravel and lignite (Johns 1968). Southward along the Torrens lineament, on the west side of the St Vincent Basin, Stuart (1970) determined from studies of cross-bedding that the transport of the middle and late Eocene quartz sandstone with subordinate conglomerate, siltstone, and clay was towards the east-south-east; and on the other side of the St Vincent Basin, Cooper (1979) reported that the basal deposit, the middle Eocene non-marine North Maslin Sand Member, was deposited from the east, as shown by studies of cross-bedding. The overlying late Eocene through middle Miocene strata comprise marine limestone that passes eastward into marginal marine to non-marine quartz sand, called the Pirramimma Sand Member. The proximity of these facies changes to the basin-bounding faults shows that the faults were active through deposition, and that the ancestral Mt. Lofty Ranges formed the upthrown block. The situation on the eastern side of the Mt. Lofty Ranges is not so clear: as already related, in the area north-west of Norwest Bend, Williams and Goode (1980) report that 'no coarse clastics indicative of a nearby early Tertiary mountain-front have yet been

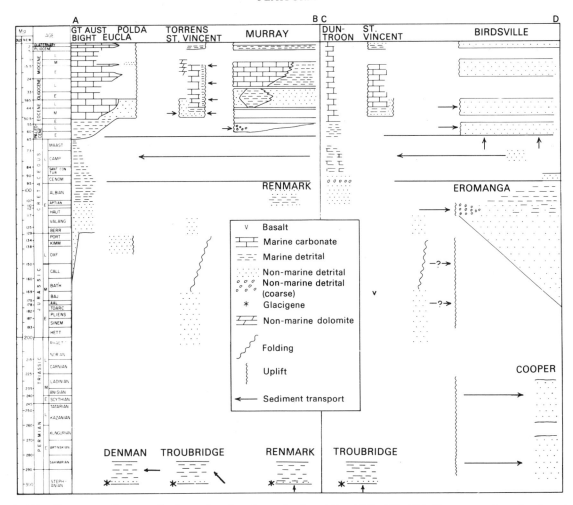

Fig. 76. Time–space diagrams of basins projected on the section lines AB and CD of Fig. 75B. Short arrows show direction of sediment transport from local sources, and the long arrows the direction from the distant source of the Eastern Highlands.

encountered in any drill hole'. Smaller Cainozoic basins (Fig. 75B, 7, 8) are shallow depressions with lignitic sands of probable middle and late Eocene age, but the information available is insufficient to indicate local contemporaneous earth movement.

In summary, the depositional patterns around the South Australian Highlands indicate uplift in the Palaeocene, and, supplemented by internal data, indicate another pulse of uplift and denudation in the middle and late Eocene. Ground movement has probably continued, with varying degrees of intensity, through to the present day, as manifested in present seismicity.

Williams and Goode (1980) argue that uplift of the Mt. Lofty–Flinders block was probably non-uniform, so that, at times and places of lesser uplift, the ancestral Murray River may have crossed the Highlands as far as Spencer Gulf to form the Broughton delta. The postulated delta has a volume of 0.002×10^6 km^3, calculated from a radius of 55 km and a generously estimated thickness of 0.2 km. In contrast is the great volume (0.520×10^6 km^3) of sediment of the Late Cretaceous Ceduna depocentre, beneath the Great Australian Bight, with no competent source other than the Eastern Highlands; this is compelling evidence for the westward extension in the Late Cretaceous of the ancestral

Murray River across the entire area of the South Australian Highlands to the Great Australian Bight, a topic that we develop later.

(iii) Later Cretaceous

In and around the South Australian Highlands, the depositional sedimentary record of the later Cretaceous, between the Cenomanian and the Palaeocene, 90 to 65 Ma ago, is negligible except offshore. This is true for the whole continent, in contrast to the immediately preceding mid-Cretaceous epoch, which saw at least half of the continental platform covered by sediment. The only deposits which might render this lacuna anything less than absolute are the scattered outcrops of possibly Late Cretaceous fluvial Mt. Howie Sandstone (Wopfner 1963; Forbes 1972), which disconformably overlie the Cenomanian Winton Formation in north-east South Australia. The absence of widespread sediment deposition during this period is coupled with intense chemical weathering to produce a thick carapace on the Cenomanian to Early Cretaceous substrate.

In terms of evidence from within the geographical frame we are considering, the depositional history of the South Australian Highlands during the later Cretaceous is a blank because the region then formed part of a land surface of continental extent (Twidale *et al.* 1976; Twidale and Harris 1977). Only from the external evidence of thick deposition in the Ceduna depocentre can we start to deduce the region's history. This also is deferred to our review in Chapter V of the history of the later Cretaceous at continental scale.

(iv) Jurassic through Cenomanian

The sediment of the Eromanga Basin laps against the northern flank of the Highlands (Fig. 76). The only part of this onlapping sequence with a clear provenance in the Highlands is the Early Cretaceous Mt. Anna Sandstone Member in the area between the Gawler Block and the Pedirka Basin (Wopfner *et al.* 1970). It consists of cross-bedded quartz sandstone with interbeds of porphyry conglomerate, the porphyry being identical with exposures of the Proterozoic Gawler Ranges Volcanics 300 km to the south-south-west (Fig. 75B), and cross-bedding directions and thickness distribution confirm this provenance. The calibre of the porphyry boulders implies considerable elevation of the ancestral Gawler Ranges. A solitary outlier of flat-lying Jurassic/Cretaceous sandstone near Leigh Creek (9 on Fig. 75B) unconformably overlies folded Late Triassic/mid-Jurassic coal measures (Johns 1975a). Folding was coeval with deposition (Townsend 1979) and was accompanied by uplift and erosional bevelling. If uplift had been prevalent over a wider area – and similar coal basins of this age 160 km southward, 10 in Fig. 75B, suggest that this may have been so – it would have provided a source for the near-by Jurassic sediments of the Eromanga Basin.

(v) Late Triassic–Early Jurassic

Little of the coeval history of the Highlands can be inferred from the tiny coal basins at Leigh Creek, Boolcunda, and Springfield, which have a maximum thickness of 1000 m and an aggregate area of 60 km^2. They are probably remnants of a much more extensive sequence (Johns 1975a) almost all of which was rapidly removed by erosion in the Middle and Late Jurassic. It is tempting to believe that the Triassic/Jurassic coal basins anticipated the Eocene ones in being deposited in intramontane sags and grabens (Townsend 1979) – the Eocene Willochra Basin overlies part of the Triassic Boolcunda Basin – but evidence is lacking.

(vi) Late Carboniferous–mid-Triassic

According to Wopfner (1981), ice issued in the Permian from the conjugate Antarctica 'towards South Australia where an elevated block formed a median ridge that divided it and provided foci for local ice accumulations. The dividing ridge remained during the later marine and fluviatile episodes of the Permian' (Fig. 77). We note, in passing, that the glacial sediments, formerly dated as Permian, are now regarded as Late Carboniferous (Cooper 1981). With respect to the postulated Highlands, the pattern of ice transport is radial outwards in the Denman and Arckaringa Basins, and cross-cutting in the Troubridge Basin, and the subsequent marginal marine sediments ring all sides except the north, which is occupied by the non-marine Cooper Basin, with sediment transport directions from the south and south-west (Thornton 1979). Cooper *et al.* (1982) note an occurrence of presumably non-marine glacigene sediments in the Polda Basin (McInerney 1979) beneath the Late Jurassic Polda Formation (Gatehouse and Cooper 1982).

(vii) Discussion

From this review, we conclude that the area of the South Australian Highlands has been high ground intermittently back to the Late Carboniferous. Evidence not reviewed here shows that Early Cambrian events in the Adelaide area, called the Cassinian Uplift (Parkin 1969, pp. 99, 100), marked the inception of the Highlands.

In the summary of events (Fig. 76), the Late Cretaceous is conspicuous for the absence of any depositional record onshore in or about the Highlands.

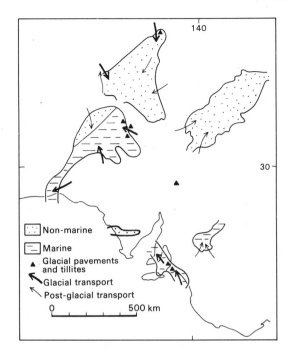

Fig. 77. Late Carboniferous and Permian basins bordering the South Australian Highlands, after Wopfner (1981), with ages from Cooper (1981).

As we relate later, this lacuna extends over the entire continental platform, and indicates a change of regime.

Another aspect of regional interest is the role of the ancestral South Australian Highlands as a source of sediment deposited far from the source. We have noted already that the Late Carboniferous to Triassic Cooper Basin had a source in the ancestral Highlands, and point out here that the Cooper Basin extends some 700 km from this source. An even wider influence is suggested by the occurrence in the Early Jurassic Precipice Sandstone of the Surat Basin, some 1000 km away, of muscovite with K-Ar dates about 500 Ma that coincide with a major period of retrogressive metamorphism in the Willyama Block (Martin 1981).

(c) History of the Great Western Plateau

The Great Western Plateau occupies all but the coastal fringe of the western two-thirds of the continental platform, and is bounded on the east by the Central-Eastern Lowlands. The higher parts of the Plateau correspond with the structural units of the Western Shield, the Kimberley Block, the Pine Creek–Arnhem Block, and the Amadeus Transverse Zone. The analysis of the Great Western Plateau is focused on these parts.

(i) Western Shield

The Western Shield (Trendall 1975; Gee 1979, 1980; Hunter 1981) comprises the Archean Pilbara and Yilgarn Blocks and the intervening and bordering Proterozoic provinces (Figs. 78, 79, 80). The Pilbara and Yilgarn Blocks are Archean (> 2.5 Ga) metamorphic and plutonic terrains intruded by mafic dykes (2.4 Ga), and overlain round their edges by sedimentary and volcanic rocks of Proterozoic and Phanerozoic age; these rocks overlie the Archean rocks at unconformities or nonconformities except along the sharp meridional line of the Darling Fault.

The Archean and Pilbara Blocks and intervening Proterozoic basins today occupy ground higher than 200 m, and the surrounding terrain of Proterozoic mobile zones and superimposed Phanerozoic basins, except the Officer Basin, occupies ground lower than 200 m (Figs. 78, 79). Thus, the Darling Mobile Zone (Glikson and Lambert 1976) and the overlying Perth Basin wrap round the western edge of the Yilgarn Block, marked by the Darling Fault, and the Albany–Fraser Province and the overlying Bremer and Eucla Basins wrap round the southern and south-eastern sides of the Yilgarn Block. The abrupt discontinuity of the structural grain of the Yilgarn Block and the encircling trends of the Proterozoic Darling and Albany–Fraser mobile zones resemble the Precambrian structural pattern of much of the rest of Gondwanaland (Johnstone et al. 1973), in particular that of East Africa (Veevers 1981), and has led to a similar pattern of location of subsequent rift structures around the Precambrian nuclei.

The correlation between the present ground surface elevation of 200 m and the boundary between Precambrian and Phanerozoic rocks breaks down in the area of the north-western Officer Basin, which occupies a topographic saddle, at an elevation of ~500 m, between the high ground of the Yilgarn Block in the Ernest Giles Range (EGR) and the Amadeus Transverse Zone in the Warburton Ranges (WR). As we shall see later, because the present 200 m contour approximates the limit of the Aptian–Albian shoreline, the Officer Basin must have risen subsequently by about 300 m relative to the other flanks of the Shield.

(1) Topography and sedimentation in the Cainozoic
The southern flank of the Western Shield is lapped by late Eocene marine and paralic sediments of the Bremer Basin and, on the south-east, by similar early and middle Miocene and Eocene sediments of the Eucla Basin. Jackson and van de Graaff (1981)

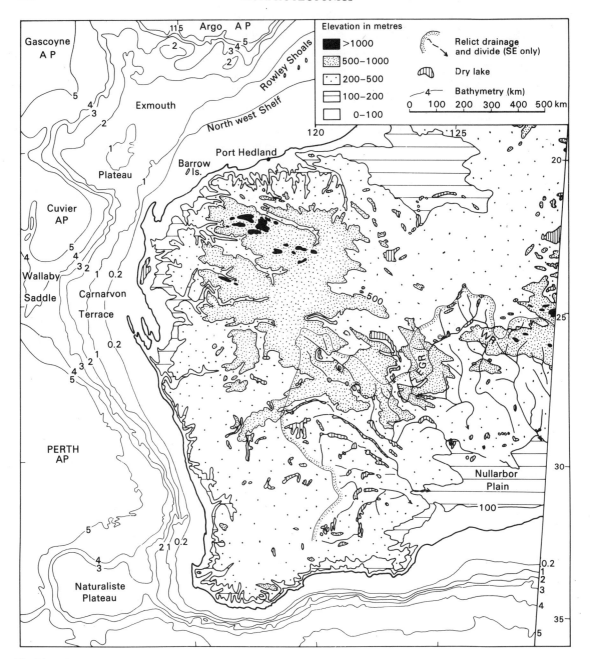

Fig. 78. Morphology of the Western Shield surrounded on the south, west, and north by a coastal plain and submarine margin, and on the east by a saddle within the Great Western Plateau between the Ernest Giles Range (EGR) and the Warburton Ranges (WR). Land contours (m) from *Times Atlas of the World* (1975), submarine contours (km) from BMR (1979). Relict divide and drainage in south-east from van de Graaff et al. (1977b).

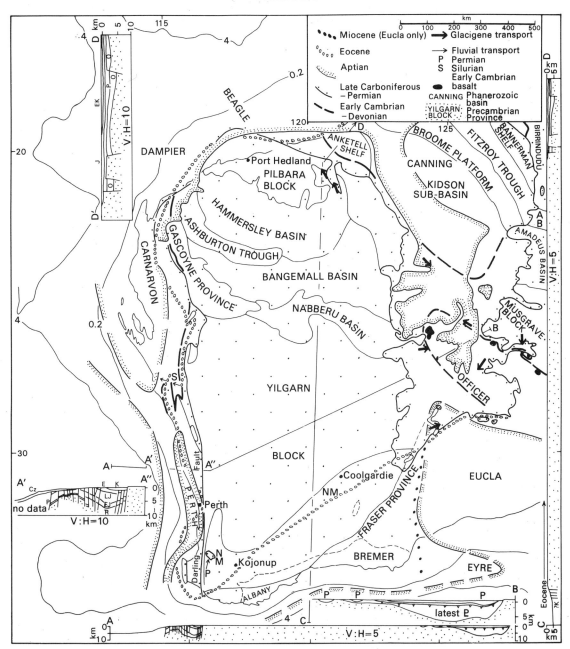

Fig. 79. Geology of the Western Shield showing Precambrian provinces (from Gee 1980), bounded by the Perth, Carnarvon, Dampier, Beagle, Canning, Officer, Eucla/Eyre, and Bremer Basins, with limits of Early Cambrian to Devonian, Late Carboniferous to Permian, Aptian, Eocene, and, in the Eucla Basin only, Miocene sediments, and outcrops of Early Cambrian basalt in the Officer Basin. Provinces of the Amadeus Transverse Zone shown on the north-east. Wide arrows show the direction of Late Carboniferous glacigene transport, and thin arrows Silurian (S) and Permian (P) fluvial transport.

Cross-section AB from Jones (1976, fig. 8, B) and Jackson and van der Graaff (1981, pl. 1, EF); CD from Cooney *et al.* (1975, fig. 4) and Playford *et al.* (1975, fig. 50, AB). Enlargements A′A″, B′B, D′D show details of Phanerozoic basins. P̄, Proterozoic; O, Ordovician; P, Permian; R̄, Triassic; EJ, Early Jurassic; LJ, Late Jurassic; EK, Early Cretaceous. AB Arunta Block.

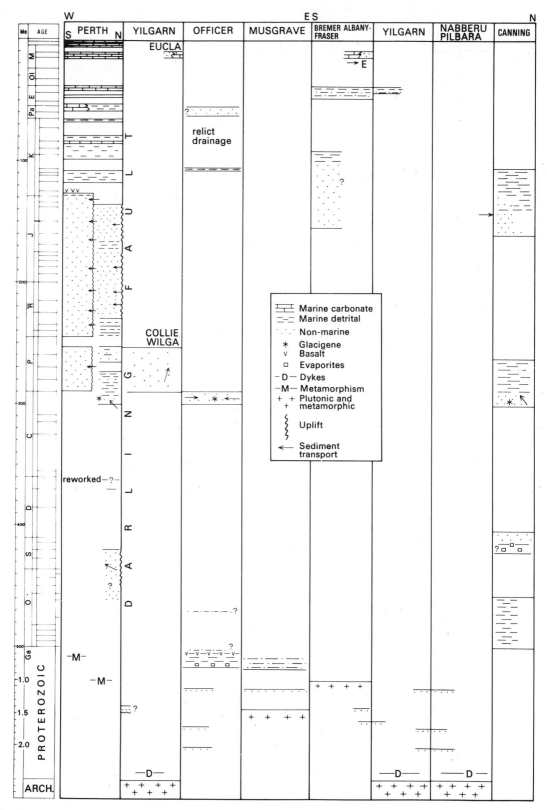

Fig. 80. Time–space diagrams of basins projected on the section lines AB and CD of Fig. 79 but without horizontal scale. Arrows show direction of sediment transport. Note change of scale (to Ga) below 500 Ma. Ages of Precambrian units, not discussed further, from Gee (1980).

interpret the Miocene feather-edge of the Eucla Basin as a transition from fluvial coastal detrital sediment in the north, including conglomerate with clasts of laterite and silcrete, to open marine limestone in the south. A similar transition is found in the underlying middle and late Eocene marine sediments in the Eucla Basin (Jackson and van de Graaff 1981) and in paralic to marine late Eocene sediments including lignite and spongolite, in the Bremer Basin (Lord 1975; Playford et al. 1975). Scattered coeval non-marine quartz sandstone and conglomerate occur at Kojonup, 250 km south-east of Perth, and shale, lignite, and conglomerate near Coolgardie (Playford et al. 1975).

On the eastern side of the Western Shield, the only Cainozoic deposits, besides the unconsolidated superficial fluvial and aeolian sediment, are thin (5 m) but widespread silicified fluvial conglomeratic sandstones called the Lampe Beds. They lack fossils, and therefore cannot be dated directly; they overlie Aptian rocks and are thus Late Cretaceous or Cainozoic, and because they are intensely ferruginized, are considered to be older than the laterite, which is pre-Miocene (Jackson and van de Graaff 1981): 'The fluviatile Lampe Beds, which are only preserved on the divides of the palaeodrainage system must have been deposited during an earlier fluvial episode. As the preserved palaeodrainage patterns in the Officer Basin area were established by the Late Eocene, the Lampe Beds must predate the Late Eocene, and a Late Cretaceous–Palaeocene age is favoured . . . The Lampe Beds represent a widespread phase of post-Early Cretaceous alluviation.' Late Cretaceous–Palaeocene encompasses the age of the lower part of the Eyre Formation of the Birdsville Basin, and the Lampe Beds possibly represent the same phase of continental alluviation. The development of the palaeodrainage mapped by van de Graaff et al. (1977b) and Jackson and van de Graaff (1981) (Fig. 78) probably started in the Late Cretaceous; it involved about 100 m of denudation (van de Graaff 1981, p. 167), and provided a minor source of the Late Cretaceous sediment in the Ceduna Depocentre to the south-east. It was then clogged, probably in the Palaeocene, by the deposition of the Lampe Beds; the surface was lightly imprinted and deeply weathered before a change from humid to arid climate and then the late Eocene marine onlap. Under aridity, the Oligocene uplift of 300 m in the area between the late Eocene and the Miocene shorelines failed to revive the drainage.

The Cainozoic record on the north-eastern flank of the Western Shield is negligible. Pisolitic ironstone and quartz sandstone (Poondano Formation) that cap mesas in the area 100 km south-east of Port Hedland, and chalcedony, marl, and limestone 200 km south-east of Port Hedland (Oakover Beds) may be Cainozoic, but evidence of their age, beyond their being Mesozoic–Cainozoic, is lacking (Playford et al. 1975). In contrast, west of Port Hedland, and along the western flank, abundant evidence from marine strata, mainly offshore, shows that the present situation of a shelf and narrow coastal plain abutting the plateau has persisted throughout the Cainozoic. Shelf carbonate with equivalent inshore to non-marine terrigenous sandstone was deposited in four transgressive–regressive cycles (Quilty 1977) separated by hiatuses. Quilty (1977, p. 338) notes that 'conditions along the Western Australian margin throughout the Cenozoic seem unusual, in that they caused such intervening regressive periods to be virtually devoid of erosion, the margin being an area of minimal drainage and sediments usually being dominantly of marine biogenic carbonate'. The only evidence of drainage from the land is in the late Palaeocene and middle/late Eocene near Perth (Playford et al. 1976b) and in the late Eocene and the Pliocene/Pleistocene in the Carnarvon Basin. Quaternary beach/dune sequences are described by Semeniuk and Johnson (1982). The modern sediments of Shark Bay have been monographed by Logan et al. (1970, 1974).

(2) Late Cretaceous

During the Late Cretaceous in the Dampier and Beagle Basins of the North-west Shelf on the north-west flank of the Western Shield, the depositional environments anticipated the present range of inner shelf to upper slope, as detailed by Apthorpe (1979). Shales were deposited on the inner shelf, and carbonates in deeper water. To the south, the shallow shelf sediments of the Carnarvon Basin are likewise mainly carbonate, whereas those of the Perth Basin are mainly terrigenous (sandstone and greensand) though thin (< 100 m onshore) so that they reflect a meagre supply of sediment from drainage off the land. The only evidence on the southern flank comes from the Eyre and Eucla Basins. At Jerboa, 350 m of marine Cenomanian siltstone, sandstone, and shale (Fig. 99: Bein and Taylor 1981), and, in the south-western Eucla Basin (Playford et al. 1975), 200 m of Cenomanian through Coniacian siltstone and greensand represent the Late Cretaceous. Further to the south-east, in the Ceduna Depocentre, the Late Cretaceous thickens to 8 km, most of it derived from Eastern Australia, as discussed later, with a smaller, but still important, contribution from the eastern flank of the Shield, which rose rapidly after the Albian and in the process was denuded during the Late Cretaceous by about a hundred metres (Van de Graaff 1981). This uplift of the south-eastern part of the Shield and of the north-western Officer Basin is the only event in the region during the last 95 Ma that has produced an important volume of sediment, all of which was deposited well outside the region. Sediments of this interval are too thin to show on the cross-sections of Fig. 79. In the preceding time interval, from the Late Carboniferous to the mid-Cretaceous, sediments are particularly thick along the western margin, and appreciably so along the other flanks.

(3) Late Carboniferous to mid-Cretaceous
This period encompasses the inception and filling of the rift-valley complex of the Perth/Carnarvon Basin and the rift-divergence zone of the Dampier and Beagle Basins (Veevers 1981), and the intermittent deposition of sediment round the other flanks of the Western Shield.

As described later, the area that was to become the western margin accumulated thick sediments in a broad downwarp along its length during the Late Carboniferous to the mid-Triassic, and then in rift valleys until continental breakup in the Middle–Late Jurassic (160 Ma) in the Dampier and Beagle Basins (Crostella and Barter 1980), and in the Early Cretaceous (128 Ma) in the Carnarvon and Perth Basins. East of the Darling Fault, the shoulders of the Western Shield towered over the rift valley of the Perth Basin, and further north, blocks on the eastern side of the Flinders and Rough Range Fault systems rose above the Carnarvon and Dampier Basins.

According to van de Graaff (1981), all available (but little published) palaeocurrent, facies, and sediment petrographic data from the Perth Basin sediments point to the very thick fluvial formations, such as the 3.6-km-thick Late Jurassic Yarragadee Formation (Veevers and Hansen 1981), being poured in from the rising shoulder of the Yilgarn Block, which was denuded by at least 500 m. The only sediment found outside the rift-valley complex is in the Early Permian Collie and Wilga Basins south-south-east of Perth. They contain fluvial sandstone deposited from palaeo-currents that flowed northward (Wilde and Walker 1978). In the Dampier and Beagle Basins of the North-west Shelf, the Triassic though Middle Jurassic sequence contains facies changes along its wedge-out near the present coast that indicate deposition along a shoreline except during Late Triassic deposition of thick fluvial sands from the east (Crostella and Barter 1980).

Along the southern flank of the Shield, a Late Jurassic to Early Cretaceous rift valley is interpreted from seismic profiles (Cooney et al. 1975) but drilling is needed to confirm the age.

The saddle between the Western Shield and the Amadeus Transverse Zone is covered by a thin (450 m) but widespread Late Carboniferous non-marine glacigene sequence of diamictite, pebbly sandstone, and siltstone (Paterson Formation) overlain on the north-east by a very thin (100 m) Aptian marine claystone, siltstone, and sandstone (Samuel Formation and Bejah Claystone), and on the south-east by marine Aptian to Senonian conglomerate, sandstone, siltstone, and shale (Jackson and van de Graaff 1981). Glacigene transport in the Paterson Formation is from the Yilgarn Block and the Amadeus Transverse Zone (Fig. 79). The Aptian marine conglomerate and sandstone of the Eucla Basin possibly reflect deposition along the subsiding coast of the south-east Shield but the other Early Cretaceous deposits, lacking sediment any coarser than medium sand, simply reflect a quiet shallow marine environment of deposition unaffected by the surrounding land except that marine microplankton are generally subordinate to spores and pollen, 'indicating that deposition was at no time far from the Cretaceous shoreline nor from the parent vegetation which provided the source of the spores and pollen' (Jackson and van de Graaff 1981). The Canning Basin that laps the north-eastern flank of the Shield has similar sequences except that deposition was continuous from the Late Carboniferous into the Permian, and the Aptian was preceded by deposition back to the Late Jurassic, with a source to the south (Forman and Wales 1981). East of the Pilbara Block, in the valley of the Oakover River, the Late Carboniferous Braeside Tillite overlies glaciated pavements that indicate transport northward and northwestward, outward from the Shield (Veevers and Wells 1962) (Fig. 79).

(4) Cambrian to Carboniferous
Evidence of regional palaeoslopes during this interval is scanty. The western flank of the Shield is indicated by west-north-west palaeocurrents in the fluvial parts of the 3-km-thick Silurian Tumblagooda Sandstone, which was probably deposited at the foot of the scarp of the Darling Fault (Hocking 1979). During the Late Devonian and Early Carboniferous, shallow marine limestone and alternating fine quartz sandstone were deposited in the eastern Carnarvon Basin alongside the Gascoyne Province, which shed boulder conglomerate along the coast (Thomas and Smith 1974). A final phase of faulting occurred in the mid-Carboniferous. In the southern Canning Basin, the Ordovician and Silurian–Devonian sequences are entirely subsurface

and too little is known about them to confirm the presence of land to the south.

In the Officer Basin, the saddle between the Western Shield and the Amadeus Transverse Zone is underlain by Early Cambrian basalt overlain by thin poorly dated early Palaeozoic sediments, which indicate nothing about the coeval landscape.

(5) Discussion
From this survey, we confirm the widely held view (Johnstone *et al.* 1973) that the area of the Western Shield has been a tract of ground higher than its surroundings, at least intermittently, since the Late Carboniferous, and along its western flank since the Silurian. Incidentally, this pattern, according to Daniels (1975, figs. 79–81), has a history stretching back into the Proterozoic. The Western Shield was the predominant, if not sole, source of the Phanerozoic detrital sediment on the west and north; a source, jointly with the Musgrave Block, of the Permian and Cretaceous sediments of the Officer Basin; and a minor source of the Late Cretaceous succession of the Ceduna Depocentre. The western flank of the Shield, the southern part of which is marked by the Darling Fault, has been a scarp since the Silurian (Playford *et al.* 1976b), and only twice, during the Early Permian and late Eocene, were sediments deposited and subsequently preserved east of the fault. The northern and eastern flanks have likewise been stable since at least the Permian, and only the southern flank has moved substantially when during the late Eocene it subsided to admit the sea and then rebounded in the Oligocene.

(ii) Kimberley Block, Pine Creek Inlier, Arnhem Block

(1) Introduction
The high ground of the northern part of the Great Western Plateau (Figs. 81, 82) is occupied by the Kimberley Plateau and Arnhem Land. The Kimberley Plateau corresponds to the Kimberley Block, rimmed by the Halls Creek Province of Early Proterozoic (~ 2 Ga) metamorphic and plutonic rocks and overlain by the Kimberley Basin of Middle Proterozoic (1.8 Ga) volcanic and sedimentary rocks (Fig. 83). The Kimberley Block is enclosed by thick Phanerozoic basins on all sides except the east. Arnhem Land occupies the Pine Creek Inlier and the Arnhem Block of Early Proterozoic (2.0 to 1.7 Ga) metamorphic and plutonic rocks separated by the McArthur Basin of Middle Proterozoic (1.7 to 1.4 Ga) volcanic and sedimentary rocks. Between the earlier Precambrian Kimberley and Pine Creek–Arnhem Blocks are the Late Proterozoic (1.2 to 0.9 Ga) Victoria River Basin and remnants of an originally extensive sheet of Early Cambrian basalt and Middle Cambrian through Early Ordovician sedimentary rocks. From the south-west to the north-east, the Precambrian blocks are bounded on the north-west by thick Phanerozoic basins: the Canning Basin onshore and offshore; the Browse Basin, wholly offshore, beneath the North-west Shelf and Scott Plateau; the Bonaparte Gulf Basin beneath the Sahul Shelf and Timor Trough, abutting to the north-west the Timor orogen, and extending southeastward onshore to a small area of Palaeozoic rocks; the Money Shoal Basin; and, beneath the Arafura Sea, the Arafura Basin, with a small area of early Palaeozoic rocks onshore.

The correlation of high ground with Precambrian terrain and low ground with Phanerozoic basins, clear in the southern part of the Great Western Plateau, is less clear here: the correlation is negative for the exposed Pine Creek Inlier and the Arnhem Block but positive if the entire Precambrian of Arnhem Land is taken into account.

The older Precambrian terrain of the Kimberley Plateau and Arnhem Land rises above and protrudes into the offshore Phanerozoic terrain. This is not so towards the interior; here the remnant early Phanerozoic basins that overlie Late Precambrian basins have much the same elevation as the older Precambrian terrain. This reflects the contrast that has pertained throughout the Phanerozoic between the stable continental platform and the subsiding continental margin. Here as elsewhere in much of Australia, the present shoreline is a feature of great antiquity. Within the offshore area, the antiquity of the modern morphology is expressed by the coincidence of the Bonaparte Depression, in the middle of the Sahul Shelf, shown in Fig. 81 by the 0.1 km isobath, with the Permian through Cainozoic depocentre of the Bonaparte Gulf Basin (Veevers and van Andel 1967). Here a present morphological relief of 0.1 km is matched by a structural relief of at least 13 km that has developed at the intersection of a Palaeozoic and early Mesozoic north-west trend (Petrel Sub-basin) and a late Mesozoic and Cainozoic trend (Malita Graben) during the past 300 Ma, from the Permian to the present day (Laws and Kraus 1974), and probably back to the Cambrian.

These long-standing morphotectonic features of north-west Australia contrast with those of the youthful orogen of the Outer Banda Arc to the north-west, with a morphological relief of 6 km, from the crest of the arc in Timor to the axis of the Timor Trough, generated during the past 3 Ma (Carter *et al.* 1976; Veevers *et al.* 1978).

In what follows, we review the evidence from

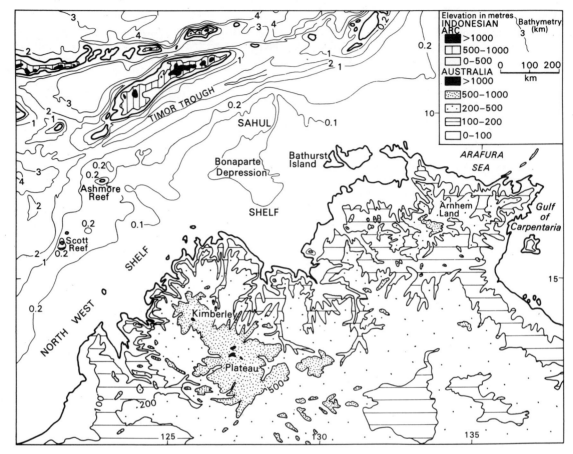

Fig. 81. Morphology of the Kimberley–Arnhem Land region and offshore areas of north-west Australia. Land contours (m) from *Times Atlas of the World* (1975), submarine contours (km) from BMR Earth Science Atlas (1979) and van Andel and Veevers (1967).

onshore and offshore concerning past uplands, but leave a fuller description of the offshore geology for a later occasion.

(2) Cainozoic patterns of erosion and deposition

The record of marine deposition during the Cainozoic is wholly offshore except for two localities onshore, described by Lloyd (1968a, b): chert and shale 100 m thick at White Mountain in the Ord Basin, and limestone 20 m thick at Rockhampton Downs in the Georgina Basin contain the marine foraminifer *Ammonia beccarii* associated with non-marine fossils. *A. beccarii* ranges back to the Miocene, and Lloyd suggests that it was deposited during a Miocene marine incursion. The present elevation of these outcrops, 250 m at White Mountain and 200 m at Rockhampton Downs, entails a post-Miocene uplift of these amounts, as discussed later.

Offshore, the modern sedimentation in the Arafura Sea (Jongsma 1974), Sahul Shelf (van Andel and Veevers 1967), and North-west Shelf (Jones 1973b) follows a pattern that can be traced back to the Late Cretaceous, with its first appearance in the Late Devonian of the Bonaparte Gulf and Canning Basins. This modern pattern, though smeared today by the effect of the rapid Holocene transgression, comprises quartzose sediment close inshore, in particular round the Kimberley Block, and calcareous sediment offshore, with a minimum $CaCO_3$ content ranging from 30 to 80 per cent, dominated by the calcarenite of the offshore shoals such as the stable Sahul Platform and Londonderry High and the calcareous silty clay of

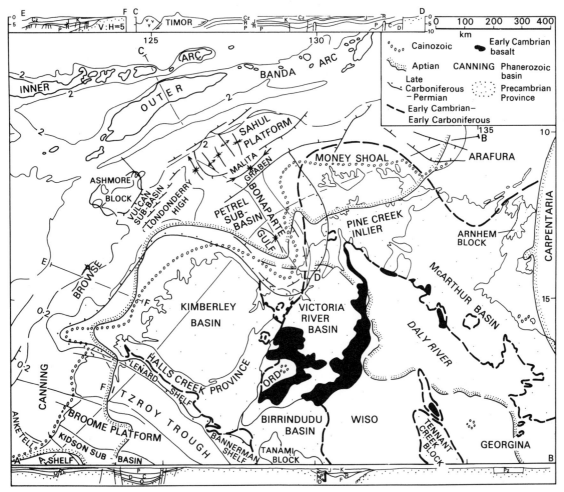

Fig. 82. Geology of the Kimberley–Arnhem Land region and offshore areas, showing Precambrian provinces (from Gee 1980, and Plumb 1979), with limits of Early Cambrian to Early Carboniferous, Late Carboniferous–Permian, Aptian, and Eocene sediments, and outcrops of Early Cambrian basalt. Age and extent of the Early Cambrian of the Arafura Basin from Plumb et al. (1976) and of the Daly River and Georgina Basins from Muir (1980).
Cross-section AB from Playford et al. (1975, fig. 50, ABC) and Laws and Kraus (1974, fig. 5, CC); CD adapted from Carter et al. (1976, fig. 8), Veevers et al. (1978), and Laws and Kraus (1974, fig. 5, AA); EF from Crostella (1976, fig. 4).

the subsiding Bonaparte Depression and the Central Arafura Shelf. The only obvious influence of the land on this pattern is to provide a narrow coastal zone of detrital sand and to fill depressions in the shelf with detrital mud; the North-west Shelf lacks extensive depressions, and carbonate sand predominates to the outer edge. Ancient counterparts of this pattern would be expected to show the proximity of any ancestral Great Western Plateau more by depositional wedge-out, as shown by isopachs, than by a facies change from offshore carbonate to inshore detrital sand. As mentioned already, the Bonaparte Gulf Basin can be traced back at least 300 Ma (Laws and Kraus 1974), and we may reasonably infer the same antiquity for the Great Western Plateau.

Offshore, the Cainozoic sediments, mainly shelf carbonate with a detrital component inshore, thicken from the depositional feather-edge near the present coastline to 2.4 km beneath the Bonaparte Depression and a maximum of 3.5 km, including 1.4 km of middle Miocene to Holocene reef (Wright 1977), at Scott Reef in the Browse Basin. Deposition at the edge of the

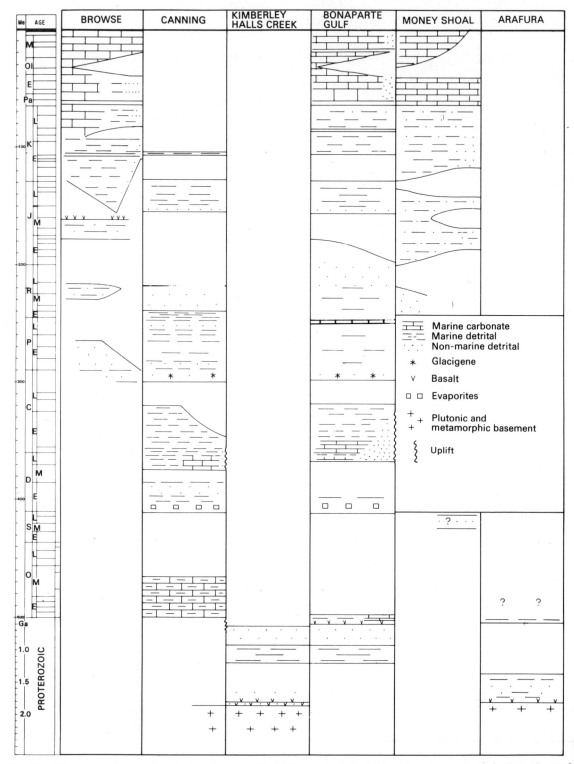

Fig. 83. Time–space diagram along the line of section AB, with Browse Basin in addition. Note change of scale (to Ga) below 500 Ma. Age of Precambrian provinces, not discussed further, from Gee (1980) and Plumb (1979). Based on information in Balke et al. (1973), Carter et al. (1976), Crostella (1976), Crowell and Frakes (1971a, b), Jones (1973b), Jongsma (1974), Laws and Kraus (1974), Lloyd (1968a, b), Muir (1980), Plumb et al. (1976), Roberts and Veevers (1973), Skwarko (1966), Turner et al. (1981), van Andel and Veevers (1967), Veevers and van Andel (1967), Veevers et al. (1967, 1978), and Wright (1977).

shelf was continuous, as at Scott Reef, with intervals that lack planktic foraminifers in the middle Eocene, middle Miocene, and late Miocene (Wright 1977). The Oligocene lacuna on the rest of the shelf expands towards the shore to encompass part of the Eocene. This pattern of deposition is traceable through the Cainozoic into the Late Cretaceous, and shows that the land has contributed only an insignificant volume of detrital sediment except in the Palaeocene and Eocene when a wedge of detrital sandstone stretched half-way across the Browse Basin (Powell 1976b), reflecting coeval uplift of the Kimberley Block.

(3) Late Jurassic to end of Cretaceous
The hinge-line of the Browse Basin has remained in much the same position from the present day back to the Late Jurassic (Powell 1976b), and, during the Late Jurassic through Neocomian, and the Campanian, was marked by the basinward passage from detrital sandstone into claystone and, in the Late Cretaceous, into carbonate, as detailed by Apthorpe (1979). During the Late Jurassic to Aptian, the shoreline advanced southeastward across the Canning Basin; fluvial quartzose sediment came from the Pilbara Block; and Aptian shale and claystone with Radiolaria were deposited in a sea flanked by basin margins that provided no coarse material.

During the Aptian, the sea swept across parts at least of the Pine Creek Inlier and Daly and McArthur Basins (Skwarko 1966) to deposit a sheet of quartzose sediment most of which was subsequently removed. For the rest of the region, the Aptian shoreline remained close to its present location.

(4) Late Carboniferous to Middle Jurassic
As in the south-west, this interval encompasses the inception and filling of a rift-valley complex, with its transformation, in the Middle Jurassic, into a juvenile ocean. The Browse Basin, opposite the Kimberley Block, extended to the south-west across the north-west-trending cross-structure of the older Canning Basin and to the north-east across the Bonaparte Gulf Basin. The hinge-line or wedge-out of sediments remained in a position similar to that of later deposition round the north-west and north-east flanks of the Kimberley Block, but for the first time enclosed the south-west flank along the Fitzroy Trough, and contracted to the western side of Bathurst Island to define the eastern side of the Petrel Sub-basin.

The Late Carboniferous and earliest Permian deposits are glacigene but the direction of ice transport is not known. The Grant Formation is 3 km thick in the Fitzroy Trough, and contains clasts with a wide variety of compositions compatible with their derivation from the Kimberley Block; its sand:shale ratio decreases southward in the Fitzroy Trough (Forman and Wales 1981), suggesting a source in the Kimberley Block. On the other side of the Kimberley Block, the onshore Bonaparte Gulf Basin contains possible tillites (Roberts and Veevers 1973). All this evidence points to a source in the Kimberley Block, which has been sketched as high land covered by an ice-cap in the Late Carboniferous and earliest Permian (Crowell and Frakes 1971a, b).

(5) Late Devonian to Early Carboniferous
Thick wedges of Late Devonian conglomerate along the south-east margin of the onshore Bonaparte Gulf Basin and along the northern edge of the onshore Canning Basin pass laterally into shallow marine sediments, including carbonate reefs, and were derived from fault scarps that separated the basins from the uplifted Victoria River Basin and the Kimberley Block. Occurrences of littoral talus breccia in the middle Tournaisian and late Visean of the south-western margin of the onshore Bonaparte Gulf Basin (Roberts and Veevers 1973) show that the north-eastern flank of the Kimberley Block at this time formed a precipitous coast. For most of the Late Devonian and Early Carboniferous, the onshore Bonaparte Gulf Basin was a marine shelf with two parts: a shallow platform with well-washed quartz sand and carbonate, including reefs, that enclosed a basinal depression that accumulated muds, a forerunner, on a smaller scale, of the modern Sahul Shelf. Along the northern Canning Basin, shallow marine carbonate, including reefs, accumulated on the Lennard Shelf and deeper water sediments in the Fitzroy Trough: towards the south-east, quartz sandstone predominates (Veevers *et al.* 1967), suggesting a source to the east. In the mid-Carboniferous, the latest phase of faulting, related to wrench faults, disrupted the Bonaparte Gulf Basin (Laws 1981) and the northern Canning Basin (Forman and Wales 1981).

(6) Silurian to Middle Devonian
Non-marine Silurian sandstone in the Money Shoal Basin (Balke *et al.* 1973) and evaporites in the Petrel Sub-basin (Laws and Kraus 1974) reveal nothing about the coeval morphology of the Kimberley Block, but in the Canning Basin similar sediments together with coeval detrital sediments on the northern margin of the basin suggest a source in the Kimberley Block. The succeeding red and brown sandstones were probably deposited in a desert (Forman and Wales 1981) but their source is unknown.

(7) Cambrian and Ordovician

The 3-km-thick Early and Middle Ordovician sediments of the Canning Basin stretch over a wide area, with carbonate over the shallow Broome Platform and shale in deeper water elsewhere, and with quartz sand, as in the Late Devonian, towards the east, suggesting a source in this direction.

Deposition in the originally continuous Bonaparte Gulf, Ord, Daly River, Wiso, and Georgina Basins began in the Early Cambrian with the subaerial eruption of flood basalt followed in the Middle Cambrian by shallow marine detrital and carbonate sediment which, except the Ord Basin, extended into the Early Ordovician. A latest Proterozoic/earliest Cambrian sandstone overlies the southern edge of the McArthur Basin (Muir 1980). Evidence of the source of the detrital sediment is lacking. In the Arafura Basin (Plumb *et al.* 1976), a 1.5-km-thick sequence of Middle Cambrian quartz sandstone, shale, and dolomite rests without intervening basalt on the Proterozoic McArthur Basin, but whether it too was part of a continuous sheet of sediment is not known. The absence of Cambrian sediment in the Canning Basin and Kimberley Block and over the main part of the Arnhem Land Block suggests that these areas were land, but facies changes indicating adjacent upland sources are not known in the surrounding basins.

(8) Conclusion

The evidence reviewed shows that the modern topography and pattern of deposition of the northern part of the Great Western Plateau can be traced without change back through the Cainozoic and Late Cretaceous. Before the Late Cretaceous, the Canning Basin occupied on the south-west side of the Kimberley Block a depression that can be traced back to the Ordovician. In the rest of the region, the ancestral Great Western Plateau can be seen to have persisted back to the late Palaeozoic and possibly to the Cambrian, with a broad depositional basin occupying the low ground between the Kimberley Block and the Pine Creek–Arnhem Block, a pattern repeated during later sea-level highstands in the Aptian and Miocene.

(iii) Central Australia: Amadeus Transverse Zone

(1) Introduction

In Central Australia, the central-eastern part of the Great Western Plateau corresponds to the complex of east–west trending Precambrian blocks and Vendian and Phanerozoic basins called the Amadeus Transverse Zone (Figs. 84, 85). Within this part of the plateau, two highland masses that reach an elevation of 1500 m are separated by a depression that contains Lake Amadeus and other dry lakes. The northern highland mass is centred on the MacDonnell Ranges, and is underlain by the Precambrian Arunta Block, the Ngalia Basin, and the northern part of the Amadeus Basin; the north-eastern part of the Plateau, the Mt. Isa highlands, is underlain by the Precambrian Mt. Isa Block; and the southern highland includes the Warburton, Petermann, Mann, and Musgrave Ranges, and is underlain by the Precambrian Musgrave Block. The monolithic dome of Uluru (Ayers Rock) and the neighbouring Mt. Olga complex of domes occupy an easterly spur off the Petermann Ranges, and are notable not only for their scenery but also for their registering in a compact area the earliest and latest phases in an eventful Phanerozoic history.

On the south, the Plateau descends across the arheic Great Victoria Desert, underlain by the Officer Basin, to the Nullarbor Plain, underlain by the Eucla Basin. A coherent pattern of relict drainage (Pitt 1979; van de Graaff *et al.* 1977b) probably dates from the Late Cretaceous, during an interval of uplift of the Musgrave–Officer region. On the east, the west-to-east trend of the Plateau is truncated by an embayment of the Central-Eastern Lowlands, underlain by Permian and younger basins. The western edge of this embayment may correspond to the eastern margin of Australia at the outset of the Phanerozoic. Coherent drainage from the highlands is concentrated on the eastern flank, and contributes mud and sand to the depocentre of Lake Eyre and the sand of the Simpson Desert. On the north, the highlands are flanked by an arheic belt of aeolian dunes between 18° and 22° S, and on the west by the arheic dunefields of the Gibson and Great Sandy Deserts.

(2) Cainozoic record

The southern perimeter of the Great Western Plateau in the Cainozoic is delineated by marine transgressions in the Palaeocene, Eocene, and Miocene, with late Eocene and middle Miocene marine sediments reaching the northern edge of the Eucla Basin. The only other marine Cainozoic sediment known in the region is a limestone in the area between the Pedirka Basin and the Mt. Isa Block, which contains the marine foraminifer *Ammonia beccarii*, thought to be a remnant of a more extensive deposit which lapped the north-eastern edge of the Great Western Plateau in the Miocene (Lloyd 1968a, b).

Within the Amadeus Transverse Zone, only three small areas of well-dated Cainozoic rocks are known (Figs. 86, 87). They are all non-marine. Near Alcoota, east of the Ngalia Basin, vertebrate fossils in a clay to sandstone sequence are tentatively assigned a late Miocene age (Woodburne 1967). Near Napperby, on the eastern side of the Ngalia Basin, Kemp (1976) found middle Eocene pollen in a carbonaceous clay at the bottom of a borehole that penetrated 100 m of sand and 45 m of clay: freshwater dinoflagellates indicate deposition in a lake. And in an 84-m-deep drillhole in the plain between Ayers Rock and the Olgas, Twidale and Harris (1977) found a middle Palaeocene palynomorph assemblage in a lignite immediately above bedrock, believed to be part of an old valley system draining to the south-west. Besides showing the antiquity of the bedrock depression and of the Olgas and Ayers Rock as upland masses, this date denotes the onset of widespread deposition, as manifested in closely similar palynomorph assemblages in the lower part of the Eyre Formation of the Birdsville Basin to the south-east (Wopfner et al. 1974), which laps the eastern flank of the Amadeus Transverse Zone, and the middle Palaeocene inception of deposition in the Murray Basin. The middle Eocene lake deposit at Napperby has the same age as the upper part of the Eyre Formation, the numerous small coal basins in South Australia, and oil-shale basins in Queensland. This was a time of basin formation, commonly accompanied by faulting, with uplift in one place attended by downfaulting in another, and the Napperby deposit is consistent with this event extending to central Australia. Neogene block-faulting along ancient, probably Precambrian, fractures, is inferred from palaeomagnetic studies (Burek et al. 1978).

In summary, the Cainozoic record of central Australia is a continuation back to 60 Ma of the present pattern of local sags, such as Lake Amadeus, within highlands flanked by broad plains. Today, little sediment is shed by the highlands, and the wholly carbonate province on the south, in the Great Australian Bight, is traceable in the Eucla Basin back to the Eocene. What little sediment escapes central Australia today, via the coherent stream system on the east and south-east – the Finke River and others to the south are the only streams that maintain flow out of the region – is concentrated in the sump of Lake Eyre and the dunefields of the Simpson Desert, and this pattern (but not facies) of deposition persists to the beginning of the Cainozoic in the form of the Eyre Formation (Wopfner et al. 1974). The aeolian sand dunes of the Great Sandy Desert of the Canning Basin and of the area to the north between 18° and 22° S, in which only a few tens of metres of sediment have accumulated, are the only other sediments derived from central Australia.

(3) Late Carboniferous to Cretaceous

A sequence of Neocomian non-marine pebbly sandstone overlain by Aptian marine shale, like that of the northern flank of the South Australian highlands, laps the eastern flank of the Amadeus Transverse Zone, and represents the widest expansion of the phase of deposition that started in the Late Triassic/Early Jurassic and produced the Eromanga Basin. Shoreline features in the marine shale are not known, and the proximity of highlands to the west is based on the coarse calibre of the Neocomian sandstone. On the southern flank, the edge of the Aptian marine sediment represents a shoreline, but on the west, the Aptian marine shale and mudstone with Radiolaria have a single facies only, which includes an outlier south of Lake Lucas, and the original extent of these deposits is unknown.

At the base of the Late Carboniferous to Triassic sequence, glaciated pavements and the direction of glacial outwash (Fig. 88A) point to high ground in part at least of central Australia, as is confirmed by the lapping of the western and eastern flanks by this sequence. The Arckaringa Basin, on the south-east flank of the Musgrave Block, formed a narrow saddle between the Musgrave and Gawler Blocks, and its southern part was filled by the sea that followed the glaciation. The outlier of Permian sediments at Lake Lucas shows that subsequent movements disrupted the originally continuous sheet of sediment, but its proximity to the Canning Basin suggests little extension further east.

(4) Devonian and Carboniferous

The earliest evidence of the ancestral MacDonnell Ranges on the northern edge of the Amadeus Basin and similar ancestral ranges on the northern edge of the Ngalia Basin is provided by the 3-km-thick non-marine Pertnjara Group and Mt. Eclipse Sandstone, which were deposited at the foot of the block mountains uplifted along thrusts and steep reverse faults during the Alice Springs Orogeny. Radiometric ages of micas in the Arunta Block show that these movements culminated about 320 Ma ago (Armstrong and Stewart 1975). Palynomorphs in the Mt. Eclipse Sandstone indicate Visean (355–325 Ma) (Kemp et al. 1977), and palynomorphs and fish plates in the preserved part of the Pertnjara Group show that it was deposited during most of the Late Devonian, that is between 360 and 370 Ma (Playford et al. 1976a). The eastern part of the Musgrave Block was uplifted at the same time and shed the Finke Group, including the Polly Conglomerate, towards the north-east. Uplift along steep reverse faults of the northern flank, at the

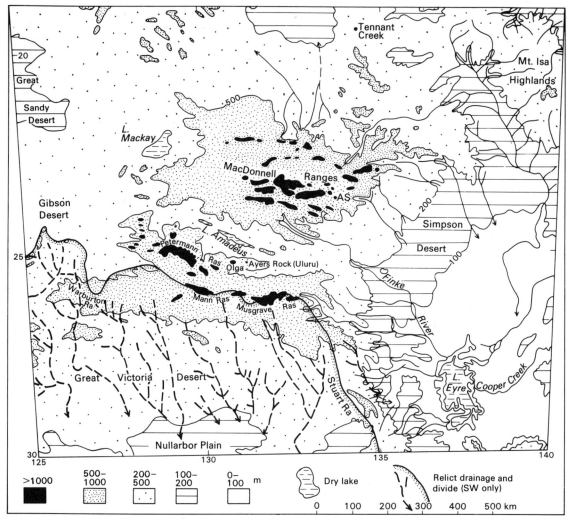

Fig. 84. Morphology of central Australia, with the highlands of the MacDonnell and Musgrave Ranges and the outlying Mt. Isa highlands occupying the central-eastern part of the Great Western Plateau, flanked on the east by the Central-Eastern Lowlands of Lake Eyre and the Simpson Desert, on the south by the Great Victoria Desert, on the west by the Gibson and Great Sandy Deserts, and on the north, at the latitude of Tennant Creek, by an unnamed sandy desert. The only coherent drainage flows from the eastern flank into the Central-Eastern Lowlands. Contours (m) from *Times Atlas of the World* (1975). Relict drainage in south-west from Pitt (1979). AS Alice Springs.

boundary of the Arunta Block and the Lander Trough of the Wiso Basin (Kennewell *et al.* 1977; Kennewell and Huleatt 1980), the Dulcie Syncline (Tucker *et al.* 1979) and Toko Syncline (Harrison 1980) of the Georgina Basin, and along the southern flank of the Musgrave Block at its boundary with the Officer Basin yielded pebbly sandstone deposited during the Devonian: at the Early/Middle Devonian boundary in the Cravens Peak Beds of the Toko Syncline (Turner *et al.* 1981) and probably also Late Devonian in the Dulcie Sandstone. The 3-km-thick sequence of shale to sandstone to conglomerate in the Officer Basin, drilled in Munyarai–1 (Fig. 88B) is dated no more closely than Devonian.

Earlier uplift of the northern edge of the Amadeus Basin in the Devonian and Silurian (Pertnjara and Rodingan Movements) was mild, and did not dismember the presumed continuous cover of sediment northward from the Musgrave Block, as the terminal Alice Springs Orogeny did.

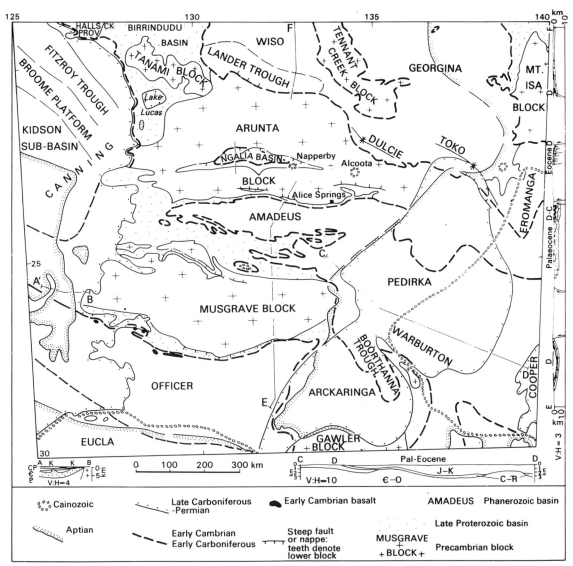

Fig. 85. Geology of central Australia (from Geological Society of Australia 1971, updated from Jackson and van de Graaff 1981; Pitt et al. 1980; Devine and Youngs 1975; Wopfner et al. 1974; Kemp 1976; Woodburne 1967; and Twidale and Harris 1977). The Amadeus Transverse Zone of east–west trending Precambrian blocks and intervening Late Proterozoic and Phanerozoic basins is bounded on the west and east by Late Carboniferous and younger sediments, and on the south by Early Cretaceous and younger sediments.

Cross-section AB (from Jackson and van de Graaff 1981) shows the onlap of strata, dating from the Late Proterozoic (dotted), latest Proterozoic/earliest Cambrian (1, Babbagoola Beds), Early Cambrian (2, Table Hill Volcanics, and 3, overlying sediments), Late Carboniferous/Permian (CP), and Aptian (K). Cross-section CD (from Devine and Youngs 1975, and Jones 1973a) shows Late Carboniferous to Triassic (C–R), Jurassic–Early Cretaceous (J–K), and Palaeocene–Eocene (Pal-Eoc) strata lapping on the eastern flank of the Amadeus Transverse Zone, and Late Devonian (D) strata lapping out. Cross-section EF (Krieg 1969; Wells et al. 1970, 1972; Kennewell et al. 1977) shows the fault-block terrain of the Amadeus Transverse Zone, with older Precambrian blocks (clear) juxtaposed against Late Proterozoic (dotted), Early Palaeozoic (clear), and Devono-Carboniferous (black) basins at steep reverse faults or at the root of nappes. The Palaeocene sediments between Ayers Rock and the Olgas, and the Eocene sediments near Napperby are too thin to be represented.

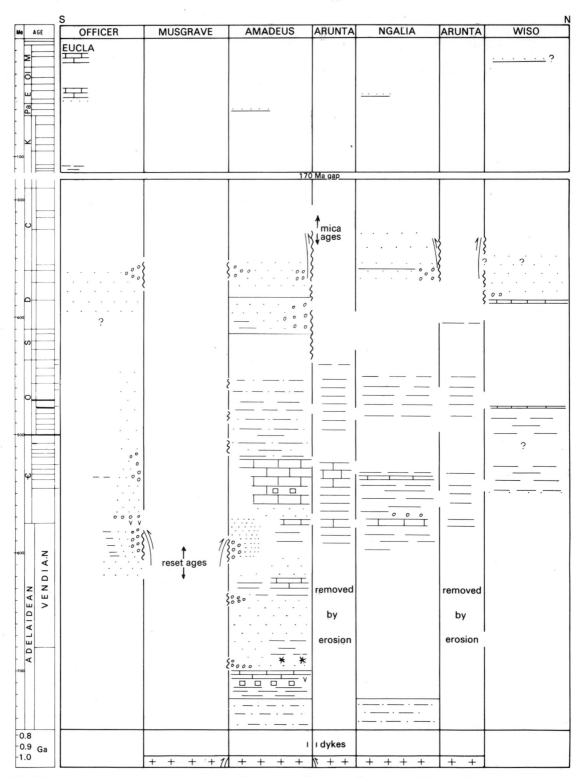

Fig. 86. Time–space diagram of the components of the Amadeus Transverse Zone arranged from south to north along section-line EF of Fig. 85.

PLATFORM

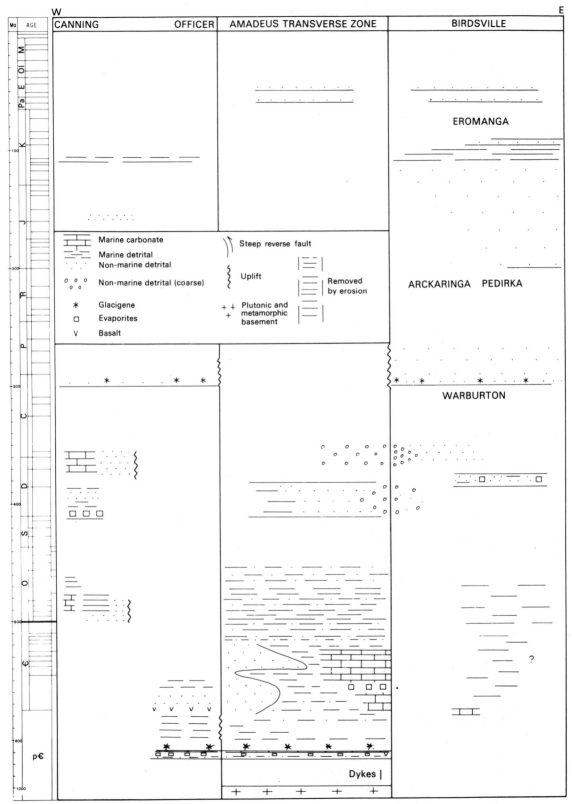

Fig. 87. Time–space diagram of the components of the Amadeus Transverse Zone arranged from west to east. Information of Figs. 86 and 87 from: Officer, Pitt *et al.* (1980), Jackson and van de Graaff (1981); Musgrave, Wells *et al.* (1970); Amadeus, Wells *et al.* (1970), Black *et al.* (1980); Ngalia, Wells *et al.* (1972); Wiso, Kennewell *et al.* (1977), Kennewell and Huleatt (1980); Canning, Forman and Wales (1981); Birdsville/Eromanga/Archaringa/Pedirka/Warburton, Jones (1972, 1973a), Devine and Youngs (1975); Arunta, Marjoribanks and Black (1974).

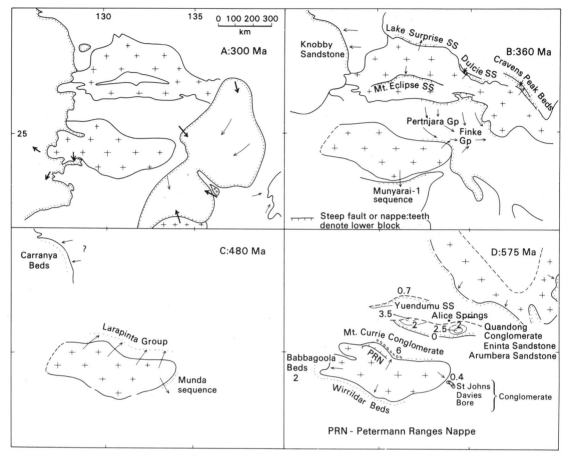

Fig. 88. Maps summarizing the changes in morphology during the Palaeozoic. Exposed older Precambrian blocks shown by crosses, Late Proterozoic and early Palaeozoic sediments clear, and sediment transport directions by arrows. In A, at 300 Ma, the Archaringa and Pedirka Basins occupy the low ground between the Gawler and Peake–Denison Blocks on the south-east and the Amadeus Transverse Zone on the west, as shown by the directions of glacigene sediment transport (broad arrows, from Wopfner 1981), and the Officer Basin occupies low ground on the west, as shown by the direction of glacigene transport (Jackson and van de Graaff 1981). The morphology of high ground underlain by the Amadeus Transverse Zone flanked by lowlands on the east and west, established by 300 Ma, has persisted to the present day. In B, at 360 Ma, during the progress of the earth movements called the Alice Springs Orogeny, the Musgrave Block was uplifted, and the area north of it was broken by faults and nappes into highlands corresponding to the Arunta Block and the northern Amadeus Basin. Coarse detritus was shed out of the highlands to the north-east (Lake Surprise Sandstone, Dulcie Sandstone, Cravens Peak Beds) and south (Pertnjara Group), and within the highlands into the Ngalia Basin (Mt. Eclipse Sandstone). The uplifted eastern part of the Musgrave Block shed sediment to the north-east (Finke Group) and south (sequence in Munyarai-1). On the north-west, the Knobby Sandstone marked the edge of the uplifted Arunta Block. In C, 480 Ma, the Musgrave Block shed detritus (Larapinta Group) northwestward into the broad sag that stretched across the entire area of the map, and southwestward into the Officer Basin (Munda sequence). The Carranya Beds of the north-east Canning Basin indicate a source to the east. At the beginning of the Palaeozoic, at 575 Ma (D), the Musgrave Block was uplifted during the formation of the Petermann Ranges Nappe, and 6 km of coarse detritus (Mt. Currie Conglomerate and the arkose at Ayers Rock) were deposited at the foot of the mountains; beyond a sill to the north-east, the Quandong Conglomerate, Eninta Sandstone, and Arumbera Sandstone were deposited in what was to become the northern Amadeus Basin, and the Yuendumu Sandstone was deposited in the future Ngalia Basin. On the east, the St Johns and Davies Bore Conglomerates are similar piedmont deposits, and on the south-west the little known Wirrildar and Babbagoola Beds lapped the Musgrave Block.

As a highland, the Musgrave Block can thus be traced back to 575 Ma, and the Arunta Block, including the Ngalia Basin, and part of the Amadeus Basin, to 360 Ma, at the end of the Devonian. This high ground persists to the present day.

(5) Latest Proterozoic/earliest Cambrian to Silurian

What is now the southern margin of the Georgina Basin was a highland underlain by the north-east part of the Arunta Block, which first appeared at the beginning of the Vendian and persisted into the Early Cambrian (Shergold and Druce 1980). From the Middle Cambrian to the Middle Ordovician, the region became the site of the deposition of the marine Georgina Basin, probably continuous with the Amadeus Basin (Henderson 1980). The Mt. Isa highlands were probably covered during the Cambrian and Early Ordovician (Henderson 1980; Shergold and Druce 1980) and did not rise above their surroundings until later.

The highlands of the Musgrave Block are denoted by piedmont fanglomerates of latest Proterozoic/earliest Cambrian age: the St Johns Conglomerate and Davies Bore Conglomerate, and the arkose at Ayers Rock on the north. The Wirrildar Beds on the south (Major 1973) contain arkosic sandstone, and the Babbagoola Beds on the south-west (Jackson and van de Graaff 1981) contain very coarse pebbly sandstone attributable to the Musgrave Block (Fig. 88D). According to Wells et al. (1970), the Mt. Currie Conglomerate and the arkose at Ayers Rock have been folded and deeply eroded since deposition; at least 2 km are exposed at Ayers Rock, and the total thickness may exceed 6 km. Both are interpreted as edges of piedmont fanglomerate deposited in front of the mountain chain formed by folding and thrusting of the Petermann Ranges, dated isotopically on biotite and microcline from the reactivated basement at about 600 Ma. Sediment, including conglomerate, from this source is inferred to have spread across a sill to the northern Amadeus Basin (Quandong Conglomerate, Eninta Sandstone, Arumbera Sandstone) and the Ngalia Basin (Yuendumu Sandstone). The sea lay to the east, and in the later Cambrian encroached from this direction so that coarse detrital sediment accumulated in the west, and carbonate in the east.

In the Ordovician, thick quartz sandstone replaced carbonate, and alternated with siltstone to form the Larapinta Group. I interpret this alternation as reflecting pulses of uplift in the Musgrave Block, which shed probably coeval sediment (Munda sequence) to the south-east also (Fig. 88C). A facies change during the Early Ordovician, from sandstone (Carranya Beds) in the north-east Canning Basin to carbonate in the west, suggests a source of detritus to the east.

During the Silurian and Early Devonian, the north-east part of the Amadeus Basin, hitherto a region of uniform deposition, was uplifted (Rodingan Movement) and was followed by the deposition of the Mereenie Sandstone in an environment that ranged from a shallow sea to an aeolian sand desert. The Mereenie Sandstone is a precursor of the Pertnjara Group, but its lack of very thick conglomerate suggests that the Rodingan Movement was mild in comparison with the violent disruption of the Alice Springs Orogeny.

The Palaeozoic history of the ancestral mountains in central Australia is summarized in Fig. 88.

(d) Summary: the antiquity of the highlands of Australia

Our review, summarized in Fig. 89, shows that most of the present highland areas of the Great Western Plateau date back to the beginning of the Phanerozoic, and the Western Shield much earlier. Only the MacDonnell Ranges, at 360 Ma, are younger. The South Australian Highlands are at least 300 Ma old, and probably date from the earliest Phanerozoic.

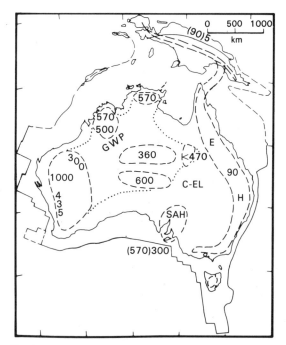

Fig. 89. The Great Western Plateau (GWP), Central-Eastern Lowlands (C-EL) with the South Australian Highlands (SAH), and the Eastern Highlands (EH) including New Guinea. The ages (in Ma) of the highlands are shown, including the individual highlands of the Western Shield, Kimberley Block, Arnhem Land, MacDonnell Ranges, Musgrave Block, and the Mt. Isa Block.

Moving from the Great Western Plateau across the Central-Eastern Lowlands, we find that the Eastern Highlands are no older than 90 Ma, and the New Guinea Highlands only 5 Ma, though both have antecedents in much older highland terrains, at least back to the Late Carboniferous.

The ancient highlands, whose record perforce has been read almost entirely from the adjacent lowland basins, were the positive elements of the landscape that, by their propensity to rise, shed sediment intermittently into the concomitantly subsiding basins. In a later section, we analyse the vertical motions of the platform, both positive and negative, during the past 100 Ma.

4. MORPHOTECTONICS OF THE DIVERGENT OR RIFTED MARGINS

The following analysis is made in terms of a concept of the development of rifted continental margins based on a tectonic interpretation of the morphological features that exist today in Africa–Arabia, as detailed by Veevers (1981).

(a) Development of rifted margins: a conceptual framework

(i) Life cycle

At the core of tectonic models of rifted continental margins is the concept, summarized by Wilson (1974), of a life cycle from an embryonic rift-valley stage through juvenile and mature oceanic stages. As in biological systems, so here the early formative stages contain the biggest changes in configuration.

The choice of the rift system for analysis – the East Africa–Arabia rift system (Fig. 90) – is easy: it is the only modern example of diverse and well-developed rift valleys that pass along strike into a juvenile ocean. Other rift valleys lack this continuity with a juvenile ocean, and, moreover, the Rhine Graben and the Lake Baikal rift probably originated from continental collision rather than plate divergence (Sengor and Burke 1978; Molnar and Tapponier 1975).

A morphotectonic analysis of the embryonic and juvenile stages of the East Africa–Arabia system is followed by a projection of the tectonics of a mature continental margin by means of the subsidence history of the ocean floor, to afford comparison with the observed history of the mature margins of the western and southern margins of Australia.

The morphological elements of the East African rift system are (1) plateaus that rise as high as 4.5 km

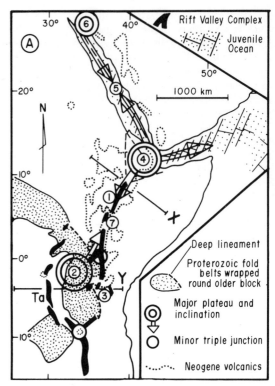

Fig. 90. Map of East Africa–Arabia showing longitudinal variation of relief. From Veevers et al. (1982). 1 Ethiopian rift valley, 2 Central Plateau about Lake Victoria, 3 Northern Tanzania: rift divergence zone, 4 Ethiopia–Yemen plateau, 5 Red Sea saddle, 6 Sinai plateau, 7 Lake Turkana saddle.

above a 'background' elevation of about 0.5 km, separated by saddles; (2) rift valleys that split the plateaus centrally, as in Ethiopia, or on the flank within individual arches, as on either side of Lake Victoria – I use the term 'arch' for the uplift on either side of a rift valley, regardless of whether it is the primary structure, as in Ethiopia, or a secondary one, as it is to the south; and (3) rift-divergence zones, with a unique example from north-east Tanzania where the eastern rift degenerates from a single rift-valley into a zone of tilted fault blocks that radiate a short distance southward. The paths of sediment transport from the Ethiopian plateau are (a) transverse from the axis of the flank; (b) longitudinal from the axial crest (the plume site of Kinsman 1975) to the saddle (the interplume site); and (c) along the depression on the side of the plateau.

With the addition to the East African rift system of the juvenile ocean between Africa and Arabia we see

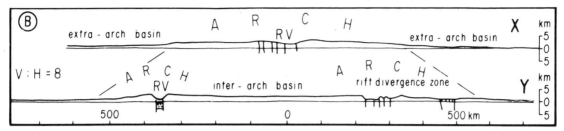

Fig. 91. Profiles across East Africa, showing the distribution of basins with respect to the arch of a single system (X) and the arches of a multiple system (Y). From Veevers *et al.* (1982).

that the Ethiopia–Yemen plateau passes to the north-north-west into the Red Sea saddle and the Sinai plateau, and to the south-south-west into the Lake Turkana saddle and the Central Plateau.

The rifted arch is the dominant morphotectonic feature (Fig. 91), and it supplies sediment to the depressions within and outside the single arch, as in section X, to form intra-arch and extra-arch basins and, additionally, between multiple arches, as in section Y, to form an inter-arch basin.

With the splitting of the arch by plate divergence, and the emplacement of oceanic lithosphere within the rift valley, a dismembered or half arch becomes a rim to the newly formed continental margin. According to the rate of subsidence of the oceanic lithosphere and the adjacent continental margin back to an average distance of 500 km, the margin subsides about a hinge, as shown from the development of a single medianly split arch in Fig. 92. The shallowest depth or greatest height at the crest of the system is given at the front of each block diagram, and the greatest depth in a saddle at the back. All values are in depths so that positive values are below sea-level, and negative values above. The diagrams contain the following quantities derived from the East African–Arabian system: (1) a 60-km-wide rift valley; (2) a distance of 120 km from crest to crest of the arch; (3) a structural relief of 4 km from the floor of the rift valley to the crest of the arch and 1.5 km at the saddle; (4) half-width of the arch of 500 km, and subsequently a 500 km distance from the COB to the hinge of the rifted margin; (5) a spacing of 1000 km between the highest part of the plateau and the lowest part of the saddle. An arbitrary half-spreading rate of 1 cm/year was used.

At breakup (0 Ma), which marks the end of the rift-valley stage and the inception by plate divergence of the juvenile ocean stage with its adjacent continental margins, the region is dominated by an arch 1000 km wide split in the middle by a rift valley 60 km wide. The structural relief from the shoulder of the arch to the axis of the rift valley is 4 km at the crest and 1.5 km at the saddle, and the arch and rift-valley axis have a longitudinal gradient of 4.5 and 2 km respectively from the crest to the saddle. Three areas of deposition are the rift valley itself within the arch (an intra-arch basin) drained longitudinally by rivers locally dammed in lakes, and on either side by an extra-arch basin. On the left-hand side is shown a river flowing down the axis of the extra-arch basin (as the Nile drains the Ethiopian and Central Plateaux) and locally accumulating thick sediment (as beneath the present Sudd; Williams and Williams 1980), and on the right-hand side the axis is occupied by a lake. The flank of the arch is lettered Shelf and the rift-valley wall Slope to show their ultimate transformation (40 Ma later) into these parts of the mature continental margin. Likewise, the subaerial canyons on either side of the crest of the arch will be transformed into submarine canyons, in a way like that described by Du Toit (1940). The deep structure on the front of the diagrams follows Roots *et al.*'s (1979) model of an 18-km-thick oceanic crust at the COB that thins towards the mid-ocean ridge.

At 20 Ma after breakup, as seen in the Red Sea–Gulf of Aden system, the juvenile ocean basin has a mid-ocean ridge with crestal depth from 0.8 to 1.5 km, and a COB depth from 2.0 to 2.8 km. The half-arch or continental rim has a depth of -2.0 to 1.8 km, and the lower parts of the rim basins are covered by the sea. By extrapolation, at 40 Ma after breakup and concomitant with the subsidence of the rims, the axis of the rim basin migrates oceanward. All but the highest parts of the rims are covered by the sea, and shortly afterwards all of the rim will be covered so that the entire margin will be open to the sea.

The main development of the margin is now complete and subsequently, with exponential decay of the subsidence rate, the margin approaches its final

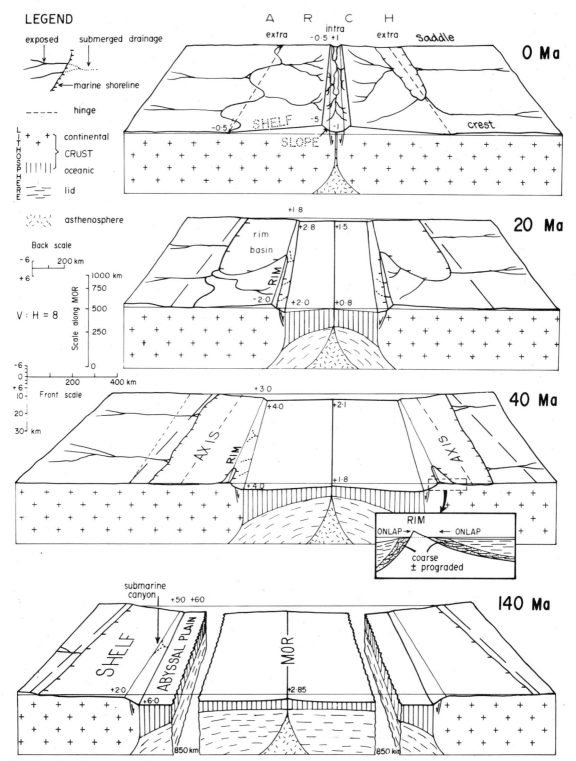

Fig. 92. Block diagrams showing the development of a rifted-arch system from continental breakup at 0 Ma, to 20 Ma, 40 Ma, and 140 Ma after breakup. The high (crestal) part of the system, with rim at −5 km at breakup, is shown at the front, and the low (saddle) part, with rim at −0.5 km, at the back. In the diagram 40 Ma after breakup, details of the stratigraphic relations at the rim are shown in an enlarged cross-section with greater vertical exaggeration. From Veevers (1981).

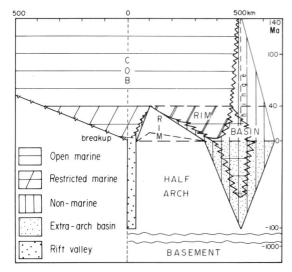

Fig. 93. Time–space diagram of a rifted margin that evolved from the median splitting of a single rifted arch, as shown in Fig. 92. The rim in the high part of the system is shown by a full line and in the low part by a broken line. The doubly restricted marine environment of the rim basin is shown by double oblique lines. From Veevers (1981).

state, shown at 140 Ma after breakup with a broad shelf and narrow slope bordered by a 6.0-km-deep abyssal plain.

These developments are shown in Fig. 93 in the form of a time–space diagram which highlights the onlap of the rim from both the landward and the oceanward sides. Evidence of this onlap on an ancient margin is likely to be the principal evidence of a continental rim (Schuepbach and Vail 1980). The main stratigraphical relation at the rim will be the lateral termination of strata at their depositional limit, called lapout. Onlap is the kind of lapout in which the deposited strata are initially horizontal and the surface on which they are deposited is inclined; the other form of basal lapout, call downlap, comes about when initially inclined strata terminate downdip against an initially horizontal or inclined surface. Mitchum et al. (1977, pp. 57, 58) discuss these terms and note that 'onlap or downlap usually can be readily identified. However, later structural movement may necessitate the reconstruction of depositional surfaces. In areas of great structural complication, the discrimination between onlap and downlap may be practically impossible, and the worker may be able to determine only that the strata are in a baselap relation.' This kind of uncertainty is inherent in Fig. 92. Whereas during the first 40 Ma the strata are deposited on either side of the rim essentially horizontally over inclined surfaces and so express onlap, subsequently by the continued subsidence of what was the rim by rotation about a hinge, these strata may become inclined while correspondingly the surface on which they were deposited becomes horizontal to simulate the relation of downlap. This is the kind of situation described above by Mitchum et al. (1977), in which the geometry alone indicates nothing more than baselap or the termination of strata at the lower boundary of a depositional sequence. This ambiguity can be resolved by seeking facies trends along a shoreline that indicates the local derivation of strata from the postulated rim. In the case of onlap (Fig. 92, box), the sediments will coarsen towards the shoreline and, if deposition were in deltas, may dip by progradation away from it. In the case of downlap, neither feature will occur. Another kind of test will be whether or not the basin fill was deposited in an open or a restricted marine environment. An open environment would tend to suggest the absence of a rim, whereas a restricted environment could point to the presence of a rim, but not uniquely because the juvenile ocean of which the sea over the margin is only a part may itself have a restricted circulation. The possibility of a doubly restricted circulation within a rim basin is shown in Fig. 93 by double oblique lines.

The onlapping facies about the rim may be prograded deltaic deposits or, given a suitable climate and reduced detrital runoff, reefs. In Fig. 93, the middle part of the extra-arch basin is shown as occupied by either marine or non-marine sediments. Which of these dominates depends on the geometry of the system – whether or not the extra-arch basin is connected to the ocean – and the eustatic level – whether high or low – of the sea. The duration of the rim above sea-level is expected to be about 40 Ma in the high part of the system (shown by a full line), and much less in the lower part (shown by a broken line).

A model of a continental margin evolved from a multiple-arch system (Fig. 94) will produce an inter-arch basin before breakup, and secondary rims (II and III) round the distal arch in addition to the primary rim (I) after breakup. A model of a continental margin evolved from a rift-divergence zone (Fig. 95) contains a rim basin that rests directly on the rift-divergence zone, without the intervention, as in the other models, of an extra-arch basin.

(ii) Falvey's (1974) model for the recognition of developmental stages

In developing a plate-tectonic model of margins, Falvey (1974), Falvey and Mutter (1981), and Falvey

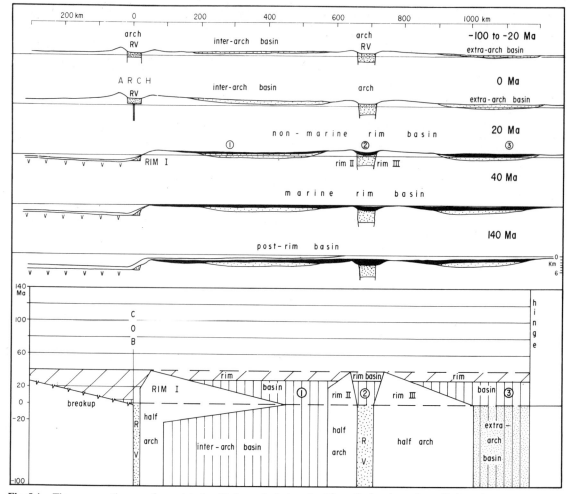

Fig. 94. Time–space diagram of a model of a rifted margin that evolved from the breakup of a multiple-arch system.

and Middleton (1981) describe the tectonic and time-stratigraphic stages in the development of rifted continental margins that can be recognized in seismic profiles controlled by drilling, as summarized in Fig. 96. According to Falvey and Middleton (1981),

1) Subsidence begins with the infrarift phase (usually about 100 m.y. before breakup) preceded by some erosion of basement or prerift sediments. The infrarift basin grows along the incipient continent–ocean boundary. It does not appear to be fault controlled and contains mostly non-marine and non-volcanic sediments;

2) from as much as 50 m.y. before breakup, through to breakup time, basin subsidence continues only in rift grabens and half grabens flanking the incipient continent-boundary. Sediments are marginal to non-marine. Some volcanism is present close to the incipient continent–ocean boundary but is absent from most major depocentres in the Australian region. Some uplift and erosion of infra-rift phase sediments is evident away from the major depocentres. There is no specific evidence that any of the en echelon rift grabens are interconnected by transform or transcurrent faults. Thus there is no direct evidence of lithosphere pull-apart.

3) From breakup time, subsidence becomes widespread. Bathyal sediments are deposited at the continent–ocean boundary and onlap progressively younger oceanic crust. A marine transgression extends shoreward. Shelf and slope deposition is commonly interrupted by massive submarine erosion caused by changing current patterns in the progressively widening and deepening ocean basin (Deighton et al., 1976).

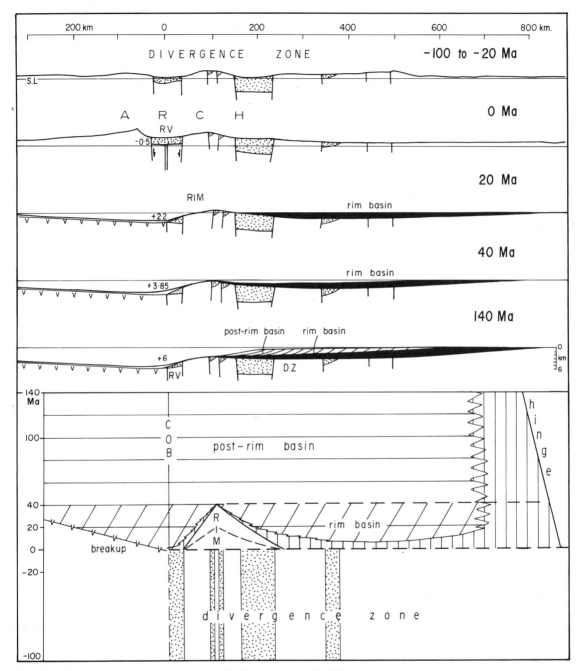

Fig. 95. Time–space diagram of a model of a rifted margin that evolved from the breakup of a rift-divergence zone.

According to Falvey (1974), the rift-valley stage is bounded by angular unconformities: a rift-onset unconformity below and a breakup unconformity above. The breakup unconformity is caused by erosion during the final uplift pulse associated with pre-breakup upwelling in the mantle. This unconformity is more localized than the rift-onset unconformity, being difficult to define in troughs. However, where it can be

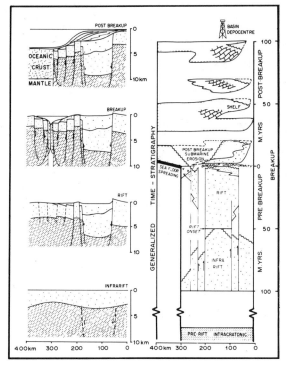

Fig. 96. Generalized development of a rifted continental margin, shown in cross-sections (left) and a time–space diagram (right), from Falvey and Middleton (1981, fig. 2).

clearly defined, it represents the youngest cycle of subaerial erosion in a rift basin which is very nearly the same age as the oldest oceanic crust in the adjacent deep ocean basin.

In recent reviews, Falvey and Mutter (1981) identified these stages in the Australian rifted margins, and Johnstone (1981) in Western Australia. These concepts, modified in the light of the latest available information, are incorporated in the analysis that follows.

(b) Southern margin

(i) Introduction

The southern margin (Fig. 97) has been much less intensively explored than the western margin, and drilling has been restricted to the central and eastern parts offshore. In compensation, two complementary surveys, by Boeuf and Doust (1975) and Talwani et al. (1979), provide excellent data, and these surveys augmented by work by the Bureau of Mineral Resources (Willcox 1978; Fraser and Tilbury 1979), and others (Griffiths 1971; Falvey 1974; Deighton et al. 1976; Denham and Brown 1976) comprise the source of information for this analysis. The margin extends eastward from the south-western tip of Australia to longitude 132° E, where it splits off the failed arm of the Polda Trough, then bears south-eastward to be offset by transform faults past Tasmania and the South Tasman Rise. Between long. 115° and 125° E, the margin has a narrow shelf, a steep ($\leq 8°$), locally terraced, continental slope which passes at 3.5 km into a wide smooth continental rise, and a rough, ridged, 5-km-deep oceanfloor, the Diamantina Zone. Further east, the rise narrows by the buildup of the Ceduna Plateau (Fraser and Tilbury 1979), and the Diamantina Zone gives way to isolated seamounts in an abyssal plain at a depth of 5.5 km. East of longitude 135° E, the slope becomes steep and is crossed by deep canyons (von der Borch 1968), and then widens again off the Otway Basin.

Talwani et al. (1979) have mapped a magnetic trough along the western half of the margin, which marks the northern boundary of a smooth magnetic field in the east and a progressively more disturbed, but not aligned, one toward the west, all called a magnetic quiet zone. This zone is bounded southward at the COB by the oldest identified seafloor-spreading magnetic anomaly, A34. The quiet zone continues southeastward seaward of the Otway Basin and western side of Tasmania (Weissel and Hayes 1972). On the basis of seismic refraction probes, including work by Hawkins et al. (1965), Talwani et al. find four different kinds of crust in the area south of the magnetic trough: (1) oceanic seaward of the quiet zone, and (2) three groups in the quiet zone: Group I (mainly in the eastern and central parts), in which the velocity of the main crustal layer ranges from 6.5 to 7.1 km/s; and Group II (mainly in the western part), 7.2 to 7.5 km/s; and Group III (central part), 5.8 to 6.2 km/s. These groups, whose distribution overlaps each other, constitute an exceptional variety of crustal layers – some typically continental, others oceanic, though much thicker than is typical, others still neither continental or oceanic – which Talwani et al. interpret as indicating a unique rift crust 'comparable with the deep structure of the Rhine Graben' generated in the 200- to 300-km-wide continental rift that was later split evenly between Australia and Antarctica. The quiet zone that extends southeastward to the west side of Tasmania is explained by Roots (1976) as due to narrow compartments of oceanic crust generated alongside a continental margin oblique to the spreading direction (Fig. 98).

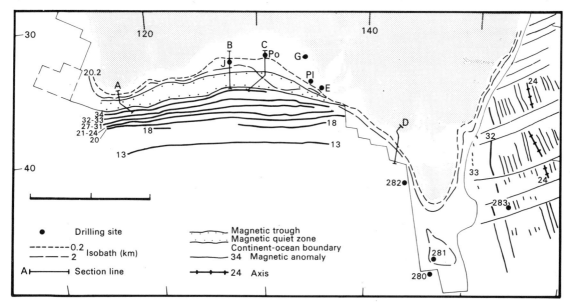

Fig. 97. Southern and south-eastern margins and adjacent oceans, showing selected drilling sites, magnetic anomalies (Cande and Mutter 1982), isobaths, sections-lines, and the magnetic quiet zone in the Great Australian Bight between the magnetic trough and the continent–ocean boundary (COB) (Cande and Mutter 1982). DSDP drilling sites shown by numerals. Other drilling sites are E Echidna, J Jerboa, Pl Platypus, Po Potoroo, G Gemini (Milnes et al. 1982).

(ii) Age of breakup of the southern and south-eastern margins

From deep-sea drilling, the age of the oldest oceanic lithosphere south of Australia is a minimum of 50 Ma (middle Eocene) at DSDP Site 280 (Kennett et al. 1975). Drilling terminated in an intrusive basalt on a basement pinnacle, so that even if the intrusive were close to oceanic basement this is only a minimum age because the oldest sediments probably lie at the bottom of the pinnacle. Site 282 reached oceanic basement beneath late Eocene (45 Ma) sediment, equivalent in age to magnetic anomaly 18; its modelled age (Fig. 33) is 66 Ma. A precise estimate of the inception of spreading or the age of breakup must depend on the dating of the magnetic anomalies closest to the continent–ocean boundary. Weissel and Hayes (1972) identified A22 (= 54 Ma, Mankinen and Dalrymple 1979) as the oldest seafloor-spreading magnetic anomaly in the south-east Indian Ocean, so that breakup of the southern margin has been taken as 54 Ma or early Eocene. Cande and Mutter (1982) have now revised the identification of the oldest anomalies. They confirm the set A1 to A18 (44 Ma) as given by Weissel and Hayes (1972), with a half spreading rate of 23 mm/year, recalculated here, from the distance between A18 north and south of the mid-ocean ridge, as 29 mm/year, and suggest that the rest of the sequence may range to A34 (82 Ma), at a half rate of 6 mm/year (Fig. 97). 'The key to this model is that at very slow spreading rates each observed anomaly corresponds to the combined effect of several polarity reversals.' A34 is the oldest anomaly of the Cainozoic–Late Cretaceous set, and is bounded by the Cretaceous long normal-polarity interval. Depending on the precise position of the COB within this interval, breakup occurred between 82 Ma and 110 Ma. This range can be narrowed by identifying in the stratigraphical section of the adjacent continental margin the major unconformity that developed during breakup – the breakup unconformity. In the Great Australian Bight, from seismic profiles tied to Potoroo well, the unconformity (horizon C of Fraser and Tilbury 1979) is dated as early to middle Cenomanian (93 Ma), and from Jerboa, some time between the Cenomanian and the Eocene (Bein and Taylor 1981). The clearer and precisely dated unconformity in the Otway Basin, between the volcanogenic Otway Group and the quartzose Sherbrook Group, lies within the *Phimopollenites pannosus* palynological zone (Douglas and Ferguson 1976, p. 169), which straddles the Albian–Cenomanian boundary at 95 Ma. Confirmation is provided by Weissel *et al.* (1982), who interpret the change at 90 ± 5 Ma from rapid to slow subsidence along the southern margin, detailed by Falvey and Mutter (1981, fig. 18), as marking the age of breakup. And this age of 95 Ma we take as the best estimate of breakup, with an assumed uniform slow half spreading rate of 4.3 mm/year for the interval 95 to 44 Ma. Thus, in the interval of slow spreading between Australia and Antarctica, the amount of separation is 440 km, and in the interval of fast spreading, from 44 Ma to the present, 2530 km.

Off the south-east margin, the oldest sediment recovered from DSDP site 283 is 66 Ma (uppermost Maastrichtian) (Partridge 1976), consistent with the determination of A30 at this site. Site 283 is 200 to 300 km from the COB, so that opening must have been much earlier. The oldest identified magnetic anomaly adjacent to the south-eastern margin is A33 (78 Ma) (Weissel and Hayes 1977). Magnetic lineations are oblique to the margin and appear to be truncated by it (Fig. 97). Hayes and Ringis (1973) explained this unusual geometry by suggesting limited subduction of Tasman Sea lithosphere beneath Australia, but Weissel and Hayes (1977) have since produced a tectonic model, from reidentification of magnetic anomalies and new finite rotations, that does not require subduction. But this does not overcome the geometrical difficulty of long coherent magnetic lineations that strike the continental margin at an angle of 45°. Roots (1976) has shown that spreading oblique to a continental margin cannot be accommodated except by initially narrow spreading segments that generate aggregate magnetic anomalies parallel to the margin (Fig. 98). Only later can the segments widen to form blocks broad enough to generate coherent magnetic anomalies. In view of this difficulty, the age of the oldest recognized magnetic anomaly in the Tasman Sea is seen to provide only a minimum age of spreading.

On the adjacent continental margin, in the Strzelecki–Gippsland Basin, the same unconformity as is found in the Otway Basin separates the deformed volcanogenic Strzelecki Group from the overlying comparatively undeformed quartzose Latrobe Group (Threlfall *et al.* 1976, p. 46), and likewise straddles the *pannosus* zone (James and Evans 1971). Consequently, we interpret the unconformity as providing a definitive age of 95 Ma for the breakup along the south-eastern margin also, and the model of seafloor spreading given in Chapter II incorporates this age of the inception of the Tasman Sea.

(iii) Ceduna Depocentre

The Ceduna Plateau is a marginal plateau in the Great Australian Bight marked by a prominent bulge in the 4 km isobath (Fig. 97), and underlain by the main depocentre of the Great Australian Bight Basin, which we call the Ceduna Depocentre. Fraser and Tilbury (1979) described the stratigraphy and structure of the depocentre, based on geophysical data collected by oil exploration companies (Boeuf and Doust 1975; Pattinson *et al.* 1976; Whyte 1978), the Bureau of Mineral Resources (Willcox 1978), and Lamont–Doherty Geological Observatory (Talwani *et al.* 1979), and offshore drilling at three sites called Echidna, Platypus, and Potoroo (Pattinson *et al.* 1976; Whyte 1978), subsequently augmented by a fourth well to the west, called Jerboa (Bein and Taylor 1981), all shown in Fig. 99. Little was known of the Ceduna Depocentre until this recent phase of exploration so that it was not recognized in previous work. During the Cretaceous the Ceduna Depocentre lay in an east–west saddle between uplifts at either end of the rift-valley system (Veevers 1981, fig. 2). On the west, the area of the Naturaliste Plateau was a triple-rift junction up to the 128 Ma breakup of India and Australia–Antarctica, and on the east the area of the Strzelecki Basin was a triple junction up to the 95 Ma breakup of Australia and the Lord Howe Rise. Occupying the ground in the rift-valley saddle at the axis of the Australian–Antarctic Depression, the Ceduna Depocentre was the focus of deposition in the late Mesozoic.

Its prominence in terms of bulk shows on the isopach maps of Jurassic, Cretaceous, and Cainozoic sediments (Figs. 149, 150). As delimited by the 2 km isopach between constrictions south of Kangaroo Island and at long. 128° E, the Ceduna Depocentre contains an estimated 0.95×10^6 km^3 of sediment. Compared with the other depocentres of the southern margin combined – the Otway, Bass, and Gippsland Basins – the Ceduna Depocentre has a slightly greater area (0.170 v. 0.135×10^6 km^2) and a much greater volume (0.950 v. 0.605×10^6 km^3) (Table 7).

The thickness and age of the sediments on the landward periphery of the Ceduna Depocentre have been found by drilling at the four well-sites (Fig. 99, located on Fig. 97), and the rest of the information on thickness comes from geophysical methods (Fig. 100). Line-drawings of some seismic profiles in Fig. 101 are located on Fig. 100. The age of breakup of the southern margin, as described above, is taken to be 95 Ma.

The Ceduna Depocentre comprises two sequences (Fig. 102). In ascending stratigraphical order, they are (1) thick (3 km) Late Jurassic and Early Cretaceous

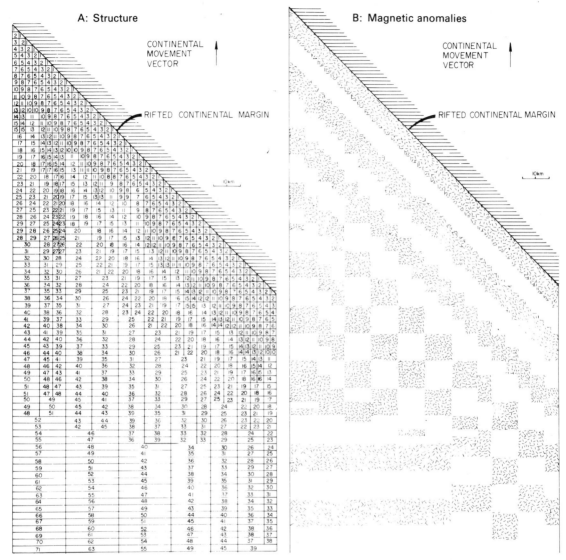

Fig. 98. A, Stages of growth of spreading segments along an oblique margin, and B, the magnetic reversal pattern produced by A, assuming a magnetic reversal every four time-units. The three distinct zones generated are (1) the zone of anomalies parallel to the margin; (2) the zone with a chess-board pattern; and (3) the zone of anomalies parallel to the mid-ocean ridge. All after Roots (1976).

detrital sediments, non-marine except in the Aptian–Albian of Potoroo and the Albian of Jerboa; these sediments are cut into fault blocks, and are interpreted (Veevers 1981) as a rift-valley divergence basin, unconformably overlain by (2) very thick (8 km) Late Cretaceous non-marine to paralic detrital sediments, cut by faults of smaller throw, and interpreted as a rim basin deposited between the continent and a postulated marginal rim, called the 'Outer Continental Margin Ridge' by Boeuf and Doust (1975); the direction of sediment transport during the Late Cretaceous is shown by foresets 1 km or so thick that dip southwestward (Figs. 100 and 101), interpreted as 'a thick southwest trending deltaic lobe of interbedded sands

TABLE 7. *Volumes and areas of Cainozoic and Mesozoic sediment*

Age	Basin						
	Ceduna Depocentre	Otway Basin	Bass Basin	Murray Basin	Gippsland Basin	Eromanga and Surat Basins	Tasman Sea
	Volume ($\times 10^6$ km^3)						
Cainozoic	0.050	0.070	0.050	0.090	0.035	0.020	?1.125
Late Cretaceous later Cretaceous Cenomanian	0.200 0.320	0.150	0.040	–	0.030	0.130	?1.125
Early Cretaceous and Jurassic	0.380	0.170	0.010	0.020	0.050	1.150	–
Total	0.950	0.390	0.100	0.110	0.115	1.300	2.250
	Area ($\times 10^6$ km^2)						
	0.170	0.070	0.025	0.300	0.040	1.500	2.750

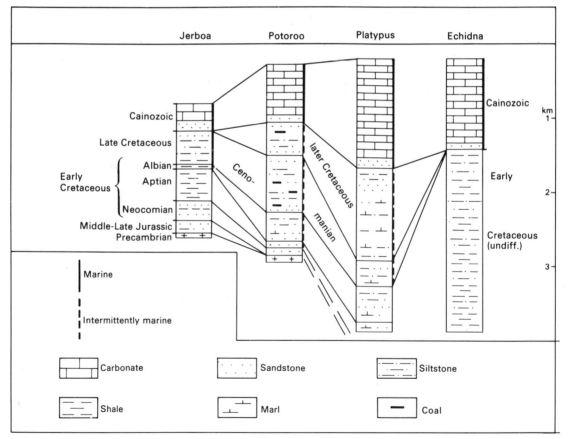

Fig. 99. Stratigraphic columns of four drillholes in the Great Australian Bight. From Bein and Taylor (1981), Fraser and Tilbury (1979), Whyte (1978), and Pattinson *et al.* (1976). The depositional environment is non-marine except where marked by vertical lines on the right-hand side of the columns.

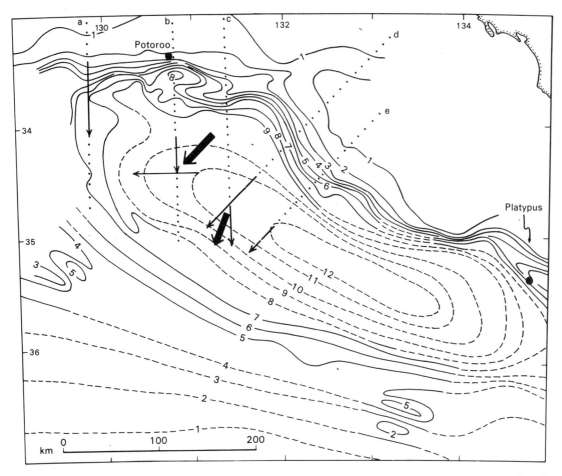

Fig. 100. Total sediment thickness (km) of the Ceduna Depocentre, from Fraser and Tilbury (1979). Continuous lines from seismic reflection measurements, broken lines from estimates of the depth of acoustic basement by sonobuoys, and of magnetic basement by interpretation of magnetic anomalies. Dotted lines show the location of profiles (Fig. 101), within which thin full lines show the location of prograded layers in the Late Cretaceous Potoroo Formation, all west of long. 132° E, and the arrow tips the direction of apparent dips of these layers. Thick arrows show the true dip, found by resolving intersecting pairs of apparent dips.

and shales built outward across a shallow sea' (Fraser and Tilbury 1979). The depocentre is overlain by a film, nowhere thicker than 0.6 km, of Cainozoic sediment, mainly carbonate, deposited on the open margin.

The Ceduna Depocentre follows the model of a margin that developed from a rift-divergence zone (Fig. 95), both in size (except for different sediment thicknesses) and in the timing of events. The 30 Ma duration of the postulated rim is longer than that envisaged in the model for a margin situated, as in the Ceduna Depocentre, low in the rift system, and is explicable by the extremely low rate of spreading and hence of migration of the heat source of the mid-ocean ridge, which involves a correspondingly low rate of cooling and hence subsidence of the adjacent rim of the continent margin.

(iv) East of the Ceduna Depocentre

The same sequences are recognized in the Otway Basin (Fig. 103), with the distinction between (1) and (2) enhanced by an abrupt change from volcanogenic to quartzose sediment.

The revision of the age of breakup from 53 to 95 Ma entails a change in my earlier designation (Veevers 1981, 1982a) of an extra-arch basin to a rim basin. This

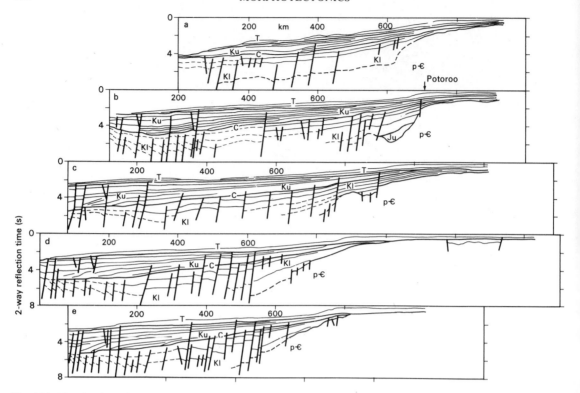

Fig. 101. Line drawings of seismic profiles across the Ceduna Plateau, from Fraser and Tilbury (1979), located in Fig. 100. Solid lines represent good or fair reflectors, broken lines poor or inferred reflectors. Reflector C lies at the base of the prograded Late Cretaceous Potoroo Formation. T Tertiary, Kl Late Cretaceous, Ke Early Cretaceous, Jl Late Jurassic, p€ Precambrian.

designation departs in some important ways from that of Falvey (1974), Deighton et al. (1976), and Falvey and Mutter (1981). In devising a model of rifted margins, Falvey (1974, fig. 11) identified the Early Cretaceous sequence of the Otway Basin as a 'pre-rift basin' and the Late Cretaceous sequence as a 'rift valley basin', so that the unconformity between these sequences was correspondingly called a 'rift onset unconformity'. In company with Boeuf and Doust (1975), I regard the Early Cretaceous sequence as the first deposit of the rift system, with the result that its boundary with basement is recognized as the rift-onset unconformity, and the no less conspicuous surface between the Early and Late Cretaceous sequences as the breakup unconformity that separates the sequence of the divergence zone from that of the rim basin. In revising part of Falvey's (1974) model, Deighton et al. (1976, fig. 7) recognized that 'the early "rift valley" section would probably have been deposited on the flank of the initial rift' but showed this flank to be below sea level before breakup, thus implying that the seaward lapout of sediment is due to downlap, and not to onlap against an outer ridge. On such a crucial point, hard evidence is not likely to accrue without deep marginal drilling.

In a novel application of the fission-track dating method to the study of rift systems, Gleadow (1978) and Gleadow and Lovering (1978) measured the fission-track ages of sphene and apatite from mid-Palaeozoic granite of King Island, off the north-west tip of Tasmania, and found that whereas the sphene ages match K–Ar ages and indicate emplacement at about 350 Ma, apatite ages are all younger by about 80 to 200 Ma. Fission tracks were not fully retained in the apatite, which has an annealing temperature about 110° C, until the Cretaceous, the youngest age being 112 Ma, or close to the boundary between the Early and Late Cretaceous, indicating the latest age that

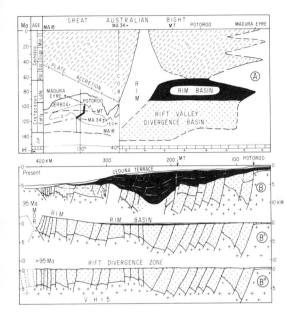

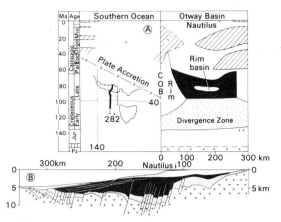

Fig. 102. A. Time–space diagram of the Ceduna Depocentre, outlined by the 2 km isopach (inset), from Veevers *et al.*(1982). MT: magnetic trough of Talwani *et al.* (1979); MA 34+: continent–ocean boundary (COB) dated by Cande *et al.* (1981) as 85 Ma at least, and taken here as 95 Ma; MA18: 43 Ma magnetic isochron at the boundary between slow and fast spreading. B. Depth section of the Ceduna Depocentre, modified from Boeuf and Doust (1975) as amended by Fraser and Tilbury (1979), located by the heavy line in A. B′. Reconstruction at breakup. B″. Before breakup, modified from Boeuf and Doust (1975).

Fig. 103. A. Time–space diagram and location map of the Otway Basin, modified from Veevers (1982a). Details out to Nautilus well from Boeuf and Doust (1975), and beyond by extrapolation of seismic profiles. B. Depth section of the Otway Basin to the continent–ocean boundary, located in A, modified from Boeuf and Doust (1975) by the work of Denham and Brown 1976).

these rocks cooled through 110 °C. These authors attribute cooling to the uplift and removal of overlying rock at a rift shoulder.

(v) West of the Ceduna Depocentre

The sediment thickness decreases as the Ceduna Plateau passes westward into the Eyre Plateau (Fig. 104). The Eyre Sub-basin (Bein and Taylor 1981) is a series of half-grabens with a maximum sediment thickness of about 6 km. The only drilling, at Jerboa, proved 1738 m of Jurassic, Cretaceous, and Cainozoic sediment above a Precambrian tilted fault-block (Fig. 99). Oceanward of a fault that displaces basement as much as 8 km downward, the broad lower slope and rise are underlain by sediment in narrow fault-blocks. The western end of the southern margin (Fig. 105) has a similar structure, with the Bremer Basin, interpreted to have a similar sequence to that of the Eyre Sub-

basin, perched on a terrace above the continental rise. The rise is underlain by locally thick sediment that overlies a rugged basement, interpreted by Falvey and Mutter (1981, fig. 9) as a 'volcanic basement complex', from which Nicholls *et al.* (1981) describe an upper-mantle nodule dredged from an exposed basement peak. Oceanwards, at A34 and beyond, is the oceanic lithosphere of the Diamantina Zone, a 300-km-wide zone of very rugged topography that stretches from south of Broken Ridge to long. 125° E. Cande and Mutter (1982) attribute this rough topography to the very slow rate of spreading from breakup to 48 Ma ago.

(vi) Summary

As shown in B′ and B″ of Fig. 102, I interpret the central part of the southern margin, at the Ceduna Plateau, as a Late Jurassic–Early Cretaceous rift-divergence zone succeeded at breakup by a Late Cretaceous rim basin, in turn succeeded by Cainozoic sediment, mainly carbonate, of an open margin. In terms of the models derived from East Africa, the Ceduna Depocentre corresponds to the configuration shown in Fig. 95. The width of 600 km of the margin, from the hinge at the landward wedge-out of the Eucla Basin to the COB, and the duration of 30 Ma of the rim basin, account being taken of the low rate of subsidence due to slow spreading, both agree with these parameters in the model. The main difference is in the sediment

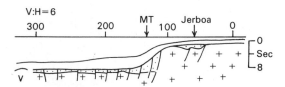

Fig. 104. Time section of the Eyre Plateau, adapted from Talwani *et al.* (1979). MT: magnetic trough.

thickness: the Ceduna Depocentre, lying in the saddle of the rift system where it corresponds to the Australian–Antarctic Depression, accumulated about 12 km of sediment on rift-attenuated crust.

To the east, in the Otway Basin, the morphotectonic sequence is the same. Both regions contain prograding clinoforms in the rim basin, which I interpret as due to onlap against an outer ridge, but direct evidence of this will come only through deep marginal drilling. Drilling would also provide a check on the ages of the sediment that laps the outermost part of the margin, estimated by extrapolation from distant drillholes on the shelf. This contrasts with the situation on the western margin where, at Scott Reef and near Perth, drillholes on the outer margin provide direct evidence of the age and geometry of the sediment that onlaps the rim. Compared with the Ceduna Depocentre, the rift-divergence and rim basins of the Otway Basin are thinner, and the open-marine basin thicker; the rift-divergence zone is mainly volcanogenic, reflecting its connection eastward with a coeval magmatic source described in our analysis of the Eastern Highlands. The width of the margin and the duration of the rim basin are again consistent with the model.

West of the Ceduna Depocentre, the margin lacks very thick sediment oceanward of the major step fault, coincident with the magnetic trough, probably because this part of the margin was starved of the great volume of sediment that focused on the Ceduna saddle.

(c) Western and north-western margins

(i) Introduction

Most of the information about the western and north-western margins has accrued during the past 25 years from petroleum and mineral exploration onshore, summarized by the Geological Survey of Western Australia (1975) and from exploration offshore by petroleum companies (Johnstone *et al.* 1973; Powell 1976b, 1982; Laws and Kraus 1974; Crostella and Barter 1980; Barber 1982) and the Bureau of Mineral Resources (Branson 1974; Symonds and Willcox 1976; Exon and Willcox 1978; 1980; Stagg and Exon 1981; and the Bundesanstalt für Geowissenschaften und Rohstoffe (Hinz *et al.* 1978; von Stackelberg *et al.* 1980; Exon *et al.* 1982; von Rad and Exon 1983). The adjacent oceanfloor has been explored by the Deep Sea Drilling Project (Davies *et al.* 1974; Veevers *et al.* 1974b; Hayes *et al.* 1975), by the Lamont–Doherty Geological Observatory (Markl 1974a; Larson 1975, 1977; Larson *et al.* 1979), and by Woods Hole Oceanographic Institution (Heirtzler *et al.* 1978). Summaries have been made by Veevers and Cotterill (1978), Falvey and Mutter (1981), Willcox (1981), and Veevers (1982a, c). In addition, accounts of the submarine physiography of various parts of the margin are given by van Andel and Veevers (1967), Falvey and Veevers (1974), Jones (1973), Jongsma (1974), Veevers (1974), and Markl (1974b).

Among passive margins, the western and north-western margins and the adjacent oceanfloor are unique in having received such a thin sediment cover since the inferred continental breakup in the Mesozoic so that the rifted-arch stage that preceded breakup is particularly well documented (Powell 1976b; Exon and Willcox 1978; Veevers and Cotterill 1978; Barber 1982).

The western and north-western margins are separated from each other by the change in trend of the coastline at North West Cape and of the 4 km isobath at the north-western tip of the Exmouth Plateau

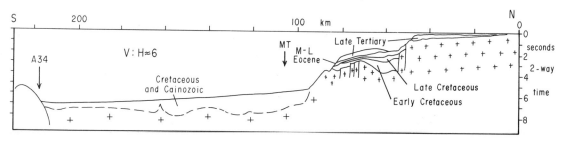

Fig. 105. Time section at long. 118° E, modified from Cooney *et al.* (1975), Talwani *et al.* (1979), and Falvey and Mutter (1981, fig. 9). MT: magnetic trough.

(Fig. 106). They are bounded on the landward side by the present shoreline and, in places, by a low coastal plain, and this boundary has occupied the same position since at least the Permian (Fig. 90), as we have seen in our analysis of the antecedents of the Great Western Plateau. On the oceanward side, the margin is bounded by the oceanic lithosphere of the Argo, Gascoyne, and Perth Abyssal Plains; off the north-west, the oceanic lithosphere dates from 160 Ma, and off the west 128 Ma (Fig. 107). Antecedents of the oceanic lithosphere were a rifted-arch complex that dates from 230 to 180 Ma (Late Triassic to Early Jurassic) and an infrarift basin that had its inception 320 Ma ago, in the Late Carboniferous. Accordingly, the margin as such can be traced back to 160 and 128 Ma, and its antecedent basins to 320 Ma. An earlier margin is postulated to have arisen from a previous cycle of breakup in the early Palaeozoic.

At the extreme north, in Timor, the margin has converged with the Sunda Arc 5 Ma ago, as is described elsewhere.

(ii) Physiography and structure

The western and north-western margins (Fig. 106) are a complex of marginal plateaux indented by abyssal plains; from south to north, these are the Naturaliste Plateau, Perth Abyssal Plain, Wallaby Plateau, Cuvier Abyssal Plain, Exmouth Plateau, Argo Abyssal Plain, and Scott Plateau; only in the extreme north-west does this pattern vary; here the margin is bounded by the Timor Trough and the island of Timor. Landward of the plateaus, whose surfaces are from 1 to 3 km below sea level, is a shelf and coastal plain ranging in width from 50 km in the south to 700 km in the north-west, and bounded by contours 0.2 km below and above sea level. The submerged or shelf part of this area is likewise narrow in the south and broad in the north-west, where it is called the North-west Shelf and Sahul Shelf. The Sahul Shelf is notable in containing a central low, the Bonaparte Depression (Fairbridge 1953; Edgerley 1974).

The land behind the margin is nowhere higher than 1250 m, and most of it, in the Great Western Plateau, lies between elevations of 200 m and 500 m. The elevated areas of the Kimberley Plateau and the Hamersley Range are separated by the broad lowland of the Great Sandy Desert. Only two areas, at either end of the margin, receive more than 1000 mm of annual precipitation, and this, together with the low relief, leads to modest sediment transport to the sea, a condition that seems to have prevailed for most of the past 100 Ma, as described already. The modern continental deposits comprise salt, silt, clay, aeolian sands, and residual deposits (laterite, caliche); offshore, carbonate dominates.

The deep structure of the Great Western Plateau comprises Archean and older Proterozoic nuclei wrapped round by younger Proterozoic mobile zones (Fig. 107), exemplified by the Darling Mobile Zone (Glikson and Lambert 1976) on the west of the Archean Yilgarn Block, and the Albany–Fraser Mobile Zone on the south-east. The major lineaments, shown in Fig. 107, are described by Plumb (1971):

> The post-Archean tectonics of the Australian Precambrian Craton are controlled by an almost rectangular pattern of major lineaments which generally trend about west to northwest and north to northeast. They are reflected in major shears, mobile belts, metamorphic belts, grabens, platform downwarps, etc. They developed early in the Proterozoic and have formed tectonic loci ever since . . . Locally, outcropping Archean shield areas have formed rigid buttresses, around which later structures have 'wrapped'.

The subsequent structure of the overlying Phanerozoic basins reflects the deep structure (Johnstone *et al.* 1973). The north-east trend is followed by the Malita Graben (2 in Fig. 107), Sahul Platform (4), Vulcan Sub-basin (26), the synclinal axis of the Browse Basin (25), Dampier and Barrow Basins (14, 15), Rankin Platform (11), and Exmouth Sub-basin (16). The north-west trend is followed by the Money Shoal Graben (1), the axis of the Palaeozoic Bonaparte Gulf Basin (5) and the Canning Basin (27), including the Fitzroy Trough (9). The north trend is followed by the Perth (21) and Carnarvon (24) Basins. Even as far offshore as the Exmouth Plateau, these trends are reflected in the pattern of faults generated during the Jurassic breakup. Barber (1982) related the north-trending faults (13) to the Proterozoic wrench faulting along the western flank of the Yilgarn Block, which later became the normally faulted Darling Fault System, and the north-north-east to north-east normal-fault trends of the Barrow–Dampier rift system to the folds of the Proterozoic mobile zones wrapped round the Pilbara Block.

(iii) Age of breakup

The age of breakup is indicated by seafloor-spreading magnetic anomalies, calibrated by deep-sea drilling at four sites, and by the breakup unconformity on the margin. Note that the ages given here are based on the new constants of the K–Ar system, and are about 2.5 per cent greater than those based on the old constants.

In the Argo Abyssal Plain, off the north-western margin (Fig. 107), Falvey (1972) described a set of magnetic anomalies which Larson (1975) and Heirtzler *et al.* (1978) showed to range back to M25 (157 Ma), in

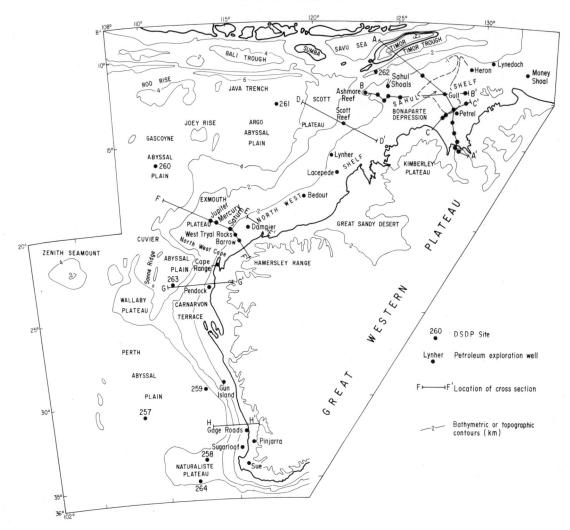

Fig. 106. Physiography, shown by the 2 and 0.2 km topographic contours, coastline, and 0.2, 2, 4, and 6 km isobaths. The 0.1 km isobath on the Sahul Shelf delineates the central Bonaparte Depression. Also shown are DSDP drilling sites and selected drilling sites on the margin and the location of cross-sections. Modified from Veevers (1982c).

agreement with DSDP Site 261 (Veevers *et al.* 1974b), which recovered late Oxfordian (145 Ma) sediment overlying oceanic basement. The uniform proximity of M25 to the continental margin indicates an age of slightly more than 157 Ma for breakup. From the spreading half-rate of 4.2 cm/year (Heirtzler *et al.* 1978) and the distance of M25 from the continent–ocean boundary of 100 km, the age of inception of spreading or breakup is calculated to be $157 + 2.4 = 159.4$ Ma, which we round to 160 Ma.

In the Cuvier Abyssal Plain, Larson (1977), Larson *et al.* (1979), and Johnson *et al.* (1980a) identified a set of anomalies, reflected about an axis along the Sonne Ridge (von Stackelberg *et al.* 1980), with the oldest anomaly being M10 (125 Ma). The oldest sediment recovered at DSDP Site 263, which did not reach oceanic basement, is of Albian age (105 Ma), though Veevers and Cotterill (1978) point out that the sediment is probably allochthonous. From M10's distance from the well-defined continent–ocean boundary (Roots *et al.* 1979) of 100 km, and a spreading half-rate of 3.2 cm/year, the inception of

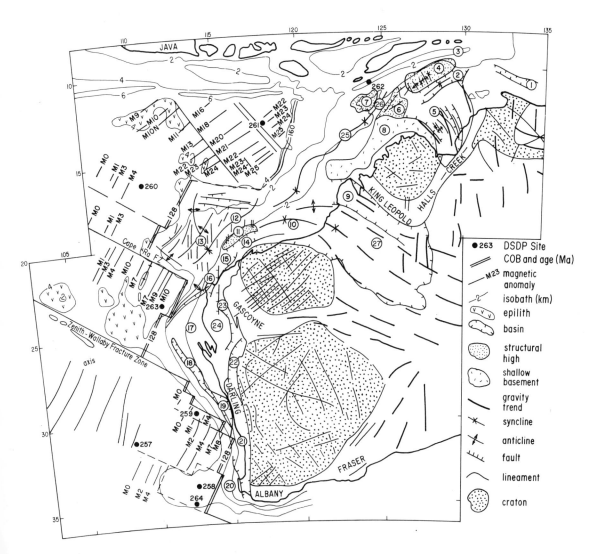

Fig. 107. Structure of the margin and adjacent oceanfloor. Onshore structure from Geological Society of Australia (1971), and gravity trends from Wellman (1976). Individual structures are: 1, Money Shoal graben; 2, Malita Graben; 3, Timor Trough; 4, Sahul Platform; 5, Petrel Sub-basin; 6, Londonderry High; 7, Ashmore Block; 8, Leveque Platform; 9, Fitzroy Graben; 10, Bedout Sub-basin; 11, Rankin Platform; 12, Kangaroo Syncline; 13, Exmouth Plateau Arch; 14, Dampier Sub-basin; 15, Barrow Sub-basin; 16, Exmouth Sub-basin; 17, Carnarvon Terrace; 18, Houtman Sub-basin; 19, Abrolhos Sub-basin; 20, Leeuwin Block; 21, Perth Basin and Perth; 22, Coolcalalaya Sub-basin; 23, Merlinleigh Sub-basin; 24, Carnarvon Basin; 25, Browse Basin; 26, Vulcan Sub-basin; 27, Canning Basin. Seafloor-spreading magnetic anomalies in the Argo Abyssal Plain and environs are from Heirtzler *et al.* (1978), in the Gascoyne Abyssal Plain from Powell (1978), in the Cuvier Abyssal Plain from Larson *et al.* (1979) and Johnson *et al.* (1980a), and in the Perth Abyssal Plain from Markl (1974a, 1978 a, b) and Johnson *et al.* (1980a). Modified from Veevers (1982c).

spreading is calculated to be $125 + 3.1 = 128.1$ Ma, rounded to 128 Ma. The same age is found off Perth, from the set of anomalies identified by Markl (1974a, 1978a, b), the oldest of which, according to Johnson *et al.* (1980a), is M8 (123 Ma). From M8's distance from the COB of 80 km, and a spreading half-rate of 3.2 cm/year, the oldest seafloor is calculated to be $123 + 2.5 = 125.5$ Ma; with an estimated 2 Ma of initial spreading transferred to India (Johnson *et al.* 1980a), the inception of spreading comes out at 127.5 Ma. The adjacent spreading compartment to the north that contains DSDP Site 259, with a basement age of Aptian (110 Ma) or, according to Morgan (1980), slightly older, has a set of anomalies (Larson *et al.* 1979; Johnson *et al.* 1980a), the oldest of which is M4 (120 Ma). From its distance of 140 km to the COB, and a half-rate of 3.2 cm/year, the oldest preserved seafloor is calculated to be $120 + 4.5 = 124.5$ Ma, adjusted as above to 126.5 Ma.

West of the Exmouth Plateau and north of the Cape Range Fracture Zone, Larson *et al.* (1979) and Powell (1978) have mapped a set of anomalies with the oldest identified one being M4 (120 Ma) near DSDP site 260, which bottomed in a basalt sill that intrudes Albian (105 Ma) sediment. Unpublished work by Veevers and L. Hansen suggests that the spreading compartment that contains Site 260 has an abandoned spreading axis, as in the Cuvier Abyssal Plain, and that spreading probably started, as it did in the south, at about 128 Ma.

The age of the breakup unconformity along the margin agrees with the magnetically determined age of the inception of spreading. The main unconformity, interpreted as reflecting break-up, in the Browse, Bonaparte, and Dampier Basins (Powell 1976b) and the Exmouth Plateau (Barber 1982) is dated closely as Callovian (160 Ma), and was accompanied in the Browse Basin and on the Ashmore Block by the extrusion of basalt (Powell 1976b). As noted by Barber (1982), seafloor spreading in the Argo Abyssal Plain affected not only the immediately adjacent margin of the Browse Basin but also the rest of the North-west Shelf to the south-west and the Exmouth Plateau, whereas the later breakup in the Early Cretaceous (128 Ma) by spreading in the Gascoyne Abyssal Plain along the western side of the Exmouth Plateau produced no unconformity. Coeval spreading in the Cuvier Abyssal Plain on the south-west side of the Exmouth Plateau, however, produced an unconformity beneath a sediment wedge that prograded from the Cape Range Fracture Zone, as described later. At the latitude of Perth, an angular unconformity separates a thick non-marine faulted and tilted sequence below from a thin marine sequence above (Veevers and Johnstone 1974, fig. 8). The lacuna at the unconformity lies within the Neocomian, probably in the Valanginian (125-117 Ma); at the unconformity in the southern Perth Basin is the Bunbury Basalt, which, as reported by Playford *et al.* (1975, p. 252; 1976b, p. 196), is enclosed by Late Jurassic strata below and Early Cretaceous strata above, dating the unconformity here as 129 Ma, in agreement with the magnetically determined age of breakup of 128 Ma. Incidentally, radiometric determinations of the age of the Bunbury Basalt (McDougall and Wellman 1976) have failed to yield anything older than a minimum age of 90 Ma.

(iv) Boundary between oceanic and continental crust

As noted already, the thin cover of post-breakup sediment over the margin and adjacent oceanfloor facilitates the discrimination of continental from oceanic crust, as exemplified by studies in the Cuvier Abyssal Plain (Roots *et al.* 1979), but the structure – whether thin continental crust or thick oceanic crust – of particular marginal plateaux as delimited by the continent–ocean boundary, is obscure. The oceanic plateaux of the Joey Rise and the Roo Rise (Fig. 106) are crossed by seafloor-spreading magnetic anomalies (Heirtzler *et al.* 1978) (Fig. 107) and accordingly are regarded as being underlain by oceanic crust. Such anomalies have not yet been reported from any part of the marginal plateaux. Veevers and Cotterill (1978) introduced the term epilith for an upgrowth of oceanic crust developed after continental breakup, and postulated that the Naturaliste Plateau, Wallaby Plateau, and adjacent Zenith Seamount, the north-west part of the Exmouth Plateau, and most of the Scott Plateau are epiliths. In reply to a discussion of the Scott Plateau by Stagg and Exon (1979), Veevers (1979) concluded that the only definitive criterion of epiliths is that the basement beneath the superficial sediment consist of volcanogenic rock with a composition appropriate to an oceanic setting *and* an age equal to or younger than breakup. The age criterion is vital: volcanic rocks are expected to be common along the rift valley that evolves into oceanic lithosphere, as borne out by the common occurrence of volcanics older than breakup along the western and north-western margins (Figs. 108, 109; Table 8). The three marginal plateaux that have been sampled for basement rock satisfy the first half only of the criterion, and lack the crucial evidence of age. The Naturaliste Plateau has been sampled at two DSDP sites, 258 (Davies *et al.* 1974) and 264 (Hayes *et al.* 1975), and by dredging (locality 13 on Fig. 108) (Coleman *et al.* 1982), and all three have yielded volcanogenic rock of appropriate oceanic composition,

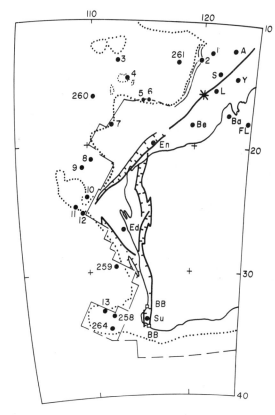

Fig. 108. Distribution of Phanerozoic volcanic and volcanogenic rocks, with respect to the continent–ocean boundary (COB), 160 Ma old on the north-west (double lines) and 128 Ma on the west (single line). Note alternative positions of COB for the Naturaliste Plateau. Also shown are the 4 km isobath (dotted lines) and the synclinal axis and grabens. Details in Table 8. From Veevers and Hansen (1981).

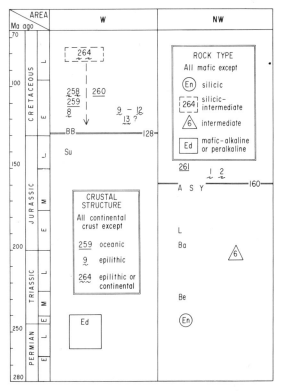

Fig. 109. Ages of volcanic and volcanogenic rock, located on Fig. 108, detailed in Table 8. The age of breakup is shown by double lines. From Veevers and Hansen (1981).

as discussed by Coleman *et al.* (1982). But the age indicated by the oldest overlying sediment – mid-Albian or 105 Ma at Site 258 – is not definitive, so that the question of the crustal structure of the Naturaliste Plateau remains open. The Wallaby Plateau contains unsampled surficial sediment draped over an unconformity cut across a dipping layered sequence at least 2 seconds' reflection time (3000 m) thick. Samples of this sequence from two areas (10 and 11 on Fig. 108) suggest that it comprises oceanic volcanics and volcanogenic sediments (von Stackelberg *et al.* 1980). The only datable, but weathered, material from the southern flank of the Wallaby Plateau is a tholeiitic basalt (11) that gave a minimum K–Ar age of 90 Ma, but this too is not definitive. Amygdaloidal basalt fragments indicate original shallow-water or subaerial extrusion. The volcaniclastic sediments were probably eroded by waves and deposited as wedges that prograded outward from an island or as deeper marine deposits on the proto-abyssal plain. The likelihood that the Wallaby Plateau is an epilith is strengthened by our unpublished work that shows the continuity in seismic and magnetic profiles of the abandoned spreading ridge in the Cuvier Abyssal Plain southward into the Wallaby Plateau.

In the Scott Plateau area, we have found definitive evidence of the location of the continent–ocean boundary in recently collected magnetic profiles. A prominent magnetic anomaly along the oceanward edge of the Scott Plateau marks the continent–ocean boundary, so that, contrary to Veevers and Cotterill (1978), the entire plateau must be underlain by continental crust, as suggested by Allen *et al.* (1978), Stagg and Exon (1979), and Falvey and Mutter (1981).

A recently published seismic profile across the

TABLE 8. *Volcanic and volcanogenic rocks of the western and north-western margins and adjacent ocean*

SAMPLE	LOCATION			ROCK TYPE			FORMATION		AGE		Radiometric Ma	REFERENCE
	Symbol on Figs. 108, 109	Name	Description	Silicic	Intermediate	Mafic	Mafic-alkaline or peralkaline		Biostratigraphic and superpositional			
DRILL HOLES	A	Ashmore	basic lava			+			Middle Jurassic		128, 143	Allen et al. (1978); Veevers (1969)
	Ba	Barlee	dolerite intrusion			+					200	Veevers & Evans (1975)
	Be	Bedout	basalt, volc. congl.			+			Middle Triassic			Powell (1976b)
	Ed	Edel	phonolite, trachyte				+				245, 266	V + E (75), Le Maitre (75)
	En	Enderby	rhyolite	+								V + E (75)
	L	Lombardina	basic lava			+			Permo-Triassic			Allen et al. (1978)
	S	Scott Reef	basic lava			+			Early Jurassic			Allen et al. (1978)
	Su	Sue	dolerite intrusion			+			Middle Jurassic			Playford et al. (1975)
	Y	Yampi	basic lava			+			Middle Jurassic			Allen et al. (1978)
	258	Deep	volcanogenic SS			+			Albian			Davies et al. (1974)
	259	Sea	tholeiitic basalt			+			Aptian			Veevers et al. (1974b)
	260	Drilling	basalt sill			+			Albian			Veevers et al. (1974b)
	261	Project	basalt sill and flows			+			Oxfordian			Veevers et al. (1974b)
	264		rhyolitic and andesitic conglomerate, tuff	+	+				Santonian or older			Ford (1975)
CORES and DREDGE HAULS	1	KD40	basalt, breccia, tuff			+			?Jurassic			Hinz et al. (1978)
	2	KD2-4, 12, 13	basalt, breccia, tuff			+			?Jurassic			Hinz et al. (1978)
	3	PC36	basaltic hyaloclastite			+						Cook et al. (1978)
	4	PC37	basaltic hyaloclastite			+						Cook et al. (1978)
	5	PC38	basalt			+						Cook et al. (1978)
	6	KD63, 65, 66	?trachyte		+						192, 213	von Stackelberg et al.
	7	KD73	?tuff						?Cretaceous			von Stackelberg et al.
	8	KD148, 149	basalt, tuff, breccia			+			?Middle Cretaceous		120	von Stackelberg et al.; Johnson et al. (1980a)
	9	KD155	basalt, tuff, breccia			+			?Middle Cretaceous			von Stackelberg et al.
	10	KD161, 162, 163	volcanic sandstone			+			?Middle Cretaceous			von Stackelberg et al.
	11	KD167, 168, 170	basalt, tuff, breccia			+			?Middle Cretaceous		>89	von Stackelberg et al.
	12	KD173	tuff, volc. claystone			+			?Middle Cretaceous			von Stackelberg et al.
	13	Naturaliste Plat.	tholeiitic basalt			+			?Middle Cretaceous		?118–128	Heezen & Tharp (1973); Coleman et al. (1982)
OUTCROP	BB	Bunbury, Black Pt.	tholeiitic basalt			+		Bunbury Basalt	Neocomian		>90	Edwards (1938); Johnstone et al. (1973); Playford et al. (1975)
	FL	Fitzroy R.	lamproite				+	Fitzroy Lamproite			17–21	Wellman (1973)

The ages of Be, 259, 261, and 264 are minima indicated by the overlying sediment. In 258, basalt clasts are Albian (the age of the enclosing sediment) or older. The ages of 8 and 13 are from seafloor-spreading magnetic determinations (Johnson et al. 1980a). The K-Ar age of 11 is a minimum as probably are those of A. The maximum age of the sill at 260 (Albian) corresponds with the magnetically determined age of the crust. Seismic interpretation indicates that the Bunbury Basalt extends northwestward from Bunbury into the Vlaming Sub-basin towards the Fremantle Canyon (P. G. Quilty, pers. comm.). K-Ar age determination of the Bunbury Basalt (McDougall & Wellman 1976) indicates a minimum age of 90 Ma. Its enclosure by Jurassic/Cretaceous sediments on either side of the breakup north-western tip of the Scott Plateau (Hinz 1981, fig. 10) shows a complex of oceanward-dipping reflectors beneath a smooth acoustic-basement reflector, interpreted as 'volcanics possibly interbedded with sediment' formed 'just before the opening of the Argo Abyssal Plain'. Hinz (1981) describes similar sequences from many other continental margins, and envisages their development in four steps: (1) extension and attenuation of continent crust accompanied by injection of dykes and extrusion of lavas under subaerial or shallow-water conditions; (2) continuing crustal extension accompanied by violent volcanic eruptions and deposition of tuff and agglomerate; gradual sagging causes the reversal of dip of the volcanics towards the active centre; (3) subsidence increases the dip of the older volcanics and leads to submarine volcanism; and (4) further subsidence to abyssal depth, with an increase in hydrostatic pressure, reduces the degassing of the lava so that normal oceanic crust is accreted. Mutter *et al.* (1982) describe oceanward-dipping reflectors from the Norwegian margin similar to those described by Hinz (1981) but within crust shown to be oceanic by seafloor-spreading magnetic anomalies; accordingly, they interpret the reflectors as produced by 'subaerial sea-floor spreading'. The discoveries of Hinz (1981) and Mutter *et al.* (1982) confirm the ideas of the development of subaerial and shallow-water facies at the ridge crust and flanks of a newly formed ocean basin advanced by Veevers (1977).

(v) Stages of development

The western and north-western margins developed through the Phanerozoic in five stages (Fig. 110): an obscure initial early Palaeozoic stage of plate divergence with associated failed arms, and four stages associated with plate divergence in the Mesozoic.

(1) Failed-arm stage
Warris (1973), Laws and Kraus (1974), Lofting *et al.* (1975), Veevers (1976), and Brown (1980) (Fig. 111) have speculated that much of what later became the north-western margin developed by plate divergence between about 575 and 510 Ma by the spreading of the Tethyan Ocean. The speculation is based on the sequence of Early Cambrian eruption of voluminous basalt (Antrim Plateau Volcanics) followed by the deposition of marine strata in the Bonaparte Gulf Basin and then, in the Ordovician, of marine strata in the Canning Basin and non-marine strata in the Carnarvon and northernmost Perth Basins. These deposits are interpreted as reflecting the subsequent fill of failed arms (Burke and Dewey 1973) that radiated from nodes initiated during the breakup of an ancestral north-western margin (Fig. 111). An intense (+80 mGal) north-trending positive gravity anomaly beneath the central axis of the Petrel Sub-basin is interpreted as reflecting crustal thinning in the failed arm of the Bonaparte Gulf Basin (Brown 1980).

Nothing is known of the sediments deposited on the original margins – they would have been obscured by deposits of the later stages – and the postulated failed arms are the sole source of information. The style of deposition is shown by the Cambrian through Carboniferous parts of the composite stratigraphical columns (Fig. 112). Deposition was wholly in shallow seas except possibly in non-marine to shallow marine situations in the Perth and Carnarvon Basins during part of the Ordovician and Silurian, and in the south Canning Basin in part of the Silurian and Devonian. Most of the Devonian carbonate rocks are reefal (Geary 1970; Playford 1980, 1982), and they grade basinwards to deeper water (but probably still neritic) shales. The only notable Palaeozoic facies omitted in the columns of Fig. 112 is the (?) Silurian to Early Devonian diapiric evaporate of the offshore Bonaparte Gulf Basin (Laws and Kraus 1974; Edgerley and Crist 1974), seen in the cross-sections of Fig. 113. The only volcanics are the Early Cambrian Antrim Plateau Volcanics and equivalents, which cover a large area of north-west Australia, and are interpreted as accompanying the initiation of the margin.

This stage of deposition extended no farther south than the northernmost part of the Perth Basin, where it is represented by the fluvial Tumblagooda Sandstone. Consistent with the idea of a failed arm, faulting accompanied deposition of this formation (Johnstone *et al.* 1973; Playford *et al.* 1975; Hocking 1979), but faulting is not known during the Ordovician in the Canning Basin or in the Cambrian of the Bonaparte Gulf Basin, and becomes obvious in these basins only in the Middle and Late Devonian. The bounding faults of the Money Shoal Graben are probably post-Silurian, and possibly older.

The postulated configuration of the ancestral north-western margin facing the open ocean of Tethys, with the site of the later western margin lying within the interior of Gondwanaland, is consistent with the ubiquitous distribution of marine Palaeozoic and pre-breakup Mesozoic sediments in the Bonaparte Gulf, Canning, and Carnarvon Basins, and the preponderance of non-marine coeval sediments in the Perth Basin (Veevers *et al.* 1971). The extreme example of this trend is shown by the 9 km of wholly non-marine Permian through Jurassic sediments of the southernmost Perth Basin (Veevers 1971a).

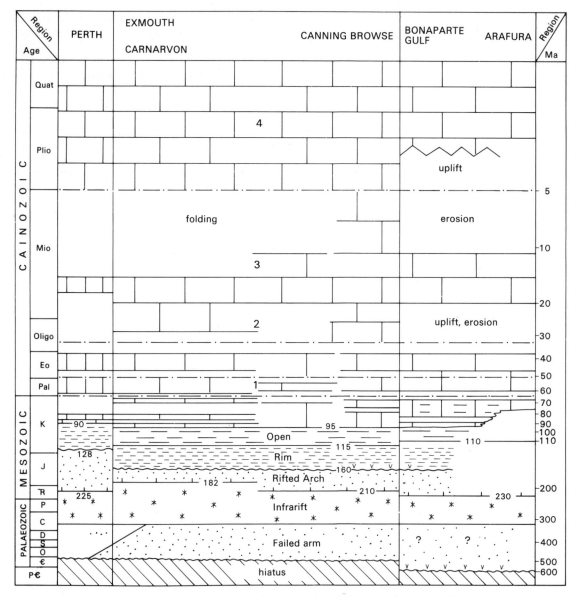

Fig. 110. Time--space diagram showing the stages of development during the Phanerozoic, including the four numbered cycles of carbonate deposition during the Cainozoic. Time shown on a logarithmic scale.

(2) Infrarift stage

As part of a stage of development that affected all Gondwanaland, and gave rise to the deposition of the Gondwana Series and its equivalents, the terrain of the western and north-western margins subsided in the Late Carboniferous and Early Permian, while the bordering shield areas were uplifted. This stage was characterized by broad downwarps, and was succeeded in the Mesozoic by a stage of rifted-arch development out of which grew the oceanic lithosphere during continental breakup in the late Mesozoic. In earlier accounts of the western margin, I classed these two stages that precede breakup as a single stage of rifting, but here I follow Falvey and Mutter's (1981) distinction

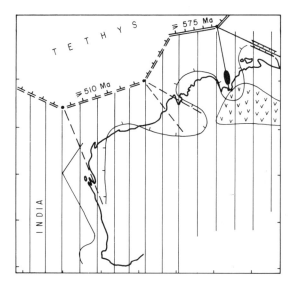

Fig. 111. Failed-arm stage, during the Cambrian and Ordovician, with failed arms radiating from nodes at re-entrants, of continental margin adjacent to Tethyan Ocean, with Money Shoal Graben and Bonaparte Gulf Basin stemming from oldest margin, and the Canning and Carnarvon Basins from the younger margin. Continental lithosphere shown by vertical lines, oceanic lithosphere clear. Solid black ellipse in the southern Bonaparte Gulf Basin represents an intense positive gravity anomaly.

of an early stage of downwarping (infrarift) from the later stage of rift-valley development.

In the Perth Basin, except the northernmost part, the Late Carboniferous–Permian deposition is the oldest to be preserved since the Precambrian, and elsewhere these deposits followed the areas of the failed arms. The basal Late Carboniferous and Sakmarian deposits are marine to non-marine, including glacigene, sediments, and are followed later in the Permian by coal measures. All were deposited in broad downwarps little affected by faults except along the Darling Fault. Even here, deposition extended eastward to the Yilgarn Block to form the Collee and Wilga Basins (Wilde and Walker 1978), as mentioned in the analysis of the Great Western Plateau. As much as 4.5 km of shallow marine to non-marine sediment accumulated in parts of the Canning and Carnarvon Basins, and as much as 6 km in the Petrel Sub-basin of the Bonaparte Gulf Basin. Volcanism, not seen since the outpouring of the Antrim Plateau Volcanics in the Early Cambrian, is represented by rhyolite at Enderby-1 and by a carbonatite-like association of phonolite and trachyte at Edel-1 (Veevers and Hansen 1981) (Figs 108, 109).

In the Early Triassic a transgressive sea covered the region, as documented for the Bonaparte Gulf Basin (Laws and Kraus 1974), the northern part of the Canning Basin and offshore (Gorter 1978), the Dampier and northern Carnarvon Basins (Crostella and Barter 1980) and probably the Exmouth Plateau (Barber 1982), and the northern Perth Basin (Jones 1976). A regional regression followed in the Middle and Late Triassic, and led to the deposition of the 3-km-thick fluvial Mungaroo Sandstone on the Exmouth Plateau (Barber 1982) (Fig. 115), with an unknown extent further west on the neighbouring part of Gondwanaland (Fig. 114). Because this thick layer of sandstone prograded westward and north-westward (Crostella and Barter 1980; Vos and McHattie 1981; Barber 1982), its source must have been the Pilbara Block, which, in the process, must have been denuded to a depth of several kilometres. The change of stage, from downwarping over broad areas (infrarift) to subsidence along narrow fault-bounded zones (rifted-arch complex), did not reach the Exmouth Plateau until the later part of the Early Jurassic (182 Ma), some 30 to 50 Ma later than elsewhere.

(3) Rifted-arch stage

In the Perth Basin, the change in stage from infrarift to rifted arch is marked by the deposition in the Late Triassic (225 Ma) of a poorly sorted conglomerate and sandstone (Lesueur Sandstone) that thins westward across growth faults from its maximum thickness of 2 km against the Darling–Urella Faults (Jones 1976) (Fig. 114). At the northern end of the system in the Bonaparte Gulf Basin, major block-faulting, from the mid-Triassic to the mid-Jurassic, superimposed north-east trends on the earlier north-west trends (Brown 1980; Laws and Kraus 1974).

Late Triassic fluvial and redbed facies in the Petrel Sub-basin pass northwestward into equivalent marine to marginal marine carbonate, sandstone, and shale, as are found in the para-autochthon of Timor (Audley-Charles *et al.* 1979). Laws and Kraus (1974) interpret a sediment source in the north-west from an increase in the concentration of sand to the north-west, the presence of a barrier-island complex in Sahul Shoals-1, and evidence from seismic profiles across the Ashmore Block of southeasterly prograding sediment. Allen *et al.* (1978) draw a similar conclusion from dip-meter evidence at Scott Reef (Fig. 114A). Fifty Ma later, at the end of the Middle Jurassic (160 Ma), the region was to develop into a new continental margin, and the shedding back of sediment towards the continent in

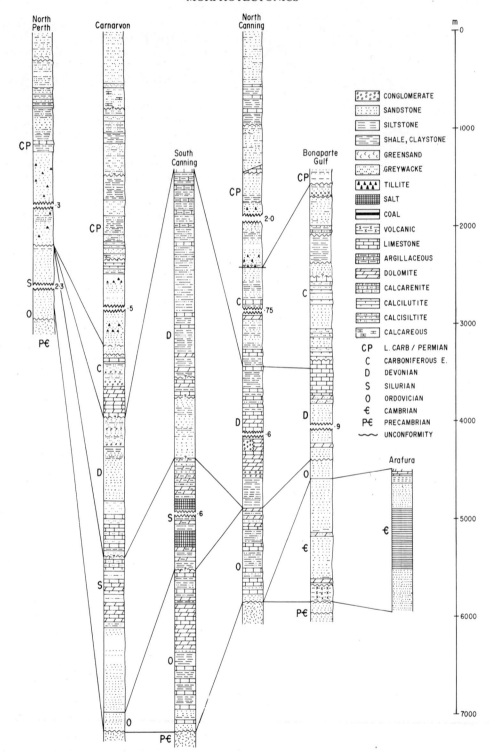

Fig. 112. Composite stratigraphical columns of the Palaeozoic parts of onshore basins. All except the Late Carboniferous–Permian are interpreted as deposits of the failed-arm stage. Numbers indicate thickness (km) of section omitted. From Veevers (1982c).

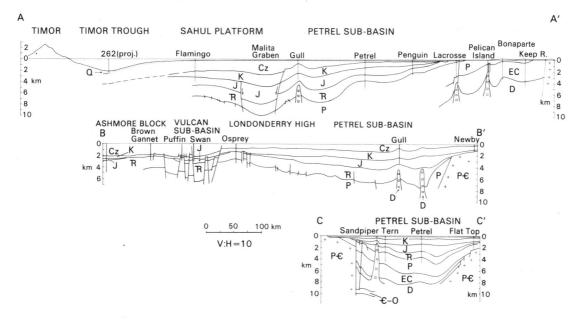

Fig. 113. Cross-sections of the Bonaparte Gulf Basin, located in Fig. 106. Modified from Brown (1980).

the Late Triassic is consistent with the development of a rifted arch to which the Browse and the Bonaparte Gulf Basins were yoked as extra-arch basins. Accompanying events about the Triassic–Jurassic boundary were the dolerite intrusion at Barlee-1 (Ba in Fig. 108), the emplacement of trachyte into what was to become the northern edge of the Exmouth Plateau (5 and 6 in Fig. 108), the folding in the Fitzroy Graben by right-lateral wrenching on the enclosing basement blocks (Rattigan 1976; Smith 1968a; Rixon 1978) (Fig. 114A), and the development of salt pillows in the Silurian–Devonian evaporites of the southern Canning Basin.

The latest onset of rift-valley development, at the Pliensbachian–Toarcian boundary (182 Ma), was in the Exmouth Plateau–Carnarvon Basin region. 'Two concurrent tectonic events occurred; as the newly created Barrow–Dampier Rift began to subside, pre-rift valley arching developed to the west of the Exmouth Plateau. This caused a regional tilt of the entire plateau area to the east, exposing all of the pre-existing Jurassic section to erosion' (Barber 1982). Thus, by the end of the Early Jurassic, the rifted-arch system formed a linear zone along the western and north-western margins – the Westralian Geosyncline of Teichert (1939).

By the mid-Jurassic (160 Ma), at the time of breakup along the north-western margin (Fig. 114B), the divergent plate boundary was marked by a postulated rifted arch on the western side of the Browse Basin, and by an oblique transform zone on the northern side of the Exmouth Plateau. A postulated rifted arch along the western side of the Exmouth Plateau and southward past several offsets to Perth did not break up until 128 Ma. On the Australian side of the rifted arch were the broad extra-arch basins of the Bonaparte Gulf Basin (Fig. 113), with the local grabens of the Malita Graben and Vulcan Sub-basin, and the Browse Basin (Fig. 116). To the south-west, the system became a narrow rift-divergence basin between the Rankin Trend and the Pilbara Block (Figs. 117, 118), which tapered out near the outer rifted arch on the western side of the Carnarvon Terrace (Fig. 119, GG'). Further south, a rift valley passed through the Abrolhos Sub-basin and the Perth Basin (Figs. 119, HH'; 120; 122).

While no analogue can be found for the infrarift stage, except perhaps for the Karroo basins of East Africa (Veevers and Cotterill 1976), the rifted-arch stage can be matched, at all scales, with the Arabian–East African rift system. At the greatest scale (Fig. 121), the two systems are the same size, as shown by Veevers and Cotterill (1976), and have the same distribution of volcanics (Veevers and Hansen 1981). The outer arm of the rift complex, along which breakup took place, has been found, where dredged

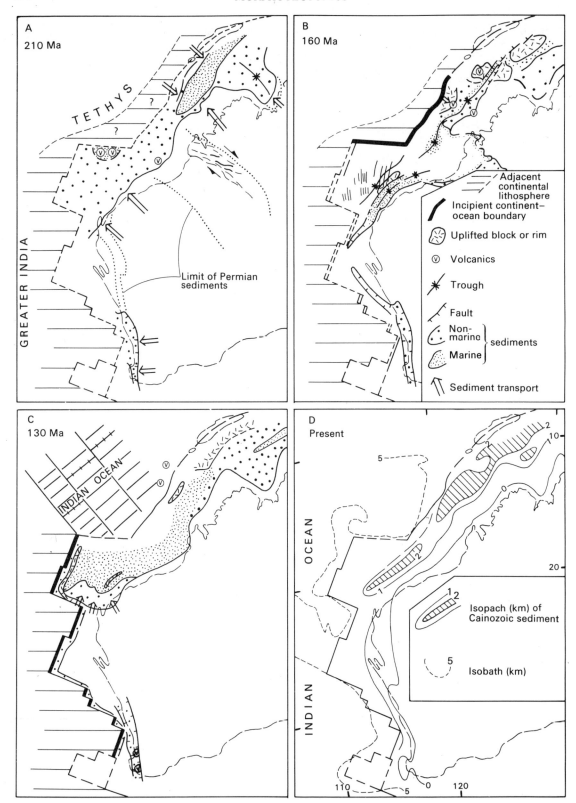

DIVERGENT MARGINS

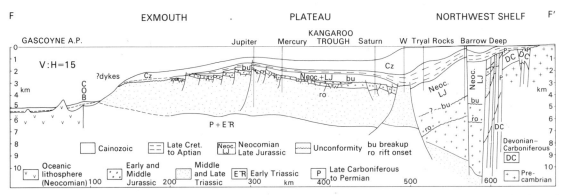

Fig. 115. Cross-section of the Gascoyne Abyssal Plain, Exmouth Plateau, and North-west Shelf (Dampier-Barrow Basin), located in Fig. 106. From Meath and Bird (1976, fig. 2), Barber (1982, fig. 16), and Veevers et al. (1974a, fig. 3, AL). Determination of COB and dykes beneath lower slope from unpublished work on magnetic anomalies, by Veevers and Hansen.

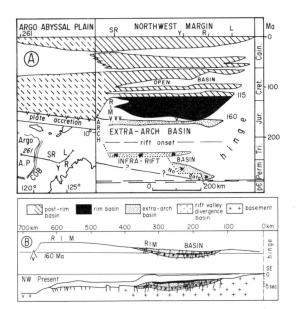

Fig. 116. A. Time–space diagram. B. Cross-section DD' (located in Fig. 106), and reconstructed cross-section at time of breakup (160 Ma). Derived from Powell (1976b) and Allen et al. (1978).

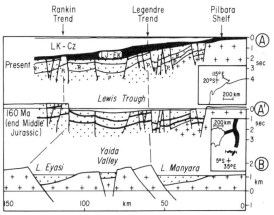

Fig. 117. Cross-section of the Dampier Basin (A), and reconstructed section at breakup (160 Ma) (A'), compared with the rift divergence zone of Tanzania (B). From Veevers et al. (1982).

and cored, to be volcanic, and matches the volcanic eastern arm of the East African rift system, which would be expected to mark the line of future breakup in Africa; and the inner arm, like the western arm of the East African rift system, is non-volcanic except at scattered centres, marked in Fig. 121 by asterisks.

The complex of diverging rift valleys in the Perth area immediately before breakup (128 Ma ago) is shown schematically in Fig. 122. The postulated outer set of rift valleys is traced back from the COB, represented off Perth (Fig. 107) by short spreading compartments offset by transform faults, and the COB is taken to mark the median line of the rift valley. The deep

Fig. 114. Palaeogeographical maps. A. Infra-rift stage, in the Late Triassic (210 Ma). B, Breakup along the north-western margin at the end of the Middle Jurassic (160 Ma), and continued rifted-arch development along the western margin. C. Rim-basin development during the Late Jurassic north of the Cape Range Fracture Zone, and breakup to the south. D. Present state, showing thickness (km) of Cainozoic sediment.

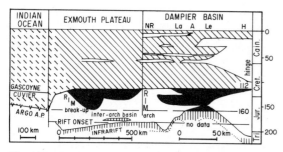

Fig. 118. Time–space diagram of the Gascoyne/Cuvier and Argo Abyssal Plains, Exmouth Plateau, and Dampier Basin, modified from Veevers and Cotterill (1978), using new drilling information from the Exmouth Plateau (Barber 1982). In the rim basin of the Exmouth Plateau, the lower part indicates an extremely condensed section, and the upper part a thick Neocomian section, whose base coincides with the start of seafloor spreading in the Cuvier Abyssal Plain.

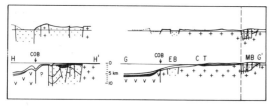

Fig. 119. Cross-sections of the Carnarvon Terrace and adjacent Cuvier Abyssal Plain (GG´) and Perth Basin and adjacent oceanfloor (HH´), with reconstructed section at breakup (128 Ma) above. From Veevers and Cotterill (1978). In GG´, the basement, denoted by crosses, includes Palaeozoic strata.

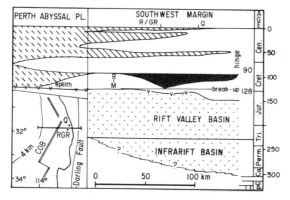

Fig. 120. Time–space diagram of the Perth Basin. From Veevers and Cotterill (1978).

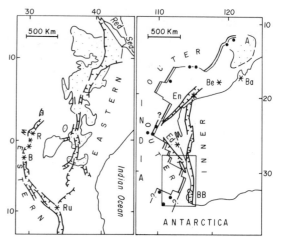

Fig. 121. Left: rift-valley faults and associated volcanic rocks in East Africa, showing the eastern arm dominated by volcanics (stippled) and the western arm non-volcanic except in local centres (asterisks) at Rungwe (Ru), Birunga (B), and Ruwenzori (R). Right: pre-breakup western and north-western margins, showing rift-valley faults, volcanic outer arm (double lines) along the incipient COB with volcanics (dots and stipple—Ashmore Volcanics (A) and Bunbury Basalt (BB)), and eastern arm non-volcanic except at local centres (asterisks) of Barlee (Ba), Bedout (Be), Enderby (En), and Edel (Ed). Modified from Veevers and Hansen (1981).

structure beneath the outer rift valleys and the western part of the inner rift (0–300 km) is hypothetical, based on Girdler's (1978, fig. 12) interpretation of extreme thinning of the lithosphere beneath the eastern volcanic rift of East Africa. The deep structure of the eastern part of the inner rift and eastward (300-500 km) is from Mathur (1974). We interpret his crust layer 3 (seismic velocity 7.49 km/sec), which wedges out beneath Coolgardie, about 500 km east of the Darling Fault (j), to be accreted asthenosphere, as suggested by Glikson and Lambert (1976), so that, at the time of breakup, it would have been continuous with the asthenosphere on the west.

The chief comparison with East Africa is the concentration of volcanism in one arm of the rift system compared to its virtual exclusion in the other (Fig. 121). With the ultimate breakup of East Africa by plate divergence along the volcanic eastern rift, as envisaged by Dietz and Holden (1970), the western rift would be preserved intact on the landward side of the continent-ocean boundary. It is this kind of configuration we postulate in the Perth region at breakup. In

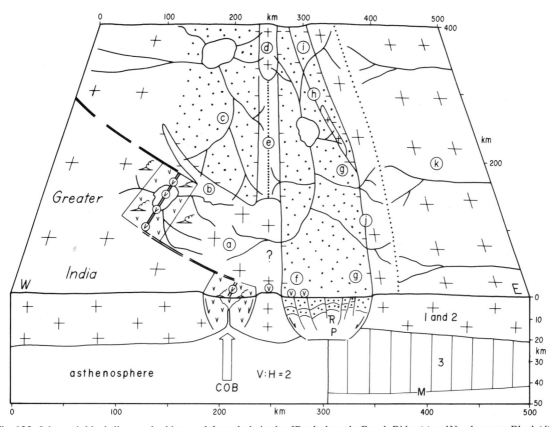

Fig. 122. Schematic block diagram, looking north from the latitude of Perth along the Beagle Ridge (e) and Northampton Block (d), of the reconstructed rift-valley system immediately before continental breakup in the Late Jurassic and Early Cretaceous (150-128 Ma ago). From Veevers and Hansen (1981). Double broken lines indicate the COB that came into being at 128 Ma ago with the separation of Greater India and Australia. Across the front of the diagram, from left to right, the asthenosphere at breakup is shown from 0-300 km; and crust layer 3 of Mathur (1974), which we presume to be accreted asthenosphere, as suggested by Glikson and Lambert (1976), is shown in its present configuration from 300– 500 km. Layer 3 wedges out beneath Coolgardie, about 500 km east of the Darling Fault (j). Mathur's (1974) crust layers 1 and 2 and outcropping Precambrian rocks are shown by crosses, and Jurassic and Early Cretaceous (up to 128 Ma ago) rift-valley fill by dots. The volcanic terrain of the postulated rift valleys that evolved into ocean is hachured with Vs and depicted by vents (smoking triangles), the emissions of which possibly included silicic pyroclastics, some of which ultimately reached the Vlaming Sub-basin (f) and Dandaragan Trough (g). Circled Vs indicate tholeiitic basalt flows that constituted the surface of the first oceanic lithosphere generated along the COB at breakup (128 Ma ago) and also the basalt that spilt over on to the continent to form the Bunbury Basalt. Watersheds (dotted lines), drainage channels, and lakes are wholly schematic. Structure of Vlaming Sub-basin (f) and Dandaragan Trough (g) shown in sections after Jones and Pearson (1972). a, Edward's Island Block; b, Turtle Dove Ridge; c, Abrolhos Sub-basin; d, Northampton Block; e, Beagle Ridge; f, Vlaming Sub-basin; g, Dandaragan Trough; h, Urella Fault; i, Irwin Sub-basin; j, Darling Fault.

the Perth Basin, the only volcanogenic rocks are the possible pyroclasts in the Yarragadee Formation described by Veevers and Hansen (1981), emplaced by air-fall or fluvial transport, or both, and the Bunbury Basalt that spilt over from the earliest emplaced oceanic lithosphere at breakup. Minor hypabyssal intrusions include dolerite in Sue–1, and are inferred to have caused the injection in places of heat-altered coal (Kantsler and Cook 1979, p. 98). Thus the Perth Basin itself was mainly non-volcanic, and probably lacked the high heat flow of the outer volcanic rift and its succeeding young ocean. This accords with Kantsler and Cook's (1979) finding, by studies of the reflectance of vitrinite clasts, of low geothermal gradients since the Permian in the Bunbury Trough and since the Jurassic in the thick Late Jurassic sections of some of the deeper sub-basins. Furthermore, the presence of very high ranks in parts of the Permian section of the Beagle Ridge suggests that a Permian to Jurassic thermal event associated with local igneous activity or the initiation of rifting (= plate divergence), or both, may also be a controlling factor (Kantsler and Cook 1979, p. 94), consistent with the location of the postulated hot volcanic rift system on the Beagle Ridge side of the Perth Basin.

The Perth Basin was generally cool during deposition, as befitted its mainly non-volcanic state; the postulated volcanic arm of the rift-valley system that lay to the west was generally hot during the Permian and part of the Mesozoic, and some of this heat may have affected the western part of the Perth Basin occupied by the Beagle Ridge. The rest of the western margin probably evolved in a similar way, as suggested by the widespread occurrence of volcanic rocks along the landward side of the continent–ocean boundary (Fig. 108).

During the rift-valley stage, the Bonaparte Gulf and Browse Basins developed as extra-arch basins to the postulated rifted arch that developed into the ocean (Fig. 114 B), the Dampier Basin developed as a narrow rift-divergence zone, and between it and the postulated outer rifted arch the Exmouth Plateau was an inter-arch basin. A direct comparison of the Lewis Trough of the Dampier Basin just before breakup with the modern rift-divergence zone of Tanzania (Fig. 117) shows the similarity of structure. South of the Exmouth Plateau, the Exmouth Sub-basin and the Abrolhos-Perth Basins developed as inner rifted-arches close to the postulated outer one (Fig. 114B, C).

(4) Rim-basin stage
Following breakup of the north-western margin at 160 Ma (Fig. 114C), the extra-arch basin of the Browse Basin was succeeded by a rim basin (Fig. 116) in which 'restricted marine claystones were initially deposited in the troughs, passing laterally into marginal marine to deltaic facies in the topographically elevated areas' (Powell 1976b, p. 21). This rim basin is no thicker than 1 km, but in the Dampier Basin the Late Jurassic–Neocomian rim basin behind the rim of the Rankin Trend (marked by West Tryal Rocks-1 in Fig. 115) is over 3 km thick and 2 km thick to the north-east (Fig. 117; Crostella and Chaney 1978, fig. 9). On the west, between the Rankin Trend and the outer part of the plateau, a broad rim basin succeeded the narrow inter-arch basin. Barber (1982) shows that at the site of Jupiter–1, only 15 m of marine clay and silt accum-

ulated as an extremely condensed sedimentary section during the interval of 30 Ma of the Late Jurassic, but 234 m accumulated during the next 20 Ma of the Neocomian. According to Barber (1982, fig. 11), as shown in Fig. 114C, Jupiter–1 lay in front of a fluvio-deltaic wedge, with sources to the south-west, south, and south-east. Veevers and Powell (1979) envisaged deposition as shown in Fig. 123, with the south-western part of the wedge prograding from the continental rim about the extensional transform-fault zone of the Cape Range Fracture Zone. From an actualistic study of the Levant transform zone, Veevers and Powell (1979) argued that the rim along the south-west flank of the Exmouth Plateau remained high only while it was a continent–continent transform fault; with the passage of the trailing edge of Greater India past the south-west tip of the Exmouth Plateau 123 Ma ago (= 120 Ma, using the old K–Ar constants), the rim would have started to subside, so that the age of the prograded wedge shed from the rim was predicted to be no greater than 128 Ma ago, or Berriasian. Barber's (1982) dating of the prodelta silts and shales in Jupiter–1 as Neocomian (Berriasian to Barremian) confirms this prediction. The wedge itself has been drilled at sites between Jupiter–1 and the Cape Range Fracture Zone but no details have yet been published, beyond a correlation diagram in Powell (1982, fig. 11), which dates the deltaic sequence as Ryazanian (= Berriasian) and Valanginian, precisely satisfying our prediction.

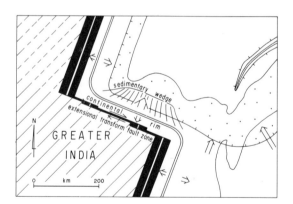

Fig. 123. Palaeogeographical reconstruction, soon after breakup at 128 Ma by seafloor spreading in the Cuvier and Gascoyne Abyssal Plains, showing sedimentary wedge prograding from the rim (double line) about the extensional fault zone of the Cape Range Fracture Zone. Coarse stipple indicates fluvio-deltaic complex, beyond which is shelf sediment, after Barber (1982). The Rankin Platform constitutes a second rim. From Veevers and Powell (1979).

In the Perth Basin, the rim basin is 1.4 km thick off Perth (Fig. 119, HH'), and succeeds, at a spectacular angular unconformity, a faulted and tilted rift-valley basin.

In the Bonaparte Gulf Basin, the Late Jurassic to Neocomian sequence above the breakup unconformity is 1.2 km thick in the Malita Graben, and thins 'northward across the Sahul Platform, suggesting the presence of a high in that direction. To the west the sequence thins over the Londonderry High, thickens erratically in the Vulcan Sub-basin, and is largely missing from the Ashmore Block due to restricted deposition and/or subsequent erosion in that region' (Laws and Kraus 1974, p. 82). I interpret the Ashmore Block and the outer part of the Sahul Platform as a rim behind which marine shale in the Malita Graben and non-marine sand elsewhere accumulated in a rim basin (Fig. 114). The postulated rim lasted for 50 Ma before it was covered by the sea in the Aptian (110 Ma).

The rims in the Browse and Dampier Basins were covered by the sea in the late Neocomian (112-115 Ma), some 45 Ma after breakup, and that in the Perth Basin in the Cenomanian, about 40 Ma after the later breakup of the western margin. Thereafter, the entire margin was fully open to the sea.

(5) Open-margin stage

By about 90 Ma ago, not only had the last rim along the margin submerged, but Greater India had cleared the epiliths of the Wallaby Plateau and Zenith Seamount that had developed along the north-east side of the Zenith–Wallaby transform fault during spreading. The synchronous replacement of detrital sediment by carbonate along the entire margin, from the Naturaliste Plateau in the south to the Bonaparte Gulf Basin in the north (Fig. 110) and at many DSDP sites (Robinson *et al.* 1974) points to the possibility that the widening strait between Greater India and the epiliths off Australia facilitated oceanic circulation that led to increased productivity and the onset of carbonate deposition on the margin.

In the aftermath of the hectic events that preceded, accompanied, and immediately followed continental breakup, the margin was indeed passive. Slow subsidence about the long-standing hinge was matched by an equivalent deposition of carbonate sediment, with little contribution of sediment from the subdued continental platform, except during the Palaeocene and Eocene when detrital sediment extended half-way across the Browse Basin as an expression of coeval uplift of the Kimberley Block. The main accumulation of detrital sediment on the margin today is in the Bonaparte Depression of the Sahul Shelf (Fig. 106), in a depocentre that can be traced back to the Permian and probably earlier, as related previously. Elsewhere the shelf slopes uniformly seaward, and, away from a narrow coastal zone of detrital sediment, is blanketed by carbonate. With the exception of Timor and the Timor Trough, the modern margin faithfully reflects the developmental stage of the past 95 Ma.

Within this interval, six cycles of carbonate deposition, two in the Late Cretaceous and four in the Cainozoic, are discernible. Apthorpe (1979) detailed the Late Cretaceous cycles, and Quilty (1980b) additionally the Cainozoic ones. A schematic curve of the transgressive/regressive cycles, extended back to the Jurassic, is shown in Fig. 124. The transgressions of the Early Cretaceous (γ to ζ of Morgan 1980, including γ and δ described by Wiseman 1979 from the Carnarvon Basin) are inferred to be eustatic, and involve detrital sediment and radiolarite only. After the regression of the Early Cretaceous shoreline, carbonate sediment predominates, with transgressive peaks in the Santonian, late Maastrichtian, late Palaeocene, middle and late Eocene, early and middle Miocene, and Pliocene. Quilty (1980b) regards these migrations of the shoreline as due to fluctuating eustatic sea-levels, and the coincidence of the Cainozoic transgressions round Australia, summarized in Fig. 151, and elsewhere (Quilty 1977), confirms this conclusion. Apthorpe (1979) does not address the question of eustatic sea-level changes, and relates differences in the Late Cretaceous transgressive–regressive history of the Dampier and Browse Basins to differential tilting.

Progradation of sediment, expressed in the Cainozoic isopachs (Figs. 114, 150), has resulted in a 2- to 3-km-thick carbonate lens along the outer part of the Northwest Shelf. North of Lat 18° S, the outer shelf is dotted with reefs that have grown since the middle Miocene at Scott Reef (Wright 1977), Ashmore Reef, and on the outer Sahul Platform (Laws and Kraus 1974). The complex of modern inshore carbonate sediments at lat. 26° S in Shark Bay has been monographed by Logan *et al.* (1970, 1974).

The only events that puncuated this quiet history are uplift and erosion in the Bonaparte Gulf Basin during the Oligocene (Laws and Kraus 1974) and late Miocene–early Pliocene (Veevers 1971b), rapid southward migration of the Timor Trough and Timor starting in the Pliocene (Veevers *et al.* 1978), and folding in the Carnarvon Basin during the late Miocene. On the platform, the Fitzroy Lamproites were erupted during the early Miocene (Wellman 1973). In Chapter III, we described the structural style in both the Carnarvon and Gippsland Basins as indicating Neogene dextral motion between the

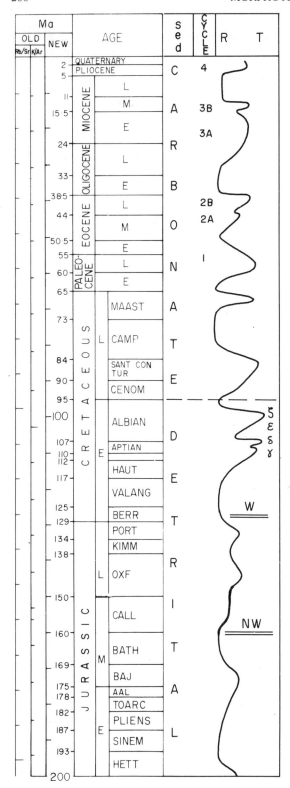

basement blocks, and a similar zone of deformation is seen in the Cooper Basin. The Pliocene continent–arc collision on the northern margin of Australia, described in Chapter II, may be reflected in the uplift of the outer part of the Bonaparte Gulf Basin, within 100 km of the collision zone.

(vi) Comparison of the rifted-arch and other stages with the conceptual morphotectonic model

The morphotectonic model, based on East Africa–Arabia, relates to continental margins that grew out of the breakup of a single rifted arch (Fig. 92, 93), of a multiple rifted-arch system (Fig. 94), and of a rift-divergence zone (Fig. 95). The north-western margin, north of the Exmouth Plateau, exemplified by the Browse Basin (Fig. 116), is compared with the single rifted-arch model (Figs. 92, 93). In the Browse Basin, the succession of basins, from the embryonic infrarift and extra-arch through to rim and open margin, their size, and their duration match the model of a single rifted-arch system (cf. Figs. 116, 92, 93). Thus, the width of the Browse Basin from the COB to the coastal hinge, the 45 Ma duration of the rim basin, and the restricted marine environment of the rim basin, all match the model.

The Exmouth Plateau–Dampier (Barrow) Basin region (Figs. 117, 118) is compared with the model of a multiple-arch system (Fig. 94). In cross-section, the Exmouth Plateau–Dampier region is only half the width of the model, and – the reverse of the model – has an arch (at Jupiter) delimited on the east by the depressed second rim of the Rankin Platform (at West Tryal Rocks) and by the collapsed first rim near the COB. Most of the depression of the second rim is attributable to isostatic loading by the lens of open-margin carbonate, and collapse of the first rim to subsidence along numerous faults instead of flexure, both secondary factors not taken into account in the model. Except the narrow rift-divergence zone of the Dampier Basin in place of the single rifted-arch basin of the model, the basins of the region, including the inter-arch basin of the Exmouth Plateau, match those of the model, except extra-arch and rim basins on the landward side of the Dampier Basin are lacking. A complication, not in the model, is the upper part of the Exmouth Plateau rim basin derived from an extensional continent–continent transform fault zone. The duration of the rims is 48 Ma compared with 40 Ma in

Fig. 124. Schematic diagram of transgressive (T)–regressive (R) cycles since the Triassic. Breakup indicated by double lines. Numbered Cainozoic cycles from Quilty (1980b).

the model. Again, the restricted marine environment of the rim basins matches the model.

The Perth Basin (Figs. 119, HH'; 120; 122) is interpreted as a complex of individual rift valleys that approach each other near Perth. Basin width and form of the individual rift valleys match those of the model – the Bunbury Trough between the Naturaliste Plateau and the Darling Fault is 50 km wide, equal to the average width of the East African rift valleys (Veevers and Cotterill 1976). A notable lack in the Perth region, shown up in the anomalously narrow 150 km width from the hinge to the COB, is any sign of an extra-arch basin. According to the model, the onset of a rifted arch in the Triassic and its persistence through the rim stage to 40 Ma after breakup entails the development of an extra-arch and then a rim basin behind the half arch of the western side of the Yilgarn Block from the Triassic to the Late Cretaceous; such a structure is patently lacking in the Yilgarn Block itself. If such sediment had bypassed a high-standing Yilgarn Block, it would be expected to lodge in the Ceduna Depocentre; the oldest known appropriate deposit here is Middle Jurassic, but reworked Triassic spores in the Cretaceous sediment in the Potoroo and Platypus wells (Shell Development (Australia) Pty Ltd 1976) may have been derived from an extensive extra-arch basin of Triassic sediments since removed by erosion.

(d) Eastern margin

(i) Introduction

The eastern margin of Australia is bounded on the west by the shoreline at the foot of the Eastern Highlands, and on the east by the oceanic basins of the Tasman Sea, Cato Trough, and Coral Sea (Fig. 125A).

As outlined in Chapter II, the ocean basins adjacent to the eastern margin were generated during three stages of spreading (Fig. 125B):

(1) in the Tasman Sea off south-east Australia, very slow spreading from 95 to 82 Ma; followed by
(2) rapid spreading from 82 to 57.5 Ma, involving asymmetrical accretion and ridge-jumping; and
(3) an extension of the second stage northward to the Coral Sea, from 63.5 to 57.5 Ma. Spreading terminated here and southward at 57.5 Ma.

There is a dearth of critical stratigraphical information compared with the western and southern margins. A gap of 1750 km separates the drillholes of the Gippsland Basin from those of the Capricorn Basin (Figs. 125A, 126), and another 1750 km thence to Anchor Cay of the Papuan Basin, with only DSDP 209, which penetrated no farther than the Eocene, in between.

Two parts of the margin are distinguished: the broad north-eastern margin, from the Gulf of Papua to lat. 24°S; and the narrow south-eastern margin, from lat. 24°S to the South Tasman Rise.

(ii) South-eastern margin

The south-eastern margin is very narrow between lat. 24° S and Bass Strait, only 100 km separating the shore from the 4 km isobath at the foot of the slope. The margin broadens at the re-entrant of Bass Strait; and south-east of Tasmania, it broadens further to encompass the marginal plateaux of the Cascade Plateau and South Tasman Rise. The margin in the north parallels the South-east Highlands, and marks the eastern limit of the zone of seismicity in south-east Australia. Except at the foot of the slope, marked by a band of sediment 1.5 km thick (Fig. 149), sediment accumulation is slight – at the shelf edge, the thickness of sediment above the Permo-Triassic Sydney Basin is estimated from seismic profiles to be no greater than 250 m (Davies 1975), increasing to a maximum of 600 m beneath the upper slope (Kamerling 1966). Representing this part of the south-eastern margin is the section off Sydney (CE of Fig. 127), with a steep slope, a thin, presumably Late Cretaceous–Cainozoic, sediment wedge beneath a narrow shelf, and elevated land adjacent to the coast, contrasted in the same figure with a regional section of the Gippsland Basin (CD).

The Gippsland Basin in Bass Strait is an oasis in this desert of information. Burke and Dewey (1973) identified the Gippsland Basin as a failed arm of the Tasman Sea Basin. The cross-section AA' in Fig. 128 shows an Early and Late Cretaceous graben, 60 km wide, succeeded by a Cainozoic basin that has expanded across the graben, probably by regional isostatic adjustment and flexure in the manner outlined by Beaumont and Sweeney (1978). The longitudinal section (BB'B") in Fig. 128 contains data of varied reliability: the boxed part of BB' is controlled by published drilling and processed seismic profiles, and what lies beneath is a simple downward projection; section B'B", derived from a single-channel seismic profile, is highly interpretative, and the only secure parts of this interpretation are the shape of the seafloor, the shape of the oceanic basement, and the dominantly westward dip of the deepest reflection. Incidentally, westward dips prevail also in the offshore Sydney Basin (Kamerling 1966). In terms of the concepts applied to the southern and western margins, the outermost Gippsland/Strzelecki Basin can be viewed

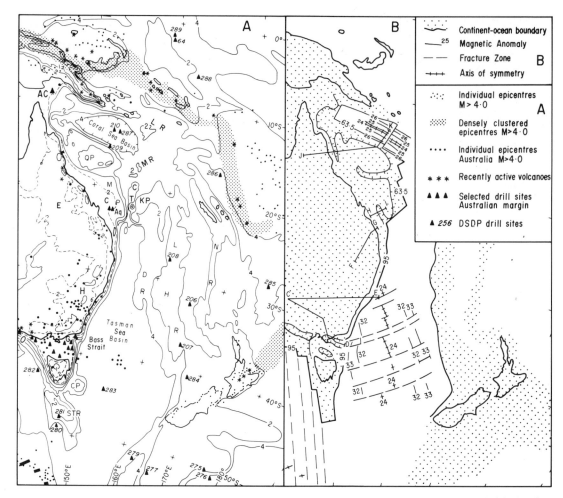

Fig. 125. Eastern margin of Australia in its setting between the Eastern Highlands and continental borderland of the South-west Pacific Ocean. A. Morphology (contours in km), seismicity (Denham *et al.* 1975; Baraganzi and Dorman 1969), modern volcanism (Douglas and Ferguson 1976; Stephenson *et al.* 1980; Holmes 1965), and location of drilling sites. C, Capricorn Basin; CP, Cascade Plateau; CT, Cato Trough; DR, Dampier Ridge; EH, Eastern Highlands; G, Gippsland Basin; KP, Kenn Plateau; LHR, Lord Howe Rise; LR, Louisiade Rise; MP, Marion Plateau; MR, Mellish Rise; NR, Norfolk Ridge; O, Otway Basin; QP, Queensland Plateau. B. Distribution of continental (stippled) and oceanic lithosphere, age of continent–ocean boundary (Ma), seafloor-spreading magnetic anomalies (Weissel and Hayes 1977; Weissel *et al.* 1977; Weissel and Watts 1979), and location of regional cross-sections.

as part of a subsided arch and successor marginal rim. To these are related an Early Cretaceous extra-arch basin containing the graben fill and a succeeding Late Cretaceous rim basin, overlain by a Cainozoic open-marine basin (Fig. 128C). In the context of the eastern margin as a whole, the pre-breakup Strzelecki Basin was an east–west rift valley filled during the Early Cretaceous by volcanogenic sediment derived from a volcanic source on a north–south trend coincident with the present eastern margin. Seafloor spreading from this boundary, starting 95 Ma ago, led to the development of a breakup unconformity between the faulted volcanogenic Strzelecki Group below, and the less intensely faulted quartzose Latrobe Group above, and converted the Strzelecki rift valley into the Gippsland failed arm.

The Latrobe Group offshore extends from the Late Cretaceous through the Eocene, and is wholly non-

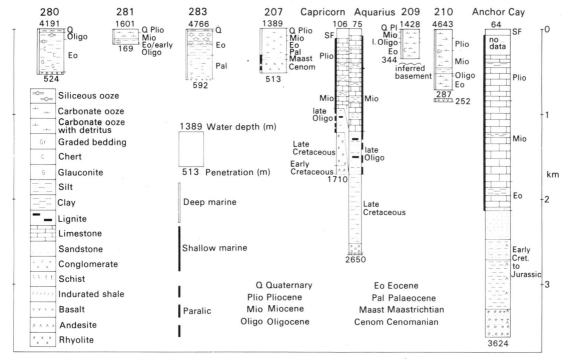

Fig. 126. Logs of drilling sites along the eastern margin (DSDP 281, Capricorn (Cap) and Aquarius (Aq)), DSDP 209, and Anchor Cay (AC)), in the adjacent ocean basins (DSDP 280, 283, 210, 287), and on Lord Howe Rise (DSDP 207), located on Fig. 125. Depositional environments, except non-marine, which is left blank, indicated on side of log. Inferred depth of seismic basement shown for DSDP 209, and basalt at base of DSDP 287 shown beneath adjacent DSDP 210. DSDP 207, 209, and 210 from Burns et al. (1973), 280, 281, and 283 from Kennett et al. (1975), and 287 from Andrews et al. (1975); Capricorn and Aquarius from Ericson (1976), and Anchor Cay from Oppel (1970).

marine until the start of a marine transgression, punctuated by several minor regressions and lacunae, that culminated in the early Miocene (Jones and Veevers 1982). Partridge (1976) shows that fine-grained terrigenous sediment bypassed the shelf during the Palaeocene and Eocene to be deposited in the Tasman Sea, as seen at DSDP 283 (Fig. 128C).

The *en échelon* anticlines in the Gippsland Basin that now serve as hydrocarbon reservoirs were generated by east–west right-lateral shear during the late Eocene and Oligocene, and again in the late Miocene (Threlfall et al. 1976), as mentioned in Chapter III.

South of the Gippsland Basin, the southern part of

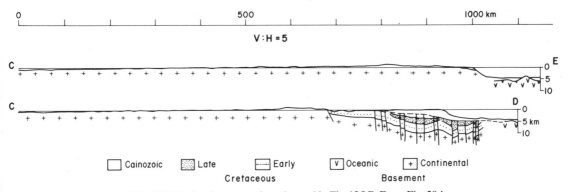

Fig. 127. Regional cross-sections, located in Fig. 125 B. From Fig. 59A.

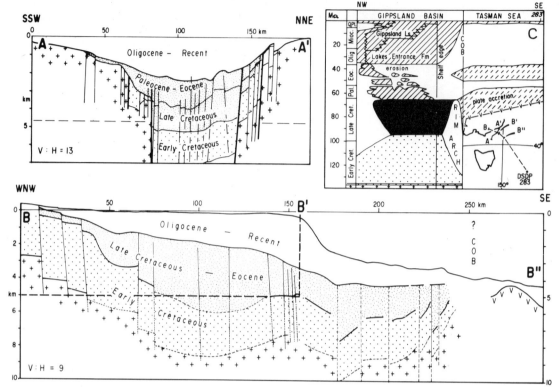

Fig. 128. Gippsland Basin. AA′: transverse section, with main faults emphasized, and BB′B″: longitudinal section, both located in inset of C. Top part of sections AA′ and BB′, above broken line, modified from James and Evans (1971); lower parts projected downward here. B′B″ is a depth section converted from a single-channel seismic profile made in 1973 by Shell International Petroleum from the MV Petrel, and available under the terms of the Submerged Lands Act. C: time–space diagram of profile BB″ extended to DSDP 283. Offshore Gippsland Basin out to shelf edge and at DSDP 283 modified from Partridge (1976). Extra-arch basin stippled, rim basin solid, and open-margin basin ruled.

the margin includes two marginal plateaux. The South Tasman Rise is underlain at DSDP 281 by Palaeozoic schist covered by a film of late Eocene and early Oligocene shallow marine sediment, and Miocene and younger deeper marine sediment (Fig. 126). Willcox (1981) interprets a seismic profile situated near the saddle that separates the rise from Tasmania as showing a central core, probably Palaeozoic, flanked by two small rift basins, containing a few kilometres of faulted fluvial-deltaic sediment (?Late Cretaceous) overlain by a sheet of undeformed marine sediment (?Cainozoic). The Cascade Plateau is poorly known. According to Willcox (1981), basement appears to be rugged, and is probably block faulted. The overlying sediments range up to 1.5 km thick. Intense magnetic anomalies on its eastern side are interpreted as due to volcanic rock, which we interpret as part of an oceanic epilith.

(iii) North-eastern margin

North of lat. 24° S, the margin widens into the broad shelf of the Great Barrier Reef and the adjacent Marion and Queensland Plateaux, bounded by the oceanic basins of the Coral Sea and Cato Trough, and the oceanic plateaux of the Kenn Plateau, Mellish Rise, and Louisiade Rise, of uncertain lithospheric affinity. A map of the region is shown in Fig. 129, cross-sections incorporating the results of the only deep drilling in the region in Fig. 130, and a time–space diagram in Fig. 131.

In examining the geology of the landward part of the region in our analysis of the Eastern Highlands earlier in this chapter, we found just offshore from the present coast the axial relics of a magmatic arc which died in the mid-Cretaceous. The north-eastern margin developed out of the continental borderland that lay north-east of this arc.

Apart from the growth of the Coral Sea Basin between Australia and the Papuan Peninsula, dated as Palaeocene by drilling at DSDP 287 (Fig. 126) and by the interpretation of magnetic anomalies (Weissel and Watts 1979), and the drilling at DSDP 209 on the north-east edge of the Queensland Plateau, which penetrated no further than middle Eocene, and at Capricorn and Aquarius in the Capricorn Basin, the only available information is from the interpretation of geophysical profiles (Mutter and Karner 1980; Falvey and Mutter 1981). Falvey and Mutter's (1981) interpretation of the structure and history of the region is shown in Figs. 129 and 131, with a cross-section (Fig. 130, JK) modified from Taylor and Falvey (1977), and a second section from our account of the Eastern Highlands (Fig. 130, FI). The Queensland Trough and the Capricorn Basin are seen as a Late Cretaceous rift-valley complex that separates the Queensland Plateau and the Marion Plateau from the mainland. In turn these plateaux are separated from each other by the Townsville Trough, and from the Coral Sea to the north-east and the Cato Trough to the east by marginal rift valleys. The onset of rift-valley faulting follows the demise in the mid-Cretaceous of the magmatic arc, shown at H in section FI of Fig. 130. The rift valleys developed until continental breakup and the inception of the margin in the Palaeocene, followed by progressive marginal subsidence and the accumulation of marine sediment. As part of an event felt throughout the South-west Pacific, much of the region was subjected to submarine erosion which produced an intra-Oligocene vacuity (Kennett *et al.* 1975), followed in the late Oligocene by the start of growth of the Great Barrier Reef, described by Fairbridge (1950), Maxwell (1968), and Hopley (1982).

(iv) Comparison with the western margin

A glance at Fig. 4 shows that the eastern margin is a crude mirror-image of the western margin. The north-west trend of the north-east margin and its considerable width are mirrored by the north-east trend of the north-west margin and its expansion in the Scott and Exmouth Plateaux, as likewise the north-west trend of the seafloor-spreading magnetic anomalies in the Coral Sea Basin is reflected by the north-east trend of the anomalies in the Argo Abyssal Plain. Further, the narrow south-east margin, at least from lat. 24° S to Bass Strait, and the north-north-west magnetic lineation of the adjacent Tasman Sea are mirrored by the narrow margin south-east of the Wallaby Plateau and by north-north-east-trending anomalies. The similarity ends here; in more important aspects, the margins differ.

(1) The eastern margin is bounded by narrow marginal oceanic basins that developed during the expansion of the Pacific margin, whereas the western margin is bounded by the broad Indian Ocean limited only to the north by convergence along the Sunda Trench.
(2) Breakup was at 66 Ma on the north-east and 95 Ma on the south-east versus 160 Ma on the north-west and 128 Ma on the south-west.
(3) Breakup in the north-east was preceded by a short (30 Ma) episode during which rift valleys developed immediately out of a terrain bordered by a magmatic arc; and in the south-east the line of breakup developed across a triple-rift junction of a failed arm filled with alkaline intermediate volcanogenic sediment; all this contrasts with the western margin development as infrarift basins and then a rift-valley complex 120 to 150 Ma before breakup.
(4) Half-arches and rims, as found on the western margin, are lacking in the east, except in the Gippsland Basin.
(5) Voluminous sediment on the western margin, in particular the 10-km-thick rift-valley fill of the Perth Basin, contrasts with the thin fill of the postulated rift valleys of the Queensland Plateau. Again the Gippsland Basin is an exception in containing thick sediment, but in trending normal to the margin it differs from its western counterparts, which parallel the margin.

In our opinion, these differences stem from the setting of the eastern margin alongside the Pacific Ocean. The eastern margin has come about by processes of marginal sea development along the South-west Pacific borderland by a change from convergence to divergence: the western margin, in contrast, has developed as the embryonic ocean stage that led to the growth of the broad Indian Ocean.

In terms of the concepts of Sengor and Burke (1978) and Baker and Morgan (1981), the eastern margin is the product of 'passive rifting, in which a preliminary cracking of the lithosphere occurs because of differential stress resulting from the interaction of lithospheric plates', with a sequence of rifting–(?uplift)–volcanism (?strongly alkaline); in contrast, the western margin is the product of 'active rifting, in which the lithosphere is cracked by asthenospheric upwelling', with a sequence of doming–volcanism (weakly alkaline to tholeiitic)–rifting.

(v) Development of the eastern margin

The development since the mid-Cretaceous of the marginal seas along the South-west Pacific borderland

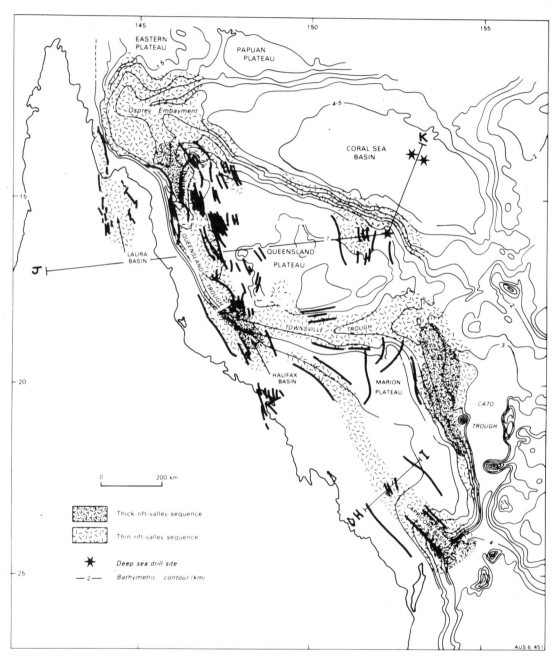

Fig. 129. North-eastern margin, showing bathymetry, and structural and depositional patterns (from Falvey and Mutter 1981), and location of sections HI and JK of Fig. 130.

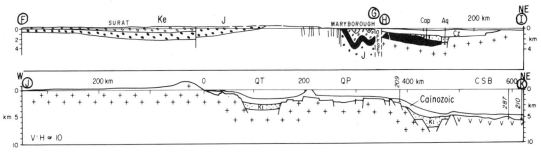

Fig. 130. Cross-sections FG (from Jones and Veevers 1983—see Fig. 68), HI, and JK (from Veevers 1982a, modified from Taylor and Falvey 1977).

and of the cognate eastern margin is sketched in Fig. 132. The south-east margin and the Tasman Sea resulted from a change of regime 95 Ma ago: a continental sliver was torn off Australia along a plate boundary, marked by volcanic centres related to the prior regime of convergence, by divergent sinistral shear (Fig. 132A). The major volcanic centre at the triple junction of the Tasman Sea and the Gippsland–Strzelecki Basin was the source of the voluminous alkaline–intermediate sediment of the Early Cretaceous Strzelecki and Otway Groups to the west; accompanying breakup, under the new regime of divergence, was the emplacement of shoshonitic magma along the south-east margin and of rhyolite on the adjacent Lord Howe Rise.

During the Cenomanian, the magmatic arc along the Queensland coast to New Guinea reached a climax of effusion of andesitic debris, some of it now preserved in the final deposits of the Great Artesian Basin. The death of the magmatic arc at 90 Ma was followed by the inception of rift valleys over the north-east margin and the rapid rise of the Eastern Highlands with a concomitant shedding of sediment southwestward (Fig. 132B). Except in the Otway Basin and in Papua New Guinea, eastern Australia was well above sea level, including the region of rift valleys of the north-east margin, where epiclastic volcanogenic conglomerate and redbeds, as seen in the Capricorn and Aquarius wells, were deposited in the intermontane troughs. Small discontinuous basins along the western margin of the Lord Howe Rise developed parallel to the trend of the short spreading compartments of the Tasman Sea (Willcox et al. 1980). In the Gippsland Basin, dominantly quartzose sediment succeeded the Strzelecki Basin of dominantly volcanogenic sediment.

By the end of the Late Cretaceous (66 Ma = A28) (Fig. 132C), a divergent plate boundary appeared between north-east Australia and the Papuan Peninsula, marking the end of rift-valley development on the adjacent north-east margin. The Eastern Highlands were lowered and narrowed to their present size. The Lord Howe Rise, isolated from New Caledonia and the Norfolk Rise by the opening of the New Caledonia Basin, was covered by a shallow sea in the south and by a deeper sea in the north. With the subsidence of its marginal rim, the Gippsland Basin started to be covered by a transgressing sea.

Towards the end of the Palaeocene (57.5 Ma = A24) (Fig. 132D), speading was terminated in the Tasman Sea, Cato Trough, and Coral Sea, and activity along the Pacific Borderland migrated eastward, as described by Coleman (1980) and Malahoff et al. (1982). The Lord Howe Rise subsided to bathyal depths, and the eastern margin continued to subside, to bathyal depths at the edge of the north-east margin no later than the mid-Eocene, to shallow shelf depths in the Gulf of Papua

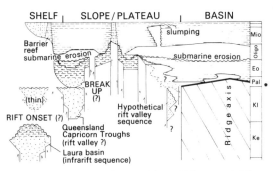

Fig. 131. Time–space diagram, from Falvey and Mutter (1981). Marine environments indicated by broken lines, carbonate by brick pattern, and marginal to non-marine by dots. Volcanics indicated by Vs, and the onset of seafloor spreading by the star.

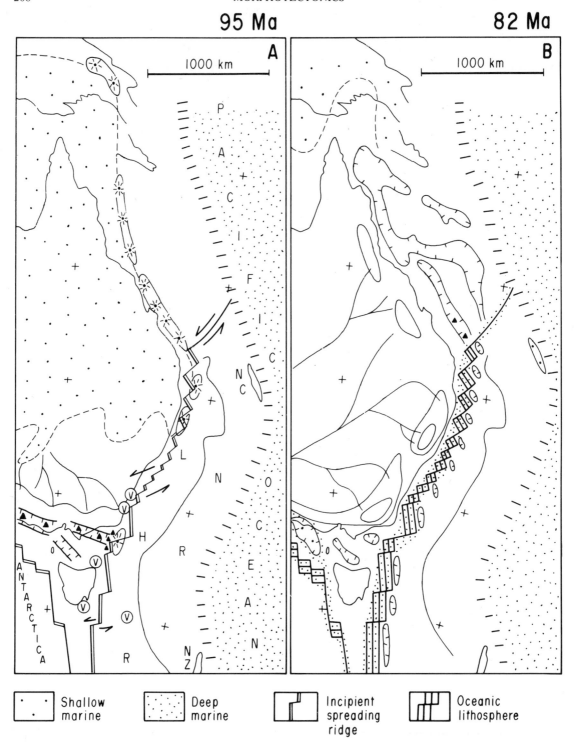

Fig. 132. Palaeogeographical sketches of the eastern margin, from its inception 95 Ma ago to the end of seafloor spreading in the adjacent ocean basins 57.5 Ma ago. A. Start of Late Cretaceous, 95 Ma ago. Inception of the south-eastern margin by breakup and subsequent eastward to northeastward rotation of New Zealand–Lord Howe Rise (LHR)–Norfolk Ridge (NR)–New Caledonia (NC). B. Campanian, 82 Ma ago (=A34). Narrow zones of newly generated oceanic lithosphere south and south-east of Australia flooded by the sea; north-eastern margin crossed by rift valleys.

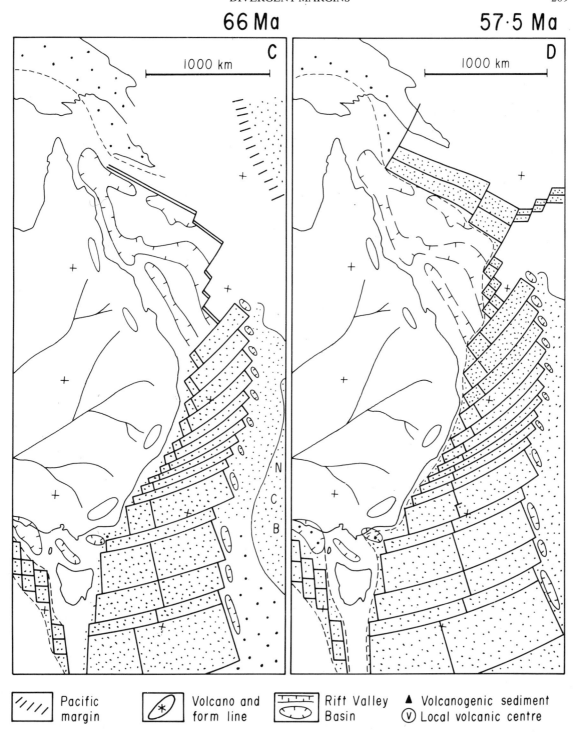

Fig. 132. C. End of the Late Cretaceous (66 Ma ago=A28). Final development of the rift valleys of the north-eastern margin before the opening of the Coral Sea Basin along a suture between the Papuan Peninsula and north-eastern Australia. D. Towards the end of the Palaeocene (57.5 Ma=A24). The end of spreading in the Tasman Sea, Cato Trough, and Coral Sea.

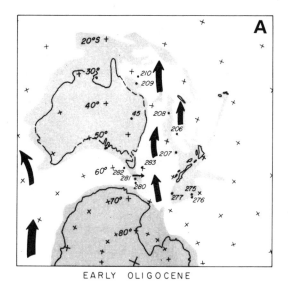 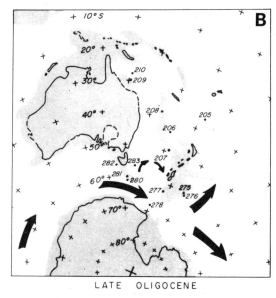

Fig. 133. A. Early Oligocene reconstruction, with suggested direction of bottom-water circulation (arrows) northward from Antarctic sources through the Tasman–Coral Sea area. No Circum-Antarctic Current had developed by this time, but currents flowed across the shallow South Tasman Rise. B. Late Oligocene reconstruction, showing the direction of the Circum-Antarctic Current south of Australia and New Zealand and a western-boundary current east of New Zealand. This direction has been retained until the present. From Kennett et al. (1975, figs. 9 and 10, p. 1166).

from the mid-Eocene onwards, and to shallow shelf depths during the marine transgression of the Gippsland Basin, which culminated in the early Miocene. Only in the late Oligocene did the inner part of the north-east margin subside below sea level to initiate growth of the Great Barrier Reef. A widespread lacuna centred on the early Oligocene is interpreted by Kennett et al. (1975) as due to marine erosion by bottom currents derived from the Ross Sea sector of Antarctica flowing northward as a broad western boundary current (Fig. 133). With the opening of an oceanic gulf between the South Tasman Rise and Antarctica in the late Oligocene, this system was replaced by the Circum-Pacific Current, and deposition resumed on the eastern margin.

5. QUANTITATIVE ESTIMATES OF THE VERTICAL MOTIONS OF THE AUSTRALIAN PLATFORM SINCE THE JURASSIC

(a) Introduction

The flooding of the Australian continental platform 110 to 95 Ma by an epeiric sea immediately followed by the isolation of Australia by the dispersal of Antarctica and New Zealand/Lord Howe Rise provides an opportunity of making quantitative estimates of the vertical motions of the platform since 95 Ma, during the Potoroo regime, and a little before. This essay is addressed to one aspect only of the tectonic history – the vertical motion of the platform – and it provides a calibration of the motions described in other parts of the book.

Lying beyond the platform, the divergent margins on the west, south, and east lie outside the scope of this analysis. Since their inception at breakup, the margins have subsided along a hinge, shown by the landward wedge-out of sediment to lie near the present shoreline, which is therefore seen to be a feature of some antiquity. This means that the landward effect of the processes connected with the development of the uplifted rifted arches and the subsequent half-arches is confined to the submergent margin. The vertical motion of the divergent margin, coupled at the continent–ocean boundary with the oceanic lithosphere, is greatest at the boundary and diminishes landward to vanishing point at the hinge. Since the hinge marks the periphery of the continental platform, the motion of the margin is expected to be independent of that of the platform, and hence lies outside the scope of this analysis.

(b) Summary

Bond's (1978, 1979) method of estimating the relative motions of the continents and sea level by the analysis of continental hypsometries is applied to Australia. Past sea levels relative to Australia and to four other continents, each assumed to have undergone simple uniform vertical motion as a block, are estimated to have been +20 m in the middle Miocene, +75 m in the Eocene, +110 m in the later Cretaceous, +75 m again in the Aptian–Albian, and −30 m in the Late Jurassic. From these datum levels, it is calculated that a surface at present sea level had a general elevation of +55 m in the Late Jurassic, −100 m in the Aptian–Albian, +180 m in the Late Cretaceous, +100 m in the Eocene, and 0 m in the Miocene. In its rapid subsidence in the Early Cretaceous, rapid uplift in the Late Cretaceous, and slow subsidence in the Cainozoic, Australia's general motion differs from that of South America, which has remained steady, and from that of Europe, North America, and Africa, which at various times since the Aptian have risen to their present elevations. Australia's contrary motion is attributed to the terminal climax in the mid-Cretaceous of the volcanic arc and the subsequent uplift of the eastern margin.

Local variations from the general vertical motions are found by a comparison of present elevation with the estimated past sea levels. Since the last continent-wide marine transgression in the Albian, the east-central part of Australia round Innamincka has subsided by about 1000 m, and the Eastern Highlands have risen by at least 400 m. The motion comprised rapid subsidence (50 m/Ma due to tectonic forces) during the Cenomanian interregnum, a reversal to slow uplift during the rest of the Late Cretaceous, and slow subsidence during the Cainozoic.

(c) Bond's method

Bond (1978, 1979) developed a method based on the hypsometric analysis of continents for distinguishing between (a) continents flooded during a eustatic rise of sea level followed by a substantial change in continental elevation, and (b) continents likewise flooded but followed by little or no change in continental elevation. The percentages of flooding of five continents for six time intervals since the Jurassic revealed substantial changes in continental hypsometries, including that of Australia, and indicated relative motions of the continents and sea level.

In this paper, the method is applied specifically to Australia. Bond's (1978, 1979) work is extended (a) by widening the frame of Australia to embrace the natural entity of the Australian continental block above the −200 m elevation contour, thus including cratonic New Guinea; (b) by using more detailed information on the extent of past seas beneath the broad shelves of north-western (Allen et al. 1978; Apthorpe 1979) and southern Australia (Fraser and Tilbury 1979; Bein and Taylor 1981); and (c) by inferring local changes of continental elevation from changes in the elevation of past shorelines.

The area of a continent flooded by the sea, expressed as a percentage of the continent's total surface area, is a value that can be plotted on the continent's hypsometric curve. The elevation of the percentage point on the curve (Fig. 134) can be regarded as equal to the rise of sea level required to flood the modern continent by the same percentage that was flooded during the past. The equivalence holds only if the hypsometry of the continent did not change significantly since the past flooding. If the percentage areas of several continents, plotted on the corresponding hypsometric curves, lie at the same elevation (Fig. 134, solid dots), this is interpreted to mean (a) that the flooding was caused by a sea-level rise equal to the elevation of the points on the hypsometric curves, and (b) the continental hypsometries have not changed significantly since the flooding. If the points do not lie at the same elevation (Fig. 134, open circles), this is interpreted to mean fairly large differential changes in continental hypsometries that post-date flooding. In view of the uncertainties in the palaeogeographical information of the real continents,

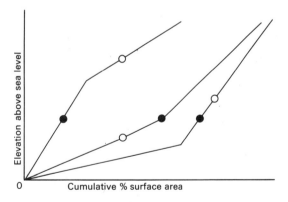

Fig. 134. Hypothetical distribution of points representing percentages of continental areas flooded by the sea. Solid dots: flooding due to sea-level rise without subsequent change in continental hypsometries. Open circles: flooding followed by substantial change in continental hypsometries. From Bond (1978, fig. 1).

and of the possibility of some continual change in hypsometry, only relatively large differences in elevation of the points should be taken as significant.

Applying this information from five continents, Bond (1979) (Fig. 135) found that the inferred past elevations of the continents tended to cluster, indicating a uniform elevation of sea level for each time interval considered. Geoid anomalies lie apparently within the measuring error. For example, in the Miocene all but Africa cluster about a sea-level elevation of $+20$ m. The high elevation of the African (Af) bar is taken to indicate that Africa, relative to the other continents, has been uplifted since the Miocene. The relative post-Miocene uplift of Africa is removed by shifting the African bar down to the inferred Miocene sea-level elevation (Fig. 135, position 1 shown by the open bar). This correction, made for Africa in the Eocene interval, is insufficient to bring Africa to the inferred Eocene sea-level elevation, and a further correction (2) is required. Other corrections in Fig. 135 are likewise shown by arrows.

(d) Past sea levels

For the interval 130 Ma to the present (Fig. 135), the points of sea-level elevation measured from the continents describe a curve which starts and finishes at 0 m, and moves to a broad maximum in the Late Cretaceous. Curve A, from Watts and Steckler (1979), is a sea-level curve derived by removing the tectonic subsidence and sediment loading effects from the total sedimentary thicknesses in offshore wells along the North American Atlantic continental margin, and it confirms the curve described from continental hypsometry. Curve B is from Pitman (1978) augmented by Vail *et al.* (1977). Changes in the volume of the mid-oceanic ridge system, due to changes in the rate of spreading, affect eustatic sea-level, which Pitman (1978) has measured as a change of continental freeboard. Pitman (1978) calculated these changes back to 85 Ma, and Vail *et al.* (1977) extended them back to 200 Ma by calibrating their curve showing relative changes of sea level from seismic stratigraphy against Pitman's curve.

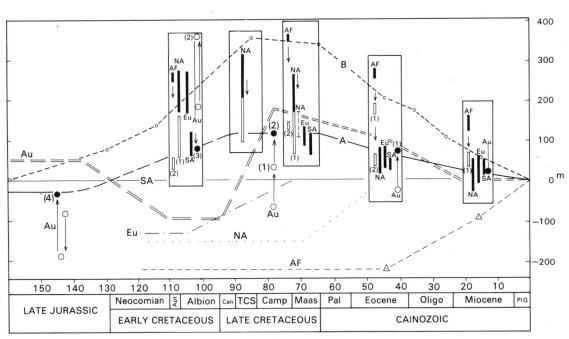

Fig. 135. Plot of elevation of sea level versus time, adapted from Bond (1979, fig. 5) and extended from 120 to 160 Ma. Elevation of flooding bars for Africa (Af), North America (NA), Europe (Eu), and South America (SA) from Bond; elevation of flooding mark for Australia (Au), (solid circle), from this paper, modified from Bond. Curve A is a sea-level curve from Watts and Steckler (1979), slightly modified from 80 to 140 Ma according to the position of Au. Individual curves for each continent integrate individual departures from curve A, and indicate the vertical motion of a point now at sea level on each continent, assuming simple uniform motion of continent. Curve B from Pitman (1978) and Vail *et al.* (1977).

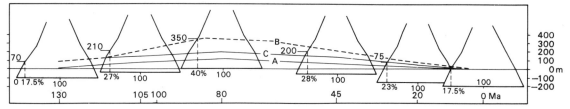

Fig. 136. A model in which the difference between curve A and curve B is due wholly to a uniform vertical motion of the continents. Continental platforms (above −200 m height contour) simulated by a two-sided hypsometric curve derived from that of South America, which from at least 120 Ma has apparently had the same hypsometry. The elevation of the continental platform was found for each of the time intervals 130, 105, 80, 45, 20, and 0 Ma by matching the apparent sea-level elevation from curve B (Pitman 1978; Vail et al. 1977) with the mean percentage flooding of the continents. Curve C denotes the apparent elevation of a point on the continent now at sea level. Curve A is the elevation of sea level derived directly from the continents.

The difference between curves A and B is due, I believe, to differences in the basic assumptions of each method: Pitman's (1978) method entails vertically fixed continents and a constant depth of the oceanfloor beyond the mid-oceanic ridge system; Bond (1979) and Watts and Steckler (1979) make their measurements on the continents themselves. The difference between the curves is thus due either to differential vertical motion between the continents (each assumed to be moving uniformly as a block) and the oceanfloor, as shown in Fig. 136, or to a change in the shape of the old (>80 Ma) and deep (>5.5 km) oceanfloor beyond the limits of the mid-oceanic ridge system, or, as Harrison et al. (1981) propose, that the hypsometry of a supercontinent is lowered by breakup. Whatever the case, curve A describes best the motion of eustatic sea-level relative to the continents, and is thus adopted here as the tentative datum against which Australia's vertical motions are gauged. Falvey and Deighton (1982) adopt the same curve.

The departures of individual continents from curve A (Fig. 135), interpreted as indicating vertical motions that post-date flooding, are integrated in the curves that lie, except for Au, wholly below A. Each curve describes the vertical motion of a point on the continent at present sea level, with the assumption that each continent has undergone simple uniform vertical motion as a block; that is, that the hypsometry does not change except to vary position vertically. The curve for SA is unchanging at present sea level, those for Eu, NA, and Af rise at various times, that is, they express uplift only, whereas that for Au falls in the Early Cretaceous, rises in the first half of the Late Cretaceous, and falls in the second half of the Late Cretaceous through the Palaeogene. We now turn to a detailed examination of Australia's vertical motions.

(e) Australian hypsometry

The hypsometry of Australia (Fig. 137) was determined by planimetry from the equal-area Geographic Map of the Circum-Pacific Region, South-west Quadrant (American Association of Petroleum Geologists 1978), and the area measured is defined by the −200 m contour, which encloses the entity of New Guinea, mainland Australia, and Tasmania. With uncertainty about the configuration of the northern half of New Guinea, which was affected by large sinistral motion in the Neogene (Powell et al. 1980b), the Australian block has existed with this outline since the dispersal from Australia of the Gondwanaland fragments of the Lord Howe Rise/New Zealand Plateau and Antarctica 95 Ma, and the rotation of the Papuan Peninsula from north-east Australia 62 Ma. The platform perimeter has thus existed during the Cainozoic, and, except off

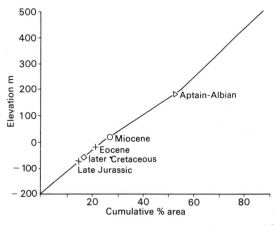

Fig. 137. Percentages of flooding plotted on the hypsometric curve of Australia.

north-east Australia, the Late Cretaceous. The changes in the hypsometry caused by the dismemberment of Gondwanaland by plate divergence are expected to be important near the original plate boundary, but diminish to vanishing-point at the edge of the platform, as shown by the shorelines in the Miocene, Eocene, and later Cretaceous, which generally parallel the present shoreline and lie close to it (Fig. 90).

In the Albian, the shoreline generally corresponds to the present +200 m height contour (Fig. 138), suggesting that since the Albian the continent has remained level; that is to say, it has not been sensibly tilted. This condition, necessary for constant hypsometry, is not sufficient, and this analysis is limited by the enabling assumption that Australia has moved up and down uniformly as a block, without significant change of hypsometry, both after breakup and dispersal of Antarctica and Lord Howe Rise/New Zealand 95 Ma and for an interval before.

The percentages of flooding for five time intervals were measured by planimetry from the palaeogeographical reconstructions of Figs. 139 to 143. Few past shorelines are known from a lateral facies change from marine to non-marine; most are indicated by the eroded feather-edge of marine strata, with an unknown loss of record by erosion after deposition, or by a vertical transition from marine to non-marine facies, so that the amount of flooding is a minimum estimate. The values were plotted on Fig. 137, and apparent sea levels, assuming constant subsequent hypsometry, were read off the ordinate. For example, the proportion that was covered by the sea in the early Miocene is estimated to be 26.5 per cent, so that the apparent sea level at this time was +20 m. Plotted on Fig. 135, the value of Au lies in the middle of a cluster of the values of SA, Eu, and NA, to which Bond (1979) adjusted Af to compensate for post-Miocene uplift, a compensation carried through to earlier time intervals. For the late Eocene, the area covered by the sea was 21 per cent, corresponding to a sea level of −25 m. This is adjusted 100 m upward to +75 m (1) to fit at the level of A within the cluster of SA, Eu, and NA, as Af is adjusted downward, implying a net subsidence of Au of 100 m between the Eocene and Miocene. In the later Cretaceous (Turonian through Maastrichtian), 15.9 per cent of the platform was covered by the sea, corresponding to a sea level of −70 m. The correction of 100 m for Eocene to Miocene subsidence is insufficient to bring Au to the cluster, and a further correction of 80 m, implying post-Cretaceous to pre-late Eocene subsidence, is required (2). In the Aptian–Albian, 52 per cent of the platform was covered, corresponding to a sea level of +180 m. A cumulative correction (1+2) of +180 m is first applied, and then a second correction (3) of −280 m, implying 280 m of post-Albian to pre-later Cretaceous uplift, is required to bring Au to the cluster. Finally, in the Late Jurassic, 15 per cent of the platform was covered, corresponding to a sea level of −80 m. A cumulative adjustment of −100 m reduces this to −180 m, and a final adjustment of +155 m, implying a net subsidence of 155 m between the Jurassic and the Aptian, is required to bring Au (4) to the level of A. These results confirm and refine those over the interval 0 to 120 Ma ago for Australia of Bond (1979), and extend them to 160 Ma ago, given that Australia has moved as a block during this time. Arranged to show the variation of general elevation of the Australian block through time, the results reveal that a surface at present sea level (double broken line in Fig. 135) had an elevation of +55 m in the Late Jurassic, −100 m in the Aptian–Albian, +180 m in the later Cretaceous, +100 m in the Eocene, and 0 m in the Miocene,

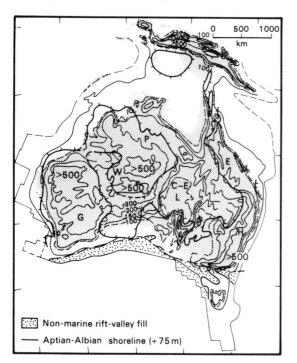

Fig. 138. Modern morphology, (on the same base as the Tectonic Map of Australia and New Guinea: Geological Society of Australia 1971) showing the morphotectonic divisions of the Great Western Plateau (GWP), Central-Eastern Lowlands (C-EL), and the Eastern Highlands (EH) (Gentilli and Browne 1963), and, in addition, the Aptian–Albian shoreline.

expressing in quantitative terms the well-known state of Australia as thalassocratic in the Aptian–Albian, and geocratic at other times in the interval 160 Ma ago to the present. These states differ from those of other continents (Fig. 135) and are interpreted as reflecting the terminal climax of the volcanic arc on the eastern margin in the mid-Cretaceous and the subsequent uplift in the later Cretaceous, as described already. Local anomalies of elevation lie outside these general values and are now examined for five time intervals.

(f) Local anomalies of continent level

(i) Since the middle Miocene (< 11 Ma) (Fig. 139)

On an Australia of constant hypsometry, the early-middle Miocene shoreline should correspond, according to curve A of Fig. 135, to the +20 m elevation contour, shown in Fig. 139 by a dotted line. The sea covered 26.5 per cent of Australia, an area barely bigger than the area of 24 per cent it covers today. The mismatch of the present +20 m contour and the Miocene shoreline (at a hypothetical elevation of +20 m) points to subsequent local changes of elevation. Individual areas, numbered on the map, are described below.

1. Most of New Guinea has risen above sea level since the middle Miocene (Harrison 1969). In the highlands orogen, middle Miocene shallow marine limestone is now at elevations exceeding 1000 m. In the depositional trough or foreland basin at the southern foot of the highlands, middle Miocene shallow marine limestone is now at a depth of 3000 m (Ridd 1976), indicating subsidence of this amount; elsewhere in southern New Guinea the shoreline has retreated in a depositional regression.

2. Around all but a few parts of the Australian coastline, the wedge-out of Miocene sediments, taken here as marking the shoreline, is estimated as lying a short distance offshore at an approximate elevation of −50 m, so that since the Miocene the Miocene shoreline apparently subsided by approximately 70 m plus the thickness of any overlying sediment. This amount is probably an overestimate because the actual Miocene shoreline probably lay closer to the present shore.

3. A north-west-trending strip across northern Australia has outcropping limestone that contains foraminifers, no older than Miocene, that indicate marine deposition (Lloyd 1968a, b). The limestone is now at elevations of about +250 m, implying an uplift of 250 − 20 = 230 m since deposition. Nearly parallel with the uplift of New Guinea, this strip possibly marked the position of the outermost downwarp connected with the New Guinea orogen.

4. In the westernmost part of Western Australia, the Cape Range contains middle Miocene shallow marine limestone now at an elevation of +300 m, indicating a local uplift of 280 m since deposition.

5. A similar arrangement at the north-west tip of Tasmania (Quilty 1972) indicates subsequent uplift of about 75 m.

6. On the Nullarbor Plain, shallow marine middle Miocene limestone slopes from an elevation of +200 m inland to about 0 m at the coast, indicating uplift since deposition of 180 m by tilting about a hinge near the coast (Cope 1975).

7. In the Murray Basin, the top of the middle Miocene shallow marine limestone and clay has an elevation

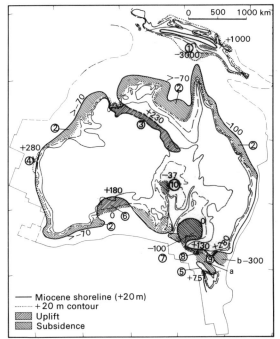

Fig. 139. (and Figs. 140–3). Palaeogeographical maps showing shorelines, and areas of subsequent vertical motion. Circled numbers indicate areas referred to in the text. The observed or inferred shoreline for the indicated age (full line) and that of the succeeding age (broken line) are shown on a base with modern contours at −200, 0, +100 (faint broken line), and +200 m. The present elevation contour equivalent to the past sea level is shown by a dotted line. Sources of general information, not cited in the text, are Ludbrook (1978), Playford et al. (1975), Harrison (1969), Exon and Senior (1976), and Leslie et al. (1976).

of 0 m in the east and −80 m in the west (Jones and Veevers 1982), indicating subsequent subsidence of at least 100 m about a hinge on the eastern side of the basin.

8. On the northern edge of the Otway Basin, middle Miocene shallow marine limestone today has an elevation of 150 m, and near the coast is close to sea level, indicating uplift by tilting in the north. To the west, Cook et al. (1977) found an uplift by tilting and faulting in sediments <0.7 Ma old of nearly 100 m at Naracoorte, 100 km from the coast, so that most of the post-middle Miocene uplift is probably fairly recent.

9. In the Bass Basin (a) and Gippsland Basin (b), the middle Miocene shallow marine limestone is several hundred metres below sea level today, indicating subsequent subsidence of this amount, but in the onshore Gippsland Basin, the limestone stands locally, in the Baragwanath Anticline, at +300 m, indicating local uplift of 280 m.

10. Lake Eyre lies today 12 m below sea level but was not covered by the Miocene sea which rose 20 m above present sea level. If Lake Eyre were open to the sea in the Miocene, presumably through Lake Torrens, then the absence of Miocene marine sediments in this low-lying area indicates subsequent subsidence of at least 37 m, from the Miocene sea level of +20 m to the present elevation of the top of the Miocene Etadunna Formation (−17 m) (Johns and Ludbrook 1962).

(ii) Between the Eocene and Miocene (38.5 to 24 Ma) (Fig. 140)

Sea level in the Eocene was at +75 m, and elevation changes between the Eocene and Miocene, that is, in the Oligocene, can be gauged by the relative positions of the Eocene shoreline (= +75 m elevation contour in the Eocene) and the Miocene shoreline (= +20 m elevation contour in the Miocene).

1. In the south-west, late Eocene shallow marine sediment now has an elevation of about +250 m near Norseman (Churchill 1973) (1a) so that the net upward tilt since the Eocene (Cope 1975) is about 175 m. At the northern edge of the Eucla Basin (1b), along the western and north-western margins, and off northeastern Queensland, the Eocene and Miocene shorelines coincide, indicating a subsidence of 55 m in the Oligocene. The wide separation of the Eocene and Miocene shorelines in southern New Guinea and the Gulf of Carpentaria indicates Oligocene subsidence greater than 55 m.

2. In the Lake Eyre area, the top of the Eocene nonmarine Eyre Formation has an elevation of −60 m, so that this area must have subsided by at least 135 m since the Eocene, comprising at least 63 m in the Oligo-

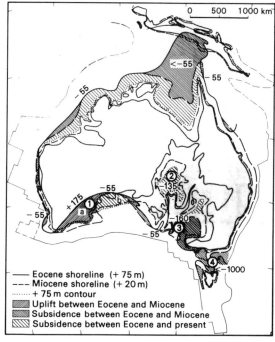

Fig. 140.

cene (4 m/Ma), 35 m in the Miocene (= thickness of Etadunna Formation) (3 m/Ma), and at least 37 m since the Miocene (3 m/Ma), as found above.

3. In the Renmark area of the Murray Basin, the top of the late Eocene shallow marine clay has an elevation of −200 m, indicating 275 m of subsidence since the Eocene, comprising at least 160 m in the late Oligocene and Miocene (= thickness of shallow marine sediments of this age), and, as found above, at least 100 m since the middle Miocene, all at rates about double those at Lake Eyre.

4. In the Bass and Gippsland Basins, late Eocene shallow marine sediment is now at depths of 1 to 2 km below sea level, implying subsidence of this amount since the Eocene, most of it during the Oligocene and Miocene, at rates an order of magnitude greater than those of the Murray Basin and Lake Eyre.

(iii) Between the later Cretaceous (90 to 65 Ma) and mid-Eocene (50 Ma) (Fig. 141)

The shoreline stood at +110 m in the later Cretaceous and at +75 m in the Eocene.

1. In the Lake Eyre region, the later Cretaceous lacuna (between the Winton and Eyre Formations) lies nearly 100 m below sea level, and must have lain >110 m above sea level during the later Cretaceous, implying

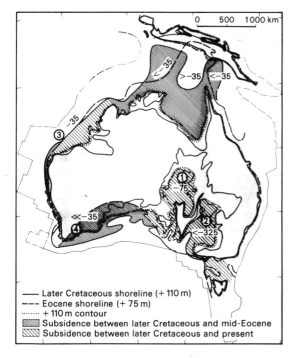
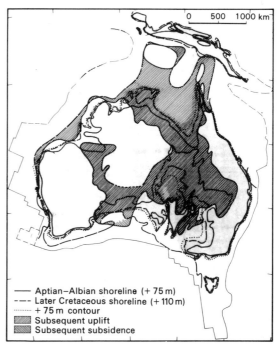

Fig. 141.

- Later Cretaceous shoreline (+110 m)
--- Eocene shoreline (+75 m)
······ +110 m contour
▨ Subsidence between later Cretaceous and mid-Eocene
▧ Subsidence between later Cretaceous and present

Fig. 142.

- Aptian–Albian shoreline (+75 m)
--- Later Cretaceous shoreline (+110 m)
······ +75 m contour
▨ Subsequent uplift
▧ Subsequent subsidence

>210 m of subsequent subsidence, including at least 135 m since the Eocene, as found above, and hence at least 75 m during the Palaeocene and Eocene ($\geqslant 3$ m/Ma).

2. The later Cretaceous lacuna in the Murray Basin is now as deep as 500 m below sea level but at the time must have been $> +110$ m. The subsequent subsidence is therefore >610 m, including at least 275 m since the Eocene, and hence >325 m in the Palaeocene and Eocene (12 m/Ma).

3. The near-coincidence of the later Cretaceous and Eocene shorelines along the western and north-eastern margins indicates a subsidence of about 35 m during the Palaeocene, and the separation of the shorelines in southern New Guinea indicates subsidence >35 m in places and <35 m in others.

4. The wide separation of the later Cretaceous and Eocene shorelines in south-west Australia implies a subsidence in the Palaeocene and early Eocene very much greater than 35 m.

(iv) Between the Aptian–Albian (95 Ma) and the later Cretaceous (90 Ma) (Fig. 142)

The feather-edge of the Aptian–Albian marine sediment, taken as indicating the minimum extent of the Aptian–Albian sea, with a sea level at +75 m, lies today, with a few exceptions, at elevations that range from +200 to +500 m (Fig. 138). A broad area of Aptian–Albian sediment in the north on either side of the Gulf of Carpentaria narrows southward to a constriction about Mt. Isa, and widens again between the eastern highlands and central Australia to terminate in the south about the Cobar Spur and the South Australian Highlands, with a narrow gulf into the Murray Basin. Past a constriction west of Lake Eyre, the area of Aptian–Albian sediment widens southwestward into the Eucla Basin and Ceduna Depocentre, possibly to link westward with the Indian Ocean, and northwestward into the Officer and Canning Basins, probably to connect across a saddle with the offshore Canning Basin. The correlation between the extent of Aptian–Albian marine sediment and the broad lowland today about the Gulf of Carpentaria and in the Lake Eyre–Surat Basin area is interpreted as showing that central-eastern Australia had a shape in the Aptian–Albian similar to that of today, but with a general elevation a few hundred metres lower, indicating this amount of subsequent uplift, probably more in central Australia, and much more, allowing for subsequent loss off the Great Dividing Range, along the western flank of the Eastern Highlands (Fig. 145). In the southeasternmost areas of Aptian–Albian sediment, in the Murray Basin, the sediment wedges out in the direction of a downslope today, indicating subsequent subsidence along the southern margin. Several hundred metres of subsequent uplift is indicated by the Aptian–Albian sediments across the high ground between the Canning and Officer Basins. With allowance for the subsided margin and uplifted Officer Basin, Eastern Highlands, and New Guinea, modern Australia imitates the shape of the continent in the Aptian–Albian, showing that the continent has since maintained this level (without tilt). And this provides a means of calibrating the net general uplift of 100 m since the Aptian–Albian, found by the hypsometric analysis (Fig. 135), with a net maximum subsidence of about 1000 m in the middle of the Eromanga/Birdsville Basin, shown by the post-Albian deposits in this area (Fig. 144), balanced by the inferred uplift of several hundred metres about the periphery of the broad Central-Eastern Lowlands (Fig. 145).

As shown by Fig. 135, the net general uplift since the Aptian–Albian comprises a rapid uplift of 280 m in the Late Cretaceous and a slow subsidence of 180 m in the Cainozoic. The distribution of the uplifted ground is shown in east-central Australia by the 2000 km long regression of the shoreline between the Aptian–Albian and the later Cretaceous (Fig. 142), during an interval when the sea level increased from $+75$ m to $+110$ m. This entire east-central region, which lay below $+75$ m in the Aptian–Albian, and above $+110$ m in the later Cretaceous, is analysed further in Figs. 144 and 145. During and after the deposition of the non-marine Winton Formation in the Cenomanian, the general uplift of 280 m was sufficient to change the former low-lying area of marine deposition into an area of no deposition, except in a few isolated pockets such as the Mt. Howie Sandstone.

On the western and north-western margins, the proximity of the Aptian–Albian and later Cretaceous shorelines indicates only a very slight uplift. In the Great Australian Bight, the overlapping shorelines indicate again very slight uplift, but, more importantly, because the Aptian–Albian shoreline transgressed from the north, and the later Cretaceous from the south, they indicate a radical change of slope from northward in the Aptian–Albian to southward in the later Cretaceous, due presumably to post-breakup subsidence of the margin.

(v) Between the Late Jurassic and the Aptian–Albian (150 Ma to 110 Ma) (Fig. 143)

In the Late Jurassic, only the north-western and northern parts of the continent were covered by a shallow sea, which stood at an elevation of -30 m. The almost static shoreline in these areas between the Late Jurassic and the Aptian–Albian suggests either a very steep coastline, with deposition keeping pace with the rise in sea level, or an uplift of about 100 m during the Neocomian. Elsewhere, the shoreline transgressed about 2500 km by the Aptian–Albian so that the sea occupied the lowland of east-central Australia and its south-western extension into the Eucla Basin. This transgression was due not only to the eustatic rise of sea level from -30 m to $+75$ m, but also, as shown by the hypsometric analysis, to a general continental subsidence of 150 m.

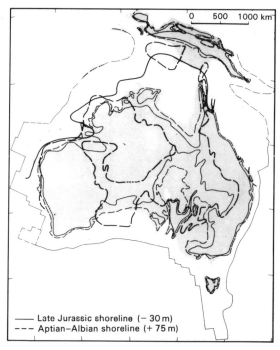

——— Late Jurassic shoreline (− 30 m)
- - - Aptian–Albian shoreline (+ 75 m)

Fig. 143.

(g) Discussion

A cumulative-subsidence diagram (Van Hinte 1978) of the Innamincka area (Fig. 144) shows the relations between eustatic sea level (dotted line), the curve of vertical motion of Australia (from Fig. 135), measured from the present elevation ($+40$ m) of the area, and the rate of accumulation of formations. For example, from the estimated net motion of $+100$ m of the land-surface since the Albian (from -60 m to $+40$ m), the present depth of the Allaru Mudstone at -1100 m,

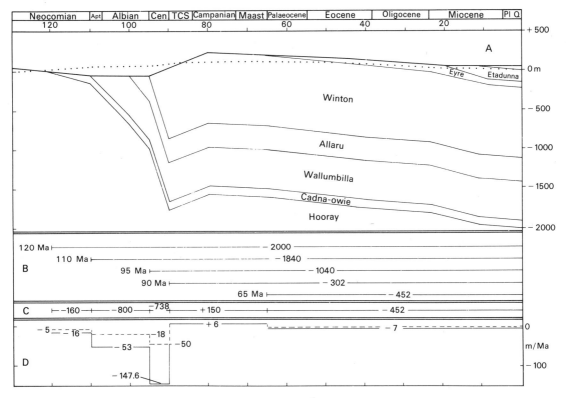

Fig. 144. A. Diagram showing changes of elevation through time of the Innamincka area of the Eromanga Basin. Details of age and thickness of formations from Wopfner et al. (1974, fig. 9) and Exon and Senior (1976, fig. 3). Dotted line is the eustatic sea-level curve, and the upper heavy line of the diagram, from which the formation thicknesses were measured, is the curve of vertical motion of Australia (double lines in Fig. 135), calibrated to the present elevation of +40 m of the Innamincka area. No allowance has been made for post-depositional compaction; that is, present compacted thicknesses are shown throughout. B. Net vertical motion (m) since 120, 110, 95, 90, and 65 Ma. C. Net vertical motion (m) between ages above. D. Rates of vertical motion (m/Ma) during age intervals. Full line is total motion, with two components: primary (tectonic), shown by the broken line, and secondary (due to isostatic sediment loading). Sediment loading is calculated as 2.2 (sediment density, from Dampney et al. 1978, fig 6)÷3.3 (asthenosphere density)=2/3 of total subsidence, and primary motion is therefore 1/3 of total motion. Effect of water displacement (≦170 m) at 95 Ma is less than 1/10 that of sediment loading, and is ignored.

and the indicated palaeobathymetry of 135 m in the Albian, within the limits found by Scheibnerova (1976) and Haig (1979), the net motion since the Albian is (− 1100 + (135−75)) = − 1040 m. These are compounded of motions, shown in Fig. 144B, over shorter time intervals, so that the motions between the various ages can be found by the difference (Fig. 144C). Finally, the rates of motion (Fig. 144D) can be found by dividing the motion by the duration of each time interval. And the motion due to tectonic forces (broken line) is found by removing the isostatic effect of sediment loading from the total motion. These rates show that tectonic subsidence increased from earlier rates of 5 to 18 m/Ma to a maximum of 50 m/Ma, 110 to 90 Ma,

changed to an apparently uniformly slow uplift during the interval 90 to 65 Ma, and to an apparently uniformly slow subsidence from 65 Ma to the present.

It is only in the Cretaceous and Cainozoic depocentre of the Innamincka area, where the depositional record is thickest and most complete, that the vertical motions of the central part of the platform can be estimated for several time intervals. Elsewhere in Australia the record is much less complete; and in intervals later than the Albian the sea covered the margins only. The transgressive shoreline of the widespread Aptian–Albian sea (Fig. 138) thus provides a unique datum, parallel to the present sea level and estimated to be 75 m above it, against which the present shape of much of

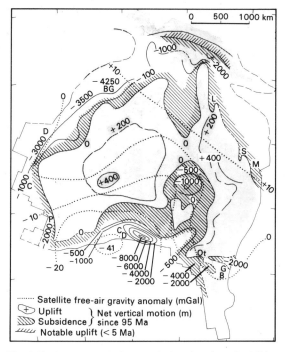

Fig. 145. Contours (m) of net vertical motion since the Albian (<95 Ma). Areas of net subsidence ruled. Dotted lines are contours (mGal) of satellite free-air gravity anomalies (Anderson et al. 1973). Sedimentary basins, clockwise from north-east, are: L Laura, S Styx, M Maryborough, G Gippsland, B Bass, O Otway, CD Ceduna Depocentre, P Perth, C Carnarvon, D Dampier, BG Bonaparte Gulf.

area of net subsidence, again relative to a fixed point at the centre of the earth, about the western, southern, and eastern coastlines and a deep re-entrant that passes through the Gulf of Carpentaria southward across the Euroka Arch to the central eastern lowland of the Innamincka area and then divides on either side of the South Australian Highlands into the Murray Basin on the east and the Nullarbor Plain on the west. The area of net uplift of the Eastern Highlands is limited to the north-east by the record of an Albian shoreline in the Laura Basin (Day 1976) and Styx Basin (de Jersey 1960), and of an Aptian shoreline in the Maryborough Basin (Ellis 1976).

The subsidence of the margin since the Albian is much greater. Deposition during the Albian in the Gippsland, Otway, and Bass Basins was in a non-marine, probably fluvial and lacustrine environment, and hence probably above sea level, so that the minimum subsidence since the Albian is measured by the thickness of the overlying sediments, which reaches 4000 m in the Otway, and 2000 m in the Bass and Gippsland Basins. Other parts of the margin have registered similarly large subsidences: on the western margin, from about 2000 m in the Perth Basin to greater than 4000 m in the Bonaparte Gulf Basin; and, on the southern margin, a maximum of 8000 m in the

Australia, and hence post-Albian vertical motions, may be gauged. The contours in Fig. 145 show the displacement from the present surface of the outcropping feather-edge of the Aptian–Albian strata or the subcrop of the shoreline in shallow-water deposits. If, for example, the Albian shoreline (= +75 m elevation in the Albian) now has an elevation of +300 m, then the net vertical motion at this spot, relative to a fixed point at the centre of the earth, since the Albian, = 300 − 75 = +225 m. If an area now at +100 m has an Albian shallow marine subcrop of −400 m, the post-Albian vertical motion would be −400 − 75 = −475 m. The greatest motion recorded onshore in Fig. 145 is in the Innamincka area, where the subcrop of the Albian shallow marine Allaru Mudstone is at −1100 m (Exon and Senior 1976), indicating a subsequent net vertical motion of −1175 m, as found above.

The indicated motions (Fig. 145), which apply mainly to the area of Albian flooding, show a broad

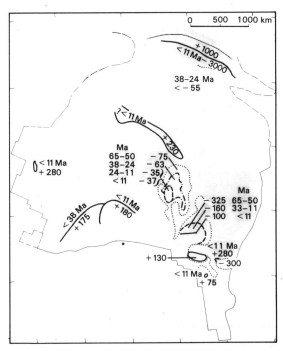

Fig. 146. Summary of vertical motions (m) in the Cainozoic.

Ceduna Depocentre (shown by the isopachs of post-Albian sediment – Fig. 145), all characteristic of the mobility of the margin. The great subsidence of the Ceduna Depocentre is attributable to its situation at the intersection of the east–west trending margin and the north- to north-east-trending Australian–Antarctic Depression, which extends landward into the Innamincka depocentre, and corresponds today with a negative free-air gravity anomaly, interpreted as indicating the dynamic influence of downwelling asthenospheric currents, as described in Chapter II.

From the datum of sea level at $+75$ m in the Aptian–Albian (Fig. 138), we see that about half the platform lay above $+75$ m and the other half below. The shape was similar to, and a few hundred metres lower than, that today except for a reverse of slope in the area of the Murray Basin. The areas of land, except the central-north-west area of the southern Northern Territory and the Kimberleys of Western Australia, were disposed symmetrically about a north–south axis, with a broad plateau of the Yilgarn–Pilbara Blocks (a half-arch) on the west, and the narrow ancestral Eastern Highlands on the east, with indentations of the Nullarbor Plain and Murray Basin in the high ground of the south, with a watershed along the northern shoulder of the rift-valley system.

In the Cenomanian, the sea withdrew to the north, and a kilometre of non-marine sediments (Winton Formation) was deposited in the Innamincka depocentre and, with a reverse of the platform slope southward, at least double this volume in the Ceduna Depocentre during a general submergence of the southern margin. As related earlier, after the Cenomanian, the Eastern Highlands expanded and rose rapidly, with the result that a great volume of sediment continued to accumulate in the Ceduna Depocentre north of a marginal rim adjacent to the slowly spreading South-east Indian Ocean.

In the Cainozoic (Fig. 146), the Eastern Highlands contracted eastward to their present width by subsidence of the Murray, Birdsville, and Karumba Basins, the depocentre migrated to the Lake Eyre region, the western part of the southern margin subsided and later rose, and New Guinea rose rapidly with a parallel but slight uplift across northern Australia. The margins continued to subside except for local uplifts at Northwest Cape, the north-west tip of Tasmania, and the northern part of the Otway Basin.

V. AUSTRALIA'S PHANEROZOIC HISTORY

1. INTRODUCTION

Phanerozoic Australia was marked by three regimes that followed the late Cryptozoic (or Precambrian) Adelaidean Regime:

(b) Late Innamincka Regime

The Innamincka Regime, which lasted from the Late Carboniferous (320 Ma) to the mid-Cretaceous (90 Ma), encompassed the development of the Gondwana

Regime	Ma ago	Stage	Ma ago
Potoroo	95 to 0	Cenomanian Interregnum	95 to 90
		Late	190 to 90
Innamincka	320 to 90	Middle	225 to 190
		Early	320 to 225
		Devonian/Carboniferous	370 to 320
		Silurian/Devonian	425 to 370
Uluru	575 to 320	Ordovician	500 to 425
		Cambrian	575 to 500
Adelaidean	c. 850 to 575	Ediacaran	c. 650 to 575

The history that follows is told in terms of the regimes, from the youngest to the oldest, and, because the Adelaidean Regime is a prelude to the succeeding Uluru Regime, these two are described together at the end.

2. POTOROO REGIME

(a) Preamble

We extract from the foregoing analyses a comprehensive history of the continent during the Potoroo Regime, which takes its name from Potoroo-1 Well in the Great Australian Bight that sampled the dominant depocentre of this regime. Much of this history is an exercise in neotectonics, in that the Potoroo Regime extends to the present, and the last part of this history is expressed in the configuration of Australia today. Having argued a history of this latest regime, we are then in a position to explore the preceding regimes. The late Innamincka Regime is outlined to provide continuity.

The history is presented in two sets of maps (Fig. 147): data maps (A, C, E, G) on which the salient evidence is summarized, and interpretative maps (B, D, F, H), supported by a correlation diagram (Fig. 148).

Series basins in Australia, and on the east was characterized by the almost continuous activity of a magmatic arc system which jumped eastward in the early Mesozoic to its Jurassic and Early Cretaceous position along the present coast. Volcanogenic sediment from the arc system interleaved with quartzose sediment dispersed centrifugally from the craton. Particularly towards the end of the Innamincka Regime, in the Aptian–Albian (Fig. 147A, B), great volumes of volcanogenic sediment were deposited westward in the foreland Surat Basin and epicratonic Eromanga Basin and in the Strzelecki–Otway rift-valley system in south-eastern Australia. Estimated sediment volumes (Table 7) (Fig. 147A) show that the Eromanga–Surat Basin overshadowed the southern rift-valley complex, and that, within this complex, the greatest volume of sediment was deposited in the Ceduna Depocentre.

(c) Cenomanian Interregnum and later Cretaceous deposition

During the transition to the Potoroo Regime from 95 to 90 Ma (Fig. 147C, D), which we call the Cenomanian Interregnum, the shoreline regressed rapidly to the north, due partly to continent-wide uplift, and partly to depositional regression. As we have seen (Fig. 144), the depocentre of the Winton Formation in the Innamincka area ran counter to the general uplift, with a rapid subsidence of 50 m/Ma (corrected for isostatic loading), but this subsidence was outpaced by non-marine deposition at a rate of 150 m/Ma of volcanogenic sediment derived from the north-east orogen in its terminal climax. The youngest known igneous rocks from the orogen are the 94 Ma old granite at Mt. Victor in New Guinea, although along the Queensland coast they are no younger than 109 Ma. However, the freshness of the andesitic debris in the Winton Formation suggests that this material is juvenile, and came from sources either unidentified or largely offshore.

With the concomitant broadening of the orogen, involving an uplift of the south-west flank of the Eastern Highlands of at least 400 m (Fig. 145), and the rise of the Euroka Arch (Senior *et al.* 1978), the former northward axial slope of Central-Eastern Australia was replaced by an extension of the southwestward slope that persists to the present. The northern limit of the depositional regression is marked on the northern side of the Euroka Arch by the paralic Normanton Formation, coinciding with the present shoreline of part of the Gulf of Carpentaria. Then, as since, the formerly important eastward drainage of the craton became incorporated in the centripetal drainage towards south-central Australia. In Fig. 147D, we speculate that this drainage terminated in the Ceduna Depocentre. Previous reconstructions (Brown *et al.* 1968, fig. 10.3; Veevers and Evans 1975, fig. 40.11) show the sea of the Great Artesian Basin withdrawing to the north, with the development of an extensive north-flowing river system. Brown *et al.* cite unpublished information of north-flowing currents in the northern part of the basin, but details are lacking. In our view, only the Carpentaria Basin drained to the north.

The three possible sources of the large volume of sediment deposited in the Ceduna Depocentre during the Cenomanian, almost as much in this 5 Ma span as in the previous 30 or 40 Ma, are:

(1) From the initial denudation of the area of the Yilgarn Block and Officer Basin formerly covered by the sea. Relict drainage in this area is dated stratigraphically as post-Aptian and pre-Eocene. Van de Graaff (1981) estimates that this area was denuded by about 100 m during this interval; even if all of this were concentrated in the Cenomanian, no more than 0.06×10^6 km^3, a relatively small volume, would have been provided.

(2) From the north-east arc, by southwestward continuation of the drainage past the area of the preserved Winton Formation. This is an imponderable in view of the scanty published knowledge of the composition of the Cenomanian fill of the Ceduna Depocentre. If a significant volume of volcanogenic sand had reached the depocentre, it should be obvious in the cores and cuttings recovered in Potoroo-1; but if the bulk of the sand had been deposited in the proximal Winton Formation, and only mud had reached the distal Ceduna Depocentre, this source would not be obvious.

(3) From the epeirogen of the ancestral South-east Highlands, which came into being with the splitting of the arc at 95 Ma, as shown in our analysis of the Eastern Highlands. Fission-track and magnetic studies indicate that the presently exposed coastal strip of south-eastern Australia (Fig. 147C) cooled through 100 °C some time in the interval 100 to 70 Ma ago. This cooling event spans (a) the cooling of shoshonitic magmas along the coast (dated at Mt. Dromedary as an average 95.5 Ma) and of the rhyolite on the conjugate Lord Howe Rise (96 Ma); and (b) the replacement of the Early Cretaceous volcanogenic sediment of the Strzelecki Basin by the Late Cretaceous quartzose sediment of the succeeding Gippsland Basin, within a microfloral zone that straddles the Albian–Cenomanian boundary, dated as 95 Ma ago. We believe that the primary cause of cooling was the cessation of orogenic magmatism 95 Ma ago, and the secondary cause was the subsequent denudation of the rapidly uplifted epeirogen. The fission-track, palaeomagnetic, and vitrinite-reflectance studies indicate that at least 1 to 2 km of cover were removed from the coastal strip, all within the interval 100 to 70 Ma ago. Over this interval, in the Cenomanian and later Cretaceous (Fig. 147C, E), the Ceduna Depocentre was filled with sediment to a thickness of 8 km. There is an eastward passage from this voluminous deposit through a lacuna to the eastern epeirogen. In our view, these are different aspects of the one event: the rapid uplift of the ancestral Eastern Highlands, their rapid downwearing, and the transport of the resulting sediment by an ancestral Murray–Darling drainage system down the western flank, represented by the Late Cretaceous lacuna beneath the Murray Basin, to the Ceduna Depocentre. Unlike in the present situation, the area of the South Australian Highlands, except perhaps the Gawler Ranges, did not deflect this drainage. Later in the Cretaceous (Fig. 147F), the thick western part of the Ceduna Depocentre that prograded south through south-west is seen to have been nourished, not by the incompetent drainage system to the north-west, but by another river system – an ancestral Cooper Creek – that debouched into the

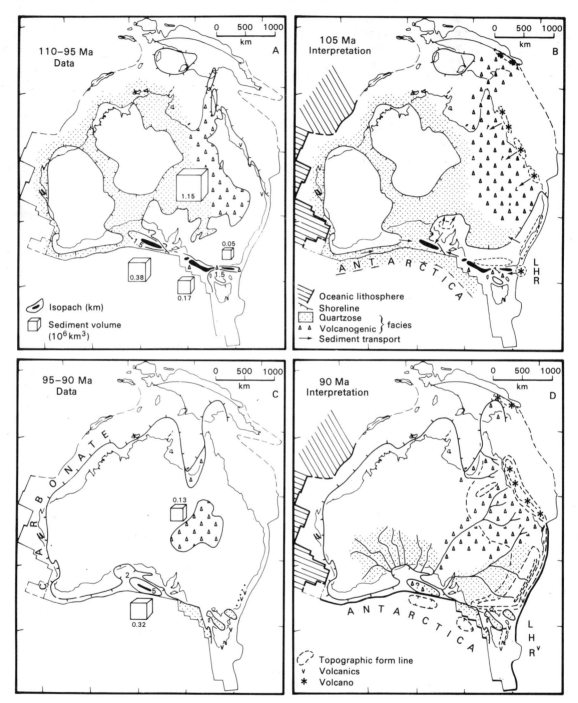

Fig. 147. A. Data map for the interval 110 to 95 Ma (Aptian–Albian). Volumes ($10^6 km^3$) are of Early Cretaceous and Jurassic sediment, and isopachs (km) of depocentres on southern margin, with the area circumscribed by the 3 km isopach shown in black. B. Interpretative map at 105 Ma (early Albian) showing additionally topographic form-lines of the eastern orogen, the direction of sediment transport, and the oceanic lithosphere on the west. C. Data map for the interval 95 to 90 Ma (Cenomanian). Shown are the volumes ($10^6 km^3$) of the Winton Formation and the Ceduna Depocentre, which is additionally outlined by isopachs (km). The Vs along the south-east coast are igneous rocks emplaced between 100 and 90 Ma, and the dotted line is the trend of cooling shown by palaeomagnetic overprints, and studies of vitrinite reflectance and apatite fission tracks. Fig. D. Interpretative map at 90 Ma (end of Cenomanian).

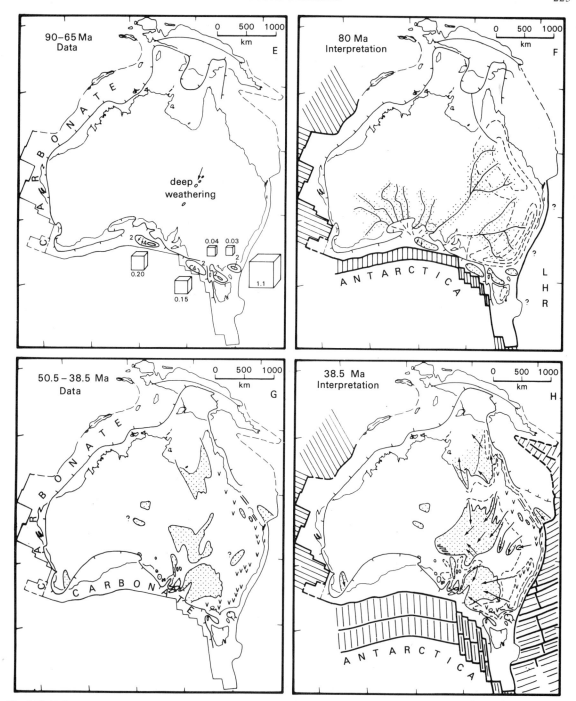

Fig. 147. E. Data map for the interval 90 to 65 Ma (post-Cenomanian Late Cretaceous), showing volumes (10^6km^3) and isopachs (km) of Ceduna Depocentre, Otway, Bass, and Gippsland Basins, and possible volume of Tasman Sea Basin. Single arrow indicates tentative direction of sediment transport of the Mt. Howie Sandstone, and arrows in the Ceduna Depocentre the direction of sediment progradation. F. Interpretative map at 80 Ma (Campanian). G. Data map for the interval 50.5 to 38.5 Ma (middle and late Eocene). H. Interpretative map for 38.5 Ma (end of Eocene). Drainage in north-east from Grimes (1980, fig. 9.2).

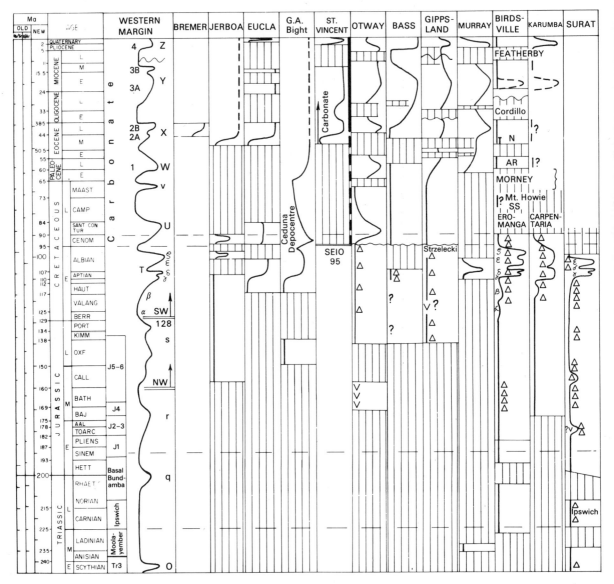

Fig. 148. Correlation diagram for the Mesozoic and Cainozoic (245 to 0 Ma). Localities are regions or sedimentary basins (Fig. 44) except Jerboa, an oil exploration well on the western side of the Great Australian Bight. Radiometric dates of volcanism from the eastern region include concordant fission-track dates on zircon, sphene, and apatite from the Otway Basin (Gleadow and Duddy 1981). The arrows indicate the best estimate of the ages of the Ferrar Supergroup of Antarctica (Kyle et al. 1981). The Cainozoic cycles are denoted by letters (from Jones and Veevers 1982) or numbers (Quilty 1977; McGowran 1979). Dating of the Early Cretaceous, including transgressions α to γ, from Morgan (1980). Sources of the individual columns are as follows.

Western margin: Apthorpe (1979), Powell (1976b), Quilty (1977, 1980b) whence the depositional cycles are taken, Wiseman (1979), Forman and Wales (1981), and Crostella and Barter (1980). Bremer: Playford et al. (1975). Jerboa: Bein and Taylor (1981). Eucla: Playford et al. (1975), Ludbrook (1978). Great Australian Bight: Fraser and Tilbury (1979), Harris and Foster (1974) (Polda Basin). St Vincent: Cooper (1979), Ludbrook (1980), McGowran (1978, 1979). South-east Indian Ocean (SEIO): Cande et al. (1981). Otway: Wopfner and Douglas (1971), Boeuf and Doust (1975), Gleadow and Duddy (1981). Bass: Brown (1976), Robinson (1974), Quilty (1972, 1980a). Gippsland: Abele et al. (1976), James and Evans (1971), Partridge (1976). Murray: Thornton (1972), Abele et al. (1976), Lawrence (1975),

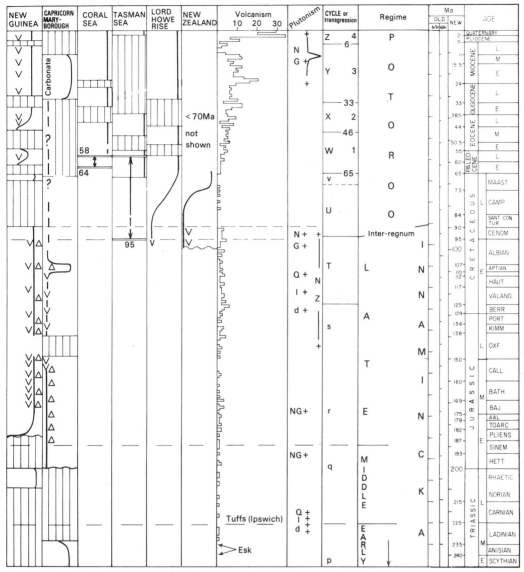

Macumber (1978), Woolley (1978). Birdsville, Eromanga: Grimes (1980), Ludbrook (1980), McGowran (1979), Wopfner et al. (1974), Wopfner (1963), Morgan (1980), Vine (1976a). N—Napperby (Kemp 1976), AR—Ayers Rock (Twidale and Harris 1977). Deep weathering profiles are the Morney Profile (Idnurm and Senior 1978), the Cordillo Surface (Wopfner 1974), and the Featherby Surface (Grimes 1980). Karumba, Carpentaria: Grimes (1980), Smart et al. (1980). Surat: Exon (1976), Allen (1976). New Guinea: Burns and Bein (1980), McGowran (1979). Capricorn, Maryborough: Day et al. (1974), Ericson (1976), Ellis (1976). Coral Sea: Weissel and Watts (1979), Andrews et al. (1975). Tasman Sea: Weissel and Hayes (1977), Partridge (1976). Lord Howe Rise: Willcox et al. (1980), McDougall and van der Lingen (1974). New Zealand: Laird (1981). Volcanism: Helby and Morgan (1979), Crawford et al. (1980), Kyle et al. (1981), Schmidt and McDougall (1977), Green and Webb (1974), Dulhunty (1972, 1976), Dulhunty and McDougall (1966), Evernden and Richards (1962), McDougall and Leggo (1965), McDougall and Wellman (1976), Wellman et al. (1970), Exon et al. (1970), Griffin and McDougall (1975), McDougall and Slessar (1972), Stephenson et al. (1980), Sutherland et al. (1978), Webb and McDougall (1967a), Webb et al. (1967), Wellman and McDougall (1974a, b), Wyatt and Webb (1970), and references in Jones and Veevers (1982, 1983). Plutonism: Page (1976), Webb and McDougall (1964, 1968), and references in Veevers and Evans (1975).

Polda Trough while the ancestral Murray–Darling system drained into the eastern (Duntroon) part of the depocentre. Regardless of the details of drainage, which can only be sketched, the Late Cretaceous part of the Ceduna Depocentre has a volume (0.5×10^6 km^3) too big to have been derived from a nearby source – the South Australian Highlands, for example, would have had to have been lowered by about 3 to 4 km if all the sediment so generated went to the Ceduna Depocentre, and double this if it were dispersed on all sides: only the orogen and the newly uplifted Eastern Highlands would suffice.

(d) Later Cretaceous lacuna on the platform

The interval from 90 to 65 Ma ago is marked by a depositional lacuna on the platform except for traces of possibly Late Cretaceous sandstone in the Innamincka area. This Mt. Howie Sandstone (Wopfner 1963; Forbes 1972), 50 m of fluvial quartz sandstone, disconformably overlies the Winton Formation, and contains plant fossils with Mesozoic affinities. Wopfner (1963) tentatively suggests that the depositing palaeocurrents were south to south-west.

With the possible exception of the Mt. Howie Sandstone, the later Cretaceous platform lacuna is absolute, in contrast to the preceding widespread marine deposition in the Aptian–Albian and non-marine Cenomanian deposition, and to the widespread fluvial deposition in the succeeding Palaeocene and Eocene (Fig. 147H). The main event registered on the platform was deep weathering to form the Morney Profile (Senior *et al.* 1977; Grimes 1980) in the interval 90 to 60 Ma, as shown by palaeomagnetic studies (Idnurm and Senior 1978). We interpret this lacuna as due to non-deposition on the south-western and western flanks of very broad Eastern Highlands; major rivers carried quartzose sediment derived from the Eastern Highlands across this terrain without deposition, except locally and intermittently, as possibly exemplified by the Mt. Howie Sandstone; not until the shoreline about the southern margin was reached did the rivers deposit their load of sediment. Most of it was deposited in the Ceduna Depocentre. The volume of sediment deposited in the Otway Basin (Table 7; Fig. 147E) is much greater than that of the adjacent Bass or Gippsland Basins, derived from local sources, and we follow Denham and Brown's (1976) interpretation that 'a proto-Murray River may have contributed large volumes of coarser clastics from a Palaeozoic hinterland', as shown in Fig. 147F, with the proto-Murray River system debouching into the Otway Basin or through Spencer Gulf into the Ceduna Depocentre, or both.

With a slow (6 mm/yr) spreading half-rate, the South-east Indian Ocean was only 180 km wide by 80 Ma ago (Fig. 147F) and 360 km wide by the end of the Cretaceous. With the change along the southern margin from rift-valley development to seafloor spreading at 95 Ma ago, a rift-divergence zone was succeeded by a rim basin, which subsided rapidly to accommodate the 8-km-thick fill of the Ceduna Depocentre (Fig. 102). On the western margin, the detrital deposition of the Early Cretaceous gave way to carbonate in the Late Cretaceous. On the southern margin, this change did not take place until the Eocene.

(e) Cainozoic resumption of platform deposition

On the southern margin, only a film less than 1 km thick of sediment, most of it carbonate, accumulated during the Cainozoic. Fig. 149 shows the isopachs of all sediments, from the Jurassic to the present, along the southern and south-eastern margins and adjacent oceans, in which the Ceduna Depocentre is shown to be 12 km thick. Fig. 150 is an isopach map of the Cainozoic sediment. The thickest Cainozoic sediment

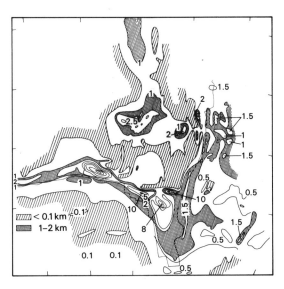

Fig. 149. Sediment isopachs of Late Jurassic, Cretaceous, and Cainozoic sediments in southern and south-eastern Australia and adjacent ocean basins. From Cooney *et al.* (1975), Denham and Brown (1976), Fraser and Tilbury (1979), Geological Society of Australia (1971), Houtz *et al.* (1973), James and Evans (1971), Robinson (1974), Symonds (1973), and Talwani *et al.* (1979).

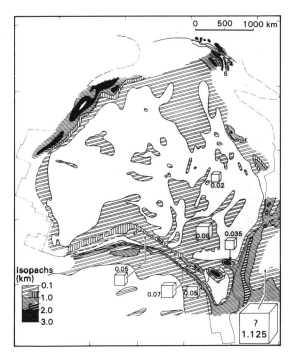

Fig. 150. Isopachs of Cainozoic sediments. Volumes of sediment ($10^6 km^3$) in the Ceduna Depocentre, Otway, Bass, Gippsland, Murray, and Birdsville Basins, and in the Tasman Sea shown diagramatically by cubes. Sediment wholly onshore, from Doutch and Nicholas (1978), is all less than 0.5 km thick, and most of it less than 0.1 km. Sources as in Fig. 149, and additionally Grund (1976), Ridd (1976), Doutch (1976), Mutter and Karner (1980), Ericson (1976), Laws and Kraus (1974), Allen *et al.* (1978), Powell (1976b), and Quilty (1978a).

on the southern margin is little more than 2 km; the Cainozoic sediment in the area of the Ceduna Depocentre, little more than 0.5 km thick, is thinner than elsewhere. The very thick Ceduna Depocentre was succeeded by a uniformly thin Cainozoic cover due to an abrupt loss of supply of detrital sediment. Tectonic subsidence in the Australian–Antarctic Depression has been maintained, but unlike the depocentres of the Otway, Bass, and Gippsland Basins to the east, the Ceduna Depocentre has been starved.

We see this as caused by (a) the contraction and lowering of the Eastern Highlands by the end of the Cretaceous so that they supplied less sediment to the westward and southwestward drainage system, and (b) the Palaeocene resurgence across the path of the drainage of the ancestral South Australian Highlands, resulting through a diminished gradient in the deposition of the Birdsville and Murray Basins within the previously erosional domain of the lower flanks of the Eastern Highlands. By the end of the Eocene (Fig. 147G, H), after a second pulse of uplift of the South Australian Highlands, broad sheets of mainly quartz sand occupied the Central-Eastern Lowlands between the Eastern Highlands and the South Australian Highlands, and this configuration has persisted to the present day.

(i) Cycles in the Cainozoic

Four Cainozoic cycles of transgression–regression–lacuna have been recognized (Fig. 151). Quilty (1977, 1980b) recognized four cycles of deposition in the Cainozoic sediment, mainly carbonate, of the western margin, and interpreted them as reflecting eustatic fluctuations of sea level. McGowran (1979) extended this scheme to the entire continent and added that 'although some form of eustasy is implied by the four main transgressions and by the intra-Miocene unconformity, eustasy alone cannot generate the Sequences; there has to be tectonism underlying the erosion-deposition relationships and permitting the accumulation of neritic carbonates'. Grimes (1980) independently found three cycles of geological activity in north-east Australia: 'Each cycle commenced with a period of epeirogenic movement which uplifted areas to become erosional sources of sediment, and downwarped basins to receive these sediments. There was accompanying volcanism in some areas. The erosion

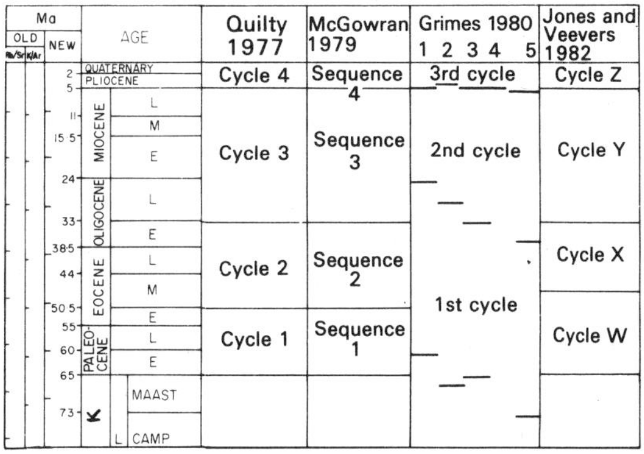

Fig. 151. Cainozoic cycles and sequences of the western margin (Quilty 1977, 1980b), Australia (McGowran 1979), north-east Australia (Grimes 1980), and the South-eastern Highlands and adjacent basins (Jones and Veevers 1982). North-east Australia comprises: 1, Inland; 2, Carpentaria; 3, North Coast; 4, Burdekin; 5, Fitzroy.

and deposition continued to the subsequent phase of tectonic stability until the relief was reduced to a planation surface which was then deeply weathered.'
We have analysed the better-known South-east Highlands and its flanking basins (Chapter IV); using the 300 or more dates of basalts of the South-east Highlands, we pointed to the correlation of the basalt chronology with the depositional cycles of the flanking basins, and advanced the hypothesis that intervals of more intense volcanism correspond with uplift of the highlands and concomitant subsidence of the flanking basins. These tectonic cycles correspond with those of Quilty (1977), McGowran (1979) and, in part, Grimes (1980), and all are probably local manifestations of global cycles.

We speculate that the South Australian Highlands followed the same uplift–subsidence cycles as the Eastern Highlands – Wopfner *et al.* (1974) indicate uplifts in the Palaeocene and Eocene – and indicate, with short arrows in the South column of Fig. 152, the most propitious times for the Murray River to have crossed the South Australian Highlands, as postulated by Williams and Goode (1978) and Goode and Williams (1980).

(f) Summary

The main events of the Potoroo Regime and of the late Innamincka Regime are summarized in the correlation diagram (Fig. 152).

(i) Innamincka prelude

The interval of the Jurassic, from 200 to 130 Ma, encompassed the breakup of the north-western margin and the development of a rim basin, the continued development of a rift-valley system in the south-west, the start of rift-valley development in the south, and the continuation of orogenic activity, after an eastward jump, in the east. In the east of the continent, the craton shed quartzose sediment to the north-east, and the orogen shed volcanogenic sediment to the south-west so that quartzose and volcanogenic sediments interleaved in the foreland and epicratonic basins (Fig. 147B). During the Early Cretaceous, volcanogenic sediment from the north-east orogen overwhelmed the quartzose sediment from the craton to reach the Innamincka depocentre. At the end of the Early Cretaceous, an epicontinental sea intermittently covered east-central Australia and parts of the north, centre, and west; the southern rifted-arch system remained non-marine during the Early Cretaceous except at the very end in the low-lying Ceduna Depocentre and to the west. The eastern part of the zone (Otway–Bass–Strzelecki), as to the north, was overwhelmed by volcanogenic sediment. The rim of the north-western margin became submerged 100–115 Ma ago, soon after the western margin broke up from India. In turn, the rim of the western margin lasted to about 90–95 Ma ago, at the same time as Greater India became detached from its former connexion along the Zenith–Wallaby transform fault, and the resulting deep oceanic circulation led to widespread deposition of carbonate.

(ii) Interregnum

The Innamincka–Potoroo Interregnum (95–90 Ma) was marked by the coexistence of the orogen in the north-east, in a climax that terminated the Innamincka Regime, and the breakup of the south-eastern and southern margins, which was the initial event of the Potoroo Regime. The joint effect of these events was the establishing of an almost continent-wide drainage to the Ceduna Depocentre, with the main sources of sediment in the newly formed epeirogen in the south-east and the terminal orogen in the north-east.

(iii) Potoroo Regime

The Potoroo Regime comprises the events that have marked the development of Australia since the mid-Cretaceous (95 Ma). The principal events were:

1. (a) At 95 Ma, the separation of the southern margin from Antarctica by a change from rift-valley development to seafloor spreading; the separation of the south-eastern margin from Lord Howe Rise/New Zealand, involving the splitting of a magmatic arc; and the detachment of Greater India from the western margin by unkeying along the Zenith–Wallaby transform fault. Hitherto a province of Gondwanaland, Australia was now a continent.
   (b) At 90 Ma, the demise of the magmatic arc in the north-east.
   (c) At 95 Ma, the uplift of Australia *en masse* so that the mid-Cretaceous epicontinental sea retreated to the margins. Within the platform, the Central-Eastern Lowlands continued to subside from 95 to 90 Ma at an accelerating rate concomitant with climactic uplift of the north-eastern magmatic arc, but sediment from the arc built up faster than the subsidence thus holding out the sea. With the demise of the arc at 90 Ma, the Central-Eastern Lowlands rose in concert with the rest of the continent.
2. (a) From 95 to 65 Ma, the Australian drainage was largely centripetal to the Ceduna Depocentre. Uplift of the South Australian Highlands at the

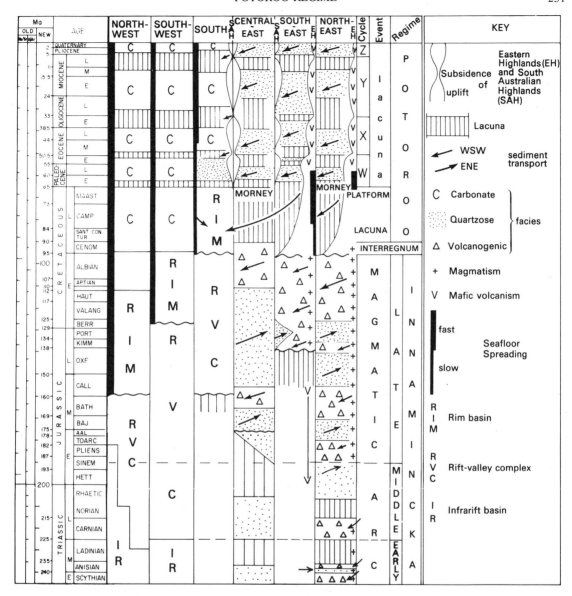

Fig. 152. Correlation diagram of the Potoroo Regime and the Mesozoic part of the Innamincka Regime.

beginning of the Cainozoic, 65 Ma ago, impeded this drainage to form the Birdsville and Murray Basins, so cutting off the supply of detrital sediment to the Ceduna Depocentre (Fig. 147H).

(b) Also at 65 Ma, the Papuan Peninsula separated from north-east Australia by seafloor spreading in the Coral Sea.

(c) During the Cainozoic, deposition on the platform and margins was dominated by four cycles of transgression–regression–lacuna that correlate with global changes of sea level, and with the waxing and waning of basaltic volcanism in the Eastern Highlands. During the Late Cretaceous, deposition on the southern margin was encompassed by a single cycle of transgression and regression, whereas two cycles operated on the west. Deposition on the western margin has been dominated by carbonate since the mid-Cretaceous (95 Ma) following the detachment of Greater India and the inception of oceanic circulation, on the southern margin since the middle Eocene (50 Ma), and on the north-east, as seen in the Great Barrier Reef, since the late Oligocene (25 Ma).

(d) At 5 Ma ago, the northern margin, in Timor and New Guinea, was uplifted as the result of convergence in the Banda Arcs and Melanesian Borderlands.

(e) During the Cainozoic, parts of the Carnarvon, Gippsland, and Birdsville Basins were folded by the lateral movement of basement blocks to provide suitable structures for petroleum reservoirs (Evans 1981, 1982).

These Cainozoic events in mainland Australia are seen as the aftermath of the shaping tectonism of the mid-Cretaceous. New Guinea, which underwent bodily rotation of its eastern tip with the spreading of the Coral Sea in the early Cainozoic, has seen a resurgence of orogenesis in the late Cainozoic with the uplift of the Central Highlands to their present elevation of 5 km.

On the other side of Australia and extending southward, the Australian–Antarctic Depression continued to develop during the Potoroo Regime. The next section describes its development in Antarctica.

(g) Australian–Antarctic Depression in Antarctica (by J. J. Veevers and P. G. Quilty)

(i) Late Cretaceous and Cainozoic history

The Transantarctic Mountains today constitute the eastern flank of the Depression in Antarctica, and are collinear in the pre-breakup reconstruction with the Eastern Highlands of Australia (Fig. 153). This collinearity is not peculiar to the reconstruction used here: it applies equally in the reconstructions of Griffiths (1974) and Weissel *et al.* (1977). The Transantarctic Mountains in the Scott Glacier area (at lat. 83° S, S in Fig. 7) existed 18 Ma ago (Stump *et al.* 1980), and Drewry (1975b) inferred uplift in the late Eocene (*c.* 40 Ma ago), but their earlier history is unknown. On the other side of the Depression, the flank of the ancestral Great Western Plateau lines up with the Antarctic flank marked by Mount Sandow (MS), but nothing of its history is known.

The late Mesozoic depositional record of East Antarctica (Table 9) comprises a single *in situ* non-

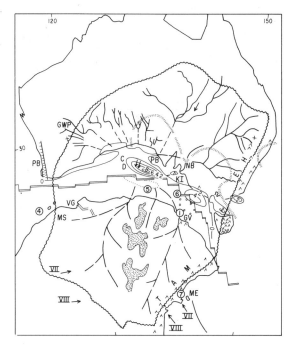

Fig. 153. Australia and Antarctica during the post-Cenomanian Late Cretaceous (90-65 Ma ago) in a pre-breakup reconstruction made by juxtaposing the continent–ocean boundaries (COBs), shown by a double line. The wiggly line shows the perimeter of the Australian–Antarctic depression, as approximated by its present position. Drainage network in Australia from Fig. 147. In Antarctica, inferred drainage shown by broken lines, inferred central highlands by stippled areas, inferred depocentre by a dotted line, and inferred basin opposite the Otway Basin by circles. Light arrows indicate direction of Late Cretaceous sediment transport into the Ceduna Depocentre (CD), shown by isopachs (km). Circled 1 off the George V Coast (GV) is an *in situ* occurrence of Aptian non-marine sediments, 4 to 6 are occurrences of reworked Late Jurassic to Eocene microfossils, and 7 opposite Mount Erebus (ME) is an occurrence of reworked Late Cretaceous fossils, all listed in Table 9. Inverted Vs indicate Early and Middle Jurassic volcanics (Dulhunty 1972; Kyle *et al.* 1981; McDougall and Wellman 1976; Schmidt and McDougall 1977; Milnes *et al.* 1982). Parallel lines indicate the Perth Basin (PB) and the trough occupied by the Vanderford Glacier (VG). Light dotted lines indicate the known extent of the Late Carboniferous sea. Broad arrows indicate location of the modern profiles VII and VIII of Fig. 6. Other abbreviations are Eastern Highlands (EH), Kangaroo Island (KI), Polda Basin (PB), Great Western Plateau (GWP), Mount Sandow (MS), and George V coast (GV). Lambert equal-area projection; grid refers to present position of Australia.

TABLE 9. *Evidence of Late Mesozoic and Early Cainozoic depositional environments, East Antarctica*

	Location (see Figs. 7 and 153) Lat. S, Long. E	Age	Depositional environment	Reference
①	67° 44′, 146° 51′ offshore George V Basin, depth 1407 m	IN SITU Aptian palynomorphs	non-marine	Domack et al. (1980)
②	66° 30′, 77° 30′ offshore Prydz Bay, water depth 320–1500 m	REWORKED Late Cretaceous–middle Eocene palynomorphs	non-marine (Eocene marine)	Kemp (1972)
		Albian palynomorphs	non-marine	Kemp (1972)
		Permian palynomorphs	non-marine	Kemp (1972)
③	55° 32′, 109° 57′ DSDP site 265, water depth 3582 m	Late Palaeocene–Eocene forams	marine	Engelhardt and Webb (in press)
		Campanian–Maastrichtian forams	marine carbonate	Engelhardt and Webb (in press)
④	65°, 96° Shackleton Ice Shelf	Late Palaeocene through Eocene dinoflagellates	marine	Truswell (1982)
		latest Cretaceous through Palaeocene pollen	non-marine	Truswell (1982)
		Late Jurassic through Early Cretaceous palynomorphs	non-marine, scantily marine	Truswell (1982)
		Late Carboniferous, Permian palynomorphs	non-marine	Truswell (1982)
⑤	65°, 134° offshore Cape Carr	latest Cretaceous through Eocene palynomorphs	marine	Truswell (1982)
		mid-Cretaceous spores and pollen	non-marine	Truswell (1982)
⑥	66°, 144° Mertz Glacier area	Eocene dinoflagellates	marine	Truswell (1982)
		Late Cretaceous through Palaeocene spores and pollen	non-marine	Truswell (1982)
		Early Cretaceous spores and pollen	non-marine	Truswell (1982)
⑦	77° 42′, 161° Taylor Valley	Late Cretaceous (Maastrichtian) forams	marine	Webb and Neall (1972)

marine Aptian deposit in the offshore George V Basin (shown by circled 1 in Fig. 153) and six deposits that contain reworked fossils (circled 4 to 7 in Fig. 153 and 2 to 7 in Fig. 7). The non-marine Aptian of the George V Basin and the reworked mainly mid-Cretaceous offshore from Prydz Bay, Cape Carr, and about the Shackleton Ice Shelf and Mertz Glacier correspond with deposits of this age in the basins of offshore southern Australia, all of which are notably non-marine, in contrast to the marine Aptian and Albian deposits of the Great Artesian Basin to the north. The reworked Campanian–Maastrichtian foraminifers at DSDP Site 265 are inferred to have been derived from west of long 110° E (Engelhardt and Webb, in press) and the Maastrichtian ones at Taylor Valley from 'somewhere in the Ross Sea–Ross Ice Shelf area or in Marie Byrd Land' (Webb and Neall 1972), all of which lie outside the region described here. The Late Cretaceous to middle Eocene non-marine palynomorphs and the Eocene microplankton off Prydz Bay and beneath the Shackleton Ice Shelf parallel the depositional sequence of events in southern Australia, though the Prydz Bay fossils have a likely source 'somewhere in the West Ice Shelf–Prydz Bay area' (Kemp 1972), again outside the region of interest here.

The occurrences of Early and Middle Jurassic volcanic rocks expressed in the Ferrar Dolerite and Kirkpatrick Volcanics of the Transantarctic Mountains (Elliot 1975b; Kyle et al. 1981), the Tasmanian Dolerite (Schmidt and McDougall 1977), and numerous minor centres in south-eastern Australia (Dulhunty 1972; McDougall and Wellman 1976; Crawford et al. 1980)

(Fig. 153) are collinear in the reconstruction along the eastern flank of the depression, and the isolated occurrences at Horn Bluff (Elliot 1975a) on the George V Coast (GV) and at Kangaroo Island (KI) (McDougall and Wellman 1976) lie nearly opposite each other.

We postulate the late Mesozoic existence of the entire eastern flank of the depression, including the Transantarctic Mountains (Fig. 153) because the terrains of the Eastern Highlands and the Transantarctic Mountains are linked today in a single morphotectonic feature; they were collinear before breakup; they were the locus of a widespread but narrow belt of Jurassic volcanism; and because the Eastern Highlands stood high in the Late Cretaceous. A critical test of this extrapolation to Antarctica would be provided by palaeomagnetic and fission-track age work in the Transantarctic Mountains.

On the western perimeter of the Depression, the only possible connection with Australia is that between the deep (> 2 km) trough occupied by the Vanderford Glacier (VG) (Allen and Whitworth 1970) and the Perth Basin (PB) of Western Australia. The surface of both features is a valley, and the Perth Basin is known additionally to be a relict rift valley filled with 10 km of sediment. The trough of the Vanderford Glacier differs from the Perth Basin in being sinuous and much narrower, and opens southward into the broad Aurora sub-ice basin. If this trough is the surface expression of thick sediments that date back to the Mesozoic, it would be on line with the Perth Basin possibly because they radiate from a common focus of the postulated triple-(hence quadruple-)rift junction (Veevers 1981) in this region.

Had the highlands of Adelie Land (Steed and Drewry 1982), with so much more relief today than their counterparts in the South Australian Highlands, existed in the Mesozoic (stippled pattern in Fig. 153), they would have divided the drainage network into three parts: one part in the ancestral Aurora basin, another in the ancestral Wilkes Basin, and the third in the intra-montane axis. All three may have drained into a depocentre that mirrored the Ceduna Depocentre, a speculation that could be tested by aeromagnetic and seismic surveys. Of the few published thicknesses of sediment off Antarctica (Houtz and Markl 1972), the area at lat. 63° S, long. 127° E, contains about 5 km of sediment, but how much of this pre-dates or post-dates breakup is unknown. Streams that flowed north-westward from the ancestral Transantarctic Mountains would have supplied most of the sediment to this depocentre, as did the south-west-flowing streams from the Eastern Highlands of Australia to the Ceduna Depocentre. In doing so, both sets of streams were diverted westward by the high ground of the former (Early Cretaceous) rift-valley shoulder, and, as in the Otway Basin of Australia, sediments may have been deposited between the shoulder and the suture. If, as in Australia, no sediment was deposited on the flanks outside the depocentre, then the sediment beneath the axis of the Wilkes sub-ice basin would be either Cainozoic, as in the Murray Basin but pre-glacial (Palaeogene), or older than Late Cretaceous, within the span of Steed and Drewry's (1982) postulated Mesozoic to early Cainozoic sediment fill over Beacon Supergroup.

(ii) Earliest history

The earliest sign of the depression is provided by the distribution of a shallow sea across part of southern Australia in the Late Carboniferous (McGowran 1973; Veevers and Evans 1975). Two embayments, an eastern one coexistent with the Cainozic Murray Basin, and a western one landward of the Great Australian Bight, are separated by a promontory on the site of the South Australian Highlands, which, immediately before this marine transgression, are inferred to have been high ground (Wopfner 1981). The morphology of the depression then may have been as it is today, with subsidiary lowlands on either side of a median axis through a highland.

(iii) Conclusions

The Australian-Antarctic Depression is coextensive with a long-wavelength negative satellite gravity anomaly that suggests that the Depression is caused by downward convection in the underlying asthenosphere. In Australia, the ancestral Depression is manifested by a thick pile of Late Cretaceous sediment deposited beneath the Great Australian Bight in a saddle in the southern margin between flanking highlands, and possibly by a wide area of Late Carboniferous marine sediments in southern and south-eastern Australia. In a reconstructed Australia and Antarctica, the eastern flanks of the Depression, represented by the Eastern Highlands of Australia and the Transantarctic Mountains, are collinear, and have a common history of widespread volcanism in the Jurassic. Because the present Eastern Highlands were high during the Late Cretaceous, we postulate that the present Transantarctic Mountains were high too, and that a thick pile of Late Cretaceous sediment, a mirror image of the Ceduna Depocentre, may exist beneath offshore Antarctica.

3. INNAMINCKA REGIME

(a) Introduction

The Innamincka Regime prevailed during the deposition of the Australian counterparts of the Indian Gondwana Series and correlatives elsewhere in Gondwanaland; it succeeded the Uluru Regime in the Late Carboniferous (320 Ma), and was succeeded by the Potoroo Regime in the mid-Cretaceous (95 Ma). It takes its name from the Innamincka region, which accumulated sediments in the depocentre of the Cooper and Eromanga Basins almost continuously during the regime.

Following the Early Carboniferous Kanimblan and Alice Springs regional deformations and uplifts, the Innamincka Regime started in the Late Carboniferous with the inception of basins across south-central Australia and the rejuvenation of old basins: in the west to form a zone of infrarift basins along which the continent split in the Jurassic and Cretaceous, and in the east, foreland and epicratonic basins linked to a magmatic arc which had jumped eastward during the Late Carboniferous. The regime concluded in the mid-Cretaceous with the effective isolation of Australia from India, its initial separation from Antarctica, and the demise of the magmatic arc in the east. Eastward jumps of the magmatic arc and related basins in the Late Triassic and Early Jurassic divided the regime into three stages.

During the Innamincka Regime, Australia occupied high southern latitudes whereas earlier (Cambrian through Devonian) and later (Cainozoic) it occupied lower latitudes (Fig. 40), and this is reflected, as elsewhere in Gondwanaland, in a distinctive style of deposition, characterized by (a) widespread glacigene sediments in the Late Carboniferous/Early Permian, lacking in the previous regime, and confined to the Quaternary in the Potoroo Regime; (b) widespread coal measures in the Permian, Late Triassic, Jurassic, and Early Cretaceous, also lacking in the earlier Phanerozoic; (c) rare limestone, in contrast to its common occurrence before and after; and (d) rare evaporites, again common before.

(b) Mid-Carboniferous lacuna (Fig. 154)

At the close of the prior regime (320 Ma), Australia was an area of non-deposition with erosion of the newly uplifted highlands in the centre and east. The mid-Carboniferous (Namurian/Westphalian, 325–310 Ma) is a lacuna in Australia except along narrow areas in the Bonaparte and Canning Basins in the north-west that continued to subside, and in the Werrie Trough and Yarrol Shelf in the central-east (Roberts and Engel 1980) (Fig. 154). In places, in particular in the south, the lacuna extends deep into the earlier Phanerozoic, and, at an extreme, beneath the Collie Basin, it extends back to the Archean, a span of 2200 Ma.

In describing the change from the Uluru Regime to the Innamincka Regime at the end of this chapter, we postulate that deposition from the uplifted areas was arrested by the formation of a continent-wide ice sheet, giving rise to the lacuna except on the margins. Only with the retreat of the ice, during the Stephanian, from 310 to 290 Ma ago, did the initial Innamincka (or Gondwana Series) basins start to fill the structural depressions generated by the mid-Carboniferous movements.

(c) Gondwanan style of deposition

The relative development of four facies characterizes the style of deposition.

(i) Glacigene sediments

The first flush of sediment intercepted by the Innamincka basins was glacigene, reflecting the combination of widespread highlands and high latitude; much of it is very coarse, and was deposited in narrow valleys on an ice-striated basement, as exemplified by the oldest known, Namurian/Westphalian, glacigene sediment of the Spion Kop Conglomerate and equivalent Mt. Johnstone Formation in the Werrie Trough (Fig. 154) (Herbert 1980b). These and other glacial occurrences in the Late Carboniferous and Permian of Australia are documented in Hambrey and Harland (1981).

(ii) Coal measures

Coal-making plants had evolved early in the Devonian, but coal does not appear in Australia until the Late Devonian, in the Amadeus Basin (Kurylowicz et al. 1976; Playford et al. 1976a), and in the mid-Visean of eastern Australia. According to J. Roberts (personal communication 1983), the first coal in eastern Australia occurs in the mid-Visean non-marine middle part of the Ararat Formation of eastern New South Wales between carbonate-bearing marine sediments. Slightly younger is the coal in the nearby Namurian Italia Road Formation (Rattigan 1966). As shown in Fig. 154, these coals are succeeded by the first glacigene deposits, and then, in the Permian, by coal in thick and widespread coal measures that continue, except in the Early and Middle Triassic, to the mid-Cretaceous, and then through the Potoroo Regime. So, like the glacigene

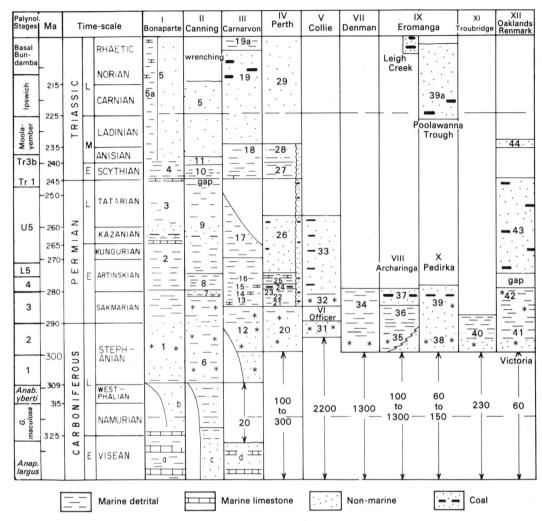

Fig. 154. Correlation diagram of the Late Carboniferous-Triassic basins from the north-west (Bonaparte) to south-east (Tasmania) and east (Clarence). General correlation and palynological stages from Kemp et al. (1977), extended and modified from Moore (1982). Additional general correlation from Dickins (1976) and Jones et al. (1973).

Bonaparte: Dickins et al. (1972), Laws and Brown (1976), Playford et al. (1975). Canning: Forman and Wales (1981), Playford et al. (1975). Carnarvon: Crostella and Barter (1980), Playford et al. (1975). Perth: Kemp et al. (1977), Playford et al. (1975). Collie: Kemp et al. (1977), Playford et al. (1975). Officer: Jackson and van de Graaff (1981). Denman: Wopfner (1981), Thornton (1979). Eromanga: Moore (1982). Arckaringa: Wopfner (1981), Thornton (1979), Cooper (1981). Pedirka: Wopfner (1981), Thornton (1979), Cooper (1981). Troubridge: Wopfner (1981), Cooper (1981). Oaklands: Yoo (1982). Renmark: Wopfner (1981), Thornton (1976). Victoria: Douglas and Ferguson (1976). Tasmania: Truswell (1978), Playford (1965), Clarke et al. (1976). Cooper: Thornton (1979). Galilee and Bowen: Fig. 156. Callide, Ipswich, Clarence: Day et al. (1974). Werrie Trough: Roberts and Engel (1980). Sydney: Herbert (1980a), McClung (1980), Runnegar (1980), Retallack (1980).

Ages of igneous rocks from text. Major marine transgressions indicated by capital letters K (Early Carboniferous) through Q (Rhaetic), and minor transgression p (Anisian). Lacuna at base of Innamincka succession shown by arrows (broad arrow indicates downward extension past the Visean), and duration in Ma. Letters identify formations before and during the lacuna, numerals after.

INNAMINCKA REGIME

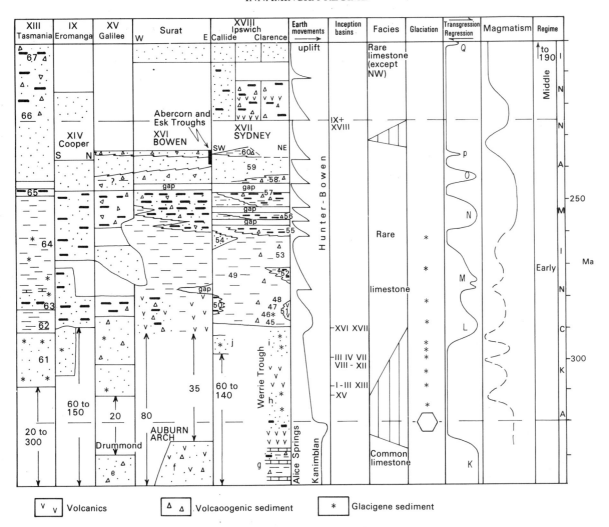

1, Kulshill Formation; 2, Fossil Head Formation; 3, Hyland Bay Formation; 4, Mt. Goodwin Shale; 5, Unnamed redbeds; 5a, Unnamed shale, sandstone, siltstone, and shale; 6, Grant Formation; 7, Poole Sandstone; 8, Noonkanbah Formation; 9, Liveringa Formation; 10, Blina Shale, 11, Erskine Sandstone; 12, Lyons Group; 13, Carrandibby Formation; 14, Callytharra Formation; 15, Wooramel Group; 16, Byro Group; 17, Kennedy Group; 18, Locker Shale; 19, Mungaroo Beds; 19a, Brigadier Beds; 20, Nangetty Formation; 21, Holmwood Shale; 22, Fossil Cliff Formation; 23, High Cliff Sandstone; 24, Irwin River Coal Measures; 25, Carynginia Formation; 26, Wagina Sandstone; 27, Kockatea Shale; 28, Woodada Formation; 29, Lesueur Sandstone; 30, Sue Coal Measures; 31, Paterson Formation; 32, Stockton Formation; 33, Collie Coal Measures; 35, Boorthanna Formation equivalent; 36, Stuart Range Formation; 37, Mt. Toondina Formation; 38, Crown Point Formation; 39, Purni Formation; 39a, Peera Peera Formation; 40, Cape Jervis Beds; 41, Unnamed; 42, Unnamed; 43, Coorabin Coal Measures; 44, Jerilderie Formation; 45, Lochinvar Formation; 46, Allandale Formation; 47, Rutherford Formation; 48, Farley Formation; 49, Shoalhaven Group; 50, Clyde Coal Measures; 51, Werrie Basalt; 52, Greta Coal Measures; 53, Maitland Group; 54, Nowra Sandstone; 55, lower Tomago Coal Measures; 56, upper Tomago Coal Measures; 57, Newcastle Coal Measures; 58, Narrabeen Group; 59, Hawkesbury Sandstone; 60, Wianamatta Group; 61, tillite; 62, Lower marine sequence; 63, Lower freshwater sequence; 64, Upper marine sequence; 65, Upper freshwater sequence; 66, Upper Parmeener Group; 67, New Town Coal Measures.

a, Milligans Formation to Tanmurra Formation; b, Unnamed formations in Pelican-1 and Lacrosse-1; c, Anderson Formation; d, Yindagindy Formation; e, Ducabrook Formation; f, Torsdale Beds; g, Namoi Formation; h, Mount Johnstone Formation and Paterson Volcanics; i, Seaham Formation; j, Tallong Conglomerate.

sediments, coal measures start in the Innamincka Regime, but are preceded by thin coal towards the end of the preceding regime.

(iii) Carbonate

In the Phanerozoic up to the Visean/Namurian (c. 325 Ma), deposits on the platform commonly contain carbonate, including oolite, last seen in the Visean/Namurian Tanmurra Formation and equivalent Medusa Beds in the Bonaparte Gulf Basin and the top of the late Visean Ararat Formation in the Hunter Valley (J. Roberts, personal communication 1983) and equivalent Namoi Formation to the north-west. In the Innamincka succession, limestone is rare, consists mainly of shell debris, and, except in a few known occurrences, lacks oolite. The known occurrences of limestone more than a few metres thick, from older to younger, are: (a) Late Sakmarian/Artinskian – a 150-m-thick lens of limestone near the top of the Carynginia Formation of the Perth Basin (Lowry 1981), limestone in the Callytharra Formation and Wooramel Group of the Carnarvon Basin, and in the Nura Nura Member at the base of the Poole Sandstone of the Canning Basin (Playford *et al.* 1975), limestone (c. 100 m thick) in the Yarrol Formation of the Yarrol Basin, Queensland (Maxwell 1960), and the Berriedale Limestone of the Tasmania Basin (Rao 1981a); the basal beds (Artinskian/Kungurian) of the Liveringa Formation of the Canning Basin contain limonitic oolite that may have been deposited as primary chamosite or goethite (Edwards 1958; Forman and Wales 1981). (b) Late Permian limestone in the Bonaparte Gulf Basin, at the Kungurian/Kazanian boundary, and at the end of the Tatarian (Laws and Kraus 1974), including the 50 m of limestone at the bottom of Sahul Shoals–1 Well (Campbell 1970); and 250 m of Late Triassic limestone in Ashmore–1 Well (Roberts and Veevers 1973); up to 250 m of limestone of Late Triassic age (Kuta Formation) occurs also in New Guinea (Skwarko *et al.* 1976). (c) 15 m of limestone in the Early Triassic Locker Shale of the Carnarvon Basin (McTavish 1970). (d) A widespread Early Jurassic (Sinemurian) limestone in the Dampier Basin, notabale for containing oolite (Crostella and Barter 1980), and an oolitic ironstone in the Evergreen Formation and equivalents of the Surat–Moreton Basin (Day *et al.* 1974). (e) The Middle Jurassic (Bajocian) Newmarracarra Limestone, only 10 m thick, of the Perth Basin (Playford *et al.* 1975) and coeval limestone in the Scott Reef area (Allen *et al.* 1978). And (f) the thin (<25 m) but very extensive coquinite in the Albian Toolebuc Formation of the Eromanga and Carpentaria Basins (Exon and Senior 1976).

Of these occurrences, only the Berriedale Limestone has been studied in detail (Rao 1981a). From its contained dropstones, the Berriedale Limestone is inferred to have been deposited in very cold water; it is an accumulation of fragments of bryozoans, brachiopods, and univalves, rare echinoderms, foraminifera, ostracods, gastropods, and very rare corals. 'Blue green algae, red algae, richtofenid brachiopods, fusulinids, goniatites, and conodonts were not observed despite a long search', nor were 'oolites, grapestones and oncolites'. Rao (1981b) described a set of criteria for the recognition of cold-water limestone related to the original calcite mineralogy and subsequent diagenetic minerals.

What little is known of the other limestones deposited in the Innamincka Regime is consistent with their origin in cold water except the Late Permian and Triassic limestones of western and north-western Australia, and the Late Triassic limestone of New Guinea. The Late Permian limestone of Sahul Shoals–1 contains oolite as well as fossils with affinities with those of Timor (Campbell 1970), regarded as indicating warm water. The Late Triassic limestone of New Guinea contains conodonts, which so far have not been found in the Permian rocks of Australia (Nicoll 1976), but are known also in the limestone from the Early Triassic Locker Shale of the Carnarvon Basin and in the Kockatea Shale of the Perth Basin (McTavish 1975). These occurrences on the west and north indicate that for the Late Permian and Triassic the western and northern parts of Australia were warmer than elsewhere, as to be expected from their lower latitude (Figs. 10, 157) and hence the relative profusion of limestone.

(iv) Evaporites

A fourth distinctive feature of the Innamincka succession is its virtual lack of evaporites. Evaporites are common in the Phanerozoic up to the mid-Carboniferous, but are represented during the Innamincka Regime only by anhydrate patches and veins in a 2 m interval of a Sakmarian sandstone in the Browse Basin and by very thin (< 5 cm) deposits of gypsum in the Early Cretaceous of the Eromanga Basin (Wells 1980). The distinction on the basis of evaporites is sharp below the Innamincka succession but weak above it, because the only widespread (but very thin) evaporites in the Potoroo succession are those in modern playa lakes.

(d) Late Carboniferous jump of the magmatic arc and related basins

Magmatism of the Devonian/Carboniferous stage of the Uluru Regime in the Werrie–Yarrol Trough and to the west (Bathurst Granite (Facer 1979), Auburn and

Urannah Granites (Webb and McDougall 1968)) continued into the Stephanian, and, after a transition marked by the Bulganunna Volcanics (297 Ma) and coeval granites (296 Ma) (Webb and McDougall 1968), jumped to an arc that remained in much the same place through the Permian and Early and Middle Triassic. The oldest dates for the Innamincka arc are 289 Ma (Hillgrove Suite) and 286 Ma (Bundarra Suite) in New England (Shaw and Flood 1981), and they coincide with the inception of the Bowen–Sydney foreland basin, marked by voluminous mafic volcanics (Camboon Andesite and Werrie Basalt) that were erupted about the Stephanian/Sakmarian boundary. These magmatic events are shown by a broken line in Fig. 154; the more fully determined part of the curve from 260 to 225 Ma ago (full line) is from Fig. 165. Later jumps of the arc south of a syntaxis at 20° S, in the Late Triassic (Fig. 173), and eastward in the Early Jurassic, separated the Innamincka Regime into early, middle, and late stages.

(e) Early Innamincka stage

In this account, a stratigraphical review of the Late Carboniferous through Middle Triassic of the continent is followed by a closer examination of the mid-Permian to mid-Triassic Bowen–Sydney Basin as a foreland basin in the style of the modern Papuan Basin, and its connection with the epicratonic basins.

(i) Review

Five depositional provinces and three highland areas on the platform are distinguished at the beginning of the Innamincka Regime (Fig. 155). (In what follows, numerals and letters refer to locations given in Fig. 155.)

1. The infrarift Perth and Carnarvon Basins on the western flank of a glaciated Great Western Plateau (I); the non-marine Collie Basin and southern Perth Basin in the south were possibly also on the northern flank of a glaciated highland in Antarctica.
2. The marine gulfs of the Bonaparte Gulf and Canning Basins, with a non-marine extension into the Officer Basin, all on the western side of high, glaciated ground in central Australia (II).
3. On the southern flank of II, the non-marine Pedirka Basin and the western part of the Arckaringa Basin.
4. The low ground of the ancestral Australian–Antarctic Depression, first covered by the distal part of a large ice-sheet centred in Victoria Land (Crowell and Frakes 1975; Wopfner 1981), then covered by a shallow sea broken into two lobes by the ancestral South Australian Highlands (III).
5. (a) An eastern magmatic arc, stretching from Cape York in the north past the eastern side of (b) the Bowen–Sydney foreland basin, and possibly to New Zealand; (c) a foreswell separating the Bowen–Sydney Basin from (d) the epicratonic Galilee and Cooper Basins, Murray infrabasins, and Tasmania Basin.

Except for continuous deposition from the Early Carboniferous in subsiding axes in the Bonaparte Basin, north Canning Basin, and Werrie Trough (Fig. 154), the basement to the Innamincka basins was an erosional land-surface carved out of the ground uplifted during the Kanimblan and Alice Springs Orogenies; the high ground of the ancestral Great Western Plateau (I and II) was covered by an ice-sheet, and the Kanimblan (5c) and New England highlands (5a) by glaciers; a large ice-sheet, probably centred in Victoria Land, flowed north into the ancestral Australian–Antarctic Depression, and a further ice-sheet lay south of southwest Australia. We regard these glaciated areas as remnants of a continent-wide ice sheet (hexagon in Fig. 154) that came into being at the climax of uplift in the mid-Carboniferous and arrested deposition, as reflected in the continent-wide lacuna. With the initial retreat of ice in the Stephanian, newly formed depressions in the surface, including rejuvenated former basins, locally with high relief, started filling with glacigene sediment, much of it of very coarse calibre.

Towards the end of the Stephanian, the ice retreated farther and the sea transgressed (transgression L – Fig. 154) the low-lying ground between the highlands, including the ancestral Australian–Antarctic Depression. In the east, this marine transgression coincided with the eastward jump of the magmatic arc, and with concomitant initial subsidence of the Bowen–Sydney foreland basin, which started filling with volcanics and volcanogenic sediment derived from the arc; in the north, volcanogenic sediment spread across the foreswell into the epicratonic Galilee Basin (Fig. 156) but in the south was confined to the Sydney Basin. The sea penetrated only a short distance into the Sydney and Bowen Basins, and coal measures were deposited on their cratonic flank and in the Galilee and Cooper Basins. In the Tasmania and Arckaringa Basins, coal measures in the late Sakmarian succeeded the marine deposits of the ancestral Australian–Antarctic Depression during a general regression (L–M) that extended into the earliest Artinskian, as indicated by the Irwin River Coal Measures of the Perth Basin. The sea soon returned to the Tasmania and Perth Basins as part of a general transgression (M), but not to the mainland part

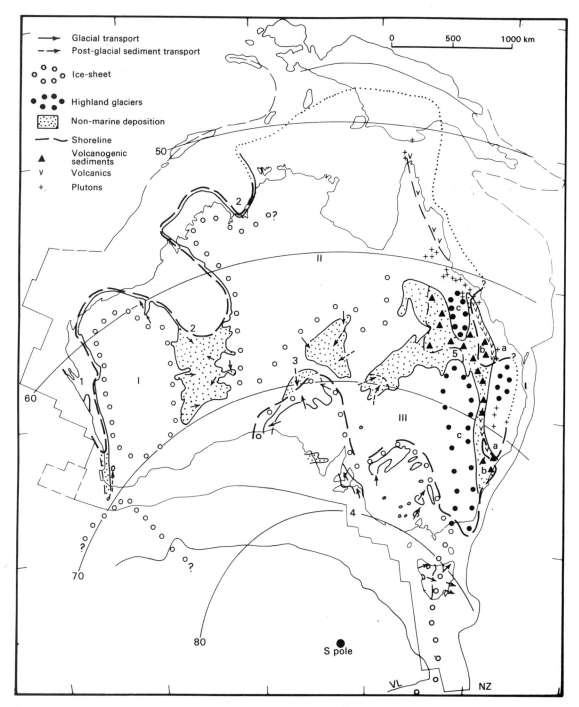

Fig.155. Palaeogeography about the Late Carboniferous/Permian boundary (300 to 280 Ma ago—palynological stages 2 and 3). Depositional provinces (1 to 5) modified from Veevers and Evans (1975, fig. 40.3), and platform uplands (I to III) described in text. Sediment-transport directions from Wopfner (1981), Crowell and Frakes (1971a, b), Jackson and van de Graaff (1981), and Wilde and Walker (1978). Distribution of ice modified from Wopfner (1981), and Crowell and

of the Australian–Antarctic Depression, which remained above the sea until the mid-Cretaceous.

The general transgression in the Artinskian (M) saw the deposition of the Berriedale Limestone in Tasmania and the limestone in the Caryginia Formation of the Perth Basin. In eastern Australia, Evans and Roberts (1980) record a short regression within this transgression, in which the orogen was uplifted along the western frontal thrust of the Hunter–Mooki Fault to shed the first pulse of volcanogenic (Greta) coal measures into the foreland basin; coal measures in the epicratonic Murray infrabasins began to be deposited at this time, and continued to be deposited in the Cooper Basin. In the Galilee Basin, a flood of quartz sand from the craton reached its peak (Fig. 156). This was the first of several cycles of transgression/regression/lacuna that punctuated the history of the Innamincka Regime, as described in detail by Jones et al. in the next section. Uplift and magmatic activity in the orogen were balanced by subsidence in the adjoining basin, which accumulated mainly volcanogenic sediment.

A general regression (M–N) is indicated by the regressive quartz sand of the Nowra Sandstone in the Sydney Basin (54) (Fig. 154), the coal measures of the Condren Member of the Liveringa Formation of the Canning Basin (9), the coal measures above the Fossil Head Formation (2) of the Bonaparte Basin, and probably by the non-marine exposed top of the Kennedy Group of the Carnarvon Basin (17) (McWhae et al. 1958). The regression was short-lived. In the Tatarian, the sea (N) re-entered the Bonaparte and Canning Basins, crossed the Sydney Basin by onlap of the Nowra Sandstone, and crossed the Springsure Shelf of the Bowen Basin (Fig. 156) to reach the eastern edge of the Galilee Basin, accompanied there and in the Sydney Basin by a depositional regression from the east by the progradation of the first of the three Late Permian cycles of volcanogenic coal measures. The third cycle (N–O) drove the sea out of the Bowen and Sydney Basins, and the same regression is seen in Tasmania (6) and in the Canning Basin at a lacuna between the Liveringa Formation (9) and the Blina Shale (10) (Balme 1969). The following Scythian transgression (O) (Fig. 157) is represented in the west by marine shales: the Blina (10), Locker (18), and Kockatea (27). In eastern Australia, the evidence suggests a much shorter transgression: acritarchs and calcareous foraminifera at the base of the Narrabeen Group (Herbert 1980a), in the Sydney Basin, and acritarchs of the same kind as those in the Kockatea Shale in Jericho-1 of the Galilee Basin (Evans 1980), located on Fig. 157 but not shown as a shoreline.

Comparison of Figs. 155 and 157 shows that in the west and east the Early Triassic shoreline lay in much the same position near the edge of the platform as it had in the latest Carboniferous/earliest Permian, though the gulf in the Canning Basin was narrower. In the south, however, there was no recurrence of the Stephanian/Sakmarian transgression. Moreover, by the Early Triassic, the only basin in central and southern Australia still receiving sediment was the Tasmania Basin. In compensation, the Early Triassic fill of the waning Bowen–Sydney foreland basin is very thick (up to 3.5 km), and labile sediment extended to the western edge of the Galilee Basin. Direct evidence suggests that the magmatic arc was restricted to the central-east. By this time, eastern Australia had migrated to the South Pole, but this did not involve glaciation, as glacigene sediments or any other sign of a frigid climate are lacking; indeed the widespread redbeds of this age would suggest the contrary.

A regression (O/p) is represented by the replacement of marine by non-marine sediments in the Bonaparte and Canning Basins in the north-west, and by the deposition of the regressive quartz sandstone of the Hawkesbury and Clematis Groups on the east. The final transgression (p) of the early Innamincka Regime is indicated by the estuarine base of the Anisian/Ladinian Wianamatta Group (60) (Herbert 1980a) and of equivalent strata in the Bowen/Surat Basin (Wiltshire 1982a). A non-marine equivalent, the Jerilderie Formation (44), terminated deposition in the Murray infrabasins. East of the Bowen Basin, the Abercorn and Esk Troughs filled with 5 km of volcanics (Day et al. 1974). The final regression of the early Innamincka Regime was registered by the replacement of the marine Locker Shale of the Carnarvon Basin by the non-marine Mungaroo Formation, and continued non-marine deposition in the Bonaparte Gulf, Canning, and Tasmania Basins. Except in these basins, the Ladinian (235 to 225 Ma) is a lacuna.

The terminal deformation of the Bowen Basin took place at much the same time, within the 30 Ma lacuna after the deposition of the Anisian (235 Ma) Moolayember Formation and before the deposition of the overlying Precipice Sandstone of the north-east Surat

---

Frakes (1971a, b). Ice was restricted to the uplands except the southern tongue in the ancestral Australian–Antarctic Depression, as shown by the sea that followed. Broken lines in the Galilee and Cooper Basins show the Late Carboniferous limits of deposition. Palaeolatitudes from Embleton (Chapter II). Igneous rocks from Richards and Willmott (1970), Richards (1980), Oversby et al. (1980), Webb and McDougall (1968), and Shaw and Flood (1981).

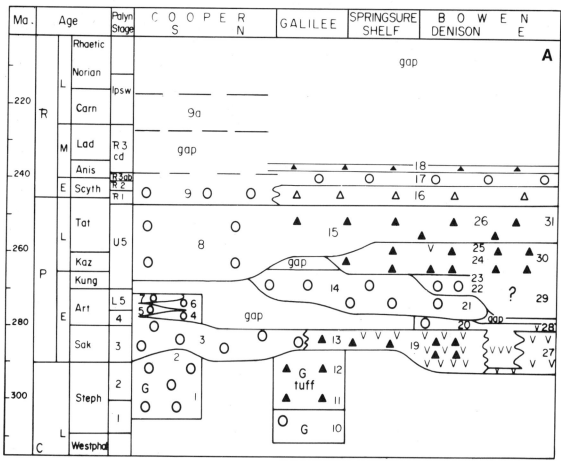

Fig. 156. A. Correlation diagram of Cooper Basin to Bowen Basin, showing compositional facies. Vs: volcanics. Solid triangles: labile sediment (< 50 per cent quartz), including tuff and coal. Open triangles: labile sediment without tuff or coal, and with redbeds. Open circles: quartzose sediment (> 50 per cent quartz). B. Environmental wetness, from fluvial (wide bars) through paludal (swamp, lake) and paralic (coastal), to shallow marine (close bars). Data from Thornton (1979, fig. 25), Battersby (1976), Papalia (1969), Martin and Hamilton (1981), Evans (1980), Paten et al. (1979). Correlation of European and Russian stages with Australian palynological stages from Moore (1982). G glacigene.

Basin, whose base is Rhaetic (205 Ma) (McKellar 1982). A shorter lacuna, spanning 10 Ma of the Norian, separates the deformed sediment of the Carnian (225 to 215 Ma) Ipswich Basin from the overlying undeformed Moreton Basin (Day et al. 1974), whose base is Rhaetic (205 Ma) (McKellar 1982). By direct evidence, the age of deformation of the Early and Middle Triassic Abercorn and Esk Troughs is no narrower than that of the Bowen Basin, though the relative age of the fault movements suggests that the Esk Trough was folded during the Ladinian (Day et al. 1974). Day et al. (1974)

regard the tentative Ladinian age of deformation of the Esk Trough as general, affecting the Bowen Basin, whereas Evans and Roberts (1980) regard the Norian deformation of the Ipswich Basin as general. These divergent views cannot be resolved without new evidence. Regardless of the precise age of deformation, the Late Triassic was a time of change: 'Commencing in the Late Triassic (Norian), the tectonic regime changed completely: new structural and sedimentary patterns accompanied initiation of the Great Artesian Basin' (Evans and Roberts 1980). Accordingly, we

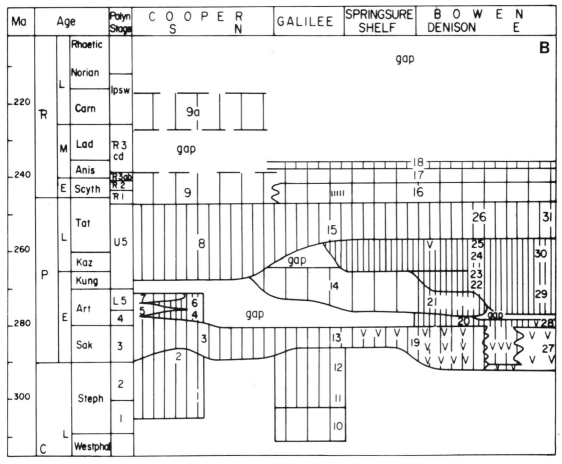

1, Merrimelia Formation; 2, Tirrawarra Sandstone; 3, Patchawarra Formation; 4, Murteree Shale; 5, Epsilon Formation; 6, Roseneath Shale; 7, Daralingie Beds; 8, Toolachee Formation; 9, Nappamerri Formation; 9a, Peera Peera Formation; 10, Lake Galilee Sandstone; 11, Jericho Formation; 12, Jochmus Formation; 13, Aramac Coal Measures; 14, Colinlea Sandstone; 15, Bandanna Formation equivalent; 16, Rewan Group; 17, Clematis Group; 18, Moolayember Formation; 19, Reids Dome Beds; 20, Cattle Creek Formation; 21, Aldebaran Sandstone; 22, Ingelara/Freitag Formation; 23, Catherine Sandstone; 24, Peawaddy Formation; 25, Black Alley Shale; 26, Bandanna Formation; 27, Camboon Andesite; 28, Buffel Formation; 29, Unnamed; 30, Boomer Formation; 31, Blackwater Group.

divide the Innamincka Regime at the Middle/Late Triassic into early and middle stages.

In the foregoing review of the early Innamincka, we have outlined the development of the rim orogen and related basins in the east. In the following section, the mid-Permian to mid-Triassic development of the Bowen–Sydney foreland basin is examined more closely.

(ii) Papuan Basin analogue and a foreland basin model for the Bowen–Sydney Basin (by J. G. Jones, P. J. Conaghan, K. L. McDonnell, R. H. Flood, and S. E. Shaw)

(1) Introduction

In its current tectonic aspect the northern margin of the Australian craton is a foreland basin yoked to the orogen of New Guinea (Figs. 53C, 158). In the Permian and Triassic, a precursor of the Papuan Foreland Basin ran along the Pacific margin of Gondwanaland in Australia (Bowen–Sydney Basin) and Antarctica (Nilsen–Mackay Basin) (Fig. 159). This section recon-

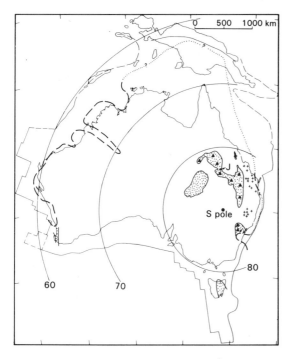

Fig. 157. Early Triassic (Scythian), 245 to 240 Ma ago. Symbols as in Fig. 155. Data from Laws and Kraus (1974), Gorter (1978), Crostella and Barter (1980), Balme (1969), Papalia (1969), Vine (1976b), Paten and McDonagh (1976), Day *et al.* (1974), Herbert (1980a), Evans and Roberts (1980), and Veevers and Evans (1975, fig. 40.6). Palaeolatitudes from Embleton (Chapter II). J: Jericho–1.

structs, by analogy with the modern Papuan Basin, the setting and development of the Bowen and Sydney Basins in the mid-Permian to mid-Triassic interval during the Early Innamincka Regime.

Carey (1938) outlined the character and affinities of the modern Papuan depositional domain on the Australian foreland adjoining the cordillera of New Guinea, likening the 'Papuan geosyncline' to the Indo-Gangetic plain. Voisey (1959), without reference to modern analogues, labelled the Permian Sydney Basin 'the Newcastle Exogeosyncline', a foredeep on the eastern margin of the end-Palaeozoic Australian craton, fed largely by sediments from a progressively

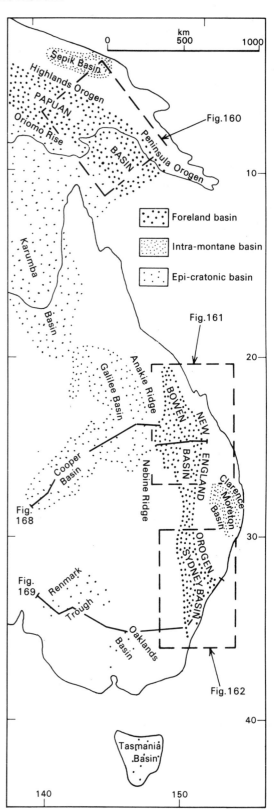

Fig. 158. Principal Permo-Triassic sedimentary basins of eastern Australia, and modern analogues in northern Australia and Papua New Guinea. Boxes indicate locations of Figs. 160, 161, and 162. Lines show location of sections of Figs. 168 and 169.

INNAMINCKA REGIME 245

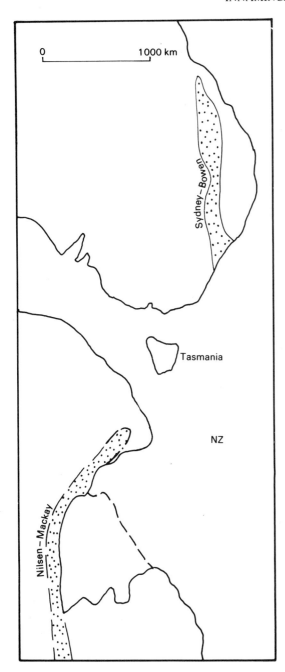

Fig. 159. Permo-Triassic foreland basin of Eastern Gondwanaland traced through the Bowen and Sydney Basins, then through a presumed inflection (not shown) east of the epicratonic Tasmania Basin to the Nilsen–Mackay Basin. Australian–Antarctic assembly from Chapter II.

deforming cordillera along its eastern edge. Jones and McDonnell (1981) linked the Papuan and Sydney Basins in the context of a focal analogy between the Papuan Peninsula segment of the New Guinea Orogen and the Tamworth Arc of the New England Orogen. In this analysis, our focus shifts to the Papuan Basin in and around the Gulf of Papua within the elbow at the junction of the Highlands and Peninsula sectors of the New Guinea Orogen, as an analogue of the Bowen and Sydney Basins in the Late Permian.

Integration of foreland basin sedimentation and orogenic magmatism is important to this study. The stratigraphic and radiometric time-scales we adopt are appended in Tables 10 and 11.

(2) Modern Papuan Basin

The Gulf of Papua is a shallow marine shelf adjoining the node between the Highlands and Peninsula Orogens (Fig. 160). To the north-west of the shelf, the sediment surface emerges as the Fly–Strickland Lowlands, a dissected piedmont alluvial plain less than 100 m a.s.l. over more than 100 000 km^2; to the south-east it submerges abruptly into the kilometres-deep Moresby Trough. These three topographic domains comprise the modern Papuan Basin in Papua New Guinea.

The Gulf of Papua is fed from streams around its Papuan perimeter, draining principally from the Highlands, but also from the Peninsula cordillera (Fig. 160). The largest of these is the Fly, which for 200 km below its junction with the Strickland River flows parallel to the Highlands on the northern flank of the mild swell of the Oriomo Plateau. A parallelism with the Peninsula cordillera is similarly manifest in the bathymetry of the Moresby Trough.

The New Guinea Highlands consist of two topographic domains: the Highlands proper and the foothills. The Highlands proper is a deeply dissected plateau, most of which is above 1500 m, with peaks commonly rising to over 3000 m (Dow 1977; Loffler 1977). The foothills are characterized by long strike-ridges of moderate relief, rarely more than 1500 m a.s.l. Rising from both plateau and foothills are a number of dormant or dead stratovolcanoes encircled by massive outwash aprons, the largest of which (Mt. Bosavi) dominates the Lowlands within the Fly–Strickland elbow from a vent located at the boundary of alluvial piedmont and foothills.

While the structure of the Papuan Basin in the depositional domains of Lowlands, Gulf, and Trough is characteristically a north-easterly homocline with dips of no more than a degree or two, the structure of the erosional domain of the foothills and Highlands proper is one of close-spaced faults and intervening

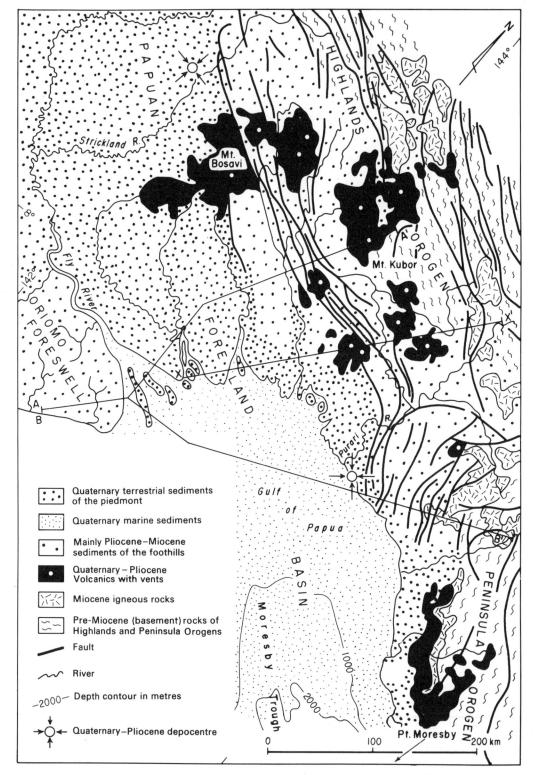

Fig. 160. Late Cainozoic geology of Papua New Guinea within the catchment of the Gulf of Papua. AA′ and BB′ are lines of cross-sections of Fig. 163; XX′ is line of time–space diagram of Fig. 164. Modified from D'Addario *et al.* (1976).

folds (Fig. 163A): open folds between north-easterly-dipping thrust faults in dominantly Cainozoic rocks of the foothills, grading to tighter folds between near-vertical strike-slip faults with associated ultramafics in the dominantly Mesozoic rocks of the Highlands proper (Dow 1977).

The Peninsula in the vicinity of the Gulf closely resembles the Highlands in morphology and structure (cf. Fig. 163A, B), though it bears only one Quaternary stratovolcano.

The boundary between highlands and lowlands, which in morphotectonic terms may be considered the boundary between orogen and craton, is by no means static. In the last 5 Ma (Pliocene–Quaternary) it has migrated cratonwards more than 100 km over much of its length, by the southwesterly tectonic transport of foreland basin sediments in folds and thrusts above surfaces of decollement (Jenkins 1974; Tallis 1975). This expansion of the orogen is concomitant with massive uplift in the same interval, which Jenkins (1974) estimated at as much as 10 km, with potent concurrent volcanism north-west of the Gulf.

The dramatic expansion of the orogen in the last 5 Ma is no less conspicuous in the sedimentary succession of the foreland basin. Thrust slices in the foothills expose Pliocene–Quaternary successions (APCP 1961; Dow 1977), recording the cratonwards advance of the piedmont in a sequence of shallow marine through paralic to paludal and fluvial sediments (Fig. 164), the same range of environments represented in and around the Gulf today. These sediments include mud, silt, sand, and gravel, substantially volcanogenic, with tuff and coal, grossly coarsening upwards. The thickness of these Pliocene–Quaternary sediments within the foothills ranges from a few hundred metres south-west of Mt. Kubor to 2-3 km along the Strickland River and beneath the alluvial embayments north and north-east of the Gulf (APCP 1961).

The Papuan Basin contiguous with the Highlands Orogen – the Papuan piedmont (essentially the Fly-Strickland Lowland) – is bounded to the south by the Oriomo Rise, a foreswell which separates the foreland basin from the epicratonic Karumba Basin (Fig. 158; Doutch 1976).

(3) Late Permian Bowen Basin

The Bowen Basin of Queensland (Fig. 158) is the remnant of a Permo-Triassic foreland basin. Figure 161 shows its present extent north of 27° S. In the context of a perceived analogy with the modern Papuan foreland basin, Figs. 160 and 161 are drawn at the same scale with common symbols denoting comparable elements.

The northern Bowen Basin (north of 24° S) expands southwards from a northern apex to a width of 200 km. To the east it is flanked by the northern extremity of the Permo-Triassic New England Orogen (Day et al. 1978). To the west it is flanked by the Anakie Ridge, a foreswell which separated the northern Bowen foreland basin from the Galilee epicratonic basin (Fig. 158).

In our view, the modern Papuan Basin adjoining the New Guinea Highlands is an illuminating analogue of the Late Permian northern Bowen Basin. Analogous morphotectonic elements are: onshore Papuan foreland basin/northern Bowen Basin; Highlands Orogen/northern New England Orogen; Oriomo Rise/Anakie Ridge. The folding which characterizes the eastern half of the northern Bowen Basin is largely Triassic (Staines and Koppe 1980), and we may envisage the basin at the end of the Permian as undeformed for at least 50 km east of the present axial zone of thrusts (Fig. 161). This gives the northern Bowen Basin at this time a width of about 200 km to the crest of the Anakie foreswell, which matches the width of the onshore Papuan Basin from frontal thrust to Oriomo foreswell. The Late Permian of the northern Bowen Basin is dominated by coal measures, predominantly lithic silts, sands, and gravels, largely volcanogenic, with a substantial component of coal and fine ashfall tuff (Dickins and Malone 1973; Jensen 1975; Koppe 1978; Staines and Koppe 1980). The same suite of sediments characterizes the Fly–Strickland Lowlands as they have developed through the Pliocene–Quaternary (APCP 1961). And as the Papuan piedmont has prograded southeasterly down the axis of the Papuan Basin, fed from the Highlands Orogen on its north-east flank, so a Late Permian piedmont prograded southeasterly down the axis of the northern Bowen Basin, fed from the northern New England Orogen along its north-east flank (Jensen 1975; Koppe 1978; Staines and Koppe 1980; Conaghan et al. 1982).

From the northern apex of the basin to about 21°30′ S, the Late Permian succession thickens from 1-2 km (Koppe 1978; comparable to Fig. 163A) to 4-5 km (Fig. 163C), with concomitant increase in coal rank (Shibaoka and Bennett 1976). Further south, deformation and poor exposure hinder measurement of thickness, but a continuing increase in coal rank to around 23° S and a decrease beyond (Shibaoka and Bennett 1976) suggest a depocentre here (Fig. 161). This depocentre has an analogue in the Gulf of Papua depocentre (Fig. 160), where as much as 10 km of sediment has accumulated in the comparable span of the Late Cainozoic (Fig. 163B), adjacent to a node between the Highlands and Peninsula Orogens (Fig.

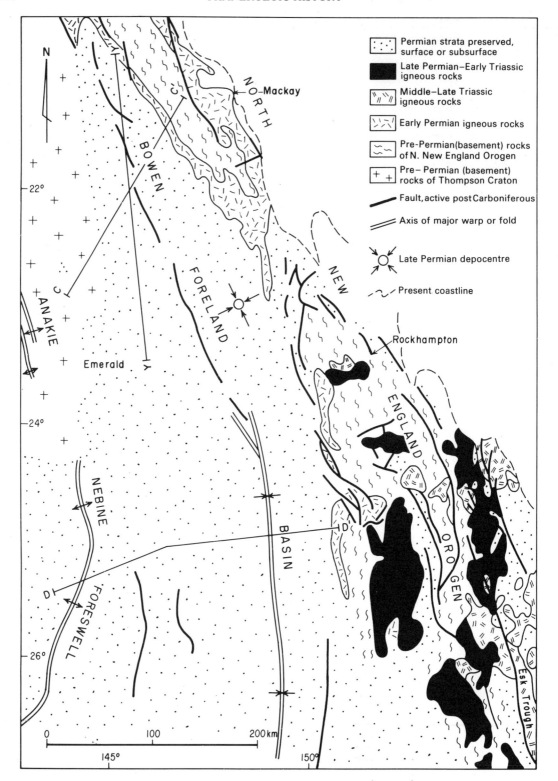

Fig. 161. Geology of the north-central Bowen Basin, focused on the Late Permian. CC' and DD' are lines of cross-sections of Fig. 163; YY' is line of time–space diagram of Fig. 165. Modified from Palfreyman *et al.* (1976).

160). The Bowen depocentre lies close to an analogous but more diffuse change in trend in the New England Orogen, most clearly reflected in a bend in the basin axis at about 24° S (Fig. 161).

In contrast to the central Bowen Basin, which is grossly a mildly asymmetric, west-facing syncline (Fig. 163D), the northern Bowen Basin is a synclinorium with a crumpled eastern limb (Fig. 163C), its axial zone of thrusts merging with the synclinal axis of the central Bowen Basin at the axial inflection at 24° S (Fig. 161). Deformation, involving steeply east-dipping thrusts and tight, west-facing folds (Malone et al. 1969), is most intense in the vicinity of the inferred depocentre. However, a seismic survey showed gentle dips below about 6 km, leading Malone et al. (1969) to propose a decollement at that depth, above which the basin fill moved westwards by folding and thrusting. The timing of this folding, beyond the western limits of orogenic igneous rock, is Middle or Late Triassic, not Late Permian (Staines and Koppe 1980).

(4) Late Permian Sydney Basin
The Sydney Basin of New South Wales is continuous with the Bowen Basin beneath an intervening cover of Jurassic–Early Cretaceous sediments. In subcrop the Bowen–Sydney Basin contracts to its narrowest across a basement saddle at about 30° S, which is taken to be the boundary between the Bowen Basin and the Sydney Basin *sensu lato* (i.e. including the Gunnedah Basin; Bembrick et al. 1980). Two sub-basins may be recognized within the latter: the Gunnedah Sub-basin north of about 32° S; and to the south what we will term the Newcastle Sub-basin (Sydney Basin of Bembrick et al. 1980). Fig. 162 depicts the Sydney Basin at the same scale as Figs. 160 and 161, with common symbols denoting comparable elements.

The Gunnedah Sub-basin (Gunnedah Basin of Bembrick et al. 1980; see Runnegar 1970; Weber et al. 1982) contains a Late Permian succession of the same facies as, and laterally continuous with, the coeval succession of the Bowen Basin (Runnegar 1970). The basin is bounded on the east, at its thickest, by the frontal (westernmost) thrust of a belt of north-westerly-striking, westerly-facing folds and east-dipping reverse faults forming the western flank of the New England Orogen (Fig. 162). To its western feather-edge on the basement rocks of the Lachlan Craton, the basin is generally a little over 100 km wide and its Late Permian succession is up to 1 km thick. Its preserved width approaches the minimum width of the onshore, presently aggrading Papuan Basin (cf. Figs. 160 and 162); and whereas the thickness of the Late Permian succession is much less than for the Late Cainozoic depocentre of the Papuan Basin, it is close to values (c. 1 km) along the relatively starved Mt. Bosavi transect (Fig. 163A). The erosional stripping of foreland basin strata which is currently active in the foothills thrust belt of the Highlands Orogen has proceeded almost to completion along the western perimeter of the southern New England Orogen (cf. Figs. 160 and 162).

Although the Sydney Basin appears to be truncated by the present coastline, there is evidence that the Newcastle Sub-basin, like the north-central Bowen Basin and, more conspicuously, the Gulf sector of the Papuan Basin, lies adjacent to a node between orogens or segments of orogens (cf. Figs. 160, 161, 162). The presently onshore southern New England Orogen trends grossly NW–SE within the frame of Figure 162. In our view, it terminates near Newcastle at a junction or node with the remnant of a SW–NE orogen underlying the present continental shelf, which we term the Currarong Orogen (after the Currarong Peninsula, 35° S, 150°50′ E; cf. 'Gerringong volcanic ridge' of Harrington 1982).

Perhaps the most direct evidence for the existence of the Currarong Orogen comes from the coastal cliffs near Newcastle, where Late Permian coal measures, indistinguishable from coeval rocks along the length of the Bowen–Sydney Basin, show dominant westerly palaeocurrents (Conaghan et al. 1982). These rocks include piedmont gravels denoting the proximity of the erosive domain. The most extensive indication lies in the grossly coast-parallel orientation of the mostly gentle warps which characterize the structure of the Newcastle Sub-basin for up to 100 km onshore between Wollongong and Newcastle (Bembrick et al. 1980, fig 1.2), and which meet the southernmost New England Orogen in a zone of sigmoid folds which Korsch and Harrington (1981) interpret as Permian 'interference structures'. Other clues include (1) warps concentric about a focus offshore of the Currarong Peninsula and intruded by dolerite no younger than Triassic (Bembrick and Holmes 1976); (2) small east-dipping thrusts intimately associated with soft-sediment folds in Early Permian strata on the coast close to the southern end of the Basin (Warden Head, 35°22′ S, 150°20′ E: Gostin and Herbert 1973); (3) a coast-parallel basement high beneath the continental shelf between Newcastle and Sydney (Fig. 162), with cover suspected to contain substantial volcanics (Palfreyman et al. 1976; Bembrick et al. 1980, fig. 1.1); and (4) a number of Late Permian–Early Triassic mafic igneous rocks south of Wollongong, confined to within 20 km of the coast (Facer and Carr 1979).

In transverse profile (Fig. 163E), the Newcastle Sub-

250 PHANEROZOIC HISTORY

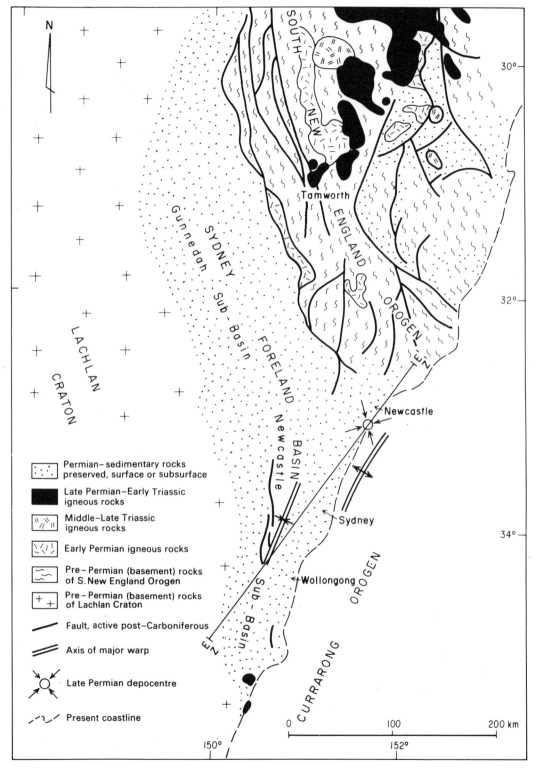

Fig. 162. Geology of the Sydney Basin, focused on the Late Permian. EZ-E″Z′ is line of cross-section of Fig. 163 and the time–space diagram of Fig. 166. Modified from Palfreyman et al. (1976).

basin is a prism, thickest in the north-east adjacent to the New England–Currarong node, and thinning stratigraphically to the south-west by an order of magnitude in a little over 200 km. The watershed of the Eastern Highlands, which runs just west of the feather-edge of the Sub-basin, may coincide with a Permo-Triassic foreswell, west of which probably lay a southern extension of the Galilee Basin, subsequently stripped to the small residual of the Oaklands Basin (Fig. 158). In its morphotectonic setting, the Permian Newcastle depocentre may be compared with the Pliocene–Quaternary (Fig. 160) and almost coincident Miocene depocentre (Aure Trough: Ridd 1976) of the Papuan Basin adjoining the Highlands–Peninsula node. There is, however, a large discrepancy in thickness per unit time: a late Cainozoic maximum of $c.$ 10 km for the Papuan depocentre versus a Late Permian maximum of little more than 2 km for the Newcastle depocentre.

Like the Late Permian Bowen Basin and the Pliocene–Quaternary of the onshore Papuan Basin, the Late Permian succession of the Sydney Basin is dominated by lithic silts, sands, and gravels, largely volcanogenic, with a substantial component of coal and fine ashfall tuff (Herbert and Helby 1980; Jones and McDonnell 1981; Conaghan *et al.* 1982; McDonnell *et al.*, in preparation). And as the Pliocene–Quaternary Papuan piedmont has prograded south-west more than 200 km to and beyond the crest of the Oriomo foreswell, driving the strandline of the Gulf to the south-east, so the Late Permian piedmont of the Newcastle Sub-basin prograded south-west more than 200 km (cf. Figs. 164 and 166), driving the strandline of an analogous Gulf to the south-east (Herbert 1980a; Jones and McDonnell 1981).

How is the spatial relationship of the Bowen and Sydney Basins to be seen in the light of the Papuan analogy? We suggest that the basement saddle which separates the basins at 30° S has a modern surface analogue in the watershed which separates the south-easterly-discharging Fly River system of Papua New Guinea (cf. Sydney Basin) from the westerly-discharging Digul River system of Irian Jaya (cf. Bowen Basin). That analogy involves a north-flowing trunk stream or streams in the southern Bowen Basin in the Late Permian, an implication supported by reconnaissance palaeocurrent data from the coal measures at the northern edge of the Jurassic cover (Jensen 1975; Conaghan *et al.* 1982). Counter-flow in the Gunnedah Sub-basin cannot be assessed for lack of palaeocurrent data, though it is established for the end-Permian Newcastle Sub-basin (Conaghan *et al.* 1982).

Perhaps the most conspicuous discrepancy between the Bowen–Sydney Basin and the Cainozoic Papuan Basin lies in the integral, though subordinate, component of craton-derived quartzose sediment in the former and its near absence in the latter. Through most of the Cainozoic, the trickle of quartz sand and silt from the Australian craton has been trapped within the epicratonic Karumba Basin (Smart *et al.* 1980), behind the sill of the Oriomo Rise (Fig. 158). This was not so throughout the Jurassic–Early Cretaceous, when voluminous cratonic quartz sand overtopped the foreswell to interleave with volcanogenic silts and sands from the orogen in an ancestral Papuan foreland basin (Fig. 70), yielding a stratigraphy still more closely comparable with that of the Bowen–Sydney Basin. In some measure, carbonate has taken the place of quartz in the Cainozoic Papuan Basin.

(5) Tectonics of the Papuan Basin
Certain relationships between erosion and sedimentation, uplift and subsidence, magmatic activity and inactivity, and deformation emerge clearly from analysis of the Pliocene–Quaternary or 'Cordilleran' phase of the late Cainozoic history of Papua New Guinea.

1. *Subsidence of the foreland is coeval with uplift of the orogen.* From around 10 to 6 Ma ago, the Papuan Basin was at sea level or slightly emergent, and a domain of nil deposition or mild erosion (Fig. 164). Around 5 to 6 Ma ago, the foreland basin began to subside, as recorded by a brief renewal of marine carbonate sedimentation. This was quickly suppressed by a flood from the rapidly rising orogen of lithic silt and sand, which accumulated in shallow marine conditions to a thickness of 3 km, denoting commensurate subsidence. The same pattern is seen at the late Oligocene onset of the largely Miocene 'Archipelagic' phase of the late Cainozoic history of Papua New Guinea. After an Oligocene interval of up to 10 Ma of non-deposition or mild subaerial erosion (Fig. 164), the foreland commenced to subside around 28 Ma ago, to accumulate a great thickness of Oligo-Miocene shallow marine carbonate (3 km) and coeval deeper marine lithic detritus (6 km) from an increasingly mountainous archipelago to the north-east.

The concurrence of subsidence of the foreland and uplift of the orogen in the Pliocene–Quaternary of Papua New Guinea has been stressed by Doutch (1976) and Lloyd (1978). The same concurrence of uplift of the highlands and subsidence of flanking basins, with the same timing, is seen in the late Cainozoic history of the Eastern Highlands epeirogen of mainland Australia,

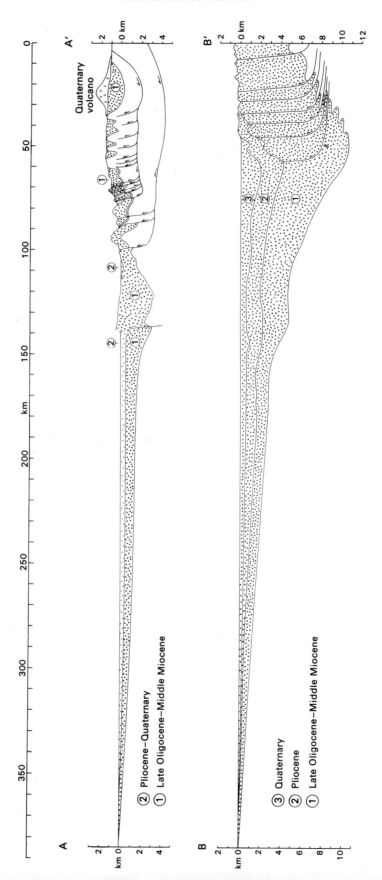

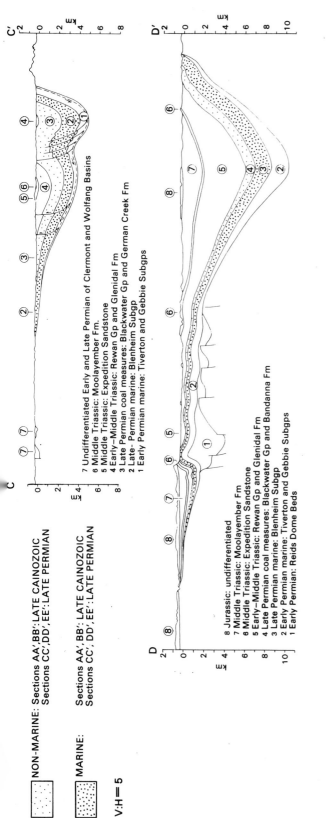

Fig. 163. Cross-sections of Papuan (AA′, BB′), Bowen (CC′, DD′), and Sydney (EE′) Basins. Lines of sections shown in Figs. 160 to 162. All sections at same scale. Note that fault attitudes and fold shapes are distorted by fivefold vertical exaggeration. Sources: AA′ modified from Jenkins (1974, fig. 8B) and Brown et al. (1979, fig. 4); BB′ modified from Brown et al. (1975, fig. 6), Brown et al. (1979, fig. 4), and incorporating data from Tallis (1975); CC′ based on 1:500 000 map of the Bowen Basin in Malone et al. (1967), augmented by data from Geological Society of Australia (1971), Cook and Taylor (1979), Dickins and Malone (1973), Goscombe and Koppe (1976), Jensen (1975), Koppe (1978), and Staines and Koppe (1980); DD′ modified from section HJK in Malone et al. (1967) on the basis of data in Brakel (1982b), Dear et al. (1981), Dixon and Bauer (1982), Exon (1976), Flood et al. (1981), Gray (1980), Gray and Heywood (1978), Herbert (1982), Jensen (1975), Paten et al. (1979), and Offshore Oil NL, pers. comm.; EE′ modified from Conaghan, in prep.

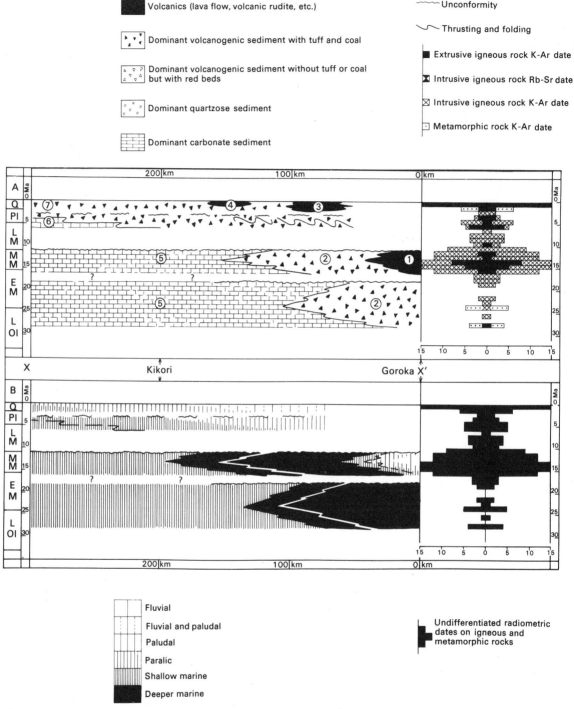

Fig. 164. Time–space transects of late Cainozoic Papuan Basin along line XX' in Fig. 160. A, rock composition; B, environmental wetness. Stratigraphic units are 1, Yaveufa Formation; 2, Aure and Movi Beds; 3, Crater Mountain and Karimui Volcanics; 4, Duau Volcanics; 5, Darai Limestone; 6, Orubadi and Era Beds; 7, Era Beds. Stratigraphic data from APCP (1961), Bain *et al.* (1975), Brown *et al.* (1975), Dow (1977), Findlay (1974), Jenkins (1974), Lloyd (1978), Ridd (1976), Tallis (1975). Histogram on right of A records K–Ar ages of single samples of late Cainozoic extrusive and intrusive igneous rocks and metamorphic rocks of the Highlands and Peninsula orogens, reflected on either side of zero. Histogram on right of B is a composite of ages from A, reflected on either side of zero. Ages expressed in new constants, and come from Davies and Smith (1971), Jenkins (1974), Loffler (1977), Page (1976), Page and Johnson (1974), Page and McDougall (1972), Ruxton and McDougall (1967), and Williams *et al.* (1972).

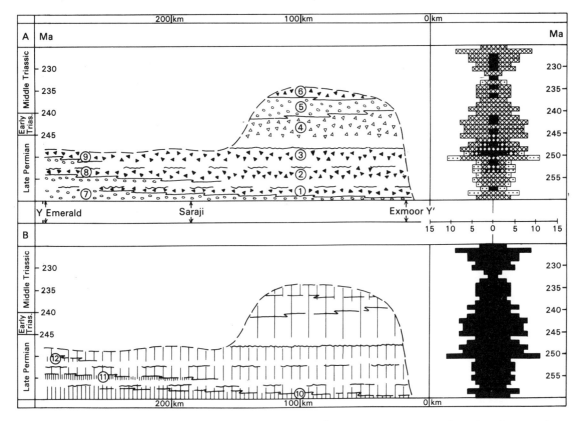

Fig. 165. Time–space transects of northern Bowen Basin along line YY' in Fig. 161. A, rock composition; B, environmental wetness. Stratigraphic units are 1, Moranbah Coal Measures; 2, Fort Cooper Coal Measures; 3, Rangal Coal Measures; 4, Rewan Group; 5, Clematis Group; 6, Moolayember Formation; 7, German Creek Formation; 8, Fair Hill Formation; 9, Rangal Coal Measures; 10, Exmoor Formation; 11, MacMillan Formation; 12, Burngrove Formation. Stratigraphic data from Brakel (1982a, b), Dickins (1982), Jensen (1975), Koppe (1978), Staines and Koppe (1980), and Waterhouse and Jell (1982). Histogram on right of A records K-Ar and Rb–Sr ages in the range 260 to 225 Ma ago of single samples of extrusive and intrusive igneous rocks and metamorphic rocks of the New England Orogen, reflected on either side of zero. Histogram on right of B is a composite of ages from A. Ages are expressed in the new constants, and come from Binns (1966), Cooper et al. (1963), Cranfield et al. (1976), Ellis and Whitaker (1976), Evernden and Richards (1962), Facer and Carr (1979), Green and Webb (1974), Kleeman (1982), Leitch and McDougall (1979), Murphy et al. (1976), Phillips (1968), Pogson and Hilyard (1981), Shaw and Flood (unpubl.), Webb and McDougall (1967b, 1968), Webb (1981), and Whitaker et al. (1974).

as described above (Figs. 57, 60: Grimes 1980; Jones and Veevers 1982).

2. *Waxing magmatism in the orogen is coeval with uplift of the orogen and subsidence of the foreland.* For the Cordilleran phase of Papua New Guinea, this is demonstrated both by the largely volcanogenic character of the foreland basin fill, including air-fall tuffs, and by the Pliocene–Quaternary cluster of magmatic ages from within the bounds of the orogen (Fig. 164), which Jenkins (1974) considers to have risen possibly as much as 10 km in this interval. This pattern is still more clearly evident in the more thoroughly dated middle Miocene pulse of the Archipelagic phase. A similar pattern is apparent in the late Cainozoic of the Eastern Highlands of mainland Australia (Figs. 57, 60).

3. *Waning magmatism in the orogen is coeval with gentle rise of the foreland.* The Papuan piedmont is presently in a state of mild uplift and shallow dissection, its surficial sediments being deeply weathered and oxidized (Blake 1971). Blake relates this dissection to glacio-eustatic regression between 27 and 17 ka ago,

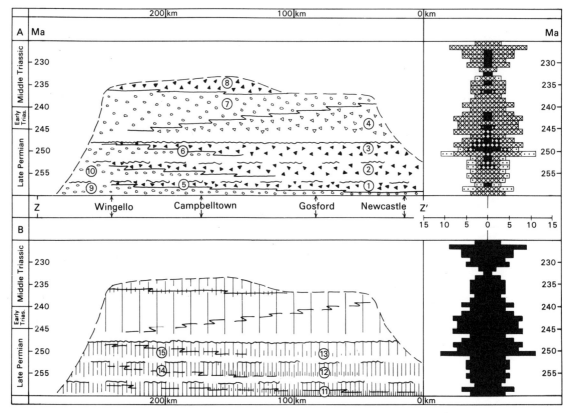

Fig. 166. Time-space transects of Sydney Basin along line ZZ' in Fig. 162. A, rock composition; B, environmental wetness. Stratigraphic units are 1, lower Tomago Coal Measures; 2, upper Tomago Coal Measures; 3, Newcastle Coal Measures; 4, Narrabeen Group excluding upper Gosford Formation; 5, Cumberland Sub-Group; 6, Sydney Sub-Group; 7, Hawkesbury Sandstone and upper Gosford Formation; 8, Wianamatta Group; 9, Nowra Sandstone; 10, Berry Formation; 11, Mulbring Formation; 12, Kulnura Marine Tongue; 13, Dempsey Formation; 14, upper Erins Vale Formation; 15, upper Wilton Formation. Stratigraphic data from Brakel (1982a), Conaghan et al. (1982), and Herbert and Helby (1980). Histograms and sources of radiometric age data as for Fig. 165.

but Loffler (1977) considers it much older. We suggest it coincides with a decline of volcanic activity in the foothills and highlands of central Papua New Guinea some time between 200 and 50 ka ago (Pain and Blong 1979), coeval with the uplift of beach ridges and the dissection of piedmonts contiguous with the Victorian Highlands of mainland Australia, as described in Chapter IV.

Both in Papua New Guinea and eastern Australia, the contemporaneity of a low level of volcanic activity in the highlands with general exposure and erosion of contiguous basins is most evident for the late Miocene (Figs. 57, 164).

4. *Folding and thrusting is coeval with volcanogenic sedimentation in the foreland basin and uplift and magmatism in the orogen.* During the Cordilleran phase, the New Guinea Orogen has greatly expanded southwards at the expense of the foreland basin, whose folded and thrust-faulted strata have been incorporated in the erosive domain of the foothills (Figs. 163A, B, 164). That this substantial deformation is coeval with volcanogenic sedimentation in the foreland basin is shown by the near or complete structural concordance between middle Miocene and latest Miocene–Pliocene strata in the fold belt, despite the gross angular unconformity of the latter with most of the younger, morphologically recognizable stratovolcanoes (Fig. 164: Jenkins 1974).

The time-pattern of cordilleran magmatism in the late Cainozoic of Papua New Guinea as it emerges from our compilation (Fig. 164) closely resembles the coeval pattern of the Eastern Australian Highlands (Fig. 57) and of the broader South-west Pacific region in which these former lie (Kennett et al. 1977, fig. 4). The late Miocene minimum evident in all three is obvious in the plot of New Guinea volcanic rocks, but is largely masked in the composite plot by dates from small, late-stage intrusions (Page 1976), which may not represent accession of new magma from depth. The paucity of dates in the interval 28 to 18 Ma ago in the Papua New Guinea plot may simply be a function of exposure, since igneous and metamorphic rocks of this age come from the northern foothills of the Highlands and isolated outcrops amid the alluvium of the Sepik-Ramu Plains (Page 1976).

In their review of circum-Pacific Cainozoic volcanism, Kennett et al. (1977) commented that 'Vail and his co-workers find that the episodes of strong volcanic activity we have observed in the Cainozoic record correspond to periods of world-wide marine transgressions, while periods of volcanic quiescence approximately coincide with times of low sealevel.' The cycles of transgression–regression linked to magmatic activity–passivity that we recognize in the late Cainozoic of Papua New Guinea correspond closely to Australasian and global patterns of transgression-regression recently summarized by Loutit and Kennett (1981, especially figs. 1 and 3).

(6) Tectonics of the Bowen–Sydney Basin

Consideration of the geology of the Papuan foreland basin and New Guinea Orogen led us to four conclusions about their tectonic relationships in the late Cainozoic. Those same relationships appear to hold for the Bowen–Sydney Basin and the linked New England Orogen.

1. *Subsidence of the foreland is coeval with uplift of the orogen*. From about 260 Ma to about 248 Ma, the Bowen–Sydney Basin accumulated as much as 5 km of coal measure sediments deposited in paludal to paralic environments (Figs. 163, 165, 166). Accumulation of these thicknesses close to sea level denotes commensurate subsidence of the foreland. That the coal measure sediments were derived largely from the orogen (Staines and Koppe 1980; Conaghan et al. 1982) confirms concurrent cordilleran uplift, for, although juvenile volcanogenic constituents may be dominant, unequivocal basement contributions are substantial (McDonnell et al., in preparation). The reality of basement uplift in the orogen is further confirmed by concurrent cratonward tectonic transport of basin fill in folds and thrusts (Staines and Koppe 1980; Cameron et al. 1982).

2. *Waxing magmatism in the orogen is coeval with subsidence of the foreland*. The subsidence of the foreland which accommodated several kilometres of coal measures in the interval 260 to 248 Ma ago was coeval with waxing magmatism in the orogen, expressed both in intrusive and extrusive activity (Figs. 165, 166). Dated extrusives in the orogen are few, though comparable in area to intrusives in New South Wales (Shaw et al. 1982b). However, the characteristic abundance of tuffs in the Late Permian coal measures of the Bowen–Sydney Basin (Koppe 1978; Jones and McDonnell 1981; McDonnell et al., in preparation) is unequivocal evidence of potent coeval volcanism in the orogen.

3. *Waning magmatism in the orogen is coeval with gentle rise of the foreland*. In the Bowen–Sydney Basin succession, the end of the Permian was a time of marked facies change from labile paludal-fluvial sediments, including coal and tuff, to labile fluvial sediments lacking coal and tuff, and including redbeds (Figs. 165, 166). We link this change to a lowering of the water-table (cf. Jensen 1975; Staines and Koppe 1980) concomitant on a gentle rise or recovery of the foreland, also expressed by a brief interlude of erosion (Uren 1980; Retallack 1980). This rise of the foreland occurred as magmatism in the orogen began to decline towards a minimum at around 232 Ma ago (Figs. 165, 166).

The disappearance of tuffs from the Bowen–Sydney Basin succession at around 248 to 246 Ma ago (Figs. 165, 166), even from floodplain and lacustrine facies, suggests an abrupt reduction of extrusive, and presumably intrusive, activity in the orogen, an abruptness not evident in the age histogram. We regard the apparent gradual decline of the next 10 Ma (Figs. 165, 166), depicted almost wholly by K–Ar dates, as recording the cooling rather than the emplacement history of the dated intrusions. This interpretation is consistent with the age distribution of igneous rocks in the southern Sydney Basin, where mafic lavas range in age from 258 to 249 Ma, whereas potentially comagmatic mafic intrusions range from 245 to 238 Ma (Facer and Carr 1979). And it is supported by the Late Permian clustering of Rb–Sr dates, which Webb (1981) regards as more likely than K–Ar dates to record the age of emplacement.

4. *Folding and thrusting are coeval with volcanogenic sedimentation in the foreland basin*. There is consensus on the contemporaneity of folding-cum-thrusting with coal measure sedimentation in the Bowen–Sydney Basin (Staines and Koppe 1980; Herbert 1980a), but pertinent data, especially stratigraphic constraints on the timing of thrust formation,

are sparse. Cameron *et al.* (1982) cite evidence on the latter, and Stuntz (1972) substantiates the concurrence of folding and coal measure sedimentation immediately south-west of the limiting thrusts of the Sydney Basin, while Dickins and Malone (1973) comment on the predominant Late Permian isotopic age of igneous rocks intruding the zone of most intense folding and thrusting in the Bowen Basin, and considered coeval with deformation.

(7) A foreland basin model

Foreland basin sedimentation has dominated the stratigraphic development of eastern Australia from the Permian to the Cretaceous, and is resurgent in the late Cainozoic of Papua New Guinea. The foreland basin model we present in this section arises primarily from our focus on the mid-Permian to mid-Triassic succession and context of the Bowen–Sydney Basin in the light of the Papuan Basin analogue (see also Jones and McDonnell 1981; Conaghan *et al.* 1982). However, a review of the Jurassic through Early Cretaceous (late Innamincka) stratigraphy of eastern Australia, including Papua New Guinea (Jones and Veevers, Chapter IV), has contributed to its formulation, and extended the range of its application.

In our model of a foreland basin fronting a volcanically active orogen (Fig. 167), the piedmont is characteristically a domain of labile sediment deposited by sediment-flows directed away from the orogen. Where sediment influx from the craton is substantial, these flows are opposed by flows of quartz sand. Sediment-flows down the basin axis may carry sediment of either provenance, or a mixture of both. Depending on location (for example, along the length of the Papuan

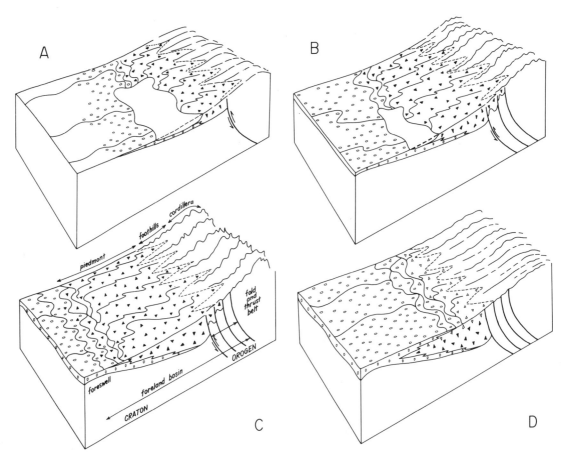

Fig. 167. Model of a foreland basin developing through a single cycle. Lettering on C identifies topographic and tectonic elements common to all block diagrams. Sediment symbols as for Figs. 164 to 166.

Basin) and on the phase of a tectonic cycle, the basin may be partially flooded or completely dry, the character of sediment-flows and sedimentary facies varying accordingly.

Fig. 167 portrays four instants in the course of a foreland basin cycle on a scale of about 5 to 25 Ma. Onset of uplift-cum-volcanism initiates growth of a mountain range and of piedmont alluvial fans, which begin to prograde into a concomitantly sagging and flooding foreland basin (Fig. 167A). With continuing uplift and outward expansion of the orogen, involving thrusting and folding of proximal outwash, the piedmont advances cratonwards into a contracting marine or lacustrine moat, pushing back the limits of craton-derived sediment (Fig. 167B). Ultimately marine and/or lacustrine conditions are eliminated from the foreland basin as the piedmont of a now mature cordillera spans most or all of the width of the basin to the limiting foreswell (Fig. 167C). With the decline of tectonism, the cordillera begins to settle, and the foreland basin rises gently, resulting in a falling water-table and an interval of non-deposition or mild erosion across the foreland while streams from the cordillera readjust to reduced gradients. Tuff and coal, which manifest the concurrent volcanism and high ground-water levels of the prograding piedmont, are no longer formed, and redbeds make their appearance as the piedmont begins to retreat. Without regeneration through uplift and volcanism, the orogen dwindles in height and extent through continued erosion and settling, and the piedmont shrinks, allowing a compensatory advance of cratonic quartz sands (Fig. 167D).

(8) Bowen–Sydney Basin: mid-Permian to mid-Triassic part of the early Innamincka Regime

Mid-Permian to mid-Triassic rocks constitute over 90 per cent of the exposure of the Bowen–Sydney Basin, and probably more than 75 per cent of the fill. Their stratigraphy is largely a record of a single gross cycle of advance and retreat of the piedmont of the New England Orogen across the proximate Australian craton, a piedmont which came into being about 260 Ma ago in response to tectonic resurgence of the orogen, and shrank to insignificance 20 Ma later.

The Late Permian is an interval, not of a simple, steady progradation of the piedmont, but of three pulses of advance, the last at least breaking across the foreswell into the Galilee Basin (Evans 1980). Each may have had a duration comparable to the Pliocene pulse of Papua New Guinea (Fig. 164), and likewise terminated in a brief interlude of foreland recovery expressed in mild uplift and erosion. Succeeding pulses opened with progressively more confined floods of the foreland basin by salt-, brackish-, and fresh-water, deepest and most persistent at nodal depocentres where subsidence and sedimentation were greatest. Recurrent prolonged rapid subsidence and sedimentation provided ideal conditions for the accumulation and preservation of peat in vast swamps and deep lakes (Conaghan 1982; Cameron et al. 1982), interleaved with mantles of ash from coeval volcanic eruptions in the cordillera.

At the end of the Permian, volcanism in the cordillera declined abruptly, though the cooling of magma emplaced in the subsurface in the Late Permian persisted for several million years into the Triassic. With the cessation of active magmatism, the orogen began to settle and the foreland rose gently, leading to mild dissection of the piedmont, and a lull in deposition while streams from the cordillera adjusted to reduced gradients. When deposition resumed across the piedmont, it was characteristically fluvial, in contrast to the paralic-paludal mode of the Late Permian. A concomitantly lower water-table was expressed in a change from deciduous swamp forests of *Glossopteris* to coniferous forests of *Voltziopsis* (Retallack 1980), an absence of significant accumulations of peat, and the widespread appearance of redbeds indicative of interludes of profound desiccation.

The Early Triassic retreat of the piedmont saw a concomitant advance of cratonic quartz sands, which, by the early Middle Triassic, blanketed the foreland basin and lapped the foothills of a degraded orogen. This retreat has been reconstructed in detail for the Sydney Basin succession where facies boundaries young from south-west to north-east through several microfloral zones (Conaghan et al. 1982).

The Late Permian advances of the piedmont were linked to expansion of the foothills by thrusting and folding of basin fill. By the end of the Permian, it seems that in the Sydney Basin the leading edge of the foothills fold and thrust belt had attained its ultimate limits: but not in the northern Bowen Basin, where the last substantial advance (beyond the present western limit of exposed Permo-Triassic igneous rocks) occurred no earlier than Middle Triassic.

The analogies drawn in earlier sections between the Late Permian Bowen–Sydney Basin and the late Cainozoic Papuan Basin are enhanced by consideration of the broader contexts. In the early Late Permian, marine sedimentation in central eastern Australia was not confined to the Bowen–Sydney Basin, but also occurred east of 152° E (Evans and Roberts 1980; Jones and McDonnell 1981) to confine the early Late Permian orogen to the dimensions of a mountainous archipelago or sea-girt cordillera analogous to the Highlands Orogen in the early Pliocene (Fig. 54: Dow 1977) or the

Peninsula Orogen today (cf. Jones and McDonnell 1981). In the process of piedmont progradation through the Late Permian, the sea was expelled beyond the limits of the present coastline (Evans and Roberts 1980), as the Pliocene–Quaternary outgrowth of the Highlands cordillera and piedmont has expelled the sea 300 km to the south and 100 km to the north (Fig. 54).

Analysis of the Early Permian stratigraphy of the Bowen–Sydney Basin is beyond the scope of this study. The mid-Permian to mid-Triassic succession on which this study is focused was attributed by Conaghan et al. (1982) to a cycle terminating in the Sydney Basin at the top of a basin-wide quartzose sheet sandstone (Fig. 166, Hawkesbury Sandstone; cf. Fig. 165, Expedition Sandstone). The top of the preceding quartzose sheet sandstone (Fig. 166, Nowra Sandstone; cf. Fig. 165, German Creek Formation) was nominated as the terminal unit of a preceding cycle. In the Sydney Basin, a still earlier cycle may close with the quartzose sheet sandstone of the Snapper Point Formation (Herbert and Helby 1980). The contrast between the predominantly marine Early Permian and predominantly paralic-to-terrestrial Late Permian of the Bowen–Sydney Basin succession recalls the contrast between the predominantly marine Miocene and conspicuously paralic-to-terrestrial Pliocene–Quaternary of the Papuan Basin, and may have the same significance: the progressive development of the orogen from an Early Permian archipelago to a Late Permian cordillera.

The Bowen–Sydney Basin succession terminates, not with the Early to early Middle Triassic quartzose sheet sandstone, but with mid-Triassic sediments of a facies most nearly equated with the Late Permian coal measures (Figs. 165, 166), but lacking workable thicknesses of coal. We regard this Middle Triassic labile facies as recording the prograde phase of a terminal cycle (Conaghan et al. 1982), linked to a glow of igneous activity in the orogen immediately preceding the Middle Triassic minimum at about 232 Ma ago (Figs. 165, 166). The applicability of the Papua New Guinea analogy to this mid-Triassic interval, as to the Late Permian, is confirmed by the character of a coeval succession within the orogen itself; this is a belt of folded subaerial volcanics and fluvial-paludal volcanogenic sediments along a major fault zone, with a preserved length of over 250 km, a width of several tens of kilometres, and a thickness of 5 km (Esk Trough: Fig. 154; Day et al. 1974; Irwin 1976). In its dimensions, lithofacies, stratigraphy, and structure, this intermontane basin closely resembles the Pliocene–Quaternary Ramu–Markham and Waria Valley fault troughs of Papua New Guinea (Grund 1976; Dow 1977), the former more closely in size, the latter in proximity to active volcanism.

Both the Bowen and Sydney Basin successions terminate in the mid-Triassic, between 235 and 230 Ma ago (Figs. 165, 166), at an apparent minimum in igneous activity in the orogen not matched at any other time in the interval 260 to 220 Ma ago. Outwash from the volcanoes which began to erupt at about 225 Ma in

TABLE 10. Dating of Late Permian and Triassic

	Time-scale	Green and Webb 1974	Armstrong 1978b[a]	Waterhouse 1978[b]	Webb 1981	Here
		—200—	—211—		—200—	—200—
L R	Rhaetic		220			
	Norian				215	215
	Carnian		229			
M R	Ladinian		234		225	225
	Anisian		238		235	235
E R	Scythian		242		240	240
		—240—	—245—	—230—	—245—	—245—
L P	Tatarian		252	256		260
	Kazanian					
			—258—	—263—		—265—

[a]Dates derived from Armstrong's (1978b) fig. 2.
[b]'High' range of values.

TABLE 11. *Eastern Australian Permo-Triassic palynostratigraphy*[a]

Age Ma	Time-scale[b]				Kemp et al. 1977	Foster 1979	Helby and Morgan 1979	Retallack 1980	Moore 1982	Here
230—	M R		Ladinian			Tr 3c–d			Tr 3c–d	
			Anisian				Dp / Ap = Tr 3b	Ap		Dp / Ap
240—	E R		Scythian	Spathian		Tr 3a–b	At = Tr 3a	At	Tr 3a–b	At
				Smithian						
				Dienerian		Tr 2	Ps = Tr 2	Ps	Tr 2	Ps
				Griesbachian	Lp	Tr 1b	Lp = Tr 1b	Lp	Tr 1b	Lp
			Tatarian				Pr = Tr 1a	Pr	Tr 1a	Pm
250—	L P	D		Do	Pr	Pm				
				Dv						
		J		Jb						
				Jv			D = upper 5	D = 5	upper 5	5
		P		Pc						
260—				Pk	D = 5					
		Z	Kazanian	Zs		5				
				Zk						

[a] Microfloral zones Tr1 to Tr3: Evans (1966); 5: Evans (1969). D *Dulhuntyispora*; Pm *Protohaploxypinus microcorpus* (= Pr *P. reticulatus*); Lp *Lunatisporites pellucidus*; Ps *Protohaploxypinus samoilovichii*: At *Aratrisporites tenuispinosus*; Ap *Aratrisporites parvispinosus*; Dp *Duplexisporites problematicus*. Dashed lines mark inferred/approximate zone boundaries.

[b] Permian time-scale includes new stage and sub-stage symbols of Waterhouse (1978).

and around the foundering Clarence–Moreton Basin (Fig. 154; Day *et al.* 1974), an intermontane basin of the character and scale of the Sepik Basin of Papua New Guinea (Grund 1976; Dow 1977), was either stripped by erosion in the Late Triassic, or more probably never lodged above the presently preserved fill of the Bowen–Sydney Basin.

(9) Appendix: Permo-Triassic time-scale and palynostratigraphy

The time-scale used in this study of the Bowen–Sydney Basin is presented in Table 10, in company with several precursors. The base of the Permian is taken to be 290 Ma, after Armstrong (1978b), close to the 'high' value of Waterhouse (1978). The placement of the Permo-Triassic boundary at 245 Ma follows Armstrong (1978b) and Webb (1981). The top of the Triassic is taken to be 220 Ma after Webb (1981).

Dates for Triassic stage boundaries follow Webb (1981). Dating of old stage boundaries within the Permian has been approached by correlating them with the new stage and sub-stage boundaries of Waterhouse (1978). Dates were assigned to these on the basis of Waterhouse's fig. 3 and Retallack's (1980)

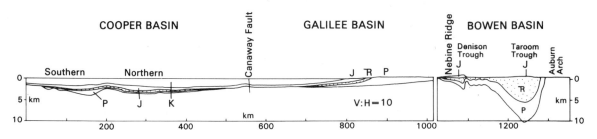

Fig. 168. Regional cross-section, located in Fig. 158, of Cooper, Galilee, and Bowen Basins. Bowen Basin from Fig. 163D, including Dickins and Malone (1973, pl. 1, H3–H4), Thomas *et al.* (1982, fig. 4, AB), and Paten *et al.* (1979, fig. 4); Galilee Basin from Evans (1980, fig. 4) and Vine (1976b), Cooper Basin from Battersby (1976, fig. 5, XX').

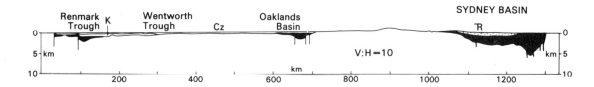

Fig. 169. Regional cross-section, located in Fig. 158, of Renmark Trough, Wentworth Trough, Oaklands Basin (all from Thornton 1976), and Sydney Basin (from Fig. 163E). Late Carboniferous and Permian shown in solid black, Triassic by stipple; Early Cretaceous (K) and Cainozoic (Cz) left clear. Triassic of Oaklands Basin, only 220 m thick, not shown.

fig. 21.2. Because the 'correlation channel' of Waterhouse (1978) is bounded by essentially straight lines, equal intervals of time (2.5 Ma) were assumed for substages from the base of the Permian (290 Ma) to the base of the Griesbachian (245 Ma). This procedure gives dates either identical with, or close to, those of Waterhouse (1978) and, in the case of the base and top of the Artinskian, those of Armstrong (1978b).

The positions we have adopted for microfloral zone boundaries in relation to stage boundaries are presented in Table 11, which summarizes some antecedent correlation profiles.

(iii) Epicratonic basins

West of the foreswell of the Nebine Ridge/Springsure Shelf/Anakie Ridge in the north are the epicratonic Galilee and Cooper Basins (Fig. 156), and west of the present South-east Highlands in the south are the epicratonic Murray infrabasins, including the Oaklands Basin and Renmark Trough (Fig. 158). As seen in regional cross-section (Figs. 168, 169), the early Innamincka epicratonic basins are broad downwarps that contain little more than 1 km of Permo-Triassic sediment that wedges out westward and eastward. In the south, the present divide of the South-east Highlands marks the general location of an early Innamincka foreswell, as shown by the westward thinning of formations in the Sydney Basin (Fig. 163E).

The modern, New Guinea analogues of the early Innamincka morphotectonic elements, as mapped in Fig. 158, are

Morphotectonic element	Modern	Permo-Triassic
Intermontane basin	Sepik	Clarence–Moreton
Orogen	New Guinea	New England
Foreland Basin	Papuan	Bowen–Sydney
Foreswell	Oriomo	Anakie–Nebine, part of South-east Highlands
Epicratonic basin	Karumba	Galilee–Cooper Murray infrabasins Tasmania

INNAMINCKA REGIME 263

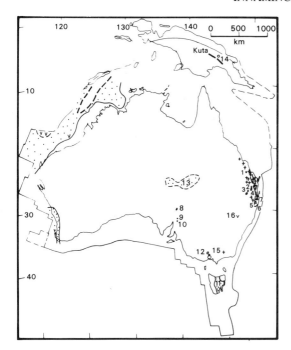

Fig. 170. Late Triassic (225 to 200 Ma), but leaving out the Rhaetic Precipice Sandstone and Bundamba Group. Data from Fig. 114 and Ludbrook (1978, fig. 4—note that Late Triassic sediment is missing in the Sydney Basin (Crawford et al. 1980) and to the north-west (Loughnan and Evans 1978)). Other sources given below. Palaeolatitudes from Embleton (Chapter II).

Coal measures, shown by solid black, are: 1 Callide, 2 Tarong, 3 Cecil Plains, 4 Ipswich, 5 Nymboida, 6 Redcliff, 7 Evans Head, 8 Leigh Creek (lower part), 9 Springfield, 10 Boolcunda, 11 Avoca, Fingal, New Town, Sandfly. Other deposits are: 12 Mudstone at Yandoit and sandstone at Bacchus Marsh (Douglas and Ferguson 1976). 13 Peera Peera Formation of the Eromanga Basin (Wiltshire 1982a, b) in the Poolawanna Trough, and above the Cooper Basin (Moore 1982). 14 Kuta Formation, with Late Triassic marine fossils (Skwarko et al. 1976). 15 Benambra Complex of granite porphyry and syenite, with an estimated age of 212 Ma (McDougall and Wellman 1976). 16 Garrawilla Volcanics and other basalts that range in age from 221 to 203 Ma (within the Late Triassic) and younger (Dulhunty 1972, 1976). Details of central-eastern coast in Fig. 171.

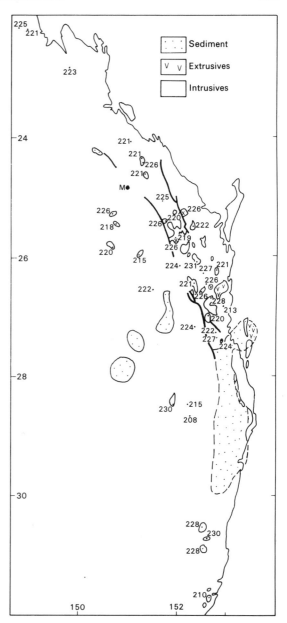

Fig. 171. Late Triassic (225 to 200 Ma) of central-eastern coast, including igneous rocks with ages in the range 230 to 200 Ma. Based on Day et al. (1974, fig. 4), with additional information from McDougall and Wellman (1976), Thomson (1975), Green and Webb (1974), Webb and McDougall (1967b), Murphy et al. (1976), Kirkegaard et al. (1970), and Leitch and McDougall (1979). M Monto.

(f) Middle Innaminka Stage: Late Triassic/earliest Jurassic (figs. 148, 154)

The change of stage of the Innamincka Regime at the Middle/Late Triassic (225 Ma) was marked by three related events: (1) the terminal deformation of the

Bowen–Sydney foreland basin; (2) the inception of the Great Artesian Basin; and (3) the eastward migration of the zone of magmatism south of a syntaxis at 20° S. We have already discussed the dating of the terminal deformation as taking place within the Ladinian-Carnian-Norian lacuna between the Bowen and Surat Basins, including the Norian lacuna between the Ipswich and Moreton Basins. The important morphotectonic change across the lacuna is the superposition of the distal Surat foreland basin on the proximal Bowen foreland basin; this is related to the eastward jump of the magmatic arc, which by the Late Triassic had come to lie from 50 km eastward of the earliest Permian position in the north to 250 km on the south (Fig. 173).

To the west, a middle and late Innamincka epicratonic basin (Eromanga) succeeded an early Innamincka one (Cooper/Galilee) after the regional lacuna of the Ladinian. Wiltshire (1982a) recognized the lacuna (or the top of the Lower to Middle Triassic unconformity) as marking the end of the 'Hunter–Bowen Orogeny'; Wiltshire (1982b) drew attention to the wide extent above the lacuna of Late Triassic sediment, shown as 13 on Fig. 170, called the Peera Peera Formation (Moore 1982), and argued for its inclusion in the Eromanga Basin on the grounds that (1) it is generally restricted to areas that underlie the main Jurassic sequence of the Eromanga Basin; (2) it is thickest in areas of thick overlying Early Jurassic sediment; and (3) it has greater affinity to the overlying Jurassic sequence than to the underlying older Triassic and Permian succession. Moore (1982) pointed to the possibility that the Peera Peera Formation continues without break from the underlying poorly dated Triassic Walkandi Formation. This would blur the distinction between the Eromanga Basin and the underlying basin, and suggest that the Late Triassic/earliest Jurassic middle Innamincka stage was an interval of transition.

The Peera Peera Formation is a sheet, no thicker than 190 m, of shale and siltstone, minor sandstone, carbonaceous shale, and rare coal. To the east, in the Late Triassic intermontane Ipswich and Clarence Basins, volcanic and volcanogenic coal measures are 1200 m thick, and correspond to an outburst of magmatism dated from 230 to 220 Ma (Fig. 171) along a diffuse zone of magmatism that had migrated eastward to the coast. The Ipswich and Clarence Basins (4-7 in Fig. 170) were deformed in the Norian, as discussed earlier, but the Callide Coal Measures (1) to the north-west were unaffected, and appear to continue upward without break into the Rhaetic–Early Jurassic Precipice Sandstone, which extends across the Surat and Ipswich/Clarence Jurassic foreland basins as the upper quartzose part of the single cycle in the middle Innamincka stage (Fig. 174). Occurrences of Rhaetic coal measures are known in South Australia in the lower part of the Leigh Creek Coal Measures (8), at Springfield (9), and Boolcunda (10), and in Tasmania (11), including the Brady Formation and 'Feldspathic Sandstone', in which Playford (1965) found palynological assemblages affiliated with those from the lower part of the Leigh Creek Coal Measures. These are the only precisely dated coal measures from the Triassic of Tasmania, which are therefore shown as young as Rhaetic on Fig. 154, but they probably started earlier, as suggested by Banks (1978), who correlated the 'Feldspathic Sandstone' at Mount Nicholas with the Carnian Ipswich Coal Measures. The Tasmanian coal measures lie above an earlier Triassic and Permian quartzose sequence, and are themselves dominated by lithic (volcanogenic) arenite and mudstone: 'All producing coal mines in Tasmania are in Triassic lithic arenites' (Hale 1962), but no volcanic rocks are known. The lower, quartzose, sandstone is interpreted by Kemp et al. (1981) as the deposit of an eastward-flowing low-sinuosity stream system, and the overlying volcaniclastic sandstone interbedded with carbonaceous shale and coal as the deposit of meandering streams with lakes and marshes. 'Abundant volcanic detritus of andesitic composition in sandstones suggests the presence of an additional source area, possibly volcanoes, along the Pacific margin of Gondwana', and direct evidence of an eastern or north-eastern source is provided by the south and south-west direction of sediment transport found by Kind (1980) at Fingal Colliery. These Late Triassic volcanogenic coal measures provide the only evidence in the Late Carboniferous through Triassic succession of Tasmania of a magmatic arc east of Tasmania, of the same general age as the exposed arc in central-eastern Australia. Much thicker Late Triassic sequences in New Zealand, with vitric tuff (Suggate 1978), are of more proximal character. The Benambra Complex (15 on Fig. 170, 212 Ma old) and the older parts of the Garrawilla Volcanics (16), 221 to 203 Ma old, are isolated occurrences of igneous rocks that lie well to the west of the presumed southern extension of the arc. On what were to become the western and north-western margins, the Late Triassic encompassed the change from infrarift to rift-valley basins, and, in the adjacent Fitzroy Trough, deformation by folding and faulting by right-lateral basement wrenching of the Fitzroy Graben (Rattigan 1967; Smith 1968a; Rixon 1978; Forman and Wales 1981) (Fig. 114A).

(g) Late Innamincka Stage

(i) Early and Middle Jurassic

In our study of the antecedents of the Eastern Highlands (Chapter IV), we traced the Jurassic/Early Cretaceous history of eastern Australia through several cycles of transgression/regression/lacuna, expressed, as in the early and middle Innamincka stages, by a pulse of volcanogenic sediment shed off the magmatic arc, followed by quartzose sediment shed off the craton. The start of the late Innamincka stage is marked by resumption of magmatic arc activity in the Early Jurassic, 190 Ma ago, in a position inferred, from the occurrence of volcanogenic sediment in the Maryborough Basin, to lie beneath the outer Barrier Reef in offshore Queensland, and northward in New Guinea where fragments of it are exposed. By about 160 Ma ago (Fig. 172), this arc shed volcanogenic sediment (Birkhead Formation) back across the craton as far west as the Birdsville Track Ridge (Moore 1982; Paton 1982). Coeval volcanic activity is registered by the youngest dated Tasmanian Dolerite, and basalts in Victoria, while on the north-west, the new oceanic lithosphere was emplaced by seafloor spreading along the newly formed north-western margin (Fig. 14), and basalt spilt into the adjacent Browse Basin and Ashmore Block. The Exmouth Plateau was cut into narrow blocks by north-trending faults, and the Perth Basin entered its final stage of rift-valley development with the very thick fill of the Yarragadee Formation. In southern Australia the first sign of the rift-valley system is Callovian sediment above basement in the Polda Trough and Jerboa-1. These events show that the Bathonian/Callovian (160 Ma ago) was a time of great geological activity: (1) on the west, breakup of the north-western margin, faulting of the Exmouth Plateau, rapid deposition in the Perth Basin, and the start of a marine transgression into the onshore Canning Basin; (2) on the south, the inception of deposition in the rift-valley system, folding and uplift in the Leigh Creek area, and the continuation of volcanism in Tasmania and Victoria; and in the east and north-east, the culmination of a phase of intense magmatic activity in the orogen and the wide westward dispersal of volcanogenic sediment. Except for the new development of the rift system along the southern margin, these events are accentuations of existing developments, as was the later breakup 128 Ma ago along the western margin and 95 Ma ago along the southern margin.

(ii) Late Jurassic (150 to 129 Ma)

By the Late Jurassic, we reach the modern or neotectonic stage of Australia's history; the north-eastern Indian Ocean and many features of the platform take their present form, so that increasing reference can be made to material given in the preceding chapters.

In the north-west, a marine transgression (s) (Fig. 148), seen in the Canning Basin at the distal part of the north-western margin that came into being 160 Ma ago, is the first expression of the eustatic rise in sea level that peaked in the Late Cretaceous (Fig. 135, curve A). Elsewhere, as in the Perth Basin, which received its final fill of thick sediment, the sea remained outside the platform. In eastern Australia, the platform accumulated quartzose sediment, the Algebuckina Sandstone (17 in Fig. 174) on the west, and the Gubberamunda (5) and Mooga (7) Sandstones on the east, split by a tongue of volcanogenic sediment (Orallo Formation, 6 in Fig. 174) that reached nearly half-way across the Eromanga Basin, derived from the source in the magmatic arc (Fig. 72C) that produced the andesite of the Grahams Creek Formation (Fig. 173; 13 in Fig. 174). In New Guinea (fig. 70), quart-

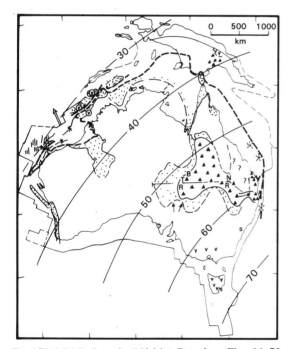

Fig. 172. Middle Jurassic (160 Ma). Data from Figs. 14, 70, 72A, and 114B, Forman and Wales (1981), Harrison (1969), Burns and Bein (1980), Doutch (1976), Paton (1982), Harris and Foster (1974), Bein and Taylor (1981), Ludbrook (1978). Line across eastern Australia indicates location of time-space diagram of Fig. 174. Palaeolatitude from Embleton (Chapter II).

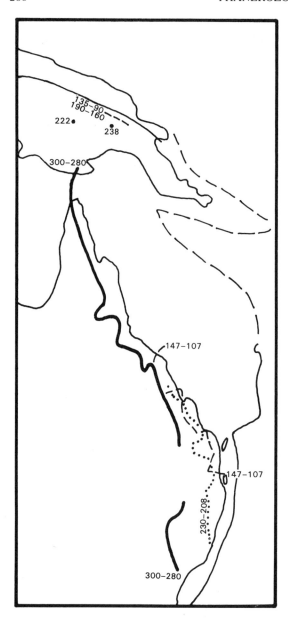

Fig. 173. Successive western limits of orogenic magmatism in eastern Australia, and southern limits in New Guinea during the Innamincka Regime. Late Carboniferous/Early Permian (300 to 280 Ma) limit (full line) from Fig. 155; Late Triassic (230 to 208 Ma) limit (dotted line) from Fig. 171, and dates of 222 and 238 Ma in New Guinea from Page (1976); Jurassic/Cretaceous (147 to 107 Ma) limit (broken line) from Fig. 66, and in New Guinea from Fig. 65B.

zose sediment covered the entire region until the resurgence of volcanism 135 Ma ago. In the south-east (Fig. 71), the oldest known volcanogenic sediment, shed from an inferred volcanic arc on the east (Fig. 72C), started filling the Strzelecki/Otway rift system, to be countered in the western Otway Basin by a tongue of quartzose sediment, the Pretty Hill Sandstone.

(iii) Early Cretaceous (129 to 95 Ma ago)

At the outset of the Cretaceous, the western margin came into being with its separation from India; at the same time, the Bunbury Basalt was erupted in the Perth Basin, and sediment prograded across the southern Exmouth Plateau from the continent–continent Cape Range Fracture Zone (Fig. 114C). The western margin was a rim basin during the Early Cretaceous, and the north-western margin passed from a rim basin to an open-marine basin half-way through the Early Cretaceous. On the southern margin, the main fill of the rift system in the Great Australian Bight and Otway–Strzelecki Basins was deposited in the Early Cretaceous. In the Otway–Strzelecki Basins, the fill was mainly volcanogenic, and, being within a rifted arch, was non-marine; only further west, in the saddle of the ancestral Australian–Antarctic Depression, did the sea enter the rift system, and this was not until the late Albian. On the platform, the first phase of the marine transgression (T) that peaked in the mid-Cretaceous is designated α (Morgan 1980), and is reflected by an expansion of sedimentation in the platform basins, including in the Eromanga Basin (Fig. 148) minor marine influence in the Hooray Sandstone (16 of Fig. 174), indicated by rare glauconie, some molluscs and foraminiferans, and very restricted assemblages of microplankton. Equivalents in the south-western part of the Eromanga Basin (Fig. 175) lack evidence of the sea. A regression in the Valanginian, between α and the next phase, β, is expressed by a lacuna, called the mid-Neocomian unconformity, by Morgan (1980) and Burns and Bein (1980), detected on the northern, north-western and western margins, as well as in the Great Artesian Basin. The overlying strata, called the Transition Beds (including the Bungil Formation, 8) in the Surat Basin and the eastern Eromanga Basin, and the Cadna–Owie Formation (18) to the west (Bowering 1982), represent marginal marine deposition during β. The Bungil Formation comprises juvenile volcanogenic material; the Cadna–Owie Formation is quartzose, though to be precise it contains fragments of volcanic rocks, but these come from the Proterozoic Gawler Ranges Volcanics (Fig. 75B), which stood high in the Early Cretaceous (see Fig. 175). By the definitive rise of sea

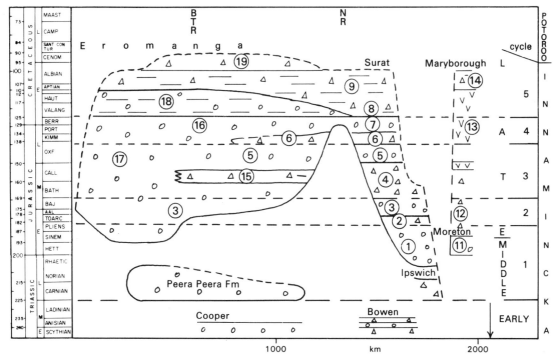

Fig. 174. Time–space diagram across the Eromanga/Cooper, Surat/Bowen, Moreton/Ipswich, and Maryborough Basins, showing the distribution of marine and non-marine environments, and volcanogenic and quartzose sediments. Location in Fig. 172. Cycles of quartzose/volcanogenic sediment in the Surat Basin indicated by numerals on right-hand side. Circled numerals are formation names, including 1-9 and 11-14, explained in Fig. 69, and: 15, Birkhead Formation; 16, Hooray Sandstone; 17, Algebuckina Sandstone; 18, Cadna–Owie Formation; 19, Winton Formation. Data from Fig. 69, Veevers et al. (1982,fig. 9b), and Moore (1982) and Wiltshire (1982a, b). BTR: Birdsville Track Ridge. NR: Nebine Ridge.

level in the Aptian (γ), volcanogenic material of the mainly shaley Rolling Downs Group had expanded across the entire Eromanga Basin to replace the westerly-derived quartzose sediment. It is this thick layer of shale that today forms the main confining bed of the quartz sandstone aquifers and volcanogenic aquitards of the Great Artesian Basin (Bowering 1982). The Eromanga and Surat Basins contain the bulk of the Jurassic-Early Cretaceous sediment deposited in Australia; much smaller amounts were deposited on the southern margin (Table 7; Fig. 147A).

The Aptian–Albian transgression (phases γ to ζ of Morgan 1980) was general (Figs. 124, 147B), and covered 52 per cent of the platform (Fig. 137) – only the ancestral Great Western Plateau, South Australian Highlands, and Eastern Highlands, and part of the southern margin remained above sea level. In the Innamincka depocentre, the rising eustatic sea-level was amplified by tectonic subsidence that reached a rate of 18 m/Ma (Fig. 144).

According to Exon and Senior (1976):

Preserved sediments fall largely into four groups: non-marine sand bodies with derived glauconie but no marine fossils; near-shore marine sand bodies with a rich molluscan benthonic fauna and abundant glauconie; thinly interbedded sands and silts with glauconie, a molluscan fauna and abundant bioturbation, deposited above wave-base; and laminated carbonaceous and pyritic silts and muds deposited offshore and largely below wave base. The latter are generally barren and non-calcareous, but contain widely separated calcareous beds with an abundant molluscan fauna. Calcareous beds are, in contrast, relatively common in the first three sediment groups. Benthonic organisms generally predominate, but pelagic organisms are abundant at some levels.

Faunal evidence (Day 1969) suggests that an easterly connection to the sea was open from late Neocomian to early Albian times, and a northerly connection from late Neocomian to late Albian times. Palaeotemperature and fossil evidence suggest a cool climate (Day, op. cit.).

A modern epicontinental sea in a cool climate, whose sedimentological features closely parallel those of the Cretaceous Eromanga and Surat Basins, is the Baltic Sea (Seibold et al.

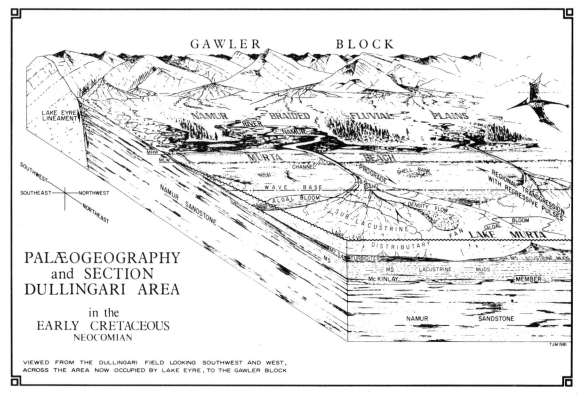

Fig. 175. Interpretation of the complex of non-marine depositional environments in the Innamincka region of the Eromanga Basin during the early part of the Early Cretaceous. From Mount (1982).

1971; Exon 1972). Within the Baltic Sea the salinity, the number of planktonic organisms, and the variety of benthonic organisms, decrease away from the ocean. Tides are small, currents are unimportant, and waves are active erosive agents. The shallowness of the sea and its entrances cause drastic changes in environment with only slight changes in sea level.

The Baltic Sea at present is a typical adjacent sea in a humid climate, with outflow of brackish surface water and inflow of saline bottom water (Seibold *et al.* 1971). The two water bodies are separated by a density barrier which approximates to wave base. Solution of carbonate and calcareous organisms is characteristic of sediments laid down below wave base, because saltwater inflow from the open sea is very limited, and the deeper waters are oxygen-poor and have a low pH (Exon 1972). In the past, when inflow was greater, the density barrier disappeared, the bottom waters were oxygenated and the pH was normal, there was no carbonate solution, and calcareous fossiliferous beds were laid down below wave base. Such fluctuations in inflow can explain the otherwise enigmatic occurrence of richly-fossiliferous beds within thick barren sequences in deeper water sediments of the Eromanga and Surat Basins.

The other major sediment groups in the Eromanga and Surat Basins also fit well with a Baltic Sea model: calcareous sand bodies containing some calcareous shelly remains near shore, and interbedded sand and silt developed offshore but above the wave base, are characteristic of the Baltic Sea. Glauconie is, however, absent.

In the Eromanga and Surat Basins, as in the Baltic Sea (Exon, 1972) the shallowness of the sea and its entrances would have meant drastic changes in environment with only slight changes in sea level. Changes in saltwater inflow through the seaways could have been related to changes in cross-section caused by eustatic sea level changes, isostatic movements or changing wind patterns. Even if the saltwater inflow did not vary, changes in the volume of the marine basin caused by sinking or infilling, or changes in freshwater inflow, could have tipped the balance of deepwater sedimentation from calcareous to non-calcareous muds or vice versa.

Another characteristic feature of the Aptian–Albian succession is radiolarite and radiolarian claystone, as, for example, in the Windalia Radiolarite of the Carnarvon Basin, in the Bejah Claystone of the southern Canning Basin and north-western Officer Basin, in the Mullaman Beds of the Darwin area, and in parts of the Great Artesian Basin succession. As noted by Brunnschweiler (1959), the varicoloured lightweight radiolarite and claystone, called porcellanite, range from almost pure kaolinite to almost pure silica. Little more than ghosts of Radiolaria are visible microscopically in thin section. Unlike the Radiolaria that accumulate today below the carbonate-compensation surface in the ocean, these accumulated on the floor of the shallow sea that covered the platform.

The interpretation of Innamincka depositional environments has developed in cycles. Many of the common non-marine parts of the Innamincka succession were originally interpreted as lake deposits (David and Browne 1950), but by the 1960s most of these had become reinterpreted as river deposits. With the wider range of conceptual models now available coupled with closer study of the outcrop or subsurface, authors have reinterpreted again some river deposits as lake deposits, in particular lake deltas in a complex of rivers, shorelines, and lakes. For example, Conaghan (1982) has put a case for regarding much of the Late Permian coal measures of the foreland basin of eastern Australia as lacustrine, with Gilbert-type deltas; Thornton (1979) interpreted the Early Permian coal measures of the Cooper Basin as deposited in an inland 'sea'; and Mount (1982) interpreted the Early Cretaceous succession in the Dullingari area, south of Innamincka, as deposited in a complex of fluvial plains on the edge of lakes (Fig. 175). In a sedimentological, but not tectonic or climatic, sense, the Central-Eastern Lowlands of Australia today (frontispiece), with long braided/meandering river systems of very low gradient ending in the sump of Lake Eyre, probably represent the same regional environment as existed during the late stage of the Innamincka Regime, as suggested by Brown et al. (1968) and Veevers and Evans (1975), qualified by Veevers and Rundle (1979).

(iv) Cenomanian Interregnum (95 to 90 Ma)

The transition to the Potoroo Regime during the Cenomanian Interregnum saw the withdrawal of the sea to the north, partly by continent-wide uplift (Fig. 135), and partly by depositional regression. Running counter to the general uplift, the Innamincka depocentre continued to subside tectonically, at an increased rate of 50 m/Ma, but the supply of volcanogenic sediment exceeded subsidence so that the same 1000 m thickness of sediment accumulated during the 5 Ma of the Cenomanian as had accumulated during the 30 Ma of the Early Cretaceous. This rapid accumulation rate does not necessarily entail greater production of volcanogenic sediment in the magmatic arc; we suggest that the arc changed at the interregnum from an archipelago, which trapped some sediment in moats, to a cordillera, from which sediment was shed down a continuous slope back across the craton. As measured by the distance travelled by the volcanogenic sediment from the arc, the terminal mid-Cretaceous spasm had the greatest effect, its products extending at least 400 km farther west than that of the Callovian Birkhead Formation, which in turn extended several hundred kilometres further than that of the Early Triassic, in turn further than that of the earliest Permian. The youngest known igneous rock from the orogen is the 94-Ma-old granite at Mt. Victor, in New Guinea; the youngest elsewhere is 109 Ma, along the Queensland coast. The freshness of the andesitic debris in the Winton Formation suggests that the material is juvenile, and came from sources either unidentified or largely offshore.

How much, if any, of the volcanogenic sediment reached the Ceduna Depocentre is unknown. In our account of the Potoroo Regime, we suggest that the northward axial slope of the Central-Eastern Lowlands was replaced in the Cenomanian by an extension of the present southwestward slope, and that the drainage may have terminated in the Ceduna Depocentre, which, like the Innamincka depocentre, accumulated about as much sediment in the Cenomanian as in the previous 30 or 40 Ma (Fig. 147A, C). A far more copious source of sediment, both in the Cenomanian and in the later Cretaceous, is the epeirogen of the South-east Highlands, which came into being 95 Ma with the splitting of the arc south of lat. 24° S (Fig. 147D); at least 1 to 2 km of cover were removed from the coastal strip of the epeirogen in the interval 100 to 70 Ma ago, and much of this must have found its way to the Ceduna Depocentre.

The terminal climax of the magmatic arc in northeastern Queensland from 95 to 90 Ma is the hold-over of the Innamincka Regime that defines the Cenomanian Interregnum. The death of the arc at 90 Ma, signified by the youngest deposits of the Winton Formation, brought to an end the magmatic arc system that had operated intermittently in eastern Australia since the early Palaeozoic. Elsewhere, the Innamincka Regime gave way to the Potoroo Regime 95 Ma ago with (a) the rearrangement of the spreading pattern in the eastern Indian Ocean (Figs. 18, 29) and the final detachment of India from its former connection along the Zenith–Wallaby transform fault, followed by widespread deposition of carbonate along the western margin from the resulting inception of deep oceanic circulation; and (b) the inception of seafloor spreading along the southern and south-eastern margins (Fig. 18).

4. ULURU AND ADELAIDEAN REGIMES

(a) Introduction

We postulate that the Phanerozoic history of Australia opened 575 Ma ago with continental breakup by plate divergence on the east and probably on the north-west, the end result of a three-stage process of extension that started in the preceding late Proterozoic or Adelaidean. The three stages are (1) from about 850 to 650 Ma ago, the Precambrian craton underwent extension concentrated along rift-valley complexes; (2) from 650 to 575 Ma ago, regional dextral shear accompanied by extension and compression led to marginal uplift; followed by (3) 575 Ma ago, continental breakup and plate divergence, which lasted until 540 Ma ago, when plate motion on the east reversed to convergence. To encompass the Phanerozoic history, it is necessary therefore to start in the Adelaidean.

Interpretation of the Adelaidean history is beset with difficulties. (a) Fossils, mainly stromatolites, provide only the grossest time-resolution. Even the Ediacara Fauna cannot resolve any interval smaller (beyond provisional early and late divisions) than its total estimated span of about 100 Ma (Cloud and Glaessner 1982); (b) a large area of Adelaidean rocks in north-eastern Australia is obscured by the overlying Georgina Basin; and (c) the effects of subsequent deformation in the Adelaide region during the Delamerian Orogeny are locally severe, and obscure the original geometry of the Adelaidean rocks.

Furthermore, whereas none of these difficulties applies to the overlying Cambrian succession west of the Tasman Line on the Precambrian craton, the history of the Cambrian of the fold belt to the east is unrecorded except for fragments along faults; and by subsequent plate motion these fragments may have come to lie considerable distances from their original location.

Our reconstruction of the Adelaidean and Cambrian is therefore provisional. It draws on four notions: (1) that the Adelaidean succession was deposited in a rift-valley setting (von der Borch 1980); (2) that shearing was important in determining structure in the late Adelaidean (a variant of Austin and Williams's 1978 view); (3) that the Tasman Line marks the suture between the Precambrian craton and the Phanerozoic fold belt to the east (Harrington 1974; Scheibner 1974a); and (4) that the present southern margin of Australia between Kangaroo Island and King Island marks the position of a past fracture zone (Crawford and Campbell 1973), which defines the pole of rotation during Early Cambrian plate divergence (Scheibner 1974a).

(b) Pre-Adelaidean: final consolidation of the Precambrian craton

By 850 Ma ago, taken to be the approximate lower limit of the Adelaidean (Cooper 1975; Black et al. 1980; Rutland et al. 1981) (Fig. 176), the Precambrian craton had taken its final form (Plumb 1979). From 1300 to 900 Ma ago (Fig. 177), intense tectonic activity, including folding, metamorphism, migmatization, and granite emplacement, had become restricted to narrow belts (Albany–Fraser, Northampton, Bangemall Basin, Musgrave, Arunta, Georgetown, Halls Creek–Fitzmaurice, Mount Painter, Houghton, western Tasmania) between or surrounding the major individual blocks. The Musgrave Block was subsequently cut by west-trending thrust faults intruded by layered mafic-ultramafic intrusions (Giles Complex) (Nesbitt et al. 1970) and granite.

(c) Adelaidean: rift-valley stage

The following Adelaidean Period saw deposition, part of it glacigene, in four main areas (Fig. 178A).

(1) Over part of the Amadeus Transverse Zone, a broad lens of shallow-water sediment, up to 5 km thick, exemplified by the succession in the Amadeus Basin (Wells et al. 1970; Preiss et al. 1978; Plumb et al. 1981) (Table 12); the diamictites are described by A. T. Wells (in Hambrey and Harland 1981);

(2) To the north-east, narrow belts of sediment, exemplified by the succession of the southern Georgina Basin (Plumb et al. 1981) (Table 13); tillites are described by M. R. Walter (in Hambrey and Harland 1981);

(3) To the south-east, the Adelaide Geosyncline (Rutland et al. 1981) or Rift (von der Borch 1980), exemplified by the succession in the Adelaide and Flinders Ranges regions. The summary given in Tables 14 and 15 cannot do justice to the complexity of the succession, details of which are given by Rutland et al. (1981). Tillites are described by R. P. Coats (in Hambrey and Harland 1981). In the northern Flinders Ranges, the Burra Group unconformably overlies the Callana Beds (Table 15);

(4) In north-west Australia, sheets of sediment, exemplified by the succession in the East Kimberley region (Plumb et al. 1981) (Table 16). Tillites are des-

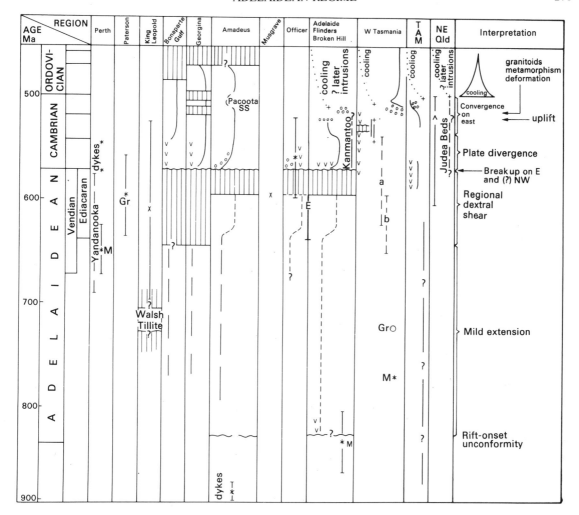

Fig. 176. Timetable of events from 900 to 450 Ma ago. Heavy lines indicate deposition, from non-marine on the left side of the column, through shallow marine in the middle to deep marine (Transantarctic Mountains, Kanmantoo Group, and ?Judea Beds only) on far right. E Ediacara Fauna, Gr granite, TAM Transantarctic Mountains. Symbols for radiometric dates in Fig. 187. In the column for western Tasmania, a=K–Ar total-rock ages of Tyennan schists and slates, and b=K–Ar slate ages in north-west Tasmania, both from Adams (1981).

cribed by K. A. Plumb (in Hambrey and Harland 1981).

Tasmania had a different history; older Adelaidean sediments, mainly interbedded siltstone and orthoquartzite and equivalent turbidite (Williams 1976), were metamorphosed about 783 Ma ago (Raheim and Compston 1977) during the Frenchman Orogeny (Spry 1962; Williams 1976; Adams 1981), and intruded by granite about 734 Ma (McDougall and Leggo 1965) and dolerite about 720 Ma (Williams 1976), and then overlain by sediment, some of which may be of glacial origin (J. B. Jago in Hambrey and Harland 1981).

South of Tasmania, along the line of the present Transantarctic Mountains, thick sediments, including turbidites (Robertson Bay Group of northern Victoria Land, with Riphean and Vendian acritarchs: Jago 1981) and the poorly dated Beardmore Group (Elliot 1975b; Stump 1981) further south probably lay on or near the continental margin of East Antarctica.

These sediments contrast with the Adelaidean sediments of Australia, which have a uniform platform

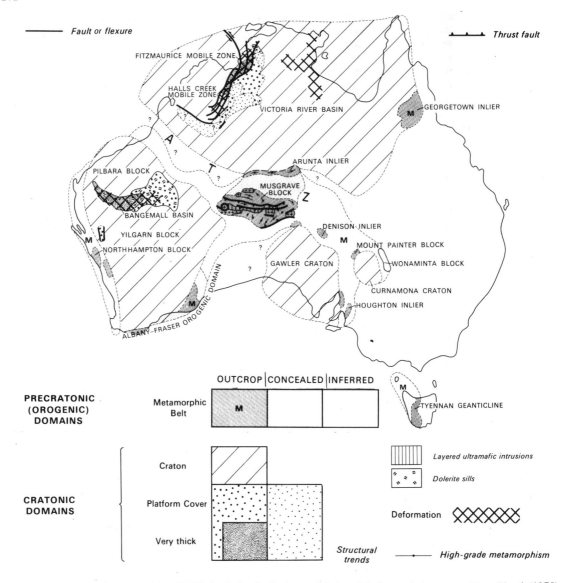

Fig. 177. Events in the interval 1300 to 900 Ma that led to the final consolidation of the Precambrian cratons. From Plumb (1979).

facies, except for local canyon-fill sediments in the Flinders Ranges (von der Borch et al. 1982). Brown et al. (1968, pp. 39–40) characterize them as 'apparently entirely devoid of greywacke-suite sediments . . . deposition took place under shallow marine or terrestrial conditions'. More specifically, Parkin (1969, p. 81) says that 'the tectonic setting of the region of the Adelaide Geosyncline is best considered as that of an area of gentle platform downwarp adjacent to rising basement areas . . . The facies changes across the Geosyncline show that the Adelaidean environment was lagoonal and shallow marine.' One of the last deposits in this sequence was the Pound Quartzite and presumed equivalents in the Officer and Amadeus Basins that contain the Ediacara fauna. Older Precambrian blocks (Tyennan–Rocky Cape, Willyama–Wonominta, Georgetown–Cape York) lay eastward. This evidence leads to the inescapable conclusion that

TABLE 12. *Adelaidean succession of the Amadeus Basin*

Unit	Thickness (m)	Description
UNCONFORMITY		
Arumbera Sandstone I		Feldspathic sandstone, Ediacara trace fossils
Julie Formation	500–2750	Dolomite, limestone, sandstone, siltstone
Pertatataka Formation		Siltstone, shale, feldspathic sandstone
Olympic Formation		Diamictite, sandstone, siltstone
UNCONFORMITY		
Aralka Formation	1800	Siltstone, shale, dolomite, calcarenite
Areyonga Formation		Diamictite, sandstone, conglomerate, dolomite
UNCONFORMITY		
Bitter Springs Formation	750–1500	Dolomite, stromatolitic limestone, chert (micro-fossils), gypsum
Heavitree Quartzite		Quartz sandstone, siltstone, conglomerate
NONCONFORMITY		
Arunta Inlier–Musgrave Block		

TABLE 13. *Adelaidean succession of the southern Georgina Basin*

Unit	Thickness (m)	Description
UNCONFORMITY		
Mopunga Group	300–1800	Arkose, shale, stromatolitic dolomite, glauconitic sandstone. More extensive than the underlying units
UNCONFORMITY		
Keepara Group	0–1100	Arkose, siltstone, shale, dolomite. Deposited only in fault troughs
UNCONFORMITY		
Yardida Tillite	0–3000	Diamictite, arkose, siltstone. Deposited only in fault troughs
Yackah Beds	0–250	Feldspathic sandstone, shale, stromatolitic dolomite. Deposited only in fault troughs
NONCONFORMITY		
Arunta Inlier		

TABLE 14. *Adelaidean succession of the Adelaide-Flinders Ranges region*

UNCONFORMITY		
Wilpena Group		
Pound Subgroup	3000	Quartzite. Contains Ediacara fauna
Wonoka Formation	460	Siltstone, silty limestone, 1.5-km-thick canyon fill (von der Borch et al. 1982)
Bunyeroo Formation	700	Shale
ABC Quartzite	1800	Quartzite
Brachina Formation	1200	Micaceous siltstone and sandstone, tuffs. The Wilpena Group is more uniform and persistent laterally than the underlying units
Umberatana Group		
Pepuarta Tillite	180	Diamictite
Elatina Formation	60	Feldspathic sandstone
Brighton Limestone	50	Oolitic limestone, micritic dolomite
Tapley Hill Formation	2000	Siltstone
Sturt Tillite	400	Diamictite
UNCONFORMITY		
Burra Group		
Belair Subgroup	3600	Feldspathic quartzite, siltstone, slate
Beaumont Dolomite	1200	Dolomite
Stonyfell Quartzite	200	Quartzite
Woolshed Flat Shale	300	Shale
Skillogalee Dolomite	4000	Dolomite, stromatolites (micro-fossils), arkose
Aldgate Sandstone	1200	Pebbly arkose, feldspathic sandstone
NONCONFORMITY		
pre-Adelaidean metamorphics		

in the Adelaidean the margin of the Australian continent lay east of these blocks, though at an unknown distance. In contrast, the eastern edge of the Antarctic platform, along the present Transantarctic Mountains, was the site of deposition of turbidite sequences of greywacke and shale, and so marked a position along or near the continental margin that faced the Pacific Ocean.

Rutland et al. (1981) draw an analogy between the evolution of the Adelaide Geosyncline and that of multiple rifted-arch systems, as interpreted from the Phanerozoic western margin of Australia (Veevers and

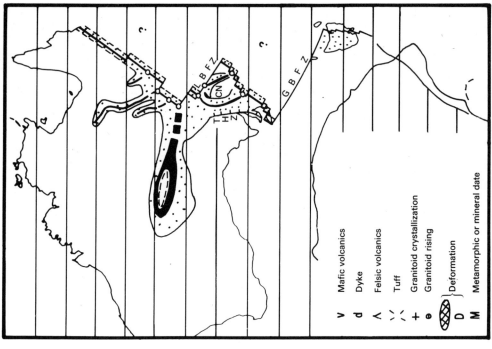

Fig. 178. A. Distribution of Adelaidean (900 to 650 Ma) rocks west of the Tasman Line (Plumb 1979; Rutland 1981; Tucker *et al.* 1979; Stump 1981) up to the epi-Adelaidean events. Symbols explained in Figs. 178, 182, and 183A, refer to the set of Figs. 178, 182–186, and 188.

Fig. 178. B. Interpretation of Adelaidean events up to the epi-Adelaidean events of 650 Ma ago. Horizontal lines (queried east of the Tasman Line): Precambrian terrain; stipple: areas of thick sediment in a broad sag over the Amadeus Transverse Zone and in rift-valley complexes to the north-east and south-east; solid black lines: axes of deposition; circles: triple rift junctions. CN: Curnamona Nucleus. GBFZ: Gambier–Beaconsfield fault zone. LBF Lake Blanche Fault. THZ Torrens Hinge Zone. Thin

TABLE 15. *Callana Beds*

Unit	Thickness (m)	Description
Wooltana Volcanics	2500	Amygdaloidal basalt, trachyte, andesite, rhyolite
Wywyana Formation	100	Calcitic and dolomitic marble
Paralana Quartzite	1100	Quartzite, minor siltstone, local talus breccia in fault-angle depressions
NONCONFORMITY		
pre-Adelaidean metamorphics		

Cotterill 1978; and Chapter IV). The Torrens Hinge Zone formed the western limit of the rifted-arch system, the Curnamona Nucleus the stable inter-arch area, and to the east an easterly rift system, as shown in Fig. 178B. Von der Borch (1980) interpreted Rutland et al.'s (1981) data in more detail, with the Adelaide Geosyncline or Rift as a rift-valley complex (Fig. 179) bounded below by a rift-onset unconformity and above by a breakup unconformity (Fig. 180), brought about by the sequence of events shown in Fig. 181. His interpretation of the Adelaide Rift follows.

The progressively evolving sedimentary facies and tectonic style of the Adelaide Rift [Fig. 179] are summarized in Figs. 3 [our Fig. 180] and 4 [our Fig. 181]. As can be seen, the evaporitic Callanna Beds at the base of the succession in the Flinders Ranges region and further north represent deposition during early rift valley stages in a tectonic, climatic and geochemical setting remarkably like that of the contemporary East African Rift Valley. The Burra Group, typified by arid playa and humid lacustrine as well as deltaic sediments, was deposited in a more continuous system of climatically oscillating lakes in growing rifts and possible extra-arch (Veevers and Cotterill, 1978) basins. The Umberatana Group, particularly at its base (Sturt Tillite and equivalents), marks a period of enhanced vertical faulting following initial rift valley formation, possibly due to accelerated uplift of hogbacks and downfaulting of adjacent basins with associated orogenic, glacigene and lacustine sedimentation. This tectonic setting is

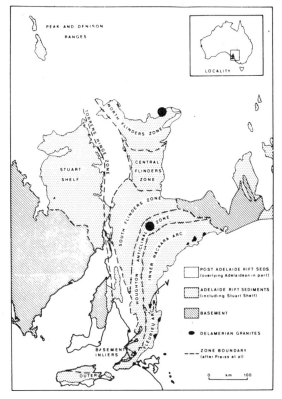

Fig. 179 (fig. 1 of von der Borch 1980). Generalized geology and structure of central southern Australia. Zone and arc nomenclature after Rutland et al. (1981); black circles refer to inferred triple junctions. Note basement inliers in the southern Houghton Anticlinal Zone which separate deep-water turbidite facies sediments of the Outer Fleurieu Arc from shelf facies sediments to the west.

TABLE 16. *Adelaidean succession in the East Kimberley region*

Unit	Thickness (m)	Description
Albert Edward Group	1450–2000+	Shale, siltstone, sandstone, dolomite
Duerdin Group	980–1450	Diamictite, sandstone, shale, siltstone

considered at least partly comparable to that of the Basin and Range province of western U.S.A. The Tapley Hill basinal siltstones reflect major subsidence and record the first significant transgression onto marginal areas of the rift (the Stuart Shelf). The Wilpena Group, with widespread deltaic sands and associated tidalites and remarkable regional uniformity, demonstrates in its basal units similar onlapping of Adelaidean sediments over the stable marginal craton. Widespread deltaic progradation which typifies this Group resembles the deltaic infilling of many younger rift systems as they evolved to maturity. The top of the Wilpena Group also contains the first tangible evidence of a marine incursion (Ediacara Fauna). This paralic fauna occurs close to the unconformable Precambrian–Cambrian boundary which everywhere is associated with sudden world-wide increases in abundance and diversity of the metazoans both as body and trace fossils (Daily, 1972). Such widespread transgressions on the stable shelf are interpreted to reflect major phases of

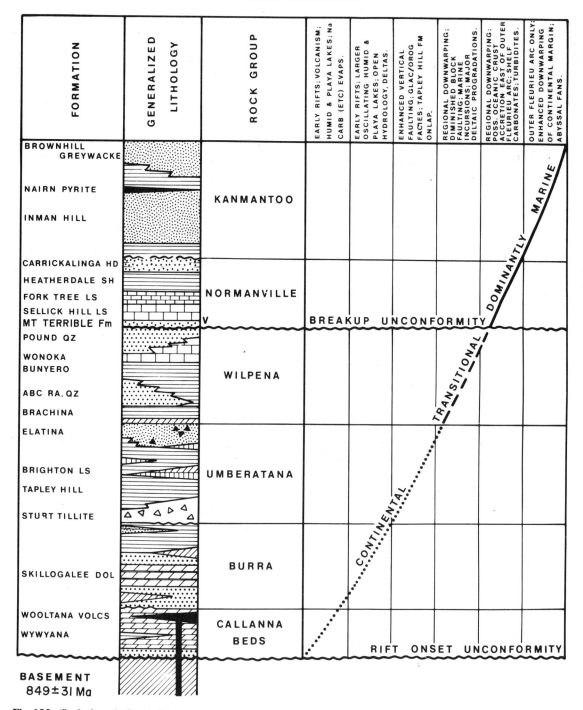

Fig. 180 (fig. 3 of von der Borch 1980). 'Composite stratigraphic column of Adelaide Rift (on left) generalized from Parkin (1969) and Preiss *et al.* (in press). Breakup unconformity separates Precambrian from Cambrian strata; rift onset unconformity separates metamorphic basement from lower Adelaidean strata; Callanna Beds and Pound Quartzite Formation occur only in Flinders Zones; Kanmantoo Group occurs only in Outer Fleurieu Arc; Normanville Group occurs only in southern Houghton Anticlinal Zone and Fleurieu Arc and has correlatives in the remainder of the rift.' Preiss *et al.* (in press) is published as part of Rutland *et al.* (1981). The basement age is from Cooper and Compston (1971).

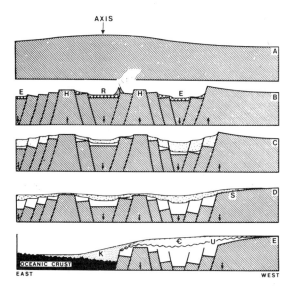

Fig. 181. (fig. 4 of von der Borch 1980). 'Palinspastic model showing sequential stages of Adelaide Rift, up to inferred oceanic crust accretion in Outer Fleurieu Arc. A. Initial arch (thermal dome) with greatest uplift at axis of future rift. B. Early rift valley stage with central rift valley complex (R), incipient marginal highs (H) and extra-arch basins (E); Callanna Beds and Burra Group time. C. Late rift valley stage illustrating widening of rift; Umberatana Group time. D. Infrabreakup stage; note downwarping involving Stuart Shelf (S) with onlap of Wilpena Group; note also continuing influence of marginal highs on sedimentation; Wilpena Group time. E. Post breakup stage for Outer Fleurieu Arc; note inferred breakup unconformity (U) developed at Cambrian–Precambrian boundary; also note marine shelf facies Cambrian sediments (Є) of Normanville Group, and continental slope and rise facies Kanmantoo Group (K).' As an indication of scale, the distance from the continent–ocean boundary to the hinge at E is about 300 km, and the Cambrian sediments are about 2 km thick and the Adelaidean sediments 10 km thick. Section oriented to facilitate comparison with Fig. 102.

regional downwarping of the rift and contiguous continental margin following earlier extensional block faulting, similar to the Hoffman et al. (1974) interpretation of the Great Slave Lake area of Canada. Finally, the Normanville and Kanmantoo Groups represent marine shoreline, shelf, and most significantly, continental slope and rise environments. It is emphasized that abyssal fan facies sediments of the Kanmantoo Group occur only in the Outer Fleurieu Arc, east of the series of basement highs described earlier within the Houghton Anticlinal Zone (Fig. 1) [our Fig. 179]. Major occurrences of deep water sediments of this type do not exist in what is interpreted as the failed arm or aulacogen (Central Flinders zone) to the north. This fact, the polarity of sediment facies in the south, and the presence of the Delamerian granite plutons of Ordovician age, strongly suggest that at least limited oceanic crust accretion occurred during Normanville and Kanmantoo Group times in the sigmoidal southern section of the rift system, in what could be interpreted as a marginal sea of the type documented by Karig (1972; see also Scheibner, 1974b). Bounding the proposed marginal sea or back-arc basin evolved a passive continental margin now represented by the Houghton Anticlinal Zone, Inner Nackara Arc and outer Fleurieu Arc. Destruction of the resulting oceanic crust in a subsequently developed active margin setting may be represented by the Delamerian granites (Fig. 1) [our Fig. 179] as roots of the inner portion of a magmatic arc, although Tertiary basin sediments obscure critical clues such as a possible ophiolite complex and forearc thrust zone to the east.

An additional significant point is the occurrence in the Outer Fleurieu Arc of the mafic Truro Volcanics immediately above the Adelaidean/Cambrian breakup unconformity. The Truro Volcanics (Forbes et al. 1972), shown by a V on Figs. 179 and 180, are interbedded with sediment in a 490-m-thick Early Cambrian sequence that unconformably overlies the Adelaidean Umberatana Group. The basal sediments contain trace fossils, and are identified as the Heatherdale Shale. The volcanics were altered during the late Cambrian–Ordovician – they have a Rb–Sr cooling date of 445 Ma (Compston et al. 1966) – and comprise metabasalt, porphyritic andesite, and amygdaloidal sodic trachyte. Whether the sodium was original or was introduced during the later Cambrian–Ordovician deformation is not known.

An interpretation more faithful to the actualistic model presented in Chapter IV would regard the Precambrian Adelaide Rift as a rift-divergence zone, consistent with its preserved width of 200 km, and the succeeding Early Cambrian marine shelf facies (Normanville Group) as a rim basin deposited behind the rim marked by the Houghton Anticlinal Zone. Von der Borch (1980) compared the Adelaide Rift with the present southern margin of Australia, an updated interpretation of which is given in Fig. 102. Comparison of Fig. 181 and Fig. 102 shows the following equivalent parts: (a) the Precambrian and Early Cretaceous rift-divergence zones, each with 5 to 10 km of block-faulted sediment; (b) the Early Cambrian and Late Cretaceous rim basins, each 300 to 400 km wide; and (c) the Early Cambrian (Kanmantoo) and Late Cretaceous abyssal sediments. Two quantitative differences – the much thicker Ceduna rim basin and the much thinner Late Cretaceous abyssal sediments – are related to local events. A qualitative difference – the absence in the Adelaide Rift of open-margin deposits equivalent to the Cainozoic shallow- to deep-water carbonate of the southern margin – reflects the reversal of subsidence after the

Early Cambrian in anticipation of the later Cambrian uplift of the Delamerian highlands, brought about by plate convergence.

The possibility that the pattern of rift-valley complexes in south-east Australia is repeated on the north-east along a similar trend (Fig. 178) is pointed to by geophysical modelling of the Adelaidean sediments beneath the Georgina Basin. Tucker et al. (1979, fig. 14) have drawn an isopach map of probable Adelaidean rocks based on gravity and magnetic modelling. From the extremely rapid thinning of sediments, for example 5 km over a distance of 15 km, they infer that the Adelaidean sediments were deposited in grabens bounded by growth faults. On the southern margin of the Georgina Basin, an unconformity separates the Adelaidean sediments from the overlying Early (but not earliest) Cambrian sediments. In places to the north and west, basalt flows of Early to early Middle Cambrian age mark the unconformity (Shergold and Druce 1980). As in the Adelaide Rift, the first marine deposition is not clearly recognized until the latest Vendian or earliest Cambrian, above the main Adelaidean sequence of glacigene sediment. Soon afterwards, in the Early Cambrian, the terrigenous sediment gave way to coastal and shallow marine carbonate, which predominated until a general regression in the late Arenigian (Shergold and Druce 1980). This chain of events is interpretable in the same way, and with the same chronology, as von der Borch (1980) interprets the Adelaide Rift, and as we interpret the overlying early Phanerozoic sequence. A big difference in the later history is that during the Late Cambrian convergence, described below, the Georgina Basin faced a marginal sea, whereas the Adelaide region faced the convergent margin. The Georgina Basin sequence is:

OPEN MARGIN	late Middle Cambrian to mid-Ordovician supratidal to shallow marine sediment
RIM BASIN	Early and early Middle Cambrian shallow marine detrital and carbonate sediment, including evaporites in the early Middle Cambrian
FLOOD BASALT	
BREAKUP UNCONFORMITY	
RIFT-VALLEY STAGE	Adelaidean detrital and dolomitic sediment, including glacigene sediment, dominantly, if not wholly, non-marine
RIFT – ONSET UNCONFORMITY	

In between the two regions of postulated rift-valley complexes lies the Amadeus Transverse Zone, dominated since 600 Ma ago by the highlands of the Musgrave Block (Fig. 88). Above the basal Heavitree/Dean Quartzites, the preserved Adelaidean sequences in the Officer Basin on the south and in the Amadeus Basin on the north thicken towards the Musgrave Block (Plumb et al. 1981), from which it is inferred (Plumb 1979, fig. 9) that at least the edges of the Musgrave Block were originally covered by thick Adelaidean sediment, all removed since uplift of the Musgrave Block 600 Ma ago, and some of it preserved as clasts in the basal part of the Mount Currie Conglomerate (Wells et al. 1970, p. 46). The pattern shown in Fig. 178A indicates a very broad region of deposition over the Amadeus Transverse Zone; the thickest deposition is on or around the region of the present Musgrave Block, interpreted in Fig. 178B as a broad sag, split along its western axis by a ridge. In the east, the axis possibly branched off the westernmost re-entrant of the eastern suture.

(d) Epi-Adelaidean: regional shear (by J. J. Veevers and C. McA. Powell)

Towards the end of the Adelaidean, in the Ediacaran (650–575 Ma), the long-standing Adelaidean development of Australia was terminated by several events (Austin and Williams 1978), all preceding breakup of the continent along the postulated eastern line of rift valleys, and in the north-west (Fig. 182), and subsequent plate divergence.

In the Amadeus Transverse Zone, the Petermann Ranges Orogeny was manifested in large thrusts and nappes, metamorphism, and northwards sliding of the Adelaidean sequence above the decollement surface of the Bitter Springs Formation during uplift of the Musgrave Block. The metamorphic cooling is dated as about 600 Ma (Forman 1972, p. 8) from Rb–Sr mineral ages on biotite and microcline, and the uplift is dated as late Vendian/Early Cambrian from piedmont deposits in the north and east (Fig. 88) and from a major lacuna in the central part of the Amadeus Basin (Wells et al. 1970). In the north-east Amadeus Basin and western Ngalia Basin (Burek et al. 1979) and in the Adelaide Rift (Rutland et al. 1981), the lacuna is shorter. The youngest Adelaidean beds in the Amadeus Basin (Arumbera Sandstone I) and Adelaide Rift (Pound Quartzite) contain the Ediacara Fauna, dated by Glaessner and Walter (1981) as within the span 640 to 580 Ma, and by Cloud and Glaessner (1982) as 670 to 550 Ma, and the oldest beds above the unconformity are near the start of the Cambrian,

ADELAIDEAN REGIME

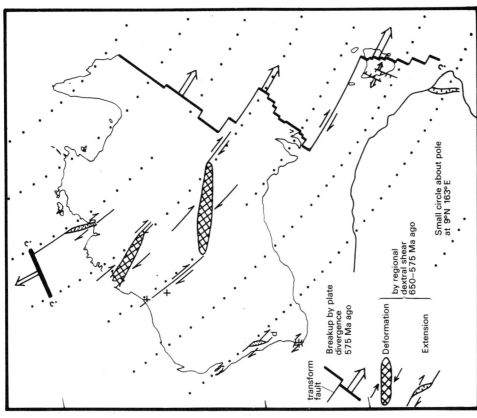

Fig. 182. B. Interpretation of epi-Adelaidean and breakup events about 650 to 575 Ma ago related to dextral shear and then plate divergence about a pole of rotation at 9° N, 163° E.

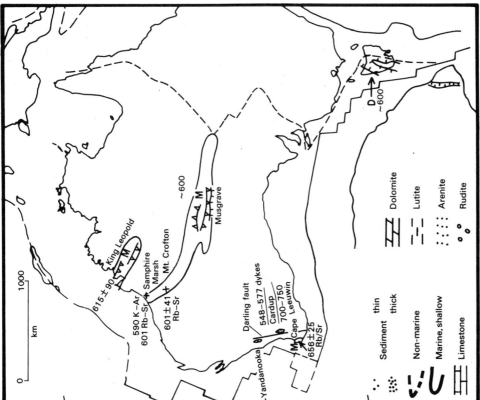

Fig. 182. A. Distribution of epi-Adelaidean and breakup events about 650 to 575 Ma ago west of the Tasman Line (Plumb 1979; Plumb and Gemuts 1976; Trendall 1974; Williams et al. 1976; Compston and Arriens 1968; Low 1975; Playford et al. 1976b; Libby and de Laeter 1979; Forman and Shaw 1973).

which we take as 575 Ma. The lacuna in these areas indicates cessation of deposition or loss of record about 600 Ma ago, which we attribute to regional uplift before breakup. Of interest is the evidence of earlier intense local uplift in the northern Flinders Ranges immediately before the deposition of the uppermost unit of the Adelaidean succession, the Pound Quartzite, about 650 Ma ago. Uplift is indicated by canyons filled with 1.5 km of sediment (Wonoka Formation) (von der Borch et al. 1982), suggesting an early phase of deformation. Incidentally, von der Borch et al. (1982) described these features as submarine canyons along a major basin slope. We suggest the alternative possibility – that the canyons were cut into the flanks of deep lakes or marine gulfs within the rift – because the Adelaide Rift, as pointed out by von der Borch (1980), did not face an ocean margin until the Cambrian.

Granite at Mount Crofton and at the bottom of Samphire Marsh Well was emplaced north-west of the Musgrave Block by 600 Ma ago (Trendall 1974; Williams et al. 1976). On the northern side of the Amadeus Transverse Zone, in the King Leopold Mobile Zone, metamorphism, intense folding, and thrusting affected the platform cover of the 1800-Ma-old Kimberley Group (Gellatly and Sofoulis 1973) and, more significantly, intensely deformed the southernmost outcrops of the late Adelaidean Walsh Tillite (Plumb and Gemuts 1976), in agreement with a 615 ± 90 Ma Rb–Sr age of metamorphic cooling at Yampi (Gellatly and Sofoulis 1973; Plumb and Gemuts 1976). In the south-west, cooling of the granulite of the Leeuwin Block at 656 ± 25 Ma (Compston and Arriens 1968) and deposition on either side of the present line of the Darling Fault, including very thick (9 km) volcanogenic sediment in the Yandanooka Group, dated within the interval 700 to 580 Ma ago (Low 1975), indicate rapid uplift and deposition. In Tasmania, a deformation, called the Penguin Orogeny, dated radiometrically as 617 to 538 Ma (Raheim and Compston 1977) and stratigraphically as older than Cambrian, is therefore within the range 617 to 575 Ma (Adams 1981), coeval with the other Petermann events.

Austin and Williams (1978) interpreted the epi-Adelaidean pattern and style of deformation as reflecting relative movement of blocks within Australia, caused by westward-directed compressive forces at the orogenic eastern margin of Antarctica. We present a different view. The epi-Adelaidean events mentioned above, plotted in Fig. 182A, can be related to a strain field defined by a pole of rotation (Scheibner 1974a) fitted to lines that delimit the boundary between the Precambrian and Phanerozoic terrains, called the Tasman Line (Harrington 1974). The model (Fig. 182B) entails two modes of strain from the same rotation pole: (a) deformation by regional dextral shear (thin arrows) during the latest Adelaidean, converted to (b) earliest Cambrian plate divergence (thick arrows) along the Tasman Line and probably also along the north-west.

The Tasman Line, the boundary between known Precambrian terrain on the west and the apparently wholly Phanerozoic terrain on the east, is constrained best by (a) the Lake Blanche Fault (see discussion by Murray and Kirkegaard 1978), parallel to O'Driscoll's (1982, fig. 25) Tethyan primary lineament, and (b) the Gambier–Beaconsfield fault zone of Crawford and Campbell (1973) and Harrington et al. (1973) – the north-western part is shown by Hills (1956) – which passes immediately south of Kangaroo Island and immediately north of King Island, and marks the boundary between the Precambrian terrain of all but the eastern part of Tasmania and the Phanerozoic terrain of Victoria. Following Scheibner (1974a), we take this boundary to be a continent–ocean transform fault, which together with the equivalent boundary along the Lake Blanche Fault defines a pole of rotation for the initial seafloor spreading on the east. The pole is located at 9° N, 163° E, and small circles about the pole, with a 5° spacing, are shown on Fig. 182B. Scheibner (1974a) found a similar, but closer, pole from the Gambier fault zone alone.

The rift valleys in the Adelaide–Broken Hill and Georgina regions intersect the small circles at an angle of about 45°, normal to the direction of extension associated with the dextral shear, but these trends may have been determined primarily by the grain of the basement. The graben of the Yandanooka Group and the complementary horst of the Leeuwin Block of south-western Australia also trend northerly, normal to the direction of extension associated with the dextral shear, but this is inherited from the northerly trend of the older Darling Mobile Zone (Glikson and Lambert 1976). The intense deformation, including thrusting, in the Musgrave Block about 600 Ma ago is explained as due to compression associated with dextral shear between subplates bounded by the Musgrave Block, with an original easterly grain, and by continent–continent transform faults trailing off its eastern and western ends. The eastern transform fault is marked by the Lake Blanche Fault, and the western fault follows the gravity trend on the north-eastern side of the Pilbara Block, along which granites are known to have been emplaced about 600 Ma ago (Trendall 1974). The few published details of the southern granite, at Mount

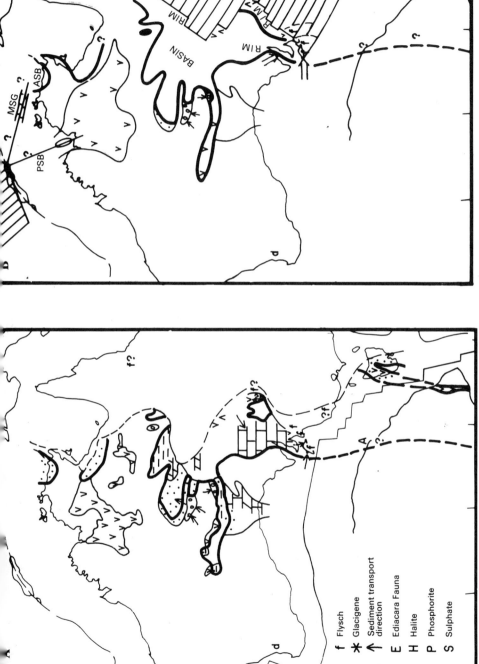

Fig. 183. A. Early Cambrian (575-545 Ma) data. From Cook (1982), and, additionally dykes (d) in the south-west, possibly Early Cambrian basalt in the Awitagoh and Kariem Formations of Irian Jaya (Visser and Hermes 1962), possibly Early Cambrian flysch (f) of the Judea Beds (Arnold and Henderson 1976) in the northeast, and of the Copper Mine Range Beds (Powell *et al.* 1982), archaeocyathid limestone belt south of the Adelaide area possibly through Cape Denison (Mawson 1940) to the Nimrod Glacier area of Antarctica (Laird *et al.* 1977), and continuation southward of the Dundas Trough and Mt. Read Volcanic Belt of Tasmania into the Bowers Trough of northern Victoria Land (Laird *et al.* 1977; Jago 1981). Incidentally, the connection of the troughs from Antarctica to Tasmania seems to rule out the possibility of the large-scale transcurrent motions between Antarctica and Tasmania advocated by Harrington *et al.* (1973) and adopted by Veevers (1976). Jago's (1981) refining of the fit of Tasmania and Antarctica by aligning the Bowers Trough west of Tasmania requires an unacceptably tight fit of Antarctica and Australia.

Fig. 183. B. Interpretation at end of Early Cambrian (575-545 Ma). Seafloor spreading pattern conjectural, with an assumed full rate of 5 cm/year. Breakup on north-west along margin of Tethys inferred from intersection of failed arms of Petrel Sub-basin (PSB), with axial positive gravity anomaly, and Money Shoal Graben (MSG) (Balke *et al.* 1973) and adjacent marine deposits of the Arafura Sea Basin (ASB) (Plumb *et al.* 1976). Shown on the platform are the Early Cambrian volcanics and marine deposits. Kanmantoo deep-sea fan, with apex at Kangaroo Island, sketched by broken lines, with main influx of sediment from the west (Flint 1978) and north (Daily *et al.* 1980).

Crofton (Trendall 1974), suggest that it may be an alkali-granite consistent with local intra-continental melting along a major shear. Thrusting in the King Leopold Mobile Zone about 600 Ma ago can be explained in the same way, but the associated transform faults are obscure.

In summary, the main effects of the dextral shear system in the epi-Adelaidean were located where lineaments determined by the older Precambrian structural grain intersected the shear system. Elsewhere, the main areas of Adelaidean deposition were uplifted and eroded to a lesser degree, as shown by the ubiquitous unconformity beneath the overlying Early Cambrian deposits, interpreted as regional doming that preceded breakup.

(e) Cambrian: plate divergence on the east and northwest, followed by plate convergence on the east (by J. J. Veevers and C. McA. Powell)

(i) Breakup and volcanism

In many areas of the platform, the first deposits above the unconformity are mafic volcanics, mainly tholeiite (Fig. 183A). Austin and Williams (1978) interpret the volcanics as indicating relief of stress. In our model, this relief of stress is by the conversion of intra-plate shear to plate divergence, as manifested by continental breakup along the Tasman Line, and the subsequent generation of oceanic lithosphere by seafloor spreading. The form of ocean generation, whether by seafloor spreading or by short segments of oceanic lithosphere in a basin-and-range terrain, is unknown.

The volcanics are attributed to four settings:

(1) The postulated newly generated oceanic lithosphere.

(2) Eruptions along the newly formed continental margin, exemplified, as noted already, by the Truro Volcanics in the Outer Fleurieu Arc.

(3) Volcanic rifts or grabens near the margin: (a) mafic and felsic volcanics in the Warburton Basin of north-east South Australia (Forbes 1969), overlain by fossiliferous Middle Cambrian tuffaceous strata, and hence probably Early Cambrian; (b) mafic and felsic volcanics interbedded with Early Cambrian sediment in the Bancannia Trough of north-west New South Wales (Cook 1982); (c) the Mt. Read Volcanic Belt (western Tasmania) of felsic to intermediate flows, breccia, and pyroclastics, deposited throughout the Cambrian, and ultramafic and mafic complexes associated with volcanics of the tholeiite suite emplaced in the adjacent Dundas Trough (Williams 1976); and (d) the Bowers Group (northern Victoria Land) of mafic volcanics interbedded with Early Cambrian (and possibly Vendian) sediment (Jago 1981).

(4) Platform tholeiite with minor tuff and agglomerate of wide extent and near-horizontal attitude on the platform, comprising the Antrim Plateau Volcanics and contiguous formations of northern Australia, and probably also the poorly dated Awitagoh Formation of West Irian, and the Table Hill Volcanics (Compston 1974) of the Officer Basin. Vents, presumably fissures, are undetected. The basalt was extruded subaerially except in a few places where its association with stromatolites suggests the possibility of subaqueous extrusion.

The platform basalt has a present area of 450 000 km^2 (400 000 km^2 in northern Australia and 50 000 km^2 in the Officer Basin) and an estimated volume (in round figures) of 100 000 km^3 (400 000 km$^2 \times 0.25$ km for northern Australia, and 50 000 km$^2 \times 0.1$ km for the Officer Basin). The poor surface exposure and the limited knowledge of the subsurface extent and thickness of the volcanics make this estimate tentative only; it is probably conservative, and, if erosion since the Early Cambrian is taken into account, then the original volume must have been greater.

The wide distribution of Early Cambrian volcanics on the Australian platform is unique in the Phanerozoic. No other Phanerozoic volcanics, except those along the western and southern margins, attributable to plate or incipient plate divergence in the Mesozoic, and the ?Miocene Fitzroy Lamproite, are found on the Australian Precambrian platform. A similar situation in the Salt Range of Pakistan, where the Khewra Trap (Gansser 1964, pp. 23, 25) lies immediately beneath the Early Cambrian sediments, hints at a still wider distribution of platform volcanics at this time in Gondwanaland.

Holmes's (1965, p. 301) list of the maximum estimated volumes of regional assemblages of continental plateau basalts has been modified by Kuno (1969), Baksi and Watkins (1973), Huber and Cornwall (1973), and Donnelly (1973), but all have overlooked the Early Cambrian volcanic province described above and also the Jurassic Ferrar Dolerite of Antarctica and the originally contiguous Tasmanian Dolerite. To provide a ready comparison, this information is consolidated in Table 17, from which it is seen that the Early Cambrian basalts of the Australian platform rank in volume with the other great plateau basalts.

A minor occurrence of mafic magma in southwestern Australia is indicated by Rb–Sr total-rock and mineral dates of 548 and 577 Ma on the sheared edges of dykes near Mundaring (Compston and

TABLE 17. *Size and age of continental plateau basalts*

Name and location	Area 10^5 km^2	Thickness km	Volume 10^5 km^3	Age
Columbia River, USA	2.00	1.0	2.00	Miocene
Deccan Traps, India	5.00	1.0	5.00	Palaeocene
		1.5	7.50	
		2.0	10.00	
Caribbean	10.00	–	–	Late Cretaceous
Parana, S. America	12.00	–	2.00	Early Cretaceous
Ferrar, Antarctica*	0.25	1.0	0.25	Middle–Late Jurassic
Tasmanian Dolerite†	0.25	0.5	0.12	Middle–Late Jurassic
Siberia	5.00	–	2.50	Early Triassic
Australian platform	4.50	0.1–0.5	1.00	Early Cambrian
Keweenawan, N. America	–	–	4.00	Proterozoic (1000–1500 Ma)

* Area measured from Craddock (1970, pl. 20). The ice cover of Antarctica makes the estimated area and volume a very low minimum.
† Calculated from Spry (1962).

Arriens 1968). Incidentally, while Libby and de Laeter (1979) acknowledge that these dates may be intrusion ages, they discount the possibility that Rb–Sr dates on biotite from Archean rocks in this region indicate a single event at this time. A further eruption of mafic (or ultramafic) magma at this time is possibly indicated by an intense gravity anomaly along the axis of the Petrel Sub-basin in the southern part of the Bonaparte Gulf (Brown 1980).

The north-western part of the Antrim Plateau Volcanics may have been continuous via the postulated failed arm of the Petrel Sub-basin with the newly generated Tethyan Ocean in the far north-west, as described later. But the main body of the platform tholeiite is widely separated from the lines of continental breakup. As we have seen in Chapter IV, the Mesozoic breakup along the western margin of Australia was accompanied by the eruption of tholeiite, but the tholeiite was confined within a distance of 300 km from the continent–ocean boundary, and the volume of magma was small. The flood of basalt across the platform in the Early Cambrian, at distances exceeding 1000 km from the nearest continent–ocean boundary, resembles the Parana basalt that accompanied the Early Cretaceous breakup of South America and Africa, and the subsequent generation of the South Atlantic Ocean.

(ii) Depositional events

Breakup and the concomitant volcanism coincide with the Precambrian/Phanerozoic boundary, shown by the first deposits of the Kanmantoo deep-water fans, the short hiatus at the breakup unconformity (late Ediacaran to earliest Cambrian), and the Early Cambrian age of the volcanism. That such events should come at the precise boundary between the Precambrian and Phanerozoic Eons is not an artefact of the time-scale but a matter of observation.

Continental breakup in the earliest Cambrian converted the intra-plate shear into plate divergence along the postulated rift valleys that marked the Tasman Line, so that the northerly trending rift valleys became failed arms, and the uplifted shoulders of the dismembered rift valleys along the Tasman Line became rims of the newly formed continental margins. The evidence for the rift-valley stage is restricted to the Adelaide–Flinders Ranges region. During the 35 Ma or so after breakup (Early and earliest Middle Cambrian), a rim along the line of the Houghton Anticlinal Zone and the Olary–Broken Hill Block shed sand northwestward into a basin whose axial part filled with shale and limestone (Parkin 1969, fig. 40; Moore 1979), in the style of the conceptual rim basin described in Figs. 92 and 93.

On the oceanward side of the rim, fans of quartzose flysch (Kanmantoo Group and possibly the sediments of the Cambrian Glenelg River Metamorphic Complex) were deposited over the newly generated oceanic lithosphere. Possible equivalents to the northeast are the poorly dated Copper Mine Range Beds (Pogson and Scheibner 1971) and the Judea Beds. 'A likely depositional model for the Judea Beds is that of deep-water flysch sedimentation on simatic crust adjacent to a quiescent cratonic margin from which mature terrestrial detritus was derived' (Arnold and Henderson 1976).

(iii) Pattern of divergence

The pattern of divergence is based on the assumption that the Lake Blanche Fault and the Gambier-Beaconsfield fault zone are transform faults. The initial segments of the oceanic lithosphere between these faults are drawn along great circles through the Outer Fleurieu Arc to east of the Wonaminta Block; and north of the Lake Blanche Fault, the ridge is located along the discontinuities of gravity trends (Wellman 1976) marked by the Diamantina Lineament and the Cork Fault/Weatherby Structure, and then the Clarke River Fault, which separate Precambrian and Phanerozoic terrain (Murray and Kirkegaard 1978). In Tasmania, the ridge follows the Precambrian/Phanerozoic boundary drawn by Williams (1976). To the west, a parallel structure, the Dundas Trough/Mt. Read Volcanic Belt, developed as a volcanic rift, in which volcanism declined during the Middle Cambrian convergence, as did its counterpart in northern Victoria Land, represented by the 3-km-thick basic volcanics of the latest Adelaidean–Early Cambrian Sledgers Group (Jago 1981). Volcanics in the Gidgealpa area (Forbes 1969) and in the Gnalta Shelf area suggest local centres of volcanism, which may have been the source of the tuffs found in the Ordian of the Flinders Ranges and Stuart Shelf (Cook 1982).

The postulated inception of the Tethyan Ocean by breakup along what ultimately became the northwestern margin, described in Chapter IV, is based on the sequence of Early Cambrian eruption of basalt in the Bonaparte Gulf Basin followed by the deposition of Middle Cambrian marine sediments in the Bonaparte and Arafura Basins, and then, in the Ordovician, of marine sediments in the Canning Basin and of possibly Ordovician non-marine sediments in the Carnarvon Basin and northernmost Perth Basin. These deposits are interpreted as reflecting the subsequent fill of failed arms that radiated from nodes initiated during the successive breakup of an ancestral northwestern margin. Details are given in Chapter IV.

(iv) Ordian (545–541 Ma) transgression

At the beginning of the Middle Cambrian, called the Ordian Stage (Öpik 1967, 1970, 1975), the sea expanded northwestward to form an epicontinental link between the Palaeo-Pacific and Tethyan Oceans (Fig. 184). The epicontinental sea was very shallow, and in places areas of emergence and hypersalinity led to the deposition of sulphate and halite evaporites (Cook 1982). The earlier detrital sediments in the Ngalia and Amadeus Basins gave way to carbonate. In the Flinders Ranges, a fan delta complex overlain by the transgressive carbonate of the Wirrealpa Formation marks the transition from a rim basin to a margin open to the ocean. According to Cook (1982), the flysch deposited in the Kanmantoo region lacked the coarse component so prominent during Early Cambrian deposition.

(v) Middle and Late Cambrian convergence

(1) Templetonian: start of convergence

During the succeeding Templetonian stage (Fig. 185A), phosphorite and other organic-rich sediments were deposited in the Georgina Basin due, according to Cook (1982), to currents sweeping nutrients into the epicontinental sea. At the same time, the sea withdrew from the Stuart Shelf, at the start of a long regression that eventually became general, so that much of northern Australia reverted to dry land (Figs. 185B, C).

The retreat of the sea is well recorded in the Flinders Ranges region and Yorke Peninsula. If, as expected, subsidence of the divergent margin had continued after the replacement of the rim basin by open-sea carbonate in the Ordian, then progressively deeper-water sediment would have overlain the Wirrealpa Limestone. But the overlying Lake Frome Group (Parkin 1969, p. 95–6) comprises 3 to 4 km of arkose that coarsens upward, from a sublittoral environment at the base to a fluvial quartz sandstone at the top (Grindstone Range Sandstone), all with a main source to the south-west (Thomson et al. 1976). The Ordian Wirrealpa Limestone thus marks the last deposit of the subsiding divergent margin, and the Lake Frome Group the first deposit of the Delamerian Orogeny (Parkin 1969, p. 96) along the rising convergent margin. Uplift and deformation in Tasmania (Tyennan Orogeny) also probably started in the Middle Cambrian, as indicated by an angular unconformity within the Mt. Read Volcanics (Williams 1976, p. 11) with a hiatus shown by fossils (Jago 1979) to span the Undillan (530–525 Ma), and possibly reflected by K-Ar mineral ages of 547 to 523 Ma from the adamellite at Murchison River (Richards and Singleton 1981).

In the Amadeus and Officer Basins, as well as in south-east South Australia, coarse detrital sediment was deposited in the later Middle Cambrian (Fig. 185).

(2) Late Cambrian deformation and plutonism

By the Late Cambrian (Fig. 186), the epicontinental sea had shrunk to indentations on the central-east and north-west, and conglomerate was deposited on the flanks of rising mountains. Idamean/post-Idamean (516 to 508 Ma) piedmont gravels in north-western New South Wales (Powell et al. 1982), Tasmania

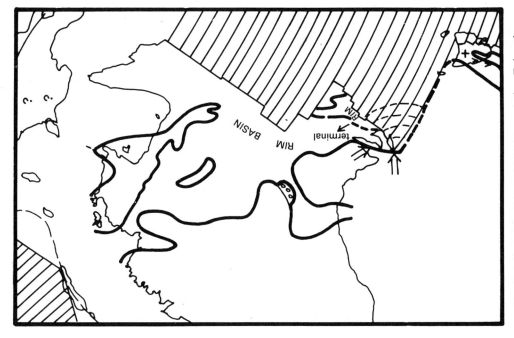

Fig. 184. B. Ordian (545-540 Ma) interpretation. Epicontinental sea links Tethyan and Paleo-Pacific Oceans. In the Flinders Ranges, open-sea carbonate (Wirrealpa Limestone) succeeds rim basin as rim submerges.

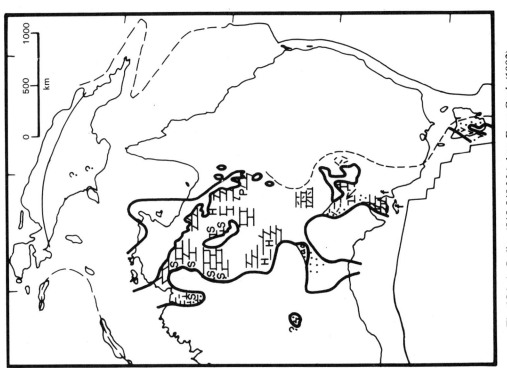

Fig. 184. A. Ordian (545-540 Ma) data. From Cook (1982), with additionally the possibly Ordian Kariem Formation of Irian Jaya. Arrow in Flinders Ranges indicates sediment-transport direction of the fan delta complex of the Eregunda Sandstone Member of the Billy Creek Formation overlain by marine-transgressive carbonate (Moore 1979).

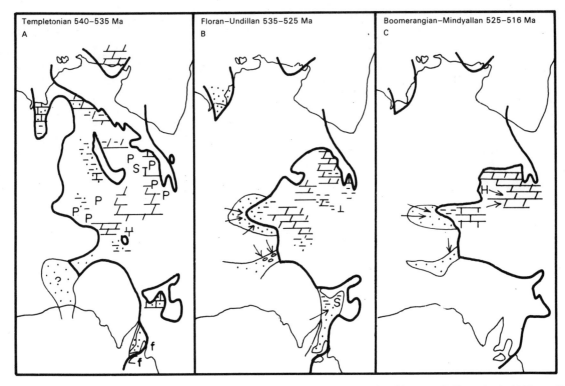

Fig. 185. Data maps for the Templetonian (A), Floran–Undillan (B), and Boomerangian–Idamean (C). From Cook (1982) modified to show the dominantly non-marine Lake Frome Group and the equivalent redbeds of Yorke Peninsula as no younger than Middle Cambrian (Klootwijk 1980).

(Webby 1978), and northern Victoria Land (Carryer Conglomerate, post-Middle Cambrian, probably Idamean/Payntonian: Laird et al. 1974; Cooper et al. 1976) reflect the main phase of uplift of the Delamerides, followed by the crystallization of granitoids and metamorphic cooling (Fig. 187). In a review of radiometric dates, Richards and Singleton (1981) note that the Glenelg River Metamorphic Complex in western Victoria is intersected by granite with K–Ar mineral ages of 484 ± 18 Ma and 476 ± 6 Ma, and a fission-track date on sphene of 470 ± 28 Ma (Gleadow and Lovering 1978). They note further that

The dates are similar in neighbouring SA . . . After correction for decay constant, granitic rocks in Kangaroo Island, Encounter Bay and the Mt. Lofty Ranges yield crystallisation ages around 495–505 Ma. The oldest K–Ar mineral ages are 475–485 Ma in these areas and E to the Victorian border; even the first K–Ar result on the Encounter Bay 'contaminated' granite (Evernden & Richards 1962) corrects to 474 ± 6 Ma.

Other mineral ages in these areas range to about 430 Ma in what has generally been interpreted as a slow-cooling pattern, which seems reasonable. Under such a regime, both the regionally-metamorphosed rocks and the granites, intruded while temperatures are still high, should yield a spread in biotite K–Ar dates and an age disparity between two minerals from a single hand specimen, reflecting differences in argon retentivity. Results for a gneiss sample from Harrow (Western Victoria) (#2: biotite, 466 ± 6 Ma; muscovite, 487 ± 6 Ma) are compatible with this concept . . . Activity at this time has been recorded even further afield. Richards & Pidgeon (1963) obtained eight Rb–Sr biotite ages from Broken Hill, which average 505 ± 10 Ma.

A mineral isochron for the Anabama Granite, 150 km SW of Broken Hill, corrects to 463 ± 3 Ma (Compston et al., 1966), a minimum age for this intrusion.

A similar history is recorded in Western Tasmania (McDougall & Leggo, 1965; Brooks, 1966). Rb–Sr on one sample of Mersey River Granite yields an essentially biotite age of 469 Ma (corrected to 477 Ma); K–Ar dates average 487 ± 13 Ma, with maximum 498 Ma for a biotite and no detectable difference between biotite and hornblende. Correlated hornblendic intrusions near the Dove River yield a mean 491 ± 15 Ma. The McIvors Hill Gabbro, near the E

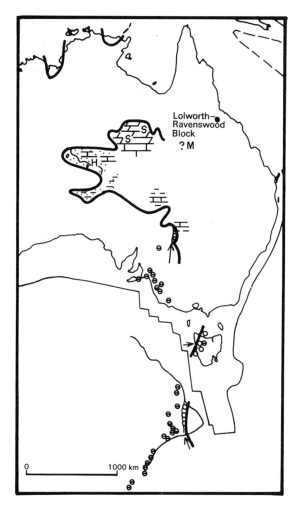
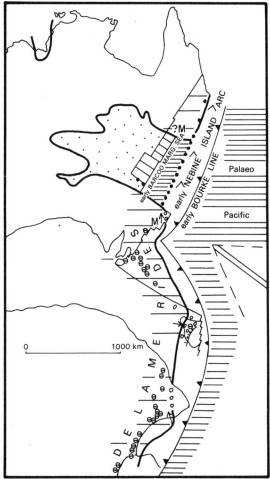

Fig. 186. A. Idamean/post-Idamean (516-508 Ma) data, from Cook (1982).

Fig. 186. B. Idamean/post-Idamean (516-508 Ma) interpretation. Filled circles: acid-intermediate volcanics (volcanic arc; Gidgealpa region; Mt. Windsor Volcanics of the Lolworth Block; Mt. Read Volcanics); ringed bars: rising granitoid plutons in Antarctica, Tasmania, western Victoria, and south-eastern South Australia; open circles: conglomerate. Convergent plate boundary at trench drawn parallel to, and less than 500 km from, magmatic arc. Uplift above magmatic arc reflected in termination of deposition of Lake Frome Group by the end of the Middle Cambrian (conglomeratic Grindstone Range Sandstone), and Idamean conglomerate in magmatic zone of western Tasmania, and on eastern side in northern Victoria Land and in Bancannia Trough. A broad marine gulf across central Australia interpreted as facing on to a marginal sea (Harrington's (1974) Barcoo Marginal Sea) behind an island arc (Harrington's Nebine Island Arc) and trench (Bourke Line). Possible direction of plate convergence shown by arrow.

boundary of the younger Heemskirk Granite, was dated by Rb–Sr at 507 ± 130 Ma. Jago et al. (1977) recorded a Rb–Sr date correcting to 480 ± 18 Ma for an andesite intruding Mid-Cambrian sediments near Ulverstone. Raheim & Compston (1977) dated the Jukesian Movement (equivalent to the SA Delamerian Movement) at 490 Ma, and reported at least three metamorphic events preceding it. Other evidence for such earlier activity includes a total-rock isochron yielding 817 ± 59 Ma for the W coast of King Island and a single K–Ar result on the Cooee Point Dolerite (725 Ma), showing that old rocks also occur on the Tasmanian N coast. Replicate measurements on one hornblende concentrate from the adamellite at Murchison River (547–523 Ma) also reveal an earlier Cambrian event.

Thus similar age patterns, leading up to the major deformation at 490–500 Ma ago, may be traced from SE

Fig. 186. C. Modern analogue: Ryukyu Trench and Island Arc, Okinawa Trough (Lee *et al.* 1980), East China Sea, and Yellow Sea, with location of active volcanoes (filled circles). Base from *Times Atlas* (1975); volcanoes from Gill (1981); plate boundaries from Hayes (1978).

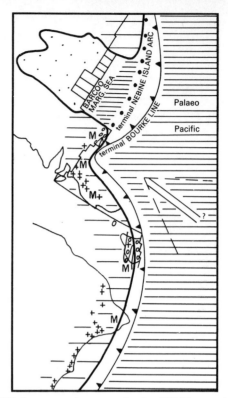

Fig. 188. Reconstruction at termination 500 Ma ago of plate convergence along Bourke Line.

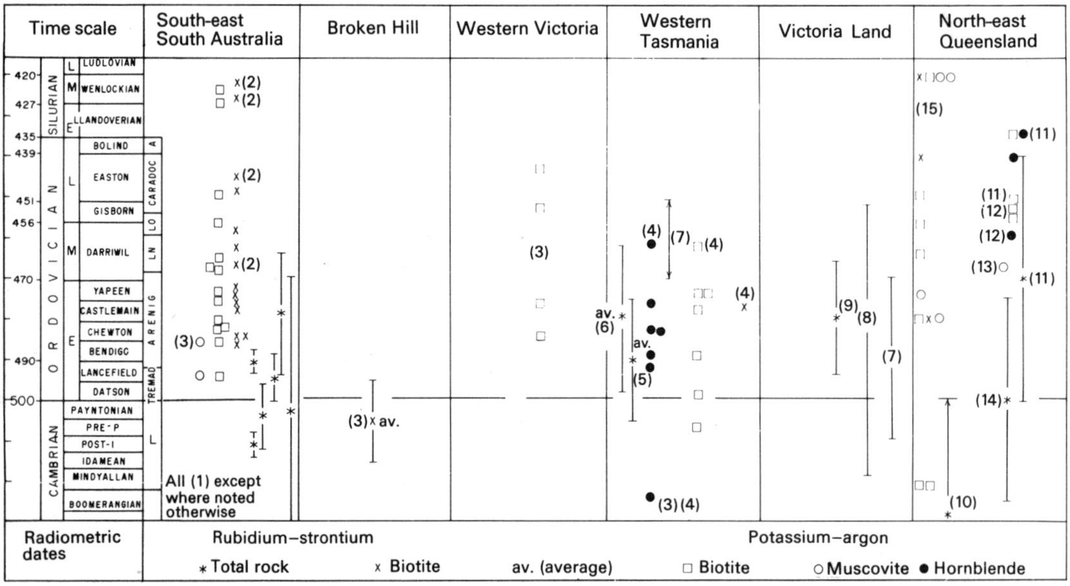

Fig. 187. Late Cambrian and Ordovician radiometric dates. 1: Milnes *et al.* (1977)—cooling dates of Encounter Bay Granites, Kangaroo Island pegmatite, Kanmantoo metasediments, Palmer Granite, Black Hill Norite. 2: Compston *et al.* (1966)—cooling dates of Anabama Granite, Mundawatana Granodiorite, Netley Gap Microtonalite, and metamorphic cooling date of the Truro Volcanics. 3: Richards and Singleton (1981)—cooling date of Broken Hill metamorphic rocks (also Richards and Pidgeon 1963; Pidgeon 1967), Dundas Tableland metamorphic rocks and granite, Murchison River adamellite. 4: McDougall and Leggo (1965)—cooling dates of Murchison River adamellite, Dove River granite. 5: Raheim and Compston (1977)—cooling dates of metasediments and eclogite, western Tasmania. 6: Jago *et al.* (1977)—andesite intrusion, Ulverstone. 7: Adams (1981)—total-rock K–Ar dates of slates in lower Mt. Read Volcanics (Tasmania) and Wilson, Robertson Bay, and Sledgers Groups (Victoria Land). 8: Jago (1981)—Transantarctic Mountains granites. 9: Jones and Faure (1969)—Granite Harbour Intrusive Complex. 10: Paine *et al.* (1974)—Mt. Windsor Volcanics (Wyatt *et al.* 1971), overlain by Early Ordovician graptolitic slate. 11: Wyatt *et al.* (1970)—Ravenswood granodiorite. 12: Mollan *et al.* (1969)—Telemon granodiorite. 13: Webb and McDougall (1968)—Anakie Metamorphics. 14: Paine *et al.* (1971)—deformation of Cape River Beds. 15: Black *et al.* (1979)—mica ages of Precambrian rocks of the Georgetown Inlier.

South Australia to the far W of NSW, S through Victoria to Tasmania. Later intrusions are also a possibility.

We interpret the pattern of dates from south-eastern Australia and the Transantarctic Mountains (Ross Orogeny) as indicating the rise of geotherms and concomitant uplift from 516 Ma ago to the end of convergence about 500 Ma ago during the peak of metamorphism and granitoid emplacement, followed by a quiescent interval of cooling, though, as noted by Richards and Singleton (1981), later intrusions cannot be ruled out.

A similar pattern is interpretable from the radiometric dates in the Ravenswood–Lolworth Block and Anakie Inlier of north-east Queensland (Fig. 187). The lower parts of the Mt. Windsor Volcanics are acid to basic calcalkaline volcanics with a Rb/Sr (corrected) age of 528 ± 100 Ma, with an upper limit set by Lancefieldian–Bendigonian (490 Ma) graptolites (Webby et al. 1981); the deformation of the Cape River Beds is dated as 500 ± 25 Ma; and part of the Ravenswood Granodiorite Complex yields a Rb–Sr date of 470 ± 30 Ma. Mica ages of Precambrian rocks of the Georgetown Inlier, biotite and hornblende dates of the Ravenswood Granodiorite and the Telemon Granodiorite, and a muscovite date from the Anakie Metamorphics range from 480 to 435 Ma: these dates may reflect cooling from a 500 Ma event or younger cooling events. A date of 520 Ma on schist at the bottom of Fermoy-1 well (Murray and Kirkegaard 1978) possibly indicates metamorphism and uplift.

On the platform, the Payntonian conglomeratic base of the Pacoota Sandstone in the Amadeus Basin succeeds finer-grained sediment, and the probable equivalent formation in the north-western part of the Georgina Basin follows a hiatus that spans the Idamean (Webby 1978; Cook 1982).

(3) Reconstructions

Data from the land and shallow sea along the eastern margin provide the basis of the palaeogeographical reconstructions shown in Figs. 186B and 188. Reconstruction of the oceanic realm on the east relies on the extrapolation of the tangible events registered on the margin. Harrington (1974) sketched the outline of an interpretation of the oceanic realm, and this interpretation, extended southward, is presented here. In the Idamean/post-Idamean (Fig. 186B), a broad epicontinental marine gulf across east-central Australia interrupted the rising highlands of the Delamerides and of north-eastern Queensland. The gulf faced on to the early stage of Harrington's (1974) Barcoo Marginal Sea, Nebine Island Arc, and trench (Bourke Line). From a node north-east of Adelaide, the trench continued south-east past Tasmania and the Transantarctic Mountains in a broad arc spaced from 200 to 400 km from the string of plutons.

Harrington (1974) envisaged the Barcoo Marginal Sea as an analogue of the Sea of Japan but we believe a closer match can be made with the Okinawa Trough and associated features south of the Sea of Japan (Fig. 186C). Analogous features, from north to south, are:

Australia	East Asia
Highlands to the north	Korea/Japan
Arc node(?) in Lolworth–Ravenswood Block	Arc node in southern Japan (Kyushu)
Epicontinental marine gulf in central Australia	Yellow/East China Sea
Barcoo Marginal Sea	Okinawa Trough
Nebine Island Arc	Ryukyu Island Arc (with volcanoes on western side)
Bourke Line	Ryukyu Trench
Highlands to south	Taiwan/east China
Node	node Taiwan
Trench south of Bourke Line	Philippines Trench

The scale of the analogous features is the same except the Barcoo Marginal Sea is much wider than its East Asian analogue. This is explained by (a) the presence of original Early Cambrian oceanic lithosphere trapped behind the arc, and (b) the longer period of dilation (possibly as much as 20 Ma by the pre-Payntonian) of the Barcoo Sea compared with the 5 Ma of the Okinawa Trough (Lee et al. 1980). By the end of convergence (500 Ma), the Barcoo Marginal Sea had dilated to its final width by eastward migration of the island arc and trench to their terminal positions, as shown in Fig. 188.

The direction of convergent plate motion shown in Figs. 186B and 188 is suggested from the direction of compressive strain found by Glen et al. (1977) from retrograde shear zones in the Adelaide Fold Belt. As it happens, the mean direction of compressive strain shown in the figures is the same as the direction of divergence but in the opposite sense, so that if it applied to plate convergence this would mean that plate divergence and convergence were along the same azimuth.

(f) Ordovician to earliest Silurian: marginal sea and island arc (by C. McA. Powell)

(i) Tasman orogen of south-eastern Australia

(1) Major facies distribution

The Cambrian cycle of divergence and convergence concluded around 500 Ma with the intrusion of the Delamerian granitoids and the accretion of the Kanmantoo flysch wedge to the continent (Fig. 188), and was followed in the Early Ordovician by an eastward jump of the island arc and marginal sea. There are four main depositional realms in south-eastern Australia in the Ordovician (Fig. 189): (1) a western belt of shallow-water marine to terrestrial clastics, which are commonly conglomeratic towards the base (Owen Conglomerate, Tasmania; Nootumbulla Sandstone, western New South Wales); (2) a central zone of deep-water terrigenous clastics of both distal and proximal turbidite facies; (3) a central-eastern belt of mafic volcanics with associated carbonates; and (4) an eastern zone of turbiditic terrigenous clastics intercalated with black shale and mafic volcanics. The western belt represents strand-line deposits, the central belt marginal-sea deposits, the central-eastern belt a volcanic island arc, and the

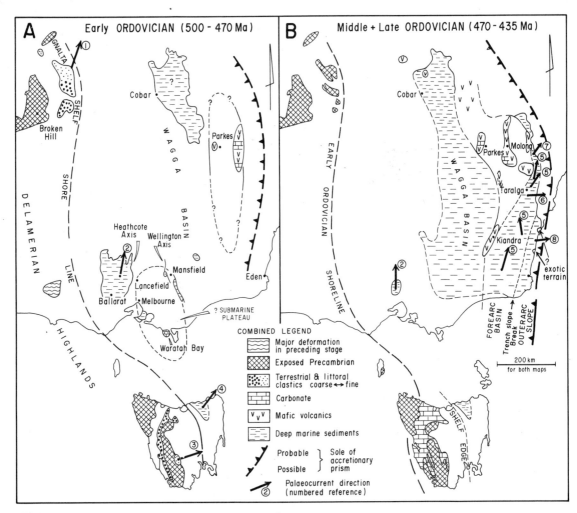

Fig. 189. Distribution of Ordovician rocks in south-eastern Australia. A. Early Ordovician. B. Middle and Late Ordovician. Based on Cas (1983, figs. 7-9), and Powell (1983b). Numbered palaeocurrent sources: 1, Webby (1976, 1978); 2, Schleiger (1968, 1969, 1974); 3, Corbett and Banks (1973); 4, Powell (unpubl. data); 5, Cas *et al.* (1980, appendix 1); 6, Powell and Conaghan (unpubl. data); 7, Fergusson (1979); 8, Powell (1983a).

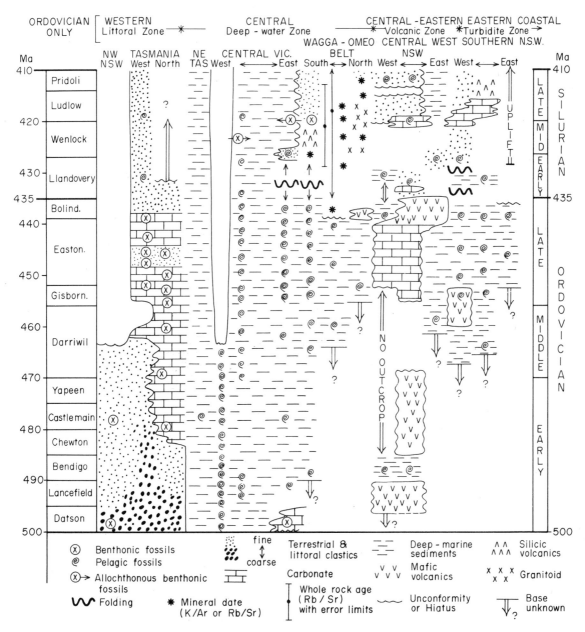

Fig. 190. Ordovician and Silurian time–space diagram west to east across south-eastern Australia. Principal data sources: Webby et al. (1981); Talent et al. (1975); Pickett (1982); Owen and Wyborn (1979).

eastern zone the accretionary prism. Summaries of the formation names and their correlation are given in Webby et al. (1981), and Cas (1983) shows their distribution. Fig. 190 is a schematic time–space diagram from west to east.

(a) Western littoral zone. The earliest Ordovician coarse clastics of the western littoral zone were derived from the epi-Cambrian Delamerian Highlands, which gradually subsided during the Early Ordovician. Thus, in Tasmania, the Owen Conglomerate and its correla-

tives pass laterally into Early Ordovician limestones of the Denison Sub-group, and upward into late Early, Middle, and Late Ordovician limestones of the Gordon Sub-group. Palaeocurrents determined from cross-bedding in the Early Ordovician Florentine Valley Formation indicate an east-north-east palaeoslope (Corbett and Banks 1973). In western New South Wales, the Nootumbulla Sandstone, conglomeratic at its base, fines upward into the Early to early Middle Ordovician siltstones and shales of the Gnalta Shelf (Webby 1978; Webby et al. 1981). The Gnalta Shelf was uplifted and being eroded in the Late Ordovician (Webby 1978).

(b) Central deep-water zone. Rocks of the central deep-water zone crop out in three main areas (Fig. 189): north-eastern Tasmania, central Victoria, and central-western New South Wales and adjacent central-eastern Victoria. The north-eastern Tasmanian localities are part of the little-studied Mathinna Beds in which Banks and Smith (1968) found graptolites, probably of Early Ordovician age. The rocks are slaty, dominantly argillaceous, and of distal turbidite facies. Palaeocurrents calculated from cross-laminations in Bouma-C horizons of graded beds indicate a north-easterly palaeoslope (Powell, unpublished data).

The central Victorian outcrops can be divided into two regions. The area west of the Cambrian Heathcote Axis is preserved in a meridional synclinorium centred on Ballarat. The rocks are mainly silty quartzose sandstone and slate, with chert bands and quartzose grits occurring locally. Sandy beds are graded, and display a variety of turbidite structures. Palaeocurrents determined from flute marks and Bouma–C cross-laminations indicate a dominantly northerly palaeoflow (Schleiger 1968, 1969, and 1974, summarized in Cas et al. 1980, fig. 2). The area south-east of Melbourne also contains graptolitic turbidite successions, but the section is condensed (Beavis 1976), and near Waratah Bay the Early Ordovician Digger Island Limestone is unusual amidst the dominantly deep-water deposits elsewhere in Victoria. The Ordovician section at Lancefield is also condensed, and at Mansfield Early Ordovician phosphate rock is overlain by Late Ordovician shale. These localities could have lain on a submarine plateau which gradually subsided through the Ordovician. The earliest Ordovician formations are generally the most sandy, with most of the Middle and Late Ordovician units being dominantly shale (Beavis 1976, table 3.5). The main turbidite fans depositing sandstones in the Ballarat area would have bypassed the submarine plateau, thus accounting for the condensed, largely pelagic sediments found.

The extensive central-western New South Wales and adjacent central-eastern Victorian outcrops comprise the deposits of the former Wagga Basin or Trough, which dates from at least the Bendigonian (Kilpatrick and Fleming 1980). The rocks in central-eastern Victoria are typically fine- to medium-grained quartzose turbidites of distal aspect. In New South Wales, Late Ordovician distal turbidites occur near Wagga Wagga, and further north, east of Cobar, the multiply-deformed Girilambone Beds (which contain thick chert intercalations – the Ballast Beds) may be even more distal equivalents. No fossils have been found in the Girilambone Beds, whose younger age limit is given by intrusions of mid-Silurian granites (Pogson and Hilyard 1981).

(c) Central-eastern mafic volcanic belt. The central-eastern belt of mafic volcanics extends from Kiandra northward toward Molong, and thence northwestward through disconnected outcrops before being lost under cover of the Great Artesian Basin. The volcanics are basaltic andesites, and date from the Early Ordovician. Near Parkes, the Nelungaloo Volcanics are overlain by beds containing Bendigonian graptolites (Sherwin 1979), and, north-east of Molong, the Fairbridge Volcanics include shales containing Bendigonian graptolites (Packham 1969). In the same region, the Walli and Cargo Andesites, themselves unfossiliferous, are unconformably overlain by Gisbornian limestone (Webby and Packham 1982), and the Mt. Pleasant Andesite by Darriwilian strata (Smith 1966). The Walli Andesite consists predominantly of porphyritic andesite and fine-grained vesicular basalt (pillowed in places) with subordinate volcanic breccia and tuff. The basalts contain phenocrysts of albite and augite, and pseudomorphs of pyroxene and olivine. Though the composition of the basalts has been extensively altered by low-grade burial metamorphism (Smith 1968b), the phenocryst population suggests they were basaltic andesites of oceanic affinity.

In the Kiandra area, Darriwilian to Gisbornian basaltic tuff, agglomerate, chert, and volcaniclastic sandstone pass upward into Gisbornian high-K porphyritic basalt of shoshonitic affinity (Owen and Wyborn 1979). The volcanism ceased by the Eastonian when distal quartzose flysch blanketed the region. Northward, in the Molong–Parkes region, the Gisbornian and Eastonian were times of extensive limestone accumulation (Webby and Packham 1982), followed in the Bolindian by a second pulse of volcanism (Fig. 190).

(d) Eastern turbidite zone. The eastern turbidite zone can be divided into two outcrop belts: (1) the coastal strip, and (2) the inland zone. The coastal strip extends up to 30 km inland, and is more highly deformed than

the inland zone, which extends across to the line of mafic volcanics. The coastal strip contains two facies associations: (a) a black shale–mafic volcanics–chert succession (the 'Wagonga Beds', Wilson 1968), and (b) a quartzose to feldspathic greywacke-and-slate succession. The 'Wagonga Beds' contain lithic greywackes, commonly with <20 per cent quartz framework grains, radiolarian cherts, and also fragmented pillow lavas. Vesicles in the pillow lava indicate extrusion in water shallower than 500 m (Cole in Powell 1983a). The 'Wagonga Beds' pass conformably upward into greywacke and slate which may have a structural thickness of more than 7 km (Cole 1982). The coastal greywacke and slate are increasingly deformed westwards where they are overprinted by a stripy, segregation cleavage that masks all but the grossest sedimentary characteristics. The stripy coastal facies abuts sharply against the inland turbidite facies, which is much less deformed. Pelites in the inland facies are slaty in appearance, and sedimentary structures are well preserved in most psammite layers.

The inland facies, like the coastal greywacke facies, comprises both distal and proximal turbidites, and in addition has extensive intercalations of chert and black shale. Sandstone composition in the inland facies varies from quartzose feldslitharenite, as in the coastal zone, to quartzarenite with >95 per cent rounded quartz grains.

The oldest graptolites in the coastal strip are Gisbornian (Late Ordovician) (Jenkins *et al.* 1982), and Darriwilian graptolites have been found in the inland facies (Jenkins 1982). In the coastal strip, fossils have been recovered only from the black shale–mafic volcanics–chert association. The youngest units in the inland facies are commonly distal quartzose turbidites in thick black-shale successions containing Eastonian or Bolindian graptolites. Thus, the oldest record of the eastern turbidite zone is significantly younger than the western zones, and this has importance in interpreting the tectonic history (see below).

(2) Palaeocurrents and provenance

The most remarkable feature of Ordovician sedimentation in south-eastern Australia is that most of the sediment is mineralogically mature quartzose sandstone and shale. The mafic-volcanic chain and associated carbonate buildups contributed only a small proportion of the total sedimentary pile. The quartz in the sandstones is commonly well rounded and well sorted, to the extent that there is commonly little grain-size grading in many beds where the full range of sedimentary structures of the Bouma sequence is present. Thin sections from coastal outcrops where the effect of grain dissolution during cleavage formation is not marked show a striking contrast between well-rounded quartz grains and angular plagioclase fragments, presumably derived from the mafic-volcanic chain. Other minor detrital components in the coastal turbidites include biotite, muscovite, sedimentary lithic fragments, and accessories such as tourmaline, epidote, and zircon. The mature quartzose sediment is evidently derived from outside the fold belt, and its high degree of rounding suggests it has been through at least one prior sedimentary cycle.

In the Early Ordovician, palaeoflow directions in the western littoral zone and adjacent central deep-water zone were towards the north and north-east (Fig. 189), away from the Gondwanaland land mass to the south. There is no record of dispersal from the central-eastern mafic-volcanic chain.

Palaeocurrents in the Middle and Late Ordovician are more informative (Figs. 189 and 191). Although there is no record in the central deep-water zone in the Melbourne area, the record in the eastern coastal zone is fairly complete (Fergusson 1979; Cas *et al.* 1980; Powell 1983a). In this zone, quartzose turbidites have a northward palaeoflow, and more feldspathic arenites an eastward palaeoflow. In the coastal strip, palaeoflow in proximal turbidites is easterly, and in intercalated distal turbidites northerly or southerly (Fig. 191; cf. Moore *et al.* 1982, fig. 4). In the inland facies close to the eastern margin of the mafic-volcanic chain, quartzarenites deposited from northward-flowing palaeocurrents are intercalated with more feldspathic arenites deposited from easterly-flowing palaeocurrents (for example Tarlo River area, Fig. 191).

In the palaeogeography responsible for this palaeoflow pattern (Fig. 192), the major sediment input is from the south parallel to the mafic-volcanic chain, with a minor amount of feldspathic sediment from the volcanic chain itself. The palaeogeography also highlights the difference between the coastal and inland facies associations in the eastern turbidite zone, the former being deposited on, or at the foot of, the outer-arc slope, and the latter in the fore-arc basin. The boundary between the fore-arc basin and outer-arc slope, commonly a high-angle reverse fault in modern arc systems (Karig *et al.* 1980a), may be indicated by the structural boundary between the coastal stripy; and inland slaty, cleavage domains (Powell 1983a).

This palaeogeography explains several otherwise perplexing features of the eastern turbidite zone. First, the model suggests intercalation in the fore-arc basin of material from two different provenances: mature quartzose arenite from the south, and immature feldspathic arenite from the west. Since each is deposited from turbidite fans that could have operated in alternating sequence or at the same time, turbidites of

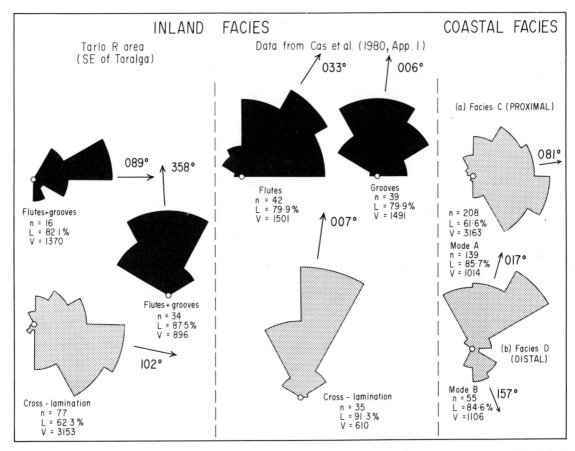

Fig. 191. Palaeocurrent summary rose diagrams for Late Ordovician turbidites in the inland and coastal facies east of the Ordovician arc. Solid-black rose diagrams are sole structures, stippled roses are Bouma–C cross-lamination. Tarlo River area is closest to the Ordovician arc, and is represented by a summary of measurements gathered by Powell and Conaghan (unpubl.). Other inland facies data from Cas *et al.* (1980, appendix 1). Coastal facies data from Powell (1983a, fig. 12); Facies C is proximal, and D is distal turbidite. In each rose diagram, n=number of measurements, L=consistency ratio (per cent), V=variance.

different provenance may be intercalated either as single flows, or as packets of flows, with little or no mixing of the two sources at the individual bed or bed-packet level.

Extremely clean, proximal quartzarenites certainly exist in the inland turbidite facies, as at Murruin Creek, north-east of Taralga. Immature feldspathic arenites intercalated with quartzose arenites are also present in the same region, north of Taralga (Packham in Webby 1976), and the easterly palaeoflow directions measured in the Tarlo River area south-east of Taralga possibly reflect the distal fringe of such arc-derived fans. To the north, in the Capertee Valley, Fergusson (1976, 1979) found immature andesitic detritus intercalated with quartzose arenites on a bed-by-bed scale, and flute marks indicate a northeasterly palaeoflow (Fig. 189). Fergusson (1979) initially thought that these intercalated quartzose and andesitic turbidites were Siluro-Devonian on the basis of tentative fossil identifications, but later work (Bischoff and Fergusson 1982) has shown that the andesitic rocks are overlain by Llandoverian/Wenlockian limey siltstone (a correlative of the Tanwarra Shale), and thus the andesitic rocks are an extension of the Late Ordovician Sofala Volcanics – a conclusion reached earlier by Packham (1969).

Two features stand out in considering the coastal outcrops. First, their palaeocurrent pattern is arc-normal, at right-angles to the main sediment-transport

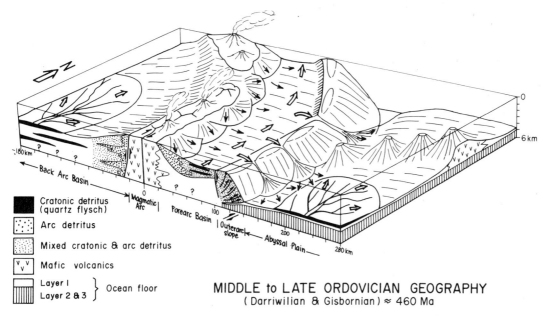

Fig. 192. Scaled block diagram of Middle to Late Ordovician geography. Arrows indicate palaeoflow direction.

direction in the inland quartzose turbidites. Second, the coastal turbidites contain a significant, though minor, proportion of angular plagioclase and volcanic-rock fragments mixed with well-rounded quartz grains. The coastal greywackes are evidently a mixture of the two provenances preserved in the inland belt, and this is explained by the palaeogeographic model if the intercalated quartzarenites and feldspathic arenites of the fore-arc basin are mixed by slumping at canyon heads before being redeposited on the fans at the foot of the outer-arc slope (Fig. 192).

The palaeogeography shown in Fig. 192 also suggests that quartzose fans prograding from the south should lie oceanward of, and be intercalated with, the mixed quartzofeldspathic detritus from the outer-arc slope. Such southerly derived quartzose material occurs at Mallacoota (Fenton et al. 1982), where it is intercalated with westerly derived turbidites (Powell 1983a).

Another feature explained by the palaeogeographic model is the apparently anomalous shallow-water pillow basalts found in the 'Wagonga Beds'. Water depths at the foot of the outer-arc slope, or on the adjacent abyssal plain are likely to have been more than 4 km, yet vesicle size indicates the pillow basalts were erupted at depths less than 500 m (Cole 1982). Even if the pillow basalts (which are mainly fragmented) were redeposited in oceanic depths by mass-flow down the flanks of a seamount, the presence of submarine volcanoes in the outer-arc region is anomalous. I consider the volcanics to be exotic terrains from former seamounts incorporated in the accretionary prism. This explanation also accounts for the fact that although the greywackes conformably overlie the black shale – mafic volcanic – chert unit (Wilson 1968; Etheridge et al. 1973; Cole 1982), the fossils in the black shale are younger (Eastonian–Bolindian, Jenkins et al. 1982) than the oldest fossils found in the greywacke (Darriwilian, Jenkins 1982). This inversion of stratigraphic facing and palaeontologic younging is typical of modern accretionary prisms where the youngest rock is emplaced at the base of an upward-facing pile (Karig et al. 1980b).

(3) Modern analogue

(a) Morphotectonic comparison. The Ordovician facies patterns in south-eastern Australia correspond to an east-facing continental margin separated from the main Palaeo-Pacific Ocean by a marginal sea rimmed by an island arc. The dimensions of the system were similar to the modern Andaman system in southeast Asia, with comparable facies and sediment-movement patterns (Fig. 193). In the Andaman Basin, a shallowly submerged continental shelf occupies the landward side of the basin, and a young, mid-Miocene to Recent, oceanfloor occupies the western half. By

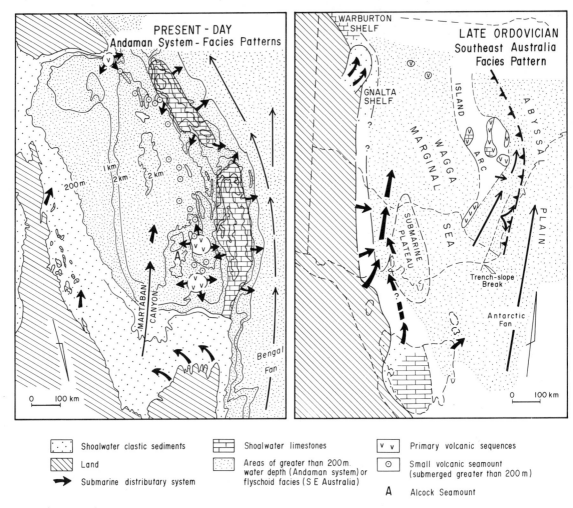

Fig. 193. Comparison of present-day facies of the Andaman–Nicobar system with the Late Ordovician facies patterns of south-eastern Australia. Modified from Cas *et al.* (1980, fig. 5), with Andaman system rotated 15° counter-clockwise to facilitate comparison with the Ordovician of south-eastern Australia.

analogy, the western part of the central deep-water zone in south-eastern Australia may be underlain by continental crust, with any young oceanic crust being confined to the Wagga Marginal Sea east of the Wellington greenstone axis. The Ordovician submarine plateau, east of Melbourne, serves the same purpose as the Alcock Seamount by diverting the major sediment input west (east) into the Ballarat/Melbourne Trough (Martaban Canyon). A difference, however, is that whereas the Ordovician submarine plateau is interpreted as a horst in continental or older island-arc basement, the Alcock Seamount is regarded as an epilith on newly formed oceanic crust (Curray *et al.* 1979).

The volcanic chain in the Andaman Basin is largely submarine, in contrast to the emergent volcanoes inferred for south-eastern Australia in the Late Ordovician. This difference may be transitory rather than substantive, because the Andaman seamounts are juvenile, and may grow to sea level and be capped by carbonate. Seafloor spreading in the Andaman Basin is less than 15 Ma old (Curray *et al.* 1979), and assuming that the Wagga Marginal Sea had begun to form by the Bendigonian (490 Ma), the comparable

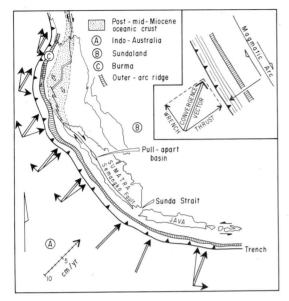

Fig. 194. Tectonic map of the western Sunda Arc (after Moore *et al.* 1980, figs. 1, 2). Inset shows decomposition of convergence vectors into thrust and wrench components.

stage in development of south-eastern Australia is still Early Ordovician (see Fig. 190). Extensive emergent volcanoes and associated carbonates did not form in south-eastern Australia until Late Ordovician, some 35 to 50 Ma after inception of the island arc.

There is no known equivalent in the Ordovician of south-eastern Australia of the emergent outer-arc ridge that forms the Andaman and Nicobar islands. Indeed, the Sunda outer arc is submerged except for only local, ephemeral islands (Moore *et al.* 1980). The Late Ordovician of south-eastern Australia has a well-developed fore-arc basin and (possibly) an outer-arc slope (Powell 1983a), which is more readily compared with the Sumatran part of the Sunda Arc system (Moore *et al.* 1980; Karig *et al.* 1980a, b). There, sediment in the fore-arc basin is transported by longitudinal and transverse currents, and sediment on the outer-arc slope is of both hemipelagic and turbiditic origin. The major source of sediment in the Sunda Arc accretionary prism appears to be the Nicobar Fan— the eastern lobe of the Bengal Fan before they were separated by the northward-converging Ninetyeast Ridge. Similarly, in south-eastern Australia, the major sediment supply is from the Gondwanaland land mass to the south.

(b) Deformation pattern. The modern deformation pattern in the Sunda Arc system can be used to interpret deformation patterns in the Ordovician of south-eastern Australia. The most important aspect of the Andaman system is that it has formed by highly oblique convergence between the Indo-Australian plate and the south-east Asian, or Sundaland plate, and as a consequence a small sliver, the 'Burma plate' of Curray *et al.* (1979), has rotated counter-clockwise from its parent, the Sundaland plate (Fig. 194). The trend of the Sunda Arc curves through 90°, and thus the azimuth of convergence goes from normal to the arc at the Sunda Strait and East Java, to nearly parallel to the arc in Burma. Earthquake first-motion studies (Fitch 1972; Verma *et al.* 1978) taken with geological observations (Hamilton 1979; Moore *et al.* 1980; Karig *et al.* 1980a, b) show that the convergence vectors between the Indo-Australian plate and the Sundaland/Burma plates can be decomposed into a thrust component normal to the arc and a wrench component parallel to the arc. The thrust motion is taken up at the foot of the outer-arc slope (Karig *et al.* 1980b) on shallowly inclined arc-dipping faults that scoop up most of the sedimentary material carried into the ocean trench and add it to the accretionary prism.

The wrench motion is accommodated mainly along the large Semangko transcurrent fault that runs along the magmatic arc into the Andaman Basin, and thence northward into Burma. Wrench motion also occurs along right-lateral faults in the accretionary prism, which trend northerly, oblique to the dominant structural and sedimentary trends. The resolution of convergence between the Indo-Australia plate and Sundaland/Burma plates shows that underthrusting is most rapid (up to 7.5 cm/yr) off Java. Conversely, the wrench component which is zero in the Sunda Strait, is dextral to the north-west along the Semangko Fault, increasing to about 4.5 cm/yr in the Andaman/Burma region.

This movement pattern has several implications for deformation in Ordovician south-eastern Australia. First, the lack of *melange* formed during Late Ordovician accretion in south-eastern Australia may be accounted for by a slow strain rate at the sole of the accretionary prism. Karig *et al.* (1980a) point out that *melange*, such as developed on Nias Island off the south-west coast of Sumatra, formed in a zone of relatively rapid strain. The strain rate depends not only on the rate of underthrusting, but also on the thickness of sediment involved; in the Andaman/Burma segment of the Sunda Arc, the combination of low underthrust rate and large thickness of subducted sediment (3 to 7 km, Moore *et al.* 1980, table 1) leads to a strain rate probably two orders of magnitude less than in the Sumatran segment. Karig et al. (1980b, p. 206) argue that under such conditions 'large tracts of trench sediments can be incorporated into the accretionary prism with surprisingly little deformation'.

Secondly, the gradual increase in offset along the arc-parallel wrench fault implies extension of the accretionary prism parallel to the arc. Some of this extension may be accomplished by flattening parallel to the arc during imbrication, and some may be accommodated by oblique dextral wrench faults which are confined to the accretionary prism.

Thirdly, the postulated dextral shear could account for the otherwise anomalous latitudinal trends of folds found in some areas of Ordovician rocks in southeastern Australia (Fig. 195C). These latitudinally trending folds (listed in Cas et al. 1980, appendix 2) are the oldest folds at each locality. They affect Late Ordovician rocks, and in places (for example localities 1, 6, and 7, Fig. 195C) are overprinted by 335°-trending folds of the Early Silurian Wagga Metamorphic Belt. Near Batemans Bay on the South Coast of New South Wales, latitudinally trending folds affect the Late Ordovician black shale – mafic volcanics – chert unit but not the conformably overlying quartzofeldspathic greywacke and slate succession (Powell 1983a). The early folds near Sofala and Batemans Bay have open profiles with little or no axial-surface cleavage, and verge slightly to the south (i.e. towards the arc) (Powell et al. 1978; Fergusson 1979; Powell 1983a, b). Although the orientation of these folds oblique to the main tectonic grain is consistent with dextral shear, no folds of a similar oblique orientation have been found in the Sunda Arc where dextral shear occurs today. Indeed, the major folds in the accretionary prism are parallel to the arc, a feature that also puzzled Karig et al. (1980b).

This last observation can be reconciled with another way of interpreting the latitudinally trending folds in south-eastern Australia. Uyeda (1981) pointed out the important distinction between two modes of subduction; (a) Chilean-type subduction, with strongly coupled plates, and horizontal compression in the arc and back-arc regions; and (b) Marianas-type subduction, with weakly coupled plates, and extension and high heat flow in the back-arc region. Uyeda also pointed out that Japan, currently undergoing Chilean-type subduction, may have been subject to Marianas-type subduction in the Miocene. By analogy, in south-eastern Australia, Marianas-type subduction in the Ordovician may have changed to Chilean-type subduction at the end of the Ordovician when compressional stress was propagated throughout the arc and back-arc regions. If this were so, the orientation of the principal direction of Early Silurian convergence between Gondwanaland and the Palaeo-Pacific plate would have been the same as in the Late Ordovician, because, as shown in Fig. 195B and C, the latitudinally trending folds are perpendicular to the same convergence vector as inferred for the formation of the Wagga Marginal Sea using the Andaman analogy. Alternation between the Chilean and Marianas modes of subduction, without the necessity for great changes in convergence vector, offers an actualistic explanation for the alternation of compressive and extensional events in the fold-belt margin of eastern Australia.

(4) Nature of the basement

(a) Distribution and composition of granitoid batholiths. The granitoid batholiths in south-eastern Australia occupy approximately 25 per cent of the present outcrop area (Fig. 196), and provide important constraints on palaeogeographic interpretations. Most of the granitoids were intruded in a 40 to 50 Ma interval from mid-Silurian to mid-Devonian, and were coeval with volcanics and volcanically derived sediments deposited in meridional grabens (Fig. 197A). The granitoids are of diverse chemistry, mineralogy, and structure, and are associated with two regional metamorphic belts – the Wagga Metamorphic Belt in the west, and the narrower Cooma Metamorphic Belt in the east (Figs. 198, 199). Both metamorphic belts are of the high-temperature, low-pressure type.

Attempts to synthesize and explain the diversity of granitoid rocks in the Lachlan Fold Belt follow three paths. First, Vallance (1969, pp. 180–200) distinguished three groups based on lithology and geological setting.

The first group, *here called the Cooma type*, includes those bodies, mainly granites and adamellites, emplaced in terrains that show the impress of medium- to high-grade regional metamorphism. Plutonic action and metamorphism appear to have been closely connected. The second group of plutonic bodies (*Murrumbidgee type*) consists chiefly of granites, adamellites, granodiorites and porphyries with some diorites and more basic rocks. These are often foliated but occur in environments displaying signs of only low-grade metamorphism. With both the first and second groups true hornfelses are rare or quite lacking in the contact areas, a feature in strong contrast to the normal pattern associated with the third group recognized. The latter, referred to as the *Bathurst type*, embraces mainly massive granites, adamellites and granodiorites, with subordinate monzonites, diorites and gabbros. These have imposed medium- to high-grade thermal metamorphic features on their contact rocks. Members of the first and second groups appear to have been closely

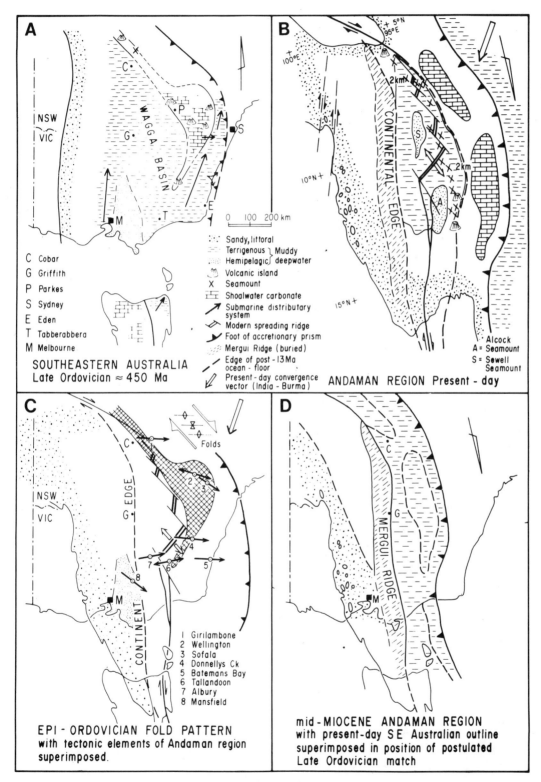

Fig. 195. Tectonic comparison between Late Ordovician south-eastern Australia and the present-day Andaman region. A. South-eastern Australia, B. Tectonic elements of Andaman region, C. Selected tectonic elements of the Andaman region superimposed on the outline of the Late Ordovician island arc in south-eastern Australia, D. Reconstructed mid-Miocene configuration of the Andaman region superimposed on south-eastern Australia.

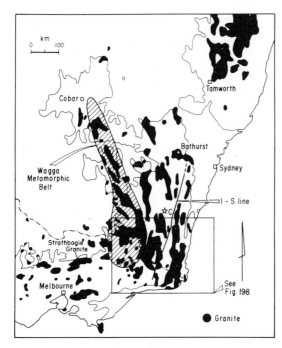

Fig. 196. Granitoids in south-eastern Australia (modified from Crook and Powell 1976, fig. 0–4). C=Canberra. I–S line interpolated from White et al. (1976) and Shaw et al. (1982a).

associated with deformation, whereas those of the third group are typically post-kinematic.

Vallance (1969) considered that the Cooma-type granites and regional metamorphism were related in space and time, and that the granites were derived in part or whole from the associated Ordovician sedimentary rocks. This view is supported by Rb–Sr geochronology of both the Cooma (Pidgeon and Compston 1965) and Corryong districts (Brooks and Leggo 1972), and by the rare-earth-element geochemistry of granite, gneiss, and migmatite in the Wagga Metamorphic Belt, which shows that muscovite-biotite granites there can be derived by partial melting of the Ordovician pelitic sediments during ultrametamorphism (Price and Taylor 1977). The relatively narrow aureoles around many of the granites can be explained by thinning during diapiric rise to their present high crustal levels (Flood and Vernon 1978).

Murrumbidgee-type granites are postulated by Vallance (1969) to have formed at depths well below those now exposed and to have reached an advanced stage of crystallization before emplacement. More recent geochemical work (Chappell and White 1974; White et al. 1976; Compston and Chappell 1979; McCulloch and Chappell 1982) has shown that the Murrumbidgee-type granites are derived from both igneous and sedimentary protoliths, most likely of Precambrian age, as discussed below.

Bathurst-type granites are postulated by Vallance to have formed from both crustal and mantle components which have reacted to approach equilibrium more closely before emplacement than the Murrumbidgee-type. Intrusion of the Bathurst-type granites as hot bodies to high crustal levels is indicated by hornfelsed contacts and inclusions.

In the second approach, White et al. (1974) subdivided granites into three types, according to their associated igneous and metamorphic rocks:

1. *Regional-aureole granites* surrounded by schists and gneisses of regional extent.

TABLE 18. *Granite types in south-eastern Australia (from Crook and Powell 1976)*

Vallance 1969	White et al. 1974		Chappell and White 1974, and pers. comm.	
	sub-volcanic			
Bathurst-type	contact aureole	II	S-type	I-type
Murrumbidgee-type		I		
Cooma-type	regional aureole		with muscovite	unknown

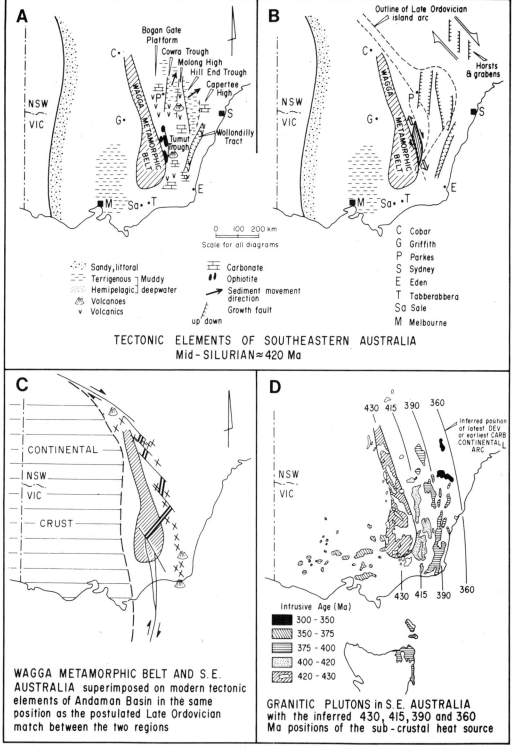

Fig. 197. Tectonic elements of south-eastern Australia in the mid-Silurian (c. 420 Ma). A. Principal palaeogeographic elements. B. Structural interpretation. C. South-eastern Australia with tectonic elements of the Andaman system superimposed on the same match as postulated for the continental edge in the Late Ordovician (Fig. 195). D. Granitic plutons in south-eastern Australia grouped according to intrusive age. Note that in the oldest group (430–420 Ma) the metamorphic aureole rather than the plutons is shaded. The axis of the 430 Ma Wagga Metamorphic Belt has been rotated eastward with an average rate of eastward movement of around 5 mm/year. The inferred 415 Ma (Late Silurian) and 390 Ma (Early Devonian) positions of the heat source are shown, and the 360 Ma (latest Devonian/Early Carboniferous) position is the inferred position of the continental arc at that time (Powell 1983a, fig. 4).

2. *Contact-aureole* granites with contact aureoles measured in hundreds of metres superimposed on all grades of any regional metamorphism which occurred prior to granite emplacement.
3. *Subvolcanic granites* with narrow contact aureoles (tens of metres) and intimately associated volcanic rocks including rhyolites, dacites, and rarer andesites, or their equivalent ashflow tuffs.

Vallance's Cooma-type are regional-aureole granites and his Murrumbidgee-type granites are contact-aureole granites. Vallance's Bathurst-type granites include both contact-aureole and subvolcanic granites, as shown in Table 18.

In the third approach, Chappell and White (1974) divided the granites into two types, S and I, according to their source material. S-type granites are derived by partial melting of sedimentary material whereas I-type granites are derived from an igneous source which has never been through a supracrustal process. Sr^{87}/Sr^{86} ratios for S-type are >0.708, and for I-type 0.704–0.706. The two types are discriminated by the following chemical and mineralogical criteria.

I-type	S-type
Relatively high sodium, Na_2O normally $>3.2\%$ in felsic varieties, decreasing to $>2.2\%$ in more mafic types	Relatively low sodium, Na_2O normally $<3.2\%$ in rocks with approx. 5% K_2O, decreasing to $<2.2\%$ in rocks with approx. 2% K_2O
Mol $Al_2O_3/(Na_2O + K_2O + CaO)<1.1$ C.I.P.W. normative diopside *or* $<1\%$ normative corundum Broad spectrum of compositions from felsic to mafic	Mol $Al_2O_3/(Na_2O + K_2O + CaO)>1.1$ $>1\%$ C.I.P.W. normative corundum Relatively restricted in composition to high SiO_2 types
Regular inter-element variation within plutons; linear or near-linear variation diagrams	Variation diagrams more irregular

Petrographic features reflect the differences in chemical composition. Hornblende is common in the more mafic I-types and is generally present in felsic varieties, whereas hornblende is absent, but muscovite is common, in the more felsic S-types; biotite may be very abundant, up to 35 per cent, in more mafic S-types. Sphene is a common accessory in the I-type granites whereas monazite may be found in S-types. Alumino-silicates, garnet, and cordierite may occur in S-type xenoliths or in the granites themselves. All of these features result from the high aluminium content relative to alkalis and calcium in S-type granites, and the converse in I-types. Apatite inclusions are common in biotite and hornblende of I-type granites whereas it occurs in larger discrete crystals in S-types.

I-type granites occur throughout south-eastern Australia, but S-type granites are restricted to west of a line that runs through the Berridale Batholith (White *et al.* 1976). This line, called the I–S line (Fig. 196), has been extended north (Shaw *et al.* 1982a), and is considered by White *et al.* (1976) to separate fundamentally different types of lower crust. Cas *et al.* (1980) consider that it marks the eastern edge of old sedimentary material that underlay the Ordovician island arc.

(b) Oceanic or continental crust? The nature of the crust beneath the Lachlan Fold Belt in early Palaeozoic times is controversial. One group (Crook 1969, 1974, 1980a, b; Scheibner 1974b; Crawford and Keays 1978) considers that the fold belt formed in an intra-oceanic setting which evolved through various thermal processes to become continental by the Early Devonian; the other group (Hills 1956; Rutland 1973, 1976; White *et al.* 1976; Compston and Chappell 1979; McCulloch and Chappell 1982) considers the fold belt to be underlain by Precambrian continental crust. An intra-oceanic origin is supported by the distribution and nature of Cambrian and Ordovician rocks, all of which indicate a setting of an island-arc and back-arc basin (data summarized in Crawford 1983).

Because no Precambrian basement is exposed in the Lachlan Fold Belt, the arguments for a continental basement are circumstantial, and fall into three categories. (1) The major-element chemistry of many of the S-type granites indicates that they are too rich in Ca, Na, Pb, and Sr to have been derived from the Ordovician quartzose clastics which they intrude (Wyborn and Chappell 1979; Phillips *et al.* 1981). Therefore, the granites must have been derived from an older, underlying, metasedimentary source. (2) From arguments based on the possible chemistry of residual and melt fractions that could have formed the granitic rocks of the south-eastern Lachlan Fold Belt, Compston and Chappell (1979) showed that the nine I-type granitoids examined first separated from the mantle about 1.1 Ga ago. Thus, newly formed oceanic crust could not have been the only source for those granites. (3) From consideration of Nd-isotopic characteristics of S- and I-type granites, McCulloch and Chappell (1982) concluded that S-type granites were derived from a *c.* 1.4 Ga sedimentary or metasedimentary source, and I-types from rocks derived pro-

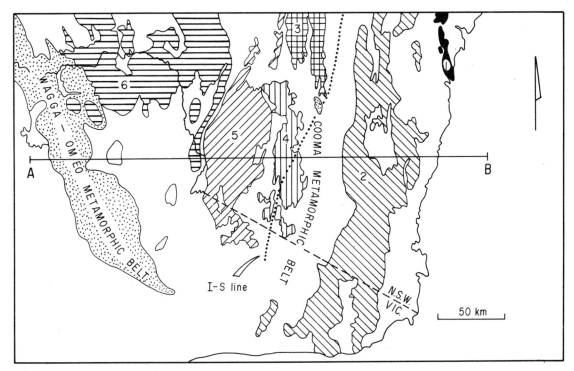

Fig. 198. South-eastern Australia showing the relationship between the two metamorphic belts, the I–S line, and the granitic batholiths (modified from Price and Taylor 1977, fig. 1). 1, Moruya Batholith; 2, Bega Batholith; 3, Murrumbidgee Batholith; 4, Berridale Batholith; 5, Kosciusko Batholith; 6, Corryong Batholith. A–B is the line of the time–space plot of Fig. 199.

gressively from the mantle over a period of 1 Ga prior to granite formation.

The evidence leads to a paradox: rocks that formed during the Ordovician suggest an intra-oceanic setting, whereas granitoids that formed immediately afterwards were derived from older, Precambrian protoliths. The paradox may be resolved by consideration of tectonic models.

(5) Tectonic Models

Recent tectonic models of Palaeozoic south-eastern Australia are of three types.

(1) Intra-oceanic fore-arc evolution. Crook (1980a, b) considered that Palaeozoic south-eastern Australia grew through fore-arc evolution with ten arc terrains being involved, most of them facing eastward. Crook suggested that most of the arcs were formed in an intra-oceanic setting similar to that of the Marianas today, and consequently the arcs would not have contained any fragments of old continental crust. Crook postulated that there is a self-sustaining cycle of fore-arc evolution in which, when the arc terrain reaches a critical thickness (12 to 20 km), internal heating causes partial melting of the subduction complex to form S-type and I-type granitic magma. Crook envisaged that no more than two of these ten arcs were active at any one time, and that the continent grew eastward by subduction of the marginal sea between the two arcs, with the landward arc being accreted on to the continent during the resulting collision. A new arc then formed oceanwards of the collision zone, and the cycle repeated itself. Crook played down the possible role of the coeval back-arc deposits because (1980a, p. 216) 'circumstances militate against the preservation of back-arc aprons as distinct strato-tectonic units'. In Crook's model the back-arc deposits are either incorporated into the subduction complex of the landward arc during collision, or become fore-arc regions by reversal of subduction polarity. After collision, the defunct landward arc is buried under 'post-arc sediments'.

(2) Early intra-oceanic arc evolution followed by

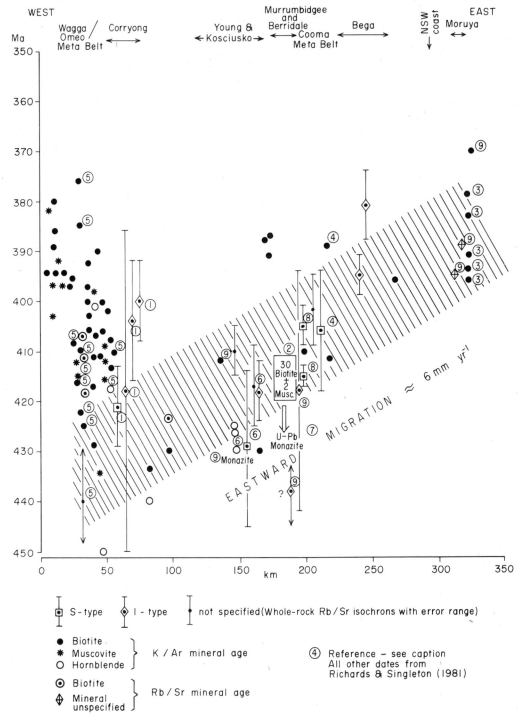

Fig. 199. East–west profile, A–B in Fig. 198, at lat 36.5° S, showing whole-rock ages of granitoid batholiths and mineral ages within the batholiths and surroundings. All ages in new constants. Principal reference is Richards and Singleton (1981), to which data have been added from (1) Brooks and Leggo (1972) as modified by Richards and Singleton (1981); (2) Williams et al. (1975), recalculated with new constants; (3) Griffin et al. (1978), recalculated with new constants; (4) Pidgeon and Compston (1965), recalculated with new constants; (5) Pogson and Hilyard (1981); (6) Owen and Wyborn (1979); (7) Williams reported in (6); (8) Roddick and Compston (1976); (9) Compston and Chappell (1979). The compilation is not exhaustive, but includes all the oldest dates in each region. The data have been projected along tectonic strike on to the east-west profile from areas in eastern Victoria and southern New South Wales. The shaded band highlights the pattern to be expected from a heat source migrating slowly eastward through the belt at 6.3 mm/year.

arc–continent collision. Crawford (1983) attempts to explain the paradox of the oceanic affinity of exposed early Palaeozoic rocks overlying an apparently continental basement by postulating an early intra-oceanic setting followed by underthrusting of a thinned Precambrian continental block in the Late Ordovician. Crawford's model accommodates both the regional facies and crustal geochemical constraints, although Cas (1983) has pointed out a number of weaknesses in the Crawford and Keays (1978) model of the intra-oceanic stage. Two difficulties with the arc–continent collision part of the model are (1) that the first regional folding on the New South Wales South Coast (the site of Crawford's postulated Late Ordovician collision) appears to be epi-Silurian (Powell 1983a, b), and (2) there are no known thrusts in the overriding block as might be expected by analogy with younger fold belts where arc–continent collisions are likely to have occurred, as for example in India–Eurasia (see Powell 1979; Tahirkheli et al. 1979; Jan and Asif 1981).

(3) Continental-margin arc-splitting with marginal-sea formation. This third grouping encompasses many specific models (for example Oversby 1971; Packham and Falvey 1971; Packham 1973; Solomon and Griffiths 1972, 1974; Scheibner 1974b; Cas et al. 1980) which have the common theme that a sliver of continent or a continental volcanic arc split to leave a passive continental margin and a young ocean-floored marginal sea in its wake. Though the details of specific models for south-eastern Australia are conflicting (see review in Packham and Leitch 1974), the back-arc model has the advantage that several modern examples have been studied (for example Japan arc and marginal sea, Ryukyu arc and marginal sea, Andaman arc and marginal sea). The main implication of this kind of model in explaining the apparent conflict between sedimentary facies distributions and basement geochemistry is that areas of both oceanic and continental crust exist in the marginal seas, and that the arcs contain slivers of older continental or island-arc material. Thus, there is no necessary conflict between the Ordovician surficial facies distribution indicating island arc – marginal sea settings and the inferred presence of old continental crust in the substrate of the arc.

In an updated version of the marginal-sea model for the Ordovician of south-eastern Australia (Fig. 200), Powell (1983b) considers that the Early Ordovician continental edge may have lain as far east as a meridional line from just west of Cobar, south along the eastern side of the Wellington greenstone axis in Victoria and thence along the Tamar Fracture in Tasmania (Fig. 195C). East of this line, the Wagga Basin was a marginal sea, probably underlain by young oceanic crust. The Ordovician andesitic arc was probably formed on a basement of pre-Ordovician continental crust or on the remnants of the Cambrian island arc that bounded the oceanward side of the Kanmantoo Basin (possibly a Cambrian equivalent of the Ordovician Wagga Basin). This model, which incorporates many aspects of previous models, accounts for the Ordovician facies distributions and the geochemistry of the Kosciusko, Young, Berridale, and Murrumbidgee batholiths, which all lie above the substrate of the Ordovician arc. The Bega and Moruya batholiths, which lie east of the I–S line, could have been generated from old oceanic crust added to the Ordovician accretionary prism during subduction.

(ii) Tasman Orogen of north-east Australia

The Ordovician record of the Tasman Orogen outside south-eastern Australia is sparse, and poorly understood. In north-eastern New South Wales, two isolated, richly fossiliferous, small bodies of Late Ordovician limestone (Uralba and Trelawney Beds, Hall 1975) lie adjacent to the Peel Fault Zone – a major serpentinite zone. Late Palaeozoic fault movements can be demonstrated along the Peel Fault Zone, but its older history is uncertain: in the Devonian it could have been part of a subduction complex. The Uralba and Trelawney Beds could thus be allochthonous material accreted during the mid-Palaeozoic, and hence cannot be used to reconstruct the Ordovician palaeogeography of the Tasman orogen.

The only other known Ordovician rocks occur in north-eastern Queensland (Murray and Kirkegaard 1978; Henderson 1980; Webby et al. 1981). Near Charters Towers, the Late Cambrian–Early Ordovician acid to intermediate Mt. Windsor Volcanics are associated with I-type granitoids of the Ravenswood Batholith, and probably represent the remnants of a volcanic island arc which was separated from the Precambrian Georgetown Inlier by a marginal sea of unknown extent (Murray and Kirkegaard 1978). Early or Middle Ordovician metamorphism also occurred in the Anakie Inlier, west of Clermont (Webb and McDougall 1968; Webb in Mollan et al. 1969; Paine et al. 1971), and Ordovician Rb–Sr mineral/whole-rock dates have been obtained from muscovite and biotite in the Einasleigh Metamorphics – a north–south metamorphic belt in the Precambrian Georgetown Inlier (Black et al. 1979). Early to Middle Ordovician K–Ar ages have also been determined from shale and altered olivine basalt to the

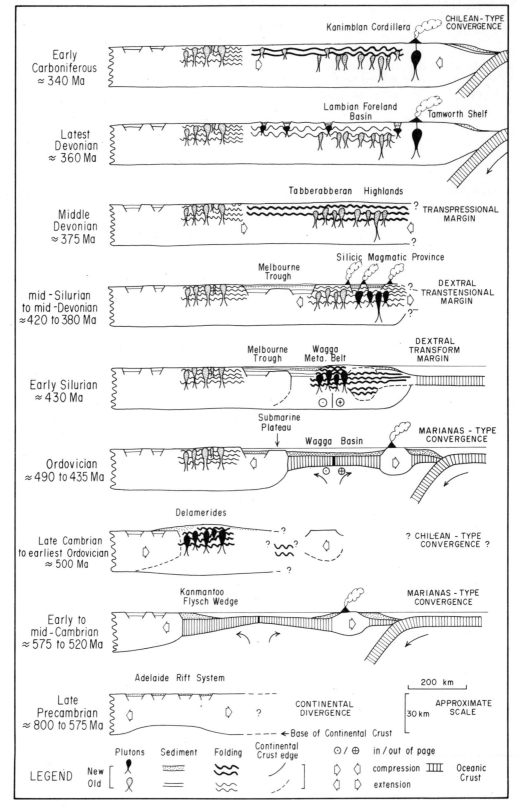

Fig. 200. Cartoon showing a plate-tectonic interpretation of the development and ultimate stabilization of the Lachlan Fold Belt. East–west sections are drawn at a latitude of about 36° S, with some elements projected on to the profile along tectonic strike. Scale is very approximate. See text for explanation.

south-west of the Anakie Inlier (Murray and Kirkegaard 1978). Even further to the south-west, just north of the Lake Blanche Fault (Fig. 178B), Early Ordovician to early Late Ordovician graptolites have been recovered from shales in deep boreholes (Murray and Kirkegaard 1978, table II).

The Ordovician sediments in Queensland have a facies, composition, and distribution similar to those in south-eastern Australia; viz. a western belt of quartzose sandstone and shale of turbidite facies, and andesitic volcanics and limestone in an eastern belt. An episode of deformation and metamorphism in the mid-Ordovician is indicated by K–Ar and Rb–Sr mineral ages and by isolated fault slices within the metamorphic belts containing non-metamorphosed Late Ordovician limestones and volcanics, for example the Fork Lagoon Beds (Anderson and Palmieri 1977), and the Carriers Well Limestone and Everett Creek Volcanics (Arnold and Fawckner 1980).

A possible palaeogeographic framework is that of a narrow marginal sea separating a north-north-east trending volcanic island arc in the east from the Precambrian shield to the west of the Tasman Line (Fig. 201). Quartzose sediment derived from the mature continental area to the west formed the main fill of the marginal sea, just as it does in the Wagga Basin to the south. The northeasterly direction of convergence between the Gondwanaland plate and the Palaeo-Pacific plate (inferred from relationships in south-eastern Australia) would have produced mainly transform motion along the southward continuation of the north-eastern Queensland island arc that terminated at a re-entrant near Charleville in southern Queensland.

Oblique convergence along the Pacific margin of Gondwanaland in the Ordovician provides a possible explanation for the apparently diachronous cessation of andesitic volcanism and the onset of metamorphism and deformation (Fig. 202). In north-eastern Queensland, island-arc magmatism, represented by the Mt. Windsor Volcanics and Ravenswood Batholith, finished by the Middle Ordovician (Webby *et al.* 1981), and was followed by mid-Ordovician metamorphism and deformation that affected the marginal sea, as well as part of the Precambrian Georgetown Inlier. In south-eastern Australia, island-arc magmatism continued throughout the Ordovician (even though it was episodic: Webby and Packham 1982) and was followed by metamorphism and deformation of the marginal-sea deposits in the Early Silurian. The situation could have been analogous to the Cainozoic history of western North America where oblique convergence has led to diachronous tectonic events (Atwater 1970).

(iii) Platform and north-western margin

In north-western Australia, the sea, hitherto confined to the Bonaparte Gulf Basin, made its way in the Early Ordovician into the Canning Basin (Legg 1976; McTavish and Legg 1976; Legg 1978; Webby 1978; Webby *et al.* 1981) in two arms separated by the emergent Broome Platform (Fig. 201). Detrital sand was deposited along a shoreline on the east, and elsewhere detrital sandstone was deposited with shale and limestone. A K–Ar date of 484 Ma on biotite from granite at the bottom of Samphire Marsh–1 Well (White 1962), mentioned by Veevers (1967) and McTavish and Legg (1976) as indicating an event immediately before deposition, is now seen (Trendall 1974) to be a minimum date of granite intruded about 600 Ma ago, as suggested by Webby (1978). We interpret the extension of the sea southwestward from the Bonaparte Gulf Basin as reflecting the further breakup along the north-western margin by divergence to extend the Tethyan Ocean (Fig. 111). The failed arms radiated from a node that developed at breakup, and were subsequently covered by the sea. Another failed arm occupied the Carnarvon/Perth Basin, as indicated by the onset of deposition of the Tumblagooda Sandstone, possibly as early as the Ordovician (Fig. 112). No direct evidence of the Tethyan Ocean is known and, like the Cambrian ocean in the east, its existence is inferred wholly from tangible data available on the platform.

A Precambrian inlier today separates the Canning and Amadeus Basins; facies trends suggest that during the Ordovician the basins were separated at least by a sill. But the distribution of fossils on either side of this division indicates interchange of marine organisms from one side to the other in a transcontinental sea called the Larapintine Sea (Webby 1978) that linked the marginal sea on the east to the failed arm on the west (Fig. 201). The southern shore of the Larapintine Sea received sediment from the high ground of the Musgrave Block, which at the same time shed non-marine sediment southeastward into the Officer Basin (Fig. 88C).

By the Late Ordovician (Fig. 202), the Larapintine Sea connection between the Tasman Orogen and north-western Australia had been broken, and local areas of uplift occurred in central Australia (Rodingan movement, Webby 1978). The Tumblagooda Sandstone may have been deposited in an arm of the sea from the Tethys Ocean that extended down the west coast of Western Australia. Possible late Middle Ordovician diastrophism in the Ballarat region of south-eastern Australia may have resulted in local uplift (VandenBerg 1978), though the age control on this deformation is poor.

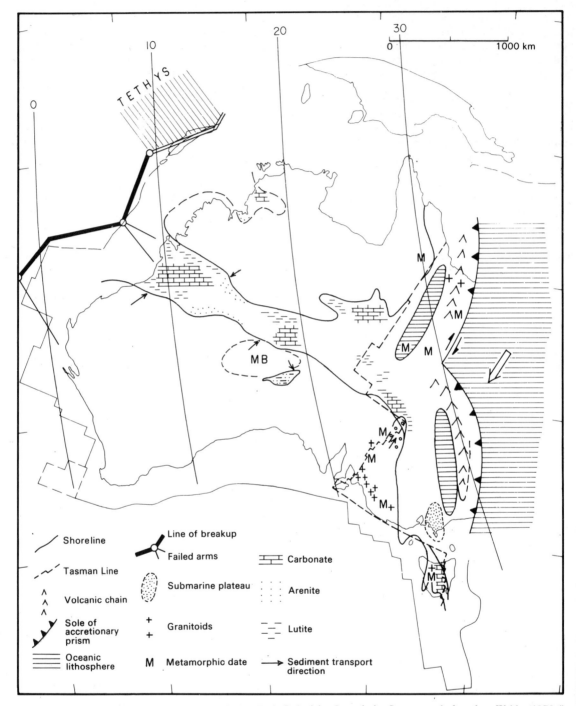

Fig. 201. Palaeogeographical reconstruction of the late Early Ordovician. Larapintine Sea across platform from Webby (1978, fig. 6B). Broad arrow is the postulated plate-convergence direction. MB—Musgrave Block.

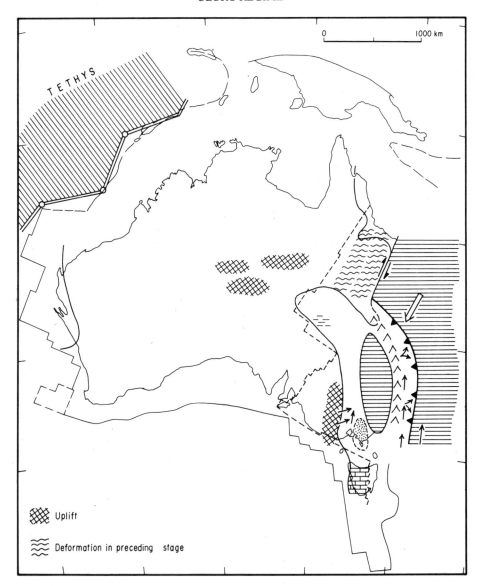

Fig. 202. Palaeogeographical reconstruction of the Late Ordovician. Shoreline in the north-eastern Tasman Orogen and western Australia is from Webby (1978, fig. 6D). Symbols, other than those explained, as for Fig. 201.

(g) Silurian to mid-Devonian—a dextral transtensional margin (by C. McA. Powell)

(i) Tasman Orogen of south-eastern Australia

(1) Major facies distribution

The volcanic island arc – marginal sea configuration of south-eastern Australia ended in the earliest Silurian with the formation of the Wagga Metamorphic Belt (WMB) over the site of the former marginal sea. Thereafter, the history of the region east of the WMB was quite different from that of the region west of the WMB until the Late Devonian when the whole region became part of a foreland and epicratonic basin west of a continental magmatic arc that lay at the eastern edge of the exposed Lachlan Fold Belt. The Silurian to mid-Devonian interval was the time of greatest facies diversity in the Lachlan Fold Belt, with many small, linear basins ('troughs' in local literature) separated by narrow uplifted areas ('highs') (Packham 1960). East of the WMB, silicic magmatism dominated the interval, with extrusive and intrusive forms, and sediment derived from this silicic magmatism was deposited in environments ranging from terrestrial to deep marine. West of the WMB, silicic magmatism was confined to a short interval (latest Silurian to Early Devonian) with the eruptive centres being located along the western margin of the WMB. The bulk of the sediment in this western belt is quartzose clastics derived from the Gondwanaland craton to the west. In regional terms, the western belt occupies the foreland-basin position between the craton to the west and a basin-and-range style silicic-magmatic province to the east.

(2) Recognition of sedimentary basins

An important aspect of the Silurian to mid-Devonian interval is the difficulty of distinguishing sedimentary basins from the present structural configurations. The Lachlan Fold Belt has been deformed into a series of meridional anticlinoria and synclinoria (Gilligan and Scheibner 1978), and many of the Silurian and Devonian sediments are preserved only in the synclinoria – Ordovician basement being exposed in the anticlinoria. There has been a tendency to regard each synclinorium as a discrete sedimentary basin with the intervening anticlinorium as a synsedimentary high. In some cases, the margins of the structural zones do lie approximately along the margins of the former sedimentary basins (for example, the Copperhannia Thrust marking the western edge of the Hill End Synclinorial Zone (Gilligan and Scheibner 1978, fig. 7) lies approximately along the western edge of the Late Silurian–Early Devonian Hill End Trough) but, in other places, the structural boundaries cut right through the sedimentary basins (for example, the western boundary of the Cowra Synclinorial Zone: Gilligan and Scheibner 1978, figs. 6 and 7).

The other problem in distinguishing sedimentary basins is the tendency for facies relationships established for one interval to be extrapolated backward in time, when a different palaeogeography may have existed. In central-western New South Wales, the Cowra Trough, Molong High, and Hill End Trough, which can be demonstrated using facies relationships to have existed in the Late Silurian and Early Devonian, have commonly been extrapolated back to the Early Silurian or even the Ordovician (for example, Brown *et al.* 1968; Webby 1978; and Scheibner 1974b; but note revised interpretation in Gilligan and Scheibner 1978).

For the geometry of a depositional basin to be extrapolated back in time, one must demonstrate that the same basin-margin facies existed at the earlier time. Unfortunately, the present state of knowledge in the Lachlan Fold Belt does not always enable this to be done.

The approach taken here is from a regional overview, and where insufficient facies relationships are available to establish the limits of a sedimentary basin, I have opted not to subdivide what may have been part of a single, more extensive, unit. Consequently, in the following outline of the evolving palaeogeography, I do not recognize any of the meridional troughs in the eastern silicic-magmatic province before the Wenlock.

A final important characteristic of this interval is the apparent change in the nature of the crust. Whereas the Ordovician configuration was largely of an ensimatic or oceanic affinity, the crust during the Silurian and thereafter has a continental character, albeit thinned continental crust in places. The continental edge must have lain east of the exposed Lachlan Fold Belt from the mid-Silurian onwards, because there is no sediment of trench or abyssal plain affinity from the Wenlock onwards. The only ocean-floor material in the Lachlan Fold Belt during this interval is in the Tumut Trough, which I argue formed as the floor of a small pull-apart basin during regional dextral shear.

(3) Main sedimentary basins

The main depositional areas west of the WMB are the Grampians, the Melbourne Trough, and the Darling Basin (Fig. 203), the latter including the latest Silurian to Early Devonian Cobar Trough along its eastern margin. The 6-km-thick section in the Grampians (Spencer-Jones 1965) is of uncertain affinity: it may be the fill of a 50-km-wide terrestrial graben, or it could be the western terrestrial equivalent of the marine Melbourne Trough. Age constraint in the Grampians is poor, but part at least of the section is coeval with the Melbourne Trough (VandenBerg *et al.* 1976). The Darling Basin extends westward across the Darling Lineament (Scheibner 1974a) where it onlaps the Precambrian Wonaminta Block (Rose and Brunker 1969; Powell *et al.* 1982). In its western two-thirds, the Darling Basin contains only shallow-marine sediments overlain by a fluviatile succession. In the eastern third, especially within 50 km of the eastern bounding fault, a thick succession of deep-marine sediments (turbidites) underlies shallow-marine sandstones and siltstones, which in turn are overlain by more than 4 km of fluviatile sediment (Glen 1982b; Glen *et al.* 1983).

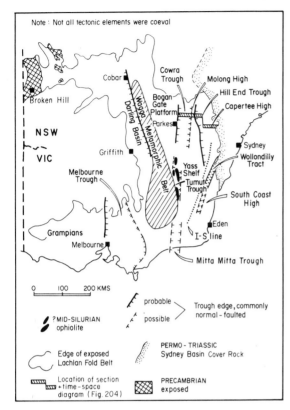

Fig. 203. Silurian to mid-Devonian tectonic elements of south-eastern Australia. Modified from Powell (1983a, fig. 3).

Siluro-Devonian felsic volcanics, possibly in a resurgent cauldron structure, occur near Mt. Hope, south of Cobar, and volcanic intercalations in the Cobar Group turbidites are associated with the base-metal deposits there. Apart from this volcanic activity along the eastern margin of the Darling Basin, there is no volcanism west of the WMB in the Silurian to mid-Devonian interval.

East of the WMB was a complete contrast. A series of meridional horsts and grabens associated with widespread silicic volcanism and subjacent granitoid intrusions occurred over a zone more than 400 km wide. The main sedimentary basins (at the latitude of Sydney) were the Cowra and Hill End Troughs, both of which were probably initiated in the Middle Silurian. Southwards (at the latitude of Canberra), the main sedimentary basins were the Tumut Trough and the Wollondilly Tract, a southward extension of which may have occupied eastern Victoria (VandenBerg et al.

1976). The more important characteristics of these sedimentary basins are summarized in Table 19.

The basins ('troughs') were separated by narrow horsts or broader platforms ('highs') on which rocks of the same age, but different facies, were deposited. The successions on the highs are thinner than those in the troughs, but contain abundant shallow-water limestone lenses which provide the main control on the age of both regions. For much of the Silurian to mid-Devonian interval, the whole region was submerged, with local volcanic islands marking the position of the highs. In the Early Silurian, most of the Lachlan Fold Belt east of the WMB was a broad shelf with carbonate patches, passing into a slope and abyssal plain in the east, but from the Wenlock onwards the palaeogeography was more complex and diverse. The horst-and-graben configuration in the eastern Lachlan Fold Belt lasted until the Middle Devonian when folding associated with the rise of the Tabberabberan Mountains destroyed the sedimentary basins, and led to the next major palaeogeographic configuration.

A time-space diagram for the northern line of basins is given in Fig. 204. A generalized time-space diagram for the Ordovician and Silurian either side of the WMB at the latitude of Canberra is given in Fig. 190. VandenBerg et al. (1976, table 4.1) have a time-space plot through the Melbourne Trough and eastern Victoria for the Silurian and Devonian at the latitude of Melbourne. Summaries of Silurian formation names, and their facies and correlation, are given in Talent et al. (1975), VandenBerg et al. (1976), Packham (1969), Pickett (1982), and Cas (1983). Devonian formations, facies, and correlations are given by Webby (1972), Roberts et al. (1972), Packham (1969), and Cas (1983). The palaeogeographic maps that follow are based on these studies, together with my own published and unpublished studies of palaeocurrents and provenance in the Devonian. Sources of palaeocurrent information are listed in the figure captions.

(4) Early Silurian

The first major deformation of south-eastern Australia took place at the end of the Ordovician, probably during the Llandovery. This deformation, called the *Benambran orogeny*, was centred over the former Wagga marginal sea, and led to the formation of the WMB – a high-temperature, low-pressure belt (Vallance 1967; 1969). The earliest folds known in the Lachlan Fold Belt are the latitudinally trending folds known from at least eight widely spaced localities (Fig. 195C). As discussed already, these early folds could have formed normal to the inferred plate-convergence

TABLE 19. *Characteristics of Silurian to Middle Devonian Basins in South-eastern Australia*

	Grampians	Melbourne Trough	Darling Basin (incl. Cobar Trough)
Extent and Width*	An arcuate line of outcrops elongate 100 km in N–S direction. Basin margins not positively determined, but may be Mt. Stavely High and Black Range, giving basin width of 50 km	Elongate NNW, more than 900 km long if extension to NE Tasmania is assumed. Preserved folded width ~ 200 km	Elongate NNW (~ 500 km) with width > 300 km. Eastern part includes narrow (~ 50 km) grabens in lowest part of section[3]
Shape in section, and palaeoslope	Not determined. Palaeoflow in part of the succession was northerly[1]	Asymmetrical with thickest part towards E edge. Palaeoslope consistently to E during sedimentation[2]	Asymmetrical with thickest part adjacent to E boundary. Palaeoslope consistently to E during sedimentation[4]
Depositional environment	Mostly deposited in fluviatile-lacustrine environment, with thin shallow-water marine intercalations	Deep-water turbidites in Early Silurian with more littoral environments in Siluro-Devonian passing into fluviatile environments in Middle Devonian[2]	Deep-water turbidites in Siluro-Devonian passing upwards through shallow marine environments to fluviatile by Middle Devonian[5]
Lithofacies and provenance	Well-rounded quartzose clastics derived from older Cambrian and ?Ordovician basement	Quartzose clastics of high mineralogical and low textural maturity derived from Gondwana craton to S + W.[1] Minor lithics and carbonate input from E	Dominantly quartzose clastics derived from craton to SW. Along the eastern edge, important silicic volcanics in the lower part of the succession
Maximum sediment thickness	6.1 km[1]	~ 10 km[1]	~ 10 km in E, of which[3,4] upper 4 km + are fluvial
Duration as a basin	?Silurian to ?early Middle or even Late Devonian (?40 Ma)	Earliest Llandovery to Eifelian[1,2] (55 Ma)	Pridolian to Tournaisian[6] (60 Ma)
Synsedimentary structure	Polymict conglomerate at base suggests active fault margin. Mafeking and Mackenzie River Granodiorites (Early to Middle Devonian) intruded half way through sedimentary history	Consistent E palaeoslope through history suggests active half-graben eastern boundary. Up-faulting occurred on the Waratah Bay axis (E margin) in Early Devonian[1]	E boundary fault active during Early Devonian. Volcanics near Mt. Hope deposited in cauldron structure.[7] Minor intrabasinal block adjustment in late Early or early Middle Devonian
Age of deformation	Broad folds, with steep dips on some marginal faults, probably formed during Middle Devonian (~ 375 Ma)[1]	Tight folding in late Middle Devonian (~ 375 Ma)[1,2]	Close (E margin) to broad folds in mid-Carboniferous[8] (~ 350 to 330 Ma)
Basement	Downfaulted continental crust formed over deformed Cambrian arc or marginal-sea material.	Thinned continental crust formed over Cambrian arc or marginal-sea material	Thinned continental crust over WMB basement in E and deformed ?Cambrian arc or marginal-sea material to W. Onlaps Precambrian Wonaminta Block NE of Broken Hill[9]

*All sedimentary basins in Lachlan Fold Belt are folded, so that width given is an estimate only.

Numbered references:

[1] Spencer-Jones (1965), VandenBerg et al. (1976).
[2] VandenBerg (1978), Garratt (1983).
[3] Glen (pers. comm.), Barron et al. (1982).
[4] Glen (pers. comm.), Glen et al. (1983), Powell et al. (1983).
[5] Ritchie (1973).
[6] Evans (1977), Sherwin (1980).
[7] Scheibner (pers. comm.), Barron et al. (1982).
[8] Glen (1982a), Powell et al. (1980a).
[9] Rose and Brunker (1969), Webby (1972), Powell et al. (1982).
[10] Killick (1982).
[11] Basden (1982), Richards et al. (1977), Lightner (1977), Owen and Wyborn (1979).
[12] Ashley et al. (1979), Scheibner and Pearce (1978), Brown (1980).
[13] Sherwin (1971).
[14] Pickett (1979), Powell and Cranc (unpub.work).
[15] Jenkins in Crook and Powell (1976, p. 62).

Tumut Trough	Cowra Trough	Hill End Trough	Wollondilly Tract †
Elongate NNW (~150 km) with width only a few 10's km	Elongate N–S, about 150 km long, up to where buried by younger cover rocks in the north. Width about 50 km	Elongate NNW, about 150 km long, up to point where buried by younger cover rocks in the north. Possible further 100 km extension south of Bathurst Granite. Width ~100 km after unfolding	Elongate NNE, about 200 km long, and ~30 km wide. May contain more than one narrow trough (e.g. Murruin and Captains Flat Troughs[23]), and may extend another 130 km SSW into Eastern Victoria
Symmetrical as far as one can tell. Axial transport of turbidites to NNW (Bumbolee Ck Fmn) or to SSE (Wyangle Formation)[10]	Symmetrical as far as one can tell. Axial transport of turbidites to N[8], and basinwards transport on eastern margin[13]	Late Silurian wedge thins to ENE. Early Devonian wedge thickens to ENE.[16] Axial transport to NNW with basinwards transport on both margins[17]	Not determined
Deep-water turbidites deposited by axial flow NNW[10]	Terrestrial to marine silicic volcanics during trough initiation in Middle Silurian, with turbidites in Late Silurian and Early Devonian. By late Early Devonian shallower marine conditions existed with bimodal volcanism	Terrestrial to marine silicic volcanics during trough initiation in Middle Silurian,[16, 18] with quartzose turbidites in Late Silurian, submarine silicic volcanics in Early Devonian,[17, 19] and volcaniclastics in late Early Devonian	Varied from terrestrial through shallow to deep marine environments, associated with dacitic to rhyolitic volcanic piles[24, 25]
A mixture of quartzose clastics off uplifted WMB to W (Bumbolee Ck Fmn), and volcanolithic sediment from E (Wyangle Formation)	Volcanolithic detritus until late Early Devonian[14] when bimodal volcanics were erupted. These are paraconformably overlain by quartzose clastics of the Hervey Group[19]	Early sediments derived from relicts of uplifted WMB to SW, late sediments dominantly volcaniclastics from coeval volcanic centres to E and SE[17, 19]	Dominantly silicic volcanic provenance for clastic sediments. Local carbonates redeposited in deeper water
Not yet determined[11]	≥3 km	5 km[16]	Not yet determined
Wenlockian to Ludlovian[11] (≤15 Ma)	Wenlock[15] to Emsian[14] (30 Ma)	Wenlock[16, 20] to Emsian[16] (30 Ma)	Ludlovian to Emsian[25] (?25 Ma)
Presumed transcurrent motion on basin margins during sedimentation. Synsedimentary recumbent folds formed by SSE to NNW gravitational gliding[10, 11]	Slump deposits[13] on eastern margin suggest active fault boundary with the Molong High. Nature of western boundary not known	Slump deposits in submarine canyon on west margin,[21] with local slumps throughout turbidite succession indicating active fault margins. Capertee High on eastern margin not established until earliest Devonian[16]	Slump deposits suggest active fault margins
End-Silurian[11] (~410 Ma)	Broad folding and uplift in Middle Devonian (380 to 370 Ma) but main folding in Early Carboniferous (350 to 340 Ma)[8]	Broad folding and uplift in Middle Devonian (380 to 370 Ma) but main folding in latest Devonian or Early Carboniferous (360 to 340 Ma)[22]	Broad folding in N becoming tight southward in mid-Devonian (?380 to 370 Ma)[26] with folding in latest Devonian or Early Carboniferous (360 to 340 Ma) decreasing in intensity south
Oceanic crust represented[12] by ophiolite in the Coolac and Goobarragandra Serpentinite Belt	Thinned continental crust formed over relicts of the deformed Ordovician volcanic island arc and accretionary prism	Thinned continental crust formed over relicts of the deformed Ordovician volcanic island arc and accretionary prism	Thinned continental crust formed over relicts of the deformed Ordovician accretionary prism

† Name suggested by J. G. Jones (unpub. MS).

[8] Packham (1966, 1968, 1969), Offenberg et al. (1971).
[10] Cas (1978a, 1979), Cas and Jones (1979).
[11] Hilyard (1981).
[12] Cas (1978b).
[13] Bischoff and Fergusson (1982).
[14] Conaghan et al. (1976), Russell (1976).
[15] Powell (1976a), Cas et al. (1976), Powell et al. (1977), Powell and Edgecombe (1978).

[23] Scheibner (1974b).
[24] Jones et al. (1977), Fergusson (1980), Cas et al. (1981), Carr et al. (1981).
[25] Sherwin in Scheibner (1973), Felton (1974), Felton and Huleatt (1975, 1976).
[26] Powell and Fergusson (1979a,b).

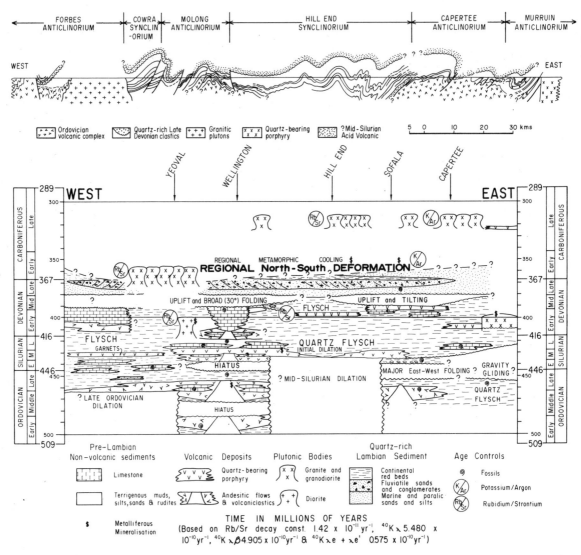

Fig. 204. Geological section and time–space diagram from Parkes to the western edge of the Sydney Basin. Location of section indicated on Fig. 203. Section has been updated and extended west from Crook and Powell (1976).

vector when the convergence changed from decoupled Marianas-type subduction to coupled Chilean-type.

The dominant structure of the WMB trends NNW, and overprints the early latitudinal folds. These NNW-trending folds, associated with metamorphism locally reaching the sillimanite isograd, did not extend west into the Melbourne Trough, or east of Tumut. In the Sofala region, just west of Sydney, the early latitudinal folds are present but the meridional overprint did not form until the latest Devonian or Early Carboniferous (Powell *et al.* 1977, 1978).

The Early Silurian palaeogeography (Figs. 205, 206) shows the distribution of NNW-trending folds as a rising fold belt (Benambran Landmass) separating the Melbourne Trough to the west from a broad platform bounded by a slope and abyssal plain in the east. In the Melbourne Trough (which had its inception as a trough from the time the Benambran Landmass began to

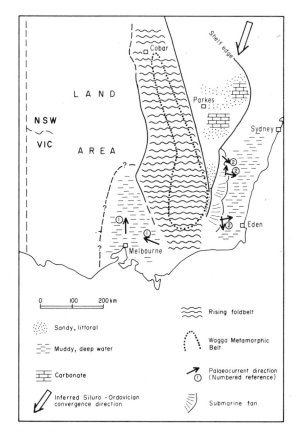

Fig. 205. Early Silurian palaeogeography of south-eastern Australia. Numbered palaeocurrent sources: 1, Garratt (1983); 2, Crook et al. (1973). The boundary between the rising fold belt and the land area to the west corresponds with that between geophysical domains 2 and 3 (Wyatt et al. 1980).

rise), sedimentation was continuous from the preceding Ordovician interval. The transition from the Ordovician to the Silurian configuration is marked only by a change from shale to siltstone deposition (VandenBerg et al. 1976), with a continuation of the mineralogically mature quartzose sediment derived largely from the south. The western edge of the Melbourne Trough is not certain at this time because of the lack of basin-margin facies (Garratt 1983). No sediments of this age are known in New South Wales west of the WMB, suggesting that western New South Wales was probably a land area at this time.

East of the WMB, shallow-water limestones, locally-derived conglomerates, sandstones, and shale occupied a broad shelf area that extended to the eastern edge of the Lachlan Fold Belt west of Sydney. The littoral late Llandovery to early Wenlock Tanwarra Shale (Packham 1968, 1969; also Bischoff and Fergusson 1982) was deposited at this time. There may have been deeper-water areas within this shelf developed over the deformed Ordovician volcanic arc (for example near Parkes: Pickett 1982), but there is no evidence of the meridional Cowra, Hill End, and Tumut Troughs at this time. In the Canberra region, Crook et al. (1973) showed that the Llandoverian facies occupied a shelf with a slope and abyssal plain further east. Quartzose clastics derived from the uplifted Benambran Landmass were deposited in submarine fans on the slope near Canberra, with the abyssal plain forming over the outer part of the former Ordovician accretionary prism (Fig. 206). Crook et al.'s (1973) 'Quidongan orogeny' is probably local movements on the eastern edge of the rising Benambran Landmass. The recumbent folds near Queanbeyan (Stauffer and Rickard 1966) may have formed at this time. The Early Silurian configuration probably lasted into the early Wenlock (c. 425 Ma). No volcanic rocks are known to have been erupted during the Early Silurian.

(5) Mid-Silurian to mid-Devonian

(a) Wenlock and Ludlow. The meridional horsts and grabens of the eastern Lachlan Fold Belt began to form during the Wenlock, and were well established by the Ludlovian (Fig. 207). The first indication of graben development was the widespread eruption of rhyolitic to rhyodacitic volcanics. In the Cowra Trough, the garnetiferous Canowindra Porphyry was emplaced; in the Hill End Trough region, the Gleneski Formation, the Mullions Range, Bells Creek, Toolamang, Dungeree, and Kangaloolah Volcanics were erupted; in the Canberra region, the silicic-volcanic Douro Group and Goobarragandra Volcanics were formed; and in eastern Victoria the Mitta Mitta Volcanics were probably erupted at this time. Many of the volcanic units were emplaced subaerially, but had lateral equivalents deposited in marine conditions, indicating that the grabens had begun to subside. In the Cowra and Hill End Troughs, these silicic volcanics are overlain by turbiditic sandstones, suggesting that subsidence outpaced sedimentation. The stratabound base-metal deposits at Woodlawn, north-east of Canberra, were formed at this time (Gilligan et al. 1979) in a Kuroko-type setting first recognized along strike by Stanton (1974).

In eastern Victoria, the Mitta Mitta Volcanics (age constrained only as post-Ordovician and pre-Ludlovian: Talent et al. 1975) form the base of the Mitta Mitta Trough, possibly a southern continuation of either the Tumut or Cowra Troughs (VandenBerg

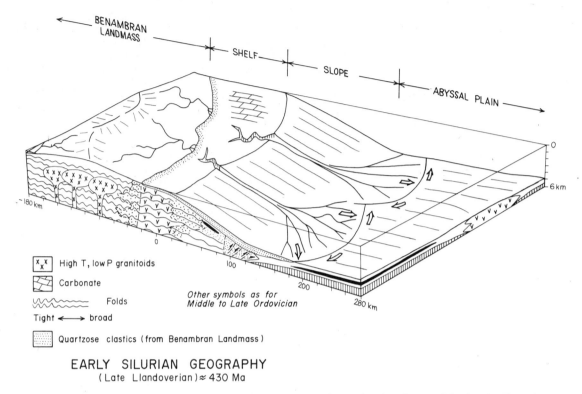

Fig. 206. Scaled block diagram of Early Silurian geography. Symbols, other than those explained, as for Fig. 192.

1978). The largely terrestrial Mitta Mitta Volcanics were overlain by a marine-shelf limestone, which was removed by erosion before deposition of a 3-km-thick terrigenous clastic succession, dominantly siltstone, of the Wombat Creek Group (VandenBerg et al. 1976).

The Tumut Trough is unique amongst these mid-Silurian grabens in that it probably had a floor of oceanic crust (Ashley et al. 1979; Brown 1980). The Coolac-Goobarragandra Serpentinite Belt, exposed along the eastern margin of the Trough, includes a succession of serpentinized harzburgite and rodingite (Coolac Serpentinite) overlain by metabasalts (Honeysuckle Beds) (see Crook and Powell 1976, fig. 4–3). A sheeted-dyke complex exists further north (Brown 1980). The ophiolite, now standing nearly vertical, is overlain conformably by quartzose turbidite of the Bumbolee Creek Formation, and andesitic volcanolithic turbidites of the Wyangle Formation. Age control within the trough succession is poor, but the presence of late Llandoverian to early Wenlockian conodonts in allochthonous limestone bodies in the Wyangle Formation (Lightner 1977) suggests that the Tumut Trough had begun to open by the Wenlockian. The trough was folded, uplifted, and eroded before the Early Devonian Gatelee Ignimbrite was deposited. I consider the Tumut Trough to be a pull-apart basin developed along the eastern margin of the cooling WMB (Fig. 208) by dextral transcurrent motion. The master fault was probably in part the Gilmore Fault Zone (Basden 1982), and stands out in regional magnetic and gravity patterns of the Lachlan Fold Belt as the discordant boundary between geophysical domains 1 and 2 (Wyatt et al. 1980).

In the Melbourne Trough, Middle and Late Silurian sedimentation was a continuation of the Early Silurian pattern. Quartzose clastics derived from the uplifted Ballarat region to the west, and from the Gondwanaland craton to the south were dispersed by turbidity currents. In the Bradford–Kilmore region, there is evidence of an ENE-draining submarine canyon and fan (Garratt 1983). Very little of the total volume of sediment was supplied from the subsiding Benambran

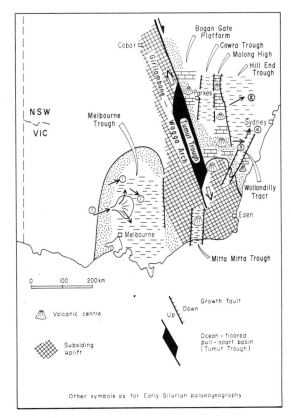

Fig. 207. Late Silurian palaeogeography of south-eastern Australia. Numbered palaeocurrent sources: 1, Garratt (1983); 2, Chesleigh Formation, in Packham (1968); 3, Argyle Formation in Cowhorn Creek, in Cas *et al.* (1981); 4, undifferentiated Taralga Group in Little Wombeyan Creek, Powell (unpubl. data). Boundary between the Girilambone–Wagga arch and the Tumut Trough and Bogan Gate Platform corresponds with that between geophysical domains 1 and 2 (Wyatt *et al.* 1980). Symbols, other than those explained, as for Fig. 205.

Landmass to the east, and there is no record of sediment of this age in New South Wales, west of the Girilambone–Wagga arch.

(b) Pridoli and Lochkovian. The sedimentary patterns established in the Late Silurian continued uninterrupted in the eastern part of the Lachlan Fold Belt into the Early Devonian (Fig. 209). The Cowra and Hill End Troughs were well established, with shallow marine to emergent conditions on the intervening Molong High. At the eastern edge of the Lachlan Fold Belt, a line of volcanoes with fringing carbonate aprons, the Capertee High, rose to provide an eastern margin to the Hill End Trough, whose basin floor, previously sloping east-northeasterly, now sloped back towards the west-south-west (Fig. 209). Southwards, the South Coast High (= Kemezys's (1978) Budawang Land) bordered the Wollondilly Tract.

West of Canberra, the Tumut Trough was folded, uplifted, and eroded in the Pridoli–Lochkovian interval, and at approximately the same time new volcanic grabens developed on the western side of the WMB in the Cobar and Mt. Hope regions (Glen 1982b; Barron *et al.* 1982). In eastern Victoria, the Mitta Mitta Trough was probably folded before intrusion of the earliest Devonian granitoids and extrusion of the Jemba Rhyolite and Snowy River Volcanics (Talent 1965; VandenBerg *et al.* 1976). The Girilambone–Wagga arch continued to subside, and shallow-marine conditions with carbonate patches, conglomerates, and sandstones formed on the shallowly submerged parts between Cobar and Parkes (Sherwin 1980).

The Darling Basin, of which the Cobar Trough and Mt. Hope volcanic-tectonic depression are early sub-basins, began to subside in the Pridoli–Lochkovian, and by the late Lochkovian (*c.* 405 Ma), a regional palaeoslope towards the east and north-east was established. There may well have been a connection between the Melbourne Trough and the Darling Basin at this time (Fig. 209), although this needs to be demonstrated. Early Devonian sediment in most of the Darling Basin west of the Cobar Trough is shallow-marine sandstone, siltstone, and shale. The marine succession is thinnest in the west (a few hundred metres in the Coppermine Range area north-east of Broken Hill), and thickens to more than 6 km in the Cobar region where most of it was deposited on turbidite fans. In view of the later regression of fluvial facies eastward across the deep marine successions, with transitional shallow-marine and littoral facies developed in between, I consider it likely that the base of the fluvial facies in the west is older than in the east, and that the facies now preserved in vertical succession (Glen 1982b; Glen *et al.* 1983) existed as lateral equivalents in the Early Devonian.

In the Melbourne Trough to the south, the first good evidence of a nearby shoreline is provided by the large-scale cross-bedded Mt. Ida Sandstone, and the continual shallowing of the Trough is suggested by the migration eastward of benthic communities (Garratt 1983). Sediment was still supplied dominantly from the west, and the basin floor sloped eastward.

The earliest Devonian is the time of the Bowning 'orogeny' (Brown 1954). In its type area, the Bowning unconformity is between mudstone of the Elmside Formation (Lochkovian) and the overlying late

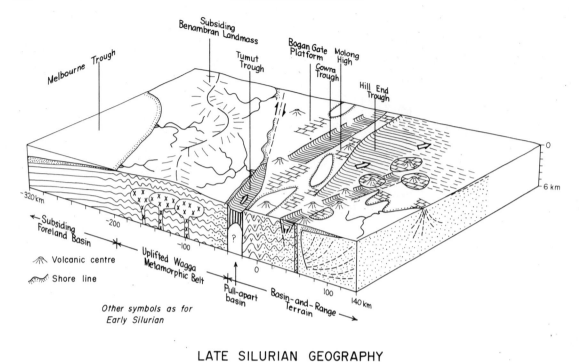

LATE SILURIAN GEOGRAPHY
(Ludlovian ≈ 415 Ma)

Fig. 208. Scaled block diagram of Late Silurian geography. Symbols, other than those explained, as for Figs. 192 and 206.

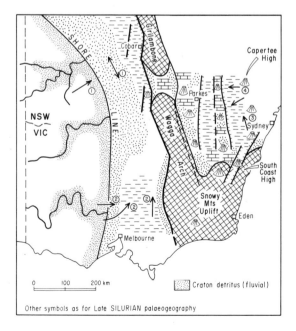

Lochkovian or Pragian Sharpeningstone Conglomerate. The unconformity is exposed in Derrengullen Creek (Stop 5–1, Crook and Powell 1976), where in the core of a small anticline, the conglomerate has gentle dips compared with the steep dips in the underlying mudstone. Near the contact, however, dips in the mudstone are very similar to those in the conglomerate, so that much of the apparent angular discordance could be caused by small-scale parasitic folding in the thinly bedded mudstone (cf. the now-discounted Ordovician unconformity on the New South Wales

Fig. 209. Earliest Devonian palaeogeography of south-eastern Australia. Based on Webby (1972, fig. 1). Numbered palaeocurrent sources: 1, Winduck Group and Meadows Tank Formation extrapolated from Glen *et al.* (1983), Powell *et al.* (1983), and Conaghan and Powell (unpubl. data); 2, Garratt (1983); 3, Kowmung Volcaniclastics, Cas *et al.* (1981); 4, Crudine Group, Packham (1968). Girilambone–Wagga arch has subsided since the Late Silurian. Symbols, other than those explained, as for Figs. 205 and 207.

South Coast: Etheridge et al. 1973). The Sharpeningstone Conglomerate cuts down only half-way through the Elmside Formation before lensing out.

The reason for mentioning these relationships is that the Bowning 'orogeny' is commonly credited with causing widespread deformation in the Lachlan Fold Belt. At Mt. Bowning, the deformation was weak, with questionable angular discordance and certainly no cleavage developed. The Tumut Trough was deformed at this time, and the first major folding of the Ordovician turbidites on the New South Wales South Coast took place between the deposition of the Ludlovian limestones and emplacement of the c. 390 Ma Bega Batholith. Similarly, in the Wollondilly Tract, Carr et al. (1981) argue that major folding occurred around this time. Away from this area, along the eastern margin of the WMB, and on the New South Wales South Coast, sedimentation continued uninterrupted. The Cowra, Hill End, and Melbourne Troughs show no sign of the Bowning orogeny, which thus may have been related to changes in the pattern of transcurrent movement that earlier had opened the Tumut Trough. Folds in the Tumut Trough associated with its Siluro-Devonian closure trend 300° (Killick 1982), markedly oblique to the north-north-west elongation of the trough. The symmetry of these folds in relation to the trough margins suggests that closure was by dextral transcurrent movement. This movement picture does not require any fundamental change to the pre-existing movement pattern: a small change in attitude of the master fault so that the Tumut Trough went from transtension to transpression is all that is required.

(c) Early Devonian. Volcanism with associated sedimentation continued in the east through the Early Devonian (Fig. 210). The 1-km-thick Merrions Tuff was emplaced by a succession of mass-flow events in the Hill End Trough (Cas 1978a; 1979), where silicic lava may have been erupted in water more than 2 km deep (Cas 1978b). To the south, the Canberra region was a series of meridional horsts and grabens in which terrestrial volcanics accumulated (Strusz 1971). Cas and Jones (1979) compared the geography of the eastern part of the Lachlan Fold Belt at this time with that of the modern Taupo Volcanic Zone of New Zealand.

In the Buchan area of eastern Victoria, the thick, silicic, mainly non-marine Snowy River Volcanics were succeeded by shallow-water limestones and intercalated terrigenous sandstones and shales (Talent 1965; Talent in VandenBerg et al. 1976). The Melbourne Trough

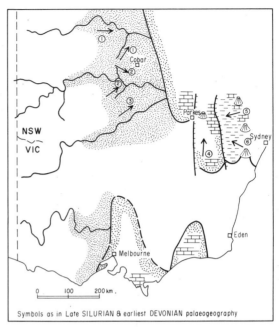

Fig. 210. Late Early Devonian palaeogeography of south-eastern Australia. Numbered palaeocurrent sources: 1, 2, and 3 =Meadows Tank and Merrimerriwa Formations in: 1, Conaghan and Powell (unpubl. data); 2, Glen et al. (1983); 3, Powell et al. (1983); 4, Moura Beds in Powell et al. (1980a); 5, Merrions Tuff and Cunningham Formation, Packham (1968); 6, Merrions Tuff, Cas (1978a). Symbols as for Figs. 205, 207, and 209.

continued to shallow, and by the end of the Early Devonian, shallow-water facies extended throughout.

In the Darling Basin, the earliest Devonian submarine volcanic deposits were buried by quartzose turbidite fans, which in turn were succeeded by the shallow-water quartzarenites of the Winduck Group. By the late Early or early Middle Devonian, the Winduck Group had been overridden by the pebbly fluvial Meadows Tank Formation at the base of the entirely fluvial Mulga Downs Group. There may have been some minor block adjustment in the late Early Devonian between the deposition of the littoral facies of the Winduck Group and the pebbly Meadows Tank Formation, but there was no regional folding or deep erosion by the fluvial facies into the Winduck Group (Glen 1982a; Glen et al. 1983; Powell et al. 1983). Regionally, the Mulga Downs and Winduck Groups are paraconformable. The fluvial sediments of the Darling Basin probably overtopped the sunken

Girilambone arch, and were deposited in a shallow sea over the former Bogan Gate Platform.

At the end of the Early Devonian, volcanism was restricted to the Capertee High in the east, and south-eastern Australia was beginning to rise above sea level.

(d) Middle Devonian. The Middle Devonian is marked by the rise of the Tabberabberan Highlands in south-eastern Australia. It was a time of regression, and, in southern New South Wales and Victoria, of deformation. In the north-west Lachlan Fold Belt, sedimentation continued, with a new supply of coarse detritus from the south.

In the Eifelian (Fig. 211), sedimentation ceased in all the eastern troughs: at least, there is no Eifelian sediment preserved except shallow-water clastics in the Mudgee district, north-west of Sydney. Terrestrial debris-flow deposits comprising the lower part of the Hatchery Creek Conglomerate west of Canberra, may be of this age: they are paraconformable above the Early Devonian Taemas Limestone, and conformably overlain by redbeds containing fossil fish probably of late Eifelian age (Young and Gorter 1981).

In the western depositional realm, the fluvial Mulga Downs Group, which contains Emsian or Eifelian fossil fish in the Merrimerriwa Formation near its base (Ritchie 1973), is conformable into the Early Carboniferous. In the Cobar region, the Merrimerriwa Formation is overlain by quartz- to sublith-arenite of the Bulgoo Formation, which was deposited on the southwesterly draining arm of a trunk stream flowing south or south-east (Glen *et al.* 1983; Powell *et al.* 1983). The lithic detritus is presumably derived from rejuvenated uplift of the Girilambone–Wagga arch (Fig. 211). In the Melbourne Trough, the fluvial Cathedral Beds overlie the Early Devonian marine succession, and may represent deposits of the trunk stream from the Darling Basin draining southwards. Alternatively, the Darling Basin trunk stream may have broken across the Girilambone–Wagga arch and drained eastward near Parkes. Palaeocurrent measurements in critical places have not yet been made.

The Givetian saw a big change in the palaeo-drainage network (Fig. 212). In the Cobar region, coarse polymictic conglomerate and pebbly sandstone were shed from the upstanding Tabberabberan Mountains to the south-east; boulders and cobbles, generally of hornfelsed fold-belt sediment, were carried in braided streams down alluvial fans that fringed the northeasterly draining trunk stream (Powell *et al.* 1983). The southward-draining trunk stream of the preceding Bulgoo Formation was evidently blocked by the rise of the Melbourne Trough as it commenced to fold. Tight folds with axial-surface slaty cleavage suggest 30 to 50 per cent east–west shortening in the former Melbourne Trough at this time.

Elsewhere in south-eastern Australia, there are two locations of Givetian sediment: (a) the redbed deposits above the late Eifelian fossil fish west of Canberra (Young and Gorter 1981) may be intra-montane fluvio-lacustrine deposits; (b) along the New South Wales South Coast, a narrow rift zone, the Eden–Comerong–Yalwal Rift Zone, extends over 300 km from Gabo Island to where it disappears under the Sydney Basin. The Eden–Comerong–Yalwal Rift contains fossil fish of late Givetian or early Frasnian age (Young in Fergusson *et al.* 1979), and genetically belongs with the Late Devonian–Early Carboniferous palaeogeographic configuration, as described below.

The Middle Devonian is the time of the Tabberabberan Orogeny, and, as with earlier orogenies in the Lachlan Fold Belt, the Tabberabberan has variable

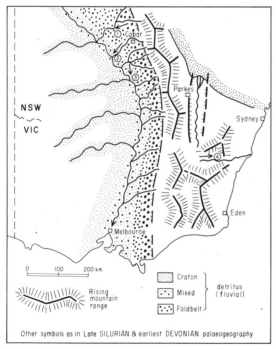

Fig. 211. Early Middle Devonian palaeogeography of south-eastern Australia. Numbered palaeocurrent sources: 1, 2, and 3=Bulgoo Formation in: 1, Conaghan and Powell (unpubl. data); 2, Glen *et al.* (1983); 3, Powell *et al.* (1983); 4, Hatchery Creek Conglomerate, Powell (unpubl. data). Symbols, other than those explained, as for Fig. 205.

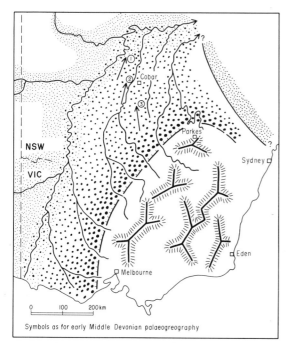

Fig. 212. Late Middle Devonian palaeogeography of south-eastern Australia. Numbered palaeocurrent sources: 1, 2, and 3=Bundycoola Formation in: 1, Conaghan and Powell (unpubl. data); 2, Glen et al. (1983); 3, Powell et al. (1983). Onset of bimodal volcanism in the Eden–Comerong-Yalwal rift system was latest Middle Devonian (Fergusson et al. 1979), and is shown in Fig. 221. Symbols as for Figs. 205 and 211.

effects throughout the region. The greatest deformation, associated with upright meridional folds and cleavage, is in the south where not only southern New South Wales and Victoria, but also Tasmania, were deformed (Williams 1978). On the New South Wales South Coast, the deformation is tightly constrained between the intrusion of the Early Devonian Bega Batholith (390 to 380 Ma) and of granites associated with the late Givetian or early Frasnian rift system (370 Ma: Powell 1983a). Similar tight age-constraint exists in the Melbourne Trough area, where the deformation is most likely Givetian (VandenBerg et al. 1976).

The Tabberabberan deformation represents an east–west shortening, and is associated with large conjugate transcurrent faults of both dextral and sinistral offset. White et al. (1976) have demonstrated an 11 km sinistral offset on the Siluro-Devonian Berridale Batholith by the northwesterly trending Berridale Wrench, and Beams (1975) has demonstrated 24 km of dextral offset of part of the Early Devonian Bega Batholith on the northeasterly trending Burragate Fault. Sixteen km of dextral offset occur on the parallel Tantawangalo Fault (Sims in Powell 1983a). VandenBerg (1978) has even suggested 150 to 180 km of dextral offset on the northeasterly trending Indi-Long Plain fault system, which, if valid, implies considerable translocation of the south-eastern third of the Lachlan Fold Belt towards the south-west. The palaeogeographic maps of south-eastern Australia presented here do not show these possible offsets, as they are not yet sufficiently well documented, but clearly they have important implications for making any truly palinspastic reconstruction.

(ii) Major facies distribution elsewhere in the Tasman Orogen

The facies contrast between north-eastern Tasmania and the western two-thirds of the island, established in the Ordovician, continued without significant change through the Silurian into the Early Devonian. In the western part, Silurian sediments were shallow-water sandstones and siltstones, and by the Devonian a few limestone lenses were also being deposited (Banks 1962). In north-eastern Tasmania, quartzose turbidites derived from the south-west (Williams 1959; Powell, unpublished data) formed the apparently conformable Early Ordovician to Early Devonian Mathinna Beds. There is no interfingering facies between the shallow-water western shelf domain and the deep-water Mathinna Beds, suggesting that the join between the two areas of sedimentation, the Tamar fracture system (Williams 1978), may be a fault of considerable lateral displacement. The sequences both east and west of the Tamar fracture system were folded in the Middle Devonian, and subsequently intruded by Middle to Late Devonian granitoids (Cocker 1982).

The pre-Devonian record in north-eastern New South Wales is sparse, with Silurian sediments preserved only in isolated fault slices from which regional facies relationships cannot be established. In the southern part of the New England Orogen (Fig. 213), patches of Silurian limestone and mafic volcanics occur as isolated bodies in a matrix of sheared, commonly younger, pelite, suggestive of olistoliths caught up in a tectonic *melange*. The general aspect of these rocks, the Woolomin Beds, is compatible with deposition at the foot of an accretionary prism, or an abyssal plain later incorporated into an accretionary prism.

By latest Silurian, there was asymmetry in the facies being deposited, with a mafic volcanic arc supplying detritus to an unstable shelf deepening progressively eastward (Tamworth Trough), and eventually passing

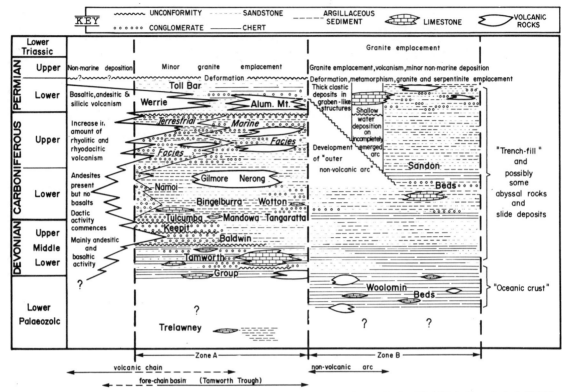

Fig. 213. Schematic representation of possible relationships in southern part of New England Fold Belt. From Leitch (1974).

into the abyssal plain or trench deposits of the Woolomin Beds (Leitch 1974). This east-facing mafic volcanic arc is also present in the Queensland part of the New England Orogen, where Day et al. (1978) identified a Late Silurian to Middle Devonian volcanic arc and unstable marine shelf on the west, facing a zone of submarine basic volcanics, pelagic sediments, and thick flysch sequences on the east (Fig. 214A). In both the New South Wales and Queensland segments of the New England Orogen, the western volcanic arc and unstable marine shelf are separated from the eastern deep-water deposits by an ultramafic belt. Their similar history lasted until the Middle Devonian, when the Calliope Island Arc was deformed and uplifted (Day et al. 1978). In contrast, the southern segment continued to shed and receive sediment until the Late Carboniferous or Early Permian (Fig. 214B), and was not deformed until the mid-Permian (Leitch 1974).

In central-western Queensland, shallow-water to terrestrial sediments accumulated in the Adavale Basin (Auchincloss 1976; Tanner 1976). The oldest unit, the Early Devonian Gumbardo Formation, is terrestrial andesitic lava flows and lithic tuffs, which grade eastward into arkosic clastics. The overlying Middle Devonian sediments contain two facies, (a) a lower sandstone and shale, and (b) an upper limestone, which together record a transgression of the sea over the volcanic Gumbardo Formation. The limestone contains an abundant selection of relatively shallow-water fossils, and becomes increasingly dolomitic and barren upwards to where it is overlain by the Givetian evaporite-bearing Etonvale Formation. The overlying Buckabie Formation was deposited under terrestrial to shallow-marine conditions, and spans the Late Devonian to Early Carboniferous. The Adavale Basin was folded in the mid-Carboniferous.

The Hodgkinson and Broken River Provinces of north Queensland (Henderson 1980) have a more complete Late Silurian to Late Devonian succession, which comprises a volcanic arc and unstable marine shelf to the west, with thick slope and basin flyschoid sequences in the east. Neritic clastic and carbonate deposits fringing the intermediate to silicic Andean-type volcanic chain to the west pass eastward into flysch that contains mafic lavas, isolated (probably

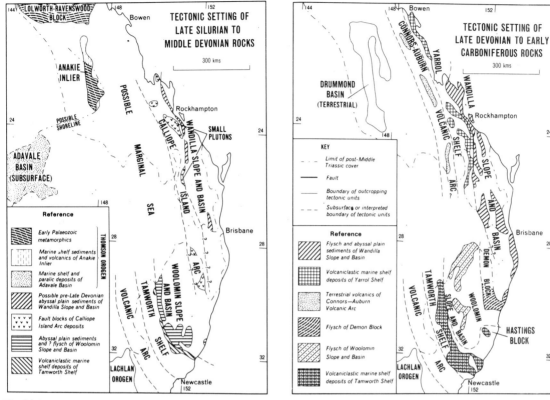

Fig. 214. A. Tectonic setting of Late Silurian to Middle Devonian rocks, New England Orogen. B. Tectonic setting of Late Devonian to Early Carboniferous rocks, New England Orogen. From Day et al. (1978).

allochthonous) limestone lenses, and abundant igneous, metamorphic, and volcanogenic detritus (Arnold and Fawckner 1980). The Hodgkinson and Broken River Provinces were folded in the latest Devonian, with another major deformation at the end of the Early Carboniferous (Day et al. 1978).

(iii) Tectonic models

(1) Early Silurian transform margin (430 ± 5 Ma)
Following the folding of the mafic volcanic arc and Wagga marginal sea at the end of the Ordovician, the Tasman Orogen was devoid of volcanics for the brief Early Silurian interval. The tectonic reconstruction (Fig. 215) shows the north-northwesterly elongate welt of the uplifted Benambran Landmass bordering the Melbourne Trough to the west, and a continental slope and abyssal plain to the east. The region may well have resembled the Californian part of the North American continent, where the uplifted Late Jurassic to Eocene accretionary prism faces on to the oceanic crust of the Pacific Plate. There are no Recent volcanoes in this part of the North American Plate boundary, and the margin is a transform zone connecting the East Pacific Rise in the Gulf of California to the Juan de Fuca Ridge off Oregon. The major plate rotations are accommodated by transcurrent offsets on the San Andreas Fault and its precursors (Atwater 1970; Barrash and Venkatakrishnan 1982). Similarly, the major plate motion between Gondwanaland and the Palaeo-Pacific Plate may have been accommodated along a north- to northeasterly trending dextral transcurrent fault system running the length of the Tasman Orogen.

A branch of this system may have run down the eastern side of the Wagga Metamorphic Belt. A small bend in the trace of this fault (Fig. 215) would have been sufficient to nucleate the Tumut pull-apart basin, and allow it to develop progressively during the offset (cf. pull-apart basins on the San Andreas fault system). Not all the plate motion need have been taken

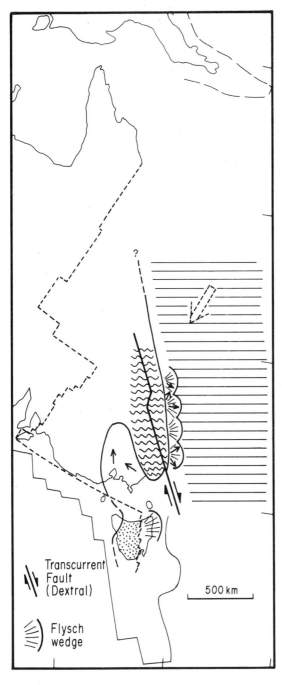

Fig. 215. Palaeogeographic reconstruction of the Tasman Orogen in the Early Silurian (c. 430 Ma). Dotted arrow is the inferred Late Ordovician convergence direction. Broken line is the Tasman Line. Other symbols as for Figs. 201 and 202.

up on a single fault that opened the Tumut Trough: the displacement could have been distributed over several faults.

(2) Mid-Silurian to mid-Devonian dextral transtensional phase (420 to 380 Ma)

One consequence of regional shear in the Early Silurian was the development by the mid-Silurian of a 400-km-wide zone of meridional horsts and grabens in south-eastern Australia. The orientation of these fault blocks is compatible with the shear being dextral, as it had been since the outset in the Ordovician, and, arguably, before that (Fig. 216A). Some grabens, such as the Wollondilly Tract, trended more easterly than others, but all were still in an orientation compatible with regional dextral shear. In the north Queensland segment of the Tasman Orogen, a continental volcanic arc faced eastward over an unstable shallow-marine shelf on to slope and abyssal plain deposits, suggesting that westward subduction was taking place in that region.

The Californian segment of the North America–Pacific Plate boundary provides a reasonable analogue for the Late Silurian–Early Devonian configuration of the Tasman Orogen (Fig. 216B). The zone of crustal extension in south-eastern Australia could have been analogous to part of the present-day Basin and Range system in the western United States of America. A tectonic sketch of the North American–Pacific Plate boundary, inverted so that north is towards the lower left, shows the San Andreas transform system passing into mid-ocean ridges both to north and south. Opposite the mid-ocean ridge segments, the boundary between the North American Plate and the oceanic plate lying immediately west changes from a transform boundary to a convergence zone. Grabens in the Basin and Range Province trend at angles of 20° (southern Basin and Range, #14) to 50° (northern Basin and Range, #6) to the transform fault system.

A possible match between the Late Silurian configuration of the Tasman Orogen and the western North American Plate is shown by the superimposed dotted outline of Australia in Fig. 216B. The directions of Late Silurian horsts and grabens in south-eastern Australia can be matched with those in the modern Basin and Range, and the trend of the Tumut Trough pull-apart basin is close to the trend of the transverse lineaments such as that of Walker Lane. However, the match is not perfect, because there is no indication of a Siluro-Devonian plate convergence zone off eastern Tasmania, and the modern Basin and Range is much larger than the maximum inferred extent of grabens in south-eastern Australia. Moreover, despite nearly 100

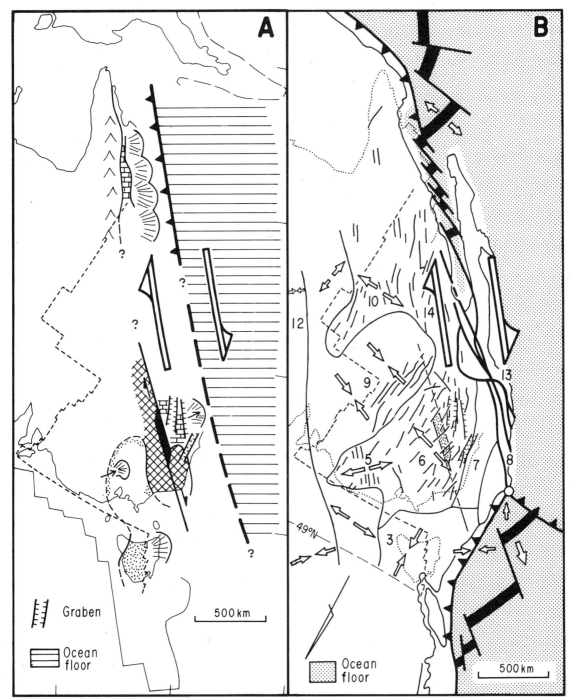

Fig. 216. A. Plate-tectonic interpretation of the Late Silurian palaeogeography of the Tasman Orogen (c.415 Ma). Dextral couple is shown along inferred transform margin. Symbols, other than those explained, as for Figs. 201, 202, and 215B. Modern elements of plate-tectonic interpretation of the Pacific coast of North America, inverted so that north points to lower left (after Barrash and Venkatakrishnan 1982). The converging arrow-heads represent the direction of regional compression, and the diverging arrows extension. The outline of Australia in the proposed match is dotted, and the graben structures of south-eastern Australia are superimposed. Numbers refer to tectonic provinces as numbered by Barrash and Venkatakrishnan (1982, fig. 1): (3) Pacific northwest, (5) Snake River Plain, (6) northern Basin and Range, (7) Sierra Nevada, (8) San Andreas Fault, (9) Colorado Plateau, (10) Rio Grande Rift, (12) Mid Continent, (13) San Andreas–Transverse Ranges, (14) southern Basin and Range. Spreading ridges are thick segments offset by thin transform faults. Screened area is oceanfloor.

per cent crustal extension in the Basin and Range (Eaton 1982), the region is still above sea level, whereas south-eastern Australia in the Siluro-Devonian was submerged.

A point of interest in this analogy is the complexity of contemporary stress directions (expressed simply in Fig. 216B as directions of compression and extension) in the North American Plate. The general north-west–south-east extension in the Basin and Range and Rio Grande Rift (#6 and #10, Fig. 216B) gives way to north-west–south-east compression in the Colorado Plateau (#9), and very complex stress distributions occur in the Pacific North-west (#3). Interesting correlations of Palaeozoic structure in the Tasman Orogen with contemporary stress in the North American Plate occur in the Snake River Plain (#5) and the Colorado Plateau (#9). In the former case, the modern extension direction transposed to the Tasman Orogen is grossly east–west, and provides a possible explanation for the meridional graben in which the ?Siluro–Devonian Grampians Group could have been deposited (VandenBerg et al. 1976). In the latter case, the transposed principal compression direction is south-east–north-west, nearly perpendicular to the trend of the Palaeozoic gravity and magnetic anomalies in this part of the Tasman Orogen (Wyatt et al. 1980). In the Palaeozoic equivalents of the Rio Grande Rift (#10) and southern Basin and Range (#14) in the proposed match, the grabens run north–south, suggesting a possible direction for the Early Devonian faults in the Adavale Basin (Tanner 1976). From a more regional viewpoint, even though the modern Basin and Range is much wider than the inferred extent of the horsts and grabens in south-eastern Australia, the overall dimensions of the transtensional North American Plate margin are almost identical to the inferred extent of the Tasman Orogen east of the Tasman Line (Fig. 216).

Our level of knowledge of the Tasman Orogen does not allow a more detailed test of the possible match with the Pacific margin of the North American Plate, but one further point stands out: the possibility, even likelihood, that terrains now preserved in the eastern part of the Tasman Orogen have undergone large lateral transport with respect to cratonic Australia. The Gulf of California is the latest in a series of slivers of continental crust carried north along the North American Plate boundary (Coney et al. 1980), and if the overall match holds, then we should expect to find evidence of such suspect terrains in the eastern half of the New England Fold Belt.

The Early Devonian configuration (Fig. 217) shows one such terrain, the Calliope Arc and Wandilla Slope,

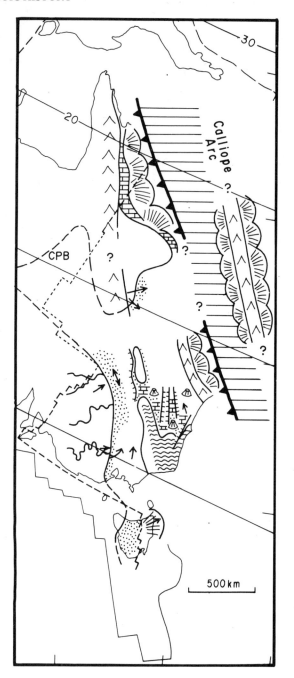

Fig. 217. Palaeogeographic reconstruction of the Tasman Orogen in the Early Devonian (c. 400 Ma). Position of the Calliope Arc is uncertain. Symbols as for Figs. 201, 202, 215, and 216A.

being rafted in from the east. Day *et al.* (1978) postulate that the Calliope Arc was separated in the Early Devonian from a continental-margin magmatic arc to the west by a marginal sea (Fig. 214A): it could well have been a wider ocean (Veevers *et al.* 1982). To the south and north, a continental volcanic arc faced east over an unstable shelf on to abyssal plain deposits. Unfortunately, the possible connection of the southern (Tamworth) and northern segments of the arc is concealed under younger rocks.

The Calliope Arc was folded in the Middle Devonian, probably as a result of collision following subduction of the marginal sea or ocean basin (Fig. 218). The Anakie Inlier in central Queensland was also folded at this time (Murray and Kirkegaard 1978). In the Tamworth Arc, there was a shallowing of the fore-arc basin in the Middle Devonian, but volcanism continued unabated. In the southern Lachlan Fold Belt, mid-Devonian was the time of major folding, uplift, and transcurrent faulting caused by east–west horizontal compression. The folding dies out northward (Powell *et al.* 1977; Powell and Edgecombe 1978), so that sedimentation continued unaffected in the Cobar region through to the Adavale Basin. In the north Queensland segment, a diachronous deformation started in the Frasnian, and became younger (Famennian) northward (Henderson 1980).

A possible explanation for the patchy distribution of zones of major deformation is that areas of deformation may correspond to promontories in the Tasman Orogen, or leading edges in the arc on the subducting plate. These salients would have been the first to collide, thus initiating strong deformation, whereas the adjacent re-entrants continued to receive sediment as if nothing had happened.

(iv) Australian platform

As shown in Fig. 219, the sea is inferred to have lapped the present western and north-western margins, and intermittently to have entered embayments in the Carnarvon/Perth, Canning, and Bonaparte Gulf Basins to deposit thick evaporites and associated redbeds. We show the central embayment of the Canning Basin continuing eastward through the Amadeus Transverse Zone as a transcontinental sea, but firm evidence of this is lacking. In central Australia, the Mereenie Sandstone (MS) of the Amadeus Basin is at least in part Devonian, as indicated by fish plates; and at least part of it is probably marine, as suggested by trace fossils (Wells *et al.* 1970) – most of it is probably aeolian. On the eastern side of the conjectured transcontinental sea are the Cravens Peak Beds (CPB). Turner *et al.* (1981) date the Cravens Peak Beds, from

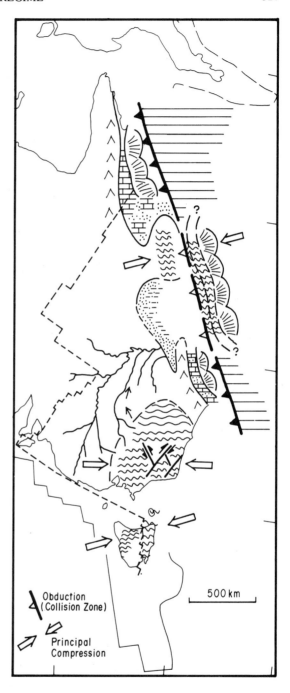

Fig. 218. Palaeogeographic reconstruction of the Tasman Orogen in the late Middle Devonian (*c.* 375 Ma). Symbols, other than those explained, as for Figs. 201, 202, and 215.

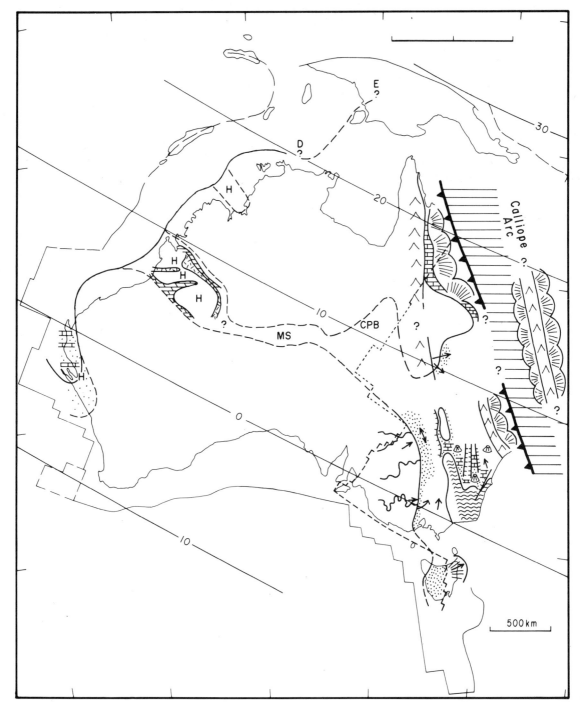

Fig. 219. Palaeogeographic reconstruction of Australia for the Siluro-Devonian (420 to 380 Ma), represented in eastern Australia by the Early Devonian (400 Ma) (Fig. 217). Shown are (a) distribution (in the Carnarvon Basin) of the Tumblagooda Sandstone (?Late Ordovician and Silurian), in part marine (Hocking 1979), and the Ludlovian Dirk Hartog Formation, including the Yaringa Evaporite Member (Playford *et al.* 1975); (b) distribution (in the Canning Basin) of evaporites in the Carribuddy Formation (?Late Silurian, Early Devonian), from Forman and Wales (1981), overlain by the Tandalgoo Redbeds (see also Wells 1980); (c) halite of the Bonaparte Gulf Basin (Edgerley and Crist 1974); (d) Silurian sediment in Money Shoal–1 (D) (Balke and Burt 1976); (e) reworked Silurian marine fossils in New Guinea (E) (Visser and Hermes 1962); (f) the Mereenie Sandstone (MS) of the Amadeus Basin (Wells *et al.* 1970); (g) the basal Cravens Peak Beds (CPB) (Turner *et al.* 1981) of Emsian/Eifelian age, and marginal marine. H=halite. Other symbols as in Figs. 201, 202, 215, and 216A.

contained fish plates (*Turinia*) as Emsian/Eifelian, and suggest from the associated ostracods and eridostracans that they were deposited at the edge of the sea that may have continued westward across Australia to the Canning Basin, coeval strata of which also contain *Turinia*.

(h) Late Devonian and Early Carboniferous: continental magmatic arc along the eastern edge of the Lachlan Fold Belt (by C. McA. Powell)

(i) Tasman Orogen of south-eastern Australia

(1) Major facies distribution in the north-eastern Lachlan Fold Belt

The Late Devonian to Early Carboniferous interval in south-eastern Australia is characterized by deposition of quartzose clastics, commonly of redbed facies and mainly in terrestrial environments. This interval is the time during which the Lambie Group was deposited in the north-eastern Lachlan Fold Belt (Conolly 1969a). Typically, the Lambie Group consists of a basal conglomeratic unit, at most a few tens of metres thick, reflecting local basement rock types, overlain by a few hundred metres of indurated fine-grained quartzarenite and shale, commonly containing the Late Devonian brachiopod *Cyrtospirifer* (Roberts et al. 1972), which, in turn, are overlain by medium-grained cross-bedded quartzarenite. In the Mount Lambie–Mount Horrible area, located in Fig. 221 (the northern #5), this succession coarsens upward into an oligomictic conglomerate composed of fine-grained indurated quartz-arenite, possibly derived mainly from extensive underlying Late Ordovician and Late Silurian quartzose turbidites (Fig. 220A). The oligomictic conglomerate is overlain, above 1082 m, by red and green lithic rudite containing a minor, but significant, proportion of silicic volcanic clasts.

The succession at Mount Horrible is 1.3 km thick, and is truncated erosionally by the overlying Permian deposits of the Sydney Basin. Similar erosionally truncated sections about 1 km thick exist in many outcrop belts of the Lambie Group and its facies equivalent, the Catombal Group, in the north-eastern Lachlan Fold Belt (Conolly 1963): the erosional top leaves open the question of original stratigraphic thickness. Near Taralga (Fig. 221, the southern #5), however, an erosionally truncated section of 4.7 km exists (Powell and Fergusson 1979a, fig. 3), and its composition above the basal 1 km of quartzarenite is so different from the lower part of the section that I conclude that it represents the only reported record of the upper 3 km of the Lambie Group in the northeastern Lachlan Fold Belt.

The Taralga section (Fig. 220B) is similar to other Lambie and Catombal Group sections in its lowest 1 km. From 1.6 km to 2.8 km, however, there is a wedge of volcanolithic sandstone and shale, with as little as 20 per cent of quartz in the framework grains of some sandstones (Powell, Conaghan and Prendergast, in preparation). The volcanolithic detritus includes devitrified microlitic rock fragments with flow-aligned plagioclase laths, euhedral volcanic β-quartz, and fragmented plagioclase grains, reflecting juvenile derivation from a dacitic or possibly andesitic volcanic source. Moreover, palaeocurrents in the volcanolithic units are directed southerly to southwesterly in contrast to the ubiquitous northeasterly to easterly palaeoflow directions obtained from the quartzarenite units. The volcanolithic facies is intercalated with tongues of the quartzarenite facies towards the top of the section (Fig. 220B), and each facies retains its integrity of provenance and palaeoflow direction, suggesting that the trunk stream drained southeasterly with juvenile volcanolithic detritus from the northeastern arm of the drainage network, and quartzarenite from the south-western arm (Powell, Conaghan and Prendergast, in preparation).

(2) Correlation of the Lambian Facies in south-eastern Australia

The Lambie Group and its correlatives in New South Wales: Merrimbula Group (Steiner 1975; Powell 1983a); Hervey Group (Conolly 1965a, b); Mulga Downs Group (Conolly 1969a, b; Glen et al. 1983; Powell et al. 1983; Powell and Conaghan, in preparation); Cocoparra Group (Conolly 1969a); Hatchery Creek Conglomerate (Owen and Wyborn 1979); and in Victoria (Marsden 1976) have a similarity in composition, facies, and tectonic position that has led previously to their being considered a single chronostratigraphic unit (for example Conolly 1969a, b). In eastern New South Wales, marine fossils near the base are of Late Devonian age (Roberts et al. 1972), no younger than late Frasnian (Pickett 1972). In the Mudgee area, late Famennian productid brachiopods have been found (N. Morris 1981, written communication), and, in far western New South Wales, Early Carboniferous spores have been recovered (Evans 1977). In eastern New South Wales and Victoria, the Lambian facies lies unconformably on folded late Early and possibly early Middle Devonian rocks (Packham 1969; Webby 1972; VandenBerg et al. 1976), so that the entire unit has commonly been regarded as Late Devonian to earliest Carboniferous (for example Conolly 1969a, b).

More recent work has shown that the base of the

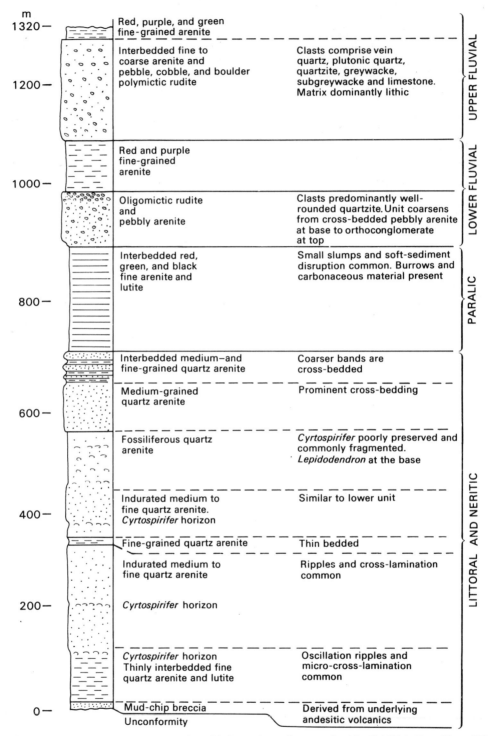

Fig. 220. A. Representative stratigraphic column of Lambie Group in north-eastern Lachlan Fold Belt (from Henry 1976).

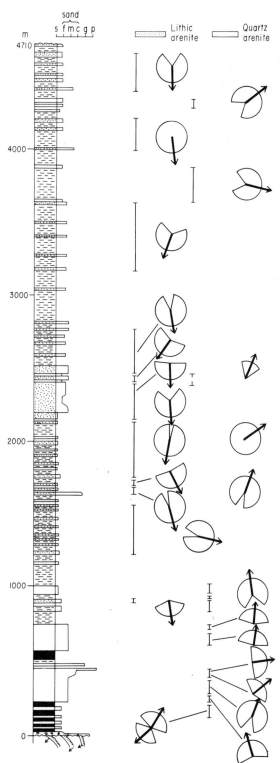

Fig. 220. B. Measured stratigraphic section with palaeocurrents in the Lambie Group near Taralga (from Powell, Conaghan and Prendergast, in prep.). Solid black denotes recessive strata. Thickness in metres.

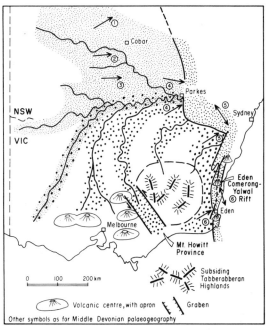

Fig. 221. Early Late Devonian palaeogeography of southeastern Australia. Numbered palaeocurrent sources: 1, 2 and 3=Crowl Creek Formation in: (1) Conaghan and Powell (unpubl. data), (2) Glen *et al.* (1983), (3) Powell *et al.* (1983); (4) Hervey Group (lower part): Powell (unpubl. data); (5) Lambie Group (lower part): Henry (1975), and Conaghan and Powell (unpubl. data); (6) Twofold Bay and Bellbird Creek Formations: Steiner (1975) and Powell (1983a). Symbols, other than those explained, as for Figs. 205 and 211.

Lambie Group is diachronous, and contains late Early or Middle Devonian fossils (Ritchie 1973; Young and Gorter 1981). The major mid-Devonian deformation and uplift centred on southern New South Wales and Victoria (Figs. 211 and 212) produced no significant break in sedimentation in the Cobar region, where over 4 km of quartzose clastics and redbeds accumulated in terrestrial environments from the late Early Devonian to the Early Carboniferous (Glen 1982; Glen *et al.* 1983; Powell *et al.* 1983). I estimate that about 2 km was deposited in the Cobar region in the Late Devonian–Early Carboniferous interval.

(3) Deposition as sheet-like bodies or in grabens?
The Lambian facies, together with the underlying older bedded rocks, was folded during the Carboniferous, and is now preserved in synclinal keels. In view

of the horst-and-graben setting for the mid-Silurian to mid-Devonian interval, the question arises as to whether the Lambian facies synclinal keels represent infilled Late Devonian grabens (Cas 1983) or the preserved remnants of more extensive sheet-like deposits.

Three lines of argument bear on the question. First, if the synclinal keels represent the approximate original extent of each graben, then one would expect local marginal facies in each synclinal keel, reflecting episodic uplift of the rift shoulders. Such facies are generally restricted to the basal 100 m of Lambian facies in each syncline, and presumably record the wearing-down of local hills at the onset of deposition. Secondly, there is no reason to expect that stratigraphic units in one graben should be readily traceable into the next. The east of correlation from one synclinal keel to the next in eastern New South Wales (Powell 1983a and unpublished work), and west of Cobar, where the same units can be identified throughout the region (Glen et al. 1983; Powell et al. 1983; Powell and Conaghan, unpublished work) indicates that the original depositional basins were of broader extent than the preserved synclinal keels. Thirdly, the palaeocurrent patterns measured in each syncline should show patterns related to the shape of the hypothetical rifts. Palaeocurrent patterns in the Cobar region and in eastern New South Wales (references given above) show remarkable consistency within correlative units, and, moreover, the inferred palaeoslope directions are commonly oblique to the graben trends postulated to parallel the synclinal keels.

These arguments show that the present distribution of Lambian facies in the northern and western Lachlan Fold Belt is the result of superimposed deformation oblique to the depositional facies trends. There are, however, two areas where volcano-tectonic grabens preceded or accompanied deposition of the Lambian facies. One area is the Eden–Comerong–Yalwal rift zone (Fig. 221) in the eastern Lachlan Fold Belt (McIlveen 1974). This meridional rift zone, of late Givetian or early Frasnian age or both, contains silicic and mafic volcanics deposited in complex, terrestrial facies intercalations (Fergusson et al. 1979). The rift may originally have been only 20 to 30 km wide, and at least 300 km long, and it formed immediately after the Middle Devonian folding and wrench faulting that gave rise to the Tabberabberan Highlands (Powell 1983a). Subvolcanic alkaline (A-type) granites (Collins et al. 1982) fed the ignimbritic rhyolite flows within the rift. The Eden–Comerong–Yalwal rift is overlain, in most places at a paraconformity and in a few places at an angular unconformity, by the Merrimbula Group (Steiner 1975; Powell 1983a), and the meridional trend of the rift affects the palaeocurrent directions in the Lambian facies for the basal 70 m only (Powell 1983a). Above this level, the regional palaeo-strandline trends north-west–south-east with a northeastward slope (Powell 1983a).

The second area where grabens may have controlled the distribution of Lambian facies is in central-eastern Victoria (Marsden 1976). The Mt. Howitt Province (Fig. 221), which lies along the line of the older Mt. Wellington axis, is a NNW-trending zone at least 100 km long and up to 50 km wide. Conglomerate, sandstone, and red shale, intercalated with rhyolite, rhyolitic tuff, and basalt, reach a thickness of 4.2 km (Marsden 1976). The silicic-volcanic intercalations appear to be related to large Late Devonian caldera complexes that developed in the newly deformed Melbourne Trough succession immediately to the west (McLaughlin 1976). Late Devonian granitoid intrusions, with or without associated volcanics, were developed throughout the former Melbourne Trough, as far west as the Ballarat region.

The silicic volcanics are restricted to the lowest kilometre or so of the Avon River Group in the Mt. Howitt Province, but sporadic basalt flows are intercalated higher in the section (Marsden 1976). The uppermost part of the section, however, is a fining-upwards redbed succession deposited in fluvial to lacustrine environments devoid of contemporaneous volcanics. The bimodal volcanic phase is probably Frasnian, and the succeeding fluvial redbed succession probably extends into the Early Carboniferous. Local facies indicate that both the eastern and western margins of the Mt. Howitt Province were faults active during part of the deposition, and at least one of the north-northeasterly trending cross-folds grew during deposition (Marsden 1976, p. 98).

No Late Devonian or Early Carboniferous sediments are preserved in Tasmania (Williams 1978), but some of the extensive granitoid plutons in north-eastern Tasmania may have been emplaced during this interval (Cocker 1982).

(4) Frasnian palaeogeography

The Frasnian palaeogeography of south-eastern Australia (Fig. 221) shows a continuation of subsidence of the Tabberabberan Highlands, and the onset of bimodal silicic and mafic volcanism in the south. The Mt. Howitt Province formed as a graben with a component of sinistral shear to account for the syn-sedimentary north-northeasterly trending cross-folds. The uplifted deformed area of central Victoria was a terrain of terrestrial silicic volcanism with cauldron subsidences, with sediment and volcanics accumulating in ephemeral lakes as well as in the Mt. Howitt Province.

On the South Coast of New South Wales, bimodal volcanism in the Eden–Comerong–Yalwal rift was short-lived, and was followed in the late Frasnian by a marine transgression (Pickett 1972; Steiner 1975). Coarse, gritty quartzarenite in the basal Merrimbula Group was derived from the south-west, mainly from uplifted and exposed Late Silurian and Early Devonian granitoid batholiths. The quartzose sediment was dispersed along a shoreline with a regional north-north-west trend.

The Frasnian transgression extended as far west as Parkes (Webby 1972; Williams 1975a), and produced a wedge of shallow-marine sediment that thickens eastward. In the Cobar region, sedimentation continued uninterrupted from the Middle Devonian, with medium- to fine-grained sandstone of the lower part of the Crowl Creek Formation being deposited in meandering streams on broad floodplains (Glen et al. 1983; Powell et al. 1983). Polymictic fold-belt detritus derived from the south failed to reach the Cobar region, which accumulated quartzarenite.

(5) Famennian–Tournaisian palaeogeography

Terrestrial silicic volcanism in central Victoria died out before the end of the Famennian, and sedimentation in the Mt. Howitt Province was of redbed facies with fluvial intercalations (Marsden 1976). On the South Coast of New South Wales, medium to coarse quartzarenite was deposited in co-sets up to 20 m thick in the channels of meandering streams, with intercalated red shale and siltstone deposited on the floodplain (Steiner 1975; Powell 1983a). The regional palaeoslope during deposition of the Worange Point Formation was to the north-east (Powell 1983a), and the upward decrease in the sand–shale ratio reflects continued wearing down of the Tabberabberan Highlands.

Sedimentation in the Cobar region was a continuation of the pattern in the Frasnian, with a few medium- to coarse-grained quartzarenite lenses representing stream deposits. The bulk of the sediment deposited, however, was fine-grained sandstone and siltstone, all formed in a meandering easterly draining system.

By the late Famennian, important changes in sedimentation occurred along the eastern edge of the Lachlan Fold Belt (Fig. 222A). Immature volcanogenic detritus derived from the north-east was deposited in the Taralga region and in the upper part of the Merrimbula Group preserved in thicker sections of the Budawang Synclinorium (Powell 1983a, and work in progress). Immediately preceding this influx of

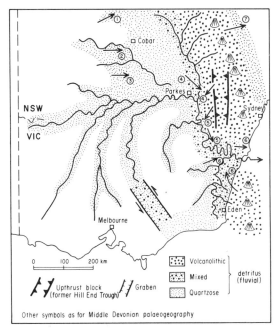

Fig. 222. A. Latest Devonian or earliest Carboniferous palaeogeography of south-eastern Australia. Numbered palaeocurrent sources: 1, 2, and 3=uppermost Crowl Creek Formation in (1) Conaghan and Powell (unpubl. data); (2) Glen et al. (1983); (3) Powell et al. (1983); (4) Hervey Group (upper part): Powell (unpubl. data); (5) middle part of Lambie Group in Taralga area (Fig. 220B); (6) Merrimbula Group: Powell (unpubl. data); (7) Keepit Conglomerate: Russell (1980). Symbols, other than those explained, as for Figs. 205, 211, and 221.

volcanogenic detritus in the eastern Lachlan Fold Belt were local movements around the margins of the former Hill End Trough. Oligomictic conglomerate, composed of fine-grained indurated quartzarenite, with clasts locally exceeding 40 cm in longest dimension, was shed away from the rising area of the former Hill End Trough, causing considerable rearrangement of the earlier palaeodrainage networks. In the Parkes area, the blocking effect of the rising Hill End region is reflected in the southerly deflection of palaeocurrents in the eastern part of the Hervey Range Group (Conolly 1965b; Powell, unpublished data). Palaeocurrent patterns in the Budawang Range area on the South Coast of New South Wales suggest that the regional east-flowing river system drained through a break in the rising eastern margin of the Lachlan Fold Belt half-way between Sydney and Eden (Fig. 222A).

The rearrangements of the palaeodrainage network appear to correspond with the establishment in the Late Devonian of a continental magmatic arc along the eastern edge of the Lachlan Fold Belt (Fig. 222A). This arc could well have been the same arc that lay to the west of the Tamworth Trough (Leitch 1974; Mory 1982), and may be reflected in a residual gravity high beneath the overlying Sydney Basin (Qureshi 1984). Subsequent deformation in the mid-Carboniferous was most intense adjacent to the continental magmatic arc, as shown by the tightness of fold profiles in the Lambie facies (Fig. 222B). The extension of this arc into Queensland, and its relationship to the New England Fold Belt that lay to the east, are discussed below.

(ii) Tasman Orogen elsewhere

(1) New England Fold Belt in New South Wales
During the Late Devonian and Early Carboniferous, the western part of the New England Fold Belt in New South Wales (i.e. west of 152° E, Fig. 214B) was an unstable marine shelf (Tamworth Trough or Shelf) that lay east of an evolving tholeiitic–calcalkaline volcanic chain and west of a zone of deep-water radiolarian chert, metabasalt, argillite, and turbidite containing various types of *melange* (Woolomin Slope and Basin in Fig. 214B) (Leitch 1974, 1975; Leitch and Cawood 1980; Cawood 1982a, b). The Tamworth Trough has a history that extends at least as far back as the Early Devonian, with continuous deposition into the Late Carboniferous (Crook 1961a, b; Leitch 1974; Roberts and Engel 1980; Mory 1982). Two general trends can be noted: first, there is a gradual change from turbidity-current deposition in the Late Devonian and Early Carboniferous to traction-current deposits advancing from the west in the upper (Late Carboniferous) part (Crook 1964). Secondly, there is a gradual change in composition of the inferred western volcanic chain from andesitic and basaltic in the Devonian to dacitic and rhyolitic in the Carboniferous (Fig. 213; Crook 1964; Leitch 1974). In the Late Devonian and Early Carboniferous part of the section, the sediments consist of westerly-derived volcanic litharenite, rudite, and mudstone (Crook 1960, 1961b) emplaced through submarine canyons, and by slumping (Crook and Powell 1976; Mory 1982).

The junction between the relatively gently folded Tamworth Trough succession and the complexly deformed Woolomin Slope and Basin is marked by a zone of serpentinite – the Peel Fault system (Corbett 1976; Cawood 1982a, b). This zone has a history at least as old as the Early Devonian (Cawood 1982a), and fragments of shallow-water limestone, probably allochthonous, as old as Cambrian (Cawood 1976) are incorporated as slivers of the fault system. In the Late Devonian and Early Carboniferous, the Peel Fault Zone was probably the non-volcanic outer arc, or trench-slope break, oceanwards of the volcanic chain west of the Tamworth Trough (Leitch 1982).

The Woolomin Association to the east of the Peel Fault Zone is generally regarded as a deep-water flysch wedge containing volcaniclastics derived from the volcanic arc to the west (Fig. 223A). Olistostromes and tectonic *melange* are present (Cawood 1982b) and there are fragments of oceanic layers 1 and 2. The Woolomin Association may have been thrust over the Tamworth Trough succession, although Cawood (1982a) notes that because the olistostromal Wisemans Arm Formation can be correlated from one side of the Peel Fault Zone to the other, with similar metamorphic grades on either side, there is unlikely to have been the 80- to 120-km of foreshortening suggested by Scheibner (1976). Scheibner and Glen (1972) consider obduction as mid-Carboniferous, whereas Leitch (1974) considers the movement to be a mid-Permian event. Subsequently, the Peel Fault Zone underwent transcurrent movement postulated to be in part dextral

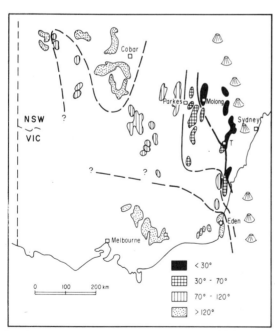

Fig. 222. B.Early Carboniferous deformation plan of southeastern Australia, as recorded by the inter-limb angle in folds in the Lambie facies and its correlatives. Postulated volcanic chain from Fig. 222A is shown for reference.

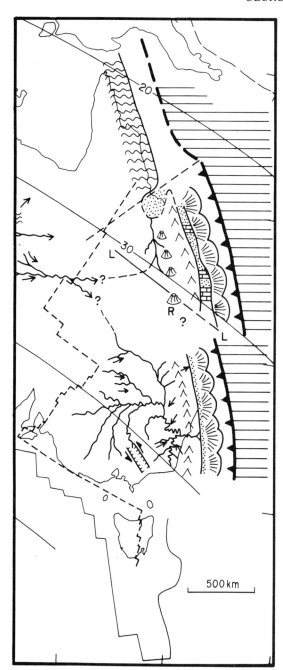

Fig. 223. A. Palaeogeographic reconstruction of the Tasman Orogen in latest Devonian–earliest Carboniferous (c. 360 Ma). L–R–L is the postulated Longreach–Roma–Lismore lineament (Evans and Roberts 1980). Other symbols as for Figs. 201, 202, 215, and 216A.

(Scheibner and Glen 1972) and in part sinistral (Corbett 1976). Movement in the Peel Fault Zone is postulated to have occurred also in the Late Permian, involving on the east the hinterland for the Hunter–Bowen foredeep (Jones and McDonnell 1981).

The Late Devonian and Early Carboniferous palaeogeography of the New England Fold Belt east of 152° E is more enigmatic (Korsch 1977; Korsch and Harrington 1981). The Demon Block (Fig. 214B) has a curved outcrop pattern convex to the south-west, and contains three main tectonostratigraphic units (Fergusson 1982a). In the north-east, the Devonian Silverwood Group and Willowie Creek Beds, which include mafic volcanics and associated sediments, are probably part of an island arc. In the middle, a *melange* unit with elongate slices of all sizes up to 15 km comprises fragments of bedded chert, greywacke, thin-bedded turbidite, argillite, bedded argillite-tuff, and greenstone. The greywacke contains microlitic and felsitic rock fragments, and oolites, the latter taken as indicating an Early Carboniferous age. In the southwest, a flysch unit containing greywacke and argillite has turbidite sequences typical of basin-plain and outer-fan depositional environments.

Fergusson (1982a, b) interprets the *melange* and flysch succession as a Carboniferous accretionary prism, now facing south-west. Flood and Fergusson (1982) have extended the units to the north-west into the Texas region, where they trace the *melange* and flysch round a megafold that appears to connect with the northern exposures of the Woolomin Slope and Basin (Fig. 214B). Fergusson (1982a) postulates that the megafold represents a Permian orocline, and that, in the Early Carboniferous, the presently south-westward-facing accretionary prism lay to the north-west and faced to the east (Fig. 223A). The Devonian Silverwood and Willowie Creek Beds may be the southern extension of the Early to mid-Devonian Calliope Island Arc (Figs. 214A, 217, 218).

(2) Queensland sector

The contrast in sedimentation style between the Lachlan and New England Fold Belts in New South Wales extends northward through Queensland. In the Adavale Basin (Fig. 214A), red and brown quartzose sandstone and variegated mudstone and shale (Buckabie Formation, 3 km thick) were deposited in the Late Devonian with no obvious structural break above the shallow-marine, partly evaporitic, Middle Devonian sediments. Whether redbed sedimentation in the Adavale Basin extended into the Early Carboniferous is debatable. Vine (1972) considered that the entire Adavale Basin succession was deposited before the first sediment of the latest Devonian–Early Carboniferous Drummond Basin to the east, but others (e.g. Auchincloss 1976; Olgers 1972; Marsden 1972) consider the two basins as overlapping in age.

The Drummond Basin (Fig. 214B) is a classical foreland basin with up to 12 km of sediment adjacent to the Anakie Inlier. After an initial burst of silicic volcanism both east and west of the Anakie Inlier, sedimentation was restricted to west of the Anakie Inlier, which constituted the rapidly uprising margin to the basin. Sedimentation was dominantly fluviatile, with a trunk stream draining to the north, parallel to the Anakie Inlier, before turning north-eastward just south of the uplifted Ravenswood Block. The fluviatile sediments were deposited in two major upward-fining cycles that reflect growing maturity in the hinterland punctuated by an episode of uplift. Tuffaceous sandstone in the upper cycle marks a renewal of volcanism in the eastern source area.

Olgers (1972) considered that the Drummond Basin was intermontane, but recent seismic work (Pinchin 1978) shows that the Drummond Basin sediments continue in the subsurface well west of the western boundary inferred by Olgers. In section, the basin is asymmetrical, with the greatest thickness deposited adjacent to the Anakie Inlier. In the mid-Carboniferous, the entire Drummond and Adavale Basin successions were deformed by east–west compression, as part of an Australia-wide orogeny.

East of the Anakie Inlier, the tectonic elements were a volcanic arc, an unstable marine shelf, and a slope and basin, arranged with the same polarity as the east-facing terrain in the western half of the New England region of New South Wales (Fig. 214B). The Connors–Auburn volcanic arc was established in the Late Devonian above, and slightly west of, the deformed Early Devonian Calliope Island Arc (Fig. 214A), and shed volcaniclastic sediment onto the shallow-marine Yarrol Shelf (a fore-arc basin) and the Wandilla Slope and Basin (the outer-arc slope and adjacent abyssal plain) (Fig. 223A). The Late Devonian and Early Carboniferous Connors–Auburn volcanic arc consisted entirely of subaerial volcanics (Murray and Whitaker 1982) ranging from andesite to dacite and rhyolite (Malone et al. 1966, 1969), suggesting that it developed on a continental Andean-type margin in contrast to the intra-oceanic setting of the Calliope Island Arc. Shallow-marine volcaniclastics and oolitic limestones deposited on the Yarrol Shelf thicken to the east (Kirkcgaard et al. 1970) where they are bounded by the Yarrol Fault Zone – an Alpine-type ultramafic belt analogous to the Peel Fault Zone in New South Wales (Day et al. 1978). The Wandilla Slope and Basin deposits contain deeper-water flyschoid equivalents of the Yarrol Shelf deposits, including oolite presumably transported eastward down slope (Day et al. 1978; Murray and Whitaker 1982).

In northern Queensland, the sedimentation pattern established in the Early Devonian continued with shallow-marine sediments being deposited above Precambrian basement on the western shelf edge, and deep-water flyschoid rocks accumulated farther east (Marsden 1972; Henderson 1980; Wyatt and Jell 1980). Upward-shallowing sedimentary facies in the Late Devonian reflect the progressive rise of the surface of the sedimentary prism, which was extensively folded in the latest Devonian (Henderson 1980). In the Broken River area, up to 4 km of lithic and arkosic arenite, siltstone, conglomerate, and shale (lower Bundock Formation) accumulated during the Frasnian and early Famennian regression, which heralded widespread late Famennian folding (Wyatt and Jell 1980). The lacuna across the unconformity in the Broken River Province is very brief, and the overlying late Famennian–Tournaisian assemblage includes lithic arkose, volcanolithic sandstone and conglomerate, and fossiliferous shale with intercalated rhyolitic ignimbrite, deposited in environments ranging from shallow marine and tidal flat to fluvial (Wyatt and Jell 1980).

Sedimentation in the Adavale and Drummond Basins, and in the western half of the north Queensland continental margin came to a close in the mid-Carboniferous (Fig. 154), with the onset of the Kanimblan deformation. This was the most widespread deformation in the orogenic stage of the eastern third of Australia, and its effects extended through central Australia and to the north-west. Thereafter, the Lachlan and Thomson Fold Belts acted as cratonic areas, and the orogenic character of the Tasman Orogenic Zone was restricted to the area east of a line connecting the Anakie Inlier south to the former volcanic arc that bounded the western edge of the Tamworth Shelf (Fig. 214B).

Little or no interruption to sedimentation occurred in the New England and Yarrol Fold Belts east of the Anakie–Tamworth line in the mid-Carboniferous, although Roberts and Engel (1980) note latest Visean and Namurian facies changes related to the major readjustments that were occurring to the west. The mid-Carboniferous (340 to 320 Ma) events produced the profoundest change in the palaeogeography of Australia since the inception of the eastern orogenic margin in the infra-Cambrian, and brought to a close the Uluru (or pre-Gondwanan) Regime.

(iii) Tectonic models

The tectonic setting of the Tasman Orogen in the Late Devonian and Early Carboniferous was that of an Andean-type continental margin facing east onto the Palaeo-Pacific Ocean (Figs. 223A, 223B). The largely terrestrial Drummond and Lambian Basins are foreland basins, and the more interior Adavale and Ravenswood Basins epicratonic basins, in relation to the magmatic arc that lay along the western edge of the New England and Yarrol Fold Belts. East of the magmatic arc the Tamworth and Yarrol Shelves were fore-arc basins separated by the Great Serpentine Belts (Peel and Yarrol Fault Zones) from the outer-arc slope and basin-plain deposits of the eastern New England and Yarrol Belts. In north Queensland, the polarity of the arc is the same but the basin arrangement is more condensed because of the eastward projection of the salient of the Precambrian Georgetown Block. This overall arrangement is consistent with a westward-dipping subduction zone beneath an Andean-type magmatic arc along the eastern edge of the Lachlan and Thomson Fold Belts (Fig. 223B).

There are, however, some problems in interpreting the palaeogeography in detail. The generally accepted connection between the Yarrol and Tamworth belts (Day *et al.* 1978; Evans and Roberts 1980; Roberts and Engel 1980) is by no means certain. The Yarrol Fault system, far from connecting across the northern extension of the Peel Fault system, appears to continue south-south-east into the eastern part of the New England belt (Murray and Whitaker 1982). Fergusson's (1982a) postulated orocline in the eastern New England belt is one way of explaining the rock distribution, and, if correct, would have formed in the Late Carboniferous or Early Permian before the more linear Late Permian Hunter–Bowen magmatic arc and foreland basin were established. Fergusson's (1982a) orocline, however, does not by itself account for another anomaly that exists if the Lambian Basin is correlated as the down-strike continuation of the Drummond Basin (Fig. 224, left side).

The distribution and nature of sediments in the Drummond Basin, where silicic volcanics and tuffs are intercalated in the latest Devonian and Early Carboniferous (Olgers 1972), suggest an active volcanic source in the bounding Anakie Inlier. The Retreat Granite (366 Ma, Webb and McDougall 1968, corrected to the new constants), is Late Devonian, non-conformably overlain by the Silver Hills Volcanics of probable Famennian age, and along strike to the south, granites in the Roma area (Fig. 224, left side) have Early Carboniferous K/Ar ages (Houston 1964). If these granitoids are the roots of a Late Devonian or Early Carboniferous volcanic arc, then they may be the northern continuation of the silicic magmatic arc inferred to bound the eastern edge of the Lambian Basin during this interval.

The Anakie–Roma line of intrusives lies 150 to 200 km west of the Connors–Auburn volcanic arc, and raises the questions of whether there was a double arc during this interval, whether the Connors–Auburn volcanic arc was more widely separated from the western arc by a marginal sea, the relicts of which now lie beneath the Bowen Basin (Veevers *et al.* 1982, fig 10b), or whether the apparent double arc is the result of large (1000 km or more) dextral transcurrent motion along a fault system now buried beneath the Permian–Triassic Bowen Foreland Basin (Fig. 224, left side). If there was a double arc, or a marginal sea between the Anakie Inlier and the Connors–Auburn arc, the Late Devonian palaeogeography of eastern Queensland could well have resembled the modern Japan–Siberia part of eastern Asia, with the marginal sea (Japan Sea) between the frontal volcanic arc (Japanese archipelago) later removed by westward subduction beneath the continental mainland (Siberia). The closure of this marginal sea in the Carboniferous may well have triggered the regional mid-Carboniferous deformation throughout eastern Australia, and marked the conversion of the Eastern Australian continental margin from a Western Pacific arc-marginal sea to a North American West Coast transform margin.

Alternatively, if the Connors–Auburn arc and Yarrol Shelf system is the northward continuation of the Tamworth arc and Shelf, large dextral displacement along a transcurrent fault zone now concealed beneath the younger Bowen Basin may have displaced the Queensland segment south-easterly, thereby forming the supposed orocline in the eastern New England region, and also accounting for the apparent double arc in central Queensland. The testing of these hypotheses awaits further work.

(iv) Australian platform

Late Devonian and Early Carboniferous deposits occur in three regions: the various structural basins of the Amadeus Transverse Zone, in the Halls Creek and King Leopold Mobile Zones around the Precambrian nucleus of the Kimberleys, and along the western margin of what subsequently became the Australian continent (Fig. 223B). In the Amadeus Basin, synorogenic conglomerates were shed southward from the rising Arunta Block (Fig. 226), with the pebble composition reflecting progressive stripping of Phanerozoic and Proterozoic sediments before high-grade

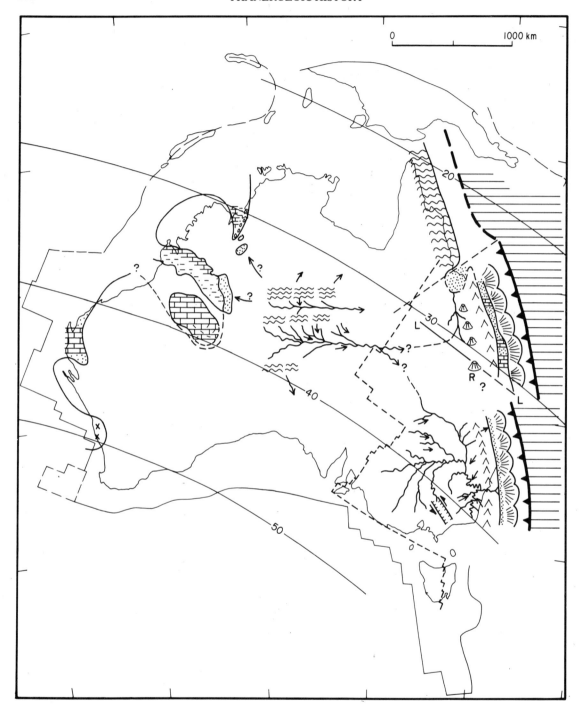

Fig. 223. B. Palaeogeographic reconstruction of the Devono-Carboniferous, including details of eastern Australia from Fig. 223A. For the rest of Australia, the entire interval of the Late Devonian and the Early Carboniferous is shown.

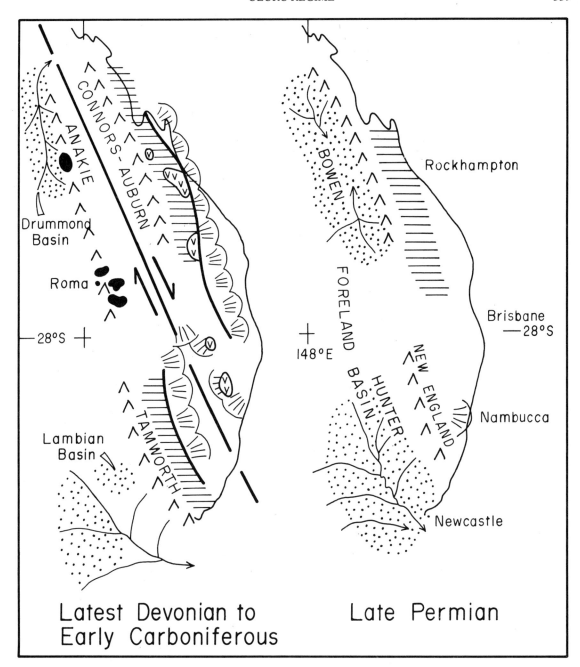

Fig. 224. Palaeogeographical comparison of the latest Devonian to Early Carboniferous and the Late Permian configurations of Eastern Australia, with palaeogeographical elements shown in their present-day position. The apparent double arc in central Queensland in the Devono-Carboniferous may be accounted for by one of at least three hypotheses, as discussed in the text. By the Late Permian, the magmatic arc and associated foreland and fore-arc basins had stepped eastward across the remains of the previous palaeogeographical system. The north-northwest trending fault in the Devono-Carboniferous is the possible transcurrent zone discussed in the text. Symbols as in previous figures, with the solid black on the left side indicating granite.

metamorphics were delivered to the basin (Jones 1972). In the Amadeus Basin, the synorogenic deposits are Late Devonian (Playford *et al.* 1976a), but in the Ngalia Basin to the north the Mount Eclipse Sandstone persisted into the Visean (Wells and Moss 1983). In the Halls Creek Mobile Zone, left-lateral wrenching opened the Bonaparte Gulf Basin (Laws 1981), with coarse clastics on the fault-bounded margins grading basinwards through carbonate into shale and siltstone.

On the west, the sea lapped the present margin with three indentations that re-occupied the failed arms (see Fig. 112 for columns): (1) in the Bonaparte Gulf Basin (Roberts and Veevers 1973), Frasnian shallow-marine quartz conglomerate and sandstone – the 1.5 km-thick Cockatoo Sandstone – was shed from the uplifted south-eastern margin of the basin, which was subjected to sinistral transcurrent movements (Laws 1981). Following uplift of the south-western margin, the depositional area in the later Devonian and Early Carboniferous was differentiated into a shallow inshore platform and an offshore basin. Carbonate, including reefs, and quartz sandstone were deposited inshore, and dark shale and siltstone offshore. Sinistral movement of blocks along the south-eastern margin continued through the Early Carboniferous up to the mid-Carboniferous lacuna that separates the Uluru Regime from the succeeding Innamincka Regime. Outliers of quartz sandstone in the Ord Basin and Ragged Range probably indicate a source of detrital sediment from the southeast; (2) in the Canning Basin (Playford *et al.* 1975), marine deposition started in the Givetian, and continued up to the mid-Carboniferous lacuna. A widespread dolomitic limestone with anhydrite and minor shale (Mellinjerrie Limestone) was deposited in the southern Canning Basin during the Givetian and Late Devonian. To the north, the sequence, which extends to the mid-Carboniferous, has three parts: (a) on the east, a belt of quartz sandstone (Knobby Hills Sandstone: Veevers and Wells 1962), probably derived from the east; and (b) on the north, a barrier reef complex (Playford 1980) that interfingers southwestward with (c) basinal shale; (3) in the Carnarvon Basin, marine platform carbonate and quartz sandstone along the edge of the basin interfinger with a reef complex (Point Maud Formation) offshore (Playford *et al.* 1975); and (4) in the Perth Basin, two occurrences of reworked Devonian microfossils point to Devonian rocks nearby (Veevers 1976): Late Devonian spores in the Jurassic/Cretaceous Yarragadee Formation, and Late Devonian conodonts in the Triassic Cockatea Shale.

Faulting that accompanied deposition in the west persisted to the general uplift in the mid-Carboniferous. The greatest uplift on the platform took place in the Amadeus Transverse Zone; its relationship with the Tasman Orogen is now examined.

(i) Comparative tectonics of the transverse structural zones of Australia and North America (by J. J. Veevers and C. McA. Powell)

(i) Introduction

The Amadeus Transverse Zone (Rutland 1973) and the Southern Oklahoma Aulacogen (Hoffman *et al.* 1974), also called the Transverse Geosynclinal Orogenic Belt (Ham and Wilson 1967), are zones of thick Palaeozoic sediment and intervening uplifts that cut across the cratonic interiors of Australia and North America (Fig. 225). These transverse zones have many features in common so that a study of their comparative tectonics, including their connection with the fold belts that developed at much the same time at convergent plate boundaries to the east, may sharpen understanding of both systems.

(ii) Amadeus Transverse Zone

Variously referred to as the Central Geosyncline (Hills 1946) and Flexible Belt (Öpik 1957), the Amadeus Transverse Zone (Rutland 1973) (Fig. 225A), described in the section on Central Australia in Chapter IV, is a major subdivision of the Australian platform which has distinctive easterly trends that strike at high angles into the northerly trends of the Tasman Fold Belt. As used here, the term encompasses all those structures that lie between the Lander Trough and the Officer Basin, and have a common set of gravity trends (Wellman 1976). The main structural element within the zone is the high-angle reverse fault of large throw (Fig. 225C), presumably with some lateral (sinistral) motion, and associated nappes that moved about 1000, 600, and 350 Ma ago (Fig. 226). The initial faulting, 1100 to 1000 Ma ago, was accompanied by migmatization and granite emplacement, and, in the western part of the Musgrave Block, by the intrusion of a stratiform gabbro (mafic to ultramafic) complex associated with basalt and rhyolite. After dolerite dykes were intruded about 900 Ma ago, sedimentary deposition started, possibly as late as 750 Ma ago, with a thin (600 m) sheet of quartz sand (Heavitree Quartzite and equivalents) that presumably covered the entire region, followed by 1000 m of a less widespread evaporitic carbonate and shale (Bitter Springs Formation and equivalents) that contain the only known volcanics (50 m of basalt) in the entire sedimentary sequence except for the 570 Ma basalt (about 100 m thick) of the Officer Basin. Uplift of the Musgrave Block is

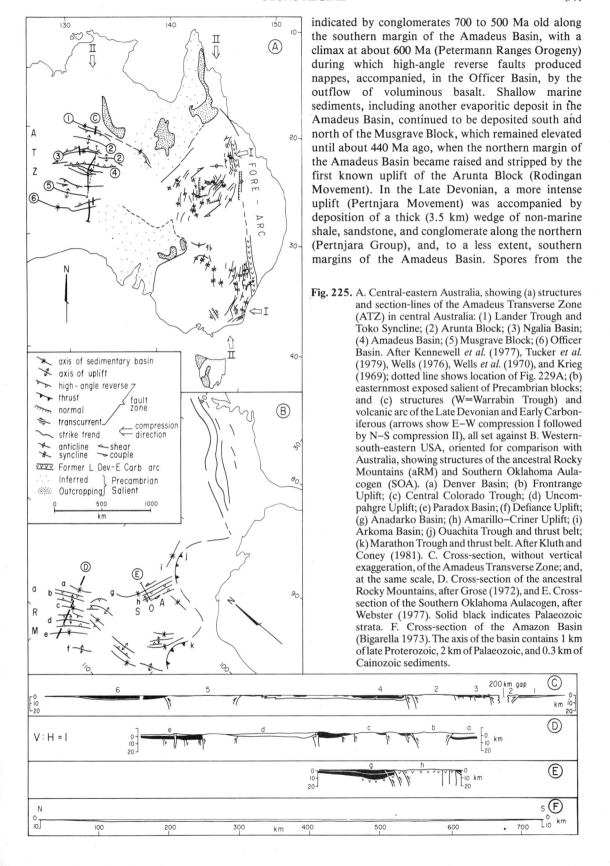

indicated by conglomerates 700 to 500 Ma old along the southern margin of the Amadeus Basin, with a climax at about 600 Ma (Petermann Ranges Orogeny) during which high-angle reverse faults produced nappes, accompanied, in the Officer Basin, by the outflow of voluminous basalt. Shallow marine sediments, including another evaporitic deposit in the Amadeus Basin, continued to be deposited south and north of the Musgrave Block, which remained elevated until about 440 Ma ago, when the northern margin of the Amadeus Basin became raised and stripped by the first known uplift of the Arunta Block (Rodingan Movement). In the Late Devonian, a more intense uplift (Pertnjara Movement) was accompanied by deposition of a thick (3.5 km) wedge of non-marine shale, sandstone, and conglomerate along the northern (Pertnjara Group), and, to a less extent, southern margins of the Amadeus Basin. Spores from the

Fig. 225. A. Central-eastern Australia, showing (a) structures and section-lines of the Amadeus Transverse Zone (ATZ) in central Australia: (1) Lander Trough and Toko Syncline; (2) Arunta Block; (3) Ngalia Basin; (4) Amadeus Basin; (5) Musgrave Block; (6) Officer Basin. After Kennewell et al. (1977), Tucker et al. (1979), Wells (1976), Wells et al. (1970), and Krieg (1969); dotted line shows location of Fig. 229A; (b) easternmost exposed salient of Precambrian blocks; and (c) structures (W=Warrabin Trough) and volcanic arc of the Late Devonian and Early Carboniferous (arrows show E–W compression I followed by N–S compression II), all set against B. Western-south-eastern USA, oriented for comparison with Australia, showing structures of the ancestral Rocky Mountains (aRM) and Southern Oklahoma Aulacogen (SOA). (a) Denver Basin; (b) Frontrange Uplift; (c) Central Colorado Trough; (d) Uncompahgre Uplift; (e) Paradox Basin; (f) Defiance Uplift; (g) Anadarko Basin; (h) Amarillo–Criner Uplift; (i) Arkoma Basin; (j) Ouachita Trough and thrust belt; (k) Marathon Trough and thrust belt. After Kluth and Coney (1981). C. Cross-section, without vertical exaggeration, of the Amadeus Transverse Zone; and, at the same scale, D. Cross-section of the ancestral Rocky Mountains, after Grose (1972), and E. Cross-section of the Southern Oklahoma Aulacogen, after Webster (1977). Solid black indicates Palaeozoic strata. F. Cross-section of the Amazon Basin (Bigarella 1973). The axis of the basin contains 1 km of late Proterozoic, 2 km of Palaeozoic, and 0.3 km of Cainozoic sediments.

342   PHANEROZOIC HISTORY

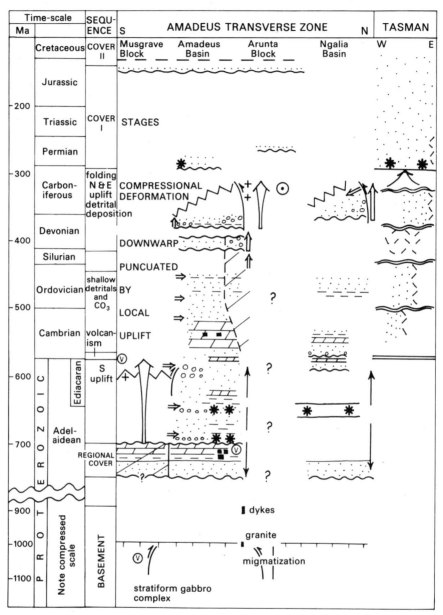

Fig. 226. Time–space diagrams of the Amadeus Transverse Zone and the Southern Oklahoma Aulacogen, and adjacent fold belts. Australian sources are Wells *et al.* (1970), Compston (1974), Wells (1976), Plumb (1979), Armstrong and Stewart

youngest preserved part of the Pertnjara Group of the northern Amadeus Basin indicate a middle Late Devonian age (Playford *et al.* 1976a). The reflectance of thin layers of vitrinite coal in the upper part of the Pertnjara Group suggests an original cover of 1 to 2 km (Kurylowicz *et al.* 1976), so that deposition may

well have extended into the Early Carboniferous, with the upper 1 or 2 km of the succession being worn off during the mid-Carboniferous lacuna or later. The Mount Eclipse Sandstone of the Ngalia Basin, 2.4 km of piedmont conglomerate, sandstone, and siltstone, contains spores that indicate a Visean age (Kemp *et al.*

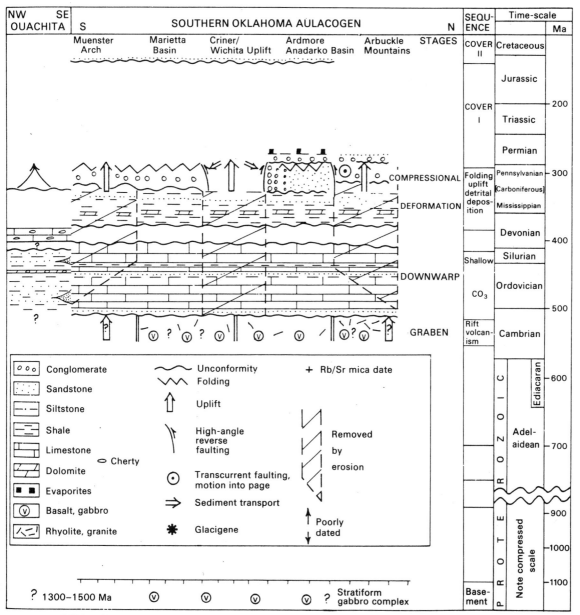

(1975), Allen and Black (1979), Black et al. (1980), Stewart (1971), Rutland (1973, 1976), and Playford et al. (1975). North American sources are Ham and Wilson (1967), Ham (1973), Webster (1977), Wickham et al. (1975), and Wickham and Denison (1978).

1977). This is the only known Early Carboniferous deposit in central Australia, and by the mid-Carboniferous, the entire region was an erosional realm. Earth movement terminated in the mid-Carboniferous, about 320 Ma ago (Alice Springs Orogeny), as indicated by Rb–Sr metamorphic mica dates (Armstrong and Stewart 1975), by complex folding and thrusting, with the formation of southward-directed nappes and thrusts that involve basement, and with the folding of the basement nappes, and of strata in the Amadeus Basin above two evaporitic layers that acted as decollement surfaces. The pattern of folds in the

Amadeus Basin (Fig. 227) suggests that some of the fault motion was left-lateral. A thin cover of Late Carboniferous/Permian (300 Ma) and Jurassic–Cretaceous (160 to 100 Ma) sediments that remain undeformed indicates quiescence during the past 300 Ma.

Most of the information about the tectonics of the Amadeus Transverse Zone comes from the Amadeus Basin (Wells et al. 1970) and the bounding Arunta and Musgrave Blocks, but what little is known of the other structures of the zone – the Officer Basin, Ngalia Basin, Lander Trough, and Toko Syncline – suggests a history of uniform deposition (in both facies and thickness) across most of the zone until dismemberment into individual blocks and basins about 320 Ma ago (Doutch and Nicholas 1978, fig. 6).

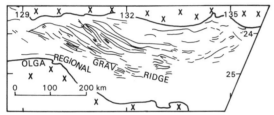

Fig. 227. Fold axes of the Amadeus Basin, from Forman and Shaw (1973).

(iii) Southern Oklahoma Aulacogen

The Southern Oklahoma Aulacogen (Figs. 225B, E; 226) cuts across Precambrian basement with radiometric ages from 1400 to 1100 Ma, and is underlain by the Layered Series of the Wichita province, interpreted by Powell and Phelps (1977) as a stratiform gabbro complex with an age of 1300 to 1500 Ma determined by the palaeomagnetic method (Roggenthen et al. 1976), supported by tectonic and petrologic constraints. Younger dates, shown in Fig. 226 by queries, are interpreted as registering a thermal overprint during emplacement 525 Ma ago of the Carlton Rhyolite and cogenetic Wichita Granite. The rhyolite, which is entirely subalkaline (Hanson and Al-Shaieb 1980), is at least 200 m thick and is restricted to a width of 90 km in the axial part of the aulacogen, presumably by emplacement along an earlier structural trend in a rift graben. Hanson and Al-Shaieb (1980) report several hundred metres of undated basalt, spilite, and minor andesite unconformably below the rhyolite, and conclude that, though not directly related to the rhyolite and granite, the more basic rocks may have been extruded during rifting. Superimposed on the graben was a broad (150 km) unfaulted downwarp in which about 9 km of Late Cambrian (520 Ma) through Pennsylvanian (290 Ma) sediments accumulated, about four times the thickness of laterally continuous sediment deposited on the adjacent stable craton. The Late Cambrian (520 Ma) to Early Devonian (410 Ma) sediments are dominantly limestones in the aulacogen and dolomite on the craton, and the younger sediments are dominantly terrigenous in the aulacogen and dominantly limestone on the craton. The Late Devonian (370 Ma) marked the earliest stage of deformation which climaxed at 300 Ma (Arbuckle Orogeny) with rapid uplift of arches bounded by high-angle reverse faults, which moved also laterally (sinistrally), accompanied by rapid deposition of conglomerate and finer detrital sediment in a dominantly marine environment and then folding in the newly defined basins. A cover of relatively lightly deformed Early Permian and Cretaceous sediments indicates quiescence for the past 290 Ma.

The three stages of development (Hoffmann et al. 1974; Wickham et al. 1975, 1976) (Fig. 226) are (1) about 525 Ma: rifting and volcanism in an 80-km-wide graben; (2) 520 to 410 Ma: subsidence in a 150-km-wide downwarp faster than in the adjacent craton; and (3) 380 to 290 Ma: compressional deformation to dismember the downwarp into individual arches and basins.

(iv) Comparisons

(1) Southern Oklahoma Aulacogen

We note some remarkable similarities or near identities between the Amadeus Transverse Zone and the Southern Oklahoma Aulacogen (Table 20): (1) the size and shape of individual basins; (2) the dominant role of high-angle reverse faults in determining structure; (3) the timing of deformation; (4) the timing of basement stabilization; and (5) the occurrence of stratiform gabbro complexes.

The essential differences are in (1) the greater width of the Australian zone (750 km versus 200 km) and its apparent lack of a volcanic graben stage; (2) the lack of a pre-breakup fill in the Southern Oklahoma Aulacogen; (3) the relations with the contemporaneously deformed fold belt; and (4) the non-marine depositional facies during compressional deformation in the Amadeus Transverse Zone versus the dominantly marine facies in the Southern Oklahoma Aulacogen.

1. Width and volcanic graben stage. The Amadeus Transverse Zone starts and persists until deformation as a broad sag, and is not centred on a volcanic graben. In so far as the graben stage is essential to the

TABLE 20. *Comparative features of the Amadeus Transverse Zone and Southern Oklahoma Aulacogen*

Feature	Amadeus Transverse Zone	Southern Oklahoma Aulacogen
SIZE (km)		
| width of zone	750	200
width of individual basins	200 (Amadeus Basin)	100 (Anadarko)
	50 (Ngalia)	25 (Marietta)
structural relief	9	8
sediment thickness	9	9 (+2 of rhyolite)
volume of basin sediment ($\times 10^3$ km^3)	800 (Amadeus Basin)	250 (Anadarko) to
	40 (Ngalia)	18 (Marietta)
STRUCTURE		
high-angle reverse fault with lateral motion suggested by oblique fold axes	✓	✓
nappes	✓	×
diapirism	✓	×
HISTORY		
| fold-belt relationship	distant	juxtaposed
timing of terminal deformation	mid-Carboniferous	Late Carboniferous
| depositional facies during compressional deformation	non-marine	dominantly marine
depositional style during downwarp stage	marine sandstone, shale, carbonate, evaporites	dominantly marine limestone
| volcanic graben stage	×	✓
| pre-breakup fill	✓	×
stratiform gabbro complex	in Musgrave Block; possibly beneath Amadeus Basin	in Wichita Uplift, presumably extends beneath aulacogen
basement stabilization	1100–1000 Ma ago	1100 Ma ago

|main differences.   ✓ present.   × absent.

definition of an aulacogen, the Amadeus Transverse Zone is strictly not an aulacogen; rather it is a broad sag, like the Amazon Basin (Figs. 225F and 228). Like an aulacogen, however, the Amadeus Transverse Zone is interpretable as having developed as the failed arm of a system that generated and then consumed the adjacent ocean, but it lacks the initial graben stage that trends to narrow the area of subsidence, so that subsidence was wide from the onset. The Olga Regional Gravity Ridge (Mathur 1976) (Fig. 227) may mark the position of an elongate body of dense rock, presumably ultramafics, beneath the sedimentary fill of the Amadeus Basin as modelled in Fig. 229A, in the same way as a corresponding gravity ridge beneath the axis of the Southern Oklahoma Aulacogen may reflect as extension of the basic intrusions exposed in the Wichita Uplift into the subsurface of the entire aulacogen (Wickham and Denison 1978, p. 21). Since in both the Amadeus Transverse Zone and the Southern Oklahoma Aulacogen the presumed sub-basin mafic and ultramafic rocks would be 1000 to 1300 Ma old and hence pre-date deposition by several hundred million years, they could not be regarded as constituting the volcanic graben. Rather, they may be viewed as being a potential basin sinker that did not cause subsidence and continuous deposition until several hundred million years after emplacement.

2. Lack or brevity of a pre-breakup fill in the Southern Oklahoma Aulacogen. The volcanic graben stage of the Southern Oklahoma Aulacogen, represented by Cambrian rhyolite and possibly by thin basalt, presumably coincided with or immediately pre-dated breakup, but lasted only a few tens of millions of years, whereas the Adelaide succession, from about 750 Ma to 570 Ma, both in the Adelaide Rift (von der Borch 1980) and the Amadeus Transverse Zone, is interpreted as a pre-breakup fill of a failed arm. While this is a notable difference between the two systems, the Southern Oklahoma Aulacogen's lack of a pre-breakup sequence, or its extreme brevity, lies within the very wide observed range of duration of such features.

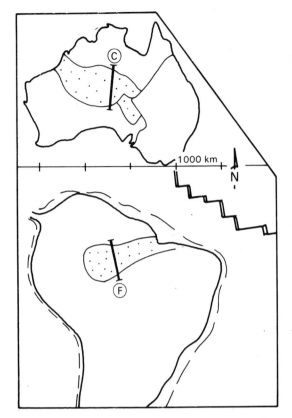

Fig. 228. Amadeus Transverse Zone and Amazon Basin (from Bigarella 1973) at the same scale. Lines show location of cross-sections C and F of Fig. 225.

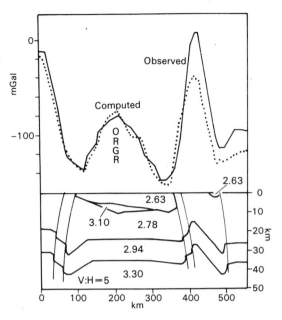

Fig. 229. A. Gravity model of part of Mathur's (1976, fig. 9) structural model, located on Fig. 225A, modified by modelling beneath the Amadeus Basin, coincident with the Olga Regional Gravity Ridge (ORGR), an ultramafic body of density $3.10\ t/m^3$ (cf. Pruatt 1975) in place of an upward bulge of the lower crust and mantle, as modelled by Mathur (1976). Gravity modelling carried out by L. Hansen.

3. Relations with contemporaneously deformed fold belt. Whereas the results of convergence in the Carboniferous produced a nearly identical set of dismembered basins in both systems, the relations of the systems with the fold belt are different: the Southern Oklahoma Aulacogen abuts the Ouachita Orogen, the Amadeus Transverse Zone is widely separated from the Tasman Fold Belt.

4. It follows that the distal Amadeus Transverse Zone was non-marine during deformation, and that the proximal Southern Oklahoma Aulacogen was marine.

(2) Ancestral Rocky Mountains

The ancestral Rocky Mountains are the same distance from the Ouachita Orogen (Fig. 225B) as is the Amadeus Transverse Zone from the Tasman Orogen. The ancestral Rocky Mountains developed out of terrain connected with the Cordilleran system on the west. As described by Kluth and Coney (1981), the structural pattern in the Mississippian/Pennsylvanian is subdued compared with the rapidly subsiding Anadarko Basin. By the middle Pennsylvanian, folding and thrusting in the Ouachita Orogen reached a climax, and deformation extended northwestward from the Anadarko Basin to affect the ancestral Rocky Mountains, where uplifts shed coarse arkose, such as the Fountain Formation, into the adjoining non-marine basins. On the west, thick marine evaporites and carbonate accumulated in the Paradox Basin. With local exceptions, deposition declined in the late Pennsylvanian and earliest Permian. According to Kluth and Coney (1981):

The component of transcurrent movement suggested on the faults bounding the foreland block uplifts probably reflects the fact that the North American craton was undergoing wrenching and some degree of internal translation due to the collision of South America–Africa with North America. We suggest that the large fault-block mountains formed when the southwestern, peninsular projection of the North American craton between the Cordilleran Geosyncline and the Ouachita–Marathon geosyncline (including the transcontinental arch) was wrenched, pushed northwestward, and deformed as the collision progressed ... Salients such as the Llano region of central Texas and recesses such as the Anadarko–Arkoma Basins region probably had an effect on the stress-strain patterns and timing of deformation within the craton ... When suturing started to affect the area south of the Ouachita–Marathon region, the peninsula of Precambrian crust of the North American craton was too weak and narrow to resist the effects of the collision. Pre-existing zones of weakness were reactivated by the collision in a complex, changing pattern that was controlled by the southwestward progressive suturing of the irregular margins ... One pre-existing zone weakness appears to have been the Anadarko Basin ... Another inherited and reactivated zone of weakness is the fault zone along the southwest margin of the Uncompaghre Uplift.

(v) North–south shortening

The deformation of the ancestral Rocky Mountains was not associated with magmatism or metamorphism. The intensity of shortening in the Amadeus Transverse Zone, expressed by thrust-faulting and the formation of large nappes, accompanied by retrogressive metamorphism, penetrative deformation, and pegmatite and vein emplacement, provides the chief contrast between the Amadeus Transverse Zone and the North American system. Parts of the Heavitree Quartzite and the Bitter Springs Formation in folded nappes on the northern edge of the Amadeus Basin underwent progressive metamorphism to sericitic quartzite and schist, while the basement rocks underwent retrogressive metamorphism of the greenschist facies (Stewart 1971). The amount of crustal shortening across the Amadeus Transverse Zone, as measured from Forman and Shaw's (1973) structural model, is at least 70 km (cf. Brewer et al.'s 1983 determination of possibly 15 ± 5 km of Pennsylvanian crustal shortening in the Southern Oklahoma Aulacogen). As in the North American system, so the Amadeus Transverse Zone was affected by transcurrent movement of unknown amount, indicated by the pattern of folds (Fig. 227), but the 70 km of crustal shortening are unlikely to be accounted for by this means alone. Kink bands in the Tasman Fold Belt, dated as mid-Carboniferous, represent about 5 per cent of north–south shortening over the measured distance of 160 km, and, if this is extrapolated over the 3000 km length of the entire Fold Belt, the total shortening is of the same order as that of the Amadeus Transverse Zone.

(vi) Discussion

North–south shortening of the Amadeus Transverse Zone during the Late Devonian and Early Carboniferous took place during east–west compression (I in Fig. 225) in the Tasman Fold Belt, from which it is inferred that part of the deformation of the Amadeus Transverse Zone may be due to the transmission of stress along transcurrent faults. Only a small part of the 70 km of shortening across the Amadeus Transverse Zone is attributable to this means, and we suggest that most of the shortening was due to superimposed north–south compression (II) such as affected the Tasman Fold Belt.

The Amadeus Transverse Zone and Southern Oklahoma Aulacogen differ in that the Amadeus Transverse Zone was not a furrow (or aulax) but a broad downwarp or syneclise (with the Musgrave Block a complementary anteclise), and this difference arose because deep-seated faults, which came into being $c.$ 1000 Ma ago, did not confine subsidence as happened in the Southern Oklahoma Aulacogen. The Amadeus Transverse Zone and the ancestral Rocky Mountains are the same distance from the conjugate fold belt, and have comparable structure, though deposition in the ancestral Rocky Mountains was connected with the western, Cordilleran, system.

If Australia contains an equivalent of the Southern Oklahoma Aulacogen, it would be expected to lie between the north-trending structures of the Tasman Fold Belt and the east-trending structures of the Amadeus Transverse Zone, in the region between lat. 135° and 140° E. That such a depression may remain undetected beneath the Eromanga Basin in this region is suggested by the discovery, made only recently by deep seismic sounding, of the Warrabin Trough (Pinchin and Senior 1982), on the western side of the Adavale Basin, within the north-trending structures of the Tasman Fold Belt.

Bally and Snelson (1980) regard the basic shape of the major Brazilian basins, including the Amazon Basin, as having been acquired during the following Mesozoic igneous events, which concluded with the inception of the South Atlantic Ocean about 130 Ma ago. The Amazon Basin 130 Ma ago thus corresponds with the Amadeus Transverse Zone 570 Ma ago (Table 21): its pre-breakup sequence stretches back some 300 Ma, compared with at least 180 Ma for the Amadeus Transverse Zone, and its stage of compressional deformation lies in the future with the demise of the South Atlantic. Compared over equivalent stages, both structures have the same width and thickness of sediment. Unknown is the effect on the Amazon Basin of the adjacent Andes. Before deformation, the Amadeus Transverse Zone was a broad sag, broken only by the Musgrave Block. It was not bounded by faults, as was the Southern Oklahoma Aulacogen, but it is possibly underlain by an old, Proterozoic, sinker. Its depositional form, before deformation, resembled the present syneclise of the Amazon Basin.

TABLE 21. *Comparative features of Amazon Basin (Bigarella 1973) and Amadeus Transverse Zone*

	Amazon Basin	Amadeus Transverse Zone
Compressional deformation (Ma ago)	≪0	325
Breakup (Ma ago)	130 ⎤	570 ⎤
	⎬ 305	⎬ ~180
Initial deposition (Ma ago)	435 ⎦	~750 ⎦
Sediment thickness (km) of comparable intervals	~5 (Silurian–present)	4.5 (Adelaidean–Ordovician)
Width (km)	800	750
Age of mafic igneous rocks (Ma ago)	120–220 (Bally and Snelson 1980, p. 40)	~700 (Amadeus Basin only) 570 (Officer Basin only)
Age of mafic igneous rocks in Parnaiba and Paraná Basins (Ma ago)	120–130	

(j) Termination of the Uluru Regime: the mid-Carboniferous lacuna (by C. McA. Powell and J. J. Veevers)

Mid-Carboniferous deformation and uplift throughout the Lachlan and Thomson Fold Belts and in the Amadeus Transverse Zone brought to an end the Uluru Regime. Sediments of the succeeding Innamincka Regime did not begin to accumulate over most of continental Australia until the latest Carboniferous or Early Permian, after the mid-Carboniferous (Namurian) lacuna (Fig. 154). Sediment accumulated only on the margins: the Yarrol–New England Fold Belt in the eastern third of the Tasman Orogen, and parts of the Bonaparte, Fitzroy, and Carnarvon Basins round the north-western margin.

Widespread mountain-building events in the east (Kanimblan orogeny) formed a meridional cordillera, and imparted the principal structural grain to the Lachlan and Thomson Fold Belts; the region changed from one of net deposition to net erosion. The inferred east–west compression that formed the eastern cordillera was followed without a break by a continent-wide north–south compression, which formed (1) megakinks in the foliated parts of the Tasman orogen (Powell, work in preparation), and (2) nappes and thrusts in the Amadeus Transverse Zone, involving an estimated 100 km of shortening. This latter mid-Carboniferous deformation must have raised the land surface, as it appears to have involved the entire continental crust (Forman and Shaw 1973; Mathur 1976). Moreover, downwearing of the highlands must have been rapid, as shown in the Amadeus Transverse Zone by the 320 Ma Rb/Sr cooling ages of metamorphic mica, and in the eastern cordillera by the removal of as much as 4.7 km of Late Devonian folded sediment, and the unroofing of mid-Carboniferous granite, such as the Bathurst Granite, before the resumption of deposition in the Late Carboniferous. This rapid uplift and concomitant rapid down-wearing must have produced a copious volume of sediment, yet the depositional record over the continent, except the eastern and north-western margins, is blank. Where could this inferred sediment have gone?

Our solution to the paradox is to postulate that the rapid uplift of much of the continent, combined with rapid movement of Australia from lower to higher latitude (Fig. 11), triggered not just an alpine glaciation in the east, as suggested by Frakes (1979, p. 135), but a continent-wide glaciation that produced a dry-based ice sheet over the entire continent except the eastern and north-western margins (Fig. 229B). The copious glacial sediment shed from the nunataks of the central uplifts and from the eastern cordillera was stored in the ice sheet. Only in the Werrie Trough of the eastern margin (Fig. 154) was glacigene sediment of this (Namurian/Westphalian) age deposited from glaciers that broke through the eastern cordillera, as

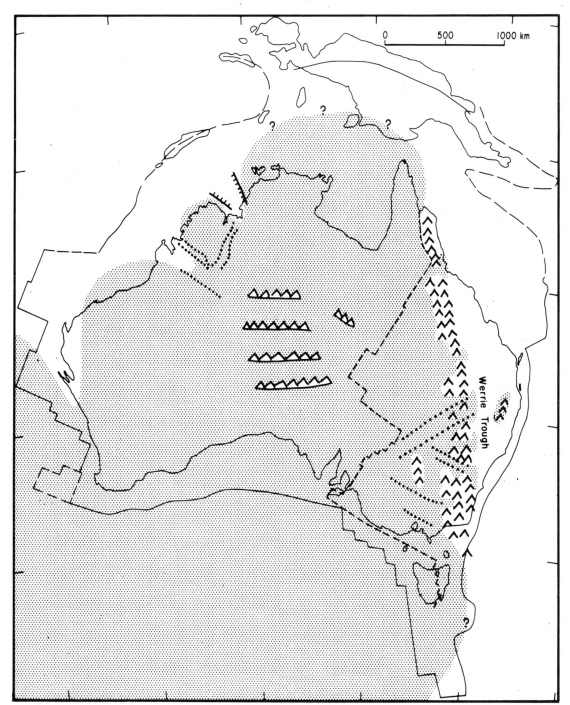

Fig. 229. B. Schematic palaeogeography of mid-Carboniferous Australia. An ice sheet (stippling) covered all the continent except the eastern and north-western margins; the nunataks of the central highlands and the eastern cordillera started being worn down by rapid wasting but the sediment released was arrested from deposition by the ice. Also shown are the structural lineaments along which movement during the Kanimblan and Alice Springs orogenies took place.

glaciers today flow through passes in the Transantarctic Mountains to deposit glacigene sediment in the Dry Valleys of the McMurdo Sound area. Not until the relaxation of intense glaciation in the Late Carboniferous/Sakmarian and the retreat of the ice sheet did the ice release its store of sediment (Fig. 155). In this view, the widespread latest Carboniferous/Sakmarian glacigene sediments at the base of the Innamincka basins reflect not the climax of glaciation, as hitherto inferred (Crowell and Frakes 1971a, b; Frakes 1979), but its decay. Furthermore, the basins that started filling at this time were generated by the terminal deformations of the Uluru Regime in the mid-Carboniferous (350 to 320 Ma), and not, as Veevers and Evans (1975), among others, have suggested, by tectonic events in the latest Carboniferous (310 Ma, Stephanian). The difference in time is short, but the distinction is important: the Innamincka basins were *generated* in the mid-Carboniferous, but did not *accumulate* sediment until the latest Carboniferous.

VI. SYNOPSIS (by J. J. Veevers, J. G. Jones, C. McA. Powell, and J. A. Talent)

1. INTRODUCTION

In this chapter, we narrate in chronological order, unfettered by documentation, the Phanerozoic and preceding Adelaidean history of Australia as detailed in the foregoing chapters. The narrative is presented as a timetable of events followed by a cinematograph of palaeogeographical reconstructions. From this history, we distil some tectonic generalizations, and conclude with a list of pressing unanswered questions.

2. TIMETABLE OF EVENTS

(a) Narrative

A timetable of events is presented in Fig. 230. The momentous events of the Phanerozoic were preceded by the 200 Ma long, comparatively quiescent, Adelaidean Regime of extension of the Precambrian craton west of the Tasman Line. Rift valleys to the north and south of a central sag all branched off the Tasman Line, itself probably marked by rift valleys. These depressions filled with thick shallow-water non-marine sandstone, shale, and carbonate; in addition, evaporites were deposited in the Amadeus Transverse Zone, and then, over the entire continent, glacigene sediment. Leading to supercontinental breakup at the start of the Phanerozoic was a 75 Ma long interval of dextral shear reflected in thrusting, the formation of nappes, and the intrusion of granite along zones across central Australia, notably in the Musgrave Block, and to the west; marine sediment, with the first traces of metazoans (the Ediacara Fauna) accumulated elsewhere.

The change to the Uluru Regime, 575 Ma ago, was reflected by breakup in the east along the Tasman Line, and in the north-west, all accompanied by the eruption of voluminous plateau basalt. Low latitude, with abundant evaporites and carbonate deposited in a broad epeiric sea that covered parts of the platform, characterized the Uluru Regime. On the east, divergence in the Early Cambrian changed to convergence in the mid-Cambrian, with the generation of a marginal sea behind a magmatic arc, and then deformation and intrusion by granitoids of the marginal sea and adjacent continental margin at the end of the Cambrian. A second stage of the Uluru Regime, in the Ordovician and Early Silurian, saw a jump of plate divergence on the north-west, and an eastward jump of the magmatic arc on the east; it too terminated with deformation. The third stage, in the Silurian/Devonian, involved the attenuation of the fold belt to a basin-and-range terrain, with a foreland basin at the edge of the newly accreted craton, all consolidated in a terminal deformation. The fourth and final stage of the Uluru Regime, in the Late Devonian/Early Carboniferous, involved an eastward jump of the reborn arc, and a mid-Carboniferous climactic terminal deformation in the east and centre, driven by compressive forces from the east and north. In the succeeding Innamincka Regime, all but the peaks of the resulting continent-wide highlands were buried by an ice sheet, and deposition was arrested until the retreat of the ice sheet in the Late Carboniferous, when the broad sags generated by the mid-Carboniferous deformation intercepted the mainly glacigene sediment released by the ice to form, as elsewhere in Gondwanaland, the initial fill of the Gondwana Series basins, including infrarift basins on the west and north-west, and a foreland basin alongside the magmatic arc that had jumped eastward. The low latitude of the Uluru Regime had given way to the high latitude of the Innamincka Regime, with a corresponding near-absence of evaporites and carbonate, and a profusion of coal measures. An initial broad marine transgression across the ancestral Australian/Antarctic Depression was followed by short marine ingressions that accompanied phases of increased tectonic activity, expressed in the arc by waxing magmatism and uplift, and in the concomitantly subsiding foreland basin by the deposition of volcanogenic sediment. Deformation that accompanied uplift extended in the mid-Triassic into the adjacent epicratonic basin to mark the end of the early Innamincka Regime. In the middle Innamincka, the arc jumped a short distance eastward to form a diffuse zone within and between uplifts; this stage terminated by deformation in the east, and in the west, where the Fitzroy Trough was folded by wrenching, and the infrarift stage of the western and north-western margins changed, by the onset of faulting, to the rift-valley stage. Then, in the late Innamincka, the arc was reborn by its sixth jump to the east, the north-western and western margins formed by seafloor spreading of the

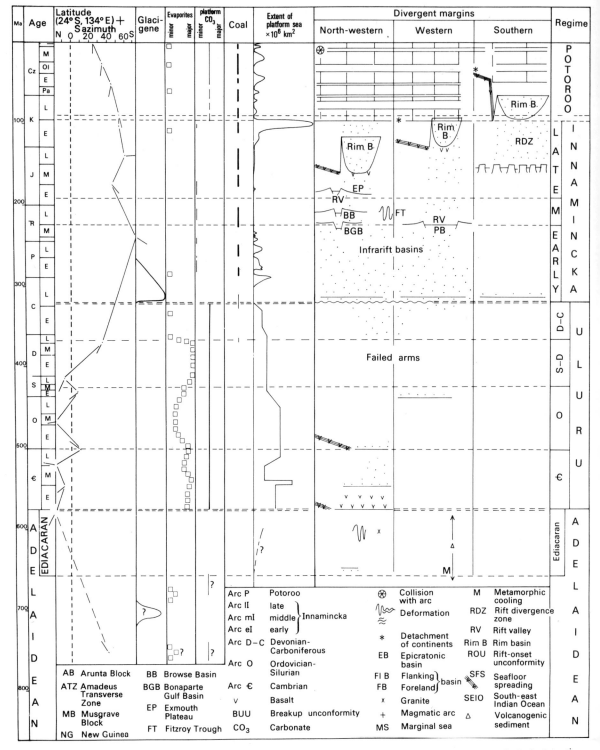

Fig. 230. Timetable of events during the Phanerozoic, and in the Adelaidean prelude. Inset map shows counter-clockwise jumping of plate divergence and plate convergence. Summarized from information given above, except extent of platform sea for the interval 575 to 100 Ma ago, from Veevers (1976) and Veevers and Evans (1973). The queries placed against some Adelaidean events indicate uncertain age determinations. Latitude and south azimuth for 750 Ma ago from Embleton

TIMETABLE OF EVENTS

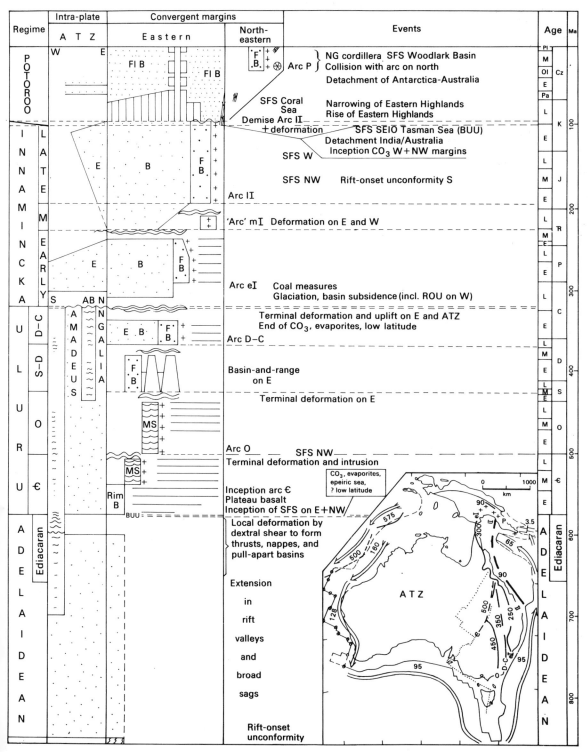

(1981, table 1); the broken line indicates the poorly determined latitude from 750 to 575 Ma ago. Embleton (Chapter II) points out the difficulty of determining palaeomagnetically the latitude of the Adelaidean glacigene sediments: low-inclined magnetizations in these sediments are possibly overprints.

eastern Indian Ocean, and the southern margin was prefigured by the inception of a rift-divergence zone. The end of the Innamincka Regime was marked by an epeiric sea that covered more than half the platform, and by a great volume of volcanogenic sediment shed from the arc across the platform that expelled the sea even from rapidly subsiding depocentres in the Innamincka area and in the ancestral Australian–Antarctic Depression. During the last 5 Ma of the arc, from 95 to 90 Ma ago, Antarctica and the Lord Howe Rise/New Zealand split off Australia, though the initial separation was very slow. The overlap of the old arc and the new marginal breakup defined the Cenomanian Interregnum. The separation of the Lord Howe Rise involved the splitting of the south-eastern part of the magmatic arc, which led to the initial uplift of the Eastern Highlands as a broad epeirogen that shed voluminous sediment into the Ceduna Depocentre during the Late Cretaceous. Also at 95 Ma ago, India cleared the epiliths off the western margin, and the deep oceanic circulation that followed led to widespread carbonate deposition, though the latitude was still fairly high. With a continuing decrease of latitude by equatorward drift, all of northern Australia and much of the south had become a carbonate province by the late Oligocene inception of the Great Barrier Reef. Cainozoic cycles of transgression–regression–lacuna in eastern Australia were matched by coeval ones in the west and south, and indeed elsewhere in the world, as expressions of global tectonic activity, expressed as eustatic changes of sea level. At the start of the Cainozoic, 65 Ma ago, the Eastern Highlands contracted to their present width, the South Australian Highlands blocked the supply of sediment to the Ceduna Depocentre, and seafloor spreading jumped to the Coral Sea. The magmatic arc was reborn (the seventh and latest generation) about 30 Ma ago in eastern New Guinea, at the same time as New Guinea collided with an external arc on the Pacific Plate. The locus of collision migrated westward to reach Timor about 5 Ma ago, by which time New Guinea had reached the cordilleran stage, and its easternmost tip, in the Woodlark Basin, had been split by seafloor spreading that started 3.5 Ma ago. Antarctica/Australia had finally separated by 37 Ma ago, and the resulting inception of the Circum-Antarctic Current succeeded the early Oligocene deep currents in the South-west Pacific that scoured sediment. The Pleistocene glaciation was manifested in Australia mainly by the onset of aridity.

(b) Pattern of plate development

Plate development was punctuated by discontinuities or jumps in supercontinental breakup by plate divergence, and in the position of the magmatic arc by plate convergence. The jumps follow a sequence that traces a counter-clockwise track, as shown in the inset of Fig. 230.

(i) Plate divergence

(1) The inception of plate divergence on the north-west 575 Ma ago was followed by (2) a jump south-westward at 500 Ma ago. A second phase of divergence in the same area started (3) 160 Ma ago in the north-west, and (4) 128 Ma on the west. Divergence continued (5) 95 Ma ago on the south and south-east to (6) the Coral Sea 65 Ma ago, and (7) finally to the Woodlark Basin 3.5 Ma ago.

(ii) Plate convergence

Convergence on the east, after the initial divergence 575 Ma ago along the Tasman Line, is shown by six positions of the magmatic arc: (1) 500 Ma ago; (2) 450 Ma; (3) 350 Ma; and (4) 250 Ma; all by counter-clockwise jumps from a syntaxis at lat. 20° S; then (5) at 190 Ma ago, the arc jumped laterally; and finally (6) by 28 Ma ago had jumped from a syntaxis at lat. 10° S to its present position.

All sequences, divergent or convergent alike, involve jumps in a counter-clockwise sense.

3. CINEMATOGRAPH

The cinematograph shown in the following pages (Figs. 231 to 246) is a time-lapse sequence of palaeogeographic reconstructions from the earliest Phanerozoic (and its late Cryptozoic forebears) to the present day. Timor, the Outer Banda Arc, the northern part of New Guinea, and the Papuan Peninsula are shown for reference only, except where shown in the present-day map of Fig. 246. The bar scale at the top right of most figures represents 1000 km.

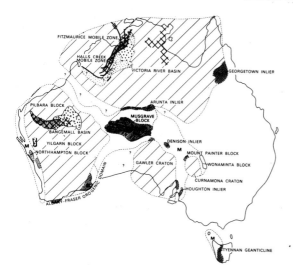

Fig. 231. Late Precambrian, c. 1000 Ma ago, during the final stage of consolidation of the Precambrian craton, which extended eastward an unknown distance east of the Tasman Line (left blank), and westward and southward into the rest of Gondwanaland. From Fig. 177 (after Plumb 1979).

From a development traceable back to 3.5 Ga ago, the separate parts of the Precambrian craton were welded together c. 1000 Ma ago by intense tectonic activity, including folding, metamorphism, migmatization, and granite emplacement in narrow belts (Albany–Fraser, Northampton, Bangemall Basin, Musgrave, Arunta, Georgetown, Halls Creek–Fitzmaurice, Mount Painter, Houghton, western Tasmania) between or surrounding the major individual blocks. Parts of the Amadeus Transverse Zone, in particular the Musgrave Block and Arunta Inlier, were consolidated by granulite metamorphism, intrusion of granites, and emplacement of layered mafic–ultramafic intrusions along west-trending faults, followed by the eruption of volcanics and granite in areas of cauldron subsidence. Similar events in the mobile zones round the Yilgarn Block consolidated it to neighbouring blocks. Internal deformation of the northern part of the craton was accompanied by the deposition of shelf carbonate and sandstone in the Victoria River Basin.

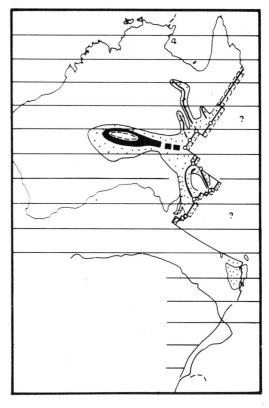

Fig. 232. Adelaidean Regime, from c. 850 to 650 Ma ago. From Fig. 178. To the south, in Antarctica, the continental margin lay along the present Transantarctic Mountains, and to the north at an unknown distance east of the Tasman Line. This interval was quiescent compared with earlier and later times. On mainland Australia, igneous activity was minimal, and shallow water, probably non-marine, sandstone, shale, and carbonate accumulated in rift-valley complexes north and south of a broad sag or syneclise across the centre. All these structures branched off the Tasman Line, the north-east-trending segments of which were probably marked by rift valleys too. Evaporites in the Amadeus Transverse Zone and subsequent glacigene sediment in all the basins suggest a change of climate within the Adelaidean Regime comparable to the subsequent change from the Uluru Regime to the Innaminka Regime. Not shown is the sheet of sediment that extended north-west of the centre into the Kimberley region. Probably reflecting its position nearer the continental margin, Tasmania had older Adelaidean sediment, including turbidites, metamorphosed and intruded by granite and dolerite by 700 Ma ago.

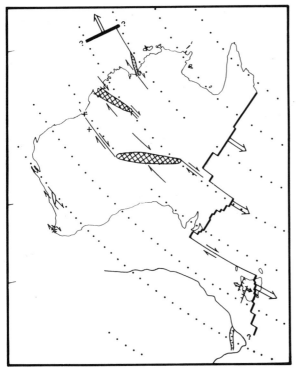

Fig. 233. Terminal (or epi-Adelaidean) phase of the Adelaidean Regime, from 650 to 575 Ma ago, during the Ediacaran, up to the time of breakup along the east and north-west at the start of the Uluru Regime. From Fig. 182.

The quiet of the Adelaidean was broken c. 650 Ma by a change from extension to dextral shear (about a pole to the north-east) that led finally to breakup by plate divergence (about the same pole) in the north-west and the east. As during the interval immediately before the Adelaidean Regime, so during the epi-Adelaidean stage, the Amadeus Transverse Zone, in particular the Musgrave Block, was the centre of activity, with thrusting, the formation of nappes, metamorphism, and, during uplift, the northward sliding of the Adelaidean sequence above a decollement of evaporites. The western part of the Amadeus Transverse Zone was intruded by granite, and its north-western part was deformed by thrusting. Outside the Amadeus Transverse Zone, in the south-west, a volcanic pull-apart basin and a deeply eroded block uplift were mirrored in the north, beneath the Bonaparte Gulf Basin, by an inferred pull-apart basin. In the Flinders Ranges, notable uplift of rift-valley shoulders and the deep erosion of canyons c. 650 Ma ago were followed by continuous quiet deposition marked by the appearance of marine metazoans of the Ediacara Fauna.

After a regional lacuna, the Adelaidean Regime passed to the Uluru Regime with a change from regional shear to plate divergence by seafloor spreading along the newly formed eastern and north-western margins.

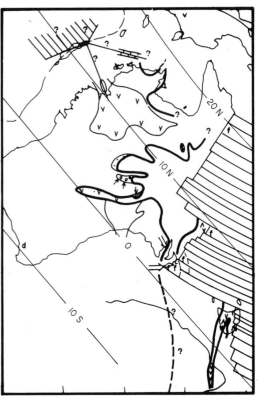

Fig. 234. End of the Early Cambrian, 545 to 540 Ma ago. From Fig. 183. Ocean basins, Tethys to the north-west, and the Palaeo-Pacific to the east, had been generated by plate divergence. The platform between the oceans was crossed in the north-west by failed arms, covered in the middle, and possibly also in New Guinea, by plateau basalt, and in the south-west intruded by dykes. The raised edge or rim of the eastern margin was backed by a marine basin with fingers that penetrated on either side of epi-Adelaidean uplifts, including the Musgrave Block, around which piedmont gravel was deposited. In the Uluru (Ayers Rock–Olgas) area of central Australia, these first deposits of the Uluru Regime, spectacularly exposed today in the monoliths of Uluru and the Olgas, give their name to the regime. Volcanic rifts dotted part of the rim, notably in Tasmania and Victoria land; in the angle formed by the rim and the Gambier–Beaconsfield fault zone, and in places to the north, wedges of quartzose flysch prograded over the newly formed oceanfloor. This was a time of evaporite deposition in the western arms of the shallow sea that faced the Palaeo-Pacific, consistent with the palaeomagnetically determined low latitude.

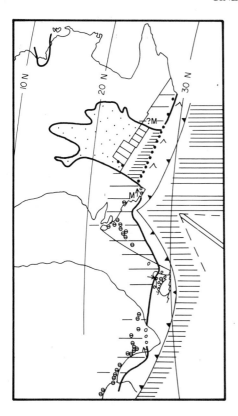

Fig. 235. Late Cambrian, 516 to 508 Ma ago. From Fig. 186B. By a reversal of plate motion on the east, from divergence to convergence, the Paleo-Pacific was subducted beneath eastern Australia. This produced different effects along strike: from north to south, (a) metamorphism and accretion to the continent of sediment deposited earlier in the Cambrian, and its intrusion by volcanics that rose into an island arc; (b) immediately southward, the island arc was backed by a marginal sea that opened by plate divergence between the arc and the continental margin or part of the old oceanfloor; the marginal sea was itself backed by an epicontinental sea; (c) between the Lake Blanche fracture zone and the Gambier–Beaconsfield fracture zone, granitoids rose through the Kanmantoo flysch wedge, and south of the Gambier–Beaconsfield fracture zone, granitoids rose into the continental margin, all of which was uplifted in the Delamerides to shed coarse detritus along the margin.

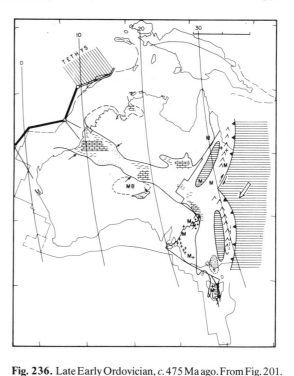

Fig. 236. Late Early Ordovician, c. 475 Ma ago. From Fig. 201. In north-western Australia, the sea, hitherto confined to the Bonaparte Gulf Basin, made its way in the Early Ordovician into the Canning Basin. Detrital sediment was deposited along a shoreline on the east, and elsewhere detrital sandstone was deposited with shale and limestone. This incursion of the sea south-westward of the Bonaparte Gulf Basin reflects the further breakup along the north-western margin by divergence to extend the Tethyan Ocean, from which failed arms radiated into the Canning and Carnarvon Basins. The sea crossed a sill south-westward of the Canning Basin to provide a link through the transcontinental Larapintine Sea with the marginal sea on the east. The southern shore of the Larapintine Sea received sediment from the high ground of the Musgrave Block, which at the same time shed non-marine sediment into the Officer Basin. The marginal sea separated a north-trending volcanic island arc in the east from the western platform, and was filled with quartzose sediment derived from the west. The north-easterly direction of convergence between the Gondwanaland plate and the Palaeo-Pacific plate produced mainly transform motion along the northern part of the island arc.

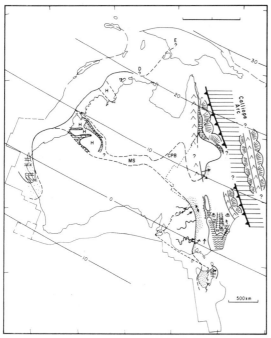
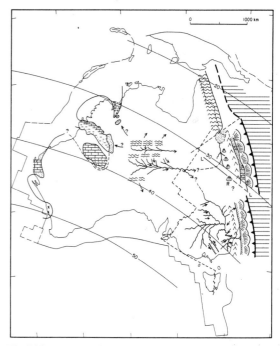

Fig. 237. Siluro-Devonian, 420 to 380 Ma ago, represented in eastern Australia by the Early Devonian (400 Ma). From Fig. 219. In the west, the arms of Tethys accumulated thick evaporites and redbeds in the Carnarvon, Canning, and Bonaparte Gulf Basins, and were possibly connected east of the Canning Basin through intermittently marine deposition across central Australia with the convergent eastern margin. Here, a transtensional basin-and-range terrain, flanked by a continental volcanic arc, faced the Palaeo-Pacific, which carried the converging Calliope Arc.

Fig. 238. Towards the end of the Uluru Regime, latest Devonian/earliest Carboniferous, 360 Ma ago. From Fig. 223B. On the west, the sea lapped again over the three failed arms. In the Bonaparte Gulf Basin, conglomerate and sandstone were shed from the uplifted south-eastern margin, which was subjected to dextral transcurrent movements. Following uplift of the south-western margin, the depositional area was differentiated into a shallow inshore platform and an offshore basin; carbonate, including reefs, and quartz sandstone were deposited inshore, and dark shale and siltstone offshore. Outliers of quartz sandstone to the south and outcrops in the north-eastern Canning Basin indicate a source of detrital sediment from the south-east and east. In the northern Canning Basin, a barrier reef complex interfingered south-westward with basinal shale, to the south of which a widespread dolomitic limestone was deposited. In the Carnarvon Basin, marine platform carbonate and quartz sandstone along the eastern edge of the basin interfinger with a carbonate reef complex offshore. The occurrence of reworked microfossils in the Perth Basin points to Late Devonian rocks nearby. The greatest uplift took place in the Amadeus Transverse Zone, where piedmont gravel, sandstone, and siltstone were deposited and then folded between rising upthrusted blocks.

On the convergent south-eastern margin, a continental magmatic arc became established along the eastern edge of the Lachlan Fold Belt, and sediment derived from the arc mingled with sediment shed from the west to fill a foreland basin. East of the magmatic arc lay a *melange* and flysch accretionary prism. To the north, drainage from central Australia led through the Adavale Basin into the foreland Drummond Basin. To the east lay a volcanic arc, fore-arc basin, and a slope. Farther north, shallow- to deep-water sediments were folded in the latest Devonian.

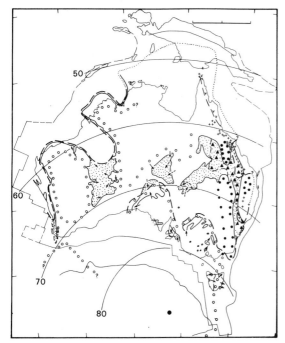

Fig. 239. Start of the Innamincka Regime, 300 to 280 Ma ago. From Fig. 155. The mid-Carboniferous deformation that terminated the Uluru Regime—east to west compressive folding on the east, and north to south compression expressed in mega-kinks on the east and transcurrent and overthrust faulting on the platform—established the salient structure of the Innamincka Regime. But deposition from the uplifted areas was arrested by the formation of a continent-wide ice sheet. Only with the retreat of the ice, towards the end of the Carboniferous, did the initial Innamincka (or Gondwana Series) basins start to fill. The Australian–Antarctic Depression was covered by an epeiric sea and then occupied by lobes of ice that issued from Victoria Land, near the South Pole, and flowed northwestward round the ancestral South Australian Highlands. Other lobes flowed out into the basins from centres on the Western Shield and in northern Australia. The basins along what was to become the western and north-western margins were infrarift basins that prefigured the margins. On the east, the magmatic arc jumped eastward, and was accompanied by the Bowen–Sydney foreland basin, and beyond a glaciated foreswell to the west, by epicratonic basins. Shortly after the glacial episode, the basins settled down to accumulate thick coal measures. In the foreland basin, cycles of transgression/regression/lacuna developed during intervals of waxing and waning activity in the magmatic arc to provide favourable conditions for the accumulation of coal enclosed by volcanogenic sediment, succeeded by barren quartzose sediment.

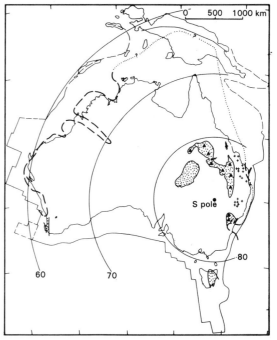

Fig. 240. Early Triassic, 245 to 240 Ma ago. From Fig. 157. By the Early Triassic, south-eastern Australia had migrated to the South Pole but did not accumulate ice. The last main pulse of activity in the early Innamincka arc generated a blanket of volcanogenic sediment that extended across the foreland basin to the westernmost limit of the Galilee epicratonic basin, but, unlike the Permian deposits of this kind, it lacks coal; its characteristic deposits are redbeds. The shoreline lay near the present coast except possibly for a brief incursion into the Galilee Basin. On the west, the shoreline lay along the present coast except for the deep embayment of the Fitzroy Trough. To the south, the sea reached the formerly non-marine part of the Perth Basin in a marine penetration that was not matched until the Jurassic/Cretaceous breakup.

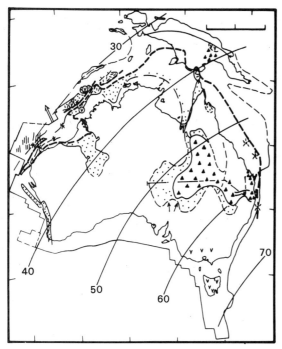 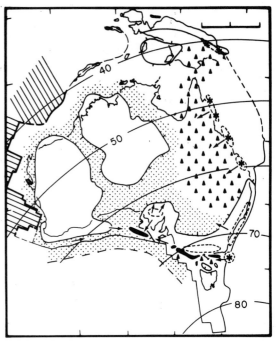

Fig. 241. Middle Jurassic, 160 Ma ago. From Fig. 172. The events of the middle Innamincka stage (225 to 190 Ma ago) encompassed the inception of rift-valley formation along the western margin, and an intermediate position of the magmatic zone on the east before its jump eastward 190 Ma ago to offshore Queensland in the late Innamincka stage. This late Innamincka arc shed volcanogenic sediment further west than any earlier arc, and at the time of this reconstruction, 160 Ma ago, the oldest volcanogenic formation of the Eromanga Basin covered all but its westernmost and southernmost parts. The New Guinea part of the arc, in contrast, shed volcanogenic sediment only a short distance back across the craton. South-eastern Australia was a region of scattered mafic volcanism; Tasmania and Antarctica saw the last pulse of intrusion of voluminous dolerite. The first signs of the rift-valley complex appeared along what was to become the southern margin, at Jerboa and in the Polda Basin. In a culmination of rifting, the north-western margin finally broke off its neighbour, basalt spilt back across the margin, the sea lapped over the Canning Basin, the Exmouth Plateau was cut into narrow blocks by north-trending faults, and the rift valley of the Perth Basin filled with thick sediment.

Fig. 242. Albian, 105 Ma ago. From Fig. 147B. The archipelagic magmatic arc had migrated back towards the Queensland coast, and extended southward to Gippsland; the volcanogenic sediment generated by the arc spread across all of sea-covered north-eastern Australia and the non-marine rifted arch of southern Australia at least as far west as the Otway Basin. The ancestral Australian–Antarctic Depression in the Great Australian Bight was covered by the epeiric sea, and accumulated the greatest volume of sediment along the southern rift system, but how much of it is volcanogenic is unknown. In the epeiric sea that extended northward round the Great Western Plateau and across its middle, thin quartzose sediment was deposited. The western and north-western margin, now fully formed after breakup in the south 128 Ma ago, was open to the ocean in the north, and rimmed by a marginal rim in the south. The epeiric sea covered about half the platform but was short-lived. Within 10 to 15 Ma, the platform rose faster than the rising eustatic sea-level, and Australia reverted to its customary geocratic state.

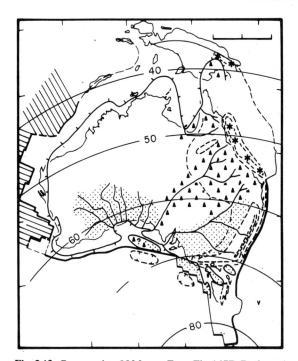

Fig. 243. Cenomanian, 90 Ma ago. From Fig. 147D. By the end of the Innamincka/Potoroo Interregnum, marked by the overlap of seafloor spreading on the south and south-east, and the terminal climax of the Innamincka arc, the platform was dry land, and the cordillera in the north-east and the epeirogen in the south-east shed copious quantities of sediment back across the craton into the Innamincka and Ceduna Depocentres through a drainage network centripetal to the Ceduna Depocentre. A subsidiary source of sediment was the area north-west of the Ceduna Depocentre. Volcanogenic sediment from the north-eastern cordillera possibly reached the Ceduna Depocentre, but the south-eastern arc was split 95 Ma ago by seafloor spreading so that the Cenomanian sediment of the previously volcanogenic Otway and Strzelecki Basins was overlain by dominantly quartzose sediment, and volcanism in the south-east was restricted to shoshonitic centres along the coast. On the west, India had cleared Australia, and the ensuing deep oceanic circulation led to carbonate being deposited over the entire, now open, margin.

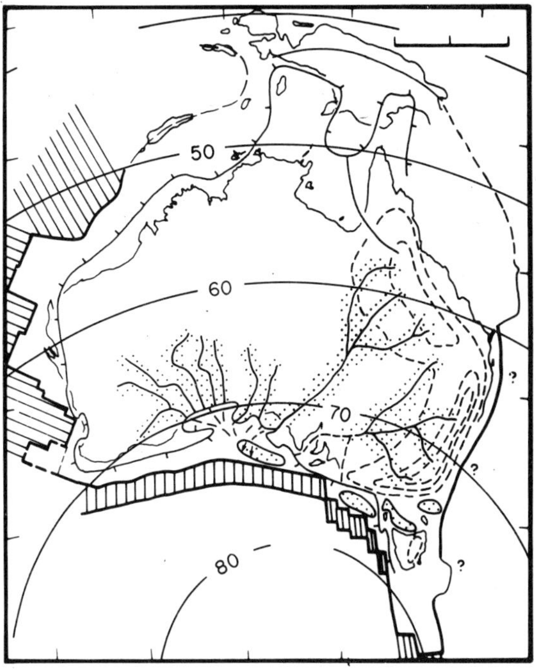

Fig. 244. Later Cretaceous, 80 Ma ago. From Fig. 147F. The death of the magmatic arc in Queensland was followed by the inception there of an epeirogen which joined with that of the South-east Highlands to form the forebears of the present Eastern Highlands. The early Eastern Highlands were much broader and higher than the present system, and were rapidly eroded to supply thick quartzose sediment through the centripetal drainage system to the Ceduna Depocentre, and smaller amounts to the Gippsland, Bass, and Otway Basins, all of which were depocentres behind the rims of the southern and south-eastern margins alongside the slowly opening ocean. Little sediment lodged on the high-standing platform, which became deeply weathered. The western and north-western margin continued its slow subsidence about a hinge along the coast, and accumulated thin carbonate sediment.

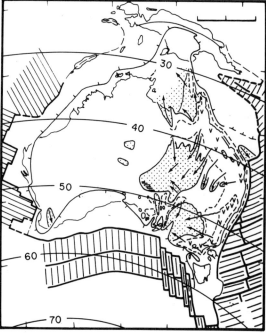

Fig. 245. Eocene/Oligocene, 38.5 Ma ago. From Fig. 147H. The uplift of the South Australian Highlands 65 Ma ago blocked the drainage to the Ceduna Depocentre, and together with the concomitant narrowing of the Eastern Highlands led to the deposition of quartzose sediment in the Birdsville and Murray Basins of the Central-Eastern Lowlands. To the north, the Karumba Basin was deposited round the Gulf of Carpentaria. By 38.5 Ma ago, parts of central Australia, hitherto a long-standing erosional domain, had also accumulated thin sediment. Rapid seafloor spreading from 82 to 57.5 Ma in the Tasman Sea and from 65 to 57.5 Ma in the Coral Sea blocked out the eastern margin. After a phase of very slow spreading from 95 to 44 Ma ago, the South-east Indian Ocean spread rapidly so that by 38.5 Ma ago Antarctica had separated 1000 km from Australia, and was on the point of complete separation along the transform faults south of Tasmania. The southern margin, open to the sea since 65 Ma ago, was now dominated by carbonate sediment, and arms of the sea penetrated the Murray Basin and part of south-western Australia. Carbonate continued to be deposited on the western and north-western margin under the influence of marine transgressions in the Palaeocene and Eocene, recorded also along the southern and eastern margins. During the transgressions, the Eastern Highlands were uplifted and, in places, covered by basalt; at the same time, the flanking basins subsided to receive sediment shed from the Highlands.

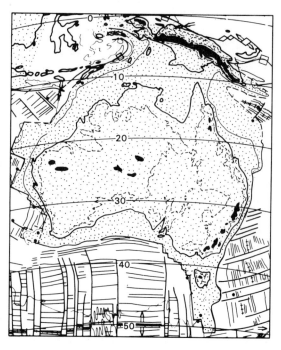

Fig. 246. Australia today. From Figs. 3 and 4. The platform remains differentiated into the highlands of the Great Western Plateau and Eastern Highlands, separated by the Central-Eastern Lowlands, co-extensive with the northern part of the Australian–Antarctic Depression, held down dynamically by asthenospheric downwelling. The Eastern Highlands and flanking basins are in the lacuna phase of a transgression/regression/lacuna cycle; that is, the highlands are low and the basins high in a phase of tectonic recovery. Tectonic activity is concentrated along the northern convergent margin, and is expressed in four features: (1) a foreland basin and frontal thrust coupled to an orogen traceable for 7000 km through the archipelago of the troughs and high islands of the Outer Banda Arc and the loop of east Sulawesi and the Vogelkop to the central New Guinea cordillera, and then south-eastward through the Owen Stanley Range of the Papuan Peninsula and the foreland basin of the Moresby Trough; (2) 100 to 200 km oceanward of the crest of the orogen, the boundary between the continental lithosphere and Cainozoic oceanic lithosphere; (3) the Northern Ranges Province in central New Guinea; and (4) volcanic arcs on the west (Inner Banda Arc) and east (Eastern Papuan Province, Bismarck Volcanic Arc).

4. DISCUSSION

(a) Australia's tectonic history back through the Phanerozoic and Adelaidean to c. 850 Ma ago is interpretable in terms of tectonic processes and rates operating in the modern world. This is exemplified by matches between modern island arcs and Australian counterparts back through the Phanerozoic, and between modern rifted-arch systems and Australian counterparts back through the Phanerozoic and Adelaidean. The depositional history of the continent in the Phanerozoic is likewise interpretable in terms of processes and rates that operate today, except for time- or climate-dependent biogenic deposits such as coal or nannofossil carbonate. (b) Two morphotectonic elements have dominated the development of Australia's Phanerozoic sedimentary basins: the rifted arch and subsequent ocean on divergent margins, and the magmatic arc paralleling the ocean margin on convergent margins. They and the intervening platform have commonly acted in concert.

Early in the Uluru Regime, at the beginning of the Phanerozoic, plate divergence on the north-west and east, part of an epoch of world-wide rifting (Bally 1980, p. 16), was followed immediately on the east by plate convergence. During the Uluru Regime, the magmatic arc jumped eastward twice, first at the end of the Cambrian, and then in the Late Devonian; between these jumps, during the mid-Silurian to mid-Devonian, the orogenic belt was occupied by a basin-and-range zone of high heat-flow. With each eastward jump of the magmatic arc, the associated sedimentary basins likewise jumped eastward; consequently, at any place in the eastern orogenic belt, basin types succeed each other vertically in the same order as they did laterally from ocean to continent, from fore-arc basin through foreland basin (or marginal sea), to epicratonic basin. The upward increase in continentality of successive basins culminated in the platform cover of the Great Artesian Basin and its successors. The mid-Carboniferous change to the Innamincka Regime was marked by a continent-wide lacuna, due to the arrest of sediment by an ice sheet, after which the Innamincka Regime followed a course resembling that of the Uluru.

In the Innamincka Regime, the convergent east and divergent west acted in concert: (1) the resumption of basin development after the mid-Carboniferous lacuna is the same age in the infrarift basins of the western and north-western margins, in the epicratonic basins of the platform, and in the foreland basin of the eastern margin; (2) waxing magmatism in the magmatic arc and concomitant subsidence in the linked foreland basin are accompanied by subsidence in the west and on the platform; (3) the Late Triassic transition from the early to late Innamincka Regime is expressed by a change from infrarift to rift-valley basins and by folding in the west and north-west, and by a coeval jump of the magmatic arc in the east; and (4), like the Uluru Regime, the Innamincka Regime was followed by a continent-wide lacuna, due, not to glaciation, but to general uplift.

Within the Potoroo Regime, greater time-resolution reveals that cycles of transgression/regression/lacuna in basins in the east, linked to waxing and waning phases of uplift and basalt magmatism in the Eastern Highlands epeirogen, correspond with similar cycles on the southern and western/north-western margins. The New Guinea Orogen and linked foreland basin also record at least the last two cycles, in the Miocene and Pliocene–Quaternary. These cycles are registered world-wide as the changes of sea level documented by Vail *et al.* (1977), and hemisphere-wide as the circum-Pacific magmatic cycles of Kennett *et al.* (1977). Thus eastern Australia did not simply register global sea-level changes passively: the epeirogen rose and sank actively in concert with the sea level, as did the New Guinea Orogen.

(c) First-order eustatic sea-level changes correlate with spreading rates of the world ocean (Pitman 1978); second-order changes in both land level and sea level also appear to be responses to global tectonic activity. Bally (1980, p. 15) suggests just this: 'In fact, Vail *et al.*'s cycles may correspond more to widespread and correlatable subsidence and uplift episodes. In this view, eustatic sea level changes may be subordinate to worldwide tectonic cycles.' An important manifestation of global tectonic cycles is the Haug Effect: 'times of orogeny are times of transgressions of epicontinental seas in the continental interiors' (Johnson 1971, 1972). That is to say, during the course of a regime, uplift and subsidence are connected: up go the rift shoulders, down goes the rift valley; up goes the orogen, down goes the foreland.

During the normal course of a regime, continental profiles are rough across the margins, whether divergent or convergent. In the interlude between regimes, the entire continental profile flattens: the sediment produced is from highlands inherited from the previous regime or generated at the transition; a negligible amount is deposited on the surface of the continent temporarily devoid of sediment traps. Thus, in the later Cretaceous lacuna that followed the Innamincka Regime, sediment bypassed the platform to terminate in the Ceduna Depocentre of the developing southern margin. During the mid-Carboniferous lacuna, the generation (except from nunataks) and deposition of sediment was arrested by an ice sheet. The Uluru Regime terminated in deformation and uplift in the east, west, and centre, whereas the termination of the Innamincka Regime was marked by deformation and uplift restricted to the northern, eastern, and southern margins.

Our synthesis of Australia's Phanerozoic history leads us to the same general conclusions as Bally (1982, p. 334): 'Sequences reflect continuing plate tectonics, while the major unconformities mark the slow-down and ultimately the cessation of one major plate tectonic regime and a plate reorganization into the new plate tectonic regime.' The Innamincka/Potoroo Interregnum (95 to 90 Ma ago) vividly exemplifies this reorganization: the death of the Innamincka arc by extinction on the north-east and by splitting on the south-east; the inception of the South-east Indian Ocean at the breakup of Antarctica and Australia; and reorganization of the spreading pattern in the eastern Indian Ocean. Bally (1980, p. 16) adds that these 'major plate reorganizations . . . are not necessarily orogenic or mountain building tectonic events'. For both the Uluru and Innamincka Regimes, the lacuna at the transition may have involved loss of record (vacuity) from uplifts inherited from the previous regime or generated at the transition, but for the rest of the continent the lacuna indicates no deposition (hiatus).

5. PRESSING UNANSWERED QUESTIONS

(a) Precisely what happened during the two continent-wide lacunae? Were these transitions between regimes mountain-building epochs? What *direct* evidence can be found for the continent-wide ice sheet postulated to account for the mid-Carboniferous lacuna? The location – in the Ceduna Depocentre – of much of the sediment produced during the later Cretaceous lacuna is suspected, but only from seismic profiles, and from drilling at its northern feather-edge. Deep drilling is indicated.

(b) What lies beneath the exposed Phanerozoic fold belt of eastern Australia: Precambrian continental crust or Palaeozoic oceanfloor? Deep seismic sounding and drilling are indicated.

(c) What were the precise connections between Australia and the continental terrains to the east? The geology of emergent New Zealand is well known but that of the submerged parts, including the Lord Howe Rise and Norfolk Ridge, is not. Further palaeomagnetic work in the intervening Tasman Sea and on the continental fragments themselves, as well as deep drilling of the Lord Howe Rise, Chesterfield Plateau, and Norfolk Ridge, are indicated. Appropriately located holes could help clarify the history of dispersal of marine and terrestrial biota across the Tasman Sea. On the Australian side of the Tasman Sea, the depositional record of the south-eastern margin, between Gippsland and the Great Barrier Reef, is unknown. Again, drilling is indicated.

(d) What is the Palaeozoic and early Mesozoic history of the northern margin (Outer Banda Arc and New Guinea), including its interaction with the Eurasian and Pacific Plates? There are suspect terrains within this region – Timor, the Vogelkop, Seram – and in eastern Australia as well. How many of these are exotic? Zoogeographic analyses to date have not rigorously confronted these issues but, clearly, more refined palaeomagnetic studies should be useful. Resolution of these questions is necessary before secure palinspastic reconstructions can be developed.

(e) What was the relative significance of gross tectonic compared with global climatic factors as causes of the rapid biogeographic differentiation of the Gondwanan marine and terrestrial realms about the beginning of the Late Carboniferous, curiously coinciding in a general way with inception of the Innamincka Regime? Likewise, the rapid decline of provinciality, commencing in the Aptian, towards the end of the Innamincka Regime.

(f) Palaeotemperature fluctuations are now well documented for the Cainozoic, but to what extent did tectonic factors (expressed in changing oceanic circulation patterns) contribute to the incursions of warm-water biota into southern Australian waters during the Cainozoic?

(g) Gondwanaland remained intact during the first 400 Ma of the Phanerozoic (McElhinny 1973; Morel and Irving 1978; Scotese *et al.* 1979; Embleton, Chapter II), so that the appropriate frame for the study of the Phanerozoic history of the daughter continents is Gondwanaland itself. A review of its Phanerozoic history would be timely.

REFERENCES

Abele, C. (1976). Tertiary, introduction. *Geol. Soc. Aust., Spec. Publ.* **5**, 177–91.
—— (1979). Geology of the Anglesea area, central coastal Victoria. *Vic., Geol. Surv., Mem.* 31.
—— and Page, R. W. (1974). Stratigraphic and isotopic ages of Tertiary basalts at Maude and Aireys Inlet, Victoria, Australia. *R. Soc. Vic., Proc.* **86**, 143–50.
——, Gloe, C. S., Hocking, J. B., Holdgate, G., Kenley, P. R., Lawrence, C. R., Ripper, D. and Threlfall, W. F. (1976). Tertiary. *Geol. Soc. Aust., Spec. Publ.* **5**, 177–274.
Adams, C. J. (1981). Geochronological correlations of Precambrian and Paleozoic orogens in New Zealand, Marie Byrd Land (West Antarctica), Northern Victoria Land (East Antarctica) and Tasmania. In *Gondwana Five* (ed. M. M. Cresswell and P. Vella), pp. 191–7. Balkema, Rotterdam.
Allen, A. R. and Black, L. P. (1979). The Harry Creek Deformed Zone, a retrograde schist zone of the Arunta Block, central Australia. *Geol. Soc. Aust., J.* **26**, 17–28.
Allen, G. A. and Whitworth, R. (1970). Ice thickness determination at Wilkes. In *International Symposium on Antarctic Exploration* (ISAGE) (ed. Gow et al.), pp. 405–14. (Hanover, New Hampshire, Sept. 1968.) Cambridge (*Int. Assoc. Hydrol. Sci., Publ.* 8607).
——, Pearce, L. G. G. and Gardner, W. E. (1978). A regional interpretation of the Browse Basin. *APEA J.* **18**, 23–33.
Allen, R. J. (1976). Surat Basin. *Australas. Inst. Min. Metall., Monogr.* **7**(3), 266–72.
—— and Houston, B. R. (1964). Petrology of Mesozoic sandstones of Carnarvon Highway section, western Bowen and Surat Basins. *Qld, Geol. Surv., Rep.* 6.
Alley, N. F. (1973). Landsurface development in the mid north of South Australia. *R. Soc. S. Aust., Trans.* **97**, 1–17.
American Association of Petroleum Geologists (AAPG) (1978). *Geographic Map of the Circum-Pacific Region, Southwest Quadrant, 1:10 000 000.* Tulsa, Oklahoma.
—— (1981). *Plate-tectonic map of the Circum-Pacific Region, Southwest Quadrant, 1:10 000 000.* Tulsa, Oklahoma.

Anderson, J. C. and Palmieri, V. (1977). The Fork Lagoons Beds, an Ordovician unit of the Anakie Inlier, central Queensland. *Qld Gov. Min. J.* **78**, 260–3.
Anderson, R. N., McKenzie, D. and Sclater, J. G. (1973). Gravity, bathymetry and convection in the earth. *Earth Planet. Sci. Lett.* **18**, 391–407.
Andrews, J. E., Packham, G. H., Eade, J. V., Holdsworth, B. K., Jones, D. L., Klein, G. D., Kroenke, L. W., Saito, T., Shafik, S., Stoeser, D. G. and van der Lingen, G. J. (1975). Southwest Pacific. *Deep Sea Dril. Proj., Initial Rep.* 30.
APCP (Australasian Petroleum Company Proprietary) (1961). Geological results of petroleum exploration in western Papua 1937–1961. *Geol. Soc. Aust., J.* **8**, 1–133.
Apthorpe, M. C. (1979). Depositional history of the Upper Cretaceous of the Northwest Shelf, based upon Foraminifera. *APEA J.* **19**, 74–89.
Arditto, P. A. (1982). Deposition and diagenesis of the Jurassic Pilliga Sandstone in the southeastern Surat Basin, New South Wales. *Geol. Soc. Aust., J.* **29**, 191–203.
Armstrong, R. L. (1978a). K–Ar dating: Late Cenozoic McMurdo Volcanic Group and dry valley glacial history, Victoria Land, Antarctica. *N.Z. J. Geol. Geophys.* **21**, 685–98.
—— (1978b). Pre-Cenozoic Phanerozoic time scale – computer file of critical dates and consequences of new and in-progress decay-constant revisions. *Am. Assoc. Pet. Geol., Stud. Geol.* **6**, 73–91.
—— and McDowell, W. G. (1975). Proposed refinement of the Phanerozoic time scale. *Geodynamic Highlights* (Inter-Union Comm. Geodynamics) **2**, 33–4.
—— and Stewart, A. J. (1975). Rubidium–strontium dates and extraneous argon in the Arltunga Nappe Complex, Northern Territory. *Geol. Soc. Aust., J.* **22**, 103–15.
Arnold, G. O. and Fawckner, J. F. (1980). The Broken River and Hodgkinson Provinces. In *The geology and geophysics of Northeastern Australia* (ed. R. A. Henderson and P. J. Stephenson), pp. 175–89. Geological Society of Australia, Queensland Division, Brisbane.
—— and Henderson, R. A. (1976). Lower Palaeozoic history of the southwestern Broken River Province,

north Queensland. *Geol. Soc. Aust., J.* **23**, 73–93.

Ashley, P. M., Brown, P. L., Franklin, B. J., Ray, A. S. and Scheibner, E. (1979). Field and geochemical characteristics of the Coolac ophiolite suite and its possible origin in a marginal sea. *Geol. Soc. Aust., J.* **26**, 45–60.

Atlas of Australian Resources (1970). *Rainfall*. Dep. National Development, Canberra.

Atwater, T. (1970). Implications of plate tectonics for the Cenozoic tectonic evolution of western North America. *Geol. Soc. Amer., Bull.* **81**, 3513–36.

Auchincloss, G. (1976). Adavale Basin. *Australas. Inst. Min. Metall., Monogr.* **7**(3), 309–15.

Audley-Charles, M. G. (1974). Banda Arcs. *Geol. Soc. Lond., Spec. Publ.* **4**, 349–63.

——, Carter, D. J., Barber, A. J., Norvick, M. S. and Tjokrosapoetro, S. (1979). Reinterpretation of geology of Ceram: implications for the Banda Arcs and northern Australia. *Geol. Soc. Lond., J.* **136**, 547–68.

Austin, P. M. and Williams, G. E. (1978). Tectonic development of Late Precambrian to Mesozoic Australia through plate motions possibly influenced by the earth's rotation. *Geol. Soc. Aust., J.* **25**, 1–21.

Aziz-ur-Rahman and McDougall, I. (1972). Potassium–argon ages on the Newer Volcanics of Victoria. *R. Soc. Vic., Proc.* **85**, 61–9.

Bain, J. H. C., Mackenzie, D. E., and Ryburn, R. J. (1975). Geology of the Kubor Anticline, Central Highlands of Papua New Guinea. *Aust., Bur. Miner. Resour., Geol. Geophys., Bull.* 155.

Baker, B. H. and Morgan, P. (1981). Continental rifting: progress and outlook. *EOS* **62**, 585–6.

Baksi, A. K. and Watkins, N. D. (1973). Volcanic production rates: comparison of oceanic ridges, islands, and the Columbia Plateau basalts. *Science* **180**, 493–96.

Balke, B. and Burt, D. (1976). Arafura Sea area. *Australas. Inst. Min. Metall., Monogr.* **7**(3), 209–12.

——, Page, C., Harrison, R. and Roussopoulos, G. (1973). Exploration in the Arafura Sea. *APEA J.* **13**, 9–12.

Bally, A. W. (1980). Basins and subsidence – a summary. In *Dynamics of the plate interiors* (ed. A. W. Bally, P. L. Bender, T. R. McGetchin, and R. I. Walcott), pp. 5–20. Geodynamics Series, 1. Am. Geophys. Union, Washington.

—— (1982). Musings over sedimentary basin evolution. *R. Soc. Lond., Philos. Trans., Ser. A* **305**, 325–38.

—— and Snelson, S. (1980). Facts and principles of world petroleum occurrence: realms of subsidence. *Canad. Soc. Pet. Geol., Mem.* **6**, 9–94.

Balme, B. E. (1969). The Triassic System in Western Australia. *APEA J.* **9**, 67–78.

Banks, M. R. (1962). Silurian and Devonian Systems. *Geol. Soc. Aust., J.* **9**, 177–87.

—— (1978). Correlation chart for the Triassic System of Australia. *Aust., Bur. Miner. Resour., Geol. Geophys., Bull.* 156 C.

—— and Smith, A. (1968). A graptolite from the Mathinna Beds, northeastern Tasmania. *Aust. J. Sci.* **31**, 118–19.

Baraganzi, M. and Dorman, J. (1969). World seismicity maps compiled by ESSA, Coast and Geodetic Survey, Epicenter data, 1961–1967. *Seismol. Soc. Am., Bull.* **59**, 369–80.

Barber, P. M. (1982). Palaeotectonic evolution and hydrocarbon genesis of the central Exmouth Plateau. *APEA J.* **22**, 131–44.

Barnes, L. C. and Pitt, G. M. (1976). The Tallaringa palaeodrainage system. *S. Aust., Geol. Surv., Q. Geol. Notes* **59**, 7–10.

Barrash, W. and Venkatakrishnan, R. (1982). Timing of late Cenozoic volcanic and tectonic events along the western margin of the North American Plate. *Geol. Soc. Am., Bull.* **93**, 977–89.

Barron, E. J. and Harrison, C. G. A. (1979). Reconstructions of the Campbell Plateau and the Lord Howe Rise. *Earth Planet. Sci. Lett.* **45**, 87–92.

Barron, L. M., Scheibner, E. and Suppel, D. W. (1982). The Mount Hope Group and its comagmatic granites on the Mount Allen 1:1 000 000 Sheet, New South Wales. *N.S.W., Geol. Surv., Q. Notes.* **47**, 1–17.

Barton, C. M. (1981). Regional stress and structure in relation to brown coal open cuts of the Latrobe Valley, Victoria. *Geol. Soc. Aust., J.* **28**, 333–9.

Basden, H. (1982). Preliminary report on the geology of the Tumut 1:1 000 000 Sheet area, southern New South Wales. *N.S.W., Geol. Surv., Q. Notes* **46**, 1–18.

Battersby, D. G. (1976). Cooper Basin gas and oil fields. *Australas. Inst. Min. Metall., Monogr.* **7**(3), 321–68.

Beams, S. D. (1975). Geology of the Wyndham–Whipstick area. B.Sc. (Hons) thesis, Aust. Nat. Univ., Canberra (unpubl.).

Beard, J. S. (1977). Tertiary evolution of the Australian flora in the light of latitudinal movements of the Continent. *J. Biogeography* **4**, 111–18.

Beaumont, C. and Sweeney, J. F. (1978). Graben generation of major sedimentary basins. *Tectonophysics* **50**, T19–23.

Beavis, F. C. (1976). Ordovician. *Geol. Soc. Aust. Spec. Publ.* **5**, 25–44.

Bein, J. and Taylor, M. L. (1981). The Eyre Sub-basin: recent exploration results. *APEA J.* **21**, 91–8.

Bembrick, C. S. (1976). Coonamble Embayment. *Australas. Inst. Min. Metall., Monogr.* **7** (3), 302–6.

—— and Holmes, G. G. (1976). An interpretation of the subsurface geology of the Nowra–Jervis Bay area. *N.S.W., Geol. Surv., Rec.* **18** (1), 5–68.

——, Herbert, C., Scheibner, E. and Stuntz, J. (1980). Structural subdivision of the Sydney Basin. *N.S.W., Geol. Surv., Bull.* **26**, 3–9.

Bentley, C. R. (1962). Glacial and sub-glacial geography of Antarctica. *Antarct. Res., Geophys. Monogr.* **7**, 11–25.

Bigarella, J. J. (1973). Geology of the Amazon and Parnaiba Basins. In *The ocean basins and margins, Vol. 1, The South Atlantic* (ed. A. E. M. Nairn and F. G. Stehli), pp. 25–86. Plenum, New York.

Binns, R. A. (1966). Granitic intrusions and regional metamorphic rocks of Permian age from the Wongwibinda district, northeastern New South Wales. *R. Soc. N.S.W., J. Proc.* **99**, 5–36.

Bischoff, G. C. O. and Fergusson, C. L. (1982). Conodont distributions, and ages of Silurian and Devonian limestones in the Palmers Oakey district, N.S.W. *Geol. Soc. Aust., J.* **29**, 469–76.

Bishop, P. (1982). Stability or change: a review of ideas on ancient drainage in eastern New South Wales. *Aust. Geographer* **15**, 219–30.

Black, L. P., Bell, T. H., Rubenach, M. J. and Withnall, I. W. (1979). Geochronology of discrete structural-metamorphic events in a multiply deformed Precambrian terrain. *Tectonophysics* **54**, 103–38.

——, Shaw, R. D. and Offe, L. A. (1980). The age of the Stuart dyke swarm and its bearing on the onset of late Precambrian sedimentation in Central Australia. *Geol. Soc. Aust., J.* **27**, 151–5.

Blake, D. H. (1971). Geology and geomorphology of the Morehead–Kiunga Area. *Aust., CSIRO, Land Res. Ser.* **29**, 56–68.

BMR (1976). *Gravity map of Australia, 1:5 000 000.* Aust., Bur. Miner. Resour., Geol. Geophys., Canberra.

—— (1979). *BMR Earth Science Atlas.* Aust., Bur. Miner. Resour., Geol. Geophys., Canberra.

Boeuf, M. G. and Doust, H. (1975). Structure and development of the southern margin of Australia. *APEA J.* **15**, 33–43.

Bolger, P. (1980). Late Cainozoic sediments in the Latrobe Valley, Victoria. In *The Cainozoic evolution of continental Southeast Australia* (ed. E. M. Truswell and R. S. Abell), p. 11. *Bur. Miner. Resour., Geol. Geophys. Aust., Rec.* 1980/67.

Bond, G. (1978). Speculations on real sea-level changes and vertical motions of continents at selected times in the Cretaceous and Tertiary Periods. *Geology* **6**, 247–50.

—— (1979). Evidence for some uplifts of large magnitude in continental platforms. *Tectonophysics* **16**, 285–305.

Bourke, D. J. (1980). Stratigraphy of the Mesozoic sequence in the Warialda–Goondiwindi area. *N.S.W., Geol. Surv., Rec.* **19** (1), 1–79.

Bowen, R. (1961). Paleotemperature analyses of Mesozoic Belemnoidea from Australia and New Guinea. *Geol. Soc. Am., Bull.* **72**, 769–74.

Bowering, O. J. W. (1982). Hydrodynamics and hydrocarbon migration – a model for the Eromanga Basin. *APEA J.* **22**, 227–36.

Bowin, C., Purdy, G. M., Johnston, C. R., Shor, G., Lawver, L., Hartono, H. M. S. and Jezek, P. (1980). Arc–continent collision in Banda Sea region. *Am. Assoc. Pet. Geol., Bull.* **64**, 868–915.

Bowler, J. M. (1973). Clay dunes: their occurrence, formation and environmental significance. *Earth-Sci. Rev.* **9**, 315–38.

—— (1976). Aridity in Australia: age, origins and expression in aeolian landforms and sediments. *Earth-Sci. Rev.* **12**, 279–310.

—— (1978). Quaternary climate and tectonics in the evolution of the Riverine Plain, southeastern Australia. In *Landform Evolution in Australasia* (ed. J. L. Davies and M. A. J. Williams), pp. 70–112. Australian National University Press, Canberra.

——, Hope, G. S., Jennings, J. N., Singh, G. and Walker, D. (1976). Late Quaternary climates of Australia and New Guinea. *Quaternary Research* **6**, 359–94.

Bowles, F. A. (1975). Paleoclimatic significance of quartz/illite variations in cores from the eastern equatorial North Atlantic. *Quaternary Research* **5**, 225–35.

Bradshaw, J. D., Adams, C. J. and Andrews, P. B. (1981). Carboniferous to Cretaceous on the Pacific margin of Gondwana: The Rangitata Phase of New Zealand. In *Gondwana Five* (ed. M. M. Cresswell and P. Vella) pp. 217–21. Balkema, Rotterdam.

Brakel, A. T. (1982a). Stratigraphic correlation of the Upper Coal Measures of the Sydney Basin with their equivalents in the Bowen Basin. *BMR J. Aust. Geol. Geophys.* **7**, 147.

—— (1982b). Correlation between the Comet Platform and the Denison Trough, Bowen Basin. *Geol. Soc. Aust., Abstr.* **8**, 31.

Branson, J. C. (1974). Structures of the Western margin of the Australian continent. *Oil and Gas (Sydney)* **20** (9), 24–35.

Brewer, J. A., Good, R., Oliver, J. E., Brown, L. D.

and Kaufman, S. (1983). COCORP profiling across the Southern Oklahoma Aulacogen: Overthrusting of the Wichita Mountains and compression within the Anadarko Basin. *Geology* **11**, 109–14.

Broecker, W. A. and Van Donk, J. (1970). Insolation changes, ice volumes, and the ^{18}O record in deep-sea cores. *Rev. Geophys. Space Physics* **8**, 169–98.

Brookfield, H. C. and Hart, D. (1966). *Rainfall in the tropical southwest Pacific.* Dept. of Geogr. Pub. G/3, Res. School Pacific Studies, Australian National University, Canberra.

Brooks, C. (1966). The rubidium–strontium ages of some Tasmanian igneous rocks. *Geol. Soc. Aust., J.* **13**, 457–69.

—— and Leggo, M. D. (1972). The local chronology and regional implications of a Rb–Sr investigation of granitic rocks from the Corryong district, South-eastern Australia. *Geol. Soc. Aust., J.* **19**, 1–19.

Brooks, J. A. (1969). Rayleigh waves in southern New Guinea. II. A shear velocity profile. *Seismol. Soc. Am., Bull.* **59**, 2017–38.

Brown, B. R. (1976). Bass Basin some aspects of the petroleum geology. *Australas. Inst. Min. Metall., Monogr.* **7** (3), 67–82.

Brown, C. M. (1980). Bonaparte Gulf Basin. *Econ. Social Comm. Asia Pacific (ESCAP), Atlas of Stratigraphy* II, VII, 42–51.

——, Pieters, P. E. and Robinson, G. P. (1975). Stratigraphic and structural development of the Aure Trough and adjacent shelf and slope areas. *APEA J.* **15**, 61–71.

——, Pigram, C. J. and Skwarko, S. K. (1979). Mesozoic stratigraphy and geological history of Papua New Guinea. *Palaeogeogr., Palaeoclimatol., Palaeoecol.* **29**, 301–22.

Brown, D. A., Campbell, K. S. W. and Crook, K. A. W. (1968). *The Geological Evolution of Australia and New Zealand.* Pergamon, Oxford.

Brown, I. (1954). A study of the Tasman Geosyncline in the region of Yass, New South Wales. *R. Soc. N.S.W., J. Proc.* **88**, 3–11.

Brown, P. F. (1980). A sheeted dyke complex within the Coolac Ophiolite, southeastern New South Wales. *Geol. Soc. Aust., J.* **26**, 411–17.

Brunnschweiler, R. O. (1957). The geology of the Dampier Peninsula, Western Australia. *Aust., Bur. Miner. Resour., Geol. Geophys., Rep.* 13.

—— (1959). New Aconeceratinae (Ammonoidea) from the Albian and Aptian of Australia. *Aust., Bur. Miner. Resour., Geol. Geophys., Bull.* 54.

—— (1978). Notes on the geology of eastern Timor. *Aust., Bur. Miner. Resour., Geol. Geophys., Bull.* **192**, 9–18.

Bullard, E. C., Everett, J. E. and Smith, A. G. (1965). The fit of the continents around the Atlantic. *Phil. Trans. R. Soc. London* **A258**, 41–51.

Burek, P. J., Wells, A. T. and Loeffler, E. (1978). Tertiary tectonomagnetism in the southern Georgina and northern Ngalia Basins, Central Australia. *Aust. National Univ., Research School of Earth Sciences, Annual Report for 1978*, 59–62.

——, Walter, M. R. and Wells, A. T. (1979). Magnetostratigraphic tests of lithostratigraphic correlations between latest Proterozoic sequences in the Ngalia, Georgina and Amadeus Basins, central Australia. *BMR J. Aust. Geol. Geophys.* **4**, 47–55.

Burger, D. and Senior, B. R. (1979). A revision of the sedimentary and palynological history of the northeastern Eromanga Basin, Queensland. *Geol. Soc. Aust., J.* **26**, 121–32.

Burke, K. and Dewey, J. F. (1973). Plume-generated triple junctions: key indicators in applying plate tectonics to old rocks. *J. Geol.* **81**, 406–33.

Burns, B. J., and Bein, J. (1980). Regional geology and hydrocarbon potential of the Mesozoic of the western Papuan Basin, Papua New Guinea. *APEA J.* **20**, 1–15.

Burns, R. E., Andrews J. E., van der Lingen, G. J., Churkin, M., Gatehouse, J. S., Packham, G. H., Davies, T. A., Kennett, J. P., Dumitrica, P., Edwards, A. R., and von Herzen, R. P. (1973). Leg 21. *Deep Sea Dril. Proj., Initial Rep.* 21.

Callen, R. A. (1977). Late Cainozoic environments of part of northeastern South Australia. *Geol. Soc. Aust., J.* **24**, 151–69.

—— and Tedford, R. H. (1976). New Late Cainozoic rock units and depositional environments, Lake Frome area, South Australia. *R. Soc. S. Aust., Trans.* **100**, 125–67.

Cameron, R. G., Conaghan, P. J. and Parbury, C. F. R. (1982). Piedmont depositional regimes within the Ravensworth alluvial embayment: Wittingham Coal Measures, northern Sydney Basin. *Sixteenth symposium: Advances in the study of the Sydney Basin*. Dep. Geol., Univ. Newcastle, 9–13.

Campbell, K. S. W. (1970). Palaeontological report. B. O. C. of Australia, completion report of Sahul Shoals No. 1 Well. *Aust., Bur. Miner. Resour., Geol. Geophys., Petroleum Search Subsidy Acts Report.*

Cande, S. C. and Mutter, J. C. (1982). A revised identification of the oldest sea floor spreading anomalies between Australia and Antarctica. *Earth Planet. Sci. Lett.* **58**, 151–60.

——, Larson, R. L. and La Brecque, J. L. (1978). Magnetic lineations in the Pacific Jurassic quiet

REFERENCES

zone. *Earth Planet. Sci. Lett.* **41**, 434–40.

——, Mutter, J., and Weissel, J. K. (1981). A revised model for the break-up of Australia and Antarctica. *EOS* **62** (17), 384.

Cardwell, R. K. and Isacks, B. L. (1978). Geometry of the subducted lithosphere beneath the Banda Sea in eastern Indonesia from seismicity and fault plane solutions. *J. Geophys. Res.* **83**, 2825–38.

Carey, S. W. (1938). The morphology of New Guinea. *Aust. Geogr.* **3**, 3–30.

—— (1958). A tectonic approach to continental drift. In *Continental Drift—a Symposium* (ed. S. W. Carey), pp. 177–355. Univ. of Tasmania, Hobart.

—— (1970). Australia, New Guinea and Melanesia in the current revolution in concepts of the evolution of the earth. *Search* **1**, 178–89.

—— (1976). *The expanding earth.* Elsevier, Amsterdam.

Carr, P. F. and Facer, R. A. (1980). Radiometric ages of some igneous rocks from the Southern and Southwestern Coalfields of New South Wales. *Search* **11**, 382–3.

——, Jones, B. G., Kanstler, A. J., Moore, P. S. and Cook, A. C. (1981). The geology of the Bungonia district, New South Wales. *Linn. Soc. N.S.W., Proc.* **104**, 229–44.

Carter, A. N. (1978). Phosphatic nodule beds in Victoria and the late Miocene–Pliocene eustatic event. *Nature* **276**, 258–9.

—— (1980). Miocene–Pliocene palaeoceanography and deposition in east Gippsland. In *The Cainozoic evolution of continental Southeast Australia* (ed. E. M. Truswell and R. S. Abell), pp. 15–16. *Aust., Bur. Miner. Resour. Geol. Geophys., Rec.* 1980/67.

Carter, D. J., Audley-Charles, M. G. and Barber, A. J. (1976). Stratigraphical analysis of island arc – continental margin collision in eastern Indonesia. *Geol. Soc. Lond., J.* **132**, 179–98.

Cas, R. A. F. (1978a). Basin characteristics of the Early Devonian part of the Hill End Trough based on stratigraphic analysis of the Merrions Tuff. *Geol. Soc. Aust., J.* **24**, 381–401.

—— (1978b). Silicic lavas in Palaeozoic flysch-like deposits in New South Wales, Australia: behaviour of deep subaqueous silicic flows. *Geol. Soc. Am., Bull.* **89**, 1708–14.

—— (1979). Mass-flow arenites from a Palaeozoic interarc basin, New South Wales, Australia: mode and environment of emplacement. *J. Sediment. Petrol.* **49**, 29–44.

—— (1983). A review of the facies patterns, palaeogeographic development and tectonic context of the Palaeozoic Lachlan Fold Belt of southeastern Australia. *Geol. Soc. Aust., Spec. Publ.* 10.

—— and Jones, J. G. (1979). Palaeozoic interarc basin in eastern Australia and a modern New Zealand analogue. *N.Z. J. Geol. Geophys.* **22**, 71–85.

——, Flood, R. H. and Shaw, S. E. (1976). Hill End Trough: new radiometric ages. *Search* **7**, 205–7.

——, Powell, C.Mc.A. and Crook, K. A. W. (1980). Ordovician palaeogeography of the Lachlan Fold Belt: modern analogue and tectonic constraints. *Geol. Soc. Aust., J.* **27**, 19–31.

——, ——, Fergusson, C. L., Jones, J. G., Roots, W. D. and Fergusson, J. (1981). The Lower Devonian Kowmung Volcaniclastics: a deep-water succession of mass-flow origin, northeastern Lachlan Fold Belt, N.S.W. *Geol. Soc. Aust., J.* **28**, 271–88.

Cawood, P. A. (1976). Cambro-Ordovician strata, northern New South Wales. *Search* **7**, 317–18.

—— (1982a). Correlation of stratigraphic units across the Peel Fault system. In *New England Geology* (ed. P. G. Flood and B. Runnegar) pp. 53–61. Univ. New England, Armidale.

—— (1982b). Structural relations in the subduction complex of the New England Fold Belt, eastern Australia. *J. Geol.* **90**, 381–92.

Chamalaun, F. H. (1977a). Palaeomagnetic evidence for the relative positions of Timor and Australia in the Permian. *Earth Planet. Sci. Lett.* **34**, 107–12.

—— (1977b). Palaeomagnetic reconnaissance result from the Maubisse Formation, East Timor, and its tectonic implications. *Tectonophysics* **42**, 17–26.

—— and Grady, A. E. (1978). The tectonic development of Timor: a new model and its implications for petroleum exploration. *APEA J.* **18**, 102–8.

Chappell, B. W. and White, A. J. (1974). Two contrasting granite types. *Pac. Geol.* **8**, 173–4.

Chappell, J. (1974). Geology of coral terraces, Huon Peninsula, New Guinea: a study of Quaternary tectonic movements and sea-level changes. *Geol. Soc. Am., Bull.* **85**, 553–70.

—— (1976). Aspects of late Quaternary palaeogeography of the Australian–East Indonesian region. In *The Origin of the Australians* (ed. R. L. Kirk and A. G. Thorne), pp. 11–22. Aust. Inst. Aborig. Studies, Canberra.

—— and Veeh, H. H. (1978). Late Quaternary tectonic movements and sea-level changes at Timor and Atauro Island. *Geol. Soc. Am., Bull.* **89**, 356–68.

Churchill, D. M. (1973). The ecological significance of tropical mangroves in the early Tertiary floras of southern Australia. *Geol. Soc. Aust., Spec. Publ.* **4**, 79–86.

Churkin, M., Carter, C. and Johnson, B. R. (1977).

Subdivision of Ordovician and Silurian time scale using accumulation rates of graptolitic shale. *Geology* **5**, 452–6.

Clarke, D. E., Paine, A. G. L. and Jensen, A. R. (1971). Geology of the Proserpine 1:250,000 Sheet area, Queensland. *Aust., Bur. Miner. Resour., Geol. Geophys., Rep.* 144.

Clarke, M. F., Wasson, R. J. and Williams, M. A. J. (1979). Point Stuart chenier plain and Holocene sea levels in Northern Australia. *Search* **10**, 90–2.

Clarke, M. J., Farmer, N. and Gulline, A. B. (1976). Tasmania Basin – Parmeener Supergroup. *Australas. Inst. Min. Metall., Monogr.* **7** (3), 438–43.

Clayton, R. N. and Stevens, G. R. (1967). Palaeotemperatures of New Zealand Jurassic and Cretaceous. *Tuatara* **16**, 3–7.

Cleary, J. R., Simpson, D. W. and Muirhead, K. J. (1972). Variations in Australian upper mantle structure from observations of the Cannikin explosion. *Nature* **236**, 111–12.

CLIMANZ (1983). *A symposium of results and discussions concerned with Late Quaternary climatic history of Australia, New Zealand and surrounding seas.* Dept. Biogeography and Geomorphology, Aust. Nat. Univ., Canberra.

Cloud, P. and Glaessner, M. F. (1982). The Ediacarian Period and System: Metazoa inherit the earth. *Science* **218**, 783–92.

Cockbain, A. E. (1967). *Asterocyclina* from the Plantagenet Beds near Esperance, W. A. *Aust. J. Sci.* **30**, 68.

—— (1969). Dasycladacean algae from the Werillup Formation, Esperance. *West. Aust., Geol. Surv., Ann. Rep.* **1968**, 52–3.

Cocker, J. D. (1982). Rb–Sr geochronology and Sr isotopic composition of Devonian granitoids, eastern Tasmania. *Geol. Soc. Aust., J.* **29**, 139–58.

Cole, J. (1982). The structure and stratigraphy of the Bodalla area, N.S.W. B.Sc. (Hons.) thesis, Macquarie Univ. (unpubl.).

Coleman, P. J. (1980). Plate tectonics background to biogeographic development in the Southwest Pacific over the last 100 million years. *Palaeogeogr., Palaeoclimatol., Palaeoecol.* **31**, 105–21.

——, Michael, P. J. and Mutter, J. C. (1982). The origin of the Naturaliste Plateau, SE Indian Ocean: implications from dredged basalts. *Geol. Soc. Aust., J.* **29**, 457–68.

Collins, W. J., Beams, S. D., White, A. J. R., and Chappell, B. W. (1982). Nature and origin of A-type granites with particular reference to southeastern Australia. *Contrib. Mineral. Petrol.* **80**, 189–200.

Collinson, D. W. (1983). *Methods in rock magnetism and palaeomagnetism: techniques and instrumentation.* Chapman and Hall, London.

——, Creer, K. M. and Runcorn, S. K. (1967). *Methods in Palaeomagnetism.* Elsevier, Amsterdam.

Commonwealth Bureau of Meteorology (1971). Seasonal rainfall zones in Australia. Melbourne.

Compston, W. (1974). The Table Hill Volcanics of the Officer Basin – Precambrian or Palaeozoic? *Geol. Soc. Aust., J.* **21**, 403–11.

—— and Arriens, P. A. (1968). The Precambrian geochronology of Australia. *Can. J. Earth Sci.* **5**, 561–83.

—— and Chappell, B. W. (1979). Sr-isotope evolution of granitoid source rocks. In *The Earth: its origin, structure and evolution* (ed. M. W. McElhinny), pp. 377–426. Academic Press, London.

——, Crawford, A. R. and Bofinger, V. M. (1966). A radiometric estimate of the duration of the sedimentation in the Adelaide Geosyncline, South Australia. *Geol. Soc. Aust., J.* **13**, 229–76.

Conaghan, P. J. (1982). Lacustrine (Gilbert) deltas in in the Permian coal measures of eastern Australia and India: implications for the widespread hydroponic origin and deep-water diagenesis of Gondwanan coals. *Sixteenth symposium: Advances in the study of the Sydney Basin.* Dept. Geol., Univ. Newcastle, 6–9.

——, Mountjoy, E. W., Edgecombe, D. R., Talent, J. A. and Owen, D. E. (1976). Nubrigyn algal reefs (Devonian), eastern Australia: allochthonous blocks and megabreccias. *Geol. Soc. Am., Bull.* **87**, 515–30.

——, Jones, J. G., McDonnell, K. L. and Royce, K. (1982). A dynamic fluvial model for the Sydney Basin. *Geol. Soc. Aust., J.* **29**, 55–65.

Coney, P. J., Jones, D. L. and Monger, J. W. H. (1980). Cordilleran suspect terrains. *Nature* **288**, 329–33.

Conolly, J. R. (1963). Upper Devonian stratigraphy and sedimentation in the Wellington–Molong district, N.S.W. *R. Soc. N.S.W., J. Proc.* **96**, 73–106.

—— (1965a). The stratigraphy of the Hervey Group in central New South Wales. *R. Soc. N.S.W., J. Proc.* **98**, 37–83.

—— (1965b). Petrology and origin of the Hervey Group, Upper Devonian, central New South Wales. *Geol. Soc. Aust., J.* **12**, 123–66.

—— (1969a). Southern and central Highlands Fold Belt: II. Upper Devonian Series. *Geol. Soc. Aust., J.* **16**, 150–78.

—— (1969b). Southern and central Highlands Fold Belt: Late Devonian sedimentation. *Geol. Soc. Aust., J.* **16**, 224–6.

Conybeare, C. E. B. and Jessop, R. G. C. (1972). Exploration for oil-bearing sand trends in the Fly River area, western Papua. *APEA J.* **12**, 69–73.

Cook, A. C. (1975). The spatial and temporal variation of the type and rank of Australian coals. In *Australian Black Coal* (ed. A. C. Cook), pp. 63–83. Australas. Inst. Min. Metall., Illawarra Branch.

Cook, F. W. and Taylor, C. P. (1979). Permian strata of the Wolfang Basin. *Qld Gov. Min. J.* **80**, 342–9.

Cook, P. J. (1982). The Cambrian palaeogeography of Australia and opportunities for petroleum exploration. *APEA J.* **22**, 42–64.

——, Veevers, J. J., Heirtzler, J. R. and Cameron, P. J. (1978). The sediments of the Argo Abyssal Plain and adjacent areas, northeast Indian Ocean. *BMR J. Geol. Geophys.* **3**, 113–24.

——, Colwell, J. B., Firman, J. B., Lindsay, J. M., Schwebel, D. A. and Von der Borch, C. C. (1977). The late Cainozoic sequence of southeast South Australia and Pleistocene sea-level changes. *BMR J. Aust. Geol. Geophys.* **2**, 81–8.

Cooney, P. M., Evans, P. R. and Eyles, D. (1975). Southern Ocean and its Margins. In *Deep Sea Drilling in Australasian Waters* (ed. J. J. Veevers), pp. 26–8. Challenger Symposium. Sydney.

Cooper, B. J. (1979). Eocene to Miocene stratigraphy of the Willunga Embayment. *S. Aust., Geol. Surv., Rep. Invest.* 50.

—— (1981). Carboniferous and Permian sediments in South Australia and their correlation. *S. Aust., Geol. Surv., Q. Geol. Notes* **79**, 2–6.

——, Harris, W. K. and Meyer, G. M. (1982). The Late Palaeozoic Coolardie Formation, Polda Basin. *S. Aust., Geol. Surv., Q. Geol. Notes* **81**, 9–13.

Cooper, J. A. (1975). Isotopic datings of the basement-cover boundaries within the Adelaide 'Geosyncline'. *Geol. Soc. Aust., 1st Geol. Convent., Abstr.*, 12.

—— and Compston, W. (1971). Rb-Sr dating within the Houghton Inlier, South Australia. *Geol. Soc. Aust., J.* **17**, 213–19.

——, Richards, J. R. and Webb, A. W. (1963). Some potassium-argon ages in New England, New South Wales. *Geol. Soc. Aust., J.* **10**, 313–16.

Cooper, R. A., Jago, J. G., MacKinnon, D. I., Simes, J. E. and Braddock, P. E. (1976). Cambrian fossils from the Bowers Group, northern Victoria Land, Antarctica. *N.Z. J. Geol. Geophys.* **19**, 283–8.

Cope, R. N. (1975). Tertiary epeirogeny in the southern part of Western Australia. *West. Aust. Geol. Surv., Ann. Rep.* **1974**, 40–6.

Corbett, G. J. (1976). A new fold structure in the Woolomin Beds suggesting a sinistral movement on the Peel Fault. *Geol. Soc. Aust., J.* **23**, 401–6.

Corbett, K. D. and Banks, M. R. (1973). Ordovician stratigraphy of the Florentine Synclinorium, southwest Tasmania. *R. Soc. Tasm., Pap. Proc.* **107**, 207–38.

Cowie, J. W. and Cribb, S. J. (1978). The Cambrian System. *Am. Assoc. Pet. Geol., Stud. Geol.* **6**, 355–62.

Craddock, C. (ed.) (1970). Geologic maps of Antarctica. *Am. Geogr. Soc., Antarctic Map Folio Ser.* 12.

Cramsie, J., Pogson, D. J. and Baker, C. J. (1975). *Yass 1:100 000 Geological Sheet 8628.* N.S.W., Geol. Surv., Sydney.

Cranfield, L. C., Schwarzbock, H. and Day, R. W. (1976). Geology of the Ipswich and Brisbane 1:250 000 Sheet areas. *Qld, Geol. Surv., Rep.* 95.

Crank, K. (1973). Geology of Barrow Island oil field. *APEA J.* **13**, 49–57.

Crawford, A. J. (1983). Construction of the continental crust of southeastern Australia and the origin of Phanerozoic Foldbelts (in preparation).

—— and Keays, R. R. (1978). Cambrian greenstone belts in Victoria: marginal sea-crust slices in the Lachlan Fold Belt of southeastern Australia. *Earth Planet. Sci. Lett.* **41**, 197–208.

Crawford, A. R. and Campbell, K. S. W. (1973). Large-scale horizontal displacement within Australo-Antarctica in the Ordovician. *Nature Phys. Sci.*, **241**, 11–14.

Crawford, E. A., Herbert, C., Taylor, G., Helby, R., Morgan, R. and Ferguson, J. (1980). Diatremes of the Sydney Basin. *N.S.W., Geol. Surv., Bull.* **26**, 294–323.

Cromer, W. C. (1980). A Late Eocene basalt date from northern Tasmania. *Search* **11**, 294–5.

Crook, K. A. W. (1960). Petrology of the Parry Group, Upper Devonian–Lower Carboniferous, Tamworth–Nundle district, N.S.W. *J. Sediment. Petrol.* **30**, 538–52.

—— (1961a). Stratigraphy of the Tamworth Group (Lower and Middle Devonian), Tamworth–Nundle district, N.S.W. *R. Soc. N.S.W., J. Proc.* **94**, 173–88.

—— (1961b). Stratigraphy of the Parry Group (Upper Devonian–Lower Carboniferous), Tamworth–Nundle district. *R. Soc. N.S.W., J. Proc.* **94**, 189–208.

—— (1964). Depositional environment and provenance of Devonian and Carboniferous sediments in the Tamworth Trough, New South Wales. *R. Soc. N.S.W. J. Proc.* **97**, 41–53.

—— (1969). Contrasts between Atlantic and Pacific geosynclines. *Earth Planet. Sci. Lett.* **5**, 429–38.

—— (1974). Kratonization of West Pacific-type geosynclines. *J. Geol.* **82,** 24–36.

—— (1980a). Fore-arc evolution in the Tasman Geosyncline. *Geol. Soc. Aust., J.* **27,** 215–32.

—— (1980b). Fore-arc evolution and continental growth: a general model. *J. Struct. Geol.* **2,** 289–303.

—— and Powell, C.McA. (1976). The evolution of the southeastern part of the Tasman Geosyncline. *Field Guide for Excursion 17A, 25th Int. Geol. Congr., Australia.*

——, Bein, J., Hughes, R. J. and Scott, P. A. (1973). Ordovician and Silurian history of the southern part of the Lachlan Geosyncline. *Geol. Soc. Aust., J.* **20,** 113–38.

Crostella, A. (1976). Browse Basin. *Australas. Inst. Min. Metall. Monogr.* **7** (3), 194–9.

—— and Barter, T. P. (1980). Triassic–Jurassic depositional history of the Dampier and Beagle Sub-basins, Northwest Shelf of Australia. *APEA J.* **20,** 25–33.

—— and Chaney, M. A. (1978). The petroleum geology of the outer Dampier Sub-basin. *APEA J.* **18,** 13–22.

—— and Powell, D. E. (1975). Geology and hydrocarbon prospects of the Timor area. *Indones. Pet. Assoc., Proc.* **2,** 149–71.

Crowell, J. C. and Frakes, L. A. (1971a). Late Paleozoic glaciation. IV. Australia. *Geol. Soc. Am., Bull.* **82,** 2515–40.

—— —— (1971b). Late Palaeozoic glaciation of Australia. *Geol. Soc. Aust., J.* **17,** 115–55.

—— —— (1975). The late Palaeozoic glaciation. In *Gondwana Geology* (ed. K. S. W. Campbell), pp. 313–31. ANU Press, Canberra.

Cull, J. P. and Denham, D. (1979). Regional variations in Australian heat flow. *BMR J. Aust. Geol. Geophys.* **4,** 1–13.

Curray, J. R., Moore, D. G., Lawver, L. A., Emmel, F. J., Raitt, R. W., Henry, M. and Kieckhefer, R. (1979). Tectonics of the Andaman Sea and Burma. *Am. Assoc. Pet. Geol., Mem.* **29,** 189–98.

——, Shor, G. C., Raitt, R. W. and Henry, M. (1977). Seismic refraction and reflection studies of crustal structure of the eastern Sunda and western Banda Arcs. *J. Geophys. Res.* **82,** 2479–89.

D'Addario, G. W., Dow, D. B. and Swoboda, R. (1976). *Map of the geology of Papua New Guinea. Scale 1:2 500 000.* Aust., Bur. Miner. Resour., Geol. Geophys., Canberra.

Daily, B. (1972). The base of the Cambrian and the first Cambrian faunas. *Centre for Precambrian Research, Univ. Adelaide, Spec. Pap.* **1,** 13–41.

——, Moore, P. S. and Rust, B. R. (1980). Terrestrial-marine transition in the Cambrian rocks of Kangaroo Island, South Australia. *Sedimentology* **27,** 379–99.

Dalrymple, G. B. (1979). Critical tables for conversion of K–Ar ages from old to new constants. *Geology* **7,** 558–60.

Dampney, C. N. G., Johnson, B. D. and Hollingsworth, R. J. S. (1978). The application of gravity interpretation to an intracratonic basin. *APEA J.* **18,** 130–6.

Daniels, J. L. (1975). Palaeogeographic development of Western Australia, Precambrian. *W. Aust., Geol. Surv., Mem.* **2,** 437–50.

David, T. W. E. and Browne, W. R. (1950). *The Geology of the Commonwealth of Australia.* Arnold, London.

Davidson, J. K. (1980). Rotational displacements in southeastern Australia and their influence on hydrocarbon occurrence. *Tectonophysics* **63,** 139–53.

Davies, H. L. and Smith, I. E. (1971). Geology of eastern Papua. *Geol. Soc. Am., Bull.* **82,** 3299–312.

Davies, J. L. (1974). Geomorphology and Quaternary environments. In *Biogeography and Ecology in Tasmania* (ed. W. D. Williams), pp. 17–27. Junk, The Hague.

—— (1977). The coast. In *Australia: a geography* (ed. D. N. Jeans), pp. 134–51. Sydney University Press.

Davies, P. J. (1975). Shallow seismic structure of the continental shelf, southeast Australia. *Geol. Soc. Aust., J.* **22,** 345–59.

—— (1979). Marine geology of the continental shelf off southeast Australia. *Aust., Bur. Miner. Resour., Geol. Geophys., Bull.* 195.

Davies, T. A., Hay, W. H., Southam, J. R. and Worsley, T. R. (1977). Estimates of Cenozoic oceanic sedimentation rates. *Science* **197,** 53–5.

——, Luyendyk, B. P., Rodolfo, K. S., Kempe, D. R. C., McKelvey, B. C., Leidy, R. D., Horvath, G. J., Hyndman, R. D., Thierstein, H. R., Boltovskoy, E. and Doyle, P. (1974). Leg 26. *Initial Rep. Deep Sea Dril. Proj.* 26.

Day, R. W. (1969). The Lower Cretaceous of the Great Artesian Basin. In *Stratigraphy and palaeontology: essays in honour of Dorothy Hill* (ed. K. S. W. Campbell), pp. 140–73. ANU Press, Canberra.

—— (1976). Laura Basin. *Australas. Inst. Min. Metall., Monogr.* **7** (3), 443–6.

——, Cranfield, L. C. and Schwarzbock, H. (1974). Stratigraphic and structural setting of Mesozoic basins in southeastern Queensland and northeastern New South Wales. In *The Tasman Geosyncline,* (ed. A. K. Denmead, G. W. Tweedale, and A. F. Wilson), pp. 319–62. Geological Society Australia, Queensland Division, Brisbane.

——, Murray, C. G. and Whitaker, W. G. (1978). The eastern part of the Tasman orogenic zone. *Tectonophysics* **48**, 327–64.

Dear, J. F., McKellar, R. G., Tucker, R. M. and Murphy, P. R. (1981). Monto, Sheet SG 56-1, Queensland. *Australia, 1:250 000 Geological Series*. Qld, Geol. Surv., Brisbane.

Deighton, I., Falvey, D. A. and Taylor, D. J. (1976). Depositional environments and geotectonic framework: southern Australian continental margin. *APEA J.* **16**, 25–36.

de Jersey, N. J. (1960). The Styx Coal Measures. *Geol. Soc. Aust., J.* **7**, 330–3.

Denham, D. (1979). Earthquakes. In *BMR Earth Science Atlas*. Aust., Bur. Miner. Resour., Geol. Geophys., Canberra.

——, Alexander, L. G. and Worotnicki, G. (1979). Stresses in the Australian crust: evidence from earthquakes and *in-situ* stress measurements. *BMR J. Aust. Geol. Geophys.* **4**, 289–95.

——, Small, G. R., Cleary, J. R., Gregson, P. J., Sutton, D. J. and Underwood, R. (1975). Australian earthquakes (1897–1972). *Search* **6**, 34–7.

——, Weekes, J. and Krayshek, C. (1981). Earthquake evidence for compressive stress in the southeast Australian crust. *Geol. Soc. Aust., J.* **28**, 323–32.

Denham, J. I. and Brown, B. R. (1976). A new look at the Otway Basin. *APEA J.* **16**, 91–8.

Denman, P. D. and van de Graaff, W. J. E. (1978). Emergent Quaternary marine deposits in the Lake Macleod area, W.A. *W. Aust., Geol. Surv., Annu. Rep.* (1976), 32–7.

Devine, S. B. and Youngs, B. C. (1975). Review of the Palaeozoic stratigraphy and petroleum potential of northern South Australia. *APEA J.* **15**, 45–54.

Dickins, J. M. (1976). Correlation chart for the Permian System of Australia. *Aust., Bur. Miner. Resour., Geol. Geophys., Bull.* 156 B.

—— (1978). Climate of the Permian in Australia: the invertebrate faunas. *Palaeogeogr., Palaeoclimatol., Palaeoecol.* **23**, 33–46.

—— (1982). The Permian Blenheim Subgroup of the Bowen Basin and its time relationships. *Geol. Soc. Aust., Abstr.* **8**, 27.

—— and Malone, E. J. (1973). Geology of the Bowen Basin, Queensland. *Aust., Bur. Miner. Resour., Geol. Geophys., Bull.* 130.

——, Roberts, J. and Veevers, J. J. (1972). Permian and Mesozoic geology of the northeastern part of the Bonaparte Gulf Basin. *Aust., Bur. Miner. Resour., Geol. Geophys., Bull.* **125**, 75–102.

Dietz, R. S. and Holden, J. C. (1970). The breakup of Pangaea. *Sci. Am.* **223**, 30–41.

Dixon, O. and Bauer, J. A. (1982). Southern Denison Trough — interpretation of seismic data from the Rolleston area. *Qld Gov. Min. J.* **83**, 122–31.

Dolby, J. H. and Balme, B. E. (1976). Triassic palynology of the Carnarvon Basin, Western Australia. *Rev. Palaeobot. Palynol.* **22**, 105–68.

Domack, E. W., Fairchild, W. W. and Anderson, J. B. (1980). Lower Cretaceous sediment from the East Antarctic continental shelf. *Nature* **287**, 625–6.

Donnelly, T. W. (1973). Late Cretaceous basalts from the Caribbean: a possible flood basalt province of vast size. *EOS (Trans. Am. Geophys. Union)* **54** (11), 1004.

Dorman, F. H. (1966). Australian Tertiary paleotemperatures. *J. Geol.* **74**, 49–61.

—— (1968). Some Australian oxygen isotope temperatures and a theory for a 30 million year world-temperature cycle. *J. Geol.* **76**, 297–313.

—— and Gill, E. D. (1959). Oxygen isotope palaeotemperature measurements on Australian fossils. *R. Soc. Vic., Proc.* **71**, 73–98.

Douglas, I. (1967). Man, vegetation and the sediment yields of rivers. *Nature* **215**, 925–8.

Douglas, J. G. and Ferguson, J. A. (eds.) (1976). Geology of Victoria. *Geol. Soc. Aust., Spec. Publ.* 5.

Doutch, H. F. (1976). The Karumba Basin, northeastern Australia and southern New Guinea. *BMR J. Aust. Geol. Geophys.* **1**, 131–40.

—— and Nicholas, E. (1978). The Phanerozoic sedimentary basins of Australia and their tectonic implication. *Tectonophysics* **48**, 365–88.

Dow, D. B. (1977). A geological synthesis of Papua New Guinea. *Aust., Bur. Miner. Resour., Geol. Geophys., Bull.* 201.

Drewry, D. J. (1975a). Radio echo sounding map of Antarctic (90° E–180°). *Polar Rec.* **17**, 359–74.

—— (1975b). Initiation and growth of the East Antarctic Ice Sheet. *Geol. Soc. Lond., J.* **131**, 255–73.

—— (1976). Sedimentary basins of the East Antarctic cratons from geophysical evidence. *Tectonophysics* **36**, 301–14.

Drummond, B. J. (1981). Crustal structure of the Precambrian terrains of northwest Australia from seismic refraction data. *BMR J. Aust. Geol. Geophys.* **6**, 123–35.

Dulhunty, J. A. (1971). Potassium–argon basalt dates and their significance in the Ilford–Mudgee–Gulgong region. *R. Soc. N.S.W., J. Proc.* **104**, 39–44.

—— (1972). Potassium–argon dating and occurrence of Tertiary and Mesozoic basalts in the Binnaway District. *R. Soc. N.S.W., J. Proc.* **105**, 71–6.

—— (1973). Potassium–argon basalt ages and their significance in the Macquarie Valley, New South Wales. *R. Soc. N.S.W., J. Proc.* **106**, 104–10.

—— (1976). Potassium–argon ages of igneous rocks in the Wollar–Rylstone region, New South Wales. *R. Soc. N.S.W., J. Proc.* **109**, 35–9.

—— (1982). Holocene sedimentary environments in Lake Eyre, South Australia. *Geol. Soc. Aust., J.* **29**, 437–42.

—— and McDougall, I. (1966). Potassium–argon dating of basalts in the Coonabarabran–Gunnedah district, New South Wales. *Aust. J. Sci.* **28**, 393–4.

Duplessy, J. C. (1982). Glacial to interglacial contrasts in the northern Indian Ocean. *Nature* **295**, 494–8.

Dury, G. H., Langford-Smith, T. and McDougall, I. (1969). A minimum age for the duricrust. *Aust. J. Sci.* **31**, 362–3.

Du Toit, A. L. (1937). *Our wandering continents.* Oliver and Boyd, London.

—— (1940). An hypothesis of submarine canyons. *Geol. Mag.* **77**, 395–404.

Eaton, G. P. (1982). The Basin and Range Province: origin and tectonic significance. *Annu. Rev. Earth Planet. Sci.* **10**, 409–40.

Edgerley, D. W. (1974). Fossil reefs of the Sahul Shelf, Timor Sea. *Proc. 2nd Int. Coral Reef Symp.* **2**, 627–37.

—— and Crist, R. P. (1974). Salt and diapiric anomalies in the southeast Bonaparte Gulf Basin. *APEA J.* **14**, 85–94.

Edwards, A. B. (1938). Tertiary tholeiite magma in Western Australia. *R. Soc. W. Aust., J.* **24**, 1–12.

—— (1958). Oolitic iron formations in northern Australia. *Geol. Rundsch.* **47**, 668–82.

—— and Baker, G. (1943). Jurassic arkose in southern Victoria. *R. Soc. Vic., Proc.* **55**, 195–226.

Edwards, A. R. (1968). Marine climates in the Oamaru district during late Kaiatan to early Whaingaroan time. *Tuatara* **16**, 75–9.

—— (1975). Southwest Pacific Neogene palaeogeography and an integrated Neogene Paleo-circulation model. *Deep Sea Dril. Proj., Initial Rep.* **30**, 667–84.

Efimova, G. P. (1974). *Bathymetric chart of Antarctica.* Dep. Geodesy and Cartography (in Russian), Moscow.

Ellenor, D. W. (1976). Otway Basin. *Australas. Inst. Min. Metall., Monogr. Ser.* **7** (3), 82–8.

Elliot, D. H. (1975a). Gondwana basins of Antarctica. In *Gondwana Geology* (ed. K. S. W. Campbell), pp. 493–536. Aust. Natl Univ. Press, Canberra.

—— (1975b). Tectonics of Antarctica: a review. *Am. J. Sci.* **275-A**, 45–106.

—— (1976). The tectonic setting of the Jurassic Ferrar Group. Antarctica. In *Andean and Antarctic Volcanology Problems* (ed. O. Gonzales Ferran), pp. 357–72. I.A.V.C.E.I., Naples.

Ellis, P. L. (1968). Geology of the Maryborough 1:250 000 Sheet area. *Qld, Geol. Surv., Rep.* 26.

—— (1976). Maryborough Basin. *Australas. Inst. Min. Metall., Monogr.* **7** (3), 447–50.

—— and Whitaker, W. G. (1976). Geology of the Bundaberg 1:250 000 Sheet area. *Qld, Geol. Surv., Rep.* 90.

Embleton, B. J. J. (1973). The palaeolatitude of Australia through Phanerozoic time. *Geol. Soc. Aust., J.* **19**, 475–82.

—— (1981). A review of the palaeomagnetism of Australia and Antarctica. In *Paleoreconstruction of the Continents* (ed. M. W. McElhinny and D. A. Valencio) pp. 77–92. Geodynamics Series 2, Am. Geophys. Union, Washington.

—— and McElhinny, M. W. (1982). Marine magnetic anomalies, palaeomagnetism and the drift history of Gondwanaland. *Earth Planet. Sci. Lett.* **58**, 141–50.

——, McElhinny, M. W., Crawford, A. R. and Luck, G. R. (1974). Palaeomagnetism and the tectonic evolution of the Tasman Orogenic zone. *Geol. Soc. Aust., J.* **21**, 187–94.

——, Veevers, J. J., Johnson, B. D. and Powell, C. McA. (1980). Palaeomagnetic comparison of a new fit of east and west Gondwanaland with the Smith and Hallam fit. *Tectonophysics* **61**, 381–90.

Emerson, D. W. and Wass, S. Y. (1980). Diatreme characteristics – evidence from the Mogo Hill Intrusion, Sydney Basin. *Aust. Soc. Expl. Geophys., Bull.* **11**, 121–33.

Engelhardt, N. L. and Webb, P. N. (in press). Recycled Foraminifera and non-biogenic material in Miocene sediments at Site 265, Southeast Indian Ocean.

Ericson, E. K. (1976). Capricorn Basin. *Australas. Inst. Min. Metall. Monogr.* **7** (3), 464–73.

Etheridge, M. A., Ransom, D. M., Williams, P. F. and Wilson, C. J. L. (1973). Structural evidence of the age of folded rocks on the south coast of New South Wales. *Geol. Soc. Aust., J.* **19**, 465–70.

Evans, P. R. (1966). Mesozoic stratigraphic palynology in Australia. *Australas. Oil Gas J.* **12** (6), 58–63.

—— (1969). Upper Carboniferous and Permian palynological stages and their distribution in eastern Australia. In *Gondwana Stratigraphy, 1st Gondwana Symposium* (ed. A. J. Amos), pp. 41–53. Unesco, Paris.

—— (1977). Petroleum geology of western New South Wales. *APEA J.* **17**, 42–9.

—— (1980). Geology of the Galilee Basin. In *The geology and geophysics of Northeastern Australia* (ed. R. A. Henderson and P. J. Stephenson), pp. 299-305. Geological Society of Australia, Queensland Division, Brisbane.

—— (1981). The petroleum potential of Australia. *Pet. Geol.* **4**, 123-46.

—— (1982). The age distribution of petroleum in Australia. *APEA J.* **22**, 301-10.

—— and Roberts, J. (1980). Evolution of central eastern Australia during the late Palaeozoic and early Mesozoic. *Geol. Soc. Aust., J.* **26**, 325-40.

Evernden, J. F. and Richards, J. R. (1962). Potassium-argon ages in eastern Australia. *Geol. Soc. Aust., J.* **9**, 1-50.

Ewart, A., Baxter, K. and Ross, J. A. (1980). The petrology and petrogenesis of the Tertiary anorogenic mafic lavas of southern and central Queensland, Australia – possible implications for crustal thickening. *Contrib. Mineral. Petrol.* **75**, 129-52.

Exon, N. F. (1972). Sedimentation in the outer Flensberg Fjord area (Baltic Sea) since the last glaciation. *Meyniana* **22**, 5-62.

—— (1976). Geology of the Surat Basin in Queensland. *Aust., Bur. Miner. Resour., Geol. Geophys. Bull.*, 166.

—— (1980). The stratigraphy of the Surat Basin, with special reference to coal deposits. *Coal Geology* **1**, 57-69.

—— and Burger, D. (1981). Sedimentary cycles in the Surat Basin and Global changes of sealevel. *BMR J. Aust. Geol. Geophys.* **6**, 153-9.

—— and Senior, B. R. (1976). The Cretaceous of the Eromanga and Surat Basins. *BMR J. Aust. Geol. Geophys.* **1**, 33-50.

—— and Willcox, J. B. (1978). The geology and petroleum potential of the Exmouth Plateau area off Western Australia. *Am. Assoc. Pet. Geol., Bull.* **62**, 40-72.

—— —— (1980). The Exmouth Plateau: stratigraphy, structure, and petroleum potential. *Aust., Bur. Miner. Resour., Geol. Geophys., Bull.* 199.

——, Langford-Smith, T. and McDougall, I. (1970). The age and geomorphic correlations of deep-weathering profiles, silcrete, and basalt in the Roma-Amby region, Queensland. *Geol. Soc. Aust., J.* **17**, 21-30.

——, von Rad, U., and von Stackelberg, U. (1982). The geological development of the passive margins of the Exmouth Plateau off northwest Australia. *Mar. Geol.* **47**, 131-52.

Facer, R. A. (1979). New and recalculated radiometric data supporting a Carboniferous age for the emplacement of the Bathurst Batholith, New South Wales. *Geol. Soc. Aust., J.* **25**, 429-32.

—— and Carr, P. F. (1979). K-Ar dating of Permian and Tertiary igneous activity in the southeastern Sydney Basin, New South Wales. *Geol. Soc. Aust., J.* **26**, 73-9.

——, Cook, A. C. and Beck, A. E. (1980). Thermal properties and coal rank in rocks and coal seams of the southern Sydney Basin, New South Wales: a palaeogeothermal explanation of coalification. *Int. J. Coal. Geol.* **1**, 1-17.

Fairbridge, R. W. (1950). Recent and Pleistocene coral reefs of Australia. *J. Geol.* **58**, 330-401.

—— (1953). The Sahul Shelf, northern Australia; its structure and geological relationships. *R. Soc. W. Aust., J.* **37**, 1-33.

Falvey, D. A. (1972). Sea-floor spreading in the Wharton Basin (northeast Indian Ocean) and the breakup of eastern Gondwanaland. *APEA J.* **12**, 86-8.

—— (1974). The development of continental margins in plate tectonic theory. *APEA J.* **14**, 95-106.

—— and Deighton, I. (1982). Recent advances in burial and thermal geohistory analysis. *APEA J.* **22**, 65-81.

—— and Middleton, M. F. (1981). Passive continental margins; evidence for a prebreakup deep crustal metamorphic subsidence mechanism. *Oceanol. Acta* **4** (supplement), 103-14.

—— and Mutter, J. C. (1981). Regional plate tectonics and the evolution of Australia's passive continental margins. *BMR J. Aust. Geol. Geophys.* **6**, 1-29.

—— and Veevers, J. J. (1974). Physiography of the Exmouth and Scott Plateaus, Western Australia, and adjacent Wharton Basin. *Mar. Geol.* **17**, 21-59.

Felton, E. A. (1974). Stratigraphic revisions in the Tarago-Woodlawn-Mount Fairy area. *N.S.W., Geol. Surv., Q. Notes* **17**, 7-12.

—— and Huleatt, M. B. (1975). *Braidwood 1:100 000 Geological Sheet 8827.* N.S.W. Geol. Surv., Sydney.

—— —— (1976). *Geology of the Braidwood 1:100 000 Geological Sheet 8827.* N.S.W., Geol. Surv., Sydney.

Fenton, M. W., Keene, J. B. and Wilson, C. J. L. (1982). The sedimentology and environment of deposition of the Mallacoota Beds, eastern Victoria. *Geol. Soc. Aust., J.* **29**, 107-14.

Ferguson, J., Ellis, D. J. and England, R. N. (1977). Unique spinel-garnet lherzolite inclusion in kimberlite from Australia. *Geology* **5**, 278-80.

——, Arculus, R. J. and Joyce, J. (1979). Kimberlite and kimberlitic intrusives of southeastern Australia: a review. *BMR J. Aust. Geol. Geophys* **4**, 227-41.

Fergusson, C. L. (1976). The structure and the stratigraphy of the Capertee-Palmers Oakey district. B.A. (Hons.) thesis, Macquarie Univ. (unpubl.).

—— (1979). Pre-cleavage folds in the mid-Palaeozoic sequence near Capertee, New South Wales. *R. Soc. N.S.W., J. Proc.* **112**, 125–32.

—— (1982a). An ancient accretionary terrain in eastern New England – evidence from the Coffs Harbour Block. In *New England Geology* (ed. P. G. Flood and B. Runnegar) pp. 63–70. Univ. New England, Armidale.

—— (1982b). Structure of the Late Palaeozoic Coffs Harbour Beds, northeastern New South Wales. *Geol. Soc. Aust., J.* **29**, 25–40.

——, Cas, R. A. F., Collins, W. J., Craig, G. Y., Crook, K. A. W., Powell, C. McA., Scott, P. A. and Young, G. C. (1979). The Upper Devonian Boyd Volcanic Complex, Eden, New South Wales. *Geol. Soc. Aust., J.* **26**, 87–105.

Fergusson, J. (1980). Yerranderie crater: a Devonian silicic eruptive centre within the Bindook Complex, New South Wales. *Geol. Soc. Aust., J.* **27**, 75–82.

Filatoff, J. (1975). Jurassic paynology of the Perth Basin, Western Australia. *Palaeontographica,* **B 154**, 1–113.

Findlay, A. L. (1974). The structure of foothills south of the Kubor Range, Papua New Guinea. *APEA J.* **14**, 14–20.

Fink, J. and Kukla, G. J. (1977). Pleistocene climates in Central Europe: at least 17 interglacials after the Olduvai Event. *Quaternary Research* **7**, 363–71.

Finlayson, D. M. (1982). Geophysical differences in the lithosphere between Phanerozoic and Precambrian Australia. *Tectonophysics* **84**, 287–312.

——, Collins, C. D. N. and Denham, D. (1980). Crustal Structure under the Lachlan Fold Belt, southeastern Australia. *Phys. Earth Planet. Int.* **21**, 321–42.

Fisher, R. A. (1953). Dispersion on a sphere. *R. Soc. London, Proc.* **A217**, 295–305.

Fitch, T. J. (1972). Plate convergence, transcurrent faults and internal deformation adjacent to southeast Asia and the western Pacific. *J. Geophys. Res.* **77**, 4432–60.

Flenley, J. (1979). *The equatorial rain forest: a geological history.* Butterworths, London.

Flint, D. J. (1978). Deep sea fan sedimentation of the Kanmantoo Group, Kangaroo Island. *R. Soc. S. Aust., Trans.* **102**, 203–22.

Flint, J. C. E., Lancaster, C. G., Gould, R. E. and Hensel, H. D. (1976). Some new stratigraphic data from the southern Clarence-Moreton Basin. *Qld Gov. Min. J.* **77**, 397–401.

Flood, P. G. and Fergusson, C. L. (1982). Tectonostratigraphic units and structure of the Texas–Coffs Harbour region. In *New England Geology* (ed. P. G. Flood and B. Runnegar) pp. 71–8. Univ. New England, Armidale.

——, Jell, J. S., and Waterhouse, J. B. (1981). Two new Early Permian stratigraphic units in the southeastern Bowen Basin, central Queensland. *Qld Gov. Min. J.* **82**, 179–84.

Flood, R. H. and Vernon, R. H. (1978). The Cooma Granodiorite, Australia: an example of in situ crustal anatexis? *Geology* **6**, 81–4.

Forbes, B. G. (1969). A review of Phanerozoic volcanism in South Australia. *Geol. Soc. Aust., Spec. Publ.* **2**, 127–32.

—— (1972). Possible post-Winton Mesozoic rocks north east of Marree, South Australia. *S. Aust., Geol. Surv., Q. Geol. Notes* **41**, 1–3.

——, Coats, R. P. and Daily, B. (1972). Truro volcanics. *S. Aust., Geol. Surv., Q. Geol. Notes* **44**, 1–5.

Ford, A. B. (1975). Volcanic rocks of Naturaliste Plateau, eastern Indian Ocean, Site 264, DSDP Leg 28. *Initial. Rep. Deep Sea Dril. Proj.* **28**, 821–33.

Forman, D. J. (1972). Petermann Ranges, Northern Territory. *Aust., Bur. Miner. Resour., Geol. Geophys., 1:250 000 Geol. Ser., Explanatory Notes,* SG/52-7.

—— and Shaw, R. D. (1973). Deformation of the crust and mantle in Central Australia. *Aust., Bur. Miner. Resour., Geol. Geophys., Bull.* 144.

—— and Wales, D. W. (1981). Geological evolution of the Canning Basin, Western Australia. *Aust., Bur. Miner. Resour., Geol. Geophys., Bull.* 210.

Foster, C. B. (1979). Permian plant microfossils of the Blair Athol Coal Measures and basal Rewan Formation of Queensland. *Qld, Geol. Surv., Palaeont. Pap.* 45.

Frakes, L. A. (1979). *Climates throughout geologic time.* Elsevier, Amsterdam.

—— and Rich, P. V. (1982). Paleoclimatic setting and paleogeographic links of Australia in the Phanerozoic. In *The fossil vertebrate record of Australasia* (ed. P. V. Rich and E. M. Thompson), pp. 28–52. Monash University Offset Printing Unit, Clayton, Victoria.

Francis, G., Douglas, I. and Walker, C. T. (1979). Landform longevity in southeastern Australia. *BGRG Conference on Longterm Landform Evolution,* Inst. Brit. Geogr., Annual Meeting, Manchester, 1–11.

Fraser, A. R. and Tilbury, L. A. (1979). Structure and stratigraphy of the Ceduna Terrace region, Great Australian Bight Basin. *APEA J.* **19**, 53–65.

Galloway, R. W. (1965). Late Quaternary climates in Australia. *J. Geol.* **73**, 603–18.

—— (1967). Pre-basalt, sub-basalt, and post-basalt surfaces of the Hunter Valley, New South Wales. In *Landform studies in Australia and New Guinea* (ed. J. N. Jennings and J. A. Mabbutt), pp. 293–414. ANU Press, Canberra.

—— and Kemp, E. M. (1980). Late Cainozoic environments in Australia. In *Ecological Biogeography in Australia* (ed. A. Keast), pp. 51–80. Junk, The Hague.

Gansser, A. (1964). *Geology of the Himalayas.* Wiley-Interscience, London.

Gaposchkin, E. M. and Lambeck, K. (1971), Earth's gravity field to the sixteenth degree and station coordinates from satellite and terrestrial data. *J. Geophys. Res.* **76**, 4855–83.

Garratt, M. J. (1983). Silurian to Early Devonian facies and biofacies patterns for the Melbourne Trough, central Victoria. *Geol. Soc. Aust., J.* **30**, 121–47.

Gatehouse, C. G. and Cooper, B. J. (1982). The Late Jurassic Polda Formation, Eyre Peninsula. *S. Aust., Geol. Surv., Q. Geol. Notes* **81**, 13–16.

Geary, J. K. (1970). Offshore exploration of the southern Carnarvon Basin. *APEA J.* **10**, 9–15.

Gee, R. D. (1979). Structure and tectonic style of the Western Australian Shield. *Tectonophysics* **58**, 327–69.

—— (1980). Summary of the Precambrian stratigraphy of Western Australia. *W. Aust., Geol., Surv., Ann. Rep.* for 1979, 85–90.

Gellatly, D. C. and Sofoulis, J. (1973). Yampi, Western Australia. *Aust., Bur. Miner. Resour., Geol. Geophys., 1:250 000 Geol. Ser., Explanatory Notes,* SE/51-3.

Gentilli, J. and Browne, W. R. (1963). Geography. In *The Australian encyclopaedia* (ed. A. Chisholm), pp. 253–61. Grolier Society, Sydney.

Geological Society of Australia (1971). *Tectonic Map of Australia and New Guinea, 1:5 000 000.* Sydney.

Geological Survey of Western Australia (1975). Geology of Western Australia. *W. Aust., Geol. Surv., Mem.* 2.

Gill, E. D. (1972). Palaeoclimatology and dinosaurs in south-east Australia. *Search* **3**, 444–6.

Gill, J. B. (1981). *Orogenic andesites and plate tectonics.* Springer-Verlag, Berlin.

Gilligan, L. B. and Scheibner, E. (1978). Lachlan Fold Belt in New South Wales. *Tectonophysics* **48**, 217–65.

——, Felton, E. A. and Olgers, F. (1979). The regional setting of the Woodlawn deposit. *Geol. Soc. Aust., J.* **26**, 135–40.

Girdler, R. W. (1978). Comparison of the East African rift system and the Permian Oslo rift. In *Petrology and geochemistry of continental rifts* (ed. E.-R. Neumann and I. B. Ramberg), pp. 329–45. Reidel, Dordrecht.

Glaessner, M. F. (1960). Upper Cretaceous larger foraminifera from New Guinea. *Tohoku Univ., Sci. Rep., Ser. 2, Special Volume* **4**, 37–44.

—— and Walter, M. R. (1981). Australian Precambrian palaeobiology. In *Precambrian of the Southern Hemisphere* (ed. D. R. Hunter), pp. 361–96. Elsevier, Amsterdam.

Gleadow, A. J. W. (1978). Fission-track evidence for the evolution of rifted continental margins. *4th Int. Conf. Geochron. Cosmochem. Isotope Geology.* U.S. Geol. Surv., Open-file Rep. 78-701, 146–8.

—— and Duddy, I. R. (1981). Early Cretaceous volcanism and the early breakup history of southeastern Australia: evidence from fission track dating of volcaniclastic sediments. In *Gondwana Five* (ed. M. M. Cresswell and P. Vella), pp. 283–7. Balkema, Rotterdam.

—— and Lovering, J. F. (1978). Fission track geochronology of King Island, Bass Strait, Australia: relationship to continental rifting. *Earth. Planet. Lett.* **37**, 429–37.

Glen, R. A. (1982a). Nature of late-Early to Middle Devonian tectonism in the Buckambool area, Cobar, New South Wales. *Geol. Soc. Aust., J.* **28**, 127–38.

—— (1982b). The Amphitheatre Group, Cobar, New South Wales: preliminary results of new mapping and implications for ore search. *N.S.W., Geol. Surv., Q. Notes* **49**, 1–14.

——, Laing, W. P., Parker, A. J. and Rutland, R. W. R. (1977). Tectonic relationships between the Proterozoic Gawler and Willyama orogenic domains, Australia. *Geol. Soc. Aust., J.* **24**, 125–50.

——, Powell, C. McA. and Khaiami, R. (1983). Mulga Downs Group. In *Geology of the Wrightville 1:100 000 Geological Sheet* (ed. R. A. Glen). N.S.W. Geol. Surv., Explan. Notes (in press).

Glikson, A. Y. and Lambert, I. B. (1976). Vertical zonation and petrogenesis of the Early Precambrian crust in Western Australia. *Tectonophysics* **30**, 55–89.

Goleby, B. R. (1980). Early Palaeozoic palaeomagnetism in south east Australia. *J. Geomag. Geoelectr.* **32**, Suppl. III, SIII 11–21.

Goode, A. D. T. and Williams, G. E. (1980). Possible western outlet for an ancient Murray River in South Australia. Reply. *Search* **11**, 227–30.

Gordon, W. A. (1973). Marine life and ocean surface currents in the Cretaceous. *J. Geol.* **81**, 269–84.

Goree, W. S. and Fuller, M. (1976). Magnetometers using R-F driven SQUIDS and their application in rock magnetism and paleomagnetism. *Rev. Geophys. Space Phys.* **14**, 591–608.

Gorter, J. D. (1978). Triassic environments in the Canning Basin, Western Australia. *BMR J. Aust. Geol. Geophys.* **3**, 25–33.

Goscombe, P. W. and Koppe, W. H. (1976). Permian coal geology – Eastern Australia; Bowen Basin, Queensland. *25th Int. Geol. Congr., Field Excursion Guidebook,* 10 A.

Gostin, V. A. and Herbert, C. (1973). Stratigraphy of the Upper Carboniferous and Lower Permian sequence, southern Sydney Basin. *Geol. Soc. Aust., J.* **20**, 49–70.

Grant, R. E. (1976). Permian brachiopods from southern Thailand. *J. Paleont.* **50**, Supplement 3, 1–269.

Gray, A. R. G. (1976). Hillsborough Basin. *Australas. Inst. Min. Metall. Monogr.* **7** (3), 460–4.

—— (1980). Stratigraphic relationships of Permian strata in the southern Taroom Trough. *Qld Gov. Min. J.* **81**, 110–30.

—— and Heywood, P. B. (1978). Stratigraphic relationships of Permian strata between Cockatoo Creek and Moura. *Qld Gov. Min. J.* **79**, 651–64.

Gray, N. M. (1982). Direction of stress, southern Sydney Basin. *Geol. Soc. Aust., J.* **29**, 277–84.

Green, D. C. and Webb, A. W. (1974). Geochronology of the northern part of the Tasman Geosyncline. In *The Tasman Geosyncline* (ed. A. K. Denmead, G. W. Tweedale, and A. F. Wilson), pp. 275–91. Geological Society of Australia, Queensland Division, Brisbane.

Green, P. W. and Bateman, R. J. (1981). The geology of the Condor oil shale deposit – onshore Hillsborough Basin. *APEA J.* **21**, 24–32.

Griffin, T. J. and McDougall, I. (1975). Geochronology of the Cainozoic McBride volcanic province, northern Queensland. *Geol. Soc. Aust., J.* **22**, 387–96.

——, White, A. J. R. and Chappell, B. W. (1978). The Moruya Batholith and geochemical contrasts between the Moruya and Jindabyne Suites. *Geol. Soc. Aust., J.* **25**, 235–47.

Griffin, W. L., Wass, S. Y. and Hollis, J. D. (1984). Ultramafic xenoliths from Bullenmerri and Gnotuk maars, Victoria, Australia: petrology of a subcontinental crust–mantle transition. *J. Petrol.* **25**, 53–87.

Griffiths, J. R. (1971). Continental margin tectonics and the evolution of Southeast Australia. *APEA J.* **11**, 75–9.

—— (1974). Revised continental fit of Australia and Antarctica. *Nature* **249**, 336–7.

Grimes, K. G. (1980). The Tertiary geology of north Queensland. In *The geology and geophysics of Northeastern Australia* (ed. R. A. Henderson and P. J. Stephenson), pp. 329–47. Geological Society of Australia, Queensland Division, Brisbane.

Grose, L. T. (1972). Tectonics. In *Geologic atlas of the Rocky Mountain region* (ed. W. W. Mallory), pp. 35–44. Rocky Mountain Association of Geologists, Denver.

Grund, R. B. (1976). North New Guinea Basin. *Australas. Inst. Min. Metall., Monogr.* **7** (3), 499–506.

Haantjens, H. A., Reynders, J. J., Mouthaan, W. L. P. J. and Van Baren, F. A. (1967). *Major soil groups of New Guinea and their distribution.* Commun. No. 55, Kon. Inst. voor de Tropen, Amsterdam.

Haig, D. W. (1979). Cretaceous foraminiferal biostratigraphy of Queensland. *Alcheringa* **3**, 171–87.

Haile, N. S. (1978). Reconnaissance palaeomagnetic results from Sulawesi, Indonesia, and their bearing on palaeomagnetic reconstructions. *Tectonophysics* **46**, 77–85.

—— (1979). Palaeomagnetic evidence for the rotation of Seram, Indonesia. In *Geodynamics of the Western Pacific* (ed. S. Uyeda, R. W. Murphy, and K. Kobayashi), pp. 191–8. Center for Academic Publications, Tokyo.

—— and Tarling, D. H. (1975). Note on reconnaissance palaeomagnetic measurements on Jurassic redbeds from Thailand. *Pac. Geol.* **10**, 101–3.

——, McElhinny, M. W. and McDougall, I. (1977). Palaeomagnetic data and radiometric ages from the Cretaceous of West Kalimantan (Borneo), and their significance in interpreting regional structure. *Geol. Soc. Lond., J.* **133**, 133–44.

Hale, G. E. (1962). Triassic System. *Geol. Soc. Aust., J.* **9**, 217–31.

Hall, R. L. (1975). Upper Ordovician coral faunas from north-eastern New South Wales. *R. Soc. N.S.W., J. Proc.* **108**, 75–93.

Hallam, A. (1975). *Jurassic environments.* Cambridge University Press, Cambridge.

Ham, W. E. (1973). *Regional geology of the Arbuckle Mountains, Oklahoma.* Guidebook Geol. Soc. Amer. Field Trip No. 5, Oklahoma Geol. Survey, Norman.

—— and Wilson, J. L. (1967). Paleozoic epeirogeny and orogeny in the central United States. *Am. J. Sci.* **265**, 332–407.

Hambrey, M. J. and Harland, W. B. (eds.) (1981). *Earth's pre-Pleistocene glacial record.* Cambridge University Press, Cambridge.

Hamilton, L. H. (1981). K–Ar age determinations of

some igneous rocks from the Newcastle and Hunter Valley coalfields. *Fifteenth symposium: Advances in the study of the Sydney Basin.* Dept Geol., Univ. Newcastle, 15-16.

Hamilton, W. (1979). Tectonics of the Indonesian region. *U.S. Geol. Surv., Prof. Pap.* 1078.

Hanson, E. R. and Al-Shaieb, Z. (1980). Voluminous subalkaline silicic magmas related to intracontinental rifting in the southern Oklahoma aulacogen. *Geology* **8**, 180-4.

Harding, R. R. (1969). Catalogue of age determinations on Australian rocks, 1962-1965. *Aust., Bur. Miner. Resour., Geol. Geophys., Rep.* 117.

Hargraves, R. B. and Onstott, T. C. (1980). Paleomagnetic results from some southern African kimberlites, and their tectonic significance. *J. Geophys. Res.* **85**, 3587-96.

Harland, W. B., Cox, A. V., Llewellyn, P. G., Pickton, C. A. G., Smith, A. G., and Walters, R. (1982). *A geologic time scale.* Cambridge University Press, Cambridge.

Harrington, H. J. (1974). The Tasman Geosyncline in Australia. In *The Tasman Geosyncline* (ed. A. K. Denmead, G. W. Tweedale, and A. F. Wilson), pp. 383-407. Geological Society of Australia, Queensland Division, Brisbane.

—— (1982). Tectonics and the Sydney Basin. *Sixteenth Symposium: Advances in the study of the Sydney Basin.* Dept. Geol., Univ. Newcastle, 15-19.

——, Burns, K. L., and Thompson, B. R. (1973). Gambier-Beaconsfield and Gambier-Sorell Fracture Zones and the movement of plates in the Australia-Antarctica-New Zealand region. *Nature* **245**, 109-12.

Harris, W. K. (1966). New and redefined names in South Australian Lower Tertiary stratigraphy. *S. Aust., Geol. Surv., Q. Geol. Notes* **20**, 1-3.

—— and Foster, C. B. (1974). Stratigraphy and palynology of the Polda Basin. *S. Aust., Miner. Resour. Rev.* **136**, 56-78.

Harrison, C. G. A., Brass, G. W., Saltzman, E., Sloan, J., Southam, J. and Whitman, J. M. (1981). Sea level variations, global sedimentation rates and the hypsographic curve. *Earth Planet. Sci. Lett.* **54**, 1-16.

Harrison, J. (1969). A review of the sedimentary history of the island of New Guinea. *APEA J.* **9**, 41-8.

Harrison, P. L. (1980). The Toomba Fault and the western margin of the Toko Syncline, Georgina Basin, Queensland and Northern Territory. *BMR J. Aust. Geol. Geophys.* **5**, 201-14.

Hawkins, L. V., Hennion, J. F., Nafe, J. E., and Doyle, H. A. (1965). Marine seismic refraction studies on the continental margin to the south of Australia. *Deep-Sea Research* **12**, 479-95.

Hayes, D. E. (ed.) (1978). Geophysical atlas of east and southeast Asian seas. *Geol. Soc. Am., Map and Chart Series* MC25.

—— and Frakes, L. A. (1975). General synthesis, Deep Sea Drilling Project Leg 28. *Deep Sea Dril. Proj., Initial Rep.* **28**, 919-42.

—— and Ringis, J. (1973). Seafloor spreading in the Tasman Sea. *Nature* **243**, 454-8.

——, Frakes, L. A., Barrett, P., Burns, D. A., Pei-Hsin Chen, Ford, A. B., Kaneps, A. G., Kemp, E. M., McCollum, D. W., Piper, D. J. W., Wall, R. E. and Webb, P. N. (1975). Leg 28. *Deep Sea Dril. Proj., Initial Rep.* 28.

Heezen, B. C. and Tharp, M. (1973). USNS *Eltanin* Cruise 55. *Antarctic J. of the U.S.* **8**, 137-41.

Heirtzler, J. R., Cameron, P., Cook, P. J., Powell, T., Roeser, H. A., Suhardi, S. and Veevers, J. J. (1978). The Argo Abyssal Plain. *Earth Planet. Sci. Lett.* **41**, 21-31.

Helby, R. and Morgan, R. (1979). Palynomorphs in Mesozoic volcanics of the Sydney Basin. *N.S.W., Geol. Surv., Q. Geol. Notes* **35**, 1-15.

Henderson, R. A. (1980). Structural outline and summary geological history for northeastern Australia. In *The geology and geophysics of Northeastern Australia* (ed. R. A. Henderson and P. J. Stephenson), pp. 1-26. Geological Society of Australia, Queensland Division, Brisbane.

Henry, N. M. (1975). Geology of the Upper Devonian between Mt Horrible and the Razorback Road: implications for the tectonics of N.S.W. B.A. (Hons) thesis, Macquarie University (unpubl.).

—— (1976). The stratigraphy and structure of the Upper Devonian rocks between Mt Horrible and the Razorback Road, and implications for the tectonics of N.S.W. *Aust. Soc. Explor. Geophys., Bull.* **7**, 29-30.

Henstridge, D. A. and Missen, D. D. (1981). *The geology of the Narrows Graben near Gladstone, Queensland, Australia.* Southern Pacific Petroleum, Central Pacific Minerals, Sydney.

Herbert, C. (1980a). Depositional development of the Sydney Basin. *N.S.W., Geol. Surv., Bull.* **26**, 11-52.

—— (1980b). Evidence for glaciation in the Sydney Basin and Tamworth synclinorial zone. *N.S.W., Geol. Surv., Bull.* **26**, 274-93.

—— (1982). Preliminary seismic interpretations in the Taroom Trough. *Geol. Soc. Aust., Abstr.* **8**, 32.

—— and Helby, R. (eds.) (1980). A guide to the Sydney Basin. *N.S.W., Geol. Surv., Bull.* 26.

Hermes, J. J. (1974). West Irian. *Geol. Soc. Lond., Spec. Publ.* **4**, 475–90.

Heusser, C. J. (1981). Palynology of the last interglacial–glacial cycle in midlatitudes of southern Chile. *Quaternary Research* **16**, 293–321.

Hills, E. S. (1938). The age and physiographic relationships of the Cainozoic volcanic rocks of Victoria. *R. Soc. Vic., Proc.* **51**, 112–39.

—— (1946). Some aspects of the tectonics of Australia. *R. Soc. N.S.W., J. Proc.* **79**, 67–91.

—— (1956). A contribution to the morphotectonics of Australia. *Geol. Soc. Aust., J.* **3**, 1–15.

—— (1975). *The physiography of Victoria; an introduction to geomorphology.* 5th edition, Whitcombe and Tombs, Melbourne.

Hilyard, D. (1981). Environmental palaeogeographic and tectonic setting of the Mullions Range Volcanics, New South Wales. *Geol. Soc. Aust., J.* **28**, 251–60.

Hinz, K. (1981). A hypothesis on terrestrial catastrophes: wedges of very thick oceanward dipping layers beneath passive continental margins. *Geol. Jahrbuch* **E22**, 3–28.

——, Beiersdorf, H., Exon, N. F., Roeser, H. A., Stagg, H. M. J. and von Stackelberg, U. (1978). Geoscientific investigations from the Scott Plateau off northwest Australia to the Java Trench. *BMR J. Aust. Geol. Geophys.* **3**, 319–40.

Hocking, J. B. (1972). Geologic evolution and hydrocarbon habitat Gippsland Basin. *APEA J.* **12**, 132–7.

Hocking, R. M. (1979). Sedimentology of the Tumblagooda Sandstone (Silurian) in the lower Murchison River area, Western Australia: a preliminary interpretation. *W. Aust., Geol. Surv., Ann. Rep.* for 1978, 40–4.

Hoffman, P., Dewey, J. F. and Burke, K. (1974). Aulacogens and their genetic relation to geosynclines with a Proterozoic example from Great Slave Lake, Canada. *Soc. Econ. Paleontol. Mineral., Spec. Publ.* **19**, 38–55.

Holcombe, C. J. (1977). How rigid are lithospheric plates? Fault and shear rotations in southeast Asia. *Geol. Soc. Lond., J.* **134**, 325–42.

Holmes, A. (1965). *Principles of physical geology.* 2nd edition, Nelson, London.

Hope, G. S. (1976). The vegetational history of Mt Wilhelm, Papua New Guinea. *J. Ecol.* **64**, 627–61.

Hope, J. H. and Hope, G. S. (1976). Palaeoenvironments for Man in New Guinea. In *The Origin of the Australians* (ed. R. L. Kirk and A. G. Thorne), pp. 29–54. Aust. Inst. Aborig. Studies, Canberra.

Hopley, D. (1982). *The geomorphology of the Great Barrier Reef: Quaternary development of coral reefs.* Wiley, New York.

Hornibrook, N. de B. (1971). New Zealand Tertiary climate. *N.Z., Geol. Surv., Rep.* **47**, 1–19.

Hos, D. (1975). Preliminary investigation of the palynology of the Upper Eocene Werillup Formation, Western Australia. *R. Soc. W. Aust., J.* **58**, 1–14.

Houston, B. R. (1964). Petrology of intrusives of the Roma Shelf. *Qld, Geol. Surv., Rep.* 7.

Houtz, R. E. and Markl, R. G. (1972). Seismic profiler data between Antarctica and Australia. *Antarct. Res. Ser.* **19**, 147–64.

——, Ewing, M., Hayes, D. and Naini, B. (1973). Sediment isopachs in the Indian and Pacific Ocean sectors (105° E to 70° W). *Am. Geogr. Soc., Antarctic Map Folio Series* **17**, 9–12.

Huber, N. K. and Cornwall, H. R. (1973). Keweenawan volcanic sequence. *Science* **182**, 1373.

Hunter, D. R. (ed.) (1981). *Precambrian of the Southern Hemisphere.* Elsevier, Amsterdam.

Idnurm, M. and Cook, P. J. (1980). Palaeomagnetism of beach ridges in South Australia and the Milankovitch theory of ice ages. *Nature* **286**, 699–702.

—— and Senior, B. R. (1978). Palaeomagnetic ages of Late Cretaceous and Tertiary weathered profiles in in the Eromanga Basin, Queensland. *Palaeogeog., Palaeoclimatol., Palaeoecol.* **24**, 263–77.

Irving, A. J. (1974). Geochemical and high-pressure experimental studies of garnet pyroxenite and pyroxene granulite xenoliths from the Delegate basaltic pipes, Australia. *J. Petrology* **15**, 1–40.

—— (1980). Petrology and geochemistry of composite ultramafic xenoliths in alkalic basalts and implications for magmatic processes within the mantle. *Am. J. Sci.* **280–A**, 389–426.

Irving, E. (1964). *Paleomagnetism and its application to geological and geophysical problems.* Wiley, N.Y.

Irwin, M. J. (1976). Aspects of the Early Triassic sedimentation in the Esk Trough, southeast Queensland. *Univ. Qld, Dept. Geol., Pap.* **7**, 46–62.

Jackson, M. J. and van de Graaff, W. J. E. (1981). Geology of the Officer Basin. *Aust., Bur. Miner. Resour., Geol. Geophys., Bull.* 206.

Jago, J. B. (1979). Tasmanian Cambrian biostratigraphy – a preliminary report. *Geol. Soc. Aust., J.* **26**, 223–30.

—— (1981). Late Precambrian–Early Palaeozoic geological relationships between Tasmania and northern Victoria Land. In *Gondwana Five* (ed. M. M. Cresswell and P. Vella), pp. 199–204. Balkema, Rotterdam.

——, Cooper, J. A. and Corbett, K. D. (1977). First evidence for Ordovician igneous activity in the Dial Range Trough, Tasmania. *Geol. Soc. Aust., J.* **24**, 81–6.

James, E. A. and Evans, P. R. (1971). The stratigraphy of the offshore Gippsland Basin. *APEA J.* **11**, 71–4.

Jan, Q. M. and Asif, M. (1981). A speculative tectonic model for the evolution of NW Himalaya and Karakoram. *Geol. Bull. Univ. Peshawar* **14**, 199–201.

Jenkin, J. J. (1968). The geomorphology and Upper Cainozoic geology of south-east Gippsland, Victoria. *Geol. Surv., Vic., Mem.* 27.

—— (1976). Geomorphology. *Geol. Soc. Aust., Spec. Publ.* **5**, 329–48.

Jenkins, C. J. (1982). Darriwilian (Middle Ordovician) graptolites from the Monaro Trough sequence, east of Braidwood, N.S.W. *Linn. Soc., N.S.W., Proc.* **106**, 173–9.

——, Kidd, R. R. and Mills, K. J. (1982). Upper Ordovician graptolites from the Wagonga Beds near Batemans Bay, New South Wales. *Geol. Soc. Aust. J.* **29**, 367–73.

Jenkins, D. A. L. (1974). Detachment tectonics in Western Papua New Guinea. *Geol. Soc. Am., Bull.* **85**, 533–48.

Jennings, J. N. (1971). Sea level changes and land links. In *Aboriginal Man and environment in Australia* (ed. D. J. Mulvaney and J. Golson), pp. 1–25, Australian National University Press, Canberra.

—— (1975). Desert dunes and estuarine fill in the Fitzroy Estuary, northwestern Australia. *Catena* **2**, 215–62.

Jensen, A. R. (1975). Permo-Triassic stratigraphy and sedimentation in the Bowen Basin, Queensland. *Aust., Bur. Miner. Resour., Geol. Geophys., Bull.* 154.

Johns, R. K. (1961). Geology and mineral resources of southern Eyre Peninsula. *S. Aust., Geol. Surv., Bull.* 37.

—— (1968). Investigation of Lakes Torrens and Gairdner. *S. Aust., Geol. Surv., Rep. Invest.* 31.

—— (1975a). Leigh Creek Coal Field, Boolcunda Basin and Springfield Basin, South Australia. *Australas. Inst. Min. Metall., Monogr.* **6** (2), 301–5.

—— (1975b). Tertiary coal in South Australia. *Australas. Inst. Min. Metall., Monogr.* **6** (2), 366–71.

—— and Ludbrook, N. H. (1962). Investigation of Lake Eyre. *S. Aust., Geol. Surv., Rep. Invest.* 24.

Johnson, B. D., Powell, C.McA. and Veevers, J. J. (1976). Spreading history of the eastern Indian Ocean and Greater India's northward flight from Antarctica and Australia. *Geol. Soc. Am. Bull.* **87**, 1560–6.

—— —— —— (1980a). Early spreading history of the Indian Ocean between India and Australia. *Earth Planet. Sci. Lett.* **47**, 131–43.

Johnson, G. L., Vanney, J.-R., Drewry, D. J. and Robin, G. de Q. (1980b). General Bathymetric Chart of the Oceans (GEBCO), Canad. Hydrographic Service, Ottawa, Sheet 5.18.

Johnson, J. G. (1971). Timing and coordination of orogenic, epeirogenic, and eustatic events. *Geol. Soc. Am., Bull.* **82**, 3263–98.

—— (1972). Antler Effect equals Haug Effect. *Geol. Soc. Am., Bull.* **83**, 2497–8.

Johnson, R. W. (1979). Geotectonics and volcanism in Papua New Guinea: a review of the late Cainozoic. *BMR J. Aust. Geol. Geophys.* **4**, 181–207.

——, Mackenzie, D. E. and Smith, I. E. M. (1978). Volcanic rock associations at convergent plate boundaries: reappraisal of the concept using plate histories from Papua New Guinea. *Geol. Soc. Am., Bull.* **89**, 96–108.

Johnston, C. R. and Bowin, C. O. (1981). Crustal reactions resulting from the mid-Pliocene to Recent continent–island arc collision in the Timor region. *BMR J. Aust. Geol. Geophys.* **6**, 223–43.

Johnstone, M. H. (1981). The importance of continental fragmentation history to petroleum accumulation. In *Gondwana Five* (ed. M. M. Cresswell and P. Vella), pp. 329–34. Balkema, Rotterdam.

——, Lowry, D. C. and Quilty, P. G. (1973). The Geology of southwestern Australia – a review. *R. Soc. W. Aust., J.* **56**, 5–15.

Jones, B. G. (1972). Upper Devonian to Lower Carboniferous stratigraphy of Pertnjara Group, Amadeus Basin, Central Australia. *Geol. Soc. Aust., J.* **19**, 229–49.

—— (1973a). Sedimentology of the Upper Devonian to Lower Carboniferous Finke Group, Amadeus and Warburton Basins, Central Australia. *Geol. Soc. Aust. J.*, **20**, 273–93.

Jones, D. K. (1976). Perth Basin. *Australas. Inst. Min. Metall., Monogr.* **7** (3), 108–26.

—— and Pearson, G. R. (1972). The tectonic elements of the Perth Basin. *APEA J.* **12**, 17–22.

Jones, H. A. (1973b). Marine geology of the North West Australian continental shelf. *Aust., Bur. Miner. Resour., Geol. Geophys., Bull.* 136.

Jones, J. G. and McDonnell, K. L. (1981). Papua New Guinea analogue for the Late Permian environment of northeastern New South Wales, Australia. *Palaeogeogr., Palaeoclimatol., Palaeoecol.* **34**, 191–205.

—— and Veevers, J. J. (1982). A Cainozoic history of Australia's Southeast Highlands. *Geol. Soc. Aust., J.* **29**, 1–12.

—— —— (1983). Mesozoic origins and antecedents of

Australia's Eastern Highlands. *Geol. Soc. Aust., J.* **30**, 305–22.

——, McPhie, J. and Roots, W. D. (1977). Devonian volcano at Yerranderie. *Search* **8**, 242–4.

Jones, L. M. and Faure, G. (1969). Age of the basement complex of Wright Valley, Antarctica. *Antarct. J. U.S.* **4**, 204–5.

Jones, P. J., Campbell, K. S. W. and Roberts, J. (1973). Correlation chart for the Carboniferous System of Australia. *Aust., Bur. Miner. Resour., Geol. Geophys., Bull.* 156 A.

Jongsma, D. (1974). Marine geology of the Arafura Sea. *Aust., Bur. Miner. Resour., Geol. Geophys., Bull.* 157.

—— and Mutter, J. C. (1978). Non-axial breaching of a rift valley: evidence from the Lord Howe Rise and the southeastern Australian margin. *Earth Planet. Sci. Lett.* **39**, 226–34.

Joyce, E. B. (1975). Quaternary volcanism and tectonics in southeastern Australia. *R. Soc. N.Z., Bull.* **13**, 169–76.

Jupp, D. L. B., Kerr, D., Lemaire, H., Milton, B. E., Moore, R. F., Nelson, R. and Vozoff, K. (1979). Joint magnetotelluric-DC resistivity survey, eastern Officer Basin. *Aust. Soc. Expl. Geophys., Bull.* **10**, 209–12.

Kalf, F. R. and Woolley, D. R. (1977). Application of mathematical modelling techniques to the alluvial aquifer system near Wagga Wagga, New South Wales. *Geol. Soc. Aust., J.* **24**, 179–94.

Kamerling, P. (1966). Sydney Basin – offshore. *APEA J.* **6**, 76–80.

Kanstler, A. J. and Cook, A. C. (1979). Maturation patterns in the Perth Basin. *APEA J.* **19**, 94–107.

Karig, D. E. (1972). Remnant arcs. *Geol. Soc. Am., Bull.* **73**, 1057–68.

—— and Jensky, W. (1972). The Proto-Gulf of California. *Earth. Planet. Sci. Lett.* **17**, 169–74.

—— Lawrence, M. B., Moore, G. F. and Curray, J. (1980a). Structural framework of the fore-arc basin, NW Sumatra. *Geol. Soc. Lond., J.* **137**, 77–91.

—— Moore, G. F., Curray, J. R. and Lawrence, M. B. (1980b). Morphology and shallow structure of the lower Trench slope off Nias Island, Sunda Arc. In *The tectonic and geologic evolution of southeast Asian seas and islands* (ed. D. E. Hayes), pp. 179–208. Amer. Geophys. Union., Monogr. 23. Washington, D.C.

Karner, G. D. and Watts, A. B. (1982). On isostasy at Atlantic-type continental margins. *J. Geophys. Res.* **87**, 2923–48.

Kemezys, K. J. (1978). Ordovician and Silurian lithofacies and base-metal deposits of the Lachlan Fold Belt. *Geol. Soc. Aust., J.* **25**, 97–107.

Kemp, E. M. (1972). Reworked palynomorphs from the West Ice Shelf area, East Antarctica, and their possible geological and palaeoclimatological significance. *Mar. Geol.* **13**, 145–57.

—— (1975). Palynology of Leg 28 drill sites. *Deep Sea Dril. Proj., Initial Rep.* **28**, 599–623.

—— (1976). Early Tertiary pollen from Napperby, Central Australia. *BMR J. Aust. Geol. Geophys.* **1**, 109–14.

—— (1978). Tertiary climatic evolution and vegetation history in the Southeast Indian Ocean region. *Palaeogeogr., Palaeoclimatol., Palaeoecol.* **24**, 169–208.

—— and Barrett, P. J. (1975). Antarctic glaciation and early Tertiary vegetation. *Nature* **258**, 507–8.

——, Balme, B. E., Helby, R. J., Kyle, R. A., Playford, G., and Price, P. L. (1977). Carboniferous and Permian palynostratigraphy in Australia and Antarctic: a Review. *BMR J. Aust. Geol. Geophys.* **2**, 177–208.

Kemp, N. R., Collinson, J. W. and Eggert, J. T. (1981). Fluvial sequence in the Triassic of Tasmania. *Geol. Soc. Aust., Abstr.* **3**, 23.

Kennett, J. P., Houtz, R. E., Andrews, P. B., Edwards, A. R., Gostin, V. A., Hajos, M., Hampton, M., Jenkins, D. G., Margolis, S. V., Ovenshine, A. T., and Perch-Nielsen, K. (1975). Leg 29. *Deep Sea Dril. Proj., Initial Rep.* 29.

——, McBirney, A. R. and Thunell, R. C. (1977). Episodes of Cenozoic volcanism in the Circum-Pacific Region. *J. Volcanol. Geotherm. Res.* **2**, 145–63.

Kennewell, P. J. and Huleatt, M. B. (1980). Geology of the Wiso Basin, Northern Territory. *Aust., Bur. Miner. Resour., Geol. Geophys., Bull.* 205.

——, Mathur, S. P. and Wilkes, P. G. (1977). The Lander Trough, Southern Wiso Basin, Northern Territory. *BMR J. Aust. Geol. Geophys.* **2**, 131–6.

Kershaw, A. P. (1978). Record of last interglacial-glacial cycle from northeastern Queensland. *Nature* **272**, 159–61.

Killick, C. L. A. (1982). Sedimentology and structural geology of the Bumbolee Creek Formation, Tumut Trough. B.Sc. (Hons) thesis, Aust. Nat. Univ., Canberra (unpubl.).

Kilpatrick, D. J. and Fleming, D. P. (1980). Lower Ordovician sediments in the Wagga Trough: discovery of early Bendigonian graptolites near Eskdale, northeast Victoria. *Geol. Soc. Aust., J.* **27**, 69–73.

Kind, H. D. (1980). Sedimentation pattern in the Triassic at Fingal Colliery, northeastern Tasmania. *Geol. Soc. Aust., 4th Geol. Convent., Abstr.*, 88.

King, L. (1967). *The morphology of the earth*. 2nd edition, Oliver and Boyd, Edinburgh.

Kinsman, D. J. J. (1975). Rift valley basins and sedimentary history of trailing continental margins. In *Petroleum and global tectonics* (ed. A. G. Fischer and S. Judson), pp. 83-127. Princeton University Press.

Kirkegaard, A. G., Shaw, R. D. and Murray, C. G. (1970). Geology of the Rockhampton and Port Clinton 1:250 000 Sheet areas. *Qld, Geol. Surv., Rep.* 38.

Kirschvink, J. L. (1980). The least-squares line and plane and the analysis of paleomagnetic data. *Geophys. J. R. Astronom. Soc.* 62, 699-718.

Kleeman, J. D. (1982). The anatomy of a tin-mineralising A-type granite. In *New England Geology* (ed. P. G. Flood and B. Runnegar), pp. 327-34. Dept Geology, University of New England, Armidale.

Klootwijk, C. T. (1980). Early Palaeozoic palaeomagnetism in Australia. *Tectonophysics* 64, 249-332.

—— and Peirce, J. W. (1979). India's and Australia's pole path since the late Mesozoic and the India-Asia collision. *Nature* 282, 605-7.

Kluth, C. F. and Coney, P. J. (1981). Plate tectonics of the Ancestral Rocky Mountains. *Geology* 9, 10-15.

Kobayashi, M. J. A. and Burton, C. K. (1971). Discovery of Ellesmereoceroid cephalopods in West Irian, New Guinea. *Japan Acad., Proc.* 47, 625-30.

Konig, M. (1981). Geophysical data from the continental margin off Wilkes Land Antarctica: implications for pre-drift fit and separation of Australia-Antarctica. *EOS* 62, 384.

Koppe, W. H. (1978). Review of the stratigraphy of the upper part of the Permian Succession in the Northern Bowen Basin. *Qld Gov. Min. J.* 79, 35-43.

Korsch, R. J. (1977). A framework for the Palaeozoic geology of the southern part of the New England Geosyncline. *Geol. Soc. Aust., J.* 24, 339-55.

Korsch, R. J. and Harrington, H. J. (1981). Stratigraphic and structural synthesis of the New England Orogen. *Geol. Soc. Aust., J.* 28, 205-26.

Krieg, G. (1969). Geological developments in the eastern Officer Basin of South Australia. *APEA J.* 9, 8-13.

Kuno, H. (1969). Plateau Basalts. In *The Earth's Crust and Upper Mantle* (ed. P. J. Hart), pp. 495-501. Am. Geophys. Union Monogr. 13.

Kurylowicz, L. E., Ozimic, S., McKirdy, D. M., Kantsler, A. J. and Cook, A. C. (1976). Reservoir and source rock potential of the Larapinta Group, Amadeus Basin, central Australia. *APEA J.* 16, 49-65.

Kushiro, I., Yoder, H. S. and Mysen, B. O. (1976). Viscosities of basalt and andesite melts at high pressures. *J. Geophys. Res.* 81, 6351-6.

Kyle, P. R., Adams, J. and Rankin, P. C. (1979). Geology and petrology of the McMurdo Volcanic Group at Rainbow Ridge, Brown Peninsula, Antarctica. *Geol. Soc. Am., Bull.* 90, 676-88.

——, Elliot, D. H. and Sutter, J. F. (1981). Jurassic Ferrar Supergroup tholeiites from the Transantarctic Mountains, Antarctica, and their relationship to the initial fragmentation of Gondwana. In *Gondwana Five* (ed. M. M. Cresswell and P. Vella), pp. 283-7. Balkema, Rotterdam.

La Brecque, J. L., Kent, D. V. and Cande, S. C. (1977). Revised magnetic polarity time scale for Late Cretaceous and Cenozoic time. *Geology* 5, 330-5.

Laird, M. G. (1981). The Late Mesozoic fragmentation of the New Zealand segment of Gondwana. In *Gondwana Five* (ed. M. M. Cresswell and P. Vella), pp. 311-18. Balkema, Rotterdam.

——, Andrews, P. B. and Kyle, P. R. (1974). Geology of northern Evans Neve, Victoria Land, Antarctica. *N.Z. J. Geol. Geophys.* 17, 587-601.

——, Cooper, R. A. and Jago, J. B. (1977). New data on the lower Palaeozoic sequence of northern Victoria Land, Antarctica, and its significance for Australian-Antarctic relations in the Palaeozoic. *Nature* 265, 107-10.

Lange, R. T. (1982). Australian Tertiary vegetation. In *A history of Australian vegetation* (ed. J. M. B. Smith), pp. 44-89. McGraw-Hill, Sydney.

Lanphere, M. A., Churkin, M. and Eberlein, G. D. (1977). Radiometric age of the *Monograptus cyphus* graptolite zone in southeast Alaska – an estimate of the age of the Ordovician-Silurian boundary. *Geol. Mag.* 114, 15-24.

Larson, R. L. (1972). Bathymetry, magnetic anomalies, and plate tectonic history of the mouth of the Gulf of California. *Geol. Soc. Am., Bull.* 83, 3345-60.

—— (1975). Late Jurassic sea-floor spreading in the Eastern Indian Ocean. *Geology* 3, 69-71.

—— (1977). Early Cretaceous breakup of Gondwanaland off Western Australia. *Geology* 5, 57-60.

—— and Hilde, T. W. C. (1975). A revised time scale of magnetic anomalies for the Early Cretaceous and Late Jurassic. *J. Geophys. Res.* 80, 2586-94.

——, Carpenter, G. B. and Diebold, J. B. (1978). A geophysical study of the Wharton Basin near the Investigator Fracture Zone. *J. Geophys. Res.* 83, 773-82.

——, Mutter, J. C., Diebold, J. B., Carpenter, G. B.

and Symonds. P. (1979). Cuvier Basin: a product of ocean crust formation by Early Cretaceous rifting off Western Australia. *Earth Planet. Sci. Lett.* **45**, 105–14.

Lawrence, C. R. (1975). Geology, hydrodynamics and hydrochemistry of the southern Murray Basin. *Vic., Geol. Surv., Mem.* 30.

——, Macumber, P. G., Kenley, P. R., Gill, E. D., Jenkin, J. J., Neilson, J. L. and McLennan, R. M. (1976). Quaternary. *Geol. Soc. Aust., Spec. Publ.* **5**, 275–327.

Laws, R. (1981). The petroleum geology of the onshore Bonaparte Basin. *APEA J.* **21**, 5–15.

—— and Brown, R. S. (1976). Bonaparte Gulf Basin – south eastern part. *Australas. Inst. Min. Metall., Monogr.* **7** (3), 200–8.

—— and Kraus, G. P. (1974). The regional geology of the Bonaparte Gulf Timor Sea area. *APEA J.* **14**, 77–84.

Leaman, D. E. (1976). Geological atlas 1:50 000 series, sheet 82 (8312S) Hobart. *Tas., Dep. Mines, Hobart.*

Lee, C., Shor, G. G., Bibee, L. D., Lu, R. S. and Hilde, T. W. C. (1980). Okinawa Trough: origin of a back-arc basin. *Mar. Geol.* **35**, 219–41.

Legg, D. P. (1976). Ordovician trilobites and graptolites from the Canning Basin, Western Australia. *Geol. et Palaeontol.* **10**, 1–58.

—— (1978). Ordovician biostratigraphy of the Canning Basin, Western Australia. *Alcheringa* **2**, 321–34.

Leitch, E. C. (1974). The geological development of the southern part of the New England Fold Belt. *Geol. Soc. Aust., J.* **21**, 133–56.

—— (1975). Plate tectonic interpretation of the Paleozoic history of the New England Fold Belt. *Geol. Soc. Am., Bull.* **86**, 141–4.

—— (1982). Crustal development in New England. In *New England Geology* (ed. P. G. Flood and B. Runnegar) pp. 9–16. Univ. New England, Armidale.

—— and Cawood, P. A. (1980). Olistoliths and debris flow deposits at ancient consuming plate margins: an eastern Australian example. *Sediment. Geol.* **25**, 5–22.

—— and McDougall, I. (1979). The age of orogenesis in the Nambucca Slate Belt: a K–Ar study of low-grade regional metamorphism. *Geol. Soc. Aust., J.* **26**, 111–19.

Le Maitre, R. W. (1975). Volcanic rocks from Edel No. 1 petroleum exploration well, offshore Carnarvon Basin, Western Australia. *Geol. Soc. Aust., J.* **22**, 167–74.

Le Pichon, X., Francheteau, J. and Bonnin, J. (1973). *Plate Tectonics.* Elsevier, Amsterdam.

Leslie, R. B., Evans, H. J. and Knight, C. L. (eds.) (1976). Economic Geology of Australia and Papua New Guinea, 3, Petroleum. *Australas. Inst. Min. Metall. Monogr.* 7.

Libby, W. G. and de Laeter, J. R. (1979). Biotite dates and cooling history at the western margin of the Yilgarn Block. *W. Aust., Geol. Surv., Annu. Rep.* for 1978, 79–87.

Lightner, J. D. (1977). The stratigraphy, structure and depositional history of the Tumut region, N.S.W. M.Sc. thesis, Aust. Nat. Univ., Canberra (unpubl.).

Lilley, F. E. M., Woods, D. V. and Sloane, M. N. (1981). Electrical conductivity profiles and implications for the absence or presence of partial melting beneath central and southeast Australia. *Phys. Earth Planet. Inter.* **25**, 419–28.

Lindsay, J. M. (1974). Marine Foraminifera from Eocene carbonaceous sediments: Polda, Cummins, and Warilla Basins, Eyre Peninsula. *S. Aust., Miner. Resour. Rev.* **136**, 46–51.

—— and Bonnett, J. E. (1973). Tertiary stratigraphy of three deep bores in the Waikerie area of the Murray Basin. *S. Aust., Geol. Surv., Rep. Invest.* 38.

—— and Harris, W. K. (1975). Fossiliferous marine and non-marine Cainozoic rocks from the eastern Eucla Basin, South Australia. *S. Aust., Miner. Resour. Rev.* **138**, 29–42.

Liu, C. S., McDonald, J. M. and Curray, J. R. (1982). A fossil spreading ridge in the northwestern Wharton Basin. *EOS* **63**, 448.

Lloyd, A. R. (1968a). Possible Miocene marine transgression in Northern Australia. *Aust., Bur. Miner. Resour., Geol. Geophys., Bull.* **80**, 87–100.

—— (1968b). Outline of the Tertiary geology of Northern Australia. *Aust., Bur. Miner. Resour., Geol. Geophys., Bull.* **80**, 87–100.

—— (1978). An outline of the Tertiary paleontology and stratigraphy of the Gulf of Papua, Papua New Guinea. In *Regional conference on the geology and mineral resources of southeast Asia, August 1975* (ed. S. Wiryosujono and A. Sudrajat), pp. 43–54. Geol. Surv. Indonesia, Bandung.

Lloyd, F. E. and Bailey, D. K. (1975). Light element metasomatism of the continental mantle; the evidence and the consequences. *Phys. Chem. of the Earth* **9**, 389–416.

Loffler, E. (1977). *Geomorphology of Papua New Guinea.* ANU Press, Canberra.

Lofting, M. J. W., Crostella, A. and Halse, J. W. (1975). Exploration results and future prospects in the northern Australasian region. *9th World Petroleum Congress, Tokyo, Panel Discussion* **7** (3), 1–18.

Logan, B. W., Davies, G. R., Read, J. F. and

Cebulski, D. E. (1970). Carbonate sedimentation and environments, Shark Bay, Western Australia. *Am. Assoc. Pet. Geol., Mem.* 13.

——, Read, J. F., Hagan, G. M., Hoffman, P., Brown, R. G., Woods, P. J. and Gebelein, C. D. (1974). Evolution and diagenesis of Quaternary carbonate sequences, Shark Bay, Western Australia. *Am. Assoc. Pet. Geol., Mem.* 22.

Lord, J. H. (1975). Bremer Basin, Western Australia. *Australas. Inst. Min. Metall., Monogr.* 6 (2), 372–3.

Loughnan, F. C. and Evans, P. R. (1978). The Permian and Mesozoic of the Merriwa–Binnaway–Ballimore area, New South Wales. *R. Soc. N.S.W., J. Proc.* 111, 107–19.

Loutit, T. S. and Kennett, J. P. (1981). New Zealand and Australian Cenozoic sedimentary cycles and global sea-level changes. *Am. Assoc. Pet. Geol., Bull.* 65, 1586–601.

Lovering, J. F. and White, A. J. R. (1969). Granulitic and eclogitic inclusions from basic pipes at Delegate, Australia. *Contrib. Mineral. Petrol.* 21, 9–52.

Low, G. H. (1975). Proterozoic rocks on or adjoining the Yilgarn Block. *West. Aust., Geol. Surv., Mem.* 2, 71–81.

Lowry, D. C. (1981). The Woodada discovery – its implications for exploration. *APEA J.* 21 (2), 30–2.

Ludbrook, N. H. (1978). Australia. In *The Phanerozoic of the World. II The Mesozoic, A* (ed. M. Moullade and A. E. M. Nairn), pp. 209–49. Elsevier, Amsterdam.

—— (1980). *A guide to the geology and mineral resources of South Australia.* South Australia, Department of Mines and Energy, Adelaide.

Mabbutt, J. A. and Sullivan, M. E. (1970). Landforms and structure. In *Australian Grasslands* (ed. R. Milton Moore), pp. 27–43.

McClenaghan, M. P., Turner, N. J., Baillie, P. W., Brown, A. V., Williams, P. R. and Moore, W. R. (1981). Geological atlas 1:50 000 series, Sheets 32 and 24 Ringarooma and Boobyalla. *Tas., Dep. Mines, Hobart.*

McClung, G. (1980). Permian marine sedimentation in the northern Sydney Basin. *N.S.W., Geol. Surv., Bull.* 26, 54–72.

McCracken, K. G. and Astley-Boden, C. E. (eds.) (1982). *Satellite images of Australia.* Harcourt Brace Jovanovich, Sydney.

McCulloch, M. T. and Chappell, B. W. (1982). Nd isotopic characteristics of S- and I-type granites. *Earth Planet. Sci. Lett.* 58, 51–64.

McDonnell, K. L., Conaghan, P. J. and Jones, J. G. (in prep.). Fluvial sandstones in a fore-deep context: Late Permian and Triassic, Sydney Basin.

McDougall, I. and Gill, E. D. (1975). Potassium–argon ages from the Quaternary succession in the Warrnambool–Port Fairy area, Victoria Australia. *R. Soc. Vic., Proc.* 87, 175–8.

—— and Leggo, P. J. (1965). Isotopic age determinations on granitic rocks from Tasmania. *Geol. Soc. Aust., J.* 12, 295–332.

—— and Roksandic, Z. (1974). Total fusion $^{40}Ar/^{39}Ar$ ages using Hifar Reactor. *Geol. Soc. Aust., J.* 21, 81–9.

—— and Slessar, G. C. (1972). Tertiary volcanism in the Cape Hillsborough area, north Queensland. *Geol. Soc. Aust. J.* 18, 401–8.

—— and van der Lingen, G. J. (1974). Age of the rhyolites of the Lord Howe Rise and the evolution of the southwest Pacific Ocean. *Earth Planet. Sci. Lett.* 21, 117–26.

—— and Wellman, P. (1976). Potassium–argon ages for some Australian Mesozoic igneous rocks. *Geol. Soc. Aust., J.* 23, 1–9.

—— and Wilkinson, J. F. G. (1967). Potassium–argon dates on some Cainozoic volcanic rocks from northeastern New South Wales. *Geol. Soc. Aust., J.* 14, 225–33.

——, Allsopp, H. L. and Chamalaun, F. H. (1966). Isotopic dating of the Newer Volcanics of Victoria, Australia, and geomagnetic polarity epochs. *J. Geophys. Res.* 71, 6170–18.

——, Embleton, B. J. J. and Stone, D. B. (1981). Origin and evolution of Lord Howe Island, southwest Pacific Ocean. *Geol. Soc. Aust., J.* 28, 155–76.

McElhinny, M. W. (1973). *Palaeomagnetism and Plate Tectonics.* Cambridge University Press, Cambridge.

—— (1978). The magnetic polarity time scale: prospects and possibilities in magnetostratigraphy. *Am. Assoc. Pet. Geol., Stud. Geol.* 6, 57–65.

——, Haile, N. S. and Crawford, A. R. (1974). Palaeomagnetic evidence shows Malay Peninsula was not a part of Gondwanaland. *Nature* 252, 641–5.

McGowran, B. (1973). Rifting and drift of Australia and the migration of mammals. *Science* 180, 759–61.

—— (1977). Maastrichtian to Eocene foraminiferal assemblages in the northern and eastern Indian Ocean region: correlations and historical patterns. In *Indian Ocean Geology and Biostratigraphy* (ed. J. R. Heirtzler *et al.*), pp. 417–58. Am. Geoph. Union, Washington, DC.

—— (1978). Stratigraphic record of Early Tertiary oceanic and continental events in the Indian Ocean region. *Mar. Geol.* 26, 1–39.

—— (1979). The Tertiary of Australia: foraminiferal overview. *Marine Micropaleontol.* 4, 235–64.

—— (1981). Palaeoenvironments of the Australasian

region – problems for resolution. In *The future of scientific ocean drilling in the Australasian region* (ed. P. J. Cook, K. A. W. Crook, and L. A. Frakes), pp. 67–79. Consortium for Ocean Geosciences (COGS), Canberra.

McIlveen, G. R. (1974). The Eden–Comerong–Yalwal rift zone and the contained gold mineralization. *N.S.W., Geol. Surv., Rec.* **16**, 245–77.

McInerney, P. M. (1979). Seismic refraction investigations in the Polda Coalfield. *S. Aust., Miner. Resour. Rev.* **147**, 9–15.

McIntyre, A. and CLIMAP project members (1976). The surface of the Ice-age earth. *Science* **191**, 1131–44.

McKellar, J. L. (1980a). Palynostratigraphy of the Burrum Coal Measures in Fairymead NS1, Maryborough Basin. *Qld Gov. Min. J.* **81**, 462–8.

—— (1980b). Palynostratigraphy of the Tiaro Coal Measures and Maryborough Formation in GSQ Maryborough 2. *Qld Gov. Min. J.* **81**, 599–604.

—— (1982). Late Triassic ('Rhaetian') and Jurassic palynostratigraphy of the Surat Basin. In *Eromanga Basin Symposium, summary papers* (ed. P. S. Moore and T. J. Mount), pp. 172–3. Geol. Soc. Aust. and Pet. Expl. Soc. Aust., Adelaide.

McKenzie, D. P. (1969). The relation between fault plane solutions for earthquakes and the directions of principal stresses. *Seismol. Soc. Am., Bull.* **59**, 591–601.

—— and Sclater, J. G. (1971). The evolution of the Indian Ocean since the Late Cretaceous. *Geophys. J. R. Astron. Soc.* **25**, 437–528.

McLaughlin, R. J. W. (1976). Upper Devonian to Lower Carboniferous provinces. *Geol. Soc. Aust., Spec. Publ.* **5**, 81–98.

McTavish, R. A. (1970). Triassic conodonts in Western Australia. *Search* **1**, 159–60.

—— (1975). Triassic conodonts and Gondwana stratigraphy. In *Gondwana Geology* (ed. K. S. W. Campbell), pp. 481–90. Australian National University Press, Canberra.

—— and Legg, D. P. (1976). The Ordovician of the Canning Basin, Western Australia. In *The Ordovician System* (ed. M. G. Bassett), pp. 447–78. Univ. Wales Press and Nat. Mus. Wales, Cardiff.

Macumber, P. G. (1978). Evolution of the Murray River during the Tertiary Period – evidence from northern Victoria. *R. Soc. Vic., Proc.* **90**, 43–52.

McWhae, J. R. H., Playford, P. E., Lindner, A. W., Glenister, B. F., and Balme, B. E. (1958). The stratigraphy of Western Australia. *Geol. Soc. Aust., J.* **4** (2).

McWilliams, M. O. and McElhinny, M. W. (1980). Late Precambrian paleomagnetism of Australia: the Adelaide Geosyncline. *J. Geol.* **88**, 1–26.

Magnetic Map of Australia (1976). Canberra, Bur. Miner. Resour., Geol. Geophys. Scale 1:2 500 000.

Major, R. B. (1973). The Wirrildar Beds. *S. Aust., Geol. Surv., Q. Geol. Notes* **45**, 8–11.

Malahoff, A., Feden, R. H. and Fleming, H. S. (1982). Magnetic anomalies and tectonic fabric of marginal basins north of New Zealand. *J. Geophys. Res.* **87**, 4109–25.

Mallett, C. W. (1978). Sea level changes in the Neogene of southern Victoria. *APEA J.* **18**, 64–9.

Malone, E. J., Jensen, A. R., Gregory, C. M. and Forbes, V. R. (1966). Geology of the southern half of the Bowen 1:250 000 Sheet area, Queensland. *Aust., Bur. Miner. Resour., Geol. Geophys., Rep.* 100.

——, Olgers, F., Mollan, R. G. and Jensen, A. R. (1967). *Geological map of the Bowen Basin, Queensland. 1:500 000.* Aust., Bur. Miner. Resour., Geol. Geophys., Canberra.

—— —— and Kirkegaard, A. G. (1969). The geology of the Duaringa and Saint Lawrence 1:250 000 Sheet areas, Queensland. *Aust., Bur. Miner. Resour., Geol. Geophys., Rep.* 121.

Mankinen, E. A. and Dalrymple, G. B. (1979). Revised geomagnetic polarity time scale for the interval 0–5 m.y. B.P. *J. Geophys. Res.* **84**, 615–26.

Marjoribanks, R. W. and Black, L. P. (1974). Geology and geochronology of the Arunta Complex north of Ormiston Gorge, central Australia. *Geol. Soc. Aust., J.* **21**, 291–9.

Markl, R. G. (1974a). Evidence for the breakup of eastern Gondwanaland by the early Cretaceous. *Nature* **251**, 196–200.

—— (1974b). Bathymetric map of the eastern Indian Ocean. *Initial Rep. Deep Sea Dril. Proj.*, **26**, 967–8.

—— (1978a). Further evidence for the Early Cretaceous breakup of Gondwanaland off southwestern Australia, *Mar. Geol.* **26**, 41–8.

—— (1978b). Basement morphology and rift geometry near the former junction of India, Australia and Antarctica. *Earth Planet. Sci. Lett.* **39**, 211–25.

Marsden, M. A. H. (1972). The Devonian history of northeastern Australia. *Geol. Soc. Aust., J.* **19**, 125–62.

—— (1976). Upper Devonian–Carboniferous. *Geol. Soc. Aust., Spec. Publ.* **5**, 77–124.

Martin, H. A. (1980). Stratigraphic palynology from shallow bores in the Namoi River and Gwydir River valleys, north-central New South Wales. *R. Soc. N.S.W., J. Proc.* **113**, 81–7.

Martin, K. R. (1981). Deposition of the Precipice

Sandstone and evolution of the Surat Basin in the Early Jurassic. *APEA J.* **21**, 16–23.
—— and Hamilton, N. J. (1981). Diagenesis and reservoir quality, Toolachee Formation, Cooper Basin. *APEA J.* **21**, 143–54.
Masolov, V. N., Kurinin, R. G. and Grikurov, G. E. (1981). Crustal structure and tectonic significance of Antarctic rift zones (from geophysical evidence). In *Gondwana Five* (ed. M. M. Cresswell and P. Vella), pp. 303–9. Balkema, Rotterdam.
Mathur, S. P. (1974). Crustal structure in southwestern Australia from seismic and gravity data. *Tectonophysics* **24**, 151–82.
—— (1976). Relation of Bouguer anomalies to crustal structure in southwestern and central Australia. *BMR J. Aust. Geol. Geophys.* **1**, 277–86.
Mawson, D. (1940). Sedimentary rocks. *Sci. Rept. Australasian Ant. Exped. 1911–14, Ser. A, Geol.*, **4** (11), 347–67.
Maxwell, W. G. H. (1960). The Yarrol Basin (southern part). *Geol. Soc. Aust., J.* **7**, 217–21.
—— (1968). *Atlas of the Great Barrier Reef.* Elsevier, Amsterdam.
Meath, J. R. and Bird, K. J. (1976). The geology of the West Trial Rocks gas field. *APEA J.* **16**, 157–63.
Mercier, J. C. C. (1979). Peridotite xenoliths and the dynamics of kimberlite intrusion. In *The mantle sample: inclusions in kimberlites and other volcanics* (ed. F. R. Boyd and H. O. A. Meyer), pp. 197–212. American Geophysical Union, Washington, DC.
Middleton, M. F. and Schmidt, P. W. (1982). Paleothermometry of the Sydney Basin. *J. Geophys. Res.* **87**, 5351–9.
Milnes, A. R., Compston, W. and Daily, B. (1977). Pre- to syn-tectonic emplacement of early Palaeozoic granites in southeastern South Australia. *Geol. Soc. Aust., J.* **24**, 87–106.
——, Cooper, B. J. and Cooper, J. A. (1982). The Jurassic Wisanger Basalt of Kangaroo Island, South Australia. *Roy. Soc. S. Aust., Trans.* **106**, 1–13.
Mitchum, R. M., Vail, P. R. and Thompson, S. (1977). The depositional sequence as a basic unit for stratigraphic analysis. *Am. Assoc. Pet. Geol., Mem.* **26**, 53–62.
Mollan, R. G., Dickins, J. M., Exon, N. F. and Kirkegaard, A. G. (1969). Geology of the Springsure 1:250 000 Sheet area, Queensland. *Aust., Bur. Miner. Resour., Geol. Geophys., Rep.* 123.
Molnar, P. and Tapponier, P. (1975). Cenozoic tectonics of Asia: effects of a continental collision. *Science* **189**, 419–26.
——, Atwater, A., Mammerickx, J. and Smith, S. M. (1975). Magnetic anomalies, bathymetry and the tectonic evolution of the South Pacific since the Late Cretaceous. *Geophys. J. R. Astron. Soc.* **40**, 383–420.
Moore, G. F., Curray, J. R., Moore, D. G. and Karig, D. E. (1980). Variations in geologic structure along the Sunda fore arc, Northeastern Indian Ocean. In *The tectonic and geologic evolution of southeast Asian seas and islands* (ed. D. E. Hayes), pp. 145–60. Amer. Geophys. Union., Monogr. 23. Washington, D.C.
Moore, J. C., Watkins, J. S., Shipley, T. H., McMillen, K. G., Bachman, S. B. and Lundberg, N. (1982). Geology and tectonic evolution of a juvenile accretionary terrane along a truncated convergent margin: synthesis of results from Leg 66 of the Deep Sea Drilling Project, southern Mexico. *Geol. Soc. Amer., Bull.* **93**, 847–61.
Moore, P. S. (1979). Deltaic sedimentation – Cambrian of South Australia. *J. Sediment. Petrol.* **49**, 1229–44.
—— (1982). Mesozoic geology of the Simpson Desert region, northern South Australia. In *Eromanga Basin Symposium, summary papers* (ed. P. S. Moore and T. J. Mount), pp. 46–57. Geol. Soc. Aust. and Pet. Expl. Soc. Aust., Adelaide.
Morel, P. and Irving, E. (1978). Tentative paleocontinental maps for the early Phanerozoic and Proterozoic. *J. Geol.* **86**, 535–61.
Morgan, R. (1980). Eustasy in the Australian Early and Middle Cretaceous. *N.S.W., Geol. Surv. Bull.* 27.
Morley, M. E., Gleadow, A. J. W. and Lovering, J. F. (1981). Evolution of the Tasman Rift: Apatite fission track dating evidence from the southeastern Australian continental margin. In *Gondwana Five* (ed. M. M. Cresswell and P. Vella), pp. 289–93. Balkema, Rotterdam.
Morris, W. A. (1977). Paleolatitude of glaciogenic upper Precambrian Rapitan group and the use of tillites as chronostratigraphic marker horizons. *Geology* **5**, 85–8.
Mory, A. J. (1982). The Early Carboniferous palaeogeography of the northern Tamworth Belt. *Geol. Soc. Aust., J.* **29**, 357–66.
Mount, T. J. (1982). Geology of the Dullingari Murta oilfield. In *Eromanga Basin symposium, summary papers* (ed. P. S. Moore and T. J. Mount), pp. 356–74. Geol. Soc. Aust. and Pet. Expl. Soc. Aust., Adelaide.
Muir, M. D. (1980). Palaeontological evidence for the Early Cambrian age of the Bukalara Sandstone, McArthur Basin, N.T. *BMR J. Aust. Geol. Geophys.* **5**, 159–60.
Murphy, P. R., Schwarzbock, H., Cranfield, L. C.,

Withnall, I. W. and Murray, C. G. (1976). Geology of the Gympie 1:250 000 Sheet area. *Qld, Geol. Surv., Rep.* 96.

Murray, C. G. (1975). Rockhampton, Queensland. *Aust., Bur. Miner. Resour. Geol. Geophys. 1:250 000 Geol. Ser., Explanatory Notes,* SF/56–13.

—— and Kirkegaard, A. G. (1978). The Thomson Orogen of the Tasman Orogenic Zone. *Tectonophysics* **48**, 299–325.

—— and Whitaker, W. G. (1982). A review of the stratigraphy, structure and regional tectonic setting of the Brisbane Metamorphics. In *New England Geology* (ed. P. G. Flood and B. Runnegar) pp. 79–94. Univ. New England, Armidale.

Mutter, J. C. and Karner, G. D. (1980). The continental margin off northeast Australia. In *The geology and geophysics of Northeastern Australia* (ed. R. A. Henderson and P. J. Stephenson), pp. 47–69. Geological Society of Australia, Queensland Division, Brisbane.

——, Talwani, M. and Stoffa, P. L. (1982). Origin of seaward-dipping reflectors in oceanic crust off the Norwegian margin by 'subaerial sea-floor spreading'. *Geology* **10**, 353–7.

Nesbitt, R. W., Goode, A. D. T., Moore A. C. and Hopwood, T. P. (1970). The Giles Complex, central Australia: a stratified sequence of mafic and ultramafic intrusions. *Geol. Soc. S. Afr., Spec. Publ.* **1**, 547–64.

Nicholls, I. A., Ferguson, J., Jones, H., Marks, G. P. and Mutter, J. C. (1981). Ultramafic blocks from the ocean floor southwest of Australia. *Earth Planet. Sci. Lett.* **56**, 362–74.

Nicoll, R. S. (1976). The effect of Late Carboniferous–Early Permian glaciation in the distribution of conodonts in Australia. *Geol. Soc. Canada, Spec. Pap.* **15**, 273–8.

Ninkovich, D. (1976). Late Cenozoic clockwise rotation of Sumatra. *Earth Planet. Sci. Lett.* **29**, 269–75.

Norton, I. O. and Molnar, P. (1977). Implications of a revised fit between Australia and Antarctica for the evolution of the eastern Indian Ocean. *Nature* **267**, 338–9.

—— and Sclater, J. G. (1979). A model of the evolution of the Indian Ocean and the breakup of Gondwanaland. *J. Geophys. Res.* **84**, 6803–30.

Norvick, M. S. (1979). The tectonic history of the Banda Arcs, eastern Indonesia: a review. *Geol. Soc. Lond., J.* **136**, 519–27.

Odin, G. S. (1978). Results of dating Cretaceous, Paleogene sediments, Europe. *Am. Assoc. Pet. Geol., Stud. Geol.* **6**, 127–41.

—— (ed.) (1982). *Numerical dating in stratigraphy.* Wiley, New York.

O'Driscoll, E. S. T. (1982). Patterns of discovery – the challenge for innovative thinking. *Pet. Expl. Soc. Aust., J.* **1**, 11–31.

Offenberg, A. C., Rose, D. M. and Packham, G. H. (1971). *Dubbo 1:250 000 Geological Series Sheet SI 55–4.* N.S.W., Geol. Surv., Sydney.

Olgers, F. (1972). Geology of the Drummond Basin, Queensland. *Aust., Bur. Miner. Resour., Geol. Geophys., Bull.* 132.

Ollier, C. D. (1979). Evolutionary geomorphology of Australia and Papua-New Guinea. *Inst. Brit. Geogr., Trans.* **4**, 516–39.

—— (1982). The Great Escarpment of eastern Australia: tectonic and geomorphic significance. *Geol. Soc. Aust., J.* **29**, 13–23.

Öpik, A. A. (1957). Cambrian palaeogeography of Australia. *Aust., Bur. Miner. Resour., Geol. Geophys., Bull.* **49**, 239–84.

—— (1967). The Ordian Stage of the Cambrian and its Australian Metadoxididae. *Aust., Bur. Miner. Resour., Geol. Geophys.* **92**, 133–70.

—— (1970). *Redlichia* of the Ordian (Cambrian) of northern Australia and New South Wales. *Aust., Bur. Miner. Resour., Geol. Geophys.* 114.

—— (1975). Templetonian and Ordian xystridurid trilobites of Australia. *Aust., Bur. Miner. Resour., Geol. Geophys.* 121.

Oppel, T. W. (1970). Exploration of the southwest flank of the Papuan Basin. *APEA J.* **10**, 62–9.

Oversby, B. (1971). Palaeozoic plate tectonics in the southern Tasman Geosyncline. *Nature Phys. Sci.* **234**, 45–8.

——, Black, L. P. and Sheraton, J. W. (1980). Late Palaeozoic continental volcanism in northeastern Queensland. In *The geology and geophysics of Northeastern Queensland* (ed. R. A. Henderson and P. J. Stephenson), pp. 247–68. Geological Society of Australia, Queensland Division, Brisbane.

Owen, M., and Wyborn, D. (1979). Geology and geochemistry of the Tantangara and Brindabella 1:100 000 Sheet Areas, New South Wales and Australian Capital Territory. *Aust., Bur. Miner. Resour., Geol. Geophys., Bull.* 204.

Packham, G. H. (1960). Sedimentary history of the Tasman Geosyncline in southeastern Australia. *Int. Geol. Congr.* **21**, 74–83.

—— (1966). *Bathurst 1:250 000 Geological Series Sheet SI 55–8.* N.S.W., Geol. Surv., Sydney.

—— (1968). The Lower and Middle Palaeozoic stratigraphy and sedimentary tectonics of the Sofala-Hill End–Euchareena region, N.S.W. *Linn. Soc. N.S.W., Proc.* **93**, 111–63.

—— (ed.) (1969). The geology of New South Wales. *Geol. Soc. Aust., J.* **16**.

—— (1973). A speculative Phanerozoic history of the south-west Pacific. In *The Western Pacific: Island Arcs, Marginal Seas, Geochemistry* (ed. P. J. Coleman), pp. 369–88. Univ. of W.A. Press, Perth.

—— (ed.) (1982). The evolution of the India–Pacific plate boundaries. *Tectonophysics* **87**, 1–397.

—— and Falvey, D. A. (1971). An hypothesis for the formation of marginal seas in the Western Pacific. *Tectonophysics* **11**, 79–109.

—— and Leitch, E. C. (1974). The role of plate tectonic theory in the interpretation of The Tasman Orogenic Zone. In *The Tasman Geosyncline* (ed. A. K. Denmead, G. W. Tweedale, and A. F. Wilson), pp. 129–54. Geological Society of Australia, Queensland Division, Brisbane.

Page, R. W. (1976). Geochronology of igneous and metamorphic rocks in the New Guinea Highlands. *Aust., Bur. Miner. Resour., Geol. Geophys., Bull.* 162.

—— and Johnson, R. W. (1974). Strontium isotope ratios of Quaternary volcanic rocks from Papua New Guinea. *Lithos* **7**, 91–100.

—— and McDougall, I. (1972). Ages of mineralisation of gold and porphyry copper deposits in the New Guinea Highlands. *Econ. Geol.* **67**, 1034–48.

Pain, C. F. and Blong, R. J. (1979). The distribution of Tephras in the Papua New Guinea Highlands. *Search* **10**, 228–30.

Paine, A. G. L., Harding, R. R. and Clarke, D. E. (1971). Geology of the northeastern part of the Hughenden 1:250 000 Sheet area, Queensland. *Aust., Bur. Miner. Resour., Geol. Geophys., Rep.* 126.

——, Clarke, D. E. and Gregory, C. M. (1974). Geology of the northern half of the Bowen 1:250 000 Sheet area, Queensland. *Aust., Bur. Miner. Resour., Geol. Geophys., Rep.* 145.

Palfreyman, W. D., D'Addario, G. W., Swoboda, R. A., Bultitude, J. M. and Lamberts, I. T. (1976). *Map of the geology of Australia*. Aust., Bur. Miner. Resour., Geol. Geophys. Scale 1:2 500 000.

Palmer, A. R. (1983). The Decade of North American Geology 1983 geologic time scale. *Geology* **11**, 503–4.

Papalia, N. (1969). The Nappamerri Formation. *APEA J.* **9**, 108–10.

Parkin, L. E. (ed.) (1969). *Handbook of South Australian geology*. Geological Survey of South Australia, Adelaide.

Partridge, A. D. (1976). The geological expression of eustacy in the Early Tertiary of the Gippsland Basin. *APEA J.* **16**, 73–9.

Paten, R. J. and McDonagh, G. P. (1976). Bowen Basin. *Australas. Inst. Min. Metall., Monogr.* **7** (3), 403–20.

——, Brown, L. N. and Groves, R. D. (1979). Stratigraphic concepts and petroleum potential of the Denison Trough, Queensland. *APEA J.* **19**, 43–52.

Paton, I. M. (1982). The Birkhead Formation – a Jurassic petroleum reservoir. In *Eromanga Basin Symposium, summary papers* (ed. P. S. Moore and T. J. Mount), pp. 346–55. Geol. Soc. Aust. and Pet. Expl. Soc. Aust., Adelaide.

Pattinson, R., Watkins, G. and van den Abeele, D. (1976). Great Australian Bight Basin, South Australia. *Australas. Inst. Min. Metall., Monogr.* **7** (3), 98–104.

Petit, J. R., Briat, M. and Royer, A. (1981). Ice age aerosol content from East Antarctic ice core samples and past wind strength. *Nature* **293**, 391–4.

Phillips, E. R. (1968). Some plutonic rocks from a northern part of the New England Batholith. *Univ. Qld, Dept. Geol., Pap.* **6** (7), 159–206.

Phillips, G. N., Wall, V. J. and Clemens, J. D. (1981). Petrology of the Strathbogie batholith: a cordierite-bearing granite. *Can. Mineral.* **19**, 47–63.

Pickett, J. W. (1972). Late Devonian (Frasnian) conodonts from Ettrema, New South Wales. *R. Soc. N.S.W., J. Proc.* **105**, 31–7.

—— (1979). Marine fossils and the age of the Dulladerry Rhyolite. *N.S.W., Geol. Surv., Palaeont. Rep.* 79-23 (unpubl.).

—— (ed.), (1982). The Silurian System in New South Wales. *N.S.W., Geol. Surv., Bull.* 29.

Pidgeon, R. T. (1967). A rubidium–strontium geochronological study of the Willyama Complex, Broken Hill, Australia. *J. Petrol.* **8**, 283–324.

—— and Compston, W. (1965). The age and origin of the Cooma Granite and its associated metamorphic zones, New South Wales. *J. Petrol.* **6**, 193–222.

Pilger, R. H. (1982). The origin of hotspot traces: evidence from eastern Australia. *J. Geophys. Res.* **87**, 1825–34.

Pinchin, J. (1978). A seismic investigation of the eastern margin of the Galilee Basin, Queensland. *BMR J. Aust. Geol. Geophys.* **3**, 193–202.

—— and Senior, B. R. (1982). The Warrabin Trough, western Adavale Basin, Queensland. *Geol. Soc. Aust., J.* **29**, 413–24.

Pitman, W. C. (1978). Relationship between eustacy and stratigraphic sequences of passive margins. *Geol. Soc. Amer., Bull.* **89**, 1389–403.

Pitt, G. M. (1979). Palaeodrainage systems in western South Australia: their detection by LANDSAT imagery, stratigraphic significance and economic potential. *S. Aust., Dept. Mines Energ., Report Book* 79/114 (unpubl.).

——, Benbow, M. C. and Youngs, B. C. (1980). A review of recent geological work in the Officer Basin, South Australia. *APEA J.* **20**, 209–20.

Pittock, A. B., Frakes, L. A., Jenssen, D., Peterson, J. A. and Zillman, J. W. (1978). *Climatic change and variability: a southern perspective.* Cambridge U.P., Cambridge.

Playford, G. (1965). Plant microfossils from Triassic sediments near Poatina, Tasmania. *Geol. Soc. Aust., J.* **12**, 173–210.

—— and Cornelius, K. D. (1967). Palynological and lithostratigraphic features of the Razorback Beds, Mount Morgan district, Queensland. *Univ. Qld, Dept Geol., Pap.* **6** (3), 81–94.

——, Jones, B. G. and Kemp, E. M. (1976a). Palynological evidence for the age of the synorogenic Brewer Conglomerate, Amadeus Basin, Central Australia. *Alcheringa* **1**, 235–43.

Playford, P. E. (1980). Devonian 'Great Barrier Reef' of Canning Basin, Western Australia. *Am. Assoc. Pet. Geol., Bull.* **64**, 814–40.

—— (1982). Devonian reef prospects in the Canning Basin: implications of the Blina oil discovery. *APEA J.* **22**, 258–71.

——, Cope, R. N., Cockbain, A. E., Low, G. H. and Lowry, D. C. (1975). Phanerozoic. *West. Aust., Geol. Surv., Mem.* **2**, 223–432.

——, Cockbain, A. E. and Low, G. H. (1976b). Geology of the Perth Basin, Western Australia. *W. Aust., Geol. Surv., Bull.* 124.

Plumb, K. A. (1971). The Archaean and Australian tectonics. *Geol. Soc. Aust., Spec. Publ.* **3**, 385.

—— (1979). The tectonic evolution of Australia. *Earth-Sci. Rev.* **14**, 205–49.

—— and Gemuts, I. (1976). Precambrian geology of the Kimberley Region, Western Australia. *Int. Geol. Congr., 25th, Excursion Guide,* 44C.

——, Shergold, J. H. and Stefanski, M. Z. (1976). Significance of Middle Cambrian trilobites from Elcho Island, Northern Territory. *BMR. J. Aust. Geol. Geophys.* **1**, 51–5.

——, Derrick, G. M., Needham, R. S. and Shaw, R. D. (1981). The Proterozoic of northern Australia. In *Precambrian of the Southern Hemisphere* (ed. D. R. Hunter), pp. 205–307. Elsevier, Amsterdam.

Pogson, D. J. and Hilyard, D. (1981). Results of isotopic age dating related to Geological Survey of N.S.W. investigations, 1974–1978. *N.S.W., Geol. Surv., Rec.* **20**, 251–72.

Pogson, D. J. and Scheibner, E. (1971). Pre-Upper Cambrian sediments east of the Copper Mine Range, New South Wales. *N.S.W., Geol. Surv., Q. Notes* **4**, 3–8.

Powell, B. N. and Phelps, D. W. (1977). Igneous cumulates of the Wichita province and their tectonic implications. *Geology* **5**, 52–6.

Powell, B. S., Gibson, D. L., Smart, J., Grimes, K. G. and Doutch, H. F. (1976). New and revised stratigraphic nomenclature, Cape York Peninsula. *Qld Gov. Min. J.* **77**, 179–88.

Powell, C. McA. (1976a). A critical appraisal of the tectonic history of the Hill End Trough and its margins. *Aust. Soc. Explor. Geophys., Bull.* **7**, 14–18.

—— (1979). A speculative tectonic history of Pakistan and surroundings: some constraints from the Indian Ocean. *Pak. Geol. Surv., Mem.* **11**, 5–24.

—— (1983a). Geology of the N.S.W. South Coast and adjacent Victoria with emphasis on the pre-Permian structural history. *Geol. Soc. Aust., Specialist Group Tectonics and Structural Geology, Field Guide* 1.

—— (1983b). Tectonic relationship between the Late Ordovician and Late Silurian palaeogeographies of southeastern Australia. *Geol. Soc. Aust., J.* **30**, 353–73.

—— and Edgecombe, D. R. (1978). Mid-Devonian movements in the northeastern Lachlan Fold Belt. *Geol. Soc. Aust., J.* **25**, 165–84.

—— and Fergusson, C. L. (1979a). The relationship of structures across the Lambian unconformity near Taralga, New South Wales. *Geol. Soc. Aust., J.* **26**, 209–19.

—— —— (1979b). Analysis of the angular discordance across the Lambian Unconformity in the Kowmung River–Murruin Creek area, eastern N.S.W. *R. Soc. N.S.W., J. Proc.* **112**, 37–42.

—— and Johnson, B. D. (1980). Constraints on the Cenozoic position of Sundaland. *Tectonophysics* **63**, 91–109.

——, Edgecombe, D. R., Henry, N. M. and Jones, J. G. (1977). Timing of regional deformation of the Hill End Trough: a reassessment. *Geol. Soc. Aust., J.* **23**, 407–21.

——, Gilfillan, M. A. and Henry, N. M. (1978). Early east-southeast trending folds in the Sofala Volcanics, New South Wales. *Roy. Soc. N.S.W., J. Proc.* **111**, 121–8.

——, Fergusson, C. L. and Williams, A. J. (1980a). Structural relationships across the Lambian Unconformity in the Hervey Range–Parkes area, N.S.W. *Linn. Soc. N.S.W., Proc.* **104**, 195–204.

——, Johnson, B. D. and Veevers, J. J. (1980b). Constraints on the positions of India, Australia and

Southeast Asia since the Late Cretaceous. *Southeast Asia Petroleum Exploration Society, Proc.* **5**, 82–9.

—— —— —— (1980c). Early Cretaceous breakup of eastern Gondwanaland, the separation of Australia and India, and their interaction with southeast Asia. In *Ecological Biogeography* (ed. A. Keast), pp. 17–29. Junk, The Hague.

—— —— —— (1980d). A revised fit of East and West Gondwanaland. *Tectonophysics* **63**, 13–29.

——, Neef, G., Crane, D., Jell, P. A. and Percival, G. (1982). Significance of Late Cambrian (Idamean) fossils in the Cupola Creek Formation, northwestern New South Wales. *Linn. Soc. N.S.W., Proc.* **106**, 127–50.

——, Scheibner, E. and Khaiami, R. (1983). Winduck Group; Mulga Downs Group. In *Geology of the Mount Allen 1:100 000 Sheet 8032.* N.S.W., Geol. Surv., Explan. Notes (In press).

Powell, D. E. (1976b). The geological evolution of the continental margin of northwest Australia. *APEA J.* **16**, 13–23.

—— (1982). The northwest Australian continental margin. *R. Soc. Lond., Philos. Trans.,* **A 305**, 45–62.

Powell, T. S. (1978). *The sea-floor spreading history of the eastern Indian Ocean.* M.A. thesis, Univ. California, Santa Barbara (unpubl.).

Preiss, W. V., Walter, M. R., Coats, R. P. and Wells, A. T. (1978). Lithological correlations of Adelaidean glaciogenic rocks in parts of the Amadeus, Ngalia, and Georgina Basins. *BMR J. Aust. Geol. Geophys.* **3**, 45–53.

Price, R. C. and Taylor, S. R. (1977). The rare-earth element geochemistry of granite, gneiss and migmatite from the western metamorphic belt of southeastern Australia. *Contrib. Mineral. Petrol.* **62**, 249–63.

Pruatt, M. A. (1975). The Southern Oklahoma Aulacogen: a geophysical and geological investigation. University of Oklahoma, M.S. thesis.

Quilty, P. G. (1972). The biostratigraphy of the Tasmanian marine Tertiary. *R. Soc. Tas., Pap. Proc.* **106**, 25–44.

—— (1973). Cenomanian–Turonian and Neogene sediments from northeast of Kerguelen Ridge, Indian Ocean. *Geol. Soc. Aust. J.* **20**, 361–71.

—— (1974). Tertiary stratigraphy of Western Australia. *Geol. Soc. Aust., J.* **21**, 301–18.

—— (1977). Cenozoic sedimentation cycles in Western Australia. *Geology* **5**, 336–40.

—— (1978a). The Late Cretaceous–Tertiary section in Challenger No. 1 (Perth Basin) – details and implications. *Aust., Bur. Miner. Resour., Geol. Geophys., Bull.* **192**, 109–35.

—— (1978b). Late Jurassic bivalves, Ellsworth Land Antarctica: their systematics and paleogeographic implications. *N.Z. J. Geol. Geophys.* **20**, 1033–80.

—— (1980a). New rotalid foraminiferids from the Oligo-Miocene of Tasmania. *Alcheringa* **4**, 299–311.

—— (1980b). Sedimentation cycles in the Cretaceous and Cenozoic of Western Australia. *Tectonophysics* **63**, 349–66.

—— (1984). Mesozoic and Cenozoic history of Australia as it affects the Australian biota. In *Arid Australia* (ed. H. G. Cogger and E. Cameron). Australian Museum, Sydney (in press).

Qureshi, I. R. (1984). Tectonic significance of the Wollondilly–Blue Mountains gravity gradient zone, N.S.W. *Geol. Soc. Aust. J.* (in press).

Raheim, A. and Compston, W. (1977). Correlations between metamorphic events and Rb/Sr ages in metasediments and eclogite from western Tasmania. *Lithos* **10**, 271–89.

Rao, C. P. (1981a). Criteria for recognition of cold-water carbonate sedimentation. Berriedale Limestone (Lower Permian), Tasmania, Australia. *J. Sediment. Petrol.* **51**, 491–506.

—— (1981b). Geochemical differences between tropical (Ordovician) and subpolar (Permian) carbonates, Tasmania, Australia. *Geology* **9**, 205–9.

—— and Green, D. C. (1982). Oxygen and carbon isotopes of Early Permian cold-water carbonates, Tasmania, Australia. *J. Sediment. Petrol.* **52**, 1111–25.

Rattigan, J. H. (1964). Occurrence and stratigraphic position of Carboniferous coals in the Hunter Valley, N.S.W. *Aust. J. Sci.* **27**, 82.

—— (1966). Cyclic sedimentation in the Carboniferous continental Kuttung Facies, New South Wales, Australia. *R. Soc. N.S.W., J. Proc.* **100**, 119–28.

—— (1967). Fold and fracture patterns resulting from basement wrenching in the Fitzroy depression, Western Australia. *Australas. Inst. Min. Metall., Proc.* **9**, 17–22.

Retallack, G. (1980). Late Carboniferous to Middle Triassic megafossil floras from the Sydney Basin. *N.S.W., Geol. Surv., Bull.* **26**, 384–430.

Richards, D. N. G. (1980). Palaeozoic granitoids of Northeastern Australia. In *The geology and geophysics of Northeastern Australia* (ed. R. A. Henderson and P. J. Stephenson), pp. 229–46. Geological Society of Australia, Queensland Division, Brisbane.

Richards, J. R. and Pidgeon, R. T. (1963). Some age measurements on micas from Broken Hill, Australia. *Geol. Soc. Aust., J.* **10**, 243–60.

—— and Singleton, O. P. (1981). Palaeozoic Victoria,

Australia: igneous rocks, ages and their interpretation. *Geol. Soe. Aust., J.* **28**, 395–421.
—— and Willmott, W. F. (1970). K–Ar age of biotites from Torres Strait. *Aust. J. Sci.* **32**, 369–70.
——, Barkas, J. P. and Vallance, T. G. (1977). A Lower Devonian point in the geological timescale. *Geochem. J.* **11**, 147–53.
Ridd, M. F. (1976). Papuan Basin – onshore. *Australas. Inst. Min. Metall. Monogr.* **7** (3), 478–94.
Ritchie, A. (1973). *Wuttagoonaspis* gen. nov., an unusual arthrodire from the Devonian of western New South Wales, Australia. *Palaeontogr., Abt. A* **143**, 58–72.
Rixon, L. K. (1978). Clay modelling of the Fitzroy Graben. *BMR J. Aust. Geol. Geophys.* **3**, 71–6.
Roberts, J. (1981). Control mechanisms of Carboniferous brachiopod zones in eastern Australia. *Lethaia* **14**, 123–34.
—— and Engel, B. A. (1980). Carboniferous palaeogeography of the Yarrol and New England Orogens, eastern Australia. *Geol. Soc. Aust., J.* **27**, 167–86.
—— and Veevers, J. J. (1973). Summary of BMR studies of the onshore Bonaparte Gulf Basin 1963–71. *Aust., Bur. Miner. Resour., Geol. Geophys., Bull.* **139**, 29–58.
——, Jones, P. J., Jell, J. S., Jenkins, T. B. H., Marsden, M. A. H., McKellar, R. G., McKelvey, B. C. and Seddon, G. (1972). Correlation of the Upper Devonian rocks of Australia. *Geol. Soc. Aust., J.* **18**, 467–90.
Robertson, A. D. (1979). Revision of the Cainozoic geology between the Kolan and Elliott Rivers. *Qld Gov. Min. J.* **80**, 350–63.
Robinson, P. T., Thayer, P. A., Cook, P. J. and McKnight, B. K. (1974). Lithology of Mesozoic and Cenozoic sediments of the Eastern Indian Ocean, Leg 27, Deep Sea Drilling Project. *Deep Sea Dril. Proj., Initial Rep.* **27**, 1001–47.
Robinson, V. A. (1974). Geologic history of the Bass Basin. *APEA J.* **14**, 45–9.
Roddick, J. C. and Compston, W. (1976). Radiometric evidence for the age of emplacement and cooling of the Murrumbidgee Batholith. *Geol. Soc. Aust., J.* **23**, 223–33.
Roggenthen, W., Fischer, A. G., Napoleone, G. and Fischer, J. F. (1976). Paleomagnetism and age of Wichita Mountains basement. *Geol. Soc. Am., Abstr. Programs* **8**, 62.
Rognon, P. and Williams, M. A. J. (1977). Late Quaternary climatic changes in Australia and North Africa: a preliminary interpretation. *Palaeogeogr., Palaeoclim., Palaeoecol.* **21**, 285–327.
Roots, W. D. (1976). Magnetic smooth zones and slope anomalies: a mechanism to explain both. *Earth Planet. Sci. Lett.* **31**, 113–18.
——, Veevers, J. J. and Clowes, D. F. (1979). Lithospheric model with thick oceanic crust at the continental boundary: a mechanism for shallow spreading ridges in young oceans. *Earth Planet. Sci. Lett.* **43**, 417–33.
Rose, G. and Brunker, R. L. (1969). The Upper Proterozoic and Phanerozoic geology of northwestern New South Wales. *Australas. Inst. Min. Metall., Proc.* **229**, 105–20.
Ross, R. J., Naeser, C. W. and Lambert, R. S. (1978). Ordovician geochronology. *Am. Assoc. Pet. Geol., Stud. Geol.* **6**, 347–54.
Runnegar, B. N. (1970). The Permian faunas of northern New South Wales and the connection between the Sydney and Bowen Basins. *Geol. Soc. Aust., J.* **16**, 697–710.
—— (1980). Biostratigraphy of the Shoalhaven Group. *N.S.W., Geol. Surv., Bull.* **26**, 376–82.
Russell, H. Y. (1976). The geometry of the Gowan Green Overfold and its relationship to the regional deformational history. *Aust. Soc. Explor. Geophys., Bull.* **7**, 22–5.
Russell, T. G. (1980). A clast fabric palaeocurrent study of the Late Devonian Keepit Conglomerate, northeastern New South Wales. *R. Soc. N.S.W., J.* **113**, 35–47.
Rust, B. R. (1981). Sedimentation in an arid-zone anastomosing fluvial system: Cooper's Creek, Central Australia. *J. Sed. Petrol.* **51**, 745–55.
Rutland, R. W. R. (1973). Tectonic evolution of the continental crust of Australia. In *Implications of continental drift to the earth sciences* (ed. D. H. Tarling and S. K. Runcorn), pp. 1011–33. Academic Press, London.
—— (1976). Orogenic evolution of Australia. *Earth-Sci. Rev.* **12**, 161–96.
—— (1981). Structural framework of the Australian Precambrian. In *Precambrian of the southern hemisphere* (ed. D. R. Hunter), pp. 1–32. Elsevier, Amsterdam.
—— (1982). On the growth and evolution of continental crust: a comparative tectonic approach. *R. Soc. N.S.W., J. Proc.* **115**, 33–60.
——, Parker, A. J., Pitt, G. M., Preiss, W. V. and Murrell, B. (1981). The Precambrian of South Australia. In *Precambrian of the Southern Hemisphere* (ed. D. R. Hunter), pp. 309–60. Elsevier, Amsterdam.
Ruxton, B. P. and McDougall, I. (1967). Denudation rates in northeast Papua from potassium–argon dating of lavas. *Am. J. Sci.* **265**, 545–61.

Salinger, M. J. (1981). Palaeoclimates north and south. *Nature* **291**, 106-7.

Sass, J. H. and Lachenbruch, A. H. (1979). Thermal regime of the Australian continental crust. In *The Earth: its origin, structure, and evolution* (ed. M. W. McElhinny), pp. 301-51. Academic Press, New York.

Scheibner, E. (1972). Tectonic concepts and tectonic mapping. *Geol. Surv. N.S.W., Rec.* **14** (1), 37-85.

—— (1973). *Geology of the Taralga 1:100 000 Sheet, 8829.* N.S.W., Geol. Surv., Sydney.

—— (1974a). Fossil fracture zones, segmentation and correlation problems in the Tasman Fold Belt System. In *The Tasman Geosyncline* (ed. A. K. Denmead, G. W. Tweedale, and A. F. Wilson), pp. 65-98. Geological Society of Australia, Queensland Division, Brisbane.

—— (1974b). A plate tectonic model of the Palaeozoic tectonic history of New South Wales. *Geol. Soc. Aust., J.* **20**, 405-26.

—— (1976). *Explanatory notes on the tectonic map of New South Wales.* N.S.W., Geological Survey, Sydney.

—— and Glen, R. A. (1972). The Peel Thrust and its tectonic history. *N.S.W., Geol. Surv., Q. Notes* **8**, 2-14.

—— and Pearce, J. A. (1978). Eruptive environments and inferred exploration potential of metabasalts from New South Wales. *J. Geochem. Explor.* **10**, 63-74.

Scheibnerova, V. (1976). Cretaceous Foraminifera of the Great Artesian Basin. *N.S.W., Geol. Surv., Pal. Mem.* 17.

Schleiger, N. W. (1968). Orientation distribution patterns of graptolitic rhabdosomes from Ordovician sediments in central Victoria, Australia. *J. Sediment. Petrol.* **38**, 462-72.

—— (1969). Problems in sampling the orientation of fossils in a graptolite band at Eaglehawk, Victoria. *R. Soc. Vic., Proc.* **82**, 161-77.

—— (1974). Statistical methods for analysis and mapping of flysch-type sediments. *Sedimentology* **21**, 223-49.

Schmidt, P. W. (1982). Linearity spectrum analysis of multicomponent magnetisations and its application to some igneous rocks from south-eastern Australia. *Geophys. J.R. Astr. Soc.* **70**, 647-65.

—— and Embleton, B. J. J. (1976). Paleomagnetic results from sediments of the Perth Basin Western Australia, and their bearing on the time of regional lateritisation. *Palaeogeog., Palaeoclimatol., Palaeoecol.* **19**, 257-73.

—— —— (1981). Magnetic overprinting in southeastern Australia and the thermal history of its rifted margin. *J. Geophys. Res.* **86**, 3998-4008.

—— and McDougall, I. (1977). Palaeomagnetic and potassium-argon dating studies of the Tasmania Dolerites. *Geol. Soc. Aust. J.* **24**, 321-8.

Schuepbach, M. A. and Vail, P. R. (1980). Evolution of outer highs on divergent continental margins. In *Continental Tectonics* (ed. B. C. Burchfiel, J. E. Oliver, and L. T. Silver), pp. 50-61. U.S. National Academy of Sciences, Washington.

Schult, A., Hussain, A. G. and Soffel, H. C. (1981). Paleomagnetism of Upper Cretaceous volcanics and Nubian Sandstones of Wadi Natash, S.E. Egypt and implications for the polar wander path for Africa in the Mesozoic. *J. Geophys.* **50**, 16-22.

Sclater, J. G. and Fisher, R. L. (1974). Evolution of the east central Indian Ocean, with emphasis on the tectonic setting of the Ninetyeast Ridge. *Geol. Soc. Am. Bull.* **85**, 683-702.

Scotese, C. R., Bamback, R. K., Barton, C., Van der Voo, R., and Ziegler, A. M. (1979). Paleozoic base maps. *J. Geol.,* **87**, 217-77.

Seibold, E., Exon, N. F., Hartmann, M., Kogler, F. C., Krumm, H., Lutze, G. F., Newton, R. S. and Werner, F. (1971). *Marine geology of Kiel Bay.* VII Int. Sedimentological Congress: guidebook to sedimentology of parts of central Europe, 209-35.

Semeniuk, V. and Johnson, D. P. (1982). Recent and Pleistocene beach/dune sequences, Western Australia. *Sedimentary Geol.* **32**, 301-28.

Sengor, A. M. C. and Burke, K. (1978). Relative timing of rifting and volcanism on earth and its tectonic implications. *Geophys. Res. Lett.* **5**, 419-21.

Senior, B. R., Mond, A. and Harrison, P. L. (1978). Geology of the Eromanga Basin. *Aust., Bur. Miner. Resour., Geol. Geophys., Bull.* 167.

——, McColl, D. H., Long, B. E. and Whiteley, R. J. (1977). The geology and magnetic characteristics of precious opal deposits, southwest Queensland. *BMR J. Aust. Geol. Geophys.* **2**, 241-51.

Shackleton, N. J. and Kennett, J. P. (1975). Paleotemperature history of the Cenozoic and the initiation of Antarctic glaciation: oxygen and carbon isotope analysis in DSDP sites 277, 279, and 281. In *Deep Sea Dril. Proj. Initial Rep.,* **29**, 743-55.

—— and Opdyke, N. D. (1977). Oxygen isotope and palaeomagnetic evidence for early Northern Hemisphere glaciation. *Nature* **270**, 216-19.

Shaw, R. D. (1978). Sea floor spreading in the Tasman Sea: a Lord Howe Rise-Eastern Australia reconstruction. *Aust. Soc. Explor. Geophys., Bull.* **9**, 75-81.

Shaw, S. E. and Flood, R. H. (1981). The New England Batholith, eastern Australia: geochemical variations

in time and space. *J. Geophys. Res.* **86,** 10530–44.
—— —— and Riley, G. H. (1982a). The Wologorong Batholith, New South Wales, and the extension of the I–S line of the Siluro-Devonian granitoids. *Geol. Soc. Aust., J.* **29,** 41–8.
—— —— and Vernon, R. H. (1982b). Permian volcanism associated with the New England Batholith. *Geol. Soc. Aust., Abstr.* **7,** 7–8.
Shell Development (Australia) Pty Ltd (1976). Potoroo No. 1 Well completion report. *Aust., Bur. Miner. Resour., Geol. Geophys., Petroleum Submerged Lands Act Report* 75/152 (unpubl.).
Shepherd, J. and Huntington, J. F. (1981). Geological fracture mapping in coalfields and the stress fields of the Sydney Basin. *Geol. Soc. Aust., J.* **28,** 299–309.
Shepherd, R. G. (1978). Underground water resources of South Australia. *S. Aust., Geol. Surv., Bull.* 48.
Sheraton, J. W. and Cundari, A. (1980). Leucitites from Gaussberg, Antarctica. *Contrib. Mineral. Petrol.* **71,** 417–27.
Shergold, J. H. and Druce, E. C. (1980). Upper Proterozoic and Lower Palaeozoic rocks of the Georgina Basin. In *The geology and geophysics of Northeastern Australia* (ed. R. A. Henderson and P. J. Stephenson), pp. 149–74. Geological Society of Australia, Queensland Division, Brisbane.
Sherwin, L. (1971). Stratigraphy of the Cheesemans Creek district, New South Wales. *N.S.W., Geol. Surv., Rec.* **13,** 199–234.
—— (1979). Age of the Nelungoloo Volcanics near Parkes. *N.S.W. Geol. Surv., Q. Notes* **35,** 15–18.
—— (1980). Faunal correlation of the Siluro-Devonian units, Mineral Hill–Trundle–Peak Hill area. *N.S.W., Geol. Surv., Q. Notes* **39,** 1–14.
Shibaoka, M. and Bennett, A. J. R. (1976). Effect of depth of burial and tectonic activity on coalification. *Nature* **259,** 385–6.
—— —— and Gould, K. W. (1973). Diagenesis of organic matter and occurrence of hydrocarbons in some Australian sedimentary basins. *APEA J.* **13,** 73–80.
Singh, G. (1981). Late Quaternary pollen records and seasonal palaeoclimates of Lake Frome, South Australia. *Hydrobiologia* **82,** 419–30.
Singleton, O. P., McDougall, I. and Mallett, C. W. (1976). The Plio-Pleistocene boundary in south-eastern Australia. *Geol. Soc. Aust., J.* **23,** 299–311.
Skwarko, W. K. (1966). Cretaceous stratigraphy and palaeontology of the Northern Territory. *Aust., Bur. Miner. Resour., Geol. Geophys., Bull.* 73.
—— (1968). Lower Cretaceous Trigoniidae from Stanwell, eastern Queensland. *Aust., Bur. Miner. Resour., Geol. Geophys., Bull.* 80.
——, Nicoll, R. S. and Campbell, K. S. W. (1976). The Late Triassic molluscs, conodonts and brachiopods of the Kuta Formation, Papua New Guinea. *BMR J. Aust. Geol. Geophys.* **1,** 219–30.
Smart, J. and Rasidi, J. S. (1979). Geology and petroleum potential of the Laura Basin, Torres Shelf and Papuan Basin, Queensland. *Qld Gov. Min. J.* **80,** 267–79.
—— and Senior, B. R. (1980). Jurassic–Cretaceous basins of northeastern Australia. In *The geology and geophysics of Northeastern Australia* (ed. R. A. Henderson and P. J. Stephenson), pp. 315–28. Geological Society of Australia, Queensland Division, Brisbane.
——, Grimes, K. G., Doutch, H. F. and Pinchin, J. (1980). The Carpentaria and Karumba Basins, north Queensland. *Aust., Bur. Miner. Resour., Geol. Geophys., Bull.* 202.
Smith, A. G. and Hallam, A. (1970). The fit of the southern continents. *Nature* **225,** 139–44.
Smith, J. G. (1968a). Tectonics of the Fitzroy wrench trough, Western Australia. *Am. J. Sci.* **266,** 766–76.
Smith, R. E. (1966). The geology of Mandurama-Panuara. *R. Soc. N.S.W., J. Proc.* **98,** 239–62.
—— (1968b). Redistribution of major elements in the alteration of some basic lavas during burial metamorphism. *J. Petrol.* **9,** 191–219.
Soil Map of New Guinea (1967). Compiled by CSIRO, Canberra, 1962–5, and by State University, Utrecht. In H. A. Haantjens *et al.*, op.cit., looseleaf folder.
Solomon, M. and Griffiths, J. R. (1972). Tectonic evolution of the Tasman Orogenic Zone, eastern Australia. *Nature Phys. Sci.* **237,** 3–6.
—— —— (1974). Aspects of the early history of the southern part of the Tasman Orogenic Zone. In *The Tasman Geosyncline* (ed. A. K. Denmead, G. W. Tweedale, and A. F. Wilson), pp. 19–46. Geological Society of Australia, Queensland Division, Brisbane.
Southgate, P. N. (1982). Cambrian skeletal halite crystals and experimental analogues. *Sedimentology* **29,** 391–407.
Spencer-Jones, D. (1965). The geology and structure of the Grampians area, western Victoria. *Vic., Geol. Surv., Mem.* 25.
Spry, A. (1962). Igneous activity. *Geol. Soc. Aust., J.* **9,** 255–85.
Stagg, H. M. J. and Exon, N. F. (1979). Western margin of Australia: evolution of a rifted arch system: Discussion. *Geol. Soc. Am., Bull.* **90,** 795–7.
—— —— (1981). Geology of the Scott Plateau and Rowley Terrace. *Aust., Bur. Miner. Resour., Geol. Geophys., Bull.* 213.

Staines, H. R. E. and Koppe, W. H. (1980). The geology of the north Bowen Basin. In *The geology and geophysics of Northeastern Australia* (ed. R. A. Henderson and P. J. Stephenson), pp. 279-98. Geological Society Australia, Queensland Division, Brisbane.

Stanton, R. L. (1974). The development of ideas on the evolution of mineralization in the Tasman Geosyncline. In *The Tasman Geosyncline* (ed. A. K. Denmead, G. W. Tweedale and A. F. Wilson), pp. 185-219. Geological Society of Australia, Queensland Division, Brisbane.

Stauffer, M. R. and Rickard, M. J. (1966). The establishment of recumbent folds in the Lower Palaeozoic near Queanbeyan, New South Wales. *Geol. Soc. Aust., J.* **13**, 419-38.

Steed, R. H. N. and Drewry, D. J. (1982). Radio echo sounding investigations of Wilkes Land, Antarctica. In *Antarctic Geoscience* (ed. C. Craddock), pp. 969-76. University of Wisconsin Press, Madison.

Steiger, R. H. and Jager, E. (1978). Subcommission on Geochronology: convention on the use of decay constants in geochronology and cosmochronology. *Am. Assoc. Pet. Geol., Stud. Geol.* **6**, 67-71.

Steiner, J. (1972). The eruptive history and geological environment of the Devonian extrusive rocks, Eden, New South Wales. *Geol. Soc. Aust., J.* **19**, 261-71.

—— (1975). The Merrimbula Group of the Eden-Merrimbula area, N.S.W. *R. Soc. N.S.W., J. Proc.* **108**, 37-51.

Stephenson, P. J., Griffin, T. J. and Sutherland, F. L. (1980). Cainozoic volcanism in northeastern Australia. In *The geology and geophysics of Northeastern Australia* (ed. R. A. Henderson and P. J. Stephenson), pp. 349-74. Geological Society of Australia, Queensland Division, Brisbane.

Stevens, G. R. and Clayton, R. N. (1971). Oxygen isotope studies on Jurassic and Cretaceous belemnites from New Zealand and their biogeographic significance. *N.Z. J. Geol. Geophys.* **14**, 829-97.

Stewart, A. J. (1971). Potassium-argon dates from the Arltunga Nappe Complex, Northern Territory. *Geol. Soc. Aust., J.* **17**, 205-11.

Stipp, J. J. and McDougall, I. (1968). Potassium-argon ages from the Nandewar volcano, near Narrabri, New South Wales. *Aust. J. Sci.* **31**, 84-5.

Stirton, R. A., Tedford, R. H. and Woodburne, M. O. (1967). A new Tertiary formation and fauna from the Tirari Desert, South Australia. *South Aust. Museum, Rec.* **15**, 427-62.

Stock, J. and Molnar, P. (1982). Uncertainties in the relative positions of the Australia, Antarctica, Lord Howe, and Pacific Plates since the Late Cretaceous. *J. Geophys. Res.* **87**, 4697-714.

Street, F. A. (1981). Tropical palaeoenvironments. *Progress in Physical Geography* **5**, 157-85.

Strusz, D. L. (1971). *Explanatory notes, Canberra 1:250 000 Geological Sheet SI 55-16.* Aust.,. Bur. Miner. Resour., Geol. Geophys., Canberra.

Stuart, W. J. (1970). The Cainozoic stratigraphy of the eastern coastal area of Yorke Peninsula, South Australia. *R. Soc. S. Aust., Trans.* **94**, 151-78.

Stump, E. (1981). Observations on the Ross Orogen, Antarctica. In *Gondwana Five* (ed. M. M. Cresswell and P. Vella), pp. 205-8. Balkema, Rotterdam.

——, Sheridan, M. F., Borg, S. G. and Sutter, J. F. (1980). Early Miocene subglacial basalts, the East Antarctic ice sheet, and uplift of the Transantarctic Mountains. *Science* **207**, 757-9.

Stuntz, J. (1972). The subsurface distribution of the Upper Coal Measures, Sydney Basin, New South Wales. *Australas. Inst. Min. Metall., Conf. Newcastle,* 1-9.

Stupavsky, M., Symons, D. T. A. and Gravenor, C. P. (1982). Caledonian remagnetization of the Dalradian tillite, Scotland: evidence against world-wide Late Precambrian glaciation. *EOS* **63** (18), 306.

Suggate, R. P. (ed.) (1978). *The Geology of New Zealand.* Government Printer, Wellington.

Sutherland, F. L. (1973). The shoshonitic association of the Upper Mesozoic of Tasmania. *Geol. Soc. Aust., J.* **19**, 487-96.

—— (1978). Mesozoic-Cainozoic volcanism of Australia. *Tectonophysics* **48**, 413-27.

—— (1981). Migration in relation to possible tectonic and regional controls in Eastern Australian volcanism. *J. Volcanol. Geotherm. Res.* **9**, 181-213.

—— and Hollis, J. D. (1982). Mantle-lower crust petrology from inclusions in basaltic rocks in eastern Australia – an outline. *Tectonophysics* **14**, 1-29.

——, Green, D. C. and Wyatt, B. W. (1973). Age of the Great Lake basalts, Tasmania, in relation to Australian Cainozoic volcanism. *Geol. Soc. Aust., J.* **20**, 85-94.

——, Stubbs, D., and Green, D. C. (1978). K-Ar ages of Cainozoic volcanic suites, Bowen-St Lawrence hinterland, North Queensland (with some implications for petrologic models). *Geol. Soc. Aust., J.* **24**, 447-60.

Swindon, V. G. (1960). Marburg Sandstone of the type area. *Geol. Soc. Aust., J.* **7**, 288-90.

Symonds, P. A. (1973). Map of sediment thickness. *Aust., Bur. Min. Resour., Geol. Geophys., Rec.* 1973/195 (unpubl.).

—— and Willcox, J. B. (1976). The gravity field of offshore Australia. *BMR J. Aust. Geol. Geophys.*

REFERENCES

1, 303–14.

Tahirkheli, R. A. K., Mattauer, M., Proust, F. and Tapponnier, P. (1979). The India Eurasia suture zone in Northern Pakistan: Synthesis and interpretation of recent data at plate scale. In *Geodynamics of Pakistan* (ed. A. Farah and K. A. De Jong), pp. 125–30. Geol. Surv. Pakistan, Quetta.

Talent, J. A. (1965). The stratigraphic and diastrophic evolution of central and eastern Victoria in Middle Palaeozoic times. *R. Soc. Vic., Proc.* **79**, 179–95.

—— (1969). Geology of east Gippsland. *R. Soc. Vic., Proc.* **82**, 37–60.

——, Berry, W. B. N. and Boucot, A. J. (1975). Correlation of the Silurian rocks of Australia, New Zealand, and New Guinea. *Geol. Soc. Am., Spec. Pap.* 150.

Tallis, N. C. (1975). Development of the Tertiary offshore Papuan Basin. *APEA J.* **15**, 55–60.

Talwani, M., Mutter, J., Houtz, R. and Konig, M. (1979). The crustal structure and evolution of the area underlying the magnetic quiet zone on the margin south of Australia. *Am. Assoc. Pet. Geol. Mem.* **19**, 151–75.

Tanner, J. J. (1976). Devonian of the Adavale Basin, Queensland. In *International symposium on the Devonian System, II* (ed. D. H. Oswald), pp. 111–16. Alberta Society of Petroleum Geologists, Calgary.

Tarling, D. H. (1972). Another Gondwanaland. *Nature* **238**, 92–3.

—— (1974). A palaeomagnetic study of Eocambrian tillites in Scotland. *Geol. Soc. Lond., J.* **130**, 163–77.

Taylor, L. and Falvey, D. (1977). Queensland Plateau and Coral Sea Basin: stratigraphy, structure and tectonics. *APEA J.* **17**, 13–29.

Taylor, T. G. (1911). A discussion of salient features in the physiography of eastern Australia. *Commonwealth Bur. Met., Bull.* 8.

Teichert, C. (1939). The Mesozoic transgressions in Western Australia. *Aust. J. Sci.* **2**, 84–6.

Thom, B. G. and Chappell, J. (1975). Holocene sea levels relative to Australia. *Search* **6**, 90–3.

Thomas, B. M. (1978). Robe River – an onshore shallow oil accumulation. *APEA J.* **18**, 3–12.

—— and Smith, D. N. (1974). A summary of the petroleum geology of the Carnarvon Basin. *APEA J.* **14**, 66–76.

——, Osborne, D. G. and Wright, A. J. (1982). Hydrocarbon habitat of the Surat/Bowen Basin. *APEA J.* **22**, 213–26.

Thomson, B. P., Daily, B., Coats, R. P. and Forbes, B. G. (1976). Late Precambrian and Cambrian geology of the Adelaide 'Geosyncline' and Stuart Shelf, South Australia. *Int. Geol. Congr.*, 25, *Excursion Guide* 33A.

Thomson, J. (1975). Results of radiometric dating programme, 1971–1973. *N.S.W., Geol. Surv., Rec.* **16**, 239–44.

Thompson, R. and Clark, R. M. (1982). A robust least squares Gondwanan apparent polar wander path and the question of palaeomagnetic assessment of Gondwanan reconstruction. *Earth Planet. Sci. Lett.* **57**, 152–8.

Thornton, R. C. N. (1972). Lower Cretaceous sedimentary units beneath the western Murray Basin. *S. Aust., Geol. Surv., Q. Geol. Notes* **44**, 5–11.

—— (1976). Murray Basin and associated infrabasins. *Australas. Inst. Min. Metall., Monogr.* **7** (3), 91–4.

—— (1979). Regional stratigraphic analysis of the Gidgealpa Group, Southern Cooper Basin, Australia. *S. Aust., Geol. Surv., Bull.* 49.

Threlfall, W. F., Brown, B. R. and Griffith, B. R. (1976). Gippsland Basin, off-shore. *Australas. Inst. Min. Metall. Monogr.* **7** (3), 41–67.

Times Concise Atlas of the World (1972). Times Newspapers Limited, London, and John Bartholomew and Sons Limited, Edinburgh.

Times Atlas of the World (1975). Comprehensive edition, 5th edition. Times Newspapers Limited, London.

Townrow, J. A. (1964). A speculation on the Rhaeto-Liassic climate of Tasmania. *R. Soc. Tas., Pap. Proc.* **98**, 113–18.

Townsend, I. J. (1979). The correlation and depositional history of the Leigh Creek coal measures. *S. Aust., Geol. Surv., Q. Geol. Notes*, **70**, 5–10.

Trendall, A. F. (1974). The age of a granite near Mount Crofton, Paterson Range Sheet. *West Aust., Geol. Surv., Annu. Rep.* for 1973, 92–6.

—— (1975). Precambrian, Introduction. *W. Aust., Geol. Surv., Mem.* **2**, 25–32.

Truswell, E. M. (1978). Palynology of the Permo-Carboniferous in Tasmania: an interim report. *Tas., Geol. Surv., Bull.* 56.

—— (1982). Palynology of seafloor samples collected by the 1911–14 Australasian Antarctic Expedition: implications for the geology of coastal East Antarctica. *Geol. Soc. Aust., J.* **29**, 343–56.

Tucker, D. H., Wyatt, B. W., Druce, E. C., Mathur, S. P. and Harrison, P. L. (1979). The upper crustal geology of the Georgina Basin region. *BMR J. Aust. Geol. Geophys.* **4**, 209–26.

Turner, S., Jones, P. J. and Draper, J. J. (1981). Early Devonian thelodonts (Agnatha) from the Toko Syncline, western Queensland, and a review of other Australian discoveries. *BMR J. Aust. Geol.*

Geophys. **6**, 51–69.

Twidale, C. R. and Harris, W. K. (1977). The age of Ayers Rock and the Olgas, central Australia. *R. Soc. S. Aust., Trans.* **101**, 45–50.

——, Bourne, J. A. and Smith, D. M. (1976). Age and origin of palaeosurfaces on Eyre Peninsula and the southern Gawler Ranges, South Australia. *Z. Geomorphol.* **20**, 28–55.

——, Lindsay, J. M. and Bourne, J. A. (1978). Age and origin of the Murray River and gorge in South Australia. *R. Soc. Vict., Proc.* **90**, 27–42.

Uren, R. (1980). Notes on the Clifton Sub-Group, northeastern Sydney Basin. *N.S.W., Geol. Surv., Bull.* **26**, 162–9.

Uyeda, S. (1981). Subduction zones and back arc basins – a review. *Geol. Rundschau* **70**, 552–69.

Vail, P. R., Mitchum, R. M. and Thompson, S. (1977). Seismic stratigraphy and global changes of sea level. *Am. Ass. Pet. Geol. Mem.* **26**, 83–97.

Vallance, T. G. (1967). Palaeozoic low-pressure regional metamorphism in south-eastern Australia. *Medd. dansk Geol. Foren.* **17**, 494–503.

—— (1969). Plutonic and metamorphic rocks. *Geol. Soc. Aust., J.* **16**, 180–200.

van Andel, T. H. and Veevers, J. J. (1967). Morphology and sediments of the Timor Sea. *Aust., Bur. Miner. Resour., Geol. Geophys., Bull.* **83**.

van de Graaff, W. J. E. (1981). Palaeogeographic evolution of a rifted cratonic margin: S.W. Australia – discussion. *Palaeogeogr., Palaeoclimatol., Palaeoecol.* **34**, 163–72.

——, Denman, P. D. and Hocking, R. M. (1977a). Emerged Pleistocene marine terraces on Cape Range, Western Australia. *W. Aust., Geol. Surv., Annu. Rep.* **(1975)**, 62–70.

——, Crowe, R. W. A., Bunting, J. A., and Jackson, M. J. (1977b). Relict Early Cainozoic drainages in arid Western Australia. *Z. Geomorphol.* **21**, 379–400.

VandenBerg, A. H. M. (1978). The Tasman Fold Belt system in Victoria. *Tectonophysics* **48**, 267–97.

——, Garratt, M. J. and Spencer-Jones, D. (1976). Silurian–Middle Devonian. *Geol. Soc. Aust., Spec. Publ.* **5**, 45–76.

Van Hinte, J. E. (1976). A Jurassic time scale. *Am. Assoc. Pet. Geol., Bull.* **60**, 489–97.

—— (1978). Geohistory analysis – application of micropaleontology in exploration geology. *Am. Ass. Pet. Geol., Bull.* **62**, 201–22.

Veevers, J. J. (1967). The Phanerozoic geological history of northwest Australia. *Geol. Soc. Aust., J.* **14**, 253–72.

—— (1969). Palaeogeography of the Timor Sea region. *Palaeogeog., Palaeoclimat., Palaeoecol.* **6**, 125–40.

—— (1971a). Phanerozoic history of Western Australia related to continental drift. *Geol. Soc. Aust., J.* **18**, 87–96.

—— (1971b). Shallow stratigraphy and structure of the Australian continental margin beneath the Timor Sea. *Mar. Geol.* **11**, 209–49.

—— (1974). Western continental margin of Australia. In *The geology of continental margins* (ed. C. A. Burk and C. L. Drake), pp. 605–16. Springer, New York.

—— (1976). Early Phanerozoic events on and alongside the Australasian–Antarctic platform. *Geol. Soc. Aust., J.* **23**, 183–206.

—— (1977). Paleobathymetry of the crest of spreading ridges related to the age of ocean basins. *Earth Planet. Sci. Lett.* **34**, 100–6.

—— (1979). Western margin of Australia: reply to discussion. *Geol. Soc. Am., Bull.* **90**, 797–8.

—— (1981). Morphotectonics of rifted continental margins in embryo (East Africa), youth (Africa–Arabia), and maturity (Australia). *J. Geol.* **89**, 57–82.

—— (1982a). Australian rifted margins. In *Dynamics of passive margins* (ed. R. A. Scrutton), pp. 72–89. Geodynamics Series 6. Am. Geophys. Union, Washington, DC.

—— (1982b). Australian–Antarctic depression from the ocean ridge to the adjacent continent. *Nature* **295**, 315–7.

—— (1982c). Western and north western margin of Australia. In *The Ocean Basins and Margins, Vol. 6, The Indian Ocean* (ed. A. E. M. Nairn and F. G. Stehli), pp. 515–44. Plenum, New York.

—— and Cotterill, D. (1976). Western margin of Australia – a Mesozoic analog of the East African rift system. *Geology* **4**, 713–17.

—— —— (1978). Western margin of Australia: evolution of a rifted arch system. *Geol. Soc. Am., Bull.* **89**, 337–55.

—— and Evans, P. R. (1973). Sedimentary and magmatic events in Australia and the mechanisms of world-wide Cretaceous transgressions. *Nature Phys. Sci.* **245**, 33–6.

—— —— (1975). Late Palaeozoic and Mesozoic history of Australia. In *Gondwana Geology* (ed. K. S. W. Campbell), pp. 579–607. ANU Press, Canberra.

—— and Hansen, L. (1981). Volcanism in the rift-valley system that evolved into the western margin of Australia. *Geol. Soc. Aust., J.* **28**, 377–84.

—— and Heirtzler, J. R. (1974). Tectonic and paleogeographic synthesis of Leg 27. *Deep Sea Dril. Proj., Initial Rep.* **27**, 1049–54.

—— and Johnstone, M. H. (1974). Comparative stratigraphy and structure of the Western Australian margin and the adjacent deep ocean floor. *Deep Sea Dril. Proj., Initial Rep.* **27**, 571–86.

—— and Powell, C.McA. (1979). Sedimentary-wedge progradation from transform-faulted continental rim: Southern Exmouth Plateau, Western Australia. *Am. Assoc. Pet. Geol., Bull.* **63**, 2088–96.

—— and Rundle, A. S. (1979). Channel Country fluvial sands and associated facies of central-eastern Australia: modern analogues of Mesozoic desert sands of South America. *Palaeogeogr. Palaeoecol., Palaeoclimatol.* **26**, 1–16.

—— and van Andel, T. H. (1967). Morphology and basement of the Sahul Shelf. *Mar. Geol.* **5**, 293–98.

——, and Wells, A. T. (1962). Geology of the Canning Basin, W.A. *Aust., Bur. Miner. Resour., Geol. Geophys., Bull.* 60.

——, Roberts, J., White, M. E. and Gemuts, I. (1967). Sandstone of probable Lower Carboniferous age in the north-eastern Canning Basin, W.A. *Aust. J. Sci.* **29**, 330–1.

——, Jones, J. G. and Talent, J. A. (1971). Indo-Australian stratigraphy and the configuration and dispersal of Gondwanaland. *Nature* **229**, 383–8.

——, Falvey, D. A., Hawkins, L. V., and Ludwig, W. J. (1974a). Seismic reflection measurements of northwest Australian margin and adjacent deeps. *Am. Assoc. Pet. Geol., Bull.* **58**, 1731–50.

——, Heirtzler, J. R., Bolli, H. M., Carter, A. N., Cook, P. J., Krasheninnikov, V. A., McKnight, B. K., Proto-Decima, F., Renz, G. W., Robinson, P. T., Rocker, K. and Thayer, P. A. (1974b). Leg 27. *Initial Rep. Deep Sea Dril. Proj.* 27.

——, Powell, C.McA. and Johnson, B. D. (1975). Greater India's place in Gondwanaland and in Asia. *Earth Planet. Sci. Lett.* **27**, 383–7.

——, Falvey, D. A. and Robins, S. (1978). Timor Trough and Australia – facies show topographic wave migrated 80 km during the past 3 m.y. *Tectonophysics* **45**, 217–27.

——, Powell, C.McA. and Johnson, B. D. (1980). Seafloor constraints on the reconstruction of Gondwanaland. *Earth Planet. Sci. Lett.* **51**, 435–44.

——, Jones, J. G. and Powell, C. McA. (1982). Tectonic framework of Australia's sedimentary basins. *APEA J.* **22**, 283–300.

Verma, R. K., Mukhopadhyay, M. and Bhuin, N. C. (1978). Seismicity, gravity and tectonics in the Andaman Sea. *J. Phys. Earth* **26**, Suppl. S, 233–48.

Vessell, R. K. and Davies, D. K. (1981). Nonmarine sedimentation in an active fore arc basin. *Soc. Econ. Paleont. Mineral., Spec. Publ.* **31**, 31–45.

Vine, R. R. (1972). Relations between the Adavale and Drummond Basins. *APEA J.* **12**, 58–61.

—— (1976a). Eromanga Basin. *Australas. Inst. Min. Metall., Monogr.* **7** (3), 306–9.

—— (1976b). Galilee Basin. *Australas. Inst. Min. Metall., Monogr.* **7** (3), 316–21.

Visser, W. A. and Hermes, J. J. (1962). Geological results of the exploration for oil in Netherlands New Guinea. *Verh. Kon. Ned. Geol. Mijnb. Gen., Geol. Ser.* 20.

Vogt, P. R., Feden, R. H. and Morgan, G. A. (1979). Project Investigator-1: a joint Australian/U.S. aeromagnetic survey of the Australia–Antarctic Discordance. *Int. Union Geodesy Geophys., 17th General Assembly, Canberra, Abstracts*, 152.

Voisey, A. H. (1959). Australian Geosynclines. *Aust. J. Sci.* **22**, 188–98.

von der Borch, C. C. (1968). Southern Australian submarine canyons; their distribution and ages. *Mar. Geol.* **6**, 267–79.

—— (1979). Continent–island arc collision in the Banda Arc. *Tectonophysics* **54**, 169–93.

—— (1980). Evolution of Late Proterozoic to Early Paleozoic Adelaide Foldbelt, Australia: comparisons with post-Permian rifts and passive margins. *Tectonophysics* **70**, 115–34.

——, Smit, C. and Grady, A. E. (1982). Late Proterozoic submarine canyons of Adelaide Geosyncline, South Australia. *Am. Assoc. Pet. Geol., Bull.* **66**, 332–47.

van Rad, V. and Exon, N. F. (1983). Mesozoic–Cenozoic sedimentary and volcanic evolution of the starved passive continental margin of northwest Australia. *Am. Assoc. Pet. Geol., Mem.* **34**, 253–81.

von Stackelberg, U., Exon, N. F., von Rad, U., Quilty, P., Shafik, S., Beiersdorf, H., Seibertz, E. and Veevers, J. J. (1980). Geology of the Exmouth and Wallaby Plateaus off northwest Australia: sampling of seismic sequences. *BMR J. Aust. Geol. Geophys.* **5**, 113–40.

Vos, R. G. and McHattie, C. M. (1981). Upper Triassic depositional environments, central Exmouth Plateau (Permit WA-84-P), northwestern Australia. *Geol. Soc. Aust., 5th Geol. Convent., Abstr.*, 16.

Vozoff, K., Kerr, D., Moore, R. F., Jupp, D. L. B. and Lewis, R. J. G. (1975). Murray Basin magnetotelluric study. *Geol. Soc. Aust., J.* **22**, 361–75.

Walcott, R. I. (1978). Present tectonics and Late Cenozoic evolution of New Zealand. *Geophys. J. R. Astron. Soc.* **52**, 137–64.

Waldman, M. (1971). Fish from the freshwater Lower Cretaceous of Victoria, Australia, with comments on the palaeo-environment. *Palaeont. Assoc.*

London, *Spec. Pap. Palaeontology* 9.

Walker, D. J. (1966). Wilkes geophysical surveys, Antarctica, 1962. *Aust., Bur. Miner. Resour., Geol. Geophys., Rec.* 1966/129 (unpubl.).

Walker, D. and Singh, G. (1981). Vegetation history. In *Australian Vegetation* (ed. R. H. Groves), pp. 26–43. Cambridge U.P., Hong Kong.

Walter, M. R. (1972). Stromatolites and the biostratigraphy of the Australian Precambrian and Cambrian. *Palaeont. Assoc. London, Spec. Pap. Palaeontology* 11.

Warren, G. (1965). Geology of Antarctica. In *Antarctica* (ed. T. Hatherton), pp. 279–320. Methuen, London.

Warris, B. J. (1973). Plate tectonics and the evolution of the Timor Sea, northwest Australia. *APEA J.* **13**, 13–18.

Wass, S. Y. (1979). Fractional crystallization in the mantle of late-stage kimberlitic liquids – evidence in xenoliths from the Kiama area, N.S.W., Australia. In *The Mantle Sample: Inclusions in Kimberlites and other Volcanics* (ed. F. R. Boyd and H. O. A. Meyer), pp. 366–73. American Geophysical Union, Washington, DC.

—— and Hollis, J. D. (1983). Crustal growth in south-eastern Australia – evidence from lower crustal eclogitic and granulitic xenoliths. *J. Met. Geology* **1**, 25–45.

—— and Irving, A. J. (1976). *XENMEG: a catalogue of occurrences of xenoliths and megacrysts in volcanic rocks of eastern Australia.* Australian Museum, Sydney.

Waterhouse, J. B. (1978). Chronostratigraphy for the World Permian. *Am. Assoc. Pet. Geol., Stud. Geol.* **6**, 299–322.

—— and Jell, J. S. (1982). The sequence of Permian rocks and faunas near Exmoor Homestead south of Collinsville, northern Bowen Basin. *Geol. Soc. Aust., Abstr.* **8**, 26.

Watts, A. B. and Daly, S. F. (1981). Long wavelength gravity and topography anomalies. *Ann. Rev. Earth Planet. Sci.* **9**, 415–48.

—— and Steckler, M. S. (1979). Subsidence and eustasy at the continent margin of eastern North America. In *Deep drilling results in the Atlantic Ocean: Continental margins and paleoenvironment,* (ed. M. Talwani, W. Hay, and W. B. F. Ryan), pp. 218–34. American Geophysical Union, Maurice Ewing Series, 3. Washington, D.C.

Watts, D. R. and Bramall, A. M. (1981). Palaeomagnetic evidence for a displaced terrain in Western Antarctica. *Nature* **293**, 638–41.

Webb, A. W. and McDougall, I. (1964). Granites of Lower Cretaceous age near Eungella, Queensland. *Geol. Soc. Aust., J.* **11**, 151–3.

—— —— (1967a). A comparison of mineral and whole rock potassium–argon ages of Tertiary volcanics from central Queensland, Australia. *Earth Planet. Sci. Lett.* **3**, 41–7.

—— —— (1967b). Isotopic dating evidence on the age of the Upper Permian and Middle Triassic. *Earth Planet. Sci. Lett.* **2**, 483–8.

—— —— (1968). The geochronology of the igneous rocks of eastern Queensland. *Geol. Soc. Aust., J.* **15**, 313–46.

——, Stevens, N. C. and McDougall, I. (1967). Isotopic age determinations on Tertiary volcanic rocks and intrusives of south-eastern Queensland. *R. Soc. Qld, Proc.* **79**, 79–92.

Webb, J. A. (1981). A radiometric time scale of the Triassic. *Geol. Soc. Aust., J.* **28**, 107–21.

Webb, P. N. and Neall, V. E. (1972). Cretaceous Foraminifera in Quaternary deposits from Taylor Valley, Victoria Land. In *Antarctic Geology and Geophysics* (ed. R. J. Adie), pp. 653–7. Universitetsforlaget, Oslo.

Webby, B. D. (1972). Devonian geological history of the Lachlan Geosyncline. *Geol. Soc. Aust., J.* **19**, 99–123.

—— (1976). The Ordovician System in southeastern Australia. In *The Ordovician System* (ed. M. G. Bassett), pp. 417–46. Univ. Wales Press, Cardiff.

—— (1978). History of the Ordovician continental platform shelf margin of Australia. *Geol. Soc. Aust., J.* **25**, 41–63.

—— and Packham, G. H. (1982). Stratigraphy and regional setting of the Cliefden Caves Limestone Group (Late Ordovician), central-western New South Wales. *Geol. Soc. Aust., J.* **19**, 297–317.

——, VandenBerg, A. H. M., Cooper, R. A., Banks, M. R., Burrett, C. F., Henderson, R. A., Clarkson, P. D., Hughes, C. P., Laurie, J., Stait, B., Thomson, M. R. A. and Webers, G. F. (1981). The Ordovician System in Australia, New Zealand and Antarctica. Correlation chart and explanatory notes. *Int. Union Geol. Sci., Publ.* 6.

Weber, C. R., Becket, J. and Hamilton, D. S. (1982). Recent exploration in the Gunnedah Basin by the Department of Mineral Resources. *Sixteenth Symposium: Advances in the study of the Sydney Basin.* Dept. Geol., Univ. Newcastle, 36–7.

Webster, R. E. (1977). Evolution of a major petroleum province: the Southern Oklahoma Aulacogen. *Compass* **54**, 59–71.

Wegener, A. (1929). *Die Entstehung der Kontinente und Ozeane.* 4th edition, Vieweg, Braunschweig.

(1966 Engl. trans. by Dover Publications, New York).

Weissel, J. K. and Hayes, D. E. (1971). Asymmetric seafloor spreading south of Australia. *Nature* **231**, 518–22.

—— —— (1972). Magnetic anomalies in the southeast Indian Ocean. *Am. Geophys. Union, Geophys. Monogr.* **19**, 165–96.

—— —— (1974). The Australian–Antarctic discordance: new results and implications. *J. Geophys. Res.* **79**, 2579–87.

—— —— (1977). Evolution of the Tasman Sea reappraised. *Earth Planet. Sci. Lett.* **36**, 77–84.

—— and Watts, A. B. (1979). Tectonic evolution of the Coral Sea Basin. *J. Geophys. Res.*, **84**, 4572–82.

——, Hayes, D. E. and Herron, E. M. (1977). Plate tectonic synthesis: the displacements between Australia, New Zealand, and Antarctica since the Late Cretaceous. *Mar. Geol.* **25**, 231–77.

——, Hegarty, K. A. and Mutter, J. (1982). The southern margin of Australia – tectonic subsidence and implications on the timing of rifting. *EOS* **63**, 444.

Wellman, P. (1973). Early Miocene potassium–argon age for the Fitzroy Lamproites of Western Australia. *Geol. Soc. Aust., J.* **19**, 471–4.

—— (1974). Potassium–argon ages on the Cainozoic volcanic rocks of eastern Victoria, Australia. *Geol. Soc. Aust., J.* **21**, 359–76.

—— (1976). Gravity trends and the growth of Australia: a tentative correlation. *Geol. Soc. Aust., J.* **23**, 11–14.

—— (1978). Potassium–argon ages of Cainozoic volcanic rocks from the Bundaberg, Rockhampton and Clermont areas of eastern Queensland. *R. Soc. Qld, Proc.* **89**, 59–64.

—— (1979a). On the Cainozoic uplift of the southeastern Australian highlands. *Geol. Soc. Aust., J.* **26**, 1–9.

—— (1979b). On the isostatic compensation of Australian topography. *BMR J. Aust. Geol. Geophys.* **4**, 373–82.

—— (1981). Crustal movement determined from repeat surveying – results from southeastern and southwestern Australia. *Geol. Soc. Aust., J.* **28**, 311–21.

—— (1982). Australian seismic refraction results, isostasy and altitude anomalies. *Nature* **298**, 838–41.

—— and McCracken, H. M. (1979). Australian region plate tectonics. *Aust., Bur. Miner. Resour., Geol. Geophys., Earth Science Atlas, 1:20 000 000.* Canberra.

—— and McDougall, I. (1974a). Potassium–argon ages on the Cainozoic volcanic rocks of New South Wales. *Geol. Soc. Aust., J.* **21**, 247–72.

—— —— (1974b). Cainozoic igneous activity in eastern Australia. *Tectonophysics* **23**, 49–65.

——, McElhinny, M. W. and McDougall, I. (1969). On the polar-wander path for Australia during the Cenozoic. *R. Astron. Soc., Geophys. J.* **18**, 371–95.

——, Cundari, A. and McDougall, I. (1970). Potassium–argon ages for leucite-bearing rocks from New South Wales, Australia. *R. Soc. N.S.W., J. Proc.* **103**, 103–7.

Wells, A. T. (1976). Ngalia Basin. *Australas. Inst. Min. Metall. Monogr.* **7** (3), 226–30.

—— (1980). Evaporites in Australia. *Aust., Bur. Miner. Resour., Geol. Geophys.* 198.

——, Forman, D. J., Ranford, L. C. and Cook, P. J. (1970). Geology of the Amadeus Basin, Central Australia. *Aust., Bur. Miner. Resour., Geol. Geophys., Bull.* 100.

——, Moss, F. J. and Sabitay, A. (1972). The Ngalia Basin, Northern Territory – recent geological and geophysical information upgrades petroleum prospects. *APEA J.* **12**, 144–51.

—— and Moss, F. J. (1983). The Ngalia Basin, Northern Territory: stratigraphy and structure. *Aust., Bur. Miner. Resour., Geol. Geophys., Bull.* **212**.

Whitaker, W. G., Murphy, P. R. and Rollason, R. G. (1974). Geology of the Mundubbera 1:250 000 Sheet area. *Qld, Geol. Surv., Rep.* 84.

White, A. J. R., Compston, W. and Kleeman, A. W. (1967). The Palmer Granite – a study of a granite within a regional metamorphic environment. *J. Petrol.* **8**, 29–50.

——, Chappell, B. W. and Cleary, J. R. (1974). Geologic setting and emplacement of some Australian Palaeozoic batholiths and implications for intrusive mechanisms. *Pac. Geol.* **8**, 159–71.

——, Williams, I. S. and Chappell, B. W. (1976). The Jindabyne Thrust and its tectonic, physiographic and petrogenetic significance. *Geol. Soc. Aust., J.* **23**, 105–12.

White, D. A. (1962). Review of age determination programme of the Bureau of Mineral Resources, Australia, 1956–1962. *Aust., Bur. Miner. Resour., Geol. Geophys. Rec.* 1962/129.

Whitehouse, F. W. (1954). The geology of the Queensland portion of the Great Artesian Basin. Appendix G. In *Artesian water supplies in Queensland*. Department of the Co-ordinator-General of Public Works, Queensland. Report A.56, 1955.

Whyte, R. (1978). Shell's offshore venture in South Australia. *APEA J.* **18**, 44–51.

Wickham, J. and Denison, R. (eds.) (1978). *Structural style of the Arbuckle region.* Geol. Soc. America, South Central Section.

——, Pruatt, M., Reiter, L. and Thompson, T. (1975). The Southern Oklahoma Aulacogen. *Geol. Soc. Am., Abstr. Programs* **7**, 1332.

——, Roeder, D. and Briggs, G. (1976). Plate tectonics models for the Ouachita foldbelt. *Geology* **4**, 173–6.

Wilde, S. A. and Walker, I. W. (1978). Palaeocurrent directions in the Permian Collie Coal Measures, Collie, Western Australia. *W. Aust. Geol. Surv., Annu. Rep.* for 1977, 41–3.

Wilkinson, J. F. G. (1974). Garnet clinopyroxenite inclusions from diatremes in the Gloucester area, New South Wales. *Contrib. Mineral. Petrol.* **46**, 275–99.

—— and Taylor, S. R. (1980). Trace element fractionation trends of tholeiitic magma at moderate pressure: evidence from an Al-spinel ultramafic-mafic inclusion suite. *Contrib. Mineral. Petrol.* **75**, 225–33.

Willcox, J. B. (1978). The Great Australian Bight: a regional interpretation of gravity, magnetic, and seismic data from the continental margin survey. *Aust., Bur. Miner. Resour., Geol. Geophys., Rep.* 201.

—— (1981). Petroleum prospectivity of Australian marginal plateaus. *Am. Assoc. Pet. Geol., Stud. Geol.* **12**, 245–72.

——, Symonds, P. A., Hinz, K. and Bennett, D. (1980). Lord Howe Rise, Tasman Sea – preliminary geophysical results and petroleum prospects. *BMR J. Geol. Geophys.* **5**, 225–36.

Williams, A. J. (1975a). Geology of the Parkes–Bumberry region, N.S.W. B.Sc. (Hons) thesis, Aust. Nat. Univ., Canberra (unpubl.).

Williams, E. (1959). The sedimentary structures of the Upper Scamander sequence and their significance. *R. Soc. Tas., Pap. Proc.* **93**, 29–32.

—— (1976). *Tasman Fold Belt System in Tasmania. Explanatory notes 1:500 000 structural map of pre-Carboniferous rocks of Tasmania.* Tas., Dep. Mines, Hobart.

—— (1978). The Tasman Fold Belt system in Tasmania. *Tectonophysics* **48**, 159–205.

Williams, G. E. (1973). Simpson Desert Sub-basin – a promising Permian target. *APEA J.* **13**, 33–40.

—— and Goode, A. D. T. (1978). Possible western outlet for an ancient Murray River in South Australia. *Search* **9**, 443–7.

Williams, I. R., Brakel, A. T., Chin, R. J. and Williams, S. J. (1976). The stratigraphy of the eastern Bangemall Basin and the Paterson Province. *West. Aust., Geol. Surv., Annu. Rep.* for 1975, 79–83.

Williams, I. S., Compston, W., Chappell, B. W. and Shirahase, T. (1975). Rubidium–strontium age determinations on micas from a geologically controlled, composite batholith. *Geol. Soc. Aust., J.* **22**, 497–505.

Williams, M. A. J. (1975b). Late Pleistocene tropical aridity synchronous in both hemispheres? *Nature* **253**, 617–18.

—— and Adamson, D. A. (1973). The physiography of the central Sudan. *Geog. J.* **139**, 498–508.

—— and Williams, F. M. (1980). Evolution of the Nile Basin. In *The Sahara and the Nile* (ed. M. A. J. Williams and H. Faure), pp. 207–24. Balkema, Rotterdam.

Williams, P. W., McDougall, I. and Powell, J. M. (1972). Aspects of the Quaternary geology of the Tari-Kobi area, Papua. *Geol. Soc. Aust., J.* **18**, 333–47.

Williams, R. M. and Drury, L. W. (1980). Cainozoic sedimentation of the eastern Murray Basin, N.S.W. In *The Cainozoic evolution of continental Southeast Australia* (ed. E. M. Truswell and R. S. Abell), pp. 77–8. *Aust., Bur. Miner. Resour., Geol. Geophys., Rec.* 1980/67.

Wilson, C. J. L. (1968). Geology of the Narooma Area, N.S.W. *R. Soc. N.S.W., J. Proc.* **101**, 147–57.

Wilson, J. T. (1974). The life cycle of ocean basins: stages of growth. In *Physics and Geology* (ed. J. A. Jacobs, R. D. Russell, and J. T. Wilson), pp. 397–425. McGraw-Hill, New York.

Wiltshire, M. J. (1982a). Late Triassic and Early Jurassic sedimentation in the Great Artesian Basin. In *Eromanga Basin Symposium, summary papers* (ed. P. S. Moore and T. J. Mount), pp. 59–67. Geol. Soc. Aust. and Pet. Expl. Soc. Aust., Adelaide.

—— (1982b). Revision of Eromanga Basin limits. In *Eromanga Basin Symposium, summary papers* (ed. P. S. Moore and T. J. Mount), pp. 69–75. Geol. Soc. Aust. and Pet. Expl. Soc. Aust., Adelaide.

Wiseman, J. F. (1979). Neocomian eustatic changes – biostratigraphic evidence from the Carnarvon Basin. *APEA J.* **19**, 66–73.

Woodburne, M. O. (1967). The Alcoota fauna, Central Australia. *Aust., Bur. Miner. Resour., Geol. Geophys., Bull.* 87.

Woolley, D. R. (1978). Cainozoic sedimentation in the Murray drainage system, New South Wales section. *R. Soc. Vict., Proc.* **90**, 61–5.

—— (1980). Geology of the eastern Murray Basin in N.S.W. In *The Cainozoic evolution of continental Southeast Australia* (ed. E. M. Truswell and R. S.

Abell), pp. 79–80. *Aust., Bur. Miner. Resour., Geol. Geophys., Rec.* 1980/67.

Wopfner, H. (1963). Post-Winton sediments of probable Upper Cretaceous age in the Central Great Artesian Basin. *R. Soc. S. Aust., Trans.* **86**, 247–53.

—— (1974). Post-Eocene history and stratigraphy of northeastern South Australia. *R. Soc. S. Aust., Trans.* **98**, 1–12.

—— (1981). Development of Permian intracratonic basins in Australia. In *Gondwana Five* (ed. M. M. Cresswell, and P. Vella), pp. 185–90. Balkema, Rotterdam.

—— and Douglas, J. G. (eds.) (1971). The Otway Basin of southeastern Australia. Spec. Publ., Geological Surveys South Australia and Victoria, Adelaide.

——, Freytag, I. B. and Heath, G. R. (1970). Basal Jurassic–Cretaceous rocks of Western Great Artesian Basin, South Australia: stratigraphy and environment. *Am. Ass. Pet. Geol., Bull.* **54**, 383–416.

——, Callen, R. and Harris, W. K. (1974). The lower Tertiary Eyre Formation of the southwestern Great Artesian Basin. *Geol. Soc. Aust., J.* **21**, 17–51.

Wright, C. A. (1977). Distribution of Cainozoic Foraminiferida in the Scott Reef No. 1 Well, Western Australia. *Geol. Soc. Aust., J.* **24**, 269–77.

Wyatt, B. W., Yeates, A. N. and Tucker, D. H. (1980). A regional review of the geological sources of magnetic and gravity anomaly fields in the Lachlan Fold Belt. *BMR J. Aust. Geol. Geophys.* **5**, 289–300.

Wyatt, D. H. and Jell, J. S. (1980). Devonian and Carboniferous stratigraphy of the northern Tasman Orogenic Zone in the Townsville hinterland, north Queensland. In *The geology and geophysics of Northeastern Australia* (ed. R. A. Henderson and P. J. Stephenson), pp. 201–28. Geological Society of Australia, Queensland Division, Brisbane.

—— and Webb, A. W. (1970). Potassium–argon ages of some northern Queensland basalts and an interpretation of Late Cainozoic history. *Geol. Soc. Aust., J.* **17**, 39–51.

——, Paine, A. G. L., Clarke, D. E. and Harding, R. R. (1970). Geology of the Townsville 1:250 000 Sheet area, Queensland. *Aust., Bur. Miner. Resour., Geol. Geophys., Rep.* 127.

—— —— ——, Gregory, C. M. and Harding, R. R. (1971). Geology of the Charters Towers 1:250 000 Sheet area, Queensland. *Aust., Bur. Miner. Resour., Geol. Geophys., Rep.* 137.

Wyborn, L. A. I. and Chappell, B. W. (1979). Geochemical evidence of a pre-Ordovician sedimentary layer in southeastern Australia. In *Crust and upper mantle of Southeast Australia* (ed. D. Denham), p. 104. *Aust., Bur. Miner. Resour., Geol. Geophys., Rec.* 1979/2.

Yoo, E. K. (1982). Geology and coal resources of the northern part of the Oaklands Basin. *N.S.W., Geol. Surv., Q. Notes* **49**, 15–27.

Young, G. C. and Gorter, J. D. (1981). A new fish fauna of Middle Devonian age from the Taemas/Wee Jasper region of New South Wales. *Aust., Bur. Miner. Resour., Geol. Geophys., Bull.* 209.

Young, R. W. (1978). The study of landform evolution in the Sydney region: a review. *Aust. Geogr.* **14**, 71–93.

—— and Bishop, P. (1980). Potassium–argon ages in Cainozoic volcanic rocks in the Crookwell–Goulburn area, New South Wales. *Search* **11**, 340–1.

Zidgerveld, J. D. A. (1967). AC demagnetisation of rocks: analysis of results. In *Methods in Palaeomagnetism* (ed. D. W. Collinson, K. M. Creer and S. K. Runcorn), pp. 254–86. Elsevier, New York.

AUTHOR INDEX

This index includes authors listed in the bibliography of biogeography on pp. 82–93.

Abele, C. 3, 57, 103, 118, 120–1, 134, 226
Adams, C. J. 20, 271, 280, 288
Adams, J. 9
Adamson, D. A. 48
Alexander, L. G. 100, 101
Allen, A. R. 343
Allen, G. A. 9, 187–8, 191, 195, 211, 229, 234, 238
Allen, R. J. 127, 134, 227
Allsopp, H. L. 118
Al-Shaieb, Z. 344
American Association of Petroleum Geologists 6, 7, 18, 100–1, 106, 108–9, 213
Anderson, J. B. 233
Anderson, J. C. 307
Anderson, R. N. 10, 220
Andrews, J. E. 37, 203, 227
Andrews, P. B. 10, 20, 38, 175, 286
APCP (Australasian Petroleum Company Proprietary) 247, 254
Apthorpe, M. C. 114, 153, 159, 199, 211, 226
Archbold, N. W. 71, 82
Archer, M. 60, 81–2
Arculus, R. J. 97, 99
Arditto, P. A. 131, 134
Arkell, W. J. 79, 82
Armstrong, R. L. 3, 9, 260, 342–3
Arnold, G. O. 281, 283, 307, 323
Arriens, P. A. 279–80, 282
Ash, S. R. 77, 86
Ashley, P. M. 312, 316
Ashlock, P. D. 58, 83
Asif, M. 305
Astley-Boden, C. E. frontispiece
Atlas of Australian Resources 44, 46
Atwater, T. 21, 307, 323
Auchincloss, G. 322, 336
Audley-Charles, M. G. 41, 67–8, 83, 108, 112–14, 155, 157–8, 191
Austin, P. M. 270, 278, 280, 282
Axelrod, D. I. 59, 68–9, 81, 89
Aziz-ur-Rahman 9, 118

Bachman, S. B. 293
Bailey, D. K. 99
Baillie, P. W. 118
Bain, J. H. C. 129, 134–6, 254
Baker, B. H. 205
Baker, G. 136
Baksi, A. K. 282

Balgooy, M. M. J. van 69, 83
Balke, B. 158, 281, 328
Bally, A. W. 347–8, 363–4
Balme, B. E. 49, 161, 236, 241, 244, 263, 342
Bambach, R. K. 73, 78, 93, 364
Banks, M. R. 3, 75, 83, 93, 264, 289–92, 305, 307, 321
Baraganzi, M. 202
Barber, A. J. 41, 112–14, 155, 157–8, 191
Barber, P. M. 182–3, 186, 193, 195–6, 198
Barkas, J. P. 312
Barlow, B. A. 59, 65, 83
Barnes, L. C. 144
Barrash, W. 323, 325
Barrett, P. 54, 182, 186
Barrett, S. F. 73, 78, 93
Barron, E. J. 21
Barron, L. M. 312, 317
Barter, T. P. 104, 154, 182, 191, 226 236, 238, 244
Bartholomai, A. 60, 82
Barton, C. 364
Barton, C. M. 100
Basden, H. 312, 316
Bateman, R. J. 132
Battersby, D. G. 242, 262
Bauer, J. A. 253
Baxter, K. 99
Beadle, N. C. W. 64–5, 83
Beams, S. D. 332
Beard, J. S. 58–9, 65, 83
Beaumont, C. 201
Beavis, F. C. 292
Beck, A. E. 136
Becket, J. 249
Beiersdorf, H. 26, 182, 184, 187–8
Bein, J. 129, 134–5, 153, 175–6, 178, 211, 226–7, 265–6, 315
Belbin, L. 80, 84
Belford, D. J. 80, 83
Bell, T. H. 288, 305
Bembrick, C. S. 134, 249
Benbow, M. C. 163, 165
Bennett, A. J. R. 132, 247
Bennett, D. 31, 34, 138, 207, 227
Bennett, E. B. 59, 93
Bennett, I. 58–9, 83
Bentley, C. R. 9
Berry, W. B. N. 48, 75, 91, 290, 311, 315
Bhuin, N. C. 297

Bibee, L. D. 288–9
Bigarella, J. J. 341, 346, 348
Binns, R. A. 255
Bird, K. J. 195
Bischoff, G. C. O. 294, 313, 315
Bishop, P. 118, 138
Black, L. P. 165, 241, 270, 288, 305, 343
Blackburn, D. T. 81, 83–4
Blake, D. H. 255
Blong, R. J. 256
BMR 116, 144, 150, 156
Boeuf, M. G. 174, 176–7, 180–1, 226
Bofinger, V. M. 277, 286, 288
Bolger, P. 121
Bolli, H. M. 182–3, 188
Boltovsky, E. 68, 83, 182, 186, 188
Bond, G. 211–13
Bonham-Carter, G. 78, 92
Bonnett, J. E. 118
Bonnin, J. 5
Borg, S. G. 232
Boucot, A. J. 48, 71, 72, 75, 83, 91, 291, 311, 315
Bourke, D. J. 134
Bourne, J. A. 146, 148
Bowen, R. 50
Bowering, O. J. W. 266–7
Bowin, C. 41–2, 112
Bowler, J. M. 42, 44–5, 47, 65, 83
Bowles, F. A. 42
Braddock, P. E. 286
Bradshaw, J. D. 20
Brakel, A. T. 253, 255–6, 279–80
Bramall, A. M. 11
Branson, J. C. 182
Brass, G. W. 213
Brenner, G. J. 80, 83
Brewer, J. A. 347
Briat, M. 45
Briden, J. C. 80, 91
Briggs, B. G. 64, 87
Briggs, G. 344
Briggs, J. C. 58, 83
Broecker, W. A. 42
Brookfield, H. C. 43–4, 46
Brooks, C. 286, 300, 304
Brooks, J. A. 96
Brown, A. V. 118
Brown, B. R. 103, 120, 123, 135, 174, 176, 181, 203, 226, 228
Brown, C. M. 135, 189, 191, 193, 253, 254

403

Brown, D. A. v, 223, 269, 272, 310
Brown, I. 317
Brown, L. D. 347
Brown, L. N. 242, 253, 262
Brown, P. F. 283, 312, 316
Brown, P. L. 312, 316
Brown, R. G. 153, 199
Brown, R. S. 236
Browne, W. R. v, 106, 107, 129, 143, 214, 269
Brundin, L. 58, 83
Brunker, R. L. 310, 312
Brunnschweiler, R. O. 79, 80, 83, 114, 188, 269
Bullard, E. C. 13
Bultitude, J. M. 248–50
Bunting, J. A. 150, 160
Burbidge, N. 59, 65, 83
Burek, P. J. 161, 278
Burger, D. 81, 89, 127, 131, 134
Burke, K. 168, 189, 201, 205, 277, 340, 344
Burns, B. J. 129, 134, 135, 227, 265, 266
Burns, D. A. 182, 186
Burns, K. L. 280, 281
Burns, R. E. 203
Burrett, C. F. 3, 75, 83, 93, 289–92, 305, 307
Burt, D. 328
Burton, C. K. 114

Callen, R. A. 54, 125, 138, 143, 146, 161, 163, 219, 227, 230
Cameron, L. M. 59, 83
Cameron, P. J. 19, 25, 26, 182–6
Cameron, R. G. 257–9
Campbell, K. S. W. v, 71, 72, 75, 76, 78, 84, 90, 91, 114, 223, 236, 238, 262, 269, 270, 272, 280, 310
Cande, S. C. 1, 16, 18, 21, 22, 32, 34, 36, 175, 181, 226
Cardwell, R. K. 101
Carey, S. W. 41, 103, 109, 244
Carolin, R. C. 65, 84
Carpenter, G. B. 25, 26, 34, 38, 182–6
Carr, P. F. 118, 131, 249, 255, 257, 313, 319
Carter, A. N. 120, 121, 123, 155, 182, 183, 188
Carter, C. 3
Carter, D. J. 41, 112–14, 157, 158, 191
Cas, R. A. F. 290, 292–4, 296, 298, 302, 305, 311, 313, 317–21, 332
Cawood, P. A. 334
Cebulski, D. E. 153, 199
Chamalaun, F. H. 41, 113, 114, 118
Chaney, M. A. 198

Chappell, B. W. 300, 302, 304, 321, 332
Chappell, J. 45, 46, 110
Chaproniere, G. C. H. 58, 84
Chatterton, B. D. E. 75, 84
Chin, R. J. 279, 280
Christofel, D. C. 81, 84
Churchill, D. M. 216
Churkin, M. 3, 203
Ciesielski, P. F. 68, 84
Clark, H. L. 58, 84
Clark, R. L. 81, 90
Clark, R. M. 13
Clarke, D. E. 131, 132, 288, 305
Clarke, M. F. 42
Clarke, M. J. 236
Clarkson, P. D. 3, 75, 93, 289–92, 305, 307
Clayton, R. N. 50
Cleary, J. R. 94, 96, 100, 202, 300
Clemens, J. D. 302
Clifford, H. T. 65, 84
CLIMANZ 44, 45, 46, 47
Cloud, P. 3, 73, 84, 270, 278
Clowes, D. F. 169, 184, 186
Coats, R. P. 270, 277, 284
Cockbain, A. E. 49, 51, 71, 84, 151, 153, 155, 157, 186, 188, 189, 215, 226, 236, 238, 279, 328, 343
Cocker, J. D. 321, 332
Cole, J. 293, 295
Coleman, P. J. 21, 72, 84, 186–8, 207
Collins, C. D. N. 98, 99
Collins, W. J. 320, 321, 332
Collinson, D. W. 11
Collinson, J. W. 264
Colwell, J. B. 216
Commonwealth Bureau of Meteorology 43, 46
Compston, W. 271, 276–7, 279–80, 282, 286–8, 300, 302, 304, 342
Conaghan, P. J. 131, 247, 249, 251, 256–60, 269, 313
Coney, P. J. 326, 341, 346
Conolly, J. R. 329, 333
Conybeare, C. E. B. 135
Cook, A. C. 132, 136, 138, 197–8, 235, 313, 319, 342
Cook, F. W. 253
Cook, P. J. 3, 19, 25–6, 48, 110, 116–17, 119, 163, 165, 167, 182, 183–6, 188, 199, 216, 270, 278, 281–2, 284–7, 289, 327, 341–2, 344
Cookson, I. C. 65, 76, 84, 87
Coomans, H. E. 72, 84
Cooney, P. M. 10, 151, 154, 182, 228
Cooper, B. J. 146, 148–9, 175, 226, 232, 236
Cooper, J. A. 118, 175, 232, 255, 270, 276, 287–8

Cooper, R. A. 3, 75, 93, 281, 286, 289–92, 305, 307
Cope, R. N. 49, 151, 153, 157, 186, 188–9, 215, 226, 236, 238, 328, 343
Corbett, G. J. 334–5
Corbett, K. D. 287–8, 290, 292
Cornelius, K. D. 131–2
Cornwall, H. R. 282
Cosgriff, J. W. 79, 84
Cotterill, D. 182, 184, 186–7, 193, 196, 201, 273, 275
Cowie, J. W. 3, 73, 84
Cox, A. V. 1
Cox, L. R. 80, 84
Cracraft, J. 59, 64, 84
Craddock, C. 283
Craig, R. Y. 320–1, 332
Cranbrook, The Earl of 67–8, 84
Crane, D. 281, 284, 310, 312
Cranfield, L. C. 127, 132–4, 227, 236, 238, 241–2, 244, 255, 260, 264
Crank, K. 103–4
Crawford, A. J. 302, 305
Crawford, A. R. 13, 38, 270, 277, 280, 286, 288
Crawford, E. A. 131–2, 227, 233, 262
Creer, K. M. 11
Crespin, I. 72, 80, 84
Cribb, S. J. 3
Crist, R. P. 189, 328
Cromer, W. C. 118
Crook, K. A. W. v, 3, 80, 84, 223, 269, 272, 290, 292–4, 296, 298, 300, 302–3, 305, 310, 312, 314–16, 318, 320–1, 332, 334
Crostella, A. 41, 104, 154, 157–8, 182, 189, 191, 198, 226, 236, 238, 244
Crowe, R. W. A. 150, 160
Cromwell, J. C. 49, 57, 158–9, 239–40, 350
Cull, J. P. 102, 104
Cundari, A. 9, 118, 131, 227
Curray, J. R. 36, 41, 293, 295–6, 298

D'Addario, G. W. 246, 248–50
Daily, B. 73, 85, 275, 277, 281, 284, 288
Dalrymple, G. B. 1, 3, 32, 118, 175
Daly, S. F. 9
Dampney, C. N. G. 219
Daniels, J. L. 155
Dartnall, A. J. 58, 84
David, T. W. E. v, 106–7, 269
Davidson, J. K. 103
Davies, G. R. 153, 199
Davies, H. L. 112, 254
Davies, J. L. 43, 45

AUTHOR INDEX

Davies, P. J. 121, 201
Davies, T. A. 53–4, 182, 186, 188, 203
Davoren, P. J. 71, 76, 91
Day, R. W. 80, 84, 127, 132, 133–4, 220, 227, 236, 238, 241–2, 244, 247, 255, 260, 264, 267, 322–3, 327, 336–7
Dear, J. F. 253
Deighton, I. 22, 38, 50, 172, 174, 180, 213
de Jersey, N. J. 220
de Laeter, J. R. 279, 283
Denham, D. 98–102, 104, 202
Denham, J. I. 174, 181, 228
Denison, R. 343, 345
Denman, P. D. 103
Derrick, G. M. 57, 270, 278
Dettmann, M. E. 81, 84
Devine, S. B. 163, 165
Dewey, J. F. 189, 201, 277, 340, 344
Diamond, J. M. 67, 85
Dickins, J. M. 49, 72, 78, 79, 85, 88, 132, 236, 247, 253, 255, 258, 262, 288, 305
Diebold, J. B. 25–6, 34, 38, 182–6
Dietz, R. S. 196
Dixon, O. 253
Dolby, J. H. 49
Domack, E. W. 233
Donnelly, T. W. 282
Dorman, F. H. 50, 53, 55
Dorman, J. 202
Douglas, I. 42, 123
Douglas, J. G. 50, 77, 81, 85, 104, 132–8, 176, 202, 226, 236, 262
Doust, H. 174, 176–7, 180–1, 226
Doutch, H. F. 129, 135, 227, 229, 251, 265, 344
Dow, D. B. 111–12, 129, 135, 139, 245–7, 254, 259–60
Doyle, H. A. 174
Doyle, P. 182, 186, 188
Draper, J. J. 158, 327–8
Drewry, D. J. 9, 232, 234
Drewry, G. E. 80, 91
Druce, E. C. 71, 75, 85, 88, 162, 167, 274, 278, 341
Drummond, B. J. 97
Drury, L. W. 119
Duddy, I. R. 131, 133, 135–6, 226
Dulhunty, J. A. 57, 118, 131, 143, 227, 232–3, 262
Dumitrica, P. 203
Duncan, P. M. 64, 91
Duplessy, J. C. 47
Dury, G. H. 118
Du Toit, A. L. 17, 169

Eade, J. V. 37, 203, 227
Eaton, G. P. 326
Eberlein, G. D. 3
Edgecombe, D. R. 313–14, 327
Edgerley, D. W. 183, 189, 328
Edwards, A. B. 136, 188, 238
Edwards, A. R. 10, 38, 50, 53–4, 175, 203
Edwards, D. 76, 85
Efimova, G. P. 9
Eggert, J. T. 264
Ekman, S. 58, 85
Ellenor, D. W. 132, 135, 137–8
Elliot, D. H. 131–2, 136, 226–7, 232–4, 271
Ellis, D. J. 99
Ellis, P. L. 127, 131–2, 134, 136–7, 220, 227, 255
Ellwood, B. B. 68, 84
Embleton, B. J. J. 11–13, 16, 34, 54, 131, 136, 139, 352
Emerson, D. W. 100
Emmel, F. J. 41, 296, 298
Engel, B. A. 72, 84, 235–6, 334, 336
Engelhardt, N. L. 233
England, R. N. 99
Ericson, E. K. 127, 132, 134, 203, 227, 229
Etheridge, M. A. 295, 319
Ettingshausen, C. von 65, 85
Evans, H. J. 215
Evans, P. R. 10, 103, 118, 131, 139, 151, 154, 176, 182, 188, 204, 223, 226, 228, 232, 234, 240–2, 244, 259–63, 269, 312, 329, 335, 337, 350, 352
Everett, J. E. 13
Evernden, J. F. 131–2, 227, 255, 286
Ewart, A. 99
Ewing, M. 228
Exon, N. F. 26, 126–7, 129, 131–2, 134, 182, 184, 186–8, 215, 219–20, 227, 253, 267–8, 288, 305
Eyles, D. 10, 151, 154, 182, 228

Facer, R. A. 118, 131, 136, 238, 249, 255, 257
Fåhraeus, L. E. 71, 85
Fairbridge, R. W. 183, 205
Fairchild, W. W. 233
Falvey, D. A. 22, 25, 37–8, 41, 50, 110, 112, 138, 155, 157–8, 171–4, 176, 181–3, 187, 190, 195, 199, 205, 213, 305
Farmer, N. 236
Faure, G. 288
Fawckner, J. F. 307, 323
Feden, R. H. 21, 34
Felton, E. A. 313, 315
Fenton, M. W. 295
Ferguson, J. 97, 99, 131–2, 181, 227, 233, 262

Ferguson, J. A. 50, 104, 132–8, 176, 202, 236, 262
Fergusson, C. L. 290, 293–4, 298, 312–13, 315, 317–21, 329, 332, 335, 337
Fergusson, J. 313, 317–18
Filatoff, J. 50
Findlay, A. L. 254
Fink, J. 42
Finlayson, D. M. 94, 96, 98–9
Firman, J. B. 216
Fischer, A. G. 344
Fischer, J. F. 344
Fisher, R. A. 12, 19
Fisher, R. L. 19
Fitch, T. J. 297
Fleming, C. A. 58, 70, 79, 85
Fleming, D. P. 292
Fleming, H. S. 21
Fleming, P. J. G. 79, 85
Flenley, J. 43, 45
Flint, D. J. 281
Flint, J. C. E. 131
Flood, P. G. 253, 335
Flood, R. H. 239, 241, 257, 300, 302, 313
Forbes, B. G. 148, 228, 277, 282, 284
Forbes, E. 58, 85
Forbes, V. R. 336
Ford, A. B. 182, 186, 188
Forman, D. J. 48, 154, 159, 163, 165, 167, 226, 236, 238, 264–5, 270, 278–9, 327–8, 341–2, 344, 347–8
Forster, R. R. 64, 85
Forsyth, D. 58, 88
Foster, C. B. 226, 263, 265
Frakes, L. A. 44, 48–9, 54–5, 57, 158–9, 182, 186, 239–40, 348, 350
Francheteau, J. 5
Francis, G. 123
Franklin, B. J. 312, 316
Fraser, A. R. 146, 174–6, 178, 179–81, 211, 226, 228
Freytag, I. B. 148
Fuller, M. 11
Furnish, W. 71–2, 85

Galehouse, J. S. 203
Galloway, R. W. 44–5, 117
Gansser, A. 282
Gaposchkin, E. M. 10
Gardner, W. E. 187–8, 191, 195, 211, 229, 238
Garratt, M. J. 310–12, 315–19, 321, 326, 329
Gatehouse, C. G. 148
Geary, J. K. 189
Gebelein, C. D. 153, 199

Gee, R. D. 149, 152, 157–8
Gehling, J. G. 73, 86
Gellatly, D. C. 280
Gemuts, I. 158–9, 279–80
Gentilli, J. 106–7, 129, 143, 214
Geological Society of Australia 106, 109, 117, 132, 145, 163, 185, 214, 228, 253
Geological Survey of Western Australia 182
Gerth, H. 71, 85
Gibson, D. L. 135
Gierlowski, E. H. 73, 78, 93
Gilbert-Tomlinson, J. 75, 89
Gilfillan, M. A. 298, 314
Gill, E. D. 50, 117–18
Gill, J. B. 136–7, 288
Gilligan, L. B. 310, 315
Girdler, R. W. 196
Glaessner, M. F. 3, 50, 73, 84–5, 270, 278
Gleadow, A. J. W. 9, 131, 133, 135–6, 139, 181, 226
Glen, R. A. 289, 310, 312, 317, 318–21, 329, 331–5
Glenister, B. F. 71–2, 75, 85, 91, 241
Glikson, A. Y. 149, 183, 196–7, 280
Gloe, C. S. 57, 103, 118, 120–1, 226
Goleby, B. R. 11, 13
Good, R. 347
Goode, A. D. T. 146–7, 230, 270
Gordon, W. A. 50
Goree, W. S. 11
Gorter, J. D. 191, 244, 320, 331
Goscombe, P. W. 253
Gostin, V. A. 10, 38, 175, 203, 249
Gould, K. W. 132
Gould, R. E. 76–8, 80, 85, 131
Grady, A. E. 41, 113–14, 272, 280
Grant, R. E. 113
Gravenor, C. P. 16
Gray, A. R. G. 132, 253
Gray, N. M. 100
Green, D. C. 49, 118, 126, 129, 132, 227, 255, 260, 264
Green, P. W. 132
Gregory, C. M. 132, 288, 336
Gregson, P. J. 100, 202
Grey, K. 72, 85
Griffin, T. J. 104, 126, 129, 202, 227, 304
Griffin, W. L. 98
Griffith, B. R. 103, 123, 176, 203
Griffiths, J. R. 22, 174, 232, 305
Grikurov, G. E. 9
Grimes, K. G. 105, 115, 124–6, 129, 132, 135, 137–8, 143, 225, 227–30, 251, 254
Grose, L. T. 341
Groves, R. D. 242, 253, 262
Grund, R. B. 103, 109–10, 229, 260
Gulline, A. B. 236

Haantjens, H. A. 43
Hagan, G. M. 153, 199
Haig, D. W. 73, 80, 85, 219
Haile, N. S. 38, 40–1, 67, 86
Hajos, M. 10, 38, 175, 203
Hakim, S. 71, 82
Hale, G. E. 264
Hall, R. L. 305
Hallam, A. 13, 17, 23, 50
Halse, J. W. 189
Ham, W. E. 340, 343
Hambrey, M. J. 235, 270–1
Hamilton, D. S. 249
Hamilton, L. H. 129, 131
Hamilton, N. J. 242
Hamilton, W. 41, 101, 112–13, 297
Hampton, M. 10, 38, 175, 203
Handby, P. L. 64, 91
Haniel, C. A. 71, 86
Hansen, L. 154, 187, 191, 193, 196–7
Hanson, E. R. 344
Harding, R. R. 118, 288, 305
Hargraves, R. B. 13
Harland, W. B. 1, 235, 270–1
Harrington, H. J. 249, 270, 280–1, 287, 289, 335
Harris, W. K. 54, 68, 88, 125, 138, 143, 146, 148, 161, 163, 219, 226–7, 230, 265
Harrison, C. G. A. 21, 213
Harrison, J. 112, 215, 265
Harrison, P. L. 162, 223, 274, 278, 341
Harrison, R. 158, 281
Hart, D. 43–4, 46
Hartmann, M. 267–8
Hartono, H. M. S. 41
Hawkins, L. V. 174, 195
Hay, W. H. 53–4
Hayes, D. E. 7, 9, 10, 16, 18, 19, 21–2, 28–9, 30, 32–5, 38, 54, 69, 80, 86, 93, 174–6, 182, 186, 202, 227–81, 232, 288
Heath, G. R. 148
Hedgpeth, J. W. 58, 72, 86
Hedley, C. 58, 86
Heezen, B. C. 188
Hegarty, K. A. 176
Hehuwat, F. 67, 90
Heirtzler, J. R. 19, 25–6, 182–6, 188
Helby, R. 131–2, 161, 227, 233, 236, 251, 256, 260, 262–3, 342
Henderson, R. A. 3, 75, 86, 93, 131, 136, 167, 281, 283, 289–92, 305, 307, 322, 327, 336
Hennig, W. 58, 86
Hennion, J. F. 174
Henry, M. 41, 296, 298
Henry, N. M. 298, 313–14, 327, 330–1
Hensel, H. D. 131

Henstridge, D. A. 132
Herbert, C. 131–2, 227, 233, 235–6, 241, 244, 249, 251, 253, 256–7, 260, 262
Hermes, J. J. 112, 114, 281, 328
Herron, E. M. 18, 19, 21–2, 28, 30, 34, 69, 80, 93, 202, 232
Heusser, C. J. 45
Heywood, P. B. 253
Hilde, T. W. C. 1, 288–9
Hill, D. 71, 73, 75–6, 86
Hills, E. S. 116–17, 119, 138, 280, 302, 340
Hilyard, D. 292, 304, 313
Hinz, K. 31, 34, 138, 188–9, 207, 227
Hirooka, K. 67, 90
Hiscock, I. D. 59, 88
Hocking, J. B. 57, 103, 118, 120–1, 135, 226
Hocking, R. M. 103, 120, 154, 189, 328
Hoffman, P. 153, 199, 277, 340, 344
Hoffstetter, R. 64, 86
Holcombe, C. J. 41
Holden, J. C. 196
Holdgate, G. 57, 103, 118, 120–1, 226
Holdsworth, B. K. 37, 203, 227
Hollingsworth, R. J. S. 219
Hollis, J. D. 98–9
Holloway, J. D. 67, 69, 86
Holmes, A. 202, 282
Holmes, G. G. 249
Holmes, W. B. K. 77, 86
Hooijer, D. A. 67–8, 83, 86
Hope, G. S. 42–5
Hope, J. H. 44
Hopley, D. 205
Hopwood, T. P. 270
Hornibrook, N. de B. 53
Horton, D. 59, 86
Horvath, G. J. 182, 186, 188
Hos, D. 51
Houston, B. R. 127, 134, 337
Houtz, R. E. 10, 20, 21, 38, 174–6, 181–2, 203, 228, 234
Huber, N. K. 282
Hughes, C. P. 3, 75, 93, 289–92, 305, 307
Hughes, R. J. 315
Huleatt, M. B. 162, 165, 313
Hunter, D. R. 149
Huntington, J. F. 100
Hussain, A. G. 13
Huxley, T. H. 67, 86
Hyndman, R. D. 182, 186, 188

Idnurm, M. 50, 110, 116–17, 119, 137, 227–8
Inger, R. F. 67, 86

Irving, A. J. 11, 98–9
Irving, E. 76, 86, 364
Irwin, M. J. 260
Isacks, B. L. 101
Ivanovsky, A. B. 75, 86

Jackson, M. J. 149–51, 153–4, 160, 163, 165–7, 236, 240
Jager, E. 3
Jago, J. B. 271, 281–2, 284–8
James, E. A. 118, 176, 204, 226, 228
Jan, Q. M. 305
Jardine, N. 67, 86
Jefferson, T. H. 80, 86
Jell, J. S. 49, 76, 86, 253, 255, 311, 329, 336
Jell, P. A. 73, 86, 281, 284, 310, 312
Jenkin, J. J. 117, 121
Jenkins, C. J. 293, 295
Jenkins, D. A. L. 110, 247, 253–6
Jenkins, D. G. 10, 38, 175, 203
Jenkins, R. J. F. 58, 73, 86
Jenkins, T. B. H. 76, 87, 311, 329
Jennings, J. N. 42, 45–7
Jensen, A. R. 131–2, 247, 251, 253, 255, 257, 336
Jensky, W. 142
Jenssen, D. 44
Jessop, R. G. C. 135
Jezek, P. 41
Johns, R. K. 146, 148, 216
Johnson, B. D. 13, 17–19, 23–7, 33–6, 38–41, 184–6, 188, 213, 219
Johnson, B. R. 3
Johnson, D. P. 153
Johnson, G. L. 9
Johnson, J. G. 71, 75, 83, 87, 363
Johnson, L. A. S. 64, 87
Johnson, R. W. 101, 104, 107, 109–10, 112, 254
Johnston, C. R. 41–2, 112
Johnstone, M. H. 54, 149, 155, 174, 182–3, 186, 188–9
Jones, B. G. 57, 161, 163, 165, 235, 313, 319, 340, 342
Jones, D. K. 151, 191
Jones, D. L. 37, 203, 227, 326
Jones, H. A. 156, 158, 181–2
Jones, J. G. 50, 95, 105, 115–16, 129, 131, 138–9, 168–9, 181, 189, 195, 203, 205, 216, 226–7, 245, 247, 249, 251, 254, 256–60, 267, 313–14, 317–18, 327, 335, 337
Jones, L. M. 288
Jones, P. J. 75, 79, 85, 87, 91, 158, 236, 311, 327–9
Jongmans, W. J. 78, 87
Jongsma, D. 138, 156, 158, 182
Joyce, E. B. 117, 123
Joyce, J. 97, 99
Jupp, D. L. B. 96

Kalf, F. R. 119
Kaljo, D. 75, 87
Kamerling, P. 201
Kaneps, A. G. 182–6
Kanstler, A. J. 197–8, 235, 313, 319, 342
Karig, D. E. 142, 277, 293, 295, 298
Karner, G. D. 31, 205, 229
Kauffman, E. G. 71–2, 80, 87
Kaufman, S. 347
Keast, A. 59, 64, 87
Keays, R. R. 302, 305
Keene, J. B. 295
Kemezys, K. J. 317
Kemp, E. M. 38, 45, 50, 53–5, 57, 65, 81, 87, 161, 163, 182, 186, 227, 233, 235–6, 263, 340, 342
Kemp, N. R. 264
Kempe, D. R. C. 182, 186, 188
Kenley, P. R. 57, 103, 117–18, 120–1, 226
Kennett, J. P. 10, 38, 42, 51, 53–5, 68, 81, 87, 175, 203, 257, 363
Kennewell, P. J. 162–3, 165, 341
Kent, D. V. 1, 32
Kerr, D. 96
Kerr, R. A. 69, 87
Kershaw, A. P. 45, 81, 90
Khaiami, R. 310, 312, 317–21, 329, 331–3.
Kidd, R. R. 293, 295
Kieckhefer, R. 41, 296, 298
Kikkawa, J. 59, 87
Killick, C. L. A. 312, 319
Kilpatrick, D. J. 292
Kind, H. D. 264
King, L. 123
Kinsman,˙D. J. J. 168
Kirkegaard, A. G. 132, 249, 264, 280, 284, 288–9, 305, 307, 327, 336
Kirschrink, J. L. 11
Klaamann, E. 75, 87
Klapper, G. 71, 87
Kleeman, J. D. 255
Klein, G. D. 37, 203, 227
Klootwijk, C. T. 11, 16, 39–40, 286
Kluth, C. F. 346
Knight, C. L. 215
Knox, G. A. 58–9, 87
Kobayashi, M. J. A. 114
Kobayashi, T. 73, 87
Kogler, F. C. 267–8
Konig, M. 20–2, 174, 176, 181–2, 228
Koppe, W. H. 132, 247, 249, 253, 255, 257
Korsch, R. J. 249, 335
Krasheninnikov, V. A. 182–3, 188
Kraus, G. P. 155–9, 182, 189, 191, 199, 229, 238, 244
Krayshek, C. 100

Krieg, G. 163, 341
Krijnen, W. F. 79, 87
Kroenke, L. W. 37, 203, 227
Krumbeck, L. 79, 87
Krumm, H. 267–8
Kukla, G. J. 42
Kummel, B. 78, 87
Kundig, E. 67, 87
Kuno, H. 282
Kurinin, R. G. 9
Kurylowicz, L. E. 235, 342
Kushiro, I. 98
Kyle, P. R. 9, 131–2, 136, 226–7, 232–3, 286
Kyle, R. A. 161, 236, 263, 342

La Brecque, J. L. 1, 32
Lacey, W. S. 78, 87
Lachenbruch, A. H. 105
Laing, W. P. 289
Laird, M. G. 227, 281, 286
Lambeck, K. 10
Lambert, I. B. 149, 183, 196–7, 280
Lambert, R. S. 3
Lamberts, I. T. 248–50
Lancaster, C. G. 131
Lang, W. H. 76, 87
Lange, R. T. 58, 88
Langford-Smith, T. 118, 126, 129, 227
Lanphere, M. A. 3
Larson, R. L. 1, 16, 25–6, 34, 38, 143, 182–6
Laurie, J. R. 3, 75, 88, 93, 289–92, 305, 307
Lawrence, C. R. 57, 103, 117–18, 120–1, 122, 226
Lawrence, M. B. 293, 295, 298
Laws, R. 155–9, 182, 189, 191, 199, 229, 236, 238, 244, 340
Lawver, L. 41, 296, 298
Leaman, D. E. 118
Ledbetter, M. T. 68, 84
Lee, C. 288–9
Legg, D. P. 75, 88, 307
Leggo, M. D. 118, 131, 227, 271, 286, 288, 300, 304
Leidy, R. D. 182, 186, 188
Leitch, E. C. 255, 264, 305, 322, 334
Leleshus, V. L. 75, 88
Lemaire, H. 96
Le Maitre, R. W. 188
Le Pichon, X. 5
Leslie, R. B. 215
Lewis, R. J. G. 96
Libby, W. G. 279, 283
Lightner, J. D. 312
Lilley, F. E. M. 94, 96
Lindner, A. W. 241
Lindsay, J. M. 68, 88, 118, 146, 216

Link, A. G. 71, 75, 88
Liu, C. S. 36
Llewellyn, P. G. 1
Lloyd, A. R. 55, 126, 156, 158, 160, 215, 251, 254
Lloyd, F. E. 99, 125
Loffler, E. 161, 245, 254, 256
Lofting, M. J. W. 189
Logan, B. W. 153, 199
Long, V. E. 228
Lord, J. H. 153
Loughnan, F. C. 262
Loutit, T. S. 257
Loveless, A. R. 64, 88
Lovering, J. F. 99, 131, 139, 180
Low, G. H. 49, 151, 153, 155, 157, 186, 188–9, 215, 226, 236, 238, 279, 328, 343
Lowry, D. C. 49, 54, 149, 151, 153, 155, 157, 182–3, 186, 188–9, 215, 226, 236, 238, 328, 343
Lu, R. S. 288–9
Luck, G. R. 13
Ludbrook, N. H. 215–16, 226–7, 263, 265
Ludwig, W. J. 195
Lundberg, N. 293
Lutze, G. F. 267–8
Luyendyk, B. P. 58, 88, 182, 186, 188
Lydekker, R. 67, 88

Mabesoone, J. M. 65, 89
Mabbutt, J. A. 43–4, 46
McBirney, A. R. 257, 363
McClenaghan, M. P. 118
McClung, G. 71, 88, 90, 236
McColl, D. H. 228
McCollum, D. W. 182, 186
McCracken, H. M. 18, 21, 36, 101
McCracken, K. G. frontispiece
McCulloch, M. T. 300, 302
McDonagh, G. P. 244
McDonald, J. M. 36
McDonnell, K. L. 131, 245, 247, 249, 251, 256–60, 335
McDougall, I. 9, 34–5, 38, 55, 65, 93, 104–5, 115, 117–18, 123, 126, 129, 131–2, 136, 138, 186, 188, 227, 232–4, 239, 241, 254–5, 262, 264, 271, 286, 288, 305, 337
McDowell, W. G. 3
McElhinny, M. W. 3, 11–13, 16, 38, 118, 364
McGowran, B. 50–1, 60, 81, 88, 118, 120, 124, 226–7, 229–30, 234
McHattie, C. M. 191
McIlveen, G. R. 332
McInerney, P. M. 148
McIntyre, A. 42, 44
McKellar, J. L. 134, 242

McKellar, R. G. 72, 76, 84, 88, 253, 311, 329
McKelvey, B. C. 72, 84, 182, 186, 188, 311, 329
McKenzie, D. 10, 34, 102, 220
Mackenzie, D. E. 110, 129, 134–6, 254
MacKinnon, D. I. 86, 286
McKirdy, D. M. 235, 342
McKnight, B. K. 182–3, 188, 199
McLaughlin, R. J. W. 332
McLean, R. A. 75, 88
McLennan, R. M. 117
McMichael, D. F. 59, 88
McMillen, K. G. 293
McPhie, J. 313
McTavish, R. A. 71, 75, 79, 85, 88, 238, 307
Macumber, P. G. 117–19, 121, 123, 227
McWhae, J. R. H. 241
McWilliams, M. O. 11, 16
Main, B. Y. 59, 64, 88
Major, R. B. 167
Malahoff, A. 21, 207
Mallett, C. W. 55, 118, 123
Malone, E. J. 132, 247, 249, 253, 258, 262, 336
Mammerickx, J. 21
Mankinen, E. A. 1, 32, 175
Margolis, S. V. 10, 38, 175, 203
Marjoribanks, R. W. 165
Markl, R. G. 16, 26, 34, 182, 185–6, 234
Marks, G. P. 181
Marsden, M. A. H. 311, 329, 332–3, 336
Martin, H. A. 65, 81, 88, 134
Martin, K. R. 131–2, 149, 242
Masolov, V. N. 9
Mathur, S. P. 162–3, 165, 196–7, 274, 278, 341, 345–8
Mattauer, M. 305
Mawson, D. 281
Maxwell, W. G. H. 72, 88, 205, 238
Mayr, E. 66–7, 88
Meath, J. R. 195
Mercer, J. H. 68, 88
Mercier, J. C. C. 98
Mertens, R. 67, 88
Meyer, G. M. 148
Michael, P. J. 186–8
Mildenhall, D. C. 69, 88
Middleton, M. F. 136, 172, 174
Mills, K. J. 293, 295
Milnes, A. R. 175, 232, 288
Milton, B. E. 96
Missen, D. D. 132
Mitchum, R. M. 57, 171, 212–3, 363
Mollan, R. G. 253, 288, 305
Molnar, P. 16, 21, 23, 168
Molnar, R. E. 80, 88

Mond, A. 223
Monger, J. W. H. 326
Moore, A. C. 270
Moore, D. G. 41, 296, 298
Moore, G. F. 293, 295, 298
Moore, J. C. 293
Moore, P. S. 236, 242, 262–3, 265, 267, 281, 285, 313, 319
Moore, R. F. 96
Moore, W. R. 118
Morel, P. 364
Morgan, G. A. 34
Morgan, P. 205
Morgan, R. 131–2, 139, 186, 199, 226–7, 233, 262–3, 266–7
Morley, M. E. 131, 139
Morris, N. 76–7, 88
Morris, W. A. 16
Mory, A. J. 334
Moss, F. J. 163, 165, 340
Mount, T. J. 268–9
Mountjoy, E. W. 313
Mouthaan, W. L. P. J. 43
Muir, M. D. 157–8, 160
Muirhead, K. J. 94, 96
Mukhopadhyay, M. 297
Müller, S. 60, 89
Murphy, P. R. 253, 255, 264
Murray, A. 60, 89
Murray, C. G. 132, 247, 255, 264, 280, 284, 289, 305, 307, 322–3, 327, 336–7
Murrell, B. 57, 270, 273, 275–6, 278
Mutter, J. C. 16, 18, 20–2, 26, 34, 36, 38, 138, 171, 174–6, 181–90, 205, 226, 228–9
Mysen, B. O. 98

Naeser, C. W. 3
Nafe, J. E. 174
Naini, B. 228
Napoleone, G. 344
Neall, V. E. 50, 233
Needham, R. S. 57, 270, 278
Neef, G. 281, 284, 310, 312
Neilson, J. L. 117
Nelson, E. C. 65, 89
Nelson, R. 96
Nesbitt, R. W. 270
Newton, R. S. 267–8
Nicholas, E. 229, 344
Nicholls, I. A. 181
Nicoll, R. S. 114, 238, 262
Ninkovich, D. 41
Nishimura, S. 67, 90
Norton, I. O. 13, 16
Norvick, M. S. 41–2, 81, 89, 191

Odin, G. S. 1, 3
O'Driscoll, E. S. T. 280

Offe, L. A. 165, 270, 343
Offenberg, A. C. 313
Olgers, F. 132, 249, 253, 315, 336-7
Oliver, J. E. 347
Oliver, W. A. 75, 89
Ollier, C. D. 124, 138
Onstott, T. C. 13
Opdyke, N. D. 42
Öpik, A. A. 73, 89, 284, 340
Oppel, T. W. 203
Osborne, D. G. 262
Otofuji, Y. 67, 90
Ovenshine, A. T. 10, 38, 175, 203
Oversby, B. 241, 305, 337
Owen, D. E. 313
Owen, M. 3, 80, 89, 291-2, 304, 312, 329
Ozimic, S. 235, 342

Packham, G. H. 3, 21, 37, 75, 92, 203, 227, 292, 294, 305, 307, 310-11, 313, 315, 317-19, 329
Page, C. 158, 281
Page, R. W. 111, 118, 136, 227, 254, 257, 266
Pain, C. F. 256
Paine, A. G. L. 131-2, 288, 305
Palfreyman, W. D. 248-50
Palmer, A. R. 1, 73, 89
Palmieri, V. 307
Papalia, N. 242, 244
Parbury, C. F. R. 257-9
Parker, A. J. 57, 270, 273, 275-6, 278, 289
Parker, W. C. 73, 78, 93
Parkin, L. E. 143, 146, 148, 272, 276, 283-4
Parrish, J. T. 73, 78, 93
Partridge, A. D. 65, 91, 118, 176, 203-4, 226-7
Paten, R. J. 242, 244, 253, 262
Paton, I. M. 265
Pattinson, R. 176, 178
Pearce, J. A. 312
Pearce, L. G. G. 187-8, 191, 195, 211, 229, 238
Pearse, K. 59, 87
Pei-Hsin Chen 182, 186
Peirce, J. W. 39-40
Perch-Nielsen, K. 10, 38, 175, 203
Percival, G. 281, 284, 310, 312
Percival, I. G. 75, 89
Petersen, M. S. 71, 76, 89
Peterson, J. A. 44
Petit, J. R. 45
Phelps, D. W. 344
Phillips, B. F. 59, 89
Phillips, G. N. 302
Phillips, J. D. 58, 88
Pickett, J. W. 71, 75-6, 89, 91, 291, 311-12, 315, 329, 333

Pickton, C. A. G. 1
Pidgeon, R. T. 286, 288, 300, 304
Pieters, P. E. 253-4
Pigram, C. J. 71, 82, 135, 253
Pilger, R. H. 104-5
Pinchin, J. 129, 135, 227, 251, 336, 347
Piper, D. J. W. 182, 186
Pitman, W. C. 118, 212-13, 363
Pitt, G. M. 57, 144, 160, 162, 163, 165, 270, 273, 275-6, 278
Pittock, A. B. 44
Platnick, N. 64, 89
Playford, G. 57, 131-2, 161, 235-6, 263-4, 340, 342
Playford, P. E. 49, 79, 83, 151, 153, 155, 157, 186, 188-9, 215, 226, 236, 238, 241, 279, 328, 340, 343
Plumb, K. A. 57, 94, 96, 157-8, 160, 270, 272, 274, 278-81, 342, 355
Plumstead, E. P. 78, 89
Pogson, D. J. 283, 292, 304
Pojeta, J. 73, 75, 89
Pope, E. C. 58-9, 83
Powell, A. W. B. 58, 89
Powell, B. N. 344
Powell, B. S. 135
Powell, D. McA. 3, 13, 17-19, 23-7, 33-6, 38-41, 95, 168-9, 181, 184-6, 188, 195, 198, 213, 267, 281, 284, 290, 292-6, 298, 300-2, 305, 310, 312-14, 316-21, 327, 329, 331-4, 337
Powell, D. E. 41, 114, 159, 182, 195, 198, 226, 229
Powell, J. M. 254
Powell, T. W. 19, 25-6, 182-6
Priess, W. V. 57, 270, 273, 275-6, 278
Price, P. L. 161, 236, 263, 342
Price, R. C. 300, 303
Proto-Decima, F. 182-3, 188
Proust, F. 305
Pruatt, M. A. 343-4, 346
Purdy, G. M. 41

Quilty, P. G. 26, 27, 34, 48, 50, 54, 55, 118, 120, 123-4, 149, 153, 155, 182-4, 187-9, 199, 200, 215, 226, 229-30
Qureshi, I. R. 334

Raheim, A. 271, 280, 287-8
Raitt, R. W. 41, 296, 298
Rand, H. M. 65, 89
Ranford, L. C. 48, 163, 165, 167, 270, 278, 327-8, 341-2, 344
Rankin, P. C. 9

Ransom, D. M. 295, 319
Rao, C. P. 49, 238
Rasidi, J. S. 129, 134-5
Ratman, N. 71, 82
Rattigan, J. H. 193, 235, 264
Raven, P. H. 59, 68-9, 80-1, 89
Ray, A. S. 312, 316
Raymond, A. 73, 78, 93
Read, J. F. 153, 199
Rieter, L. 343-4
Renz, G. W. 182-3, 188
Retallack, G. J. 77-9, 89, 236, 257, 259-60, 263
Reynders, J. J. 43
Rich, P. V. 48-9, 55, 57, 59, 64, 81, 89
Richards, D. N. G. 241
Richards, J. R. 118, 131-2, 227, 241, 255, 284, 286, 288-9, 304, 312
Rickard, M. J. 315
Ridd, M. F. 57, 109, 129, 137, 215, 229, 251, 254
Riek, E. F. 79, 89
Rigby, J. F. 78, 89
Riley, G. H. 300, 302
Ringis, J. 28-30, 69, 80, 86, 176
Ripper, D. 57, 103, 118, 120-1, 226
Ritchie, A. 312, 320, 331
Rixon, L. K. 193, 264
Roberts, J. 3, 49, 72, 76, 89, 158-9, 235-6, 238, 241-2, 244, 259-60, 311, 329, 334-7, 340
Robertson, A. D. 134
Robin, G. de Q. 9
Robins, S. 41, 110, 112, 155, 157-8, 199
Robinson, G. P. 253-4
Robinson, P. T. 182-3, 188, 199
Robinson, V. A. 120, 226, 228
Rochford, D. J. 59, 93
Rocker, K. 182-3, 188
Roddick, J. C. 304
Rodolfo, K. S. 182, 186, 188
Roeder, D. 344
Roeser, H. A. 19, 25-6, 182-6, 188
Roggenthen, W. 344
Rognon, P. 45
Roksandic, Z. 131
Rollason, R. G. 255
Roots, W. D. 30, 169, 174, 176-7, 184, 186, 313, 317-8
Rose, D. M. 313
Rose, G. 310, 312
Ross, C. A. 70, 76, 90
Ross, J. A. 99
Ross, R. J. 3
Roussopoulos, G. 158, 281
Royce, K. 131, 247, 249, 251, 256-60
Royer, A. 45
Rubenach, M. J. 288, 305
Runcorn, S. K. 11

Rundle, A. S. 45, 47, 143, 269
Runnegar, B. 71-3, 76, 78-9, 90, 236, 249
Russell, H. Y. 313
Russell, T. G. 333
Rust, B. R. 45, 281
Rutland, R. W. R. 57, 94-5, 97, 270, 273-6, 278, 289, 302, 340, 343
Ruxton, B. P. 254
Ryburn, R. J. 129, 134-6, 254

Sabitay, A. 163, 165
Saito, T. 37, 203, 227
Salinger, M. J. 45
Saltzman, E. 213
Sasajima, S. 67, 90
Sass, J. H. 104
Scheibner, E. 249, 270, 277, 280, 283, 302, 305, 310, 312-13, 316-21, 329, 331-5
Scheibnerova, V. 50, 72-3, 80, 90, 219
Schleiger, N. W. 290, 292
Schlinger, E. I. 58, 90
Schmidt, P. W. 11, 54, 131-2, 136, 139, 229, 232-3
Schopf, T. J. M. 78, 90
Schuepbach, M. A. 171
Schult, A. 13
Schwarzbock, H. 127, 132-4, 227, 236, 238, 241-2, 244, 255, 260, 264
Schwebel, D. A. 216
Sclater, J. G. 10, 13, 19, 35, 220
Sclater, P. L. 60, 90
Sclater, W. L. 60, 90
Scotese, C. R. 364
Scott, P. A. 315, 320-1, 332
Seddon, G. 311, 329
Seibertz, E. 26, 182, 184, 187-8
Seibold, E. 267-8
Semeniuk, V. 153
Sengor, A. M. C. 168, 205
Senior, B. R. 50, 127, 131-2, 134, 137, 215, 219-20, 223, 227-8, 238, 267, 347
Sepkoski, J. J. 73, 78, 93
Shackleton, N. J. 42, 51, 53-5, 81, 87
Shafik, S. 26, 37, 182, 184, 187-8, 203, 227
Shaw, R. D. 30-1, 34, 57, 132, 165, 264, 270, 278-9, 336, 343-4, 347-8
Shaw, S. E. 239, 241, 257, 300, 302, 313
Shell Development (Australia) Pty Ltd 201
Shepherd, J. 100
Sheraton, J. W. 9, 241
Shergold, J. H. 73, 75, 90, 157-8, 160, 167, 278, 281

Sheridan, M. F. 232
Sherwin, L. 75, 90, 292, 312, 317
Shibaoka, M. 132, 247
Shipley, T. H. 293
Shirahase, T. 304
Shor, G. 41, 288-9
Simberloff, D. S. 78, 90
Simes, J. E. 286
Simon, B. K. 65, 84
Simpson, D. W. 94, 96
Simpson, G. G. 60, 66-7, 90
Singh, G. 42, 45-7, 81, 90
Singleton, O. P. 55, 118, 284, 286, 288-9, 304
Skwarko, W. K. 79-80, 90, 114, 132, 135, 158-9, 238, 254, 262
Slessar, G. C. 126, 129, 227
Sloan, J. 213
Sloane, M. N. 94, 96
Small, G. R. 100, 202
Smart, J. 129, 134-5, 227, 251
Smit, C. 272, 280
Smith, A. 292
Smith, A. G. 1, 13, 17, 23, 80, 91
Smith, D. N. 103, 148, 154
Smith, I. E. 110, 112, 254
Smith, J. G. 193, 264
Smith, J. P. 71, 91
Smith, R. E. 292
Smith, S. M. 21
Snelson, S. 347-8
Soffel, H. C. 13
Sofoulis, J. 280
Soil Map of New Guinea 43
Solomon, M. 305
Southam, J. R. 53-4, 213
Southgate, P. N. 48
Specht, R. L. 65, 91
Spencer, B. 59, 91
Spencer-Jones, D. 310-12, 315-17, 319, 321, 326, 329
Sprigg, R. C. 73, 91
Spry, A. 271, 283
Stagg, H. M. J. 182, 186-8
Staines, H. R. E. 132, 247, 249, 253, 255, 257
Stait, B. 3, 75, 93, 289-92, 305, 307
Stanton, R. L. 315
Stauffer, M. R. 315
Steckler, M. S. 212-13
Steed, R. H. N. 9, 234
Steenis, G. G. G. J. van 64, 91
Stefanski, M. Z. 157-8, 160, 281
Steiger, R. H. 3
Steiner, J. 329, 331-3
Stephenson, P. J. 104, 129, 202, 227
Stevens, G. R. 50, 79, 91
Stevens, N. C. 126, 227
Stewart, A. J. 342-3, 347
Stipp, J. J. 118
Stirton, R. A. 54
Stock, J. 21, 23

Stoeser, D. G. 37, 203, 227
Stoffa, P. L. 189
Stone, D. B. 34
Stover, L. E. 65, 91
Street, F. A. 47
Strusz, D. L. 75, 91, 319
Stuart, W. J. 146
Stubbs, D. 126, 129, 227
Stump, E. 232, 271, 274
Stuntz, J. 249, 258
Stupavsky, M. 16
Suggate, R. P. 264
Suhardi, S. 19, 25-6, 182-6
Sullivan, M. E. 43-4, 46
Suppel, D. W. 312, 317
Sutherland, F. L. 98, 104, 117-18, 126, 129, 202, 227
Sutter, J. F. 68, 88, 131-2, 136, 226-7, 232-3
Sutton, D. J. 100, 202
Sweeney, J. F. 201
Swindon, V. G. 131-2
Swoboda, R. 246, 248-50
Symons, D. T. A. 16
Symonds, P. A. 26, 31, 34, 38, 138, 182-6, 207, 227-8

Tahirkheli, R. A. K. 305
Talent, J. A. 48, 58, 64, 67, 69, 71, 75, 83, 91, 121, 189, 291, 311, 313, 315, 317, 319
Tallis, N. C. 110, 247, 253-4
Talwani, M. 20-1, 174, 176, 181-2, 189, 228
Tanner, J. J. 322, 326
Tapponier, P. 168, 305
Tarling, D. H. 16-17, 38
Tasch, P. 79, 91
Tate, R. 59, 91
Taylor, C. P. 253
Taylor, D. J. 22, 38, 50, 68, 80, 91, 172, 176, 180
Taylor, G. 131-2, 227, 233, 262
Taylor, L. 37, 205
Taylor, M. L. 153, 175-6, 178, 211, 226, 265
Taylor, S. R. 100, 300, 303
Taylor, T. G. 138
Tedford, R. H. 54, 59, 91
Teichert, C. 71-2, 75, 91, 193
Telford, P. G. 71, 76, 91
Tharp, M. 188
Thayer, P. A. 182-3, 188, 199
Thierstein, H. R. 182, 186, 188
Thom, B. G. 45
Thomas, B. M. 103-4, 154, 262
Thomas, D. E. 75, 91
Thomas, G. A. 72, 91
Thomson, B. P. 284
Thomson, J. 264
Thompson, B. R. 280-1

AUTHOR INDEX

Thompson, E. M. 59, 81, 89
Thompson, R. 13
Thompson, S. 57, 171, 212-13, 363
Thompson, T. 343-4
Thomson, M. R. A. 3, 75, 93, 289-92, 305, 307
Thornton, R. C. N. 148, 226, 236, 242, 262, 269
Threlfall, W. F. 57, 103, 118, 120-1, 123, 176, 203, 226
Thunell, R. C. 257, 363
Tilbury, L. A. 146, 174-6, 178-81, 211, 226, 228
Times Atlas of the World 106, 116, 129, 144, 150, 156, 162, 288
Times Concise Atlas of the World 43, 46
Tjokrosapoetro, S. 41, 191
Townley, K. A. 72, 92
Townrow, J. A. 50, 80, 92
Townsend, I. J. 148
Trendall, A. F. 149, 279-80, 282, 307
Truswell, E. M. 233, 236
Tucker, D. H. 162, 274, 278, 315-17, 326, 341
Tucker, R. M. 253
Turner, N. J. 118
Turner, S. 158, 327-8
Twidale, C. R. 146, 148, 161, 163, 227

Underwood, R. 100, 202
Uren, R. 257
Uyeda, S. 298

Vail, P. R. 57, 171, 212-13, 363
Vakhrameev, V. A. 80, 92
Valentine, J. W. 58, 92
Vallance, T. G. 298, 300, 311-12
van Andel, T. H. 155-6, 158, 182
Van Baren, F. A. 43
van Bemmelen, R. W. 67, 83
van de Graaff, W. J. E. 103, 149-51, 153-4, 160, 165-7, 223, 236, 240
van den Abeele, D. 176, 178
VandenBerg, A. H. M. 3, 75, 93, 289-92, 305, 307, 310-12, 315-17, 319, 321, 326, 329
van der Lingen, G. J. 35, 37, 131, 203, 227
Van der Voo, R. 364
Van Donk, J. 42
Van Hinte, J. E. 3, 218
van Leeuwen, T. 67, 90
Vanney, J.-R. 9

Veeh, H. H. 110
Veevers, J. J. 6, 10, 13, 17-19, 22-7, 33-6, 38, 40-1, 45, 47, 49-50, 72, 76, 92, 95, 105, 110, 112, 115-16, 120, 129, 131-2, 134, 138-9, 142-3, 149, 154-9, 168-71, 177, 179, 181-9, 191-3, 195-9, 201, 203, 205, 213, 216, 223, 226-7, 234, 236, 238, 240, 244, 254, 258, 267, 269, 273, 275, 281, 307, 327, 337, 340, 350, 352
Venkatakrishnan, R. 323, 325
Verma, R. K. 297
Vernon, R. H. 257, 300
Vine, R. R. 227, 244, 262, 335
Visser, W. A. 112, 114, 281, 328
Vogt, P. R. 34
Voisey, A. H. 244
von der Borch, C. C. 42, 174, 216, 270, 272, 275-8, 280, 345
von Herzen, R. P. 203
von Rad, U. 26, 182, 184, 187-8
von Stackelberg, U. 26, 182, 184, 187-8
Vos, R. G. 191
Vozoff, K. 96

Wade, M. 73, 75, 85, 92
Wade, R. T. 79, 92
Walcott, R. I. 20, 103
Waldman, M. 50
Wales, D. W. 154, 159, 165, 226, 236, 238, 264-5, 328
Walker, C. T. 123
Walker, D. J. 9, 42, 45-6
Walker, I. W. 154, 191, 240
Walkom, J. B. 81, 92
Wall, R. E. 182, 186
Wall, V. J. 302
Wallace, A. R. 67, 92
Walter, M. R. 3, 270, 278
Walters, R. 1
Warren, A. 79, 92
Warren, G. 9
Warris, B. J. 189
Wass, R. E. 72, 92
Wass, S. Y. 98-100
Wasson, R. J. 42
Waterhouse, J. B. 3, 71, 78, 92, 253, 255, 260-1, 263
Watkins, G. 176, 178
Watkins, J. S. 293
Watkins, N. D. 282
Watts, A. B. 9, 18-19, 28-9, 31, 37, 202, 205, 212-13, 227
Watts, D. R. 11
Webb, A. W. 118, 126, 129, 132, 136, 227, 239, 241, 255, 260, 264, 288, 305, 337

Webb, J. A. 3, 257, 260-1
Webb, P. N. 50, 182, 186, 233
Webby, B. D. 3, 48, 75, 92-3, 286, 289-92, 294, 305, 307-12, 318, 329, 333
Weber, C. R. 249
Weber, M. 60, 93
Webers, G. F. 3, 75, 93, 289-92, 305, 307
Webster, R. E. 341, 343
Weekes, J. 100
Wegener, A. 78, 93, 107-8
Weissel, J. K. 7, 9-10, 16, 18, 19, 21-2, 28-30, 32-5, 37-8, 69, 80, 93, 174-6, 181, 202, 205, 226-7, 232
Wellman, P. 9, 18, 21, 36, 65, 93-4, 96, 101-2, 104-5, 115, 117-18, 123, 126, 129, 131-2, 136, 138, 185-6, 188, 199, 227, 232-4, 262, 264, 284, 340
Wells, A. T. 48-9, 56-7, 154, 161, 163, 165, 167, 238, 270, 278, 327-8, 340-2, 344
Werner, F. 267-8
Whitaker, W. G. 127, 134, 136, 247, 255, 322-3, 327, 336-7
White, A. H. 78, 93
White, A. J. R. 99, 300, 302, 304, 321, 332
White, D. A. 307
White, E. I. 76, 93
White, M. E. 158-9
Whitehouse, F. W. 131
Whiteley, R. J. 228
Whitley, G. P. 58, 93
Whitman, J. M. 213
Whitmore, T. C. 59, 67, 69, 93
Whittington, H. B. 75, 93
Whitworth, R. 9, 234
Whyte, R. 176, 178
Wickham, J. 343-5
Wilde, S. A. 154, 191, 240
Wilkes, P. G. 162-3, 165, 341
Wilkinson, J. F. G. 99-100, 118
Willcox, J. B. 31, 34, 138, 174, 176, 182, 204, 207, 227
Williams, A. 75, 93
Williams, A. J. 312-13, 319, 333
Williams, E. 271, 282, 284, 321, 332
Williams, F. M. 169
Williams, G. E. 146-7, 230, 270, 278, 280, 282
Williams, I. R. 279-80
Williams, I. S. 300, 302, 304, 321
Williams, M. A. J. 42, 45, 48, 169
Williams, P. F. 295, 319
Williams, P. R. 118
Williams, P. W. 254
Williams, R. M. 119
Williams, S. J. 279-80
Willmott, W. F. 241

Wilson, C. J. L. 295, 319
Wilson, J. L. 340, 343
Wilson, J. T. 168
Wiltshire, M. J. 241, 262, 267
Windle, D. L. 71–2, 85
Wiseman, J. F. 199, 226
Withnall, I. W. 255, 264, 288, 305
Woodburne, M. O. 54, 63, 93, 161, 163
Woods, D. V. 94, 96
Woods, P. J. 153, 199
Woodward, S. P. 72, 93
Woolley, D. R. 118–19, 122–3, 227
Wopfner, H. 54, 125, 132–5, 137–8, 143, 146, 148–9, 161, 163, 166, 219, 226–8, 230, 234, 236, 239–40
Worotnicki, G. 100–1
Worsley, T. R. 53–4
Wright, A. J. 262
Wright, C. A. 157–9, 199
Wright, C. W. 80, 93
Wyatt, B. W. 49, 118, 162, 274, 278, 315–17, 326, 341
Wyatt, D. H. 126, 129, 227, 288, 336
Wyborn, D. 3, 291–2, 304, 312, 329
Wyborn, L. A. I. 302

Wyrtki, K. 59, 93

Yeates, A. N. 315–17, 326
Yoder, H. S. 98
Yoo, E. K. 236
Young, G. C. 76, 93, 320–1, 332
Young, R. W. 118, 123
Youngs, B. C. 163, 165

Zhuravleva, I. T. 73, 93
Zidgerveld, J. D. A. 11
Ziegler, A. M. 73, 78, 93, 364
Zillman, J. W. 44
Zinsmeister, W. J. 63, 93

SUBJECT INDEX

ABC Quartzite 273
Abercorn Trough 241
Abrolhos Sub-basin 185, 193, 197–8
Adavale Basin 49, 56, 322, 326–7, 335–7, 347
Adelaide Fold Belt 143, 270, 272–3, 275, 289
Adelaide Rift 270, 275–8, 280, 345
Albany-Fraser Province 149, 183, 270
Albert Edward Group 275
Aldebaran Sandstone 243
Aldgate Sandstone 273
Algebuckina Sandstone 265, 267
Allandale Formation 237
Allaru Mudstone 218, 220
Alpine Fault Zone 20
Amadeus Basin 48–9, 56–7, 160, 162, 165–7, 235, 270, 272–3, 278, 284, 289, 307, 327–8, 337, 340–5, 347–8
Amadeus Transverse Zone 149, 154–5, 160–5, 166, 270, 274, 278, 280, 327, 337, 340–2, 344–5, 347–8, 351
Anabama Granite 286, 288
Anakie Metamorphics 288–9
Anakie Ridge 247, 261, 289, 305, 307, 336–7
Anchor Cay-1 124, 201, 203
Anderson Formation 56, 237
Antarctic Plate 37
Antrim Plateau Volcanics 189, 191, 282–3
Aquarius-1 124, 203, 205, 207
Arafura Basin 155, 157, 160, 281, 284
Arafura Sea 106–7, 110
Aralka Formation 273
Aramac Coal Measures 243
Ararat Formation 235, 238
Arckaringa Basin 143, 148, 161, 165–6, 236, 239
Areyonga Formation 273
Argo Abyssal Plain 21–2, 26, 183, 185–6, 189, 196, 205
Argyle Formation 317
arkose at Ayers Rock 166–7
Arumbera Sandstone 166–7, 273, 278
Arunta Block 95, 151, 160, 162, 166–7, 185, 270, 273, 337, 341, 344
Ashmore Block 186, 191, 199, 265
Ashmore Reef-1 188, 196, 199, 238

Auburn Granite 238
Aure Beds 254
Aure Trough 111
Aurora sub-ice basin 9, 234
Austral Downs Limestone 126
Australian–Antarctic Depression 6, 9–10, 110, 143–4, 176, 182, 221, 229, 232, 234, 239, 241, 266, 351, 354
Australian–Antarctic discordance 7, 9
Avoca Coal Measures 263
Avon River Group 332
Awitagoh Formation 281–2

Babbagoola Beds 163, 166–7
Balimbu Greywacke 135
Ballast Beds 292
Bancannia Trough 282, 287
Bandanna Formation 243
Bangemall Basin 270
Baragwanath Anticline 216
Barcoo Marginal Sea 287, 289
Barlee-1 188, 193, 196
Barrow Sub-basin 183, 185, 193, 195, 200
Basleo Beds 114
Bass Basin 50, 117, 120, 178, 216, 220, 225–6, 229
Bathurst Granite 238, 313, 348
Bayford Granite 132
Beagle Basin 151, 153–4
Beagle Ridge 197–8
Beardmore Group 271
Beaumont Dolomite 273
Bedout-1 188, 196
Bedout Sub-basin 185
Bega Batholith 303, 305, 319, 321
Bejah Claystone 154, 268
Belair Subgroup 273
Bellbird Formation 331
Bells Creek Volcanics 315
Benambra Complex 263–4
Berridale Batholith 302–3, 305, 321
Berridale Wrench 321
Berriedale Limestone 238, 241
Berry Formation 256
biogeography 57
Billy Creek Formation 285
Birdsville Basin 57, 125, 127, 138, 143, 145–6, 153, 161, 165, 218, 221, 227, 229, 231–2
Birdsville Track Ridge 265, 267
Birkhead Formation 265, 267, 269
Bismarck Granite 135

Bitter Springs Formation 273, 278, 340, 347
Black Alley Shale 243
Black Hill Norite 288
Blackwater Group 243
Blina Shale 237, 241
Bogan Gate Platform 317, 320
Bonaparte Gulf Basin 49, 155–7, 159–60, 183, 186, 189, 191, 193, 198–9, 200, 220, 236, 239, 241, 283–4, 307, 327–8, 340, 348
Bookpurnong Beds 119
Boolcunda Basin 145, 148
Boolcunda Coal Measures 263–4
Boomer Formation 243
Boorthanna Formation 237
Boree-1 56
Boree Salt Member 56
Bourke Line 287–9
Bowen Basin 112, 127, 131–2, 138–9, 236, 239, 241–3, 260–2, 267, 337
Bowers Group 282
Brachina Formation 273
Brady Formation 264
Braeside Tillite 154
Bremer Basin 53, 149, 151, 153, 181, 226
Brigadier Beds 237
Brighton Limestone 273
Broken Ridge 37
Broken River Province 322–3, 336
Broome Platform 307
Broome Sandstone 188
Browse Basin 56–7, 155, 157–9, 183, 185–6, 193, 198–9, 200, 265
Buchan Group 49
Buchan Trough 317
Buckabie Formation 322, 335
Budawang Land 317
Buffel Formation 243
Bulganunna Volcanics 239
Bulgoo Formation 320
Bulimba Formation 135
Bumbolee Creek Formation 313, 316
Bunbury Basalt 186, 188, 196–7, 266
Bunbury Trough 198, 201
Bundamba Group 263
Bundarra Suite 239
Bundock Formation 336
Bundycoola Formation 321

413

SUBJECT INDEX

Bungil Formation 134, 266
Bunyeroo Formation 273
Burngrove Formation 255
Burragate Fault 321
Burra Group 270, 273, 275, 277
Burrum Coal Measures 132, 134
Byro Group 237

Cadna-owie Formation 266-7
Calivil Sand 119-20
Callana Beds 270, 275-7
Callide Basin 236
Callide Coal Measures 262-3
Calliope Arc 322, 326-7, 335-6
Callytharra Formation 237-8
Camboon Andesite 239, 243
Canning Basin 48-9, 56, 71, 100, 102, 106, 151, 154, 155-6, 159, 160-1, 165-7, 183, 185, 188-9, 191, 193, 217-18, 236, 239, 241, 265, 268, 284, 307, 327-9, 340
Canowindra Porphyry 315
Cape Jervis Beds 237
Cape Range Fracture Zone 23-4, 33, 38, 186, 195, 198, 266
Cape River Beds 288-9
Capertee High 313, 317, 320
Capricorn-1 124, 203, 205, 207
Capricorn Basin 57, 126, 132, 201-2, 205, 227
Captains Flat Trough 313
Cargo Andesite 292
Carnarvon Basin 49, 53, 56, 71, 102-4, 151, 153-4, 183, 185, 189, 191, 193, 199, 220, 232, 236, 239, 241, 268, 284, 307, 327-8, 340, 348
Carnarvon Terrace 38, 185, 193, 196
Carpentaria Basin 127, 129, 134-5, 137, 223, 227
Carrandibby Formation 237
Carranya Beds 166-7
Carribuddy Formation 49, 56, 328
Carriers Well Limestone 307
Carryer Conglomerate 286
Caryneginia Formation 237-8, 241
Cascade Plateau 201-2, 204
Casterton Beds 135
Cathedral Beds 320
Catherine Sandstone 243
Catomball Group 329
Cato Trough 201-2, 204-5, 207, 209
Cattle Creek Formation 243
Cecil Plains Coal Measures 263
Ceduna Depocentre 138, 142, 147-8, 153, 155, 176-9, 181-2, 201, 217, 220-5, 228-32, 234, 269, 354, 363-4
Ceduna Plateau 174, 180

Central–Eastern Lowlands 106-7, 144, 149, 160, 162, 167-8, 214, 218, 230, 269
Central Flinders Zone 277
Channel Country 45, 47, 143
Chesleigh Formation 317
Childers Formation 121
Chim Formation 135
Circum-Antarctic Current 38, 52, 54, 84, 210, 354
Clarence-Moreton Basin 133, 236, 260, 262-4
Clarke River Fault 284
Clematis Group 241, 243, 255
Clyde Coal Measures 237
Cobar Trough 310-12, 317
Cockatea Shale 237, 340
Cockatoo Sandstone 340
Cocoparra Group 329
Coen Inlier 95
Colinlea Sandstone 243
Collie Basin 154, 191, 235-6, 239
Collie Coal Measures 237
Connors–Auburn volcanic arc 336-7
Condren Member 241
Cooee Point Dolerite 287
Coolac Serpentinite 313, 316
Coolcalalaya Sub-basin 185
Cooma Metamorphic Belt 298
Cooper Basin 103, 143, 145, 149, 200, 235-6, 239, 241-2, 261-2, 267, 269
Coorabin Coal Measures 237
Copperhannia Thrust 310
Copper Mine Range Beds 281, 283
Coral Sea 28-9, 37, 80, 108, 112, 114, 137, 201, 204-5, 207, 209-10, 227, 231-2, 354
Cordillo Surface 227
Cork Fault/Weatherby Structure 284
Corryong Batholith 303
Cowra Trough 310-11, 313, 315, 317, 319
Crater Mountain Volcanics 254
Craven's Peak Beds 49, 162, 166, 327-8
Crowl Creek Formation 331, 333
Crown Point Formation 237
Crudine Group 318
Cumberland Sub-Group 256
Cummins Basin 145
Cunningham Formation 319
Curnamona Craton 94, 96, 274-5
Currarong Orogen 249
Cuvier Abyssal Plain 26, 33, 183-7, 196, 198

Daly River Basin 48, 56, 157, 159, 160
Dampier Basin 102, 104, 151, 153-4, 183, 185-6, 191, 193, 195-6, 198-9, 200, 220

Dampier Ridge 34-5, 202
Danai Limestone 135
Dandaragan Trough 197
Darai Limestone 254
Daralingie Beds 243
Darling Basin 310, 312, 317, 319
Darling Fault 149, 154-5, 183, 191, 196-7, 201, 280
Darling Lineament 310
Darling Mobile Zone 149, 183
Davies Bore Conglomerate 166-7
Dean Quartzite 278
Demon Block 335
Dempsey Formation 256
Denison Sub-group 292
Denman Basin 143, 148, 236
Diamantina Lineament 284
Diamantina Zone 174, 181
Digger Island Limestone 292
Dirk Hartog Limestone 49, 328
Double Mountain Volcanics 132
Douro Group 315
Drummond Basin 336-7
DSDP sites:
 207 35, 131, 203
 209 201, 203, 205
 210 203
 257 27
 258 186, 188
 259 188
 260 186, 188
 261 183, 188
 262 110
 263 184
 264 186, 188
 265 35, 233
 266 35
 267 B 35
 268 35
 269 35
 277 57
 279 57
 280 35, 175, 203
 281 57, 203
 282 35, 175
 283 35, 176, 203-4
 287 37, 203, 205
Duau Volcanics 254
Duaringa Basin 132
Ducabrook Formation 237
Duck Bay Basalt 135
Duddo Limestone 121
Duerdin Group 275
Dulcie Sandstone 166
Dulcie Syncline 162
Dundas Trough 281-2, 284
Dungeree Volcanics 315
Duntroon Basin 143, 145-6, 228

Eastern Gondwanaland 17-18, 20, 23, 37

Eastern Highlands 6–7, 10, 45, 47, 105–7, 112, 114–15, 125–7, 137–8, 142, 144–5, 147, 167–8, 182, 202, 204–5, 207, 211, 214, 217–18, 220–1, 223, 228–9, 230–2, 234, 251, 265, 267, 354, 363
eastern margin 201
Echidna-1 175–6
Edel-1 188, 191, 196
Eden–Comerong–Yalwal Rift 320–1, 332–3
Edward's Island Block 197
Einasleigh Metamorphics 305
Elatina Formation 273
Elliott Formation 134
Elmside Formation 317, 319
Encounter Bay Granites 288
Enderby-1 188, 191, 196
Eninta Sandstone 166–7
Epsilon Formation 243
Era Beds 254
Eregunda Sandstone Member 285
Erins Vale Formation 256
Eromanga Basin 57, 96, 103, 129, 131–2, 138–9, 143, 145, 148, 161, 165, 178, 218–19, 222, 227, 235–6, 262–3, 265–8, 347
Erskine Sandstone 237
Esk Trough 241, 260
Etadunna Formation 216
Etonvale Formation 322
Ettrick Marl 121
Eucla Basin 53, 143, 145–6, 149, 151, 153, 160–1, 181, 216–17
Eumeralla Formation 135
Euroka Arch 129, 220, 223
Evans Head Coal Measures 263
Everett Creek Volcanics 307
Evergreen Formation 134, 238
Exmoor Formation 255
Exmouth Plateau 38, 182–3, 186, 191, 193, 195, 196, 198, 200, 205, 265
Exmouth Plateau Arch 185
Exmouth Sub-basin 183, 185, 198
Expedition Sandstone 260
Eyre Formation 125, 143, 153, 161, 216
Eyre Sub-basin 151, 153, 181–2

Fairbridge Volcanics 292
Fair Hill Formation 255
Fairymead Beds 134
Farley Formation 237
Featherby Surface 227
'Feldspathic Sandstone' 264
Fermoy-1 289
Ferrar Dolerite 131, 226, 233, 282–3
Fingal Coal Measures 263

Finke Group 161, 166
Fitzroy Lamproite 188, 199, 282
Fitzroy Trough 159, 183, 185, 193, 264, 348, 351
Florentine Valley Formation 292
Fork Lagoon Beds 307
Fort Cooper Coal Measures 255
Fossil Cliff Formation 237
Fossil Head Formation 237, 241
Freitag Formation 243

Galilee Basin 236, 239, 241–2, 247, 251, 259–62
Gambier–Beaconsfield fault zone 274, 280, 284
Garra Formation 49
Garrawilla Volcanics 263–4
Gascoyne Abyssal Plain 183, 185–6, 195–6, 198
Gatelee Ignimbrite 316
Gawler Block 94–6, 143, 146, 148, 161, 166
Gawler Ranges Volcanics 148, 266
Geera Clay 121
Gemini-1 175
Georgetown Inlier 95, 99, 270, 272, 288–9, 305, 307, 337
Georgina Basin 48, 157, 160, 162, 167, 270, 273, 278, 284, 289
German Creek Formation 255, 260
Gilbert River Formation 135
Gilmore Fault Zone 316
Gippsland Basin 50, 55, 102–3, 115, 117–21, 132–5, 137, 139, 178, 199, 201–5, 207, 210, 216, 220, 225–6, 228–9, 232
Gippsland Limestone 118, 121
Girilambone Beds 292
Girilambone–Wagga arch 317–18
Glenelg River Metamorphic Complex 283, 286
Gleneski Formation 315
Gnalta Shelf 284, 292
Goobarragandra Volcanics 313, 315–16
Gordon Sub-Group 291
Gosford Formation 256
Grahams Creek Formation 134, 265
Grampians Group 49, 312, 326
Granite Harbour Intrusive Complex 288
Grant Formation 159, 237
Great Artesian Basin 127, 143, 207, 223, 262, 266, 268, 292, 363
Great Australian Bight Basin 50, 145–6, 176, 178, 226, 266
Great Barrier Reef 47, 204–5, 210, 232, 265, 354
Greater India 16–17, 20–4, 38–9, 41, 197–9, 232

Great Western Plateau 7, 106–7, 144, 149–50, 155, 157, 160, 162, 167–8, 183, 191, 214, 232, 239, 267
Greta Coal Measures 237, 241
Grindstone Range Sandstone 284, 287
Gubberamunda Sandstone 134, 265
Gulf of Carpentaria 47, 106–7, 110, 129, 216–17, 220, 223
Gumbardo Formation 322
Gunnedah Basin 249, 251
Gurnard Formation 121

Halls Creek Province 155, 270, 337, 340
Hatchery Creek Conglomerate 320, 329
Haunted Hill Gravels 118, 120
Hawkesbury Sandstone 237, 241, 256, 260
Heathcote Axis 292
Heatherdale Shale 277
Heavitree Quartzite 273, 278, 340, 347
Heemskirk Granite 287
Helby Beds 135
Hervey Group 313, 329, 331, 333
Heytesbury Group 135
High Cliff Sandstone 237
Hill End Synclinorial Zone 310
Hill End Trough 100, 310–11, 313, 315, 317, 319, 333
Hillgrove Suite 239
Hindmarsh Clay 143
Hodgkinson Province 322–3
Holmwood Shale 237
Honeysuckle Beds 316
Hooray Sandstone 266–7
Houghton Anticlinal Zone 270, 275–6, 283
Houtman Sub-basin 185
Hunter–Mooki Fault 241
Hutton Sandstone 134
Hyland Bay Formation 237

Indian Plate 37
Indo-Australian Plate 5, 17, 23, 37, 100, 102–4, 106, 109, 113–14
Ingelara Formation 243
Injune Creek Group 134
Innamincka depocentre 110, 219–21, 230, 267, 269
Inner Nackara Arc 277
Investigator Fracture Zone 34, 38
Ipswich Basin 236, 262–4, 267
Ipswich Coal Measures 263–4
Irwin River Coal Measures 237, 239
Irwin Sub-basin 197
Italia Road Formation 235

Java Trench 32, 38
Jemba Rhyolite 317
Jemmy's Point Formation 120
Jerboa-1 153, 175-7, 181, 226, 265
Jericho-1 241, 244
Jericho Formation 243
Jerilderie Formation 237, 241
Jochmus Formation 243
Joey Rise 21, 26, 186
Judea Beds 271, 281, 283
Julie Formation 273
Jupiter-1 198, 200

Kangaloolah Volcanics 315
Kangaroo Island pegmatite 288
Kangaroo Syncline 185
Kanmantoo Basin 281, 283-4, 290, 305
Kanmantoo Group 271, 276-7, 288
Kariem Formation 281, 285
Karimui Volcanics 254
Karumba Basin 110, 127, 129, 221, 227, 247, 251, 262
Keepara Group 273
Keepit Conglomerate 333
Kendall Surface 126
Kenn Plateau 202
Kennedy Group 237, 241
Kerguelen Plateau 34, 37
Kimberley Basin 155
Kimberley Block 94-6, 149, 155-6, 159-60, 167, 199
Kimberley Group 280
King Leopold Mobile Zone 280, 282, 337
Kirkpatrick Volcanics 233
Knobby Hills Sandstone 166, 340
Kockatea Shale 238, 241
Kondaku Tuff 135
Kosciusko Batholith 303, 305
Kowmung Volcaniclastics 318
Kubor Anticline 129
Kubor Block 110, 136, 247
Kulnura Marine Tongue 256
Kulshill Formation 237
Kumbruf Volcanics 135
Kuta Formation 238, 263

Lacrosse-1 237
Lake Blanche Fault 274, 280, 284, 307
Lake Frome Group 284, 286-7
Lake Galilee Sandstone 243
Lakes Entrance Formation 118, 121
Lake Surprise Sandstone 166
Lake Torrens Basin 145
Lambian Facies 329, 333-4
Lambie Group 329-31, 333
Lampe Beds 153
Lander Trough 162, 340-1, 344
Larapinta Group 166-7

Latrobe Group 118, 123, 135, 176, 202
Laura Basin 220, 251
Leeuwin Block 185, 280
Leigh Creek Basin 145, 148
Leigh Creek Coal Measures 263-4
Lesueur Sandstone 191, 237
Leveque Platform 185
Lewis Trough 198
lithospheric structure 94
Liveringa Formation 237-8, 241
Lochinvar Formation 237
Locker Shale 237-8, 241
Lombardina-1 188
Londonderry High 185, 199
Lord Howe Rise 17-20, 23-24, 27-8, 31, 33-5, 37, 50, 69, 80, 131, 138, 142, 176, 202-3, 207-8, 210, 213-14, 223, 227, 230, 354, 364
Louisiade Rise 202
Lyons Group 237

McArthur Basin 155, 159-60
McIvors Hill Gabbro 286
MacMillan Formation 255
McLarty-1 56
Mafeking and Mackenzie River Granodiorites 312
Maitland Group 237
Malita Graben 155, 183-4, 193, 199
Maril Shale 135
Marion Plateau 202, 204-5
Maryborough Basin 127, 129, 132, 133-4, 136-8, 220, 227, 265, 267
Maryborough Formation 134
Mathinna Beds 292, 321
Meadows Tank Formation 318-19
Medusa Beds 238
Melbourne Trough 296, 310-12, 314-17, 319-21, 323, 332
Mellinjerrie Limestone 340
Mellish Rise 202
Mereenie Sandstone 48-9, 167, 327-8
Merlinleigh Sub-basin 185
Merrimbula Group 329, 332-3
Merrimelia Formation 243
Merrimerriwa Formation 319-20
Merrions Tuff 319
Mersey River Granite 286
Milligans Formation 237
Mitta Mitta Volcanics 315-16
Mologa Surface 119, 121-2
Molong High 310, 313, 317
Money Shoal-1 328
Money Shoal Basin 155, 159, 183, 185, 189, 191, 281
Mongum Volcanics 135
Mooga Sandstone 134, 265

Moolayember Formation 241, 243, 255
Mopunga Group 273
Moranbah Coal Measures 255
Moresby Trough 37, 108, 245
Moreton Basin 127, 132-3, 267
Morney Profile 125, 137, 227-8
Moruya Batholith 303, 305
Movi Beds 254
Mt Anna Sandstone Member 148
Mt Currie Conglomerate 166-7, 278
Mt Eclipse Sandstone 161, 166, 340, 342
Mt Goodwin Shale 237
Mt Howie Sandstone 148, 218, 225, 228
Mt Howitt Province 332-3
Mt Ida Sandstone 317
Mt Isa Block 95, 160, 167
Mt Johnstone Formation 235, 237
Mt Painter Block 270
Mt Pleasant Andesite 292
Mt Read Volcanics 284, 287-8
Mt Read Volcanic Belt 281-2, 284
Mt Toondina Formation 237
Mt Wellington axis 332
Mt Windsor Volcanics 287-9, 305, 307
Moura Beds 319
Mulbring Formation 256
Mulga Downs Group 319-20, 329
Mullaman Beds 268
Mullions Range Volcanics 315
Munda sequence 166-7
Mundawatana Granodiorite 288
Mungaroo Sandstone 191, 237, 241
Munyarai-1 162, 166
Murray Basin 9, 42, 55, 96, 115-23, 127, 137-8, 143, 145-6, 161, 178, 215-18, 220-1, 223, 226, 228-9, 231, 234
Murray infrabasins 239, 241, 260, 262
Murruin Trough 313
Murrumbidgee Batholith 303, 305
Murteree Shale 243
Musgrave Block 155, 160-2, 165-7, 270, 273, 278, 280, 307-8, 340-1, 344-5, 347, 351
Myrtle Creek Sandstone 134

Nambour Basin 127
Namoi Formation 237-8
Nangetty Formation 237
Nappamerri Formation 243
Narrabeen Group 237, 241, 256
Naturaliste Plateau 6, 176, 183, 186-8, 199, 201
Nautilus-1 181
Nebine Island Arc 287, 289
Nebine Ridge 261, 267

SUBJECT INDEX

Nelungaloo Volcanics 292
Netley Microtonalite 288
New Caledonia 69, 208
Newcastle Coal Measures 237, 256
New Guinea Highlands 10, 106–7, 110, 112, 115, 136, 139, 168, 215
Newmarracarra Limestone 50, 238
New Town Coal Measures 237, 263
New Zealand Plateau 17, 20, 213
Ngalia Basin 160–1, 165–7, 278, 284, 340–4
Nilsen–Mackay Basin 243, 245
Ninetyeast Ridge 36–7
Nirranda Group 135
Noonkanbah Formation 237
Nootumbulla Sandstone 290, 292
Norfolk Ridge 69, 202, 208, 364
Normanton Formation 223
Normanville Group 276–7
Northampton Block 197, 270
North Australian Craton 94, 96
Northeast Australian Orogens 94, 96
North Maslin Sand Member 146
North New Guinea Basin 108, 110, 129, 134–5
North Tryal Rocks-1 55
North West Shelf 114, 153–7, 183, 186, 195, 199
Nowra Sandstone 237, 241, 256, 260
Nura Nura Member 238
Nymboida Coal Measures 263

Oaklands Basin 236, 251, 260–2
Oakover Beds 153
Officer Basin 96, 149, 151, 153–60, 162, 165–6, 217–18, 223, 236, 239, 268, 272, 278, 282, 284, 307, 340–1, 344, 348
Olary–Broken Hill Block 283
Olga Regional Gravity Ridge 345–6
Olympic Formation 273
Orallo Formation 134, 265
Ord Basin 160, 340
Oriomo Rise 247, 251, 262
Orubadi Beds 254
Otway Basin 50, 117, 120, 131–5, 137–9, 174–6, 178–9, 181–2, 202, 207, 216, 220–1, 225–6, 228–9, 232, 234, 266
Otway Group 175, 207
Outer Fleurieu Arc 275–7, 282, 284
Owen Conglomerate 290–1

Pacific Plate 6, 23, 100, 102, 104, 109
Pacoota Sandstone 289
palaeomagnetism
  continental 11
  oceanic 17

Palmer Granite 288
Papuan Basin 57, 110, 126–7, 129, 134–5, 137, 201, 239, 243–60, 262
Papuan Peninsula 5, 17–20, 27, 37, 108, 112, 114, 207, 209, 213, 231, 245
Papuan Ultramafic Belt 111–12
Parlana Quartzite 275
Parilla Sand 119
Patchawarra Formation 243
Paterson Formation 154, 237
Paterson Volcanics 237
Peake–Denison Block 166
Peawaddy Formation 243
Pedirka Basin 103, 143, 148, 160, 165–6, 236, 239
Peel Fault Zone 305, 334, 336–7
Pelican-1 237
Peninsula Range Volcanics 132
Pepuarta Tillite 273
Pertataka Formation 273
Perth Abyssal Plain 33, 183, 185
Perth Basin 50, 71, 151, 153, 154, 183, 185–6, 189, 191, 193, 196–9, 201, 205, 220, 232, 234, 236, 239, 265–6, 284, 307, 327
Pertnjara Group 161, 166–7, 341–2
Petrel Sub-basin 155, 159, 185, 189, 191, 281, 283
Pidinga Formation 146
Pilbara Block 94, 96–7, 149, 154, 159, 183, 191, 193, 221, 280
Pilliga Sandstone 134
Pine Creek–Arnhem Block 149, 155, 159–60, 167
Pirramimma Sand Member 146
Platypus-1 175–6, 201
Point Maud Formation 340
Polda Basin 145, 148, 174, 226, 228, 232, 265
Polda Formation 148
Polly Conglomerate 161
Poolawanna Trough 263
Poole Sandstone 237–8
Poondano Formation 153
Potoroo-1 175–7, 201, 222–3
Potoroo Formation 179–80
Pound Quartzite 73, 272–3, 276, 278, 280
Precipice Sandstone 132, 134, 149, 241, 262–3
Pretty Hill Sandstone 135, 266
Purni Formation 237
Pyri Pyri Granite 132

Quandong Conglomerate 166–7
Queensland Plateau 202, 204–5

Rangal Coal Measures 255
Rankin Platform 183, 185, 193, 198, 200
Ravenswood Granodiorite Complex 288–9, 305, 307
Ravenswood–Lolworth Block 287, 289
Razorback Beds 132
Redcliff Coal Measures 263
Reids Dome Beds 243
Renmark Basin 143, 145, 236, 260–2
Retreat Granite 337
Rewan Group 243, 255
Robertson Bay Group 271, 288
Robinson Basin 145
Rolling Downs Group 57, 134–5, 267
Roo Rise 22, 26, 186
Roseneath Shale 243
Rutherford Formation 237

Sahul Platform 183, 185, 199
Sahul Shelf 45, 47, 107, 109–10, 114, 155–6, 159, 183–4, 199
Sahul Shoals-1 113–14, 191, 238
Samphire Marsh-1 280, 307
Samuel Formation 154
Sandfly Coal Measures 263
Scott Plateau 155, 183, 186–7, 189, 205
Scott Reef-1 157, 159, 182, 188, 191, 199, 238
Seaham Formation 237
Seaspray Group 135
Sharpeningstone Conglomerate 318–19
Shepparton Formation 120
Sherbrook Group 135, 175
Shoalhaven Group 237
Silver Hills Volcanics 337
Silverwood Group 335
Skillogalee Dolomite 273
Sledgers Group 284, 288
Snapper Point Formation 260
Snowy River Volcanics 317, 319
Sofala Volcanics 294
Sonne Ridge 22, 26, 184
South Australian Highlands 9, 107, 143–9, 167, 217, 220, 223, 228–30, 234, 267, 354
South Coast High 317
Southern margin 174–6
South Tasman Rise 6, 7, 22, 31, 38, 53–5, 57, 68, 174, 201, 204, 210
Spencer Gulf Basin 145
Spion Kop Conglomerate 235
Springfield Basin 145, 148
Springfield Coal Measures 263–4
Springsure Shelf 241–2, 261
Stanwell Coal Measures 132

SUBJECT INDEX

St Johns Conglomerate 166-7
Stockton Formation 237
Stonyfell Quartzite 273
Strzelecki/Gippsland Basin 32, 176, 201-2, 207, 223, 266
Strzelecki Group 50, 135, 176, 202, 207
Stuart Range Formation 237
Stuart Shelf 143, 275, 277, 284
Stuckeys Creek Formation 316
Sturt Tillite 273, 275
St Vincent Basin 145-6, 226
Styx Basin 132, 220
Sue-1 188, 197
Sue Coal Measures 237
Sunda Arc 38-40, 183
Sundaland 31, 38-41, 109
Surat Basin 127, 129, 131-4, 136-7, 139, 149, 178, 217, 222, 227, 236, 241, 262, 264, 266-8
Sydney Basin 79, 100, 112, 123, 127, 131, 133, 138-9, 201, 239, 243-62, 314, 329, 334
Sydney Sub-Group 256

Table Hill Volcanics 163, 282
Taemas Group 49, 320
Tallong Conglomerate 237
Tamar Fracture 305, 321
Tamworth Trough 321, 334, 337
Tandalgoo Red Beds 328
Tanmurra Formation 237-8
Tantawangelo Fault 321
Tanwarra Shale 294, 315
Tapley Hill Formation 273
Taralga Group 317
Tarong Coal Measures 263
Tasmania Basin 35, 236, 239, 241, 245, 262, 264
Tasmanian Dolerite 136, 233, 265, 282-3
Tasman Line 94-6, 270, 274, 279-80, 282, 307, 326, 351
Tasman Sea 21, 26, 28-9, 31, 34, 69, 80, 100, 137, 139, 176, 178, 201, 203, 205, 207, 209-10, 225, 227, 229, 364
Telemon granodiorite 288-9
Tethys 26, 113-14, 189, 191, 281, 283-5, 307
Tiaro Sandstone 132, 134
time scale 1, 260-1, 263
Timor Trough 41, 110, 112-13, 155, 183, 185, 199
Tirrawarra Sandstone 243
Toko Syncline 162, 341, 344
Tomago Coal Measures 237, 256
Toolachee Formation 243
Toolamang Volcanics 315

Toolebuc Formation 238
Torrens Basin 145-6
Torrens Fault 274-5
Torrens lineament 145, 274-5
Torsdale Beds 237
Transantarctic Mountains 9-10, 131-2, 136, 232-3, 271, 273, 288, 350
Transition Beds 266
Trelawney Beds 305
Troubridge Basin 143, 145, 148, 236
Truro Volcanics 277, 282, 288
Tumblagooda Sandstone 154, 189, 307, 328
Tumut Trough 310-11, 313, 315-17, 319, 323-4
Turtle Dove Ridge 197
Twofold Bay Formation 331
Tyennan Geanticline 95, 272

Umberatana Group 273, 275, 277
Upper Parmeener Group 237
Uralba Beds 305
Urella Fault 197
Urannah Granite 239

Victoria River Basin 155, 159
Vlaming Sub-basin 188, 197
Vulcan Sub-basin 183, 185, 193, 199

Wagga Basin 292, 305, 307
Wagga Marginal Sea 296, 298, 311, 323
Wagga Metamorphic Belt 298, 300-1, 309-15, 317, 319, 323
Wagina Sandstone 237
Wagonga Beds 293, 295
Walkandi Formation 262
Wallaby Plateau 22, 38, 183, 186-7, 199, 205
Walli Andesite 292
Walloway Basin 145
Walsh Tillite 280
Wandilla Slope 326, 336
Wangerrip Group 135
Warburton Basin 165, 282
Warrabin Trough 341, 347
Wellington greenstone axis 305
Wentworth Trough 262
Werrie Basalt 237, 239
Werrie Trough 235-6, 239, 348
West Tyral Rocks-1 198, 200
western margin 24, 37, 182, 226, 229
Western Shield 95, 149-51, 153-5, 167
Whitula Formation 57

Wianamatta Group 237, 241, 256
Wilga Basin 154, 191
Wilkes sub-ice basin 9, 234
Willochra Basin 145, 148
Willowie Creek Beds 335
Willyama Block 95, 143, 149, 272
Wilpena Group 273, 275, 277
Wilson Group 288
Wilton Formation 256
Windalia Radiolarite 268
Winduck Group 318-19
Winnambool Formation 121
Winton Formation 148, 216, 218, 221, 223-4, 228, 267, 269
Wirrealpa Formation 284-5
Wirrildar Beds 166-7
Wisemans Arm Formation 334
Wiso Basin 160, 162, 165
Wollondilly Tract 311, 313, 317, 319, 324
Wombat Creek Group 316
Wonaminta Block 272, 284, 310, 312
Wonoka Formation 273, 280
Woodada Formation 237
Woodlark Basin 354
Woolomin Beds 321-2
Woolomin Slope 334-5
Woolshed Flat Shale 273
Wooltana Volcanics 275
Wooramel Group 237-8
Worange Point Formation 333
Wurruk Sand 120
Wyaaba Formation 135
Wywyana Formation 275
Wyangle Formation 313

Yackah Beds 273
Yampi-1 188
Yandanooka Group 280
Yardida Tillite 273
Yaringa Evaporite Member 328
Yarragadee Formation 154, 197, 265, 340
Yarrol Basin 336-7
Yarrol Fault zone 336-7
Yarrol Formation 238
Yaveufa Formation 254
Yilgarn Block 94-7, 101, 149, 154, 183, 191, 201, 221, 223
Yindagindy Formation 237
Young Batholith 305
Yuendumu Sandstone 166-7

Zenith Seamount 23, 38, 186, 199
Zenith-Wallaby Fracture Zone 24, 33-4, 38, 199, 230, 269